# GAS TABLES

BOOKS BY J. H. KEENAN
    Thermodynamics
BY J. H. KEENAN, J. CHAO, AND J. KAYE
    Gas Tables
BY J. H. KEENAN AND F. G. KEYES
    Mollier Chart
BY J. H. KEENAN, F. G. KEYES, P. G. HILL, AND J. G. MOORE
    Steam Tables
BY G. N. HATSOPOULOS AND J. H. KEENAN
    Principles of General Thermodynamics

# GAS TABLES
## International Version

## THERMODYNAMIC PROPERTIES OF AIR
## PRODUCTS OF COMBUSTION
## AND
## COMPONENT GASES
## COMPRESSIBLE FLOW FUNCTIONS

Including those of Ascher H. Shapiro
and Gilbert M. Edelman

## SECOND EDITION
## (SI UNITS)

BY

### JOSEPH H. KEENAN

Professor of Mechanical Engineering
Massachusetts Institute of Technology

### JING CHAO

Research Scientist
Thermodynamics Research Center
Texas A & M University

AND

### JOSEPH KAYE

Professor of Mechanical Engineering
Massachusetts Institute of Technology

**A WILEY-INTERSCIENCE PUBLICATION**

## JOHN WILEY AND SONS

**NEW YORK      CHICHESTER      BRISBANE      TORONTO      SINGAPORE**

Published by John Wiley & Sons, Inc.
Copyright, 1945 by Joseph H. Keenan and Joseph Kaye under the title
*Thermodynamic Properties of Air, Including Polytropic Functions*
Copyright, 1948 by Joseph H. Keenan and Joseph Kaye
Copyright © 1983 by Joseph H. Keenan Trusts, Mrs. Joseph Kaye, and Jing Chao

*Library of Congress Cataloging in Publication Data:*

Keenan, Joseph Henry, 1900–1977
  Gas tables.

  "A Wiley-Interscience publication."
  Bibliography: p.

  1. Air—Tables. 2. Thermodynamics—Tables
I. Chao, Jing. II. Kaye, Joseph, 1912–1961
III. Title.

QC161.5.K43     1983     621.43'021'2     82-24779
ISBN 0-471-08874-9

Printed in the United States of America

10 9 8 7 6 5 4 3 2

# IN MEMORIAM
## Joseph H. Keenan

Professor Joseph H. Keenan, who was the motivating force behind these tables from the beginning, died at the age of 77 on July 17, 1977. Although his health gradually declined during the last two years of his life, he remained active and managed to complete the preliminary arrangements which eventually led to the publication of the two versions of the second edition.

As coauthor, Dr. Jing Chao shouldered a greater share of the responsibility for the 1980 revision in English Units and for the most recent tables in SI Units. His son, William C. H. Chao, played a crucial role in devising the computer program which made possible a calculation of the values and the tabulation of them automatically for the printed page. Other individuals, including many who had been friends of Professor Keenan, were more than generous with their assistance. Dr. George N. Hatsopoulos in particular never failed to provide the benefits of his judgement and his technical talents.

The origins of these tables date back more than 30 years to a collaboration between Professor Keenan and Professor Joseph Kaye, who died in 1961. They developed the basic concepts and methodology used in formulating much of the data for the early editions. For this reason it was considered appropriate here to include their acknowledgments and the contents of each preface from the 1945 and 1948 printings. The wives of these men deserve recognition as well—Mrs. Ida Kaye for her understanding cooperation and Mrs. Isabel M. Keenan for her encouragement and support.

I know that my father would have wanted me to thank on his behalf all of the people who contributed their time and effort in making the revisions a success. I know too how deeply he would have appreciated the readiness and willingness shown by those among them who knew him and were close to him.

MATTHEW A. KEENAN

*Belmont, Massachusetts*
*February 1983*

# PREFACE TO THE SECOND EDITION OF
# THE GAS TABLES IN SI UNITS

When the authors decided to embark upon a revision of the *Gas Tables*, they planned to publish a new edition using English Units as before and then to follow it later with another edition of these tables converted into International Units.

Professor Joseph H. Keenan directed the initial preparations, but died in 1977 before he could participate further in the project. After the unfortunate loss of his presence, it wasn't until 1980 that the revised *Gas Tables* in English Units (second edition) reached a successful conclusion. Now the final half of the overall project has become a reality with the arrival of these revised tables in International Units.

In this latest edition, the sources of data, methods of calculation, and formats for tabulating the evaluated results were the same as before. The property values given in Tables 1 through 24 were reevaluated at selected temperatures in Kelvin (K) and at P = 101.325 kPa (1 atm.). The explanatory text for "Sources of Data and Methods of Calculation," and for the chapter on "Examples," were essentially unchanged with minimum adjustment for employing different units for recalculation.

To provide the corrected entropy at 1 bar pressure for thermodynamic calculations, the constant $0.1094$ $JK^{-1}$ $mol^{-1}$ should be added to each entropy value listed in Tables 3, 5, 7, 11, 13, 15, 17, 19 and 21. In computing entropy change for a constant pressure process the listed entropy values can be used without corrections, because the same constants added to both quantities cancel each other.

The entropy values presented in Table 1 should be augmented by $4.7767$ $JK^{-1}$ $mol^{-1}$ to yield the corrected entropy for one kilogram at one bar pressure. This correction is not necessary for calculations of entropy changes at constant pressures.

SI is the official abbreviation for the International System of Units (Le Systeme International d'Unites). It is a rationalization of the cgs and MKS systems. The seven base units of meter (m), kilogram (kg), second (s), Kelvin (K), mole (mol), ampere (A), and candela (cd) correspond respectively to length, mass, time, temperature, amount of substance, electric current, and luminous intensity. Obtained therefrom are other derived and supplementary units. Some allowable units, not a part of SI, may also be used in conjunction with SI.

SI was adopted by the General Conference of Weights and Measures in 1960 as recommended units for use in Science and Technology. In 1964 the National Bureau of Standards adopted SI for use by its staff and issued a statement interpreting this policy in 1968. In recent years, scientific and technical communities in the United States also accepted the mandatory application of SI in all publications and all papers presented at national meetings.

The publication of this most recent edition of *Gas Tables* would not have been possible without the invaluable cooperation of numerous individuals. Throughout much of the undertaking, it was reassuring to have available the experienced counsel of Dr. George N. Hatsopoulos, who enjoyed a long association and close friendship with the late Professor Keenan. Another friend and colleague of the professor was Dr. Elias P. Gyftopoulos, whose helpful advice was required on several occasions. And the professor's son, Matthew A. Keenan, lent his support by coordinating the various phases of the project and by editing the manuscript.

A key contributor was my son, William C. H. Chao, who brought his knowledge of computers to the considerable task of recalculating the revised numerical tables and to the direct generation of photocomposition tapes, thereby automating the printing of the tables and eliminating the potential for typesetting errors. Sincere gratitude also belongs to the people at Inforonics, Inc., for their kind services and efficient handling of the photocomposition galleys.

The author particularly wishes to thank Dr. Lester Haar at the National Bureau of Standards for recommending that Dr. Harold W. Wooley's recently reevaluated values of ideal gas thermodynamic properties for steam be adopted in this book. I am also grateful to David T. Barr for reworking in detail the illustrative problems in the chapter on "Examples."

On behalf of all the participants, it is hoped that Professor Keenan would have been pleased with the published results and have considered them worthy of the high standards he maintained throughout a long and productive career in the field of thermodynamics.

Finally, any comments or observations that are forwarded by the reader of this revision will be greatly appreciated.

JING CHAO

*College Station, Texas*
*February 1983*

# PREFACE TO THE FIRST EDITION
# OF THE GAS TABLES

THIS book supersedes a previous one entitled *Thermodynamic Properties of Air*. The properties of air have been reexamined and recalculated, the properties of products of combustion of hydrocarbons and of the constituent gases have been added, and tables of functions useful in analysis of the flow of compressible fluids have been greatly extended.

All values of the thermodynamic properties in this book are based on the examination of data from spectroscopic sources which was published by F. D. Rossini and his coworkers of the National Bureau of Standards in 1945. This was a revision of the calculated values of Johnston, Giauque, Gordon, and Kassel, and of the tabular values of Heck. The revision consisted mainly of application of new values of the fundamental constants.

The present status of these thermodynamic properties is relatively satisfactory. The uncertainties remaining in the interpretation of spectra appear to be small for monatomic and diatomic gases, and of no great order for $CO_2$ and $H_2O$. For engineering application to the design of power apparatus the precision of the present tables is of the same order as that of modern steam tables. The convenience is of a higher order.

The base temperature, for which enthalpy is zero, is not the same here as in *Thermodynamic Properties of Air*. It is taken to be zero on the absolute Fahrenheit scale in accordance with the data of Rossini. Negative values of properties are thereby avoided.

The molal unit is employed for all tables of thermodynamic properties except air. As compared with the mass unit, the molal unit makes the range of values between tables very much smaller. It also permits the use of a single table for the products of combustion corresponding to a wide range of carbon-hydrogen ratios, provided only that the "percentage of theoretical air" is held fixed. Certain other interpretations of the same table in terms of mixtures of air and fuel vapor and of air and water vapor are also valid to good precision.

Although the major tables of thermodynamic properties can be used in the analysis of the flow of compressible fluids, several excellent tabulations which have recently appeared are of greater specific convenience. The more generally useful of these have been included here. Some of the most valuable portions of this material prepared by Ascher H. Shapiro and Gilbert M. Edelman have not been previously published except in reports of limited circulation.

Dependable values of viscosity and thermal conductivity for air cover a limited range of temperature—for other gases an entirely inadequate range. Values of these properties are tabulated here only for air. Correlation of existing data and some measurements to obtain new data are in progress for other gases. As the results of this work become available they will be incorporated into the tables. The scope of the investigations in progress, however, is entirely too limited to fill the urgent need for knowledge of these two properties.

JOSEPH H. KEENAN
JOSEPH KAYE

*Cambridge, May 1948*

# PREFACE TO
# THERMODYNAMIC PROPERTIES OF
# AIR

THE need for a working table of the thermodynamic properties of air has been emphasized recently by the rapidly growing interest in the gas turbine. The compression of atmospheric air which occurs in a gas turbine can be computed with reasonable convenience without the aid of a table. Computations of certain other processes, however—such as the heating of air in a regenerator or the expansion of air and similar gases from states of high temperature—involve laborious integrations if tabulated properties are not available.

For such computations air may be considered to be a relatively simple substance because the expression

$$pv = RT$$

is an adequate equation of state. By virtue of this fact the table of properties for air is simpler than the corresponding tables for vapors in that a single independent argument serves in place of two. Within the same space, therefore, a far more detailed table is possible for air than for steam. In the present table, as contrasted with existing tables for vapors, interpolation can often be dispensed with, and where it must be used it is a single interpolation which may be done by inspection.

Since the value of the gas constant $R$ is known with great precision, the degree of uncertainty in the properties tabulated here depends only upon the state of knowledge of specific heats. The development of spectroscopic methods of determining the specific heats of the simpler gases at zero pressure has reduced this degree of uncertainty to such an extent that there is now little room for disagreement. In general, the effects of departures in composition and in pressure from those stipulated for this table will be greater than the effects of uncertainties in the data upon which the table is based. For this good fortune we are indebted to those who have labored to perfect our knowledge of the specific heats of gases.

Table 1 appeared in an abbreviated form in the *Journal of Applied Mechanics*. Since that publication the composition assumed and the values of the specific heat for air have been revised, and the entire table has been recomputed, interpolated to smaller intervals, and extended to higher temperatures.

In addition to the usual thermodynamic properties, values are given for viscosity, thermal conductivity, and Prandtl number because of their great utility in engineering computations. Certain commonly used functions of the pressure ratio are also given in some detail for a range of values of the polytropic exponent.

The present data will satisfy all but the most exacting requirements for calculations of gas-turbine processes. In order to satisfy certain other requirements work is in progress on the extension of the present data to cover the properties of air-fuel mixtures for hydrocarbon fuels and their products of combustion.

JOSEPH H. KEENAN
JOSEPH KAYE

*Cambridge, March 1945*

# CONTENTS

$a$  velocity of sound $= \sqrt{kRT}$, m s$^{-1}$

$c_P$  specific heat at constant pressure, J K$^{-1}$ kg$^{-1}$

$c_v$  specific heat at constant volume, J K$^{-1}$ kg$^{-1}$

$G$  flow per unit area or mass velocity, kg m$^{-2}$ s$^{-1}$

$h$  enthalpy per unit mass, kJ kg$^{-1}$*

$\bar{h}$  enthalpy per mole, J g-mol$^{-1}$

$k$  $c_p/c_v$

$p$  pressure, kPa

$Pr$  Prandtl number $= c_p u / \lambda$

$p_r$  relative pressure$^\dagger$

$R$  gas constant for air $= 0.287031$ kJ K$^{-1}$ kg$^{-1}$

$T$  temperature, K

$t$  temperature, °C

$u$  internal energy per unit mass, kJ kg$^{-1}$

$\bar{u}$  internal energy per mole, J g-mol$^{-1}$

$v_r$  relative volume$^\dagger$

$\lambda$  thermal conductivity, W m$^{-1}$ K$^{-1}$

$\mu$  viscosity, NR m$^{-2}$

$\phi$  $\int_{T_0}^{T} c_p \, dT$, kJ K$^{-1}$ kg$^{-1}$

$\bar{\phi}$  $\int_{T_0}^{T} \dfrac{\bar{c}_p}{T} \, dT$, J K$^{-1}$ g-mol$^{-1}$$^\ddagger$

*To obtain values in Joules per gram-mole multiply tabulated values by molecular weight of air, namely 28.9669.

$^\dagger$The ratio of the pressures $p_a$ and $p_b$ corresponding to the temperatures $T_a$ and $T_b$, respectively, along a given isentropic is equal to the ratio of the relative pressures $p_{ra}$ and $p_{rb}$ as tabulated for $T_a$ and $T_b$, respectively. Thus

$$\left( \frac{p_a}{p_b} \right)_{s = \text{constant}} = \frac{p_{ra}}{p_{rb}}$$

Similarly

$$\left( \frac{v_a}{v_b} \right)_{s = \text{constant}} = \frac{v_{ra}}{v_{rb}}$$

$^\ddagger$For interpolation between Table 1 and Table 3 the value of $\phi$ from Table 1 should be augmented by unity and multiplied by the molecular weight of air, namely 28.9669. For interpolation between Table 1 and Table 3 the value of $V_r$ from Table 1 should be multiplied by 0.0289669.

# Table 1 Air at Low Pressures (for One Kilogram)

| T | t | h | $p_r$ | u | $v_r$ | φ | T | t | h | $p_r$ | u | $v_r$ | φ |
|---|---|---|---|---|---|---|---|---|---|---|---|---|---|
| 100 | −173.15 | 99.93 | .02977 | 71.23 | 964.19 | 4.6004 | 150 | −123.15 | 150.03 | .12259 | 106.97 | 351.20 | 5.0067 |
| 101 | | 100.93 | .03082 | 71.94 | 940.59 | 4.6104 | 151 | | 151.03 | .12547 | 107.69 | 345.44 | 5.0133 |
| 102 | | 101.93 | .03190 | 72.66 | 917.79 | 4.6203 | 152 | | 152.03 | .12839 | 108.40 | 339.81 | 5.0200 |
| 103 | | 102.94 | .03300 | 73.37 | 895.76 | 4.6300 | 153 | | 153.03 | .13137 | 109.12 | 334.30 | 5.0265 |
| 104 | | 103.94 | .03414 | 74.09 | 874.46 | 4.6397 | 154 | | 154.04 | .13439 | 109.83 | 328.92 | 5.0331 |
| 105 | −168.15 | 104.94 | .03530 | 74.80 | 853.86 | 4.6493 | 155 | −118.15 | 155.04 | .13746 | 110.55 | 323.66 | 5.0395 |
| 106 | | 105.94 | .03648 | 75.52 | 833.94 | 4.6588 | 156 | | 156.04 | .14058 | 111.26 | 318.52 | 5.0460 |
| 107 | | 106.94 | .03770 | 76.23 | 814.66 | 4.6682 | 157 | | 157.04 | .14375 | 111.98 | 313.49 | 5.0524 |
| 108 | | 107.94 | .03894 | 76.95 | 796.00 | 4.6775 | 158 | | 158.04 | .14697 | 112.69 | 308.57 | 5.0588 |
| 109 | | 108.95 | .04022 | 77.66 | 777.94 | 4.6868 | 159 | | 159.05 | .15025 | 113.41 | 303.76 | 5.0651 |
| 110 | −163.15 | 109.95 | .04152 | 78.38 | 760.44 | 4.6959 | 160 | −113.15 | 160.05 | .15357 | 114.12 | 299.05 | 5.0714 |
| 111 | | 110.95 | .04285 | 79.09 | 743.49 | 4.7050 | 161 | | 161.05 | .15695 | 114.84 | 294.44 | 5.0776 |
| 112 | | 111.95 | .04422 | 79.81 | 727.07 | 4.7140 | 162 | | 162.05 | .16038 | 115.55 | 289.94 | 5.0838 |
| 113 | | 112.95 | .04561 | 80.52 | 711.15 | 4.7229 | 163 | | 163.05 | .16386 | 116.27 | 285.53 | 5.0900 |
| 114 | | 113.96 | .04703 | 81.23 | 695.71 | 4.7317 | 164 | | 164.06 | .16740 | 116.98 | 281.21 | 5.0961 |
| 115 | −158.15 | 114.96 | .04849 | 81.95 | 680.74 | 4.7405 | 165 | −108.15 | 165.06 | .17099 | 117.70 | 276.98 | 5.1022 |
| 116 | | 115.96 | .04998 | 82.66 | 666.22 | 4.7491 | 166 | | 166.06 | .17463 | 118.41 | 272.85 | 5.1082 |
| 117 | | 116.96 | .05150 | 83.38 | 652.13 | 4.7577 | 167 | | 167.06 | .17833 | 119.13 | 268.79 | 5.1143 |
| 118 | | 117.96 | .05305 | 84.09 | 638.45 | 4.7663 | 168 | | 168.06 | .18209 | 119.84 | 264.83 | 5.1202 |
| 119 | | 118.97 | .05464 | 84.81 | 625.17 | 4.7747 | 169 | | 169.07 | .18590 | 120.56 | 260.94 | 5.1262 |
| 120 | −153.15 | 119.97 | .05626 | 85.52 | 612.27 | 4.7831 | 170 | −103.15 | 170.07 | .18977 | 121.27 | 257.13 | 5.1321 |
| 121 | | 120.97 | .05791 | 86.24 | 599.75 | 4.7914 | 171 | | 171.07 | .19369 | 121.99 | 253.40 | 5.1380 |
| 122 | | 121.97 | .05960 | 86.95 | 587.58 | 4.7997 | 172 | | 172.07 | .19768 | 122.70 | 249.75 | 5.1438 |
| 123 | | 122.97 | .06132 | 87.67 | 575.75 | 4.8078 | 173 | | 173.07 | .20172 | 123.42 | 246.17 | 5.1496 |
| 124 | | 123.98 | .06308 | 88.38 | 564.26 | 4.8160 | 174 | | 174.08 | .20582 | 124.13 | 242.66 | 5.1554 |
| 125 | −148.15 | 124.98 | .06487 | 89.10 | 553.08 | 4.8240 | 175 | −98.15 | 175.08 | .20998 | 124.85 | 239.22 | 5.1612 |
| 126 | | 125.98 | .06670 | 89.81 | 542.21 | 4.8320 | 176 | | 176.08 | .21420 | 125.56 | 235.85 | 5.1669 |
| 127 | | 126.98 | .06857 | 90.53 | 531.64 | 4.8399 | 177 | | 177.08 | .21847 | 126.28 | 232.54 | 5.1725 |
| 128 | | 127.98 | .07047 | 91.24 | 521.35 | 4.8478 | 178 | | 178.08 | .22281 | 126.99 | 229.30 | 5.1782 |
| 129 | | 128.99 | .07241 | 91.96 | 511.35 | 4.8556 | 179 | | 179.09 | .22721 | 127.71 | 226.12 | 5.1838 |
| 130 | −143.15 | 129.99 | .07439 | 92.67 | 501.60 | 4.8633 | 180 | −93.15 | 180.09 | .23168 | 128.42 | 223.01 | 5.1894 |
| 131 | | 130.99 | .07641 | 93.39 | 492.12 | 4.8710 | 181 | | 181.09 | .23620 | 129.14 | 219.95 | 5.1949 |
| 132 | | 131.99 | .07846 | 94.10 | 482.89 | 4.8786 | 182 | | 182.09 | .24079 | 129.85 | 216.95 | 5.2005 |
| 133 | | 132.99 | .08056 | 94.82 | 473.89 | 4.8862 | 183 | | 183.09 | .24544 | 130.57 | 214.01 | 5.2059 |
| 134 | | 134.00 | .08269 | 95.53 | 465.13 | 4.8937 | 184 | | 184.10 | .25015 | 131.28 | 211.13 | 5.2114 |
| 135 | −138.15 | 135.00 | .08486 | 96.25 | 456.60 | 4.9011 | 185 | −88.15 | 185.10 | .25493 | 132.00 | 208.29 | 5.2168 |
| 136 | | 136.00 | .08708 | 96.96 | 448.28 | 4.9085 | 186 | | 186.10 | .25978 | 132.71 | 205.51 | 5.2222 |
| 137 | | 137.00 | .08934 | 97.68 | 440.18 | 4.9159 | 187 | | 187.10 | .26468 | 133.43 | 202.79 | 5.2276 |
| 138 | | 138.00 | .09163 | 98.39 | 432.27 | 4.9231 | 188 | | 188.11 | .26966 | 134.14 | 200.11 | 5.2330 |
| 139 | | 139.01 | .09397 | 99.11 | 424.57 | 4.9304 | 189 | | 189.11 | .27470 | 134.86 | 197.48 | 5.2383 |
| 140 | −133.15 | 140.01 | .09635 | 99.82 | 417.06 | 4.9376 | 190 | −83.15 | 190.11 | .27981 | 135.57 | 194.90 | 5.2436 |
| 141 | | 141.01 | .09878 | 100.54 | 409.73 | 4.9447 | 191 | | 191.11 | .28498 | 136.29 | 192.37 | 5.2488 |
| 142 | | 142.01 | .10124 | 101.25 | 402.58 | 4.9518 | 192 | | 192.11 | .29023 | 137.00 | 189.89 | 5.2541 |
| 143 | | 143.01 | .10375 | 101.97 | 395.60 | 4.9588 | 193 | | 193.12 | .29554 | 137.72 | 187.44 | 5.2593 |
| 144 | | 144.02 | .10631 | 102.68 | 388.79 | 4.9658 | 194 | | 194.12 | .30092 | 138.43 | 185.05 | 5.2644 |
| 145 | −128.15 | 145.02 | .10891 | 103.40 | 382.15 | 4.9727 | 195 | −78.15 | 195.12 | .30637 | 139.15 | 182.69 | 5.2696 |
| 146 | | 146.02 | .11155 | 104.11 | 375.66 | 4.9796 | 196 | | 196.12 | .31189 | 139.86 | 180.38 | 5.2747 |
| 147 | | 147.02 | .11424 | 104.83 | 369.33 | 4.9864 | 197 | | 197.12 | .31748 | 140.58 | 178.11 | 5.2798 |
| 148 | | 148.02 | .11698 | 105.54 | 363.15 | 4.9932 | 198 | | 198.13 | .32314 | 141.29 | 175.87 | 5.2849 |
| 149 | | 149.03 | .11976 | 106.26 | 357.11 | 5.0000 | 199 | | 199.13 | .32888 | 142.01 | 173.68 | 5.2899 |

# Table 1 Air at Low Pressures (for One Kilogram)

| T | t | h | $p_r$ | u | $v_r$ | φ | T | t | h | $p_r$ | u | $v_r$ | φ |
|---|---|---|---|---|---|---|---|---|---|---|---|---|---|
| 200 | −73.15 | 200.13 | .33468 | 142.72 | 171.52 | 5.2950 | 250 | −23.15 | 250.25 | .7296 | 178.49 | 98.353 | 5.5186 |
| 201 | | 201.13 | .34056 | 143.44 | 169.41 | 5.3000 | 251 | | 251.25 | .7398 | 179.21 | 97.379 | 5.5226 |
| 202 | | 202.13 | .34652 | 144.15 | 167.32 | 5.3049 | 252 | | 252.26 | .7502 | 179.93 | 96.418 | 5.5266 |
| 203 | | 203.14 | .35254 | 144.87 | 165.28 | 5.3099 | 253 | | 253.26 | .7606 | 180.64 | 95.471 | 5.5306 |
| 204 | | 204.14 | .35864 | 145.58 | 163.27 | 5.3148 | 254 | | 254.26 | .7712 | 181.36 | 94.536 | 5.5346 |
| 205 | −68.15 | 205.14 | .36482 | 146.30 | 161.29 | 5.3197 | 255 | −18.15 | 255.27 | .7819 | 182.07 | 93.614 | 5.5385 |
| 206 | | 206.14 | .37107 | 147.02 | 159.35 | 5.3246 | 256 | | 256.27 | .7926 | 182.79 | 92.705 | 5.5424 |
| 207 | | 207.15 | .37740 | 147.73 | 157.43 | 5.3294 | 257 | | 257.27 | .8035 | 183.51 | 91.808 | 5.5463 |
| 208 | | 208.15 | .38380 | 148.45 | 155.55 | 5.3343 | 258 | | 258.28 | .8145 | 184.22 | 90.923 | 5.5502 |
| 209 | | 209.15 | .39029 | 149.16 | 153.71 | 5.3391 | 259 | | 259.28 | .8256 | 184.94 | 90.049 | 5.5541 |
| 210 | −63.15 | 210.15 | .39684 | 149.88 | 151.89 | 5.3439 | 260 | −13.15 | 260.28 | .8368 | 185.65 | 89.188 | 5.5580 |
| 211 | | 211.15 | .40348 | 150.59 | 150.10 | 5.3486 | 261 | | 261.28 | .8481 | 186.37 | 88.338 | 5.5618 |
| 212 | | 212.16 | .41020 | 151.31 | 148.34 | 5.3534 | 262 | | 262.29 | .8595 | 187.09 | 87.499 | 5.5657 |
| 213 | | 213.16 | .41700 | 152.02 | 146.61 | 5.3581 | 263 | | 263.29 | .8710 | 187.80 | 86.671 | 5.5695 |
| 214 | | 214.16 | .42387 | 152.74 | 144.91 | 5.3628 | 264 | | 264.29 | .8826 | 188.52 | 85.855 | 5.5733 |
| 215 | −58.15 | 215.16 | .43083 | 153.45 | 143.24 | 5.3674 | 265 | −8.15 | 265.30 | .8944 | 189.23 | 85.049 | 5.5771 |
| 216 | | 216.17 | .43787 | 154.17 | 141.59 | 5.3721 | 266 | | 266.30 | .9062 | 189.95 | 84.253 | 5.5809 |
| 217 | | 217.17 | .44499 | 154.88 | 139.97 | 5.3767 | 267 | | 267.30 | .9182 | 190.67 | 83.468 | 5.5846 |
| 218 | | 218.17 | .45219 | 155.60 | 138.38 | 5.3813 | 268 | | 268.31 | .9302 | 191.38 | 82.693 | 5.5884 |
| 219 | | 219.17 | .45947 | 156.31 | 136.81 | 5.3859 | 269 | | 269.31 | .9424 | 192.10 | 81.928 | 5.5921 |
| 220 | −53.15 | 220.18 | .46684 | 157.03 | 135.26 | 5.3905 | 270 | −3.15 | 270.31 | .9547 | 192.81 | 81.173 | 5.5958 |
| 221 | | 221.18 | .47429 | 157.74 | 133.74 | 5.3950 | 271 | | 271.32 | .9672 | 193.53 | 80.427 | 5.5996 |
| 222 | | 222.18 | .48183 | 158.46 | 132.25 | 5.3996 | 272 | | 272.32 | .9797 | 194.25 | 79.691 | 5.6032 |
| 223 | | 223.18 | .48945 | 159.17 | 130.77 | 5.4041 | 273 | | 273.32 | .9923 | 194.96 | 78.965 | 5.6069 |
| 224 | | 224.18 | .49716 | 159.89 | 129.32 | 5.4085 | 274 | | 274.33 | 1.0051 | 195.68 | 78.248 | 5.6106 |
| 225 | −48.15 | 225.19 | .50495 | 160.61 | 127.90 | 5.4130 | 275 | 1.85 | 275.33 | 1.0180 | 196.40 | 77.539 | 5.6143 |
| 226 | | 226.19 | .51284 | 161.32 | 126.49 | 5.4175 | 276 | | 276.33 | 1.0310 | 197.11 | 76.840 | 5.6179 |
| 227 | | 227.19 | .52080 | 162.04 | 125.11 | 5.4219 | 277 | | 277.34 | 1.0441 | 197.83 | 76.149 | 5.6215 |
| 228 | | 228.19 | .52886 | 162.75 | 123.74 | 5.4263 | 278 | | 278.34 | 1.0573 | 198.55 | 75.467 | 5.6251 |
| 229 | | 229.20 | .53701 | 163.47 | 122.40 | 5.4307 | 279 | | 279.34 | 1.0707 | 199.26 | 74.794 | 5.6287 |
| 230 | −43.15 | 230.20 | .54524 | 164.18 | 121.08 | 5.4350 | 280 | 6.85 | 280.35 | 1.0842 | 199.98 | 74.129 | 5.6323 |
| 231 | | 231.20 | .55357 | 164.90 | 119.78 | 5.4394 | 281 | | 281.35 | 1.0978 | 200.70 | 73.472 | 5.6359 |
| 232 | | 232.20 | .56198 | 165.61 | 118.49 | 5.4437 | 282 | | 282.36 | 1.1115 | 201.41 | 72.823 | 5.6395 |
| 233 | | 233.21 | .57049 | 166.33 | 117.23 | 5.4480 | 283 | | 283.36 | 1.1253 | 202.13 | 72.182 | 5.6430 |
| 234 | | 234.21 | .57908 | 167.04 | 115.99 | 5.4523 | 284 | | 284.36 | 1.1393 | 202.85 | 71.549 | 5.6466 |
| 235 | −38.15 | 235.21 | .58777 | 167.76 | 114.76 | 5.4566 | 285 | 11.85 | 285.37 | 1.1534 | 203.56 | 70.924 | 5.6501 |
| 236 | | 236.21 | .59656 | 168.48 | 113.55 | 5.4609 | 286 | | 286.37 | 1.1676 | 204.28 | 70.306 | 5.6536 |
| 237 | | 237.22 | .60543 | 169.19 | 112.36 | 5.4651 | 287 | | 287.37 | 1.1820 | 205.00 | 69.696 | 5.6571 |
| 238 | | 238.22 | .61440 | 169.91 | 111.19 | 5.4693 | 288 | | 288.38 | 1.1964 | 205.71 | 69.093 | 5.6606 |
| 239 | | 239.22 | .62347 | 170.62 | 110.03 | 5.4735 | 289 | | 289.38 | 1.2110 | 206.43 | 68.497 | 5.6641 |
| 240 | −33.15 | 240.22 | .63263 | 171.34 | 108.89 | 5.4777 | 290 | 16.85 | 290.39 | 1.2258 | 207.15 | 67.909 | 5.6676 |
| 241 | | 241.23 | .64188 | 172.05 | 107.77 | 5.4819 | 291 | | 291.39 | 1.2406 | 207.86 | 67.327 | 5.6710 |
| 242 | | 242.23 | .65123 | 172.77 | 106.66 | 5.4860 | 292 | | 292.40 | 1.2556 | 208.58 | 66.753 | 5.6745 |
| 243 | | 243.23 | .66068 | 173.48 | 105.57 | 5.4902 | 293 | | 293.40 | 1.2707 | 209.30 | 66.185 | 5.6779 |
| 244 | | 244.24 | .67023 | 174.20 | 104.50 | 5.4943 | 294 | | 294.40 | 1.2859 | 210.02 | 65.624 | 5.6813 |
| 245 | −28.15 | 245.24 | .67987 | 174.92 | 103.44 | 5.4984 | 295 | 21.85 | 295.41 | 1.3013 | 210.73 | 65.069 | 5.6847 |
| 246 | | 246.24 | .68962 | 175.63 | 102.39 | 5.5025 | 296 | | 296.41 | 1.3168 | 211.45 | 64.521 | 5.6881 |
| 247 | | 247.24 | .69946 | 176.35 | 101.36 | 5.5065 | 297 | | 297.42 | 1.3324 | 212.17 | 63.980 | 5.6915 |
| 248 | | 248.25 | .70940 | 177.06 | 100.34 | 5.5106 | 298 | | 298.42 | 1.3482 | 212.89 | 63.445 | 5.6949 |
| 249 | | 249.25 | .71945 | 177.78 | 99.34 | 5.5146 | 299 | | 299.43 | 1.3641 | 213.60 | 62.916 | 5.6983 |

# Table 1  Air at Low Pressures (for One Kilogram)

| T | t | h | p_r | u | v_r | φ | T | t | h | p_r | u | v_r | φ |
|---|---|---|---|---|---|---|---|---|---|---|---|---|---|
| 300 | 26.85 | 300.43 | 1.3801 | 214.32 | 62.393 | 5.7016 | 350 | 76.85 | 350.73 | 2.3689 | 250.27 | 42.407 | 5.8567 |
| 301 | | 301.43 | 1.3963 | 215.04 | 61.876 | 5.7050 | 351 | | 351.74 | 2.3928 | 250.99 | 42.105 | 5.8596 |
| 302 | | 302.44 | 1.4126 | 215.76 | 61.365 | 5.7083 | 352 | | 352.75 | 2.4168 | 251.71 | 41.805 | 5.8624 |
| 303 | | 303.44 | 1.4290 | 216.47 | 60.860 | 5.7116 | 353 | | 353.76 | 2.4410 | 252.43 | 41.508 | 5.8653 |
| 304 | | 304.45 | 1.4456 | 217.19 | 60.360 | 5.7149 | 354 | | 354.76 | 2.4654 | 253.16 | 41.214 | 5.8681 |
| 305 | 31.85 | 305.45 | 1.4623 | 217.91 | 59.867 | 5.7182 | 355 | 81.85 | 355.77 | 2.4900 | 253.88 | 40.923 | 5.8710 |
| 306 | | 306.46 | 1.4792 | 218.63 | 59.379 | 5.7215 | 356 | | 356.78 | 2.5147 | 254.60 | 40.635 | 5.8738 |
| 307 | | 307.46 | 1.4962 | 219.34 | 58.896 | 5.7248 | 357 | | 357.79 | 2.5396 | 255.32 | 40.349 | 5.8767 |
| 308 | | 308.47 | 1.5133 | 220.06 | 58.419 | 5.7281 | 358 | | 358.80 | 2.5647 | 256.04 | 40.067 | 5.8795 |
| 309 | | 309.47 | 1.5306 | 220.78 | 57.947 | 5.7313 | 359 | | 359.81 | 2.5899 | 256.76 | 39.787 | 5.8823 |
| 310 | 36.85 | 310.48 | 1.5480 | 221.50 | 57.481 | 5.7346 | 360 | 86.85 | 360.81 | 2.6154 | 257.48 | 39.509 | 5.8851 |
| 311 | | 311.48 | 1.5655 | 222.22 | 57.020 | 5.7378 | 361 | | 361.82 | 2.6410 | 258.21 | 39.235 | 5.8879 |
| 312 | | 312.49 | 1.5832 | 222.93 | 56.564 | 5.7410 | 362 | | 362.83 | 2.6668 | 258.93 | 38.963 | 5.8907 |
| 313 | | 313.49 | 1.6011 | 223.65 | 56.112 | 5.7442 | 363 | | 363.84 | 2.6928 | 259.65 | 38.693 | 5.8935 |
| 314 | | 314.50 | 1.6191 | 224.37 | 55.666 | 5.7474 | 364 | | 364.85 | 2.7190 | 260.37 | 38.426 | 5.8962 |
| 315 | 41.85 | 315.50 | 1.6372 | 225.09 | 55.225 | 5.7506 | 365 | 91.85 | 365.86 | 2.7453 | 261.09 | 38.162 | 5.8990 |
| 316 | | 316.51 | 1.6555 | 225.81 | 54.789 | 5.7538 | 366 | | 366.87 | 2.7718 | 261.82 | 37.900 | 5.9018 |
| 317 | | 317.51 | 1.6739 | 226.53 | 54.357 | 5.7570 | 367 | | 367.88 | 2.7986 | 262.54 | 37.641 | 5.9045 |
| 318 | | 318.52 | 1.6925 | 227.24 | 53.930 | 5.7602 | 368 | | 368.89 | 2.8255 | 263.26 | 37.384 | 5.9073 |
| 319 | | 319.53 | 1.7112 | 227.96 | 53.508 | 5.7633 | 369 | | 369.90 | 2.8526 | 263.98 | 37.129 | 5.9100 |
| 320 | 46.85 | 320.53 | 1.7301 | 228.68 | 53.091 | 5.7665 | 370 | 96.85 | 370.91 | 2.8799 | 264.71 | 36.877 | 5.9127 |
| 321 | | 321.54 | 1.7491 | 229.40 | 52.677 | 5.7696 | 371 | | 371.92 | 2.9073 | 265.43 | 36.628 | 5.9155 |
| 322 | | 322.54 | 1.7682 | 230.12 | 52.269 | 5.7727 | 372 | | 372.93 | 2.9350 | 266.15 | 36.380 | 5.9182 |
| 323 | | 323.55 | 1.7876 | 230.84 | 51.865 | 5.7759 | 373 | | 373.94 | 2.9629 | 266.87 | 36.135 | 5.9209 |
| 324 | | 324.55 | 1.8070 | 231.56 | 51.465 | 5.7790 | 374 | | 374.95 | 2.9909 | 267.60 | 35.892 | 5.9236 |
| 325 | 51.85 | 325.56 | 1.8267 | 232.27 | 51.069 | 5.7821 | 375 | 101.85 | 375.96 | 3.0191 | 268.32 | 35.651 | 5.9263 |
| 326 | | 326.57 | 1.8464 | 232.99 | 50.677 | 5.7852 | 376 | | 376.97 | 3.0476 | 269.04 | 35.413 | 5.9290 |
| 327 | | 327.57 | 1.8664 | 233.71 | 50.290 | 5.7882 | 377 | | 377.98 | 3.0762 | 269.77 | 35.177 | 5.9317 |
| 328 | | 328.58 | 1.8864 | 234.43 | 49.907 | 5.7913 | 378 | | 378.99 | 3.1050 | 270.49 | 34.943 | 5.9344 |
| 329 | | 329.58 | 1.9067 | 235.15 | 49.528 | 5.7944 | 379 | | 380.00 | 3.1340 | 271.21 | 34.711 | 5.9370 |
| 330 | 56.85 | 330.59 | 1.9271 | 235.87 | 49.152 | 5.7974 | 380 | 106.85 | 381.01 | 3.1633 | 271.94 | 34.481 | 5.9397 |
| 331 | | 331.60 | 1.9476 | 236.59 | 48.781 | 5.8005 | 381 | | 382.02 | 3.1927 | 272.66 | 34.253 | 5.9423 |
| 332 | | 332.60 | 1.9683 | 237.31 | 48.414 | 5.8035 | 382 | | 383.03 | 3.2223 | 273.39 | 34.027 | 5.9450 |
| 333 | | 333.61 | 1.9892 | 238.03 | 48.050 | 5.8065 | 383 | | 384.04 | 3.2521 | 274.11 | 33.804 | 5.9476 |
| 334 | | 334.62 | 2.0102 | 238.75 | 47.690 | 5.8096 | 384 | | 385.05 | 3.2821 | 274.83 | 33.582 | 5.9503 |
| 335 | 61.85 | 335.62 | 2.0314 | 239.47 | 47.334 | 5.8126 | 385 | 111.85 | 386.06 | 3.3123 | 275.56 | 33.362 | 5.9529 |
| 336 | | 336.63 | 2.0528 | 240.19 | 46.982 | 5.8156 | 386 | | 387.08 | 3.3427 | 276.28 | 33.145 | 5.9555 |
| 337 | | 337.64 | 2.0743 | 240.91 | 46.633 | 5.8186 | 387 | | 388.09 | 3.3733 | 277.01 | 32.929 | 5.9581 |
| 338 | | 338.64 | 2.0959 | 241.63 | 46.288 | 5.8215 | 388 | | 389.10 | 3.4042 | 277.73 | 32.715 | 5.9608 |
| 339 | | 339.65 | 2.1178 | 242.35 | 45.946 | 5.8245 | 389 | | 390.11 | 3.4352 | 278.46 | 32.503 | 5.9634 |
| 340 | 66.85 | 340.66 | 2.1398 | 243.07 | 45.608 | 5.8275 | 390 | 116.85 | 391.12 | 3.4664 | 279.18 | 32.293 | 5.9660 |
| 341 | | 341.66 | 2.1619 | 243.79 | 45.273 | 5.8304 | 391 | | 392.13 | 3.4979 | 279.91 | 32.085 | 5.9685 |
| 342 | | 342.67 | 2.1843 | 244.51 | 44.942 | 5.8334 | 392 | | 393.15 | 3.5295 | 280.63 | 31.879 | 5.9711 |
| 343 | | 343.68 | 2.2068 | 245.23 | 44.614 | 5.8363 | 393 | | 394.16 | 3.5614 | 281.36 | 31.674 | 5.9737 |
| 344 | | 344.69 | 2.2294 | 245.95 | 44.289 | 5.8393 | 394 | | 395.17 | 3.5934 | 282.08 | 31.471 | 5.9763 |
| 345 | 71.85 | 345.69 | 2.2523 | 246.67 | 43.967 | 5.8422 | 395 | 121.85 | 396.18 | 3.6257 | 282.81 | 31.270 | 5.9788 |
| 346 | | 346.70 | 2.2753 | 247.39 | 43.649 | 5.8451 | 396 | | 397.20 | 3.6582 | 283.53 | 31.071 | 5.9814 |
| 347 | | 347.71 | 2.2984 | 248.11 | 43.334 | 5.8480 | 397 | | 398.21 | 3.6909 | 284.26 | 30.874 | 5.9840 |
| 348 | | 348.72 | 2.3218 | 248.83 | 43.022 | 5.8509 | 398 | | 399.22 | 3.7238 | 284.98 | 30.678 | 5.9865 |
| 349 | | 349.72 | 2.3453 | 249.55 | 42.713 | 5.8538 | 399 | | 400.23 | 3.7569 | 285.71 | 30.484 | 5.9891 |

# Table 1   Air at Low Pressures (for One Kilogram)

| T | t | h | $p_r$ | u | $v_r$ | $\phi$ | T | t | h | $p_r$ | u | $v_r$ | $\phi$ |
|---|---|---|---|---|---|---|---|---|---|---|---|---|---|
| 400 | 126.85 | 401.25 | 3.7902 | 286.43 | 30.292 | 5.9916 | 450 | 176.85 | 452.07 | 5.7519 | 322.91 | 22.456 | 6.1113 |
| 401 | | 402.26 | 3.8238 | 287.16 | 30.101 | 5.9941 | 451 | | 453.09 | 5.7975 | 323.64 | 22.329 | 6.1136 |
| 402 | | 403.27 | 3.8576 | 287.89 | 29.912 | 5.9966 | 452 | | 454.11 | 5.8433 | 324.38 | 22.203 | 6.1158 |
| 403 | | 404.29 | 3.8916 | 288.61 | 29.724 | 5.9992 | 453 | | 455.14 | 5.8894 | 325.11 | 22.078 | 6.1181 |
| 404 | | 405.30 | 3.9258 | 289.34 | 29.538 | 6.0017 | 454 | | 456.16 | 5.9358 | 325.84 | 21.954 | 6.1203 |
| 405 | 131.85 | 406.31 | 3.9602 | 290.07 | 29.354 | 6.0042 | 455 | 181.85 | 457.18 | 5.9824 | 326.58 | 21.831 | 6.1226 |
| 406 | | 407.33 | 3.9948 | 290.79 | 29.171 | 6.0067 | 456 | | 458.20 | 6.0293 | 327.31 | 21.708 | 6.1248 |
| 407 | | 408.34 | 4.0297 | 291.52 | 28.990 | 6.0092 | 457 | | 459.22 | 6.0765 | 328.05 | 21.587 | 6.1271 |
| 408 | | 409.36 | 4.0648 | 292.25 | 28.811 | 6.0117 | 458 | | 460.24 | 6.1240 | 328.78 | 21.467 | 6.1293 |
| 409 | | 410.37 | 4.1001 | 292.97 | 28.632 | 6.0141 | 459 | | 461.26 | 6.1717 | 329.52 | 21.347 | 6.1315 |
| 410 | 136.85 | 411.38 | 4.1356 | 293.70 | 28.456 | 6.0166 | 460 | 186.85 | 462.28 | 6.2197 | 330.25 | 21.228 | 6.1338 |
| 411 | | 412.40 | 4.1714 | 294.43 | 28.281 | 6.0191 | 461 | | 463.31 | 6.2680 | 330.99 | 21.111 | 6.1360 |
| 412 | | 413.41 | 4.2074 | 295.16 | 28.107 | 6.0216 | 462 | | 464.33 | 6.3165 | 331.72 | 20.994 | 6.1382 |
| 413 | | 414.43 | 4.2436 | 295.88 | 27.935 | 6.0240 | 463 | | 465.35 | 6.3653 | 332.46 | 20.878 | 6.1404 |
| 414 | | 415.44 | 4.2800 | 296.61 | 27.764 | 6.0265 | 464 | | 466.37 | 6.4145 | 333.19 | 20.763 | 6.1426 |
| 415 | 141.85 | 416.46 | 4.3167 | 297.34 | 27.595 | 6.0289 | 465 | 191.85 | 467.40 | 6.4638 | 333.93 | 20.649 | 6.1448 |
| 416 | | 417.47 | 4.3536 | 298.07 | 27.427 | 6.0314 | 466 | | 468.42 | 6.5135 | 334.66 | 20.535 | 6.1470 |
| 417 | | 418.49 | 4.3907 | 298.80 | 27.260 | 6.0338 | 467 | | 469.44 | 6.5635 | 335.40 | 20.423 | 6.1492 |
| 418 | | 419.50 | 4.4281 | 299.53 | 27.095 | 6.0362 | 468 | | 470.47 | 6.6137 | 336.13 | 20.311 | 6.1514 |
| 419 | | 420.52 | 4.4657 | 300.25 | 26.931 | 6.0387 | 469 | | 471.49 | 6.6642 | 336.87 | 20.200 | 6.1536 |
| 420 | 146.85 | 421.54 | 4.5035 | 300.98 | 26.769 | 6.0411 | 470 | 196.85 | 472.51 | 6.7150 | 337.61 | 20.090 | 6.1557 |
| 421 | | 422.55 | 4.5416 | 301.71 | 26.608 | 6.0435 | 471 | | 473.54 | 6.7661 | 338.34 | 19.981 | 6.1579 |
| 422 | | 423.57 | 4.5799 | 302.44 | 26.448 | 6.0459 | 472 | | 474.56 | 6.8175 | 339.08 | 19.872 | 6.1601 |
| 423 | | 424.58 | 4.6184 | 303.17 | 26.289 | 6.0483 | 473 | | 475.58 | 6.8692 | 339.82 | 19.764 | 6.1623 |
| 424 | | 425.60 | 4.6572 | 303.90 | 26.132 | 6.0507 | 474 | | 476.61 | 6.9212 | 340.56 | 19.658 | 6.1644 |
| 425 | 151.85 | 426.62 | 4.6962 | 304.63 | 25.976 | 6.0531 | 475 | 201.85 | 477.63 | 6.9734 | 341.29 | 19.551 | 6.1666 |
| 426 | | 427.63 | 4.7354 | 305.36 | 25.821 | 6.0555 | 476 | | 478.66 | 7.0260 | 342.03 | 19.446 | 6.1687 |
| 427 | | 428.65 | 4.7749 | 306.09 | 25.668 | 6.0579 | 477 | | 479.68 | 7.0788 | 342.77 | 19.341 | 6.1709 |
| 428 | | 429.67 | 4.8147 | 306.82 | 25.516 | 6.0603 | 478 | | 480.71 | 7.1319 | 343.51 | 19.238 | 6.1730 |
| 429 | | 430.68 | 4.8546 | 307.55 | 25.365 | 6.0626 | 479 | | 481.73 | 7.1854 | 344.24 | 19.134 | 6.1752 |
| 430 | 156.85 | 431.70 | 4.8949 | 308.28 | 25.215 | 6.0650 | 480 | 206.85 | 482.76 | 7.2391 | 344.98 | 19.032 | 6.1773 |
| 431 | | 432.72 | 4.9353 | 309.01 | 25.066 | 6.0674 | 481 | | 483.78 | 7.2931 | 345.72 | 18.930 | 6.1794 |
| 432 | | 433.73 | 4.9760 | 309.74 | 24.919 | 6.0697 | 482 | | 484.81 | 7.3475 | 346.46 | 18.830 | 6.1816 |
| 433 | | 434.75 | 5.0170 | 310.47 | 24.773 | 6.0721 | 483 | | 485.83 | 7.4021 | 347.20 | 18.729 | 6.1837 |
| 434 | | 435.77 | 5.0582 | 311.20 | 24.628 | 6.0744 | 484 | | 486.86 | 7.4570 | 347.94 | 18.630 | 6.1858 |
| 435 | 161.85 | 436.79 | 5.0996 | 311.93 | 24.484 | 6.0768 | 485 | 211.85 | 487.89 | 7.5123 | 348.68 | 18.531 | 6.1879 |
| 436 | | 437.81 | 5.1413 | 312.66 | 24.341 | 6.0791 | 486 | | 488.91 | 7.5678 | 349.42 | 18.433 | 6.1901 |
| 437 | | 438.82 | 5.1833 | 313.39 | 24.199 | 6.0814 | 487 | | 489.94 | 7.6236 | 350.16 | 18.336 | 6.1922 |
| 438 | | 439.84 | 5.2255 | 314.12 | 24.059 | 6.0838 | 488 | | 490.97 | 7.6798 | 350.90 | 18.239 | 6.1943 |
| 439 | | 440.86 | 5.2679 | 314.85 | 23.920 | 6.0861 | 489 | | 491.99 | 7.7362 | 351.64 | 18.143 | 6.1964 |
| 440 | 166.85 | 441.88 | 5.3106 | 315.59 | 23.781 | 6.0884 | 490 | 216.85 | 493.02 | 7.7930 | 352.38 | 18.048 | 6.1985 |
| 441 | | 442.90 | 5.3536 | 316.32 | 23.644 | 6.0907 | 491 | | 494.05 | 7.8501 | 353.12 | 17.953 | 6.2006 |
| 442 | | 443.92 | 5.3968 | 317.05 | 23.508 | 6.0930 | 492 | | 495.08 | 7.9075 | 353.86 | 17.859 | 6.2027 |
| 443 | | 444.94 | 5.4403 | 317.78 | 23.373 | 6.0953 | 493 | | 496.10 | 7.9652 | 354.60 | 17.766 | 6.2047 |
| 444 | | 445.95 | 5.4840 | 318.51 | 23.239 | 6.0976 | 494 | | 497.13 | 8.0232 | 355.34 | 17.673 | 6.2068 |
| 445 | 171.85 | 446.97 | 5.5280 | 319.25 | 23.106 | 6.0999 | 495 | 221.85 | 498.16 | 8.0815 | 356.08 | 17.581 | 6.2089 |
| 446 | | 447.99 | 5.5723 | 319.98 | 22.974 | 6.1022 | 496 | | 499.19 | 8.1401 | 356.82 | 17.490 | 6.2110 |
| 447 | | 449.01 | 5.6168 | 320.71 | 22.843 | 6.1045 | 497 | | 500.22 | 8.1991 | 357.56 | 17.399 | 6.2131 |
| 448 | | 450.03 | 5.6616 | 321.44 | 22.713 | 6.1068 | 498 | | 501.24 | 8.2584 | 358.30 | 17.309 | 6.2151 |
| 449 | | 451.05 | 5.7066 | 322.18 | 22.584 | 6.1090 | 499 | | 502.27 | 8.3180 | 359.04 | 17.219 | 6.2172 |

# Table 1    Air at Low Pressures (for One Kilogram)

| T | t | h | $p_r$ | u | $v_r$ | $\phi$ | T | t | h | $p_r$ | u | $v_r$ | $\phi$ |
|---|---|---|---|---|---|---|---|---|---|---|---|---|---|
| 500 | 226.85 | 503.30 | 8.378 | 359.79 | 17.130 | 6.2193 | 550 | 276.85 | 555.01 | 11.810 | 397.15 | 13.367 | 6.3178 |
| 501 | | 504.33 | 8.438 | 360.53 | 17.042 | 6.2213 | 551 | | 556.05 | 11.888 | 397.90 | 13.303 | 6.3197 |
| 502 | | 505.36 | 8.499 | 361.27 | 16.954 | 6.2234 | 552 | | 557.09 | 11.967 | 398.65 | 13.240 | 6.3216 |
| 503 | | 506.39 | 8.560 | 362.01 | 16.867 | 6.2254 | 553 | | 558.13 | 12.045 | 399.40 | 13.178 | 6.3235 |
| 504 | | 507.42 | 8.621 | 362.76 | 16.781 | 6.2275 | 554 | | 559.17 | 12.125 | 400.16 | 13.115 | 6.3253 |
| 505 | 231.85 | 508.45 | 8.682 | 363.50 | 16.695 | 6.2295 | 555 | 281.85 | 560.21 | 12.204 | 400.91 | 13.053 | 6.3272 |
| 506 | | 509.48 | 8.744 | 364.24 | 16.610 | 6.2315 | 556 | | 561.25 | 12.284 | 401.66 | 12.992 | 6.3291 |
| 507 | | 510.51 | 8.806 | 364.99 | 16.525 | 6.2336 | 557 | | 562.29 | 12.364 | 402.42 | 12.931 | 6.3310 |
| 508 | | 511.54 | 8.869 | 365.73 | 16.441 | 6.2356 | 558 | | 563.33 | 12.445 | 403.17 | 12.870 | 6.3328 |
| 509 | | 512.57 | 8.932 | 366.47 | 16.357 | 6.2376 | 559 | | 564.38 | 12.526 | 403.93 | 12.809 | 6.3347 |
| 510 | 236.85 | 513.60 | 8.995 | 367.22 | 16.274 | 6.2396 | 560 | 286.85 | 565.42 | 12.608 | 404.68 | 12.749 | 6.3366 |
| 511 | | 514.64 | 9.058 | 367.96 | 16.192 | 6.2417 | 561 | | 566.46 | 12.689 | 405.43 | 12.690 | 6.3384 |
| 512 | | 515.67 | 9.122 | 368.71 | 16.110 | 6.2437 | 562 | | 567.50 | 12.772 | 406.19 | 12.630 | 6.3403 |
| 513 | | 516.70 | 9.187 | 369.45 | 16.029 | 6.2457 | 563 | | 568.54 | 12.854 | 406.94 | 12.571 | 6.3421 |
| 514 | | 517.73 | 9.251 | 370.20 | 15.948 | 6.2477 | 564 | | 569.59 | 12.938 | 407.70 | 12.513 | 6.3440 |
| 515 | 241.85 | 518.76 | 9.316 | 370.94 | 15.868 | 6.2497 | 565 | 291.85 | 570.63 | 13.021 | 408.46 | 12.455 | 6.3458 |
| 516 | | 519.79 | 9.381 | 371.69 | 15.788 | 6.2517 | 566 | | 571.67 | 13.105 | 409.21 | 12.397 | 6.3477 |
| 517 | | 520.83 | 9.447 | 372.43 | 15.709 | 6.2537 | 567 | | 572.71 | 13.189 | 409.97 | 12.339 | 6.3495 |
| 518 | | 521.86 | 9.513 | 373.18 | 15.630 | 6.2557 | 568 | | 573.76 | 13.274 | 410.72 | 12.282 | 6.3513 |
| 519 | | 522.89 | 9.579 | 373.92 | 15.552 | 6.2577 | 569 | | 574.80 | 13.359 | 411.48 | 12.225 | 6.3532 |
| 520 | 246.85 | 523.93 | 9.645 | 374.67 | 15.474 | 6.2597 | 570 | 296.85 | 575.84 | 13.445 | 412.24 | 12.169 | 6.3550 |
| 521 | | 524.96 | 9.712 | 375.42 | 15.397 | 6.2617 | 571 | | 576.89 | 13.531 | 412.99 | 12.113 | 6.3568 |
| 522 | | 525.99 | 9.780 | 376.16 | 15.321 | 6.2637 | 572 | | 577.93 | 13.617 | 413.75 | 12.057 | 6.3587 |
| 523 | | 527.03 | 9.847 | 376.91 | 15.244 | 6.2656 | 573 | | 578.98 | 13.704 | 414.51 | 12.002 | 6.3605 |
| 524 | | 528.06 | 9.915 | 377.65 | 15.169 | 6.2676 | 574 | | 580.02 | 13.791 | 415.27 | 11.946 | 6.3623 |
| 525 | 251.85 | 529.09 | 9.984 | 378.40 | 15.094 | 6.2696 | 575 | 301.85 | 581.07 | 13.879 | 416.02 | 11.892 | 6.3641 |
| 526 | | 530.13 | 10.052 | 379.15 | 15.019 | 6.2716 | 576 | | 582.11 | 13.967 | 416.78 | 11.837 | 6.3660 |
| 527 | | 531.16 | 10.121 | 379.90 | 14.945 | 6.2735 | 577 | | 583.16 | 14.056 | 417.54 | 11.783 | 6.3678 |
| 528 | | 532.20 | 10.191 | 380.64 | 14.872 | 6.2755 | 578 | | 584.20 | 14.144 | 418.30 | 11.729 | 6.3696 |
| 529 | | 533.23 | 10.261 | 381.39 | 14.798 | 6.2774 | 579 | | 585.25 | 14.234 | 419.06 | 11.676 | 6.3714 |
| 530 | 256.85 | 534.27 | 10.331 | 382.14 | 14.726 | 6.2794 | 580 | 306.85 | 586.29 | 14.324 | 419.82 | 11.623 | 6.3732 |
| 531 | | 535.30 | 10.401 | 382.89 | 14.654 | 6.2813 | 581 | | 587.34 | 14.414 | 420.58 | 11.570 | 6.3750 |
| 532 | | 536.34 | 10.472 | 383.64 | 14.582 | 6.2833 | 582 | | 588.39 | 14.504 | 421.33 | 11.517 | 6.3768 |
| 533 | | 537.37 | 10.543 | 384.39 | 14.511 | 6.2852 | 583 | | 589.43 | 14.596 | 422.09 | 11.465 | 6.3786 |
| 534 | | 538.41 | 10.615 | 385.13 | 14.440 | 6.2872 | 584 | | 590.48 | 14.687 | 422.85 | 11.413 | 6.3804 |
| 535 | 261.85 | 539.45 | 10.687 | 385.88 | 14.369 | 6.2891 | 585 | 311.85 | 591.53 | 14.779 | 423.61 | 11.362 | 6.3822 |
| 536 | | 540.48 | 10.759 | 386.63 | 14.300 | 6.2910 | 586 | | 592.57 | 14.871 | 424.37 | 11.310 | 6.3840 |
| 537 | | 541.52 | 10.832 | 387.38 | 14.230 | 6.2930 | 587 | | 593.62 | 14.964 | 425.13 | 11.259 | 6.3857 |
| 538 | | 542.55 | 10.905 | 388.13 | 14.161 | 6.2949 | 588 | | 594.67 | 15.058 | 425.90 | 11.209 | 6.3875 |
| 539 | | 543.59 | 10.978 | 388.88 | 14.093 | 6.2968 | 589 | | 595.72 | 15.151 | 426.66 | 11.158 | 6.3893 |
| 540 | 266.85 | 544.63 | 11.052 | 389.63 | 14.025 | 6.2988 | 590 | 316.85 | 596.77 | 15.245 | 427.42 | 11.108 | 6.3911 |
| 541 | | 545.67 | 11.126 | 390.38 | 13.957 | 6.3007 | 591 | | 597.81 | 15.340 | 428.18 | 11.058 | 6.3929 |
| 542 | | 546.70 | 11.201 | 391.13 | 13.890 | 6.3026 | 592 | | 598.86 | 15.435 | 428.94 | 11.009 | 6.3946 |
| 543 | | 547.74 | 11.275 | 391.88 | 13.823 | 6.3045 | 593 | | 599.91 | 15.531 | 429.70 | 10.960 | 6.3964 |
| 544 | | 548.78 | 11.351 | 392.63 | 13.756 | 6.3064 | 594 | | 600.96 | 15.626 | 430.46 | 10.911 | 6.3982 |
| 545 | 271.85 | 549.82 | 11.426 | 393.39 | 13.690 | 6.3083 | 595 | 321.85 | 602.01 | 15.723 | 431.23 | 10.862 | 6.3999 |
| 546 | | 550.86 | 11.502 | 394.14 | 13.625 | 6.3102 | 596 | | 603.06 | 15.820 | 431.99 | 10.814 | 6.4017 |
| 547 | | 551.89 | 11.579 | 394.89 | 13.560 | 6.3121 | 597 | | 604.11 | 15.917 | 432.75 | 10.766 | 6.4035 |
| 548 | | 552.93 | 11.656 | 395.64 | 13.495 | 6.3140 | 598 | | 605.16 | 16.015 | 433.52 | 10.718 | 6.4052 |
| 549 | | 553.97 | 11.733 | 396.39 | 13.431 | 6.3159 | 599 | | 606.21 | 16.113 | 434.28 | 10.670 | 6.4070 |

## Table 1 Air at Low Pressures (for One Kilogram)

| T | t | h | $p_r$ | u | $v_r$ | $\phi$ | T | t | h | $p_r$ | u | $v_r$ | $\phi$ |
|---|---|---|---|---|---|---|---|---|---|---|---|---|---|
| 600 | 326.85 | 607.26 | 16.212 | 435.04 | 10.623 | 6.4087 | 650 | 376.85 | 660.09 | 21.766 | 473.52 | 8.5718 | 6.4933 |
| 601 | | 608.31 | 16.311 | 435.81 | 10.576 | 6.4105 | 651 | | 661.15 | 21.890 | 474.29 | 8.5363 | 6.4949 |
| 602 | | 609.36 | 16.410 | 436.57 | 10.530 | 6.4122 | 652 | | 662.21 | 22.015 | 475.07 | 8.5009 | 6.4966 |
| 603 | | 610.41 | 16.510 | 437.33 | 10.483 | 6.4140 | 653 | | 663.28 | 22.140 | 475.84 | 8.4658 | 6.4982 |
| 604 | | 611.47 | 16.611 | 438.10 | 10.437 | 6.4157 | 654 | | 664.34 | 22.266 | 476.62 | 8.4308 | 6.4998 |
| 605 | 331.85 | 612.52 | 16.712 | 438.86 | 10.391 | 6.4175 | 655 | 381.85 | 665.40 | 22.392 | 477.40 | 8.3960 | 6.5014 |
| 606 | | 613.57 | 16.813 | 439.63 | 10.345 | 6.4192 | 656 | | 666.47 | 22.519 | 478.17 | 8.3614 | 6.5031 |
| 607 | | 614.62 | 16.915 | 440.39 | 10.300 | 6.4209 | 657 | | 667.53 | 22.647 | 478.95 | 8.3270 | 6.5047 |
| 608 | | 615.67 | 17.018 | 441.16 | 10.255 | 6.4227 | 658 | | 668.59 | 22.775 | 479.73 | 8.2928 | 6.5063 |
| 609 | | 616.73 | 17.121 | 441.92 | 10.210 | 6.4244 | 659 | | 669.66 | 22.903 | 480.51 | 8.2588 | 6.5079 |
| 610 | 336.85 | 617.78 | 17.224 | 442.69 | 10.165 | 6.4261 | 660 | 386.85 | 670.72 | 23.033 | 481.28 | 8.2249 | 6.5095 |
| 611 | | 618.83 | 17.328 | 443.46 | 10.121 | 6.4278 | 661 | | 671.79 | 23.162 | 482.06 | 8.1912 | 6.5111 |
| 612 | | 619.89 | 17.432 | 444.22 | 10.077 | 6.4296 | 662 | | 672.85 | 23.293 | 482.84 | 8.1577 | 6.5128 |
| 613 | | 620.94 | 17.537 | 444.99 | 10.033 | 6.4313 | 663 | | 673.92 | 23.423 | 483.62 | 8.1244 | 6.5144 |
| 614 | | 621.99 | 17.642 | 445.76 | 9.990 | 6.4330 | 664 | | 674.98 | 23.555 | 484.40 | 8.0913 | 6.5160 |
| 615 | 341.85 | 623.05 | 17.748 | 446.52 | 9.946 | 6.4347 | 665 | 391.85 | 676.05 | 23.687 | 485.17 | 8.0583 | 6.5176 |
| 616 | | 624.10 | 17.854 | 447.29 | 9.903 | 6.4364 | 666 | | 677.12 | 23.820 | 485.95 | 8.0255 | 6.5192 |
| 617 | | 625.16 | 17.961 | 448.06 | 9.860 | 6.4381 | 667 | | 678.18 | 23.953 | 486.73 | 7.9928 | 6.5208 |
| 618 | | 626.21 | 18.068 | 448.83 | 9.818 | 6.4398 | 668 | | 679.25 | 24.086 | 487.51 | 7.9604 | 6.5224 |
| 619 | | 627.27 | 18.176 | 449.59 | 9.775 | 6.4416 | 669 | | 680.32 | 24.221 | 488.29 | 7.9281 | 6.5240 |
| 620 | 346.85 | 628.32 | 18.284 | 450.36 | 9.733 | 6.4433 | 670 | 396.85 | 681.38 | 24.356 | 489.07 | 7.8960 | 6.5256 |
| 621 | | 629.38 | 18.393 | 451.13 | 9.691 | 6.4450 | 671 | | 682.45 | 24.491 | 489.85 | 7.8640 | 6.5272 |
| 622 | | 630.43 | 18.502 | 451.90 | 9.650 | 6.4467 | 672 | | 683.52 | 24.627 | 490.63 | 7.8322 | 6.5287 |
| 623 | | 631.49 | 18.611 | 452.67 | 9.608 | 6.4484 | 673 | | 684.59 | 24.764 | 491.41 | 7.8006 | 6.5303 |
| 624 | | 632.54 | 18.722 | 453.44 | 9.567 | 6.4500 | 674 | | 685.65 | 24.901 | 492.20 | 7.7692 | 6.5319 |
| 625 | 351.85 | 633.60 | 18.832 | 454.21 | 9.526 | 6.4517 | 675 | 401.85 | 686.72 | 25.039 | 492.98 | 7.7379 | 6.5335 |
| 626 | | 634.66 | 18.943 | 454.98 | 9.485 | 6.4534 | 676 | | 687.79 | 25.177 | 493.76 | 7.7068 | 6.5351 |
| 627 | | 635.71 | 19.055 | 455.75 | 9.445 | 6.4551 | 677 | | 688.86 | 25.316 | 494.54 | 7.6758 | 6.5367 |
| 628 | | 636.77 | 19.167 | 456.52 | 9.404 | 6.4568 | 678 | | 689.93 | 25.456 | 495.32 | 7.6450 | 6.5382 |
| 629 | | 637.83 | 19.280 | 457.29 | 9.364 | 6.4585 | 679 | | 691.00 | 25.596 | 496.10 | 7.6143 | 6.5398 |
| 630 | 356.85 | 638.89 | 19.393 | 458.06 | 9.324 | 6.4602 | 680 | 406.85 | 692.07 | 25.736 | 496.89 | 7.5839 | 6.5414 |
| 631 | | 639.94 | 19.507 | 458.83 | 9.285 | 6.4618 | 681 | | 693.14 | 25.878 | 497.67 | 7.5535 | 6.5430 |
| 632 | | 641.00 | 19.621 | 459.60 | 9.245 | 6.4635 | 682 | | 694.21 | 26.020 | 498.45 | 7.5233 | 6.5445 |
| 633 | | 642.06 | 19.736 | 460.37 | 9.206 | 6.4652 | 683 | | 695.28 | 26.162 | 499.24 | 7.4933 | 6.5461 |
| 634 | | 643.12 | 19.851 | 461.14 | 9.167 | 6.4669 | 684 | | 696.35 | 26.305 | 500.02 | 7.4635 | 6.5477 |
| 635 | 361.85 | 644.18 | 19.967 | 461.91 | 9.129 | 6.4685 | 685 | 411.85 | 697.42 | 26.449 | 500.80 | 7.4337 | 6.5492 |
| 636 | | 645.24 | 20.083 | 462.68 | 9.090 | 6.4702 | 686 | | 698.49 | 26.594 | 501.59 | 7.4042 | 6.5508 |
| 637 | | 646.30 | 20.200 | 463.46 | 9.052 | 6.4719 | 687 | | 699.56 | 26.739 | 502.37 | 7.3748 | 6.5524 |
| 638 | | 647.35 | 20.317 | 464.23 | 9.013 | 6.4735 | 688 | | 700.63 | 26.884 | 503.16 | 7.3455 | 6.5539 |
| 639 | | 648.41 | 20.435 | 465.00 | 8.976 | 6.4752 | 689 | | 701.71 | 27.030 | 503.94 | 7.3164 | 6.5555 |
| 640 | 366.85 | 649.47 | 20.553 | 465.77 | 8.938 | 6.4768 | 690 | 416.85 | 702.78 | 27.177 | 504.73 | 7.2875 | 6.5570 |
| 641 | | 650.53 | 20.672 | 466.55 | 8.900 | 6.4785 | 691 | | 703.85 | 27.324 | 505.51 | 7.2586 | 6.5586 |
| 642 | | 651.60 | 20.791 | 467.32 | 8.863 | 6.4801 | 692 | | 704.92 | 27.473 | 506.30 | 7.2300 | 6.5601 |
| 643 | | 652.66 | 20.911 | 468.09 | 8.826 | 6.4818 | 693 | | 705.99 | 27.621 | 507.08 | 7.2015 | 6.5617 |
| 644 | | 653.72 | 21.032 | 468.87 | 8.789 | 6.4834 | 694 | | 707.07 | 27.770 | 507.87 | 7.1731 | 6.5632 |
| 645 | 371.85 | 654.78 | 21.153 | 469.64 | 8.752 | 6.4851 | 695 | 421.85 | 708.14 | 27.920 | 508.65 | 7.1448 | 6.5648 |
| 646 | | 655.84 | 21.274 | 470.42 | 8.716 | 6.4867 | 696 | | 709.21 | 28.071 | 509.44 | 7.1168 | 6.5663 |
| 647 | | 656.90 | 21.396 | 471.19 | 8.680 | 6.4884 | 697 | | 710.29 | 28.222 | 510.23 | 7.0888 | 6.5679 |
| 648 | | 657.96 | 21.519 | 471.97 | 8.643 | 6.4900 | 698 | | 711.36 | 28.374 | 511.01 | 7.0610 | 6.5694 |
| 649 | | 659.02 | 21.642 | 472.74 | 8.608 | 6.4917 | 699 | | 712.44 | 28.526 | 511.80 | 7.0333 | 6.5709 |

# Table 1  Air at Low Pressures (for One Kilogram)

| T | t | h | $p_r$ | u | $v_r$ | φ | T | t | h | $p_r$ | u | $v_r$ | φ |
|---|---|---|---|---|---|---|---|---|---|---|---|---|---|
| 700 | 426.85 | 713.51 | 28.679 | 512.59 | 7.0058 | 6.5725 | 750 | 476.85 | 767.53 | 37.184 | 552.26 | 5.7894 | 6.6470 |
| 701 |  | 714.58 | 28.833 | 513.38 | 6.9784 | 6.5740 | 751 |  | 768.62 | 37.372 | 553.06 | 5.7680 | 6.6485 |
| 702 |  | 715.66 | 28.987 | 514.16 | 6.9512 | 6.5755 | 752 |  | 769.71 | 37.561 | 553.86 | 5.7467 | 6.6499 |
| 703 |  | 716.73 | 29.142 | 514.95 | 6.9241 | 6.5771 | 753 |  | 770.79 | 37.750 | 554.66 | 5.7254 | 6.6513 |
| 704 |  | 717.81 | 29.298 | 515.74 | 6.8971 | 6.5786 | 754 |  | 771.88 | 37.940 | 555.46 | 5.7043 | 6.6528 |
| 705 | 431.85 | 718.89 | 29.454 | 516.53 | 6.8702 | 6.5801 | 755 | 481.85 | 772.97 | 38.131 | 556.26 | 5.6832 | 6.6542 |
| 706 |  | 719.96 | 29.611 | 517.32 | 6.8435 | 6.5816 | 756 |  | 774.06 | 38.323 | 557.06 | 5.6623 | 6.6557 |
| 707 |  | 721.04 | 29.769 | 518.11 | 6.8170 | 6.5832 | 757 |  | 775.15 | 38.516 | 557.86 | 5.6414 | 6.6571 |
| 708 |  | 722.11 | 29.927 | 518.90 | 6.7905 | 6.5847 | 758 |  | 776.23 | 38.709 | 558.66 | 5.6207 | 6.6585 |
| 709 |  | 723.19 | 30.086 | 519.68 | 6.7642 | 6.5862 | 759 |  | 777.32 | 38.903 | 559.47 | 5.6000 | 6.6600 |
| 710 | 436.85 | 724.27 | 30.245 | 520.47 | 6.7380 | 6.5877 | 760 | 486.85 | 778.41 | 39.098 | 560.27 | 5.5795 | 6.6614 |
| 711 |  | 725.34 | 30.405 | 521.26 | 6.7120 | 6.5892 | 761 |  | 779.50 | 39.293 | 561.07 | 5.5590 | 6.6628 |
| 712 |  | 726.42 | 30.566 | 522.05 | 6.6861 | 6.5908 | 762 |  | 780.59 | 39.489 | 561.87 | 5.5386 | 6.6643 |
| 713 |  | 727.50 | 30.728 | 522.85 | 6.6603 | 6.5923 | 763 |  | 781.68 | 39.687 | 562.67 | 5.5184 | 6.6657 |
| 714 |  | 728.58 | 30.890 | 523.64 | 6.6346 | 6.5938 | 764 |  | 782.77 | 39.884 | 563.48 | 5.4982 | 6.6671 |
| 715 | 441.85 | 729.65 | 31.052 | 524.43 | 6.6091 | 6.5953 | 765 | 491.85 | 783.86 | 40.083 | 564.28 | 5.4781 | 6.6686 |
| 716 |  | 730.73 | 31.216 | 525.22 | 6.5836 | 6.5968 | 766 |  | 784.95 | 40.282 | 565.08 | 5.4581 | 6.6700 |
| 717 |  | 731.81 | 31.380 | 526.01 | 6.5584 | 6.5983 | 767 |  | 786.04 | 40.482 | 565.89 | 5.4382 | 6.6714 |
| 718 |  | 732.89 | 31.545 | 526.80 | 6.5332 | 6.5998 | 768 |  | 787.13 | 40.683 | 566.69 | 5.4184 | 6.6728 |
| 719 |  | 733.97 | 31.710 | 527.59 | 6.5082 | 6.6013 | 769 |  | 788.22 | 40.885 | 567.49 | 5.3987 | 6.6742 |
| 720 | 446.85 | 735.05 | 31.876 | 528.39 | 6.4832 | 6.6028 | 770 | 496.85 | 789.31 | 41.088 | 568.30 | 5.3791 | 6.6757 |
| 721 |  | 736.13 | 32.043 | 529.18 | 6.4584 | 6.6043 | 771 |  | 790.40 | 41.291 | 569.10 | 5.3596 | 6.6771 |
| 722 |  | 737.21 | 32.211 | 529.97 | 6.4338 | 6.6058 | 772 |  | 791.49 | 41.495 | 569.91 | 5.3401 | 6.6785 |
| 723 |  | 738.29 | 32.379 | 530.76 | 6.4092 | 6.6073 | 773 |  | 792.59 | 41.700 | 570.71 | 5.3208 | 6.6799 |
| 724 |  | 739.37 | 32.548 | 531.56 | 6.3848 | 6.6088 | 774 |  | 793.68 | 41.905 | 571.52 | 5.3015 | 6.6813 |
| 725 | 451.85 | 740.45 | 32.717 | 532.35 | 6.3605 | 6.6103 | 775 | 501.85 | 794.77 | 42.112 | 572.32 | 5.2824 | 6.6827 |
| 726 |  | 741.53 | 32.888 | 533.14 | 6.3363 | 6.6118 | 776 |  | 795.86 | 42.319 | 573.13 | 5.2633 | 6.6841 |
| 727 |  | 742.61 | 33.058 | 533.94 | 6.3122 | 6.6133 | 777 |  | 796.95 | 42.527 | 573.93 | 5.2443 | 6.6855 |
| 728 |  | 743.69 | 33.230 | 534.73 | 6.2883 | 6.6147 | 778 |  | 798.05 | 42.736 | 574.74 | 5.2254 | 6.6869 |
| 729 |  | 744.77 | 33.402 | 535.53 | 6.2644 | 6.6162 | 779 |  | 799.14 | 42.945 | 575.54 | 5.2066 | 6.6884 |
| 730 | 456.85 | 745.85 | 33.575 | 536.32 | 6.2407 | 6.6177 | 780 | 506.85 | 800.23 | 43.156 | 576.35 | 5.1878 | 6.6898 |
| 731 |  | 746.93 | 33.749 | 537.11 | 6.2171 | 6.6192 | 781 |  | 801.33 | 43.367 | 577.16 | 5.1692 | 6.6912 |
| 732 |  | 748.02 | 33.923 | 537.91 | 6.1936 | 6.6207 | 782 |  | 802.42 | 43.579 | 577.96 | 5.1506 | 6.6926 |
| 733 |  | 749.10 | 34.098 | 538.70 | 6.1702 | 6.6221 | 783 |  | 803.52 | 43.792 | 578.77 | 5.1322 | 6.6940 |
| 734 |  | 750.18 | 34.274 | 539.50 | 6.1469 | 6.6236 | 784 |  | 804.61 | 44.005 | 579.58 | 5.1138 | 6.6954 |
| 735 | 461.85 | 751.26 | 34.451 | 540.30 | 6.1238 | 6.6251 | 785 | 511.85 | 805.71 | 44.220 | 580.39 | 5.0955 | 6.6967 |
| 736 |  | 752.35 | 34.628 | 541.09 | 6.1007 | 6.6266 | 786 |  | 806.80 | 44.435 | 581.19 | 5.0772 | 6.6981 |
| 737 |  | 753.43 | 34.806 | 541.89 | 6.0778 | 6.6280 | 787 |  | 807.90 | 44.651 | 582.00 | 5.0591 | 6.6995 |
| 738 |  | 754.51 | 34.984 | 542.68 | 6.0550 | 6.6295 | 788 |  | 808.99 | 44.868 | 582.81 | 5.0410 | 6.7009 |
| 739 |  | 755.60 | 35.164 | 543.48 | 6.0323 | 6.6310 | 789 |  | 810.09 | 45.086 | 583.62 | 5.0231 | 6.7023 |
| 740 | 466.85 | 756.68 | 35.344 | 544.28 | 6.0097 | 6.6324 | 790 | 516.85 | 811.18 | 45.304 | 584.43 | 5.0052 | 6.7037 |
| 741 |  | 757.77 | 35.524 | 545.08 | 5.9872 | 6.6339 | 791 |  | 812.28 | 45.523 | 585.24 | 4.9874 | 6.7051 |
| 742 |  | 758.85 | 35.706 | 545.87 | 5.9648 | 6.6354 | 792 |  | 813.37 | 45.744 | 586.05 | 4.9696 | 6.7065 |
| 743 |  | 759.93 | 35.888 | 546.67 | 5.9425 | 6.6368 | 793 |  | 814.47 | 45.965 | 586.85 | 4.9520 | 6.7079 |
| 744 |  | 761.02 | 36.071 | 547.47 | 5.9203 | 6.6383 | 794 |  | 815.57 | 46.187 | 587.66 | 4.9344 | 6.7092 |
| 745 | 471.85 | 762.10 | 36.255 | 548.27 | 5.8982 | 6.6397 | 795 | 521.85 | 816.66 | 46.409 | 588.47 | 4.9169 | 6.7106 |
| 746 |  | 763.19 | 36.439 | 549.06 | 5.8763 | 6.6412 | 796 |  | 817.76 | 46.633 | 589.28 | 4.8995 | 6.7120 |
| 747 |  | 764.28 | 36.624 | 549.86 | 5.8544 | 6.6427 | 797 |  | 818.86 | 46.857 | 590.09 | 4.8822 | 6.7134 |
| 748 |  | 765.36 | 36.810 | 550.66 | 5.8327 | 6.6441 | 798 |  | 819.96 | 47.082 | 590.91 | 4.8649 | 6.7148 |
| 749 |  | 766.45 | 36.996 | 551.46 | 5.8110 | 6.6456 | 799 |  | 821.05 | 47.308 | 591.72 | 4.8477 | 6.7161 |

## Table 1 Air at Low Pressures (for One Kilogram)

| T | t | h | $p_r$ | u | $v_r$ | $\phi$ | T | t | h | $p_r$ | u | $v_r$ | $\phi$ |
|---|---|---|---|---|---|---|---|---|---|---|---|---|---|
| 800 | 526.85 | 822.15 | 47.535 | 592.53 | 4.8306 | 6.7175 | 850 | 576.85 | 877.35 | 60.017 | 633.37 | 4.0652 | 6.7844 |
| 801 | | 823.25 | 47.763 | 593.34 | 4.8136 | 6.7189 | 851 | | 878.46 | 60.290 | 634.19 | 4.0515 | 6.7857 |
| 802 | | 824.35 | 47.992 | 594.15 | 4.7966 | 6.7202 | 852 | | 879.57 | 60.564 | 635.02 | 4.0379 | 6.7870 |
| 803 | | 825.45 | 48.221 | 594.96 | 4.7798 | 6.7216 | 853 | | 880.68 | 60.840 | 635.84 | 4.0243 | 6.7883 |
| 804 | | 826.55 | 48.452 | 595.77 | 4.7630 | 6.7230 | 854 | | 881.79 | 61.116 | 636.66 | 4.0108 | 6.7896 |
| 805 | 531.85 | 827.65 | 48.683 | 596.59 | 4.7462 | 6.7243 | 855 | 581.85 | 882.90 | 61.394 | 637.49 | 3.9974 | 6.7909 |
| 806 | | 828.75 | 48.915 | 597.40 | 4.7296 | 6.7257 | 856 | | 884.01 | 61.672 | 638.31 | 3.9840 | 6.7922 |
| 807 | | 829.85 | 49.148 | 598.21 | 4.7130 | 6.7271 | 857 | | 885.12 | 61.951 | 639.13 | 3.9706 | 6.7935 |
| 808 | | 830.95 | 49.382 | 599.02 | 4.6965 | 6.7284 | 858 | | 886.23 | 62.232 | 639.96 | 3.9574 | 6.7948 |
| 809 | | 832.05 | 49.616 | 599.84 | 4.6801 | 6.7298 | 859 | | 887.34 | 62.513 | 640.78 | 3.9441 | 6.7961 |
| 810 | 536.85 | 833.15 | 49.852 | 600.65 | 4.6637 | 6.7312 | 860 | 586.85 | 888.45 | 62.795 | 641.61 | 3.9310 | 6.7974 |
| 811 | | 834.25 | 50.088 | 601.46 | 4.6475 | 6.7325 | 861 | | 889.57 | 63.078 | 642.43 | 3.9179 | 6.7987 |
| 812 | | 835.35 | 50.325 | 602.28 | 4.6312 | 6.7339 | 862 | | 890.68 | 63.363 | 643.26 | 3.9048 | 6.8000 |
| 813 | | 836.45 | 50.564 | 603.09 | 4.6151 | 6.7352 | 863 | | 891.79 | 63.648 | 644.08 | 3.8918 | 6.8013 |
| 814 | | 837.55 | 50.803 | 603.91 | 4.5990 | 6.7366 | 864 | | 892.90 | 63.934 | 644.91 | 3.8789 | 6.8026 |
| 815 | 541.85 | 838.65 | 51.043 | 604.72 | 4.5831 | 6.7379 | 865 | 591.85 | 894.02 | 64.222 | 645.73 | 3.8660 | 6.8039 |
| 816 | | 839.75 | 51.283 | 605.54 | 4.5671 | 6.7393 | 866 | | 895.13 | 64.510 | 646.56 | 3.8532 | 6.8051 |
| 817 | | 840.85 | 51.525 | 606.35 | 4.5513 | 6.7406 | 867 | | 896.24 | 64.800 | 647.39 | 3.8404 | 6.8064 |
| 818 | | 841.96 | 51.768 | 607.17 | 4.5355 | 6.7420 | 868 | | 897.36 | 65.090 | 648.21 | 3.8277 | 6.8077 |
| 819 | | 843.06 | 52.011 | 607.98 | 4.5198 | 6.7433 | 869 | | 898.47 | 65.381 | 649.04 | 3.8150 | 6.8090 |
| 820 | 546.85 | 844.16 | 52.256 | 608.80 | 4.5041 | 6.7447 | 870 | 596.85 | 899.58 | 65.674 | 649.87 | 3.8024 | 6.8103 |
| 821 | | 845.26 | 52.501 | 609.61 | 4.4886 | 6.7460 | 871 | | 900.70 | 65.967 | 650.69 | 3.7898 | 6.8116 |
| 822 | | 846.37 | 52.747 | 610.43 | 4.4730 | 6.7474 | 872 | | 901.81 | 66.262 | 651.52 | 3.7773 | 6.8128 |
| 823 | | 847.47 | 52.994 | 611.24 | 4.4576 | 6.7487 | 873 | | 902.93 | 66.557 | 652.35 | 3.7649 | 6.8141 |
| 824 | | 848.58 | 53.242 | 612.06 | 4.4422 | 6.7500 | 874 | | 904.04 | 66.854 | 653.17 | 3.7524 | 6.8154 |
| 825 | 551.85 | 849.68 | 53.491 | 612.88 | 4.4269 | 6.7514 | 875 | 601.85 | 905.15 | 67.152 | 654.00 | 3.7401 | 6.8167 |
| 826 | | 850.78 | 53.741 | 613.70 | 4.4117 | 6.7527 | 876 | | 906.27 | 67.450 | 654.83 | 3.7278 | 6.8179 |
| 827 | | 851.89 | 53.992 | 614.51 | 4.3965 | 6.7541 | 877 | | 907.39 | 67.750 | 655.66 | 3.7155 | 6.8192 |
| 828 | | 852.99 | 54.243 | 615.33 | 4.3814 | 6.7554 | 878 | | 908.50 | 68.051 | 656.49 | 3.7033 | 6.8205 |
| 829 | | 854.10 | 54.496 | 616.15 | 4.3664 | 6.7567 | 879 | | 909.62 | 68.352 | 657.32 | 3.6912 | 6.8218 |
| 830 | 556.85 | 855.20 | 54.749 | 616.97 | 4.3514 | 6.7581 | 880 | 606.85 | 910.73 | 68.655 | 658.15 | 3.6791 | 6.8230 |
| 831 | | 856.31 | 55.004 | 617.78 | 4.3365 | 6.7594 | 881 | | 911.85 | 68.959 | 658.97 | 3.6670 | 6.8243 |
| 832 | | 857.41 | 55.259 | 618.60 | 4.3216 | 6.7607 | 882 | | 912.97 | 69.264 | 659.80 | 3.6550 | 6.8256 |
| 833 | | 858.52 | 55.515 | 619.42 | 4.3069 | 6.7620 | 883 | | 914.08 | 69.570 | 660.63 | 3.6431 | 6.8268 |
| 834 | | 859.62 | 55.773 | 620.24 | 4.2921 | 6.7634 | 884 | | 915.20 | 69.877 | 661.46 | 3.6312 | 6.8281 |
| 835 | 561.85 | 860.73 | 56.031 | 621.06 | 4.2775 | 6.7647 | 885 | 611.85 | 916.32 | 70.185 | 662.29 | 3.6193 | 6.8293 |
| 836 | | 861.84 | 56.290 | 621.88 | 4.2629 | 6.7660 | 886 | | 917.43 | 70.495 | 663.12 | 3.6075 | 6.8306 |
| 837 | | 862.94 | 56.550 | 622.70 | 4.2484 | 6.7673 | 887 | | 918.55 | 70.805 | 663.95 | 3.5958 | 6.8319 |
| 838 | | 864.05 | 56.811 | 623.52 | 4.2339 | 6.7687 | 888 | | 919.67 | 71.116 | 664.78 | 3.5841 | 6.8331 |
| 839 | | 865.16 | 57.073 | 624.34 | 4.2195 | 6.7700 | 889 | | 920.79 | 71.429 | 665.62 | 3.5724 | 6.8344 |
| 840 | 566.85 | 866.26 | 57.336 | 625.16 | 4.2052 | 6.7713 | 890 | 616.85 | 921.90 | 71.742 | 666.45 | 3.5608 | 6.8356 |
| 841 | | 867.37 | 57.599 | 625.98 | 4.1909 | 6.7726 | 891 | | 923.02 | 72.057 | 667.28 | 3.5492 | 6.8369 |
| 842 | | 868.48 | 57.864 | 626.80 | 4.1767 | 6.7739 | 892 | | 924.14 | 72.372 | 668.11 | 3.5377 | 6.8382 |
| 843 | | 869.59 | 58.130 | 627.62 | 4.1625 | 6.7753 | 893 | | 925.26 | 72.689 | 668.94 | 3.5262 | 6.8394 |
| 844 | | 870.69 | 58.396 | 628.44 | 4.1484 | 6.7766 | 894 | | 926.38 | 73.007 | 669.77 | 3.5148 | 6.8407 |
| 845 | 571.85 | 871.80 | 58.664 | 629.26 | 4.1344 | 6.7779 | 895 | 621.85 | 927.50 | 73.326 | 670.61 | 3.5034 | 6.8419 |
| 846 | | 872.91 | 58.933 | 630.08 | 4.1204 | 6.7792 | 896 | | 928.62 | 73.646 | 671.44 | 3.4921 | 6.8432 |
| 847 | | 874.02 | 59.202 | 630.90 | 4.1065 | 6.7805 | 897 | | 929.74 | 73.967 | 672.27 | 3.4808 | 6.8444 |
| 848 | | 875.13 | 59.473 | 631.73 | 4.0927 | 6.7818 | 898 | | 930.86 | 74.289 | 673.10 | 3.4696 | 6.8457 |
| 849 | | 876.24 | 59.744 | 632.55 | 4.0789 | 6.7831 | 899 | | 931.98 | 74.613 | 673.94 | 3.4584 | 6.8469 |

# Table 1    Air at Low Pressures (for One Kilogram)

| T | t | h | $p_r$ | u | $v_r$ | $\phi$ | T | t | h | $p_r$ | u | $v_r$ | $\phi$ |
|---|---|---|---|---|---|---|---|---|---|---|---|---|---|
| 900 | 626.85 | 933.10 | 74.937 | 674.77 | 3.4473 | 6.8482 | 950 | 676.85 | 989.38 | 92.63 | 716.70 | 2.9436 | 6.9090 |
| 901 | | 934.22 | 75.263 | 675.60 | 3.4362 | 6.8494 | 951 | | 990.51 | 93.02 | 717.54 | 2.9345 | 6.9102 |
| 902 | | 935.34 | 75.589 | 676.44 | 3.4251 | 6.8506 | 952 | | 991.64 | 93.41 | 718.39 | 2.9255 | 6.9114 |
| 903 | | 936.46 | 75.917 | 677.27 | 3.4141 | 6.8519 | 953 | | 992.77 | 93.79 | 719.23 | 2.9164 | 6.9126 |
| 904 | | 937.58 | 76.246 | 678.11 | 3.4031 | 6.8531 | 954 | | 993.90 | 94.18 | 720.07 | 2.9074 | 6.9138 |
| 905 | 631.85 | 938.70 | 76.576 | 678.94 | 3.3922 | 6.8544 | 955 | 681.85 | 995.03 | 94.57 | 720.92 | 2.8985 | 6.9149 |
| 906 | | 939.82 | 76.907 | 679.77 | 3.3813 | 6.8556 | 956 | | 996.17 | 94.96 | 721.76 | 2.8896 | 6.9161 |
| 907 | | 940.95 | 77.240 | 680.61 | 3.3705 | 6.8568 | 957 | | 997.30 | 95.35 | 722.61 | 2.8807 | 6.9173 |
| 908 | | 942.07 | 77.573 | 681.44 | 3.3597 | 6.8581 | 958 | | 998.43 | 95.75 | 723.45 | 2.8719 | 6.9185 |
| 909 | | 943.19 | 77.908 | 682.28 | 3.3490 | 6.8593 | 959 | | 999.56 | 96.14 | 724.30 | 2.8631 | 6.9197 |
| 910 | 636.85 | 944.31 | 78.243 | 683.11 | 3.3383 | 6.8605 | 960 | 686.85 | 1000.69 | 96.54 | 725.14 | 2.8543 | 6.9209 |
| 911 | | 945.44 | 78.580 | 683.95 | 3.3276 | 6.8618 | 961 | | 1001.83 | 96.94 | 725.99 | 2.8455 | 6.9220 |
| 912 | | 946.56 | 78.918 | 684.79 | 3.3170 | 6.8630 | 962 | | 1002.96 | 97.34 | 726.84 | 2.8368 | 6.9232 |
| 913 | | 947.68 | 79.257 | 685.62 | 3.3064 | 6.8642 | 963 | | 1004.09 | 97.74 | 727.68 | 2.8281 | 6.9244 |
| 914 | | 948.80 | 79.597 | 686.46 | 3.2959 | 6.8655 | 964 | | 1005.23 | 98.14 | 728.53 | 2.8195 | 6.9256 |
| 915 | 641.85 | 949.93 | 79.939 | 687.29 | 3.2854 | 6.8667 | 965 | 691.85 | 1006.36 | 98.54 | 729.38 | 2.8109 | 6.9267 |
| 916 | | 951.05 | 80.281 | 688.13 | 3.2750 | 6.8679 | 966 | | 1007.49 | 98.94 | 730.22 | 2.8023 | 6.9279 |
| 917 | | 952.18 | 80.625 | 688.97 | 3.2646 | 6.8692 | 967 | | 1008.63 | 99.35 | 731.07 | 2.7938 | 6.9291 |
| 918 | | 953.30 | 80.970 | 689.80 | 3.2542 | 6.8704 | 968 | | 1009.76 | 99.76 | 731.92 | 2.7853 | 6.9303 |
| 919 | | 954.42 | 81.316 | 690.64 | 3.2439 | 6.8716 | 969 | | 1010.90 | 100.16 | 732.76 | 2.7768 | 6.9314 |
| 920 | 646.85 | 955.55 | 81.663 | 691.48 | 3.2336 | 6.8728 | 970 | 696.85 | 1012.03 | 100.57 | 733.61 | 2.7683 | 6.9326 |
| 921 | | 956.67 | 82.012 | 692.32 | 3.2234 | 6.8740 | 971 | | 1013.17 | 100.98 | 734.46 | 2.7599 | 6.9338 |
| 922 | | 957.80 | 82.361 | 693.15 | 3.2132 | 6.8753 | 972 | | 1014.30 | 101.40 | 735.31 | 2.7516 | 6.9349 |
| 923 | | 958.92 | 82.712 | 693.99 | 3.2031 | 6.8765 | 973 | | 1015.44 | 101.81 | 736.16 | 2.7432 | 6.9361 |
| 924 | | 960.05 | 83.064 | 694.83 | 3.1929 | 6.8777 | 974 | | 1016.57 | 102.22 | 737.00 | 2.7349 | 6.9373 |
| 925 | 651.85 | 961.17 | 83.417 | 695.67 | 3.1829 | 6.8789 | 975 | 701.85 | 1017.71 | 102.64 | 737.85 | 2.7266 | 6.9384 |
| 926 | | 962.30 | 83.771 | 696.51 | 3.1728 | 6.8801 | 976 | | 1018.84 | 103.06 | 738.70 | 2.7184 | 6.9396 |
| 927 | | 963.43 | 84.126 | 697.35 | 3.1628 | 6.8814 | 977 | | 1019.98 | 103.47 | 739.55 | 2.7101 | 6.9408 |
| 928 | | 964.55 | 84.483 | 698.19 | 3.1529 | 6.8826 | 978 | | 1021.12 | 103.89 | 740.40 | 2.7019 | 6.9419 |
| 929 | | 965.68 | 84.841 | 699.03 | 3.1430 | 6.8838 | 979 | | 1022.25 | 104.32 | 741.25 | 2.6938 | 6.9431 |
| 930 | 656.85 | 966.80 | 85.200 | 699.87 | 3.1331 | 6.8850 | 980 | 706.85 | 1023.39 | 104.74 | 742.10 | 2.6857 | 6.9443 |
| 931 | | 967.93 | 85.560 | 700.70 | 3.1233 | 6.8862 | 981 | | 1024.52 | 105.16 | 742.95 | 2.6776 | 6.9454 |
| 932 | | 969.06 | 85.921 | 701.54 | 3.1135 | 6.8874 | 982 | | 1025.66 | 105.59 | 743.80 | 2.6695 | 6.9466 |
| 933 | | 970.18 | 86.284 | 702.38 | 3.1037 | 6.8886 | 983 | | 1026.80 | 106.01 | 744.65 | 2.6615 | 6.9477 |
| 934 | | 971.31 | 86.648 | 703.23 | 3.0940 | 6.8898 | 984 | | 1027.94 | 106.44 | 745.50 | 2.6535 | 6.9489 |
| 935 | 661.85 | 972.44 | 87.013 | 704.07 | 3.0843 | 6.8910 | 985 | 711.85 | 1029.07 | 106.87 | 746.35 | 2.6455 | 6.9500 |
| 936 | | 973.57 | 87.379 | 704.91 | 3.0747 | 6.8922 | 986 | | 1030.21 | 107.30 | 747.20 | 2.6375 | 6.9512 |
| 937 | | 974.70 | 87.746 | 705.75 | 3.0651 | 6.8934 | 987 | | 1031.35 | 107.73 | 748.05 | 2.6296 | 6.9523 |
| 938 | | 975.82 | 88.115 | 706.59 | 3.0555 | 6.8946 | 988 | | 1032.49 | 108.17 | 748.90 | 2.6217 | 6.9535 |
| 939 | | 976.95 | 88.485 | 707.43 | 3.0460 | 6.8959 | 989 | | 1033.63 | 108.60 | 749.75 | 2.6139 | 6.9547 |
| 940 | 666.85 | 978.08 | 88.856 | 708.27 | 3.0365 | 6.8971 | 990 | 716.85 | 1034.76 | 109.04 | 750.60 | 2.6061 | 6.9558 |
| 941 | | 979.21 | 89.228 | 709.11 | 3.0270 | 6.8983 | 991 | | 1035.90 | 109.48 | 751.45 | 2.5983 | 6.9570 |
| 942 | | 980.34 | 89.602 | 709.96 | 3.0176 | 6.8995 | 992 | | 1037.04 | 109.91 | 752.31 | 2.5905 | 6.9581 |
| 943 | | 981.47 | 89.977 | 710.80 | 3.0082 | 6.9006 | 993 | | 1038.18 | 110.36 | 753.16 | 2.5828 | 6.9592 |
| 944 | | 982.60 | 90.353 | 711.64 | 2.9989 | 6.9018 | 994 | | 1039.32 | 110.80 | 754.01 | 2.5751 | 6.9604 |
| 945 | 671.85 | 983.73 | 90.730 | 712.48 | 2.9896 | 6.9030 | 995 | 721.85 | 1040.46 | 111.24 | 754.86 | 2.5674 | 6.9615 |
| 946 | | 984.86 | 91.108 | 713.33 | 2.9803 | 6.9042 | 996 | | 1041.60 | 111.68 | 755.72 | 2.5597 | 6.9627 |
| 947 | | 985.99 | 91.488 | 714.17 | 2.9711 | 6.9054 | 997 | | 1042.74 | 112.13 | 756.57 | 2.5521 | 6.9638 |
| 948 | | 987.12 | 91.869 | 715.01 | 2.9619 | 6.9066 | 998 | | 1043.88 | 112.58 | 757.42 | 2.5445 | 6.9650 |
| 949 | | 988.25 | 92.251 | 715.85 | 2.9527 | 6.9078 | 999 | | 1045.02 | 113.03 | 758.27 | 2.5370 | 6.9661 |

# Table 1 Air at Low Pressures (for One Kilogram)

| T | t | h | $p_r$ | u | $v_r$ | $\phi$ | T | t | h | $p_r$ | u | $v_r$ | $\phi$ |
|---|---|---|---|---|---|---|---|---|---|---|---|---|---|
| 1000 | 726.85 | 1046.16 | 113.48 | 759.13 | 2.5294 | 6.9673 | 1050 | 776.85 | 1103.41 | 137.86 | 802.03 | 2.1862 | 7.0231 |
| 1001 | | 1047.30 | 113.93 | 759.98 | 2.5219 | 6.9684 | 1051 | | 1104.56 | 138.39 | 802.89 | 2.1799 | 7.0242 |
| 1002 | | 1048.44 | 114.38 | 760.83 | 2.5144 | 6.9695 | 1052 | | 1105.71 | 138.91 | 803.76 | 2.1737 | 7.0253 |
| 1003 | | 1049.58 | 114.84 | 761.69 | 2.5070 | 6.9707 | 1053 | | 1106.86 | 139.44 | 804.62 | 2.1675 | 7.0264 |
| 1004 | | 1050.72 | 115.29 | 762.54 | 2.4996 | 6.9718 | 1054 | | 1108.01 | 139.98 | 805.48 | 2.1613 | 7.0275 |
| 1005 | 731.85 | 1051.86 | 115.75 | 763.40 | 2.4922 | 6.9729 | 1055 | 781.85 | 1109.16 | 140.51 | 806.34 | 2.1552 | 7.0286 |
| 1006 | | 1053.00 | 116.21 | 764.25 | 2.4848 | 6.9741 | 1056 | | 1110.31 | 141.04 | 807.21 | 2.1490 | 7.0297 |
| 1007 | | 1054.15 | 116.67 | 765.11 | 2.4775 | 6.9752 | 1057 | | 1111.46 | 141.58 | 808.07 | 2.1429 | 7.0308 |
| 1008 | | 1055.29 | 117.13 | 765.96 | 2.4702 | 6.9763 | 1058 | | 1112.62 | 142.12 | 808.94 | 2.1368 | 7.0319 |
| 1009 | | 1056.43 | 117.59 | 766.82 | 2.4629 | 6.9775 | 1059 | | 1113.77 | 142.66 | 809.80 | 2.1308 | 7.0329 |
| 1010 | 736.85 | 1057.57 | 118.06 | 767.67 | 2.4556 | 6.9786 | 1060 | 786.85 | 1114.92 | 143.20 | 810.66 | 2.1247 | 7.0340 |
| 1011 | | 1058.72 | 118.52 | 768.53 | 2.4484 | 6.9797 | 1061 | | 1116.07 | 143.74 | 811.53 | 2.1187 | 7.0351 |
| 1012 | | 1059.86 | 118.99 | 769.38 | 2.4412 | 6.9809 | 1062 | | 1117.22 | 144.28 | 812.39 | 2.1127 | 7.0362 |
| 1013 | | 1061.00 | 119.46 | 770.24 | 2.4340 | 6.9820 | 1063 | | 1118.37 | 144.83 | 813.26 | 2.1067 | 7.0373 |
| 1014 | | 1062.14 | 119.93 | 771.09 | 2.4268 | 6.9831 | 1064 | | 1119.52 | 145.38 | 814.12 | 2.1007 | 7.0384 |
| 1015 | 741.85 | 1063.29 | 120.40 | 771.95 | 2.4197 | 6.9843 | 1065 | 791.85 | 1120.68 | 145.93 | 814.99 | 2.0948 | 7.0394 |
| 1016 | | 1064.43 | 120.87 | 772.81 | 2.4126 | 6.9854 | 1066 | | 1121.83 | 146.48 | 815.85 | 2.0889 | 7.0405 |
| 1017 | | 1065.57 | 121.35 | 773.66 | 2.4056 | 6.9865 | 1067 | | 1122.98 | 147.03 | 816.72 | 2.0830 | 7.0416 |
| 1018 | | 1066.72 | 121.82 | 774.52 | 2.3985 | 6.9876 | 1068 | | 1124.13 | 147.59 | 817.58 | 2.0771 | 7.0427 |
| 1019 | | 1067.86 | 122.30 | 775.38 | 2.3915 | 6.9888 | 1069 | | 1125.29 | 148.14 | 818.45 | 2.0712 | 7.0438 |
| 1020 | 746.85 | 1069.01 | 122.78 | 776.23 | 2.3845 | 6.9899 | 1070 | 796.85 | 1126.44 | 148.70 | 819.32 | 2.0654 | 7.0448 |
| 1021 | | 1070.15 | 123.26 | 777.09 | 2.3775 | 6.9910 | 1071 | | 1127.59 | 149.26 | 820.18 | 2.0596 | 7.0459 |
| 1022 | | 1071.29 | 123.74 | 777.95 | 2.3706 | 6.9921 | 1072 | | 1128.75 | 149.82 | 821.05 | 2.0538 | 7.0470 |
| 1023 | | 1072.44 | 124.23 | 778.81 | 2.3637 | 6.9932 | 1073 | | 1129.90 | 150.38 | 821.92 | 2.0480 | 7.0481 |
| 1024 | | 1073.58 | 124.71 | 779.66 | 2.3568 | 6.9944 | 1074 | | 1131.05 | 150.94 | 822.78 | 2.0423 | 7.0491 |
| 1025 | 751.85 | 1074.73 | 125.20 | 780.52 | 2.3499 | 6.9955 | 1075 | 801.85 | 1132.21 | 151.51 | 823.65 | 2.0365 | 7.0502 |
| 1026 | | 1075.87 | 125.69 | 781.38 | 2.3431 | 6.9966 | 1076 | | 1133.36 | 152.08 | 824.52 | 2.0308 | 7.0513 |
| 1027 | | 1077.02 | 126.18 | 782.24 | 2.3363 | 6.9977 | 1077 | | 1134.52 | 152.65 | 825.38 | 2.0251 | 7.0524 |
| 1028 | | 1078.16 | 126.67 | 783.10 | 2.3295 | 6.9988 | 1078 | | 1135.67 | 153.22 | 826.25 | 2.0195 | 7.0534 |
| 1029 | | 1079.31 | 127.16 | 783.96 | 2.3227 | 6.9999 | 1079 | | 1136.83 | 153.79 | 827.12 | 2.0138 | 7.0545 |
| 1030 | 756.85 | 1080.46 | 127.65 | 784.81 | 2.3160 | 7.0010 | 1080 | 806.85 | 1137.98 | 154.37 | 827.99 | 2.0082 | 7.0556 |
| 1031 | | 1081.60 | 128.15 | 785.67 | 2.3092 | 7.0022 | 1081 | | 1139.13 | 154.94 | 828.85 | 2.0026 | 7.0566 |
| 1032 | | 1082.75 | 128.65 | 786.53 | 2.3025 | 7.0033 | 1082 | | 1140.29 | 155.52 | 829.72 | 1.9970 | 7.0577 |
| 1033 | | 1083.90 | 129.15 | 787.39 | 2.2959 | 7.0044 | 1083 | | 1141.45 | 156.10 | 830.59 | 1.9914 | 7.0588 |
| 1034 | | 1085.04 | 129.65 | 788.25 | 2.2892 | 7.0055 | 1084 | | 1142.60 | 156.68 | 831.46 | 1.9859 | 7.0599 |
| 1035 | 761.85 | 1086.19 | 130.15 | 789.11 | 2.2826 | 7.0066 | 1085 | 811.85 | 1143.76 | 157.26 | 832.33 | 1.9803 | 7.0609 |
| 1036 | | 1087.34 | 130.65 | 789.97 | 2.2760 | 7.0077 | 1086 | | 1144.91 | 157.85 | 833.20 | 1.9748 | 7.0620 |
| 1037 | | 1088.48 | 131.16 | 790.83 | 2.2694 | 7.0088 | 1087 | | 1146.07 | 158.43 | 834.06 | 1.9693 | 7.0630 |
| 1038 | | 1089.63 | 131.66 | 791.69 | 2.2629 | 7.0099 | 1088 | | 1147.22 | 159.02 | 834.93 | 1.9638 | 7.0641 |
| 1039 | | 1090.78 | 132.17 | 792.55 | 2.2564 | 7.0110 | 1089 | | 1148.38 | 159.61 | 835.80 | 1.9584 | 7.0652 |
| 1040 | 766.85 | 1091.93 | 132.68 | 793.41 | 2.2499 | 7.0121 | 1090 | 816.85 | 1149.54 | 160.20 | 836.67 | 1.9529 | 7.0662 |
| 1041 | | 1093.07 | 133.19 | 794.27 | 2.2434 | 7.0132 | 1091 | | 1150.69 | 160.79 | 837.54 | 1.9475 | 7.0673 |
| 1042 | | 1094.22 | 133.70 | 795.13 | 2.2370 | 7.0143 | 1092 | | 1151.85 | 161.39 | 838.41 | 1.9421 | 7.0684 |
| 1043 | | 1095.37 | 134.22 | 796.00 | 2.2305 | 7.0154 | 1093 | | 1153.01 | 161.99 | 839.28 | 1.9368 | 7.0694 |
| 1044 | | 1096.52 | 134.73 | 796.86 | 2.2241 | 7.0165 | 1094 | | 1154.16 | 162.58 | 840.15 | 1.9314 | 7.0705 |
| 1045 | 771.85 | 1097.67 | 135.25 | 797.72 | 2.2177 | 7.0176 | 1095 | 821.85 | 1155.32 | 163.18 | 841.02 | 1.9260 | 7.0715 |
| 1046 | | 1098.82 | 135.77 | 798.58 | 2.2114 | 7.0187 | 1096 | | 1156.48 | 163.79 | 841.89 | 1.9207 | 7.0726 |
| 1047 | | 1099.96 | 136.29 | 799.44 | 2.2050 | 7.0198 | 1097 | | 1157.64 | 164.39 | 842.76 | 1.9154 | 7.0736 |
| 1048 | | 1101.11 | 136.81 | 800.30 | 2.1987 | 7.0209 | 1098 | | 1158.79 | 164.99 | 843.63 | 1.9101 | 7.0747 |
| 1049 | | 1102.26 | 137.33 | 801.17 | 2.1924 | 7.0220 | 1099 | | 1159.95 | 165.60 | 844.51 | 1.9049 | 7.0757 |

# Table 1    Air at Low Pressures (for One Kilogram)

| T | t | h | $p_r$ | u | $v_r$ | $\phi$ | T | t | h | $p_r$ | u | $v_r$ | $\phi$ |
|---|---|---|---|---|---|---|---|---|---|---|---|---|---|
| 1100 | 826.85 | 1161.11 | 166.21 | 845.38 | 1.8996 | 7.0768 | 1150 | 876.85 | 1219.23 | 198.99 | 889.14 | 1.6588 | 7.1285 |
| 1101 | | 1162.27 | 166.82 | 846.25 | 1.8944 | 7.0779 | 1151 | | 1220.39 | 199.69 | 890.02 | 1.6544 | 7.1295 |
| 1102 | | 1163.43 | 167.43 | 847.12 | 1.8892 | 7.0789 | 1152 | | 1221.56 | 200.40 | 890.90 | 1.6500 | 7.1305 |
| 1103 | | 1164.59 | 168.05 | 847.99 | 1.8840 | 7.0800 | 1153 | | 1222.72 | 201.11 | 891.78 | 1.6456 | 7.1315 |
| 1104 | | 1165.74 | 168.66 | 848.86 | 1.8788 | 7.0810 | 1154 | | 1223.89 | 201.82 | 892.66 | 1.6413 | 7.1325 |
| 1105 | 831.85 | 1166.90 | 169.28 | 849.73 | 1.8736 | 7.0821 | 1155 | 881.85 | 1225.06 | 202.53 | 893.54 | 1.6369 | 7.1335 |
| 1106 | | 1168.06 | 169.90 | 850.61 | 1.8685 | 7.0831 | 1156 | | 1226.23 | 203.24 | 894.42 | 1.6326 | 7.1345 |
| 1107 | | 1169.22 | 170.52 | 851.48 | 1.8634 | 7.0842 | 1157 | | 1227.39 | 203.96 | 895.30 | 1.6282 | 7.1355 |
| 1108 | | 1170.38 | 171.14 | 852.35 | 1.8583 | 7.0852 | 1158 | | 1228.56 | 204.68 | 896.18 | 1.6239 | 7.1366 |
| 1109 | | 1171.54 | 171.77 | 853.22 | 1.8532 | 7.0862 | 1159 | | 1229.73 | 205.40 | 897.06 | 1.6196 | 7.1376 |
| 1110 | 836.85 | 1172.70 | 172.40 | 854.10 | 1.8481 | 7.0873 | 1160 | 886.85 | 1230.90 | 206.12 | 897.94 | 1.6154 | 7.1386 |
| 1111 | | 1173.86 | 173.03 | 854.97 | 1.8430 | 7.0883 | 1161 | | 1232.06 | 206.84 | 898.82 | 1.6111 | 7.1396 |
| 1112 | | 1175.02 | 173.66 | 855.84 | 1.8380 | 7.0894 | 1162 | | 1233.23 | 207.57 | 899.70 | 1.6068 | 7.1406 |
| 1113 | | 1176.18 | 174.29 | 856.72 | 1.8330 | 7.0904 | 1163 | | 1234.40 | 208.30 | 900.58 | 1.6026 | 7.1416 |
| 1114 | | 1177.34 | 174.92 | 857.59 | 1.8280 | 7.0915 | 1164 | | 1235.57 | 209.03 | 901.46 | 1.5984 | 7.1426 |
| 1115 | 841.85 | 1178.50 | 175.56 | 858.46 | 1.8230 | 7.0925 | 1165 | 891.85 | 1236.74 | 209.76 | 902.35 | 1.5942 | 7.1436 |
| 1116 | | 1179.66 | 176.19 | 859.34 | 1.8180 | 7.0935 | 1166 | | 1237.91 | 210.49 | 903.23 | 1.5900 | 7.1446 |
| 1117 | | 1180.82 | 176.83 | 860.21 | 1.8131 | 7.0946 | 1167 | | 1239.07 | 211.23 | 904.11 | 1.5858 | 7.1456 |
| 1118 | | 1181.99 | 177.48 | 861.08 | 1.8081 | 7.0956 | 1168 | | 1240.24 | 211.97 | 904.99 | 1.5816 | 7.1466 |
| 1119 | | 1183.15 | 178.12 | 861.96 | 1.8032 | 7.0967 | 1169 | | 1241.41 | 212.71 | 905.87 | 1.5775 | 7.1476 |
| 1120 | 846.85 | 1184.31 | 178.76 | 862.83 | 1.7983 | 7.0977 | 1170 | 896.85 | 1242.58 | 213.45 | 906.76 | 1.5733 | 7.1486 |
| 1121 | | 1185.47 | 179.41 | 863.71 | 1.7934 | 7.0987 | 1171 | | 1243.75 | 214.19 | 907.64 | 1.5692 | 7.1496 |
| 1122 | | 1186.63 | 180.06 | 864.58 | 1.7886 | 7.0998 | 1172 | | 1244.92 | 214.94 | 908.52 | 1.5651 | 7.1506 |
| 1123 | | 1187.79 | 180.71 | 865.46 | 1.7837 | 7.1008 | 1173 | | 1246.09 | 215.69 | 909.40 | 1.5610 | 7.1516 |
| 1124 | | 1188.96 | 181.36 | 866.33 | 1.7789 | 7.1018 | 1174 | | 1247.26 | 216.44 | 910.29 | 1.5569 | 7.1526 |
| 1125 | 851.85 | 1190.12 | 182.02 | 867.21 | 1.7741 | 7.1029 | 1175 | 901.85 | 1248.43 | 217.19 | 911.17 | 1.5528 | 7.1536 |
| 1126 | | 1191.28 | 182.67 | 868.08 | 1.7693 | 7.1039 | 1176 | | 1249.60 | 217.95 | 912.05 | 1.5488 | 7.1546 |
| 1127 | | 1192.44 | 183.33 | 868.96 | 1.7645 | 7.1049 | 1177 | | 1250.77 | 218.70 | 912.94 | 1.5447 | 7.1556 |
| 1128 | | 1193.61 | 183.99 | 869.83 | 1.7597 | 7.1060 | 1178 | | 1251.94 | 219.46 | 913.82 | 1.5407 | 7.1566 |
| 1129 | | 1194.77 | 184.65 | 870.71 | 1.7550 | 7.1070 | 1179 | | 1253.11 | 220.22 | 914.70 | 1.5367 | 7.1576 |
| 1130 | 856.85 | 1195.93 | 185.32 | 871.59 | 1.7502 | 7.1080 | 1180 | 906.85 | 1254.28 | 220.99 | 915.59 | 1.5327 | 7.1586 |
| 1131 | | 1197.09 | 185.98 | 872.46 | 1.7455 | 7.1091 | 1181 | | 1255.45 | 221.75 | 916.47 | 1.5287 | 7.1596 |
| 1132 | | 1198.26 | 186.65 | 873.34 | 1.7408 | 7.1101 | 1182 | | 1256.63 | 222.52 | 917.35 | 1.5247 | 7.1605 |
| 1133 | | 1199.42 | 187.32 | 874.21 | 1.7361 | 7.1111 | 1183 | | 1257.80 | 223.29 | 918.24 | 1.5207 | 7.1615 |
| 1134 | | 1200.58 | 187.99 | 875.09 | 1.7315 | 7.1121 | 1184 | | 1258.97 | 224.06 | 919.12 | 1.5168 | 7.1625 |
| 1135 | 861.85 | 1201.75 | 188.66 | 875.97 | 1.7268 | 7.1132 | 1185 | 911.85 | 1260.14 | 224.83 | 920.01 | 1.5128 | 7.1635 |
| 1136 | | 1202.91 | 189.34 | 876.85 | 1.7222 | 7.1142 | 1186 | | 1261.31 | 225.61 | 920.89 | 1.5089 | 7.1645 |
| 1137 | | 1204.08 | 190.01 | 877.72 | 1.7175 | 7.1152 | 1187 | | 1262.48 | 226.38 | 921.78 | 1.5050 | 7.1655 |
| 1138 | | 1205.24 | 190.69 | 878.60 | 1.7129 | 7.1162 | 1188 | | 1263.65 | 227.16 | 922.66 | 1.5011 | 7.1665 |
| 1139 | | 1206.41 | 191.37 | 879.48 | 1.7083 | 7.1173 | 1189 | | 1264.83 | 227.95 | 923.55 | 1.4972 | 7.1675 |
| 1140 | 866.85 | 1207.57 | 192.06 | 880.35 | 1.7038 | 7.1183 | 1190 | 916.85 | 1266.00 | 228.73 | 924.43 | 1.4933 | 7.1684 |
| 1141 | | 1208.74 | 192.74 | 881.23 | 1.6992 | 7.1193 | 1191 | | 1267.17 | 229.52 | 925.32 | 1.4895 | 7.1694 |
| 1142 | | 1209.90 | 193.43 | 882.11 | 1.6946 | 7.1203 | 1192 | | 1268.34 | 230.30 | 926.20 | 1.4856 | 7.1704 |
| 1143 | | 1211.07 | 194.12 | 882.99 | 1.6901 | 7.1213 | 1193 | | 1269.52 | 231.09 | 927.09 | 1.4818 | 7.1714 |
| 1144 | | 1212.23 | 194.81 | 883.87 | 1.6856 | 7.1224 | 1194 | | 1270.69 | 231.89 | 927.97 | 1.4779 | 7.1724 |
| 1145 | 871.85 | 1213.40 | 195.50 | 884.75 | 1.6811 | 7.1234 | 1195 | 921.85 | 1271.86 | 232.68 | 928.86 | 1.4741 | 7.1734 |
| 1146 | | 1214.56 | 196.19 | 885.62 | 1.6766 | 7.1244 | 1196 | | 1273.04 | 233.48 | 929.75 | 1.4703 | 7.1743 |
| 1147 | | 1215.73 | 196.89 | 886.50 | 1.6721 | 7.1254 | 1197 | | 1274.21 | 234.28 | 930.63 | 1.4665 | 7.1753 |
| 1148 | | 1216.89 | 197.59 | 887.38 | 1.6677 | 7.1264 | 1198 | | 1275.38 | 235.08 | 931.52 | 1.4628 | 7.1763 |
| 1149 | | 1218.06 | 198.29 | 888.26 | 1.6632 | 7.1275 | 1199 | | 1276.56 | 235.88 | 932.41 | 1.4590 | 7.1773 |

## Table 1   Air at Low Pressures (for One Kilogram)

| T | t | h | $p_r$ | u | $v_r$ | $\phi$ | T | t | h | $p_r$ | u | $v_r$ | $\phi$ |
|---|---|---|---|---|---|---|---|---|---|---|---|---|---|
| 1200 | 926.85 | 1277.73 | 236.69 | 933.29 | 1.4552 | 7.1783 | 1250 | 976.85 | 1336.60 | 279.83 | 977.81 | 1.2822 | 7.2263 |
| 1201 | | 1278.90 | 237.50 | 934.18 | 1.4515 | 7.1792 | 1251 | | 1337.78 | 280.76 | 978.71 | 1.2790 | 7.2273 |
| 1202 | | 1280.08 | 238.31 | 935.07 | 1.4478 | 7.1802 | 1252 | | 1338.96 | 281.68 | 979.60 | 1.2758 | 7.2282 |
| 1203 | | 1281.25 | 239.12 | 935.95 | 1.4441 | 7.1812 | 1253 | | 1340.14 | 282.61 | 980.49 | 1.2726 | 7.2292 |
| 1204 | | 1282.43 | 239.93 | 936.84 | 1.4403 | 7.1822 | 1254 | | 1341.33 | 283.54 | 981.39 | 1.2695 | 7.2301 |
| 1205 | 931.85 | 1283.60 | 240.75 | 937.73 | 1.4367 | 7.1831 | 1255 | 981.85 | 1342.51 | 284.47 | 982.28 | 1.2663 | 7.2310 |
| 1206 | | 1284.78 | 241.57 | 938.62 | 1.4330 | 7.1841 | 1256 | | 1343.69 | 285.40 | 983.18 | 1.2632 | 7.2320 |
| 1207 | | 1285.95 | 242.39 | 939.50 | 1.4293 | 7.1851 | 1257 | | 1344.87 | 286.34 | 984.07 | 1.2600 | 7.2329 |
| 1208 | | 1287.13 | 243.21 | 940.39 | 1.4256 | 7.1861 | 1258 | | 1346.05 | 287.28 | 984.97 | 1.2569 | 7.2339 |
| 1209 | | 1288.30 | 244.04 | 941.28 | 1.4220 | 7.1870 | 1259 | | 1347.23 | 288.22 | 985.86 | 1.2538 | 7.2348 |
| 1210 | 936.85 | 1289.48 | 244.86 | 942.17 | 1.4184 | 7.1880 | 1260 | 986.85 | 1348.42 | 289.16 | 986.76 | 1.2507 | 7.2357 |
| 1211 | | 1290.65 | 245.69 | 943.06 | 1.4147 | 7.1890 | 1261 | | 1349.60 | 290.11 | 987.65 | 1.2476 | 7.2367 |
| 1212 | | 1291.83 | 246.53 | 943.95 | 1.4111 | 7.1900 | 1262 | | 1350.78 | 291.06 | 988.55 | 1.2445 | 7.2376 |
| 1213 | | 1293.00 | 247.36 | 944.83 | 1.4075 | 7.1909 | 1263 | | 1351.96 | 292.01 | 989.44 | 1.2415 | 7.2386 |
| 1214 | | 1294.18 | 248.20 | 945.72 | 1.4040 | 7.1919 | 1264 | | 1353.15 | 292.97 | 990.34 | 1.2384 | 7.2395 |
| 1215 | 941.85 | 1295.35 | 249.04 | 946.61 | 1.4004 | 7.1929 | 1265 | 991.85 | 1354.33 | 293.92 | 991.24 | 1.2353 | 7.2404 |
| 1216 | | 1296.53 | 249.88 | 947.50 | 1.3968 | 7.1938 | 1266 | | 1355.51 | 294.88 | 992.13 | 1.2323 | 7.2414 |
| 1217 | | 1297.71 | 250.72 | 948.39 | 1.3933 | 7.1948 | 1267 | | 1356.70 | 295.84 | 993.03 | 1.2293 | 7.2423 |
| 1218 | | 1298.88 | 251.56 | 949.28 | 1.3897 | 7.1958 | 1268 | | 1357.88 | 296.81 | 993.92 | 1.2262 | 7.2432 |
| 1219 | | 1300.06 | 252.41 | 950.17 | 1.3862 | 7.1967 | 1269 | | 1359.06 | 297.77 | 994.82 | 1.2232 | 7.2442 |
| 1220 | 946.85 | 1301.24 | 253.26 | 951.06 | 1.3827 | 7.1977 | 1270 | 996.85 | 1360.25 | 298.74 | 995.72 | 1.2202 | 7.2451 |
| 1221 | | 1302.41 | 254.11 | 951.95 | 1.3792 | 7.1987 | 1271 | | 1361.43 | 299.71 | 996.61 | 1.2172 | 7.2460 |
| 1222 | | 1303.59 | 254.97 | 952.84 | 1.3757 | 7.1996 | 1272 | | 1362.61 | 300.69 | 997.51 | 1.2142 | 7.2470 |
| 1223 | | 1304.77 | 255.83 | 953.73 | 1.3722 | 7.2006 | 1273 | | 1363.80 | 301.66 | 998.41 | 1.2113 | 7.2479 |
| 1224 | | 1305.94 | 256.68 | 954.62 | 1.3687 | 7.2015 | 1274 | | 1364.98 | 302.64 | 999.30 | 1.2083 | 7.2488 |
| 1225 | 951.85 | 1307.12 | 257.55 | 955.51 | 1.3652 | 7.2025 | 1275 | 1001.85 | 1366.17 | 303.62 | 1000.20 | 1.2053 | 7.2497 |
| 1226 | | 1308.30 | 258.41 | 956.40 | 1.3618 | 7.2035 | 1276 | | 1367.35 | 304.61 | 1001.10 | 1.2024 | 7.2507 |
| 1227 | | 1309.48 | 259.28 | 957.29 | 1.3584 | 7.2044 | 1277 | | 1368.54 | 305.59 | 1002.00 | 1.1994 | 7.2516 |
| 1228 | | 1310.65 | 260.14 | 958.18 | 1.3549 | 7.2054 | 1278 | | 1369.72 | 306.58 | 1002.89 | 1.1965 | 7.2525 |
| 1229 | | 1311.83 | 261.01 | 959.07 | 1.3515 | 7.2063 | 1279 | | 1370.90 | 307.57 | 1003.79 | 1.1936 | 7.2535 |
| 1230 | 956.85 | 1313.01 | 261.89 | 959.96 | 1.3481 | 7.2073 | 1280 | 1006.85 | 1372.09 | 308.57 | 1004.69 | 1.1907 | 7.2544 |
| 1231 | | 1314.19 | 262.76 | 960.85 | 1.3447 | 7.2083 | 1281 | | 1373.27 | 309.56 | 1005.59 | 1.1878 | 7.2553 |
| 1232 | | 1315.37 | 263.64 | 961.74 | 1.3413 | 7.2092 | 1282 | | 1374.46 | 310.56 | 1006.49 | 1.1849 | 7.2562 |
| 1233 | | 1316.55 | 264.52 | 962.64 | 1.3379 | 7.2102 | 1283 | | 1375.64 | 311.56 | 1007.38 | 1.1820 | 7.2572 |
| 1234 | | 1317.72 | 265.40 | 963.53 | 1.3346 | 7.2111 | 1284 | | 1376.83 | 312.57 | 1008.28 | 1.1791 | 7.2581 |
| 1235 | 961.85 | 1318.90 | 266.29 | 964.42 | 1.3312 | 7.2121 | 1285 | 1011.85 | 1378.02 | 313.57 | 1009.18 | 1.1762 | 7.2590 |
| 1236 | | 1320.08 | 267.17 | 965.31 | 1.3279 | 7.2130 | 1286 | | 1379.20 | 314.58 | 1010.08 | 1.1734 | 7.2599 |
| 1237 | | 1321.26 | 268.06 | 966.20 | 1.3245 | 7.2140 | 1287 | | 1380.39 | 315.60 | 1010.98 | 1.1705 | 7.2608 |
| 1238 | | 1322.44 | 268.95 | 967.10 | 1.3212 | 7.2149 | 1288 | | 1381.57 | 316.61 | 1011.88 | 1.1677 | 7.2618 |
| 1239 | | 1323.62 | 269.85 | 967.99 | 1.3179 | 7.2159 | 1289 | | 1382.76 | 317.63 | 1012.78 | 1.1648 | 7.2627 |
| 1240 | 966.85 | 1324.80 | 270.74 | 968.88 | 1.3146 | 7.2168 | 1290 | 1016.85 | 1383.95 | 318.65 | 1013.68 | 1.1620 | 7.2636 |
| 1241 | | 1325.98 | 271.64 | 969.77 | 1.3113 | 7.2178 | 1291 | | 1385.13 | 319.67 | 1014.57 | 1.1592 | 7.2645 |
| 1242 | | 1327.16 | 272.54 | 970.67 | 1.3080 | 7.2187 | 1292 | | 1386.32 | 320.69 | 1015.47 | 1.1564 | 7.2654 |
| 1243 | | 1328.34 | 273.44 | 971.56 | 1.3048 | 7.2197 | 1293 | | 1387.50 | 321.72 | 1016.37 | 1.1536 | 7.2664 |
| 1244 | | 1329.52 | 274.35 | 972.45 | 1.3015 | 7.2206 | 1294 | | 1388.69 | 322.75 | 1017.27 | 1.1508 | 7.2673 |
| 1245 | 971.85 | 1330.70 | 275.26 | 973.34 | 1.2983 | 7.2216 | 1295 | 1021.85 | 1389.88 | 323.78 | 1018.17 | 1.1480 | 7.2682 |
| 1246 | | 1331.88 | 276.17 | 974.24 | 1.2950 | 7.2225 | 1296 | | 1391.07 | 324.82 | 1019.07 | 1.1452 | 7.2691 |
| 1247 | | 1333.06 | 277.08 | 975.13 | 1.2918 | 7.2235 | 1297 | | 1392.25 | 325.86 | 1019.97 | 1.1425 | 7.2700 |
| 1248 | | 1334.24 | 278.00 | 976.02 | 1.2886 | 7.2244 | 1298 | | 1393.44 | 326.90 | 1020.87 | 1.1397 | 7.2709 |
| 1249 | | 1335.42 | 278.91 | 976.92 | 1.2854 | 7.2254 | 1299 | | 1394.63 | 327.94 | 1021.77 | 1.1370 | 7.2719 |

# Table 1   Air at Low Pressures (for One Kilogram)

| T | t | h | $p_r$ | u | $v_r$ | $\phi$ | T | t | h | $p_r$ | u | $v_r$ | $\phi$ |
|---|---|---|---|---|---|---|---|---|---|---|---|---|---|
| 1300 | 1026.85 | 1395.81 | 328.98 | 1022.67 | 1.1342 | 7.2728 | 1350 | 1076.85 | 1455.35 | 384.74 | 1067.86 | 1.00716 | 7.3177 |
| 1301 | | 1397.00 | 330.03 | 1023.57 | 1.1315 | 7.2737 | 1351 | | 1456.54 | 385.93 | 1068.76 | 1.00480 | 7.3186 |
| 1302 | | 1398.19 | 331.08 | 1024.47 | 1.1288 | 7.2746 | 1352 | | 1457.74 | 387.12 | 1069.67 | 1.00246 | 7.3195 |
| 1303 | | 1399.38 | 332.14 | 1025.38 | 1.1260 | 7.2755 | 1353 | | 1458.93 | 388.31 | 1070.58 | 1.00012 | 7.3204 |
| 1304 | | 1400.57 | 333.19 | 1026.28 | 1.1233 | 7.2764 | 1354 | | 1460.12 | 389.50 | 1071.48 | .99778 | 7.3212 |
| 1305 | 1031.85 | 1401.75 | 334.25 | 1027.18 | 1.1206 | 7.2773 | 1355 | 1081.85 | 1461.32 | 390.70 | 1072.39 | .99546 | 7.3221 |
| 1306 | | 1402.94 | 335.31 | 1028.08 | 1.1179 | 7.2782 | 1356 | | 1462.51 | 391.90 | 1073.30 | .99314 | 7.3230 |
| 1307 | | 1404.13 | 336.38 | 1028.98 | 1.1153 | 7.2792 | 1357 | | 1463.71 | 393.11 | 1074.21 | .99083 | 7.3239 |
| 1308 | | 1405.32 | 337.45 | 1029.88 | 1.1126 | 7.2801 | 1358 | | 1464.90 | 394.31 | 1075.11 | .98852 | 7.3248 |
| 1309 | | 1406.51 | 338.52 | 1030.78 | 1.1099 | 7.2810 | 1359 | | 1466.10 | 395.52 | 1076.02 | .98622 | 7.3256 |
| 1310 | 1036.85 | 1407.70 | 339.59 | 1031.69 | 1.1073 | 7.2819 | 1360 | 1086.85 | 1467.29 | 396.74 | 1076.93 | .98393 | 7.3265 |
| 1311 | | 1408.88 | 340.66 | 1032.59 | 1.1046 | 7.2828 | 1361 | | 1468.49 | 397.95 | 1077.84 | .98165 | 7.3274 |
| 1312 | | 1410.07 | 341.74 | 1033.49 | 1.1020 | 7.2837 | 1362 | | 1469.68 | 399.17 | 1078.75 | .97937 | 7.3283 |
| 1313 | | 1411.26 | 342.82 | 1034.39 | 1.0993 | 7.2846 | 1363 | | 1470.88 | 400.39 | 1079.65 | .97710 | 7.3292 |
| 1314 | | 1412.45 | 343.90 | 1035.29 | 1.0967 | 7.2855 | 1364 | | 1472.07 | 401.62 | 1080.56 | .97483 | 7.3300 |
| 1315 | 1041.85 | 1413.64 | 344.99 | 1036.20 | 1.0941 | 7.2864 | 1365 | 1091.85 | 1473.27 | 402.85 | 1081.47 | .97257 | 7.3309 |
| 1316 | | 1414.83 | 346.08 | 1037.10 | 1.0915 | 7.2873 | 1366 | | 1474.46 | 404.08 | 1082.38 | .97032 | 7.3318 |
| 1317 | | 1416.02 | 347.17 | 1038.00 | 1.0889 | 7.2882 | 1367 | | 1475.66 | 405.31 | 1083.29 | .96808 | 7.3327 |
| 1318 | | 1417.21 | 348.26 | 1038.90 | 1.0863 | 7.2891 | 1368 | | 1476.86 | 406.55 | 1084.20 | .96584 | 7.3335 |
| 1319 | | 1418.40 | 349.36 | 1039.81 | 1.0837 | 7.2900 | 1369 | | 1478.05 | 407.79 | 1085.11 | .96360 | 7.3344 |
| 1320 | 1046.85 | 1419.59 | 350.46 | 1040.71 | 1.0811 | 7.2909 | 1370 | 1096.85 | 1479.25 | 409.03 | 1086.01 | .96138 | 7.3353 |
| 1321 | | 1420.78 | 351.56 | 1041.61 | 1.0785 | 7.2918 | 1371 | | 1480.44 | 410.28 | 1086.92 | .95916 | 7.3362 |
| 1322 | | 1421.97 | 352.67 | 1042.52 | 1.0760 | 7.2927 | 1372 | | 1481.64 | 411.52 | 1087.83 | .95695 | 7.3370 |
| 1323 | | 1423.16 | 353.77 | 1043.42 | 1.0734 | 7.2936 | 1373 | | 1482.84 | 412.78 | 1088.74 | .95474 | 7.3379 |
| 1324 | | 1424.35 | 354.88 | 1044.32 | 1.0709 | 7.2945 | 1374 | | 1484.03 | 414.03 | 1089.65 | .95254 | 7.3388 |
| 1325 | 1051.85 | 1425.54 | 356.00 | 1045.23 | 1.0683 | 7.2954 | 1375 | 1101.85 | 1485.23 | 415.29 | 1090.56 | .95035 | 7.3396 |
| 1326 | | 1426.73 | 357.11 | 1046.13 | 1.0658 | 7.2963 | 1376 | | 1486.43 | 416.55 | 1091.47 | .94816 | 7.3405 |
| 1327 | | 1427.92 | 358.23 | 1047.03 | 1.0632 | 7.2972 | 1377 | | 1487.62 | 417.81 | 1092.38 | .94598 | 7.3414 |
| 1328 | | 1429.12 | 359.35 | 1047.94 | 1.0607 | 7.2981 | 1378 | | 1488.82 | 419.08 | 1093.29 | .94380 | 7.3423 |
| 1329 | | 1430.31 | 360.48 | 1048.84 | 1.0582 | 7.2990 | 1379 | | 1490.02 | 420.35 | 1094.20 | .94164 | 7.3431 |
| 1330 | 1056.85 | 1431.50 | 361.61 | 1049.75 | 1.0557 | 7.2999 | 1380 | 1106.85 | 1491.21 | 421.62 | 1095.11 | .93947 | 7.3440 |
| 1331 | | 1432.69 | 362.74 | 1050.65 | 1.0532 | 7.3008 | 1381 | | 1492.41 | 422.90 | 1096.02 | .93732 | 7.3449 |
| 1332 | | 1433.88 | 363.87 | 1051.55 | 1.0507 | 7.3017 | 1382 | | 1493.61 | 424.18 | 1096.93 | .93517 | 7.3457 |
| 1333 | | 1435.07 | 365.00 | 1052.46 | 1.0482 | 7.3026 | 1383 | | 1494.81 | 425.46 | 1097.84 | .93302 | 7.3466 |
| 1334 | | 1436.26 | 366.14 | 1053.36 | 1.0458 | 7.3035 | 1384 | | 1496.00 | 426.75 | 1098.75 | .93089 | 7.3475 |
| 1335 | 1061.85 | 1437.46 | 367.28 | 1054.27 | 1.0433 | 7.3044 | 1385 | 1111.85 | 1497.20 | 428.03 | 1099.66 | .92876 | 7.3483 |
| 1336 | | 1438.65 | 368.43 | 1055.17 | 1.0408 | 7.3053 | 1386 | | 1498.40 | 429.32 | 1100.57 | .92663 | 7.3492 |
| 1337 | | 1439.84 | 369.57 | 1056.08 | 1.0384 | 7.3062 | 1387 | | 1499.60 | 430.62 | 1101.49 | .92451 | 7.3500 |
| 1338 | | 1441.03 | 370.72 | 1056.98 | 1.0359 | 7.3071 | 1388 | | 1500.80 | 431.92 | 1102.40 | .92240 | 7.3509 |
| 1339 | | 1442.22 | 371.88 | 1057.89 | 1.0335 | 7.3079 | 1389 | | 1501.99 | 433.22 | 1103.31 | .92029 | 7.3518 |
| 1340 | 1066.85 | 1443.42 | 373.03 | 1058.79 | 1.0311 | 7.3088 | 1390 | 1116.85 | 1503.19 | 434.52 | 1104.22 | .91819 | 7.3526 |
| 1341 | | 1444.61 | 374.19 | 1059.70 | 1.0287 | 7.3097 | 1391 | | 1504.39 | 435.83 | 1105.13 | .91610 | 7.3535 |
| 1342 | | 1445.80 | 375.35 | 1060.61 | 1.0262 | 7.3106 | 1392 | | 1505.59 | 437.14 | 1106.04 | .91401 | 7.3544 |
| 1343 | | 1446.99 | 376.51 | 1061.51 | 1.0238 | 7.3115 | 1393 | | 1506.79 | 438.45 | 1106.95 | .91193 | 7.3552 |
| 1344 | | 1448.19 | 377.68 | 1062.42 | 1.0214 | 7.3124 | 1394 | | 1507.99 | 439.77 | 1107.87 | .90985 | 7.3561 |
| 1345 | 1071.85 | 1449.38 | 378.85 | 1063.32 | 1.0190 | 7.3133 | 1395 | 1121.85 | 1509.19 | 441.09 | 1108.78 | .90778 | 7.3569 |
| 1346 | | 1450.57 | 380.02 | 1064.23 | 1.0166 | 7.3142 | 1396 | | 1510.39 | 442.41 | 1109.69 | .90572 | 7 3578 |
| 1347 | | 1451.77 | 381.20 | 1065.14 | 1.0143 | 7.3151 | 1397 | | 1511.58 | 443.73 | 1110.60 | .90366 | 7.3587 |
| 1348 | | 1452.96 | 382.37 | 1066.04 | 1.0119 | 7.3159 | 1398 | | 1512.78 | 445.06 | 1111.51 | .90160 | 7.3595 |
| 1349 | | 1454.15 | 383.56 | 1066.95 | 1.0095 | 7.3168 | 1399 | | 1513.98 | 446.39 | 1112.43 | .89956 | 7.3604 |

# Table 1  Air at Low Pressures (for One Kilogram)

| T | t | h | $p_r$ | u | $v_r$ | $\phi$ | T | t | h | $p_r$ | u | $v_r$ | $\phi$ |
|---|---|---|---|---|---|---|---|---|---|---|---|---|---|
| 1400 | 1126.85 | 1515.18 | 447.73 | 1113.34 | .89752 | 7.3612 | 1450 | 1176.85 | 1575.30 | 518.63 | 1159.11 | .80250 | 7.4034 |
| 1401 | | 1516.38 | 449.07 | 1114.25 | .89548 | 7.3621 | 1451 | | 1576.51 | 520.13 | 1160.02 | .80073 | 7.4043 |
| 1402 | | 1517.58 | 450.41 | 1115.16 | .89345 | 7.3629 | 1452 | | 1577.71 | 521.64 | 1160.94 | .79897 | 7.4051 |
| 1403 | | 1518.78 | 451.75 | 1116.08 | .89143 | 7.3638 | 1453 | | 1578.92 | 523.15 | 1161.86 | .79721 | 7.4059 |
| 1404 | | 1519.98 | 453.10 | 1116.99 | .88941 | 7.3647 | 1454 | | 1580.12 | 524.66 | 1162.78 | .79546 | 7.4067 |
| 1405 | 1131.85 | 1521.18 | 454.45 | 1117.90 | .88740 | 7.3655 | 1455 | 1181.85 | 1581.33 | 526.18 | 1163.70 | .79371 | 7.4076 |
| 1406 | | 1522.38 | 455.81 | 1118.82 | .88539 | 7.3664 | 1456 | | 1582.53 | 527.70 | 1164.62 | .79197 | 7.4084 |
| 1407 | | 1523.58 | 457.16 | 1119.73 | .88339 | 7.3672 | 1457 | | 1583.74 | 529.22 | 1165.53 | .79023 | 7.4092 |
| 1408 | | 1524.78 | 458.52 | 1120.64 | .88140 | 7.3681 | 1458 | | 1584.94 | 530.75 | 1166.45 | .78849 | 7.4101 |
| 1409 | | 1525.98 | 459.89 | 1121.56 | .87941 | 7.3689 | 1459 | | 1586.15 | 532.28 | 1167.37 | .78676 | 7.4109 |
| 1410 | 1136.85 | 1527.18 | 461.25 | 1122.47 | .87742 | 7.3698 | 1460 | 1186.85 | 1587.36 | 533.81 | 1168.29 | .78504 | 7.4117 |
| 1411 | | 1528.39 | 462.62 | 1123.38 | .87544 | 7.3706 | 1461 | | 1588.56 | 535.35 | 1169.21 | .78332 | 7.4125 |
| 1412 | | 1529.59 | 464.00 | 1124.30 | .87347 | 7.3715 | 1462 | | 1589.77 | 536.89 | 1170.13 | .78161 | 7.4134 |
| 1413 | | 1530.79 | 465.37 | 1125.21 | .87150 | 7.3723 | 1463 | | 1590.98 | 538.44 | 1171.05 | .77990 | 7.4142 |
| 1414 | | 1531.99 | 466.75 | 1126.13 | .86954 | 7.3732 | 1464 | | 1592.18 | 539.99 | 1171.97 | .77819 | 7.4150 |
| 1415 | 1141.85 | 1533.19 | 468.14 | 1127.04 | .86759 | 7.3740 | 1465 | 1191.85 | 1593.39 | 541.54 | 1172.89 | .77649 | 7.4158 |
| 1416 | | 1534.39 | 469.52 | 1127.95 | .86564 | 7.3749 | 1466 | | 1594.60 | 543.10 | 1173.81 | .77479 | 7.4167 |
| 1417 | | 1535.59 | 470.91 | 1128.87 | .86369 | 7.3757 | 1467 | | 1595.80 | 544.66 | 1174.73 | .77310 | 7.4175 |
| 1418 | | 1536.79 | 472.30 | 1129.78 | .86175 | 7.3766 | 1468 | | 1597.01 | 546.22 | 1175.65 | .77142 | 7.4183 |
| 1419 | | 1538.00 | 473.70 | 1130.70 | .85982 | 7.3774 | 1469 | | 1598.22 | 547.78 | 1176.57 | .76974 | 7.4191 |
| 1420 | 1146.85 | 1539.20 | 475.10 | 1131.61 | .85789 | 7.3783 | 1470 | 1196.85 | 1599.42 | 549.35 | 1177.49 | .76806 | 7.4199 |
| 1421 | | 1540.40 | 476.50 | 1132.53 | .85597 | 7.3791 | 1471 | | 1600.63 | 550.93 | 1178.41 | .76639 | 7.4208 |
| 1422 | | 1541.60 | 477.91 | 1133.44 | .85405 | 7.3800 | 1472 | | 1601.84 | 552.51 | 1179.33 | .76472 | 7.4216 |
| 1423 | | 1542.80 | 479.32 | 1134.36 | .85214 | 7.3808 | 1473 | | 1603.04 | 554.09 | 1180.25 | .76305 | 7.4224 |
| 1424 | | 1544.01 | 480.73 | 1135.27 | .85023 | 7.3816 | 1474 | | 1604.25 | 555.67 | 1181.17 | .76140 | 7.4232 |
| 1425 | 1151.85 | 1545.21 | 482.15 | 1136.19 | .84833 | 7.3825 | 1475 | 1201.85 | 1605.46 | 557.26 | 1182.09 | .75974 | 7.4240 |
| 1426 | | 1546.41 | 483.56 | 1137.10 | .84644 | 7.3833 | 1476 | | 1606.67 | 558.85 | 1183.01 | .75809 | 7.4249 |
| 1427 | | 1547.61 | 484.99 | 1138.02 | .84455 | 7.3842 | 1477 | | 1607.88 | 560.44 | 1183.93 | .75645 | 7.4257 |
| 1428 | | 1548.82 | 486.41 | 1138.93 | .84266 | 7.3850 | 1478 | | 1609.08 | 562.04 | 1184.85 | .75480 | 7.4265 |
| 1429 | | 1550.02 | 487.84 | 1139.85 | .84078 | 7.3859 | 1479 | | 1610.29 | 563.64 | 1185.77 | .75317 | 7.4273 |
| 1430 | 1156.85 | 1551.22 | 489.27 | 1140.77 | .83891 | 7.3867 | 1480 | 1206.85 | 1611.50 | 565.25 | 1186.69 | .75154 | 7.4281 |
| 1431 | | 1552.42 | 490.71 | 1141.68 | .83704 | 7.3875 | 1481 | | 1612.71 | 566.86 | 1187.61 | .74991 | 7.4289 |
| 1432 | | 1553.63 | 492.15 | 1142.60 | .83517 | 7.3884 | 1482 | | 1613.92 | 568.47 | 1188.54 | .74829 | 7.4298 |
| 1433 | | 1554.83 | 493.59 | 1143.51 | .83331 | 7.3892 | 1483 | | 1615.12 | 570.09 | 1189.46 | .74667 | 7.4306 |
| 1434 | | 1556.03 | 495.04 | 1144.43 | .83146 | 7.3901 | 1484 | | 1616.33 | 571.71 | 1190.38 | .74505 | 7.4314 |
| 1435 | 1161.85 | 1557.24 | 496.48 | 1145.35 | .82961 | 7.3909 | 1485 | 1211.85 | 1617.54 | 573.33 | 1191.30 | .74344 | 7.4322 |
| 1436 | | 1558.44 | 497.94 | 1146.26 | .82777 | 7.3917 | 1486 | | 1618.75 | 574.96 | 1192.22 | .74184 | 7.4330 |
| 1437 | | 1559.64 | 499.39 | 1147.18 | .82593 | 7.3926 | 1487 | | 1619.96 | 576.59 | 1193.14 | .74024 | 7.4338 |
| 1438 | | 1560.85 | 500.85 | 1148.10 | .82410 | 7.3934 | 1488 | | 1621.17 | 578.23 | 1194.07 | .73864 | 7.4346 |
| 1439 | | 1562.05 | 502.31 | 1149.01 | .82227 | 7.3942 | 1489 | | 1622.38 | 579.87 | 1194.99 | .73705 | 7.4355 |
| 1440 | 1166.85 | 1563.26 | 503.78 | 1149.93 | .82045 | 7.3951 | 1490 | 1216.85 | 1623.59 | 581.51 | 1195.91 | .73546 | 7.4363 |
| 1441 | | 1564.46 | 505.25 | 1150.85 | .81863 | 7.3959 | 1491 | | 1624.80 | 583.16 | 1196.83 | .73388 | 7.4371 |
| 1442 | | 1565.66 | 506.72 | 1151.76 | .81682 | 7.3968 | 1492 | | 1626.01 | 584.80 | 1197.75 | .73230 | 7.4379 |
| 1443 | | 1566.87 | 508.20 | 1152.68 | .81501 | 7.3976 | 1493 | | 1627.21 | 586.46 | 1198.68 | .73072 | 7.4387 |
| 1444 | | 1568.07 | 509.68 | 1153.60 | .81321 | 7.3984 | 1494 | | 1628.42 | 588.12 | 1199.60 | .72915 | 7.4395 |
| 1445 | 1171.85 | 1569.28 | 511.16 | 1154.52 | .81141 | 7.3993 | 1495 | 1221.85 | 1629.63 | 589.78 | 1200.52 | .72758 | 7.4403 |
| 1446 | | 1570.48 | 512.65 | 1155.43 | .80962 | 7.4001 | 1496 | | 1630.84 | 591.44 | 1201.44 | .72602 | 7.4411 |
| 1447 | | 1571.69 | 514.14 | 1156.35 | .80783 | 7.4009 | 1497 | | 1632.05 | 593.11 | 1202.37 | .72446 | 7.4419 |
| 1448 | | 1572.89 | 515.63 | 1157.27 | .80605 | 7.4018 | 1498 | | 1633.26 | 594.78 | 1203.29 | .72291 | 7.4427 |
| 1449 | | 1574.10 | 517.13 | 1158.19 | .80427 | 7.4026 | 1499 | | 1634.47 | 596.46 | 1204.21 | .72136 | 7.4436 |

# Table 1 Air at Low Pressures (for One Kilogram)

| T | t | h | $p_r$ | u | $v_r$ | $\phi$ | T | t | h | $p_r$ | u | $v_r$ | $\phi$ |
|---|---|---|---|---|---|---|---|---|---|---|---|---|---|
| 1500 | 1226.85 | 1635.68 | 598.14 | 1205.14 | .71982 | 7.4444 | 1550 | 1276.85 | 1696.32 | 687.01 | 1251.42 | .64759 | 7.4841 |
| 1501 | | 1636.89 | 599.82 | 1206.06 | .71827 | 7.4452 | 1551 | | 1697.53 | 688.89 | 1252.35 | .64624 | 7.4849 |
| 1502 | | 1638.10 | 601.51 | 1206.98 | .71674 | 7.4460 | 1552 | | 1698.75 | 690.77 | 1253.27 | .64489 | 7.4857 |
| 1503 | | 1639.31 | 603.20 | 1207.91 | .71520 | 7.4468 | 1553 | | 1699.96 | 692.66 | 1254.20 | .64355 | 7.4865 |
| 1504 | | 1640.53 | 604.89 | 1208.83 | .71367 | 7.4476 | 1554 | | 1701.18 | 694.55 | 1255.13 | .64221 | 7.4873 |
| 1505 | 1231.85 | 1641.74 | 606.59 | 1209.75 | .71215 | 7.4484 | 1555 | 1281.85 | 1702.39 | 696.44 | 1256.06 | .64088 | 7.4880 |
| 1506 | | 1642.95 | 608.29 | 1210.68 | .71063 | 7.4492 | 1556 | | 1703.61 | 698.34 | 1256.99 | .63955 | 7.4888 |
| 1507 | | 1644.16 | 610.00 | 1211.60 | .70911 | 7.4500 | 1557 | | 1704.82 | 700.24 | 1257.92 | .63822 | 7.4896 |
| 1508 | | 1645.37 | 611.71 | 1212.53 | .70760 | 7.4508 | 1558 | | 1706.04 | 702.15 | 1258.85 | .63689 | 7.4904 |
| 1509 | | 1646.58 | 613.42 | 1213.45 | .70609 | 7.4516 | 1559 | | 1707.26 | 704.06 | 1259.77 | .63557 | 7.4912 |
| 1510 | 1236.85 | 1647.79 | 615.14 | 1214.37 | .70459 | 7.4524 | 1560 | 1286.85 | 1708.47 | 705.98 | 1260.70 | .63425 | 7.4919 |
| 1511 | | 1649.00 | 616.86 | 1215.30 | .70309 | 7.4532 | 1561 | | 1709.69 | 707.90 | 1261.63 | .63294 | 7.4927 |
| 1512 | | 1650.21 | 618.58 | 1216.22 | .70159 | 7.4540 | 1562 | | 1710.90 | 709.82 | 1262.56 | .63163 | 7.4935 |
| 1513 | | 1651.42 | 620.31 | 1217.15 | .70010 | 7.4548 | 1563 | | 1712.12 | 711.75 | 1263.49 | .63032 | 7.4943 |
| 1514 | | 1652.64 | 622.04 | 1218.07 | .69861 | 7.4556 | 1564 | | 1713.34 | 713.68 | 1264.42 | .62902 | 7.4951 |
| 1515 | 1241.85 | 1653.85 | 623.78 | 1219.00 | .69713 | 7.4564 | 1565 | 1291.85 | 1714.55 | 715.61 | 1265.35 | .62772 | 7.4958 |
| 1516 | | 1655.06 | 625.52 | 1219.92 | .69565 | 7.4572 | 1566 | | 1715.77 | 717.55 | 1266.28 | .62642 | 7.4966 |
| 1517 | | 1656.27 | 627.26 | 1220.85 | .69417 | 7.4580 | 1567 | | 1716.99 | 719.50 | 1267.21 | .62513 | 7.4974 |
| 1518 | | 1657.48 | 629.01 | 1221.77 | .69270 | 7.4588 | 1568 | | 1718.20 | 721.45 | 1268.14 | .62384 | 7.4982 |
| 1519 | | 1658.70 | 630.76 | 1222.70 | .69123 | 7.4596 | 1569 | | 1719.42 | 723.40 | 1269.07 | .62255 | 7.4989 |
| 1520 | 1246.85 | 1659.91 | 632.52 | 1223.62 | .68976 | 7.4604 | 1570 | 1296.85 | 1720.64 | 725.36 | 1270.00 | .62127 | 7.4997 |
| 1521 | | 1661.12 | 634.28 | 1224.55 | .68830 | 7.4612 | 1571 | | 1721.85 | 727.32 | 1270.93 | .61999 | 7.5005 |
| 1522 | | 1662.33 | 636.04 | 1225.47 | .68685 | 7.4620 | 1572 | | 1723.07 | 729.28 | 1271.86 | .61871 | 7.5013 |
| 1523 | | 1663.54 | 637.81 | 1226.40 | .68539 | 7.4628 | 1573 | | 1724.29 | 731.25 | 1272.79 | .61744 | 7.5020 |
| 1524 | | 1664.76 | 639.58 | 1227.32 | .68394 | 7.4636 | 1574 | | 1725.51 | 733.23 | 1273.72 | .61616 | 7.5028 |
| 1525 | 1251.85 | 1665.97 | 641.35 | 1228.25 | .68250 | 7.4644 | 1575 | 1301.85 | 1726.72 | 735.20 | 1274.65 | .61490 | 7.5036 |
| 1526 | | 1667.18 | 643.13 | 1229.17 | .68106 | 7.4652 | 1576 | | 1727.94 | 737.19 | 1275.58 | .61363 | 7.5044 |
| 1527 | | 1668.40 | 644.92 | 1230.10 | .67962 | 7.4660 | 1577 | | 1729.16 | 739.17 | 1276.51 | .61237 | 7.5051 |
| 1528 | | 1669.61 | 646.70 | 1231.02 | .67819 | 7.4668 | 1578 | | 1730.38 | 741.16 | 1277.44 | .61112 | 7.5059 |
| 1529 | | 1670.82 | 648.49 | 1231.95 | .67676 | 7.4676 | 1579 | | 1731.59 | 743.16 | 1278.37 | .60986 | 7.5067 |
| 1530 | 1256.85 | 1672.03 | 650.29 | 1232.88 | .67533 | 7.4684 | 1580 | 1306.85 | 1732.81 | 745.16 | 1279.30 | .60861 | 7.5074 |
| 1531 | | 1673.25 | 652.09 | 1233.80 | .67391 | 7.4692 | 1581 | | 1734.03 | 747.16 | 1280.23 | .60736 | 7.5082 |
| 1532 | | 1674.46 | 653.89 | 1234.73 | .67249 | 7.4699 | 1582 | | 1735.25 | 749.17 | 1281.16 | .60612 | 7.5090 |
| 1533 | | 1675.67 | 655.69 | 1235.66 | .67107 | 7.4707 | 1583 | | 1736.47 | 751.18 | 1282.09 | .60488 | 7.5098 |
| 1534 | | 1676.89 | 657.50 | 1236.58 | .66966 | 7.4715 | 1584 | | 1737.68 | 753.19 | 1283.03 | .60364 | 7.5105 |
| 1535 | 1261.85 | 1678.10 | 659.32 | 1237.51 | .66826 | 7.4723 | 1585 | 1311.85 | 1738.90 | 755.21 | 1283.96 | .60241 | 7.5113 |
| 1536 | | 1679.32 | 661.14 | 1238.44 | .66685 | 7.4731 | 1586 | | 1740.12 | 757.24 | 1284.89 | .60117 | 7.5121 |
| 1537 | | 1680.53 | 662.96 | 1239.36 | .66545 | 7.4739 | 1587 | | 1741.34 | 759.27 | 1285.82 | .59995 | 7.5128 |
| 1538 | | 1681.74 | 664.78 | 1240.29 | .66406 | 7.4747 | 1588 | | 1742.56 | 761.30 | 1286.75 | .59872 | 7.5136 |
| 1539 | | 1682.96 | 666.61 | 1241.22 | .66266 | 7.4755 | 1589 | | 1743.78 | 763.34 | 1287.68 | .59750 | 7.5144 |
| 1540 | 1266.85 | 1684.17 | 668.45 | 1242.14 | .66128 | 7.4763 | 1590 | 1316.85 | 1744.99 | 765.38 | 1288.61 | .59628 | 7.5151 |
| 1541 | | 1685.39 | 670.29 | 1243.07 | .65989 | 7.4771 | 1591 | | 1746.21 | 767.43 | 1289.55 | .59506 | 7.5159 |
| 1542 | | 1686.60 | 672.13 | 1244.00 | .65851 | 7.4778 | 1592 | | 1747.43 | 769.48 | 1290.48 | .59385 | 7.5167 |
| 1543 | | 1687.81 | 673.97 | 1244.92 | .65713 | 7.4786 | 1593 | | 1748.65 | 771.53 | 1291.41 | .59264 | 7.5174 |
| 1544 | | 1689.03 | 675.82 | 1245.85 | .65576 | 7.4794 | 1594 | | 1749.87 | 773.59 | 1292.34 | .59143 | 7.5182 |
| 1545 | 1271.85 | 1690.24 | 677.68 | 1246.78 | .65439 | 7.4802 | 1595 | 1321.85 | 1751.09 | 775.65 | 1293.27 | .59023 | 7.5190 |
| 1546 | | 1691.46 | 679.54 | 1247.71 | .65302 | 7.4810 | 1596 | | 1752.31 | 777.72 | 1294.21 | .58903 | 7.5197 |
| 1547 | | 1692.67 | 681.40 | 1248.63 | .65166 | 7.4818 | 1597 | | 1753.53 | 779.79 | 1295.14 | .58783 | 7.5205 |
| 1548 | | 1693.89 | 683.27 | 1249.56 | .65030 | 7.4826 | 1598 | | 1754.75 | 781.87 | 1296.07 | .58664 | 7.5213 |
| 1549 | | 1695.10 | 685.14 | 1250.49 | .64894 | 7.4833 | 1599 | | 1755.97 | 783.95 | 1297.00 | .58545 | 7.5220 |

## Table 1   Air at Low Pressures (for One Kilogram)

| T | t | h | $p_r$ | u | $v_r$ | $\phi$ | T | t | h | $p_r$ | u | $v_r$ | $\phi$ |
|---|---|---|---|---|---|---|---|---|---|---|---|---|---|
| 1600 | 1326.85 | 1757.19 | 786.04 | 1297.94 | .58426 | 7.5228 | 1650 | 1376.85 | 1818.28 | 896.05 | 1344.68 | .52855 | 7.5604 |
| 1601 | | 1758.41 | 788.13 | 1298.87 | .58308 | 7.5235 | 1651 | | 1819.50 | 898.36 | 1345.61 | .52750 | 7.5611 |
| 1602 | | 1759.63 | 790.22 | 1299.80 | .58189 | 7.5243 | 1652 | | 1820.73 | 900.69 | 1346.55 | .52646 | 7.5619 |
| 1603 | | 1760.85 | 792.32 | 1300.73 | .58072 | 7.5251 | 1653 | | 1821.95 | 903.01 | 1347.49 | .52542 | 7.5626 |
| 1604 | | 1762.07 | 794.42 | 1301.67 | .57954 | 7.5258 | 1654 | | 1823.18 | 905.35 | 1348.43 | .52438 | 7.5633 |
| 1605 | 1331.85 | 1763.29 | 796.53 | 1302.60 | .57837 | 7.5266 | 1655 | 1381.85 | 1824.40 | 907.68 | 1349.36 | .52335 | 7.5641 |
| 1606 | | 1764.51 | 798.64 | 1303.53 | .57720 | 7.5273 | 1656 | | 1825.62 | 910.03 | 1350.30 | .52232 | 7.5648 |
| 1607 | | 1765.73 | 800.76 | 1304.47 | .57603 | 7.5281 | 1657 | | 1826.85 | 912.37 | 1351.24 | .52129 | 7.5656 |
| 1608 | | 1766.95 | 802.88 | 1305.40 | .57487 | 7.5289 | 1658 | | 1828.07 | 914.72 | 1352.18 | .52026 | 7.5663 |
| 1609 | | 1768.17 | 805.00 | 1306.33 | .57370 | 7.5296 | 1659 | | 1829.30 | 917.08 | 1353.11 | .51924 | 7.5670 |
| 1610 | 1336.85 | 1769.39 | 807.13 | 1307.27 | .57255 | 7.5304 | 1660 | 1386.85 | 1830.52 | 919.44 | 1354.05 | .51822 | 7.5678 |
| 1611 | | 1770.61 | 809.27 | 1308.20 | .57139 | 7.5311 | 1661 | | 1831.75 | 921.81 | 1354.99 | .51720 | 7.5685 |
| 1612 | | 1771.83 | 811.40 | 1309.13 | .57024 | 7.5319 | 1662 | | 1832.97 | 924.18 | 1355.93 | .51618 | 7.5692 |
| 1613 | | 1773.05 | 813.55 | 1310.07 | .56909 | 7.5327 | 1663 | | 1834.20 | 926.55 | 1356.86 | .51517 | 7.5700 |
| 1614 | | 1774.27 | 815.70 | 1311.00 | .56794 | 7.5334 | 1664 | | 1835.42 | 928.93 | 1357.80 | .51416 | 7.5707 |
| 1615 | 1341.85 | 1775.49 | 817.85 | 1311.94 | .56680 | 7.5342 | 1665 | 1391.85 | 1836.65 | 931.32 | 1358.74 | .51315 | 7.5715 |
| 1616 | | 1776.71 | 820.00 | 1312.87 | .56566 | 7.5349 | 1666 | | 1837.87 | 933.71 | 1359.68 | .51214 | 7.5722 |
| 1617 | | 1777.93 | 822.16 | 1313.80 | .56452 | 7.5357 | 1667 | | 1839.10 | 936.11 | 1360.62 | .51114 | 7.5729 |
| 1618 | | 1779.15 | 824.33 | 1314.74 | .56339 | 7.5364 | 1668 | | 1840.32 | 938.51 | 1361.56 | .51014 | 7.5737 |
| 1619 | | 1780.38 | 826.50 | 1315.67 | .56226 | 7.5372 | 1669 | | 1841.55 | 940.91 | 1362.49 | .50914 | 7.5744 |
| 1620 | 1346.85 | 1781.60 | 828.67 | 1316.61 | .56113 | 7.5379 | 1670 | 1396.85 | 1842.78 | 943.32 | 1363.43 | .50814 | 7.5751 |
| 1621 | | 1782.82 | 830.85 | 1317.54 | .56000 | 7.5387 | 1671 | | 1844.00 | 945.73 | 1364.37 | .50715 | 7.5759 |
| 1622 | | 1784.04 | 833.04 | 1318.48 | .55888 | 7.5394 | 1672 | | 1845.23 | 948.15 | 1365.31 | .50616 | 7.5766 |
| 1623 | | 1785.26 | 835.22 | 1319.41 | .55776 | 7.5402 | 1673 | | 1846.45 | 950.58 | 1366.25 | .50517 | 7.5773 |
| 1624 | | 1786.48 | 837.42 | 1320.34 | .55664 | 7.5410 | 1674 | | 1847.68 | 953.01 | 1367.19 | .50418 | 7.5781 |
| 1625 | 1351.85 | 1787.71 | 839.61 | 1321.28 | .55552 | 7.5417 | 1675 | 1401.85 | 1848.90 | 955.44 | 1368.13 | .50320 | 7.5788 |
| 1626 | | 1788.93 | 841.82 | 1322.21 | .55441 | 7.5425 | 1676 | | 1850.13 | 957.88 | 1369.07 | .50222 | 7.5795 |
| 1627 | | 1790.15 | 844.02 | 1323.15 | .55330 | 7.5432 | 1677 | | 1851.36 | 960.32 | 1370.01 | .50124 | 7.5803 |
| 1628 | | 1791.37 | 846.23 | 1324.08 | .55220 | 7.5440 | 1678 | | 1852.58 | 962.77 | 1370.94 | .50026 | 7.5810 |
| 1629 | | 1792.59 | 848.45 | 1325.02 | .55109 | 7.5447 | 1679 | | 1853.81 | 965.23 | 1371.88 | .49929 | 7.5817 |
| 1630 | 1356.85 | 1793.82 | 850.67 | 1325.95 | .54999 | 7.5455 | 1680 | 1406.85 | 1855.04 | 967.69 | 1372.82 | .49832 | 7.5825 |
| 1631 | | 1795.04 | 852.89 | 1326.89 | .54889 | 7.5462 | 1681 | | 1856.26 | 970.15 | 1373.76 | .49735 | 7.5832 |
| 1632 | | 1796.26 | 855.12 | 1327.83 | .54780 | 7.5470 | 1682 | | 1857.49 | 972.62 | 1374.70 | .49638 | 7.5839 |
| 1633 | | 1797.48 | 857.36 | 1328.76 | .54671 | 7.5477 | 1683 | | 1858.72 | 975.09 | 1375.64 | .49541 | 7.5846 |
| 1634 | | 1798.71 | 859.60 | 1329.70 | .54562 | 7.5485 | 1684 | | 1859.94 | 977.57 | 1376.58 | .49445 | 7.5854 |
| 1635 | 1361.85 | 1799.93 | 861.84 | 1330.63 | .54453 | 7.5492 | 1685 | 1411.85 | 1861.17 | 980.05 | 1377.52 | .49349 | 7.5861 |
| 1636 | | 1801.15 | 864.09 | 1331.57 | .54344 | 7.5499 | 1686 | | 1862.40 | 982.54 | 1378.46 | .49253 | 7.5868 |
| 1637 | | 1802.37 | 866.34 | 1332.50 | .54236 | 7.5507 | 1687 | | 1863.62 | 985.04 | 1379.40 | .49158 | 7.5876 |
| 1638 | | 1803.60 | 868.60 | 1333.44 | .54128 | 7.5514 | 1688 | | 1864.85 | 987.53 | 1380.34 | .49062 | 7.5883 |
| 1639 | | 1804.82 | 870.86 | 1334.38 | .54021 | 7.5522 | 1689 | | 1866.08 | 990.04 | 1381.28 | .48967 | 7.5890 |
| 1640 | 1366.85 | 1806.04 | 873.12 | 1335.31 | .53913 | 7.5529 | 1690 | 1416.85 | 1867.30 | 992.55 | 1382.22 | .48873 | 7.5897 |
| 1641 | | 1807.27 | 875.40 | 1336.25 | .53806 | 7.5537 | 1691 | | 1868.53 | 995.06 | 1383.16 | .48778 | 7.5905 |
| 1642 | | 1808.49 | 877.67 | 1337.18 | .53700 | 7.5544 | 1692 | | 1869.76 | 997.58 | 1384.10 | .48684 | 7.5912 |
| 1643 | | 1809.71 | 879.95 | 1338.12 | .53593 | 7.5552 | 1693 | | 1870.99 | 1000.10 | 1385.04 | .48589 | 7.5919 |
| 1644 | | 1810.94 | 882.24 | 1339.06 | .53487 | 7.5559 | 1694 | | 1872.21 | 1002.63 | 1385.98 | .48495 | 7.5926 |
| 1645 | 1371.85 | 1812.16 | 884.53 | 1339.99 | .53381 | 7.5567 | 1695 | 1421.85 | 1873.44 | 1005.17 | 1386.92 | .48402 | 7.5934 |
| 1646 | | 1813.38 | 886.82 | 1340.93 | .53275 | 7.5574 | 1696 | | 1874.67 | 1007.71 | 1387.87 | .48308 | 7.5941 |
| 1647 | | 1814.61 | 889.12 | 1341.87 | .53170 | 7.5581 | 1697 | | 1875.90 | 1010.25 | 1388.81 | .48215 | 7.5948 |
| 1648 | | 1815.83 | 891.42 | 1342.80 | .53064 | 7.5589 | 1698 | | 1877.13 | 1012.80 | 1389.75 | .48122 | 7.5955 |
| 1649 | | 1817.05 | 893.73 | 1343.74 | .52959 | 7.5596 | 1699 | | 1878.35 | 1015.35 | 1390.69 | .48029 | 7.5963 |

# Table 1  Air at Low Pressures (for One Kilogram)

| T | t | h | $p_r$ | u | $v_r$ | $\phi$ | T | t | h | $p_r$ | u | $v_r$ | $\phi$ |
|---|---|---|---|---|---|---|---|---|---|---|---|---|---|
| 1700 | 1426.85 | 1879.58 | 1017.9 | 1391.63 | .47937 | 7.5970 | 1750 | 1476.85 | 1941.09 | 1152.6 | 1438.78 | .43582 | 7.6326 |
| 1701 | | 1880.81 | 1020.5 | 1392.57 | .47844 | 7.5977 | 1751 | | 1942.32 | 1155.4 | 1439.73 | .43500 | 7.6333 |
| 1702 | | 1882.04 | 1023.0 | 1393.51 | .47752 | 7.5984 | 1752 | | 1943.55 | 1158.2 | 1440.67 | .43418 | 7.6340 |
| 1703 | | 1883.27 | 1025.6 | 1394.45 | .47660 | 7.5991 | 1753 | | 1944.78 | 1161.1 | 1441.62 | .43337 | 7.6347 |
| 1704 | | 1884.49 | 1028.2 | 1395.39 | .47569 | 7.5999 | 1754 | | 1946.01 | 1163.9 | 1442.56 | .43256 | 7.6354 |
| 1705 | 1431.85 | 1885.72 | 1030.8 | 1396.34 | .47477 | 7.6006 | 1755 | 1481.85 | 1947.25 | 1166.8 | 1443.51 | .43174 | 7.6361 |
| 1706 | | 1886.95 | 1033.4 | 1397.28 | .47386 | 7.6013 | 1756 | | 1948.48 | 1169.6 | 1444.45 | .43094 | 7.6368 |
| 1707 | | 1888.18 | 1036.0 | 1398.22 | .47295 | 7.6020 | 1757 | | 1949.71 | 1172.5 | 1445.40 | .43013 | 7.6376 |
| 1708 | | 1889.41 | 1038.6 | 1399.16 | .47204 | 7.6027 | 1758 | | 1950.94 | 1175.3 | 1446.34 | .42932 | 7.6383 |
| 1709 | | 1890.64 | 1041.2 | 1400.10 | .47114 | 7.6035 | 1759 | | 1952.18 | 1178.2 | 1447.29 | .42852 | 7.6390 |
| 1710 | 1436.85 | 1891.87 | 1043.8 | 1401.04 | .47023 | 7.6042 | 1760 | 1486.85 | 1953.41 | 1181.1 | 1448.23 | .42772 | 7.6397 |
| 1711 | | 1893.10 | 1046.4 | 1401.99 | .46933 | 7.6049 | 1761 | | 1954.64 | 1184.0 | 1449.18 | .42692 | 7.6404 |
| 1712 | | 1894.32 | 1049.0 | 1402.93 | .46843 | 7.6056 | 1762 | | 1955.87 | 1186.9 | 1450.13 | .42612 | 7.6411 |
| 1713 | | 1895.55 | 1051.7 | 1403.87 | .46754 | 7.6063 | 1763 | | 1957.11 | 1189.8 | 1451.07 | .42532 | 7.6418 |
| 1714 | | 1896.78 | 1054.3 | 1404.81 | .46664 | 7.6071 | 1764 | | 1958.34 | 1192.7 | 1452.02 | .42453 | 7.6425 |
| 1715 | 1441.85 | 1898.01 | 1056.9 | 1405.75 | .46575 | 7.6078 | 1765 | 1491.85 | 1959.57 | 1195.6 | 1452.96 | .42374 | 7.6432 |
| 1716 | | 1899.24 | 1059.6 | 1406.70 | .46486 | 7.6085 | 1766 | | 1960.81 | 1198.5 | 1453.91 | .42295 | 7.6439 |
| 1717 | | 1900.47 | 1062.2 | 1407.64 | .46397 | 7.6092 | 1767 | | 1962.04 | 1201.4 | 1454.86 | .42216 | 7.6445 |
| 1718 | | 1901.70 | 1064.9 | 1408.58 | .46308 | 7.6099 | 1768 | | 1963.27 | 1204.3 | 1455.80 | .42137 | 7.6452 |
| 1719 | | 1902.93 | 1067.5 | 1409.52 | .46220 | 7.6106 | 1769 | | 1964.51 | 1207.3 | 1456.75 | .42059 | 7.6459 |
| 1720 | 1446.85 | 1904.16 | 1070.2 | 1410.47 | .46132 | 7.6113 | 1770 | 1496.85 | 1965.74 | 1210.2 | 1457.70 | .41980 | 7.6466 |
| 1721 | | 1905.39 | 1072.9 | 1411.41 | .46044 | 7.6121 | 1771 | | 1966.97 | 1213.1 | 1458.64 | .41902 | 7.6473 |
| 1722 | | 1906.62 | 1075.5 | 1412.35 | .45956 | 7.6128 | 1772 | | 1968.21 | 1216.1 | 1459.59 | .41824 | 7.6480 |
| 1723 | | 1907.85 | 1078.2 | 1413.29 | .45868 | 7.6135 | 1773 | | 1969.44 | 1219.0 | 1460.54 | .41747 | 7.6487 |
| 1724 | | 1909.08 | 1080.9 | 1414.24 | .45781 | 7.6142 | 1774 | | 1970.68 | 1222.0 | 1461.48 | .41669 | 7.6494 |
| 1725 | 1451.85 | 1910.31 | 1083.6 | 1415.18 | .45694 | 7.6149 | 1775 | 1501.85 | 1971.91 | 1225.0 | 1462.43 | .41592 | 7.6501 |
| 1726 | | 1911.54 | 1086.3 | 1416.12 | .45607 | 7.6156 | 1776 | | 1973.14 | 1227.9 | 1463.38 | .41514 | 7.6508 |
| 1727 | | 1912.77 | 1089.0 | 1417.07 | .45520 | 7.6163 | 1777 | | 1974.38 | 1230.9 | 1464.32 | .41437 | 7.6515 |
| 1728 | | 1914.00 | 1091.7 | 1418.01 | .45434 | 7.6171 | 1778 | | 1975.61 | 1233.9 | 1465.27 | .41361 | 7.6522 |
| 1729 | | 1915.23 | 1094.4 | 1418.95 | .45348 | 7.6178 | 1779 | | 1976.85 | 1236.9 | 1466.22 | .41284 | 7.6529 |
| 1730 | 1456.85 | 1916.46 | 1097.1 | 1419.90 | .45261 | 7.6185 | 1780 | 1506.85 | 1978.08 | 1239.9 | 1467.16 | .41207 | 7.6536 |
| 1731 | | 1917.69 | 1099.8 | 1420.84 | .45176 | 7.6192 | 1781 | | 1979.31 | 1242.9 | 1468.11 | .41131 | 7.6543 |
| 1732 | | 1918.92 | 1102.6 | 1421.78 | .45090 | 7.6199 | 1782 | | 1980.55 | 1245.9 | 1469.06 | .41055 | 7.6550 |
| 1733 | | 1920.15 | 1105.3 | 1422.73 | .45004 | 7.6206 | 1783 | | 1981.78 | 1248.9 | 1470.01 | .40979 | 7.6557 |
| 1734 | | 1921.38 | 1108.0 | 1423.67 | .44919 | 7.6213 | 1784 | | 1983.02 | 1251.9 | 1470.95 | .40903 | 7.6564 |
| 1735 | 1461.85 | 1922.61 | 1110.8 | 1424.61 | .44834 | 7.6220 | 1785 | 1511.85 | 1984.25 | 1254.9 | 1471.90 | .40828 | 7.6571 |
| 1736 | | 1923.84 | 1113.5 | 1425.56 | .44749 | 7.6227 | 1786 | | 1985.49 | 1257.9 | 1472.85 | .40752 | 7.6577 |
| 1737 | | 1925.08 | 1116.3 | 1426.50 | .44664 | 7.6234 | 1787 | | 1986.72 | 1261.0 | 1473.80 | .40677 | 7.6584 |
| 1738 | | 1926.31 | 1119.0 | 1427.45 | .44580 | 7.6242 | 1788 | | 1987.96 | 1264.0 | 1474.74 | .40602 | 7.6591 |
| 1739 | | 1927.54 | 1121.8 | 1428.39 | .44496 | 7.6249 | 1789 | | 1989.19 | 1267.1 | 1475.69 | .40527 | 7.6598 |
| 1740 | 1466.85 | 1928.77 | 1124.6 | 1429.33 | .44412 | 7.6256 | 1790 | 1516.85 | 1990.43 | 1270.1 | 1476.64 | .40452 | 7.6605 |
| 1741 | | 1930.00 | 1127.3 | 1430.28 | .44328 | 7.6263 | 1791 | | 1991.66 | 1273.2 | 1477.59 | .40378 | 7.6612 |
| 1742 | | 1931.23 | 1130.1 | 1431.22 | .44244 | 7.6270 | 1792 | | 1992.90 | 1276.2 | 1478.54 | .40303 | 7.6619 |
| 1743 | | 1932.46 | 1132.9 | 1432.17 | .44161 | 7.6277 | 1793 | | 1994.13 | 1279.3 | 1479.48 | .40229 | 7.6626 |
| 1744 | | 1933.69 | 1135.7 | 1433.11 | .44077 | 7.6284 | 1794 | | 1995.37 | 1282.4 | 1480.43 | .40155 | 7.6633 |
| 1745 | 1471.85 | 1934.93 | 1138.5 | 1434.06 | .43994 | 7.6291 | 1795 | 1521.85 | 1996.60 | 1285.4 | 1481.38 | .40081 | 7.6640 |
| 1746 | | 1936.16 | 1141.3 | 1435.00 | .43911 | 7.6298 | 1796 | | 1997.84 | 1288.5 | 1482.33 | .40008 | 7.6646 |
| 1747 | | 1937.39 | 1144.1 | 1435.95 | .43829 | 7.6305 | 1797 | | 1999.07 | 1291.6 | 1483.28 | .39934 | 7.6653 |
| 1748 | | 1938.62 | 1146.9 | 1436.89 | .43746 | 7.6312 | 1798 | | 2000.31 | 1294.7 | 1484.23 | .39861 | 7.6660 |
| 1749 | | 1939.85 | 1149.7 | 1437.84 | .43664 | 7.6319 | 1799 | | 2001.54 | 1297.8 | 1485.17 | .39788 | 7.6667 |

# Table 1   Air at Low Pressures (for One Kilogram)

| T | t | h | $p_r$ | u | $v_r$ | $\phi$ | T | t | h | $p_r$ | u | $v_r$ | $\phi$ |
|---|---|---|---|---|---|---|---|---|---|---|---|---|---|
| **1800** | 1526.85 | 2002.78 | 1300.9 | 1486.12 | .39714 | 7.6674 | **1850** | 1576.85 | 2064.65 | 1464.0 | 1533.65 | .36270 | 7.7013 |
| **1801** |  | 2004.01 | 1304.0 | 1487.07 | .39642 | 7.6681 | **1851** |  | 2065.89 | 1467.5 | 1534.60 | .36205 | 7.7020 |
| **1802** |  | 2005.25 | 1307.2 | 1488.02 | .39569 | 7.6688 | **1852** |  | 2067.13 | 1470.9 | 1535.55 | .36140 | 7.7026 |
| **1803** |  | 2006.49 | 1310.3 | 1488.97 | .39496 | 7.6694 | **1853** |  | 2068.37 | 1474.3 | 1536.50 | .36076 | 7.7033 |
| **1804** |  | 2007.72 | 1313.4 | 1489.92 | .39424 | 7.6701 | **1854** |  | 2069.61 | 1477.8 | 1537.46 | .36011 | 7.7040 |
| **1805** | 1531.85 | 2008.96 | 1316.6 | 1490.87 | .39352 | 7.6708 | **1855** | 1581.85 | 2070.85 | 1481.2 | 1538.41 | .35947 | 7.7046 |
| **1806** |  | 2010.19 | 1319.7 | 1491.82 | .39280 | 7.6715 | **1856** |  | 2072.09 | 1484.7 | 1539.36 | .35882 | 7.7053 |
| **1807** |  | 2011.43 | 1322.9 | 1492.77 | .39208 | 7.6722 | **1857** |  | 2073.33 | 1488.1 | 1540.31 | .35818 | 7.7060 |
| **1808** |  | 2012.67 | 1326.0 | 1493.71 | .39136 | 7.6729 | **1858** |  | 2074.57 | 1491.6 | 1541.27 | .35754 | 7.7066 |
| **1809** |  | 2013.90 | 1329.2 | 1494.66 | .39065 | 7.6736 | **1859** |  | 2075.81 | 1495.1 | 1542.22 | .35691 | 7.7073 |
| **1810** | 1536.85 | 2015.14 | 1332.3 | 1495.61 | .38994 | 7.6742 | **1860** | 1586.85 | 2077.05 | 1498.5 | 1543.17 | .35627 | 7.7080 |
| **1811** |  | 2016.38 | 1335.5 | 1496.56 | .38922 | 7.6749 | **1861** |  | 2078.29 | 1502.0 | 1544.12 | .35563 | 7.7086 |
| **1812** |  | 2017.61 | 1338.7 | 1497.51 | .38851 | 7.6756 | **1862** |  | 2079.53 | 1505.5 | 1545.08 | .35500 | 7.7093 |
| **1813** |  | 2018.85 | 1341.9 | 1498.46 | .38781 | 7.6763 | **1863** |  | 2080.77 | 1509.0 | 1546.03 | .35437 | 7.7100 |
| **1814** |  | 2020.09 | 1345.1 | 1499.41 | .38710 | 7.6770 | **1864** |  | 2082.01 | 1512.5 | 1546.98 | .35374 | 7.7106 |
| **1815** | 1541.85 | 2021.32 | 1348.3 | 1500.36 | .38639 | 7.6777 | **1865** | 1591.85 | 2083.25 | 1516.0 | 1547.94 | .35311 | 7.7113 |
| **1816** |  | 2022.56 | 1351.5 | 1501.31 | .38569 | 7.6783 | **1866** |  | 2084.49 | 1519.5 | 1548.89 | .35248 | 7.7120 |
| **1817** |  | 2023.80 | 1354.7 | 1502.26 | .38499 | 7.6790 | **1867** |  | 2085.73 | 1523.0 | 1549.84 | .35185 | 7.7126 |
| **1818** |  | 2025.03 | 1357.9 | 1503.21 | .38429 | 7.6797 | **1868** |  | 2086.97 | 1526.6 | 1550.80 | .35123 | 7.7133 |
| **1819** |  | 2026.27 | 1361.1 | 1504.16 | .38359 | 7.6804 | **1869** |  | 2088.21 | 1530.1 | 1551.75 | .35060 | 7.7140 |
| **1820** | 1546.85 | 2027.51 | 1364.3 | 1505.11 | .38289 | 7.6811 | **1870** | 1596.85 | 2089.45 | 1533.7 | 1552.70 | .34998 | 7.7146 |
| **1821** |  | 2028.74 | 1367.6 | 1506.06 | .38220 | 7.6817 | **1871** |  | 2090.69 | 1537.2 | 1553.66 | .34936 | 7.7153 |
| **1822** |  | 2029.98 | 1370.8 | 1507.01 | .38150 | 7.6824 | **1872** |  | 2091.93 | 1540.8 | 1554.61 | .34874 | 7.7160 |
| **1823** |  | 2031.22 | 1374.1 | 1507.96 | .38081 | 7.6831 | **1873** |  | 2093.17 | 1544.3 | 1555.56 | .34812 | 7.7166 |
| **1824** |  | 2032.46 | 1377.3 | 1508.91 | .38012 | 7.6838 | **1874** |  | 2094.41 | 1547.9 | 1556.52 | .34750 | 7.7173 |
| **1825** | 1551.85 | 2033.69 | 1380.6 | 1509.86 | .37943 | 7.6844 | **1875** | 1601.85 | 2095.66 | 1551.5 | 1557.47 | .34689 | 7.7179 |
| **1826** |  | 2034.93 | 1383.8 | 1510.81 | .37874 | 7.6851 | **1876** |  | 2096.90 | 1555.0 | 1558.43 | .34627 | 7.7186 |
| **1827** |  | 2036.17 | 1387.1 | 1511.76 | .37806 | 7.6858 | **1877** |  | 2098.14 | 1558.6 | 1559.38 | .34566 | 7.7193 |
| **1828** |  | 2037.41 | 1390.4 | 1512.71 | .37737 | 7.6865 | **1878** |  | 2099.38 | 1562.2 | 1560.33 | .34505 | 7.7199 |
| **1829** |  | 2038.64 | 1393.7 | 1513.66 | .37669 | 7.6872 | **1879** |  | 2100.62 | 1565.8 | 1561.29 | .34444 | 7.7206 |
| **1830** | 1556.85 | 2039.88 | 1397.0 | 1514.62 | .37601 | 7.6878 | **1880** | 1606.85 | 2101.86 | 1569.4 | 1562.24 | .34383 | 7.7212 |
| **1831** |  | 2041.12 | 1400.3 | 1515.57 | .37533 | 7.6885 | **1881** |  | 2103.10 | 1573.0 | 1563.20 | .34322 | 7.7219 |
| **1832** |  | 2042.36 | 1403.6 | 1516.52 | .37465 | 7.6892 | **1882** |  | 2104.34 | 1576.7 | 1564.15 | .34262 | 7.7226 |
| **1833** |  | 2043.60 | 1406.9 | 1517.47 | .37397 | 7.6899 | **1883** |  | 2105.59 | 1580.3 | 1565.11 | .34201 | 7.7232 |
| **1834** |  | 2044.83 | 1410.2 | 1518.42 | .37330 | 7.6905 | **1884** |  | 2106.83 | 1583.9 | 1566.06 | .34141 | 7.7239 |
| **1835** | 1561.85 | 2046.07 | 1413.5 | 1519.37 | .37262 | 7.6912 | **1885** | 1611.85 | 2108.07 | 1587.6 | 1567.01 | .34081 | 7.7245 |
| **1836** |  | 2047.31 | 1416.8 | 1520.32 | .37195 | 7.6919 | **1886** |  | 2109.31 | 1591.2 | 1567.97 | .34021 | 7.7252 |
| **1837** |  | 2048.55 | 1420.2 | 1521.27 | .37128 | 7.6926 | **1887** |  | 2110.55 | 1594.9 | 1568.92 | .33961 | 7.7259 |
| **1838** |  | 2049.79 | 1423.5 | 1522.22 | .37061 | 7.6932 | **1888** |  | 2111.79 | 1598.5 | 1569.88 | .33901 | 7.7265 |
| **1839** |  | 2051.03 | 1426.8 | 1523.18 | .36994 | 7.6939 | **1889** |  | 2113.04 | 1602.2 | 1570.83 | .33841 | 7.7272 |
| **1840** | 1566.85 | 2052.26 | 1430.2 | 1524.13 | .36928 | 7.6946 | **1890** | 1616.85 | 2114.28 | 1605.9 | 1571.79 | .33782 | 7.7278 |
| **1841** |  | 2053.50 | 1433.5 | 1525.08 | .36861 | 7.6953 | **1891** |  | 2115.52 | 1609.5 | 1572.74 | .33722 | 7.7285 |
| **1842** |  | 2054.74 | 1436.9 | 1526.03 | .36795 | 7.6959 | **1892** |  | 2116.76 | 1613.2 | 1573.70 | .33663 | 7.7291 |
| **1843** |  | 2055.98 | 1440.3 | 1526.98 | .36729 | 7.6966 | **1893** |  | 2118.00 | 1616.9 | 1574.65 | .33604 | 7.7298 |
| **1844** |  | 2057.22 | 1443.7 | 1527.93 | .36663 | 7.6973 | **1894** |  | 2119.25 | 1620.6 | 1575.61 | .33545 | 7.7305 |
| **1845** | 1571.85 | 2058.46 | 1447.0 | 1528.89 | .36597 | 7.6979 | **1895** | 1621.85 | 2120.49 | 1624.3 | 1576.56 | .33486 | 7.7311 |
| **1846** |  | 2059.70 | 1450.4 | 1529.84 | .36531 | 7.6986 | **1896** |  | 2121.73 | 1628.0 | 1577.52 | .33427 | 7.7318 |
| **1847** |  | 2060.94 | 1453.8 | 1530.79 | .36466 | 7.6993 | **1897** |  | 2122.97 | 1631.8 | 1578.47 | .33369 | 7.7324 |
| **1848** |  | 2062.18 | 1457.2 | 1531.74 | .36400 | 7.7000 | **1898** |  | 2124.22 | 1635.5 | 1579.43 | .33310 | 7.7331 |
| **1849** |  | 2063.41 | 1460.6 | 1532.69 | .36335 | 7.7006 | **1899** |  | 2125.46 | 1639.2 | 1580.39 | .33252 | 7.7337 |

# Table 1   Air at Low Pressures (for One Kilogram)

| T | t | h | p_r | u | v_r | φ | T | t | h | p_r | u | v_r | φ |
|---|---|---|---|---|---|---|---|---|---|---|---|---|---|
| **1900** | 1626.85 | 2126.70 | 1643.0 | 1581.34 | .33194 | 7.7344 | **1950** | 1676.85 | 2188.91 | 1838.8 | 1629.20 | .30439 | 7.7667 |
| **1901** | | 2127.94 | 1646.7 | 1582.30 | .33136 | 7.7350 | **1951** | | 2190.16 | 1842.9 | 1630.16 | .30387 | 7.7673 |
| **1902** | | 2129.19 | 1650.5 | 1583.25 | .33078 | 7.7357 | **1952** | | 2191.40 | 1847.0 | 1631.12 | .30335 | 7.7680 |
| **1903** | | 2130.43 | 1654.2 | 1584.21 | .33020 | 7.7364 | **1953** | | 2192.65 | 1851.1 | 1632.08 | .30284 | 7.7686 |
| **1904** | | 2131.67 | 1658.0 | 1585.16 | .32962 | 7.7370 | **1954** | | 2193.90 | 1855.2 | 1633.04 | .30232 | 7.7693 |
| **1905** | 1631.85 | 2132.91 | 1661.8 | 1586.12 | .32905 | 7.7377 | **1955** | 1681.85 | 2195.14 | 1859.3 | 1634.00 | .30180 | 7.7699 |
| **1906** | | 2134.16 | 1665.5 | 1587.08 | .32847 | 7.7383 | **1956** | | 2196.39 | 1863.5 | 1634.95 | .30129 | 7.7705 |
| **1907** | | 2135.40 | 1669.3 | 1588.03 | .32790 | 7.7390 | **1957** | | 2197.63 | 1867.6 | 1635.91 | .30077 | 7.7712 |
| **1908** | | 2136.64 | 1673.1 | 1588.99 | .32733 | 7.7396 | **1958** | | 2198.88 | 1871.7 | 1636.87 | .30026 | 7.7718 |
| **1909** | | 2137.89 | 1676.9 | 1589.94 | .32675 | 7.7403 | **1959** | | 2200.13 | 1875.9 | 1637.83 | .29975 | 7.7724 |
| **1910** | 1636.85 | 2139.13 | 1680.7 | 1590.90 | .32619 | 7.7409 | **1960** | 1686.85 | 2201.37 | 1880.1 | 1638.79 | .29924 | 7.7731 |
| **1911** | | 2140.37 | 1684.5 | 1591.86 | .32562 | 7.7416 | **1961** | | 2202.62 | 1884.2 | 1639.75 | .29873 | 7.7737 |
| **1912** | | 2141.62 | 1688.4 | 1592.81 | .32505 | 7.7422 | **1962** | | 2203.87 | 1888.4 | 1640.71 | .29822 | 7.7744 |
| **1913** | | 2142.86 | 1692.2 | 1593.77 | .32448 | 7.7429 | **1963** | | 2205.11 | 1892.6 | 1641.67 | .29771 | 7.7750 |
| **1914** | | 2144.10 | 1696.0 | 1594.73 | .32392 | 7.7435 | **1964** | | 2206.36 | 1896.8 | 1642.63 | .29720 | 7.7756 |
| **1915** | 1641.85 | 2145.35 | 1699.9 | 1595.68 | .32336 | 7.7442 | **1965** | 1691.85 | 2207.61 | 1901.0 | 1643.59 | .29670 | 7.7763 |
| **1916** | | 2146.59 | 1703.7 | 1596.64 | .32279 | 7.7448 | **1966** | | 2208.85 | 1905.2 | 1644.55 | .29619 | 7.7769 |
| **1917** | | 2147.83 | 1707.6 | 1597.60 | .32223 | 7.7455 | **1967** | | 2210.10 | 1909.4 | 1645.51 | .29569 | 7.7775 |
| **1918** | | 2149.08 | 1711.4 | 1598.55 | .32167 | 7.7461 | **1968** | | 2211.35 | 1913.6 | 1646.47 | .29519 | 7.7782 |
| **1919** | | 2150.32 | 1715.3 | 1599.51 | .32112 | 7.7468 | **1969** | | 2212.59 | 1917.8 | 1647.43 | .29469 | 7.7788 |
| **1920** | 1646.85 | 2151.57 | 1719.2 | 1600.47 | .32056 | 7.7474 | **1970** | 1696.85 | 2213.84 | 1922.1 | 1648.39 | .29419 | 7.7794 |
| **1921** | | 2152.81 | 1723.1 | 1601.42 | .32000 | 7.7481 | **1971** | | 2215.09 | 1926.3 | 1649.35 | .29369 | 7.7801 |
| **1922** | | 2154.05 | 1727.0 | 1602.38 | .31945 | 7.7487 | **1972** | | 2216.33 | 1930.6 | 1650.31 | .29319 | 7.7807 |
| **1923** | | 2155.30 | 1730.9 | 1603.34 | .31889 | 7.7494 | **1973** | | 2217.58 | 1934.8 | 1651.27 | .29269 | 7.7813 |
| **1924** | | 2156.54 | 1734.8 | 1604.29 | .31834 | 7.7500 | **1974** | | 2218.83 | 1939.1 | 1652.23 | .29220 | 7.7820 |
| **1925** | 1651.85 | 2157.79 | 1738.7 | 1605.25 | .31779 | 7.7506 | **1975** | 1701.85 | 2220.08 | 1943.4 | 1653.19 | .29170 | 7.7826 |
| **1926** | | 2159.03 | 1742.6 | 1606.21 | .31724 | 7.7513 | **1976** | | 2221.32 | 1947.6 | 1654.15 | .29121 | 7.7832 |
| **1927** | | 2160.27 | 1746.5 | 1607.17 | .31669 | 7.7519 | **1977** | | 2222.57 | 1951.9 | 1655.11 | .29072 | 7.7839 |
| **1928** | | 2161.52 | 1750.5 | 1608.12 | .31614 | 7.7526 | **1978** | | 2223.82 | 1956.2 | 1656.07 | .29023 | 7.7845 |
| **1929** | | 2162.76 | 1754.4 | 1609.08 | .31560 | 7.7532 | **1979** | | 2225.07 | 1960.5 | 1657.03 | .28974 | 7.7851 |
| **1930** | 1656.85 | 2164.01 | 1758.3 | 1610.04 | .31505 | 7.7539 | **1980** | 1706.85 | 2226.31 | 1964.8 | 1657.99 | .28925 | 7.7857 |
| **1931** | | 2165.25 | 1762.3 | 1610.99 | .31451 | 7.7545 | **1981** | | 2227.56 | 1969.2 | 1658.95 | .28876 | 7.7864 |
| **1932** | | 2166.50 | 1766.3 | 1611.95 | .31397 | 7.7552 | **1982** | | 2228.81 | 1973.5 | 1659.91 | .28827 | 7.7870 |
| **1933** | | 2167.74 | 1770.2 | 1612.91 | .31342 | 7.7558 | **1983** | | 2230.06 | 1977.8 | 1660.87 | .28778 | 7.7876 |
| **1934** | | 2168.99 | 1774.2 | 1613.87 | .31288 | 7.7564 | **1984** | | 2231.31 | 1982.2 | 1661.84 | .28730 | 7.7883 |
| **1935** | 1661.85 | 2170.23 | 1778.2 | 1614.83 | .31234 | 7.7571 | **1985** | 1711.85 | 2232.55 | 1986.5 | 1662.80 | .28681 | 7.7889 |
| **1936** | | 2171.48 | 1782.2 | 1615.78 | .31181 | 7.7577 | **1986** | | 2233.80 | 1990.9 | 1663.76 | .28633 | 7.7895 |
| **1937** | | 2172.72 | 1786.2 | 1616.74 | .31127 | 7.7584 | **1987** | | 2235.05 | 1995.2 | 1664.72 | .28585 | 7.7901 |
| **1938** | | 2173.97 | 1790.2 | 1617.70 | .31073 | 7.7590 | **1988** | | 2236.30 | 1999.6 | 1665.68 | .28537 | 7.7908 |
| **1939** | | 2175.21 | 1794.2 | 1618.66 | .31020 | 7.7597 | **1989** | | 2237.55 | 2004.0 | 1666.64 | .28489 | 7.7914 |
| **1940** | 1666.85 | 2176.46 | 1798.2 | 1619.62 | .30967 | 7.7603 | **1990** | 1716.85 | 2238.79 | 2008.4 | 1667.60 | .28441 | 7.7920 |
| **1941** | | 2177.70 | 1802.2 | 1620.57 | .30913 | 7.7609 | **1991** | | 2240.04 | 2012.7 | 1668.56 | .28393 | 7.7927 |
| **1942** | | 2178.95 | 1806.3 | 1621.53 | .30860 | 7.7616 | **1992** | | 2241.29 | 2017.1 | 1669.52 | .28345 | 7.7933 |
| **1943** | | 2180.19 | 1810.3 | 1622.49 | .30807 | 7.7622 | **1993** | | 2242.54 | 2021.6 | 1670.49 | .28298 | 7.7939 |
| **1944** | | 2181.44 | 1814.3 | 1623.45 | .30754 | 7.7629 | **1994** | | 2243.79 | 2026.0 | 1671.45 | .28250 | 7.7945 |
| **1945** | 1671.85 | 2182.68 | 1818.4 | 1624.41 | .30702 | 7.7635 | **1995** | 1721.85 | 2245.04 | 2030.4 | 1672.41 | .28203 | 7.7952 |
| **1946** | | 2183.93 | 1822.5 | 1625.37 | .30649 | 7.7642 | **1996** | | 2246.28 | 2034.8 | 1673.37 | .28155 | 7.7958 |
| **1947** | | 2185.17 | 1826.5 | 1626.32 | .30596 | 7.7648 | **1997** | | 2247.53 | 2039.3 | 1674.33 | .28108 | 7.7964 |
| **1948** | | 2186.42 | 1830.6 | 1627.28 | .30544 | 7.7654 | **1998** | | 2248.78 | 2043.7 | 1675.29 | .28061 | 7.7970 |
| **1949** | | 2187.67 | 1834.7 | 1628.24 | .30492 | 7.7661 | **1999** | | 2250.03 | 2048.2 | 1676.26 | .28014 | 7.7977 |

# Table 1   Air at Low Pressures (for One Kilogram)

| T | t | h | $p_r$ | u | $v_r$ | φ | T | t | h | $p_r$ | u | $v_r$ | φ |
|---|---|---|---|---|---|---|---|---|---|---|---|---|---|
| 2000 | 1726.85 | 2251.28 | 2052.6 | 1677.22 | .27967 | 7.7983 | 2050 | 1776.85 | 2313.80 | 2285.7 | 1725.38 | .25743 | 7.8292 |
| 2001 | | 2252.53 | 2057.1 | 1678.18 | .27920 | 7.7989 | 2051 | | 2315.05 | 2290.6 | 1726.35 | .25701 | 7.8298 |
| 2002 | | 2253.78 | 2061.6 | 1679.14 | .27874 | 7.7995 | 2052 | | 2316.30 | 2295.5 | 1727.31 | .25659 | 7.8304 |
| 2003 | | 2255.03 | 2066.1 | 1680.10 | .27827 | 7.8002 | 2053 | | 2317.55 | 2300.4 | 1728.28 | .25617 | 7.8310 |
| 2004 | | 2256.28 | 2070.6 | 1681.07 | .27781 | 7.8008 | 2054 | | 2318.81 | 2305.3 | 1729.24 | .25575 | 7.8316 |
| 2005 | 1731.85 | 2257.52 | 2075.1 | 1682.03 | .27734 | 7.8014 | 2055 | 1781.85 | 2320.06 | 2310.1 | 1730.21 | .25533 | 7.8322 |
| 2006 | | 2258.77 | 2079.6 | 1682.99 | .27688 | 7.8020 | 2056 | | 2321.31 | 2315.1 | 1731.17 | .25491 | 7.8328 |
| 2007 | | 2260.02 | 2084.1 | 1683.95 | .27642 | 7.8027 | 2057 | | 2322.56 | 2320.0 | 1732.14 | .25450 | 7.8334 |
| 2008 | | 2261.27 | 2088.6 | 1684.91 | .27596 | 7.8033 | 2058 | | 2323.82 | 2324.9 | 1733.10 | .25408 | 7.8340 |
| 2009 | | 2262.52 | 2093.1 | 1685.88 | .27549 | 7.8039 | 2059 | | 2325.07 | 2329.8 | 1734.07 | .25367 | 7.8346 |
| 2010 | 1736.85 | 2263.77 | 2097.7 | 1686.84 | .27504 | 7.8045 | 2060 | 1786.85 | 2326.32 | 2334.8 | 1735.04 | .25325 | 7.8353 |
| 2011 | | 2265.02 | 2102.2 | 1687.80 | .27458 | 7.8051 | 2061 | | 2327.57 | 2339.7 | 1736.00 | .25284 | 7.8359 |
| 2012 | | 2266.27 | 2106.8 | 1688.76 | .27412 | 7.8058 | 2062 | | 2328.82 | 2344.7 | 1736.97 | .25243 | 7.8365 |
| 2013 | | 2267.52 | 2111.3 | 1689.73 | .27366 | 7.8064 | 2063 | | 2330.08 | 2349.6 | 1737.93 | .25201 | 7.8371 |
| 2014 | | 2268.77 | 2115.9 | 1690.69 | .27321 | 7.8070 | 2064 | | 2331.33 | 2354.6 | 1738.90 | .25160 | 7.8377 |
| 2015 | 1741.85 | 2270.02 | 2120.5 | 1691.65 | .27275 | 7.8076 | 2065 | 1791.85 | 2332.58 | 2359.6 | 1739.86 | .25119 | 7.8383 |
| 2016 | | 2271.27 | 2125.1 | 1692.61 | .27230 | 7.8082 | 2066 | | 2333.84 | 2364.6 | 1740.83 | .25079 | 7.8389 |
| 2017 | | 2272.52 | 2129.7 | 1693.58 | .27185 | 7.8089 | 2067 | | 2335.09 | 2369.6 | 1741.79 | .25038 | 7.8395 |
| 2018 | | 2273.77 | 2134.3 | 1694.54 | .27140 | 7.8095 | 2068 | | 2336.34 | 2374.6 | 1742.76 | .24997 | 7.8401 |
| 2019 | | 2275.02 | 2138.9 | 1695.50 | .27094 | 7.8101 | 2069 | | 2337.59 | 2379.6 | 1743.73 | .24956 | 7.8407 |
| 2020 | 1746.85 | 2276.27 | 2143.5 | 1696.47 | .27050 | 7.8107 | 2070 | 1796.85 | 2338.85 | 2384.6 | 1744.69 | .24916 | 7.8413 |
| 2021 | | 2277.52 | 2148.1 | 1697.43 | .27005 | 7.8113 | 2071 | | 2340.10 | 2389.7 | 1745.66 | .24875 | 7.8419 |
| 2022 | | 2278.77 | 2152.7 | 1698.39 | .26960 | 7.8120 | 2072 | | 2341.35 | 2394.7 | 1746.62 | .24835 | 7.8425 |
| 2023 | | 2280.02 | 2157.4 | 1699.36 | .26915 | 7.8126 | 2073 | | 2342.61 | 2399.8 | 1747.59 | .24795 | 7.8431 |
| 2024 | | 2281.27 | 2162.0 | 1700.32 | .26871 | 7.8132 | 2074 | | 2343.86 | 2404.8 | 1748.56 | .24755 | 7.8437 |
| 2025 | 1751.85 | 2282.52 | 2166.7 | 1701.28 | .26826 | 7.8138 | 2075 | 1801.85 | 2345.11 | 2409.9 | 1749.52 | .24714 | 7.8443 |
| 2026 | | 2283.77 | 2171.4 | 1702.25 | .26782 | 7.8144 | 2076 | | 2346.37 | 2415.0 | 1750.49 | .24674 | 7.8450 |
| 2027 | | 2285.02 | 2176.0 | 1703.21 | .26737 | 7.8150 | 2077 | | 2347.62 | 2420.1 | 1751.46 | .24634 | 7.8456 |
| 2028 | | 2286.27 | 2180.7 | 1704.17 | .26693 | 7.8157 | 2078 | | 2348.87 | 2425.1 | 1752.42 | .24594 | 7.8462 |
| 2029 | | 2287.52 | 2185.4 | 1705.14 | .26649 | 7.8163 | 2079 | | 2350.13 | 2430.2 | 1753.39 | .24555 | 7.8468 |
| 2030 | 1756.85 | 2288.77 | 2190.1 | 1706.10 | .26605 | 7.8169 | 2080 | 1806.85 | 2351.38 | 2435.4 | 1754.35 | .24515 | 7.8474 |
| 2031 | | 2290.02 | 2194.8 | 1707.06 | .26561 | 7.8175 | 2081 | | 2352.63 | 2440.5 | 1755.32 | .24475 | 7.8480 |
| 2032 | | 2291.27 | 2199.5 | 1708.03 | .26517 | 7.8181 | 2082 | | 2353.89 | 2445.6 | 1756.29 | .24436 | 7.8486 |
| 2033 | | 2292.53 | 2204.2 | 1708.99 | .26473 | 7.8187 | 2083 | | 2355.14 | 2450.7 | 1757.25 | .24396 | 7.8492 |
| 2034 | | 2293.78 | 2209.0 | 1709.96 | .26430 | 7.8194 | 2084 | | 2356.39 | 2455.9 | 1758.22 | .24357 | 7.8498 |
| 2035 | 1761.85 | 2295.03 | 2213.7 | 1710.92 | .26386 | 7.8200 | 2085 | 1811.85 | 2357.65 | 2461.0 | 1759.19 | .24318 | 7.8504 |
| 2036 | | 2296.28 | 2218.5 | 1711.88 | .26343 | 7.8206 | 2086 | | 2358.90 | 2466.2 | 1760.15 | .24278 | 7.8510 |
| 2037 | | 2297.53 | 2223.2 | 1712.85 | .26299 | 7.8212 | 2087 | | 2360.16 | 2471.4 | 1761.12 | .24239 | 7.8516 |
| 2038 | | 2298.78 | 2228.0 | 1713.81 | .26256 | 7.8218 | 2088 | | 2361.41 | 2476.5 | 1762.09 | .24200 | 7.8522 |
| 2039 | | 2300.03 | 2232.7 | 1714.78 | .26213 | 7.8224 | 2089 | | 2362.66 | 2481.7 | 1763.06 | .24161 | 7.8528 |
| 2040 | 1766.85 | 2301.28 | 2237.5 | 1715.74 | .26169 | 7.8230 | 2090 | 1816.85 | 2363.92 | 2486.9 | 1764.02 | .24122 | 7.8534 |
| 2041 | | 2302.53 | 2242.3 | 1716.70 | .26126 | 7.8237 | 2091 | | 2365.17 | 2492.1 | 1764.99 | .24083 | 7.8540 |
| 2042 | | 2303.79 | 2247.1 | 1717.67 | .26083 | 7.8243 | 2092 | | 2366.43 | 2497.3 | 1765.96 | .24044 | 7.8546 |
| 2043 | | 2305.04 | 2251.9 | 1718.63 | .26041 | 7.8249 | 2093 | | 2367.68 | 2502.6 | 1766.92 | .24006 | 7.8552 |
| 2044 | | 2306.29 | 2256.7 | 1719.60 | .25998 | 7.8255 | 2094 | | 2368.93 | 2507.8 | 1767.89 | .23967 | 7.8558 |
| 2045 | 1771.85 | 2307.54 | 2261.5 | 1720.56 | .25955 | 7.8261 | 2095 | 1821.85 | 2370.19 | 2513.0 | 1768.86 | .23929 | 7.8564 |
| 2046 | | 2308.79 | 2266.3 | 1721.53 | .25912 | 7.8267 | 2096 | | 2371.44 | 2518.3 | 1769.83 | .23890 | 7.8570 |
| 2047 | | 2310.04 | 2271.2 | 1722.49 | .25870 | 7.8273 | 2097 | | 2372.70 | 2523.5 | 1770.79 | .23852 | 7.8576 |
| 2048 | | 2311.30 | 2276.0 | 1723.46 | .25828 | 7.8279 | 2098 | | 2373.95 | 2528.8 | 1771.76 | .23814 | 7.8582 |
| 2049 | | 2312.55 | 2280.9 | 1724.42 | .25785 | 7.8286 | 2099 | | 2375.21 | 2534.1 | 1772.73 | .23775 | 7.8588 |

# Table 1  Air at Low Pressures (for One Kilogram)

| T | t | h | $p_r$ | u | $v_r$ | $\phi$ | T | t | h | $p_r$ | u | $v_r$ | $\phi$ |
|---|---|---|---|---|---|---|---|---|---|---|---|---|---|
| 2100 | 1826.85 | 2376.46 | 2539.3 | 1773.70 | .23737 | 7.8594 | 2350 | 2076.85 | 2691.75 | 4162.3 | 2017.23 | .16206 | 8.0012 |
| 2105 | | 2382.74 | 2565.9 | 1778.54 | .23548 | 7.8623 | 2355 | | 2698.09 | 4201.5 | 2022.13 | .16088 | 8.0039 |
| 2110 | | 2389.01 | 2592.6 | 1783.38 | .23360 | 7.8653 | 2360 | | 2704.43 | 4241.1 | 2027.03 | .15972 | 8.0066 |
| 2115 | | 2395.29 | 2619.6 | 1788.22 | .23174 | 7.8683 | 2365 | | 2710.76 | 4280.9 | 2031.94 | .15857 | 8.0093 |
| 2120 | | 2401.57 | 2646.8 | 1793.06 | .22990 | 7.8713 | 2370 | | 2717.10 | 4321.0 | 2036.84 | .15743 | 8.0119 |
| 2125 | 1851.85 | 2407.85 | 2674.2 | 1797.90 | .22808 | 7.8742 | 2375 | 2101.85 | 2723.45 | 4361.4 | 2041.75 | .15630 | 8.0146 |
| 2130 | | 2414.13 | 2701.9 | 1802.75 | .22628 | 7.8772 | 2380 | | 2729.79 | 4402.2 | 2046.65 | .15518 | 8.0173 |
| 2135 | | 2420.41 | 2729.8 | 1807.60 | .22449 | 7.8801 | 2385 | | 2736.13 | 4443.2 | 2051.56 | .15407 | 8.0199 |
| 2140 | | 2426.69 | 2757.9 | 1812.45 | .22273 | 7.8831 | 2390 | | 2742.48 | 4484.5 | 2056.47 | .15297 | 8.0226 |
| 2145 | | 2432.98 | 2786.2 | 1817.30 | .22098 | 7.8860 | 2395 | | 2748.82 | 4526.2 | 2061.38 | .15188 | 8.0253 |
| 2150 | 1876.85 | 2439.26 | 2814.7 | 1822.15 | .21924 | 7.8889 | 2400 | 2126.85 | 2755.17 | 4568.1 | 2066.29 | .15080 | 8.0279 |
| 2155 | | 2445.55 | 2843.5 | 1827.00 | .21753 | 7.8918 | 2405 | | 2761.52 | 4610.3 | 2071.21 | .14973 | 8.0305 |
| 2160 | | 2451.84 | 2872.6 | 1831.85 | .21583 | 7.8948 | 2410 | | 2767.87 | 4652.9 | 2076.12 | .14867 | 8.0332 |
| 2165 | | 2458.13 | 2901.8 | 1836.71 | .21415 | 7.8977 | 2415 | | 2774.22 | 4695.8 | 2081.04 | .14762 | 8.0358 |
| 2170 | | 2464.42 | 2931.3 | 1841.56 | .21248 | 7.9006 | 2420 | | 2780.57 | 4738.9 | 2085.95 | .14658 | 8.0384 |
| 2175 | 1901.85 | 2470.71 | 2961.0 | 1846.42 | .21084 | 7.9035 | 2425 | 2151.85 | 2786.92 | 4782.4 | 2090.87 | .14554 | 8.0411 |
| 2180 | | 2477.01 | 2991.0 | 1851.28 | .20920 | 7.9064 | 2430 | | 2793.27 | 4826.2 | 2095.79 | .14452 | 8.0437 |
| 2185 | | 2483.30 | 3021.2 | 1856.14 | .20759 | 7.9092 | 2435 | | 2799.63 | 4870.4 | 2100.71 | .14351 | 8.0463 |
| 2190 | | 2489.60 | 3051.7 | 1861.00 | .20599 | 7.9121 | 2440 | | 2805.98 | 4914.8 | 2105.63 | .14250 | 8.0489 |
| 2195 | | 2495.90 | 3082.4 | 1865.86 | .20440 | 7.9150 | 2445 | | 2812.34 | 4959.6 | 2110.55 | .14150 | 8.0515 |
| 2200 | 1926.85 | 2502.20 | 3113.3 | 1870.73 | .20283 | 7.9179 | 2450 | 2176.85 | 2818.70 | 5004.7 | 2115.47 | .14051 | 8.0541 |
| 2205 | | 2508.50 | 3144.5 | 1875.59 | .20127 | 7.9207 | 2455 | | 2825.06 | 5050.1 | 2120.40 | .13954 | 8.0567 |
| 2210 | | 2514.80 | 3175.9 | 1880.46 | .19973 | 7.9236 | 2460 | | 2831.42 | 5095.8 | 2125.32 | .13856 | 8.0593 |
| 2215 | | 2521.10 | 3207.6 | 1885.33 | .19821 | 7.9264 | 2465 | | 2837.78 | 5141.9 | 2130.25 | .13760 | 8.0619 |
| 2220 | | 2527.41 | 3239.5 | 1890.20 | .19670 | 7.9293 | 2470 | | 2844.14 | 5188.3 | 2135.17 | .13665 | 8.0644 |
| 2225 | 1951.85 | 2533.71 | 3271.7 | 1895.07 | .19520 | 7.9321 | 2475 | 2201.85 | 2850.50 | 5235.0 | 2140.10 | .13570 | 8.0670 |
| 2230 | | 2540.02 | 3304.1 | 1899.94 | .19372 | 7.9349 | 2480 | | 2856.87 | 5282.1 | 2145.03 | .13477 | 8.0696 |
| 2235 | | 2546.33 | 3336.8 | 1904.81 | .19225 | 7.9378 | 2485 | | 2863.23 | 5329.5 | 2149.96 | .13384 | 8.0722 |
| 2240 | | 2552.64 | 3369.8 | 1909.69 | .19080 | 7.9406 | 2490 | | 2869.60 | 5377.2 | 2154.89 | .13291 | 8.0747 |
| 2245 | | 2558.95 | 3403.0 | 1914.56 | .18936 | 7.9434 | 2495 | | 2875.97 | 5425.3 | 2159.82 | .13200 | 8.0773 |
| 2250 | 1976.85 | 2565.26 | 3436.4 | 1919.44 | .18793 | 7.9462 | 2500 | 2226.85 | 2882.34 | 5473.7 | 2164.76 | .13110 | 8.0798 |
| 2255 | | 2571.57 | 3470.1 | 1924.32 | .18652 | 7.9490 | 2505 | | 2888.70 | 5522.4 | 2169.69 | .13020 | 8.0824 |
| 2260 | | 2577.89 | 3504.1 | 1929.20 | .18512 | 7.9518 | 2510 | | 2895.07 | 5571.5 | 2174.63 | .12931 | 8.0849 |
| 2265 | | 2584.20 | 3538.4 | 1934.08 | .18374 | 7.9546 | 2515 | | 2901.45 | 5621.0 | 2179.56 | .12843 | 8.0874 |
| 2270 | | 2590.52 | 3572.9 | 1938.96 | .18236 | 7.9574 | 2520 | | 2907.82 | 5670.8 | 2184.50 | .12755 | 8.0900 |
| 2275 | 2001.85 | 2596.84 | 3607.7 | 1943.84 | .18100 | 7.9602 | 2525 | 2251.85 | 2914.19 | 5720.9 | 2189.44 | .12669 | 8.0925 |
| 2280 | | 2603.16 | 3642.7 | 1948.73 | .17966 | 7.9629 | 2530 | | 2920.57 | 5771.4 | 2194.38 | .12583 | 8.0950 |
| 2285 | | 2609.48 | 3678.0 | 1953.61 | .17832 | 7.9657 | 2535 | | 2926.94 | 5822.2 | 2199.32 | .12497 | 8.0975 |
| 2290 | | 2615.80 | 3713.6 | 1958.50 | .17700 | 7.9685 | 2540 | | 2933.32 | 5873.4 | 2204.26 | .12413 | 8.1001 |
| 2295 | | 2622.12 | 3749.4 | 1963.39 | .17569 | 7.9712 | 2545 | | 2939.70 | 5925.0 | 2209.20 | .12329 | 8.1026 |
| 2300 | 2026.85 | 2628.45 | 3785.6 | 1968.27 | .17439 | 7.9740 | 2550 | 2276.85 | 2946.07 | 5976.9 | 2214.15 | .12246 | 8.1051 |
| 2305 | | 2634.77 | 3822.0 | 1973.16 | .17311 | 7.9767 | 2555 | | 2952.45 | 6029.2 | 2219.09 | .12164 | 8.1076 |
| 2310 | | 2641.10 | 3858.7 | 1978.06 | .17183 | 7.9795 | 2560 | | 2958.83 | 6081.8 | 2224.03 | .12082 | 8.1101 |
| 2315 | | 2647.42 | 3895.6 | 1982.95 | .17057 | 7.9822 | 2565 | | 2965.22 | 6134.8 | 2228.98 | .12001 | 8.1125 |
| 2320 | | 2653.75 | 3932.9 | 1987.84 | .16932 | 7.9849 | 2570 | | 2971.60 | 6188.1 | 2233.93 | .11921 | 8.1150 |
| 2325 | 2051.85 | 2660.08 | 3970.4 | 1992.74 | .16808 | 7.9877 | 2575 | 2301.85 | 2977.98 | 6241.9 | 2238.88 | .11841 | 8.1175 |
| 2330 | | 2666.41 | 4008.2 | 1997.63 | .16685 | 7.9904 | 2580 | | 2984.37 | 6296.0 | 2243.83 | .11762 | 8.1200 |
| 2335 | | 2672.75 | 4046.3 | 2002.53 | .16564 | 7.9931 | 2585 | | 2990.75 | 6350.4 | 2248.78 | .11684 | 8.1225 |
| 2340 | | 2679.08 | 4084.7 | 2007.43 | .16443 | 7.9958 | 2590 | | 2997.14 | 6405.3 | 2253.73 | .11606 | 8.1249 |
| 2345 | | 2685.41 | 4123.3 | 2012.33 | .16324 | 7.9985 | 2595 | | 3003.53 | 6460.5 | 2258.68 | .11529 | 8.1274 |

# Table 1   Air at Low Pressures (for One Kilogram)

| T | t | h | $p_r$ | u | $v_r$ | $\phi$ | T | t | h | $p_r$ | u | $v_r$ | $\phi$ |
|---|---|---|---|---|---|---|---|---|---|---|---|---|---|
| 2600 | 2326.85 | 3009.91 | 6516.1 | 2263.63 | .11453 | 8.1299 | 2850 | 2576.85 | 3330.50 | 9820 | 2512.46 | .08330 | 8.2476 |
| 2605 |  | 3016.30 | 6572.1 | 2268.59 | .11377 | 8.1323 | 2855 |  | 3336.94 | 9897 | 2517.46 | .08280 | 8.2498 |
| 2610 |  | 3022.69 | 6628.4 | 2273.54 | .11302 | 8.1348 | 2860 |  | 3343.37 | 9975 | 2522.46 | .08229 | 8.2521 |
| 2615 |  | 3029.09 | 6685.2 | 2278.50 | .11228 | 8.1372 | 2865 |  | 3349.81 | 10054 | 2527.46 | .08179 | 8.2543 |
| 2620 |  | 3035.48 | 6742.3 | 2283.46 | .11154 | 8.1396 | 2870 |  | 3356.24 | 10133 | 2532.47 | .08130 | 8.2566 |
| 2625 | 2351.85 | 3041.87 | 6799.8 | 2288.41 | .11081 | 8.1421 | 2875 | 2601.85 | 3362.68 | 10212 | 2537.47 | .08081 | 8.2588 |
| 2630 |  | 3048.26 | 6857.7 | 2293.37 | .11008 | 8.1445 | 2880 |  | 3369.12 | 10292 | 2542.47 | .08032 | 8.2611 |
| 2635 |  | 3054.66 | 6916.0 | 2298.33 | .10936 | 8.1470 | 2885 |  | 3375.56 | 10373 | 2547.47 | .07983 | 8.2633 |
| 2640 |  | 3061.06 | 6974.7 | 2303.29 | .10865 | 8.1494 | 2890 |  | 3382.00 | 10453 | 2552.48 | .07935 | 8.2655 |
| 2645 |  | 3067.45 | 7033.7 | 2308.26 | .10794 | 8.1518 | 2895 |  | 3388.44 | 10535 | 2557.48 | .07888 | 8.2677 |
| 2650 | 2376.85 | 3073.85 | 7093.2 | 2313.22 | .10723 | 8.1542 | 2900 | 2626.85 | 3394.88 | 10617 | 2562.49 | .07840 | 8.2700 |
| 2655 |  | 3080.25 | 7153.1 | 2318.18 | .10654 | 8.1566 | 2905 |  | 3401.32 | 10699 | 2567.50 | .07793 | 8.2722 |
| 2660 |  | 3086.65 | 7213.3 | 2323.15 | .10585 | 8.1590 | 2910 |  | 3407.77 | 10782 | 2572.50 | .07747 | 8.2744 |
| 2665 |  | 3093.05 | 7274.0 | 2328.11 | .10516 | 8.1614 | 2915 |  | 3414.21 | 10866 | 2577.51 | .07700 | 8.2766 |
| 2670 |  | 3099.45 | 7335.1 | 2333.08 | .10448 | 8.1638 | 2920 |  | 3420.65 | 10949 | 2582.52 | .07655 | 8.2788 |
| 2675 | 2401.85 | 3105.85 | 7396.5 | 2338.05 | .10381 | 8.1662 | 2925 | 2651.85 | 3427.10 | 11034 | 2587.53 | .07609 | 8.2810 |
| 2680 |  | 3112.26 | 7458.4 | 2343.01 | .10314 | 8.1686 | 2930 |  | 3433.55 | 11119 | 2592.54 | .07564 | 8.2832 |
| 2685 |  | 3118.66 | 7520.7 | 2347.98 | .10247 | 8.1710 | 2935 |  | 3439.99 | 11204 | 2597.56 | .07519 | 8.2854 |
| 2690 |  | 3125.07 | 7583.4 | 2352.95 | .10182 | 8.1734 | 2940 |  | 3446.44 | 11290 | 2602.57 | .07474 | 8.2876 |
| 2695 |  | 3131.47 | 7646.6 | 2357.92 | .10116 | 8.1758 | 2945 |  | 3452.89 | 11377 | 2607.58 | .07430 | 8.2898 |
| 2700 | 2426.85 | 3137.88 | 7710.1 | 2362.90 | .10052 | 8.1781 | 2950 | 2676.85 | 3459.34 | 11464 | 2612.60 | .07386 | 8.2920 |
| 2705 |  | 3144.29 | 7774.1 | 2367.87 | .09987 | 8.1805 | 2955 |  | 3465.79 | 11552 | 2617.61 | .07343 | 8.2942 |
| 2710 |  | 3150.70 | 7838.4 | 2372.84 | .09924 | 8.1829 | 2960 |  | 3472.24 | 11640 | 2622.63 | .07299 | 8.2964 |
| 2715 |  | 3157.11 | 7903.2 | 2377.82 | .09860 | 8.1853 | 2965 |  | 3478.69 | 11728 | 2627.64 | .07256 | 8.2986 |
| 2720 |  | 3163.52 | 7968.5 | 2382.79 | .09798 | 8.1876 | 2970 |  | 3485.14 | 11818 | 2632.66 | .07214 | 8.3007 |
| 2725 | 2451.85 | 3169.93 | 8034.1 | 2387.77 | .09736 | 8.1900 | 2975 | 2701.85 | 3491.60 | 11907 | 2637.68 | .07171 | 8.3029 |
| 2730 |  | 3176.34 | 8100.2 | 2392.75 | .09674 | 8.1923 | 2980 |  | 3498.05 | 11997 | 2642.70 | .07129 | 8.3051 |
| 2735 |  | 3182.76 | 8166.7 | 2397.73 | .09613 | 8.1947 | 2985 |  | 3504.50 | 12088 | 2647.72 | .07088 | 8.3072 |
| 2740 |  | 3189.17 | 8233.6 | 2402.70 | .09552 | 8.1970 | 2990 |  | 3510.96 | 12180 | 2652.74 | .07046 | 8.3094 |
| 2745 |  | 3195.59 | 8301.0 | 2407.68 | .09492 | 8.1993 | 2995 |  | 3517.42 | 12272 | 2657.76 | .07005 | 8.3115 |
| 2750 | 2476.85 | 3202.00 | 8368.8 | 2412.67 | .09432 | 8.2017 | 3000 | 2726.85 | 3523.87 | 12364 | 2662.78 | .06965 | 8.3137 |
| 2755 |  | 3208.42 | 8437.1 | 2417.65 | .09373 | 8.2040 | 3005 |  | 3530.33 | 12457 | 2667.80 | .06924 | 8.3159 |
| 2760 |  | 3214.84 | 8505.8 | 2422.63 | .09314 | 8.2063 | 3010 |  | 3536.79 | 12551 | 2672.83 | .06884 | 8.3180 |
| 2765 |  | 3221.25 | 8574.9 | 2427.61 | .09255 | 8.2087 | 3015 |  | 3543.25 | 12645 | 2677.85 | .06844 | 8.3201 |
| 2770 |  | 3227.67 | 8644.5 | 2432.60 | .09198 | 8.2110 | 3020 |  | 3549.71 | 12739 | 2682.87 | .06804 | 8.3223 |
| 2775 | 2501.85 | 3234.09 | 8714.5 | 2437.58 | .09140 | 8.2133 | 3025 | 2751.85 | 3556.17 | 12834 | 2687.90 | .06765 | 8.3244 |
| 2780 |  | 3240.52 | 8785.0 | 2442.57 | .09083 | 8.2156 | 3030 |  | 3562.63 | 12930 | 2692.93 | .06726 | 8.3266 |
| 2785 |  | 3246.94 | 8855.9 | 2447.56 | .09027 | 8.2179 | 3035 |  | 3569.09 | 13027 | 2697.95 | .06687 | 8.3287 |
| 2790 |  | 3253.36 | 8927.3 | 2452.54 | .08970 | 8.2202 | 3040 |  | 3575.56 | 13124 | 2702.98 | .06649 | 8.3308 |
| 2795 |  | 3259.79 | 8999.1 | 2457.53 | .08915 | 8.2225 | 3045 |  | 3582.02 | 13221 | 2708.01 | .06611 | 8.3329 |
| 2800 | 2526.85 | 3266.21 | 9071.4 | 2462.52 | .08860 | 8.2248 | 3050 | 2776.85 | 3588.48 | 13319 | 2713.04 | .06573 | 8.3351 |
| 2805 |  | 3272.64 | 9144.2 | 2467.51 | .08805 | 8.2271 | 3055 |  | 3594.95 | 13418 | 2718.07 | .06535 | 8.3372 |
| 2810 |  | 3279.06 | 9217.4 | 2472.50 | .08750 | 8.2294 | 3060 |  | 3601.41 | 13517 | 2723.10 | .06498 | 8.3393 |
| 2815 |  | 3285.49 | 9291.0 | 2477.50 | .08696 | 8.2317 | 3065 |  | 3607.88 | 13617 | 2728.13 | .06461 | 8.3414 |
| 2820 |  | 3291.92 | 9365.2 | 2482.49 | .08643 | 8.2340 | 3070 |  | 3614.35 | 13717 | 2733.16 | .06424 | 8.3435 |
| 2825 | 2551.85 | 3298.35 | 9439.8 | 2487.48 | .08590 | 8.2362 | 3075 | 2801.85 | 3620.82 | 13818 | 2738.20 | .06387 | 8.3456 |
| 2830 |  | 3304.78 | 9514.9 | 2492.48 | .08537 | 8.2385 | 3080 |  | 3627.29 | 13920 | 2743.23 | .06351 | 8.3477 |
| 2835 |  | 3311.21 | 9590.4 | 2497.47 | .08485 | 8.2408 | 3085 |  | 3633.76 | 14022 | 2748.26 | .06315 | 8.3498 |
| 2840 |  | 3317.64 | 9666.5 | 2502.47 | .08433 | 8.2431 | 3090 |  | 3640.23 | 14125 | 2753.30 | .06279 | 8.3519 |
| 2845 |  | 3324.07 | 9743.0 | 2507.47 | .08381 | 8.2453 | 3095 |  | 3646.70 | 14228 | 2758.33 | .06244 | 8.3540 |

# Table 1   Air at Low Pressures (for One Kilogram)

| T | t | h | p_r | u | v_r | φ | T | t | h | p_r | u | v_r | φ |
|---|---|---|---|---|---|---|---|---|---|---|---|---|---|
| 3100 | 2826.85 | 3653.17 | 14332 | 2763.37 | .06208 | 8.3561 | 3350 | 3076.85 | 3977.62 | 20352 | 3016.07 | .04725 | 8.4568 |
| 3105 | | 3659.64 | 14436 | 2768.41 | .06173 | 8.3582 | 3355 | | 3984.13 | 20490 | 3021.14 | .04700 | 8.4587 |
| 3110 | | 3666.11 | 14542 | 2773.45 | .06139 | 8.3603 | 3360 | | 3990.63 | 20628 | 3026.21 | .04675 | 8.4606 |
| 3115 | | 3672.59 | 14647 | 2778.49 | .06104 | 8.3623 | 3365 | | 3997.14 | 20768 | 3031.28 | .04651 | 8.4626 |
| 3120 | | 3679.06 | 14754 | 2783.52 | .06070 | 8.3644 | 3370 | | 4003.65 | 20908 | 3036.35 | .04626 | 8.4645 |
| 3125 | 2851.85 | 3685.54 | 14861 | 2788.56 | .06036 | 8.3665 | 3375 | 3101.85 | 4010.16 | 21049 | 3041.43 | .04602 | 8.4664 |
| 3130 | | 3692.01 | 14968 | 2793.61 | .06002 | 8.3686 | 3380 | | 4016.67 | 21191 | 3046.50 | .04578 | 8.4684 |
| 3135 | | 3698.49 | 15077 | 2798.65 | .05969 | 8.3706 | 3385 | | 4023.18 | 21334 | 3051.57 | .04554 | 8.4703 |
| 3140 | | 3704.97 | 15185 | 2803.69 | .05935 | 8.3727 | 3390 | | 4029.69 | 21477 | 3056.65 | .04531 | 8.4722 |
| 3145 | | 3711.45 | 15295 | 2808.73 | .05902 | 8.3748 | 3395 | | 4036.20 | 21621 | 3061.72 | .04507 | 8.4741 |
| 3150 | 2876.85 | 3717.92 | 15405 | 2813.78 | .05869 | 8.3768 | 3400 | 3126.85 | 4042.71 | 21766 | 3066.80 | .04484 | 8.4760 |
| 3155 | | 3724.40 | 15516 | 2818.82 | .05837 | 8.3789 | 3405 | | 4049.22 | 21912 | 3071.88 | .04460 | 8.4779 |
| 3160 | | 3730.88 | 15627 | 2823.86 | .05804 | 8.3809 | 3410 | | 4055.73 | 22058 | 3076.95 | .04437 | 8.4799 |
| 3165 | | 3737.36 | 15739 | 2828.91 | .05772 | 8.3830 | 3415 | | 4062.24 | 22205 | 3082.03 | .04414 | 8.4818 |
| 3170 | | 3743.85 | 15851 | 2833.96 | .05740 | 8.3850 | 3420 | | 4068.76 | 22353 | 3087.11 | .04392 | 8.4837 |
| 3175 | 2901.85 | 3750.33 | 15965 | 2839.00 | .05708 | 8.3871 | 3425 | 3151.85 | 4075.27 | 22502 | 3092.19 | .04369 | 8.4856 |
| 3180 | | 3756.81 | 16079 | 2844.05 | .05677 | 8.3891 | 3430 | | 4081.79 | 22651 | 3097.27 | .04346 | 8.4875 |
| 3185 | | 3763.29 | 16193 | 2849.10 | .05646 | 8.3911 | 3435 | | 4088.30 | 22802 | 3102.35 | .04324 | 8.4894 |
| 3190 | | 3769.78 | 16308 | 2854.15 | .05614 | 8.3932 | 3440 | | 4094.82 | 22953 | 3107.43 | .04302 | 8.4913 |
| 3195 | | 3776.26 | 16424 | 2859.20 | .05584 | 8.3952 | 3445 | | 4101.33 | 23104 | 3112.51 | .04280 | 8.4932 |
| 3200 | 2926.85 | 3782.75 | 16541 | 2864.25 | .05553 | 8.3972 | 3450 | 3176.85 | 4107.85 | 23257 | 3117.59 | .04258 | 8.4951 |
| 3205 | | 3789.24 | 16658 | 2869.30 | .05523 | 8.3993 | 3455 | | 4114.37 | 23411 | 3122.68 | .04236 | 8.4969 |
| 3210 | | 3795.72 | 16776 | 2874.35 | .05492 | 8.4013 | 3460 | | 4120.89 | 23565 | 3127.76 | .04214 | 8.4988 |
| 3215 | | 3802.21 | 16894 | 2879.40 | .05462 | 8.4033 | 3465 | | 4127.41 | 23720 | 3132.84 | .04193 | 8.5007 |
| 3220 | | 3808.70 | 17013 | 2884.46 | .05433 | 8.4053 | 3470 | | 4133.93 | 23876 | 3137.93 | .04172 | 8.5026 |
| 3225 | 2951.85 | 3815.19 | 17133 | 2889.51 | .05403 | 8.4073 | 3475 | 3201.85 | 4140.45 | 24033 | 3143.01 | .04150 | 8.5045 |
| 3230 | | 3821.68 | 17253 | 2894.57 | .05374 | 8.4093 | 3480 | | 4146.97 | 24190 | 3148.10 | .04129 | 8.5063 |
| 3235 | | 3828.17 | 17374 | 2899.62 | .05344 | 8.4114 | 3485 | | 4153.49 | 24348 | 3153.18 | .04108 | 8.5082 |
| 3240 | | 3834.66 | 17496 | 2904.68 | .05315 | 8.4134 | 3490 | | 4160.01 | 24508 | 3158.27 | .04087 | 8.5101 |
| 3245 | | 3841.15 | 17619 | 2909.73 | .05287 | 8.4154 | 3495 | | 4166.53 | 24668 | 3163.36 | .04067 | 8.5120 |
| 3250 | 2976.85 | 3847.64 | 17742 | 2914.79 | .05258 | 8.4174 | 3500 | 3226.85 | 4173.05 | 24828 | 3168.45 | .04046 | 8.5138 |
| 3255 | | 3854.14 | 17866 | 2919.85 | .05230 | 8.4194 | 3505 | | 4179.58 | 24990 | 3173.53 | .04026 | 8.5157 |
| 3260 | | 3860.63 | 17990 | 2924.91 | .05201 | 8.4213 | 3510 | | 4186.10 | 25152 | 3178.62 | .04006 | 8.5175 |
| 3265 | | 3867.12 | 18115 | 2929.97 | .05173 | 8.4233 | 3515 | | 4192.63 | 25316 | 3183.71 | .03985 | 8.5194 |
| 3270 | | 3873.62 | 18241 | 2935.03 | .05145 | 8.4253 | 3520 | | 4199.15 | 25480 | 3188.80 | .03965 | 8.5213 |
| 3275 | 3001.85 | 3880.11 | 18368 | 2940.09 | .05118 | 8.4273 | 3525 | 3251.85 | 4205.68 | 25645 | 3193.89 | .03945 | 8.5231 |
| 3280 | | 3886.61 | 18495 | 2945.15 | .05090 | 8.4293 | 3530 | | 4212.20 | 25811 | 3198.98 | .03926 | 8.5250 |
| 3285 | | 3893.11 | 18623 | 2950.21 | .05063 | 8.4313 | 3535 | | 4218.73 | 25977 | 3204.08 | .03906 | 8.5268 |
| 3290 | | 3899.60 | 18752 | 2955.27 | .05036 | 8.4332 | 3540 | | 4225.26 | 26145 | 3209.17 | .03886 | 8.5286 |
| 3295 | | 3906.10 | 18881 | 2960.33 | .05009 | 8.4352 | 3545 | | 4231.79 | 26313 | 3214.26 | .03867 | 8.5305 |
| 3300 | 3026.85 | 3912.60 | 19011 | 2965.40 | .04982 | 8.4372 | 3550 | 3276.85 | 4238.32 | 26483 | 3219.35 | .03848 | 8.5323 |
| 3305 | | 3919.10 | 19142 | 2970.46 | .04956 | 8.4392 | 3555 | | 4244.84 | 26653 | 3224.45 | .03828 | 8.5342 |
| 3310 | | 3925.60 | 19273 | 2975.53 | .04929 | 8.4411 | 3560 | | 4251.37 | 26824 | 3229.54 | .03809 | 8.5360 |
| 3315 | | 3932.10 | 19406 | 2980.59 | .04903 | 8.4431 | 3565 | | 4257.90 | 26995 | 3234.64 | .03791 | 8.5378 |
| 3320 | | 3938.60 | 19539 | 2985.66 | .04877 | 8.4450 | 3570 | | 4264.44 | 27168 | 3239.73 | .03772 | 8.5397 |
| 3325 | 3051.85 | 3945.10 | 19672 | 2990.72 | .04851 | 8.4470 | 3575 | 3301.85 | 4270.97 | 27342 | 3244.83 | .03753 | 8.5415 |
| 3330 | | 3951.61 | 19807 | 2995.79 | .04826 | 8.4490 | 3580 | | 4277.50 | 27516 | 3249.93 | .03734 | 8.5433 |
| 3335 | | 3958.11 | 19942 | 3000.86 | .04800 | 8.4509 | 3585 | | 4284.03 | 27692 | 3255.02 | .03716 | 8.5451 |
| 3340 | | 3964.61 | 20078 | 3005.93 | .04775 | 8.4529 | 3590 | | 4290.56 | 27868 | 3260.12 | .03698 | 8.5470 |
| 3345 | | 3971.12 | 20214 | 3011.00 | .04750 | 8.4548 | 3595 | | 4297.10 | 28045 | 3265.22 | .03679 | 8.5488 |

# Table 2  Air at Low Pressures

| T | t | $c_p$ | $c_v$ | $k = \dfrac{c_p}{c_v}$ | a | $\dfrac{G_{max}}{p_i}$ | $\mu \times 10^7$ | $\lambda$ | $Pr = $ |
|---|---|---|---|---|---|---|---|---|---|
| K | C | $\dfrac{J}{kg\ K}$ | $\dfrac{J}{kg\ K}$ | | $\dfrac{m}{s}$ | $\dfrac{kg}{s\ m^2}\Big/ kPa$ | $\dfrac{N\ s}{m^2}$ | $\dfrac{W}{m\ K}$ | $\dfrac{c_p \mu}{\lambda}$ |
| 50 | −223.15 | 1001.9 | 714.8 | 1.402 | 141.8 | 5.7179 | | | |
| 75 | −198.15 | 1001.9 | 714.9 | 1.402 | 173.7 | 4.6686 | | | |
| 100 | −173.15 | 1001.9 | 714.9 | 1.401 | 200.6 | 4.0431 | 70.6 | .0092 | .767 |
| 125 | −148.15 | 1002.0 | 714.9 | 1.401 | 224.2 | 3.6163 | 87.6 | .0115 | .763 |
| 150 | −123.15 | 1002.0 | 715.0 | 1.401 | 245.6 | 3.3012 | 103.8 | .0137 | .756 |
| 175 | −98.15 | 1002.1 | 715.0 | 1.401 | 265.3 | 3.0563 | 119.1 | .0159 | .749 |
| 200 | −73.15 | 1002.2 | 715.1 | 1.401 | 283.6 | 2.8589 | 133.6 | .0181 | .740 |
| 225 | −48.15 | 1002.4 | 715.4 | 1.401 | 300.8 | 2.6953 | 147.4 | .0202 | .731 |
| 250 | −23.15 | 1002.8 | 715.8 | 1.401 | 317.1 | 2.5569 | 160.6 | .0223 | .724 |
| 275 | 1.85 | 1003.5 | 716.5 | 1.401 | 332.5 | 2.4378 | 173.2 | .0242 | .717 |
| 300 | 26.85 | 1004.5 | 717.5 | 1.400 | 347.2 | 2.3338 | 185.3 | .0261 | .712 |
| 325 | 51.85 | 1006.0 | 718.9 | 1.399 | 361.3 | 2.2420 | 196.9 | .0279 | .709 |
| 350 | 76.85 | 1007.9 | 720.8 | 1.398 | 374.8 | 2.1601 | 208.1 | .0297 | .706 |
| 375 | 101.85 | 1010.2 | 723.2 | 1.397 | 387.8 | 2.0864 | 219.0 | .0314 | .705 |
| 400 | 126.85 | 1013.1 | 726.1 | 1.395 | 400.3 | 2.0196 | 229.4 | .0330 | .703 |
| 450 | 176.85 | 1020.3 | 733.2 | 1.391 | 423.9 | 1.9027 | 249.3 | .0363 | .700 |
| 500 | 226.85 | 1029.2 | 742.1 | 1.387 | 446.1 | 1.8035 | 268.2 | .0395 | .699 |
| 550 | 276.85 | 1039.4 | 752.4 | 1.381 | 467.0 | 1.7177 | 286.0 | .0426 | .698 |
| 600 | 326.85 | 1050.7 | 763.6 | 1.376 | 486.8 | 1.6426 | 303.0 | .0456 | .698 |
| 650 | 376.85 | 1062.5 | 775.4 | 1.370 | 505.6 | 1.5761 | 319.3 | .0484 | .701 |
| 700 | 426.85 | 1074.5 | 787.4 | 1.365 | 523.6 | 1.5168 | 334.9 | .0513 | .701 |
| 750 | 476.85 | 1086.5 | 799.4 | 1.359 | 540.9 | 1.4634 | 349.8 | .0541 | .702 |
| 800 | 526.85 | 1098.2 | 811.2 | 1.354 | 557.6 | 1.4151 | 364.3 | .0569 | .703 |
| 850 | 576.85 | 1109.5 | 822.5 | 1.349 | 573.7 | 1.3711 | 378.3 | .0597 | .703 |
| 900 | 626.85 | 1120.4 | 833.4 | 1.344 | 589.3 | 1.3309 | 391.8 | .0625 | .702 |
| 1000 | 726.85 | 1140.4 | 853.4 | 1.336 | 619.3 | 1.2598 | 417.7 | .0672 | .709 |
| 1100 | 826.85 | 1158.2 | 871.2 | 1.329 | 647.9 | 1.1988 | 442.0 | .0717 | .714 |
| 1200 | 926.85 | 1173.8 | 886.8 | 1.324 | 675.2 | 1.1459 | 465.0 | .0759 | .719 |
| 1300 | 1026.85 | 1187.5 | 900.5 | 1.319 | 701.5 | 1.0994 | 488.0 | .0797 | .727 |
| 1400 | 1126.85 | 1199.6 | 912.6 | 1.315 | 726.8 | 1.0581 | 509.0 | .0835 | .731 |
| 1500 | 1226.85 | 1210.2 | 923.2 | 1.311 | 751.3 | 1.0211 | | | |
| 1600 | 1326.85 | 1219.7 | 932.6 | 1.308 | 775.0 | .9877 | | | |
| 1700 | 1426.85 | 1228.1 | 941.1 | 1.305 | 798.0 | .9575 | | | |
| 1800 | 1526.85 | 1235.7 | 948.7 | 1.303 | 820.4 | .9298 | | | |
| 1900 | 1626.85 | 1242.6 | 955.6 | 1.300 | 842.1 | .9044 | | | |
| 2000 | 1726.85 | 1248.9 | 961.9 | 1.298 | 863.3 | .8810 | | | |
| 2100 | 1826.85 | 1254.7 | 967.6 | 1.297 | 884.1 | .8593 | | | |
| 2200 | 1926.85 | 1260.0 | 973.0 | 1.295 | 904.3 | .8392 | | | |
| 2300 | 2026.85 | 1264.9 | 977.9 | 1.294 | 924.1 | .8204 | | | |
| 2400 | 2126.85 | 1269.5 | 982.5 | 1.292 | 943.5 | .8028 | | | |
| 2500 | 2226.85 | 1273.8 | 986.7 | 1.291 | 962.5 | .7863 | | | |
| 2600 | 2326.85 | 1277.8 | 990.7 | 1.290 | 981.1 | .7707 | | | |
| 2700 | 2426.85 | 1281.5 | 994.5 | 1.289 | 999.3 | .7561 | | | |
| 2800 | 2526.85 | 1285.0 | 998.0 | 1.288 | 1017.3 | .7422 | | | |
| 2900 | 2626.85 | 1288.3 | 1001.3 | 1.287 | 1034.9 | .7291 | | | |
| 3000 | 2726.85 | 1291.5 | 1004.4 | 1.286 | 1052.2 | .7167 | | | |
| 3100 | 2826.85 | 1294.4 | 1007.4 | 1.285 | 1069.3 | .7049 | | | |
| 3200 | 2926.85 | 1297.2 | 1010.2 | 1.284 | 1086.0 | .6936 | | | |
| 3300 | 3026.85 | 1299.8 | 1012.8 | 1.283 | 1102.6 | .6828 | | | |
| 3400 | 3126.85 | 1302.3 | 1015.3 | 1.283 | 1118.8 | .6726 | | | |

# Table 3 Products—400% Theoretical Air (for One Gram-Mole)

| T | t | $\bar{h}$ | $p_r$ | $\bar{u}$ | $v_r$ | $\bar{\phi}$ | T | t | $\bar{h}$ | $p_r$ | $\bar{u}$ | $v_r$ | $\bar{\phi}$ |
|---|---|---|---|---|---|---|---|---|---|---|---|---|---|
| 100 | −173.15 | 2908.2 | .02860 | 2076.8 | 29.070 | 161.89 | 150 | −123.15 | 4367.6 | .11872 | 3120.4 | 10.5052 | 173.73 |
| 101 | | 2937.4 | .02962 | 2097.6 | 28.353 | 162.18 | 151 | | 4396.8 | .12152 | 3141.3 | 10.3312 | 173.92 |
| 102 | | 2966.6 | .03066 | 2118.5 | 27.661 | 162.47 | 152 | | 4426.0 | .12437 | 3162.2 | 10.1612 | 174.12 |
| 103 | | 2995.7 | .03173 | 2139.4 | 26.992 | 162.76 | 153 | | 4455.2 | .12727 | 3183.1 | 9.9951 | 174.31 |
| 104 | | 3024.9 | .03282 | 2160.2 | 26.346 | 163.04 | 154 | | 4484.4 | .13022 | 3204.0 | 9.8328 | 174.50 |
| 105 | −168.15 | 3054.1 | .03394 | 2181.1 | 25.721 | 163.32 | 155 | −118.15 | 4513.7 | .13322 | 3224.9 | 9.6741 | 174.69 |
| 106 | | 3083.3 | .03509 | 2201.9 | 25.116 | 163.59 | 156 | | 4542.9 | .13626 | 3245.8 | 9.5189 | 174.87 |
| 107 | | 3112.4 | .03627 | 2222.8 | 24.532 | 163.87 | 157 | | 4572.1 | .13935 | 3266.7 | 9.3672 | 175.06 |
| 108 | | 3141.6 | .03747 | 2243.7 | 23.966 | 164.14 | 158 | | 4601.3 | .14250 | 3287.6 | 9.2188 | 175.25 |
| 109 | | 3170.8 | .03870 | 2264.5 | 23.418 | 164.41 | 159 | | 4630.5 | .14569 | 3308.5 | 9.0738 | 175.43 |
| 110 | −163.15 | 3200.0 | .03996 | 2285.4 | 22.887 | 164.67 | 160 | −113.15 | 4659.7 | .14894 | 3329.4 | 8.9318 | 175.61 |
| 111 | | 3229.1 | .04125 | 2306.2 | 22.373 | 164.94 | 161 | | 4689.0 | .15224 | 3350.4 | 8.7929 | 175.80 |
| 112 | | 3258.3 | .04257 | 2327.1 | 21.876 | 165.20 | 162 | | 4718.2 | .15559 | 3371.3 | 8.6571 | 175.98 |
| 113 | | 3287.5 | .04392 | 2348.0 | 21.393 | 165.46 | 163 | | 4747.4 | .15899 | 3392.2 | 8.5241 | 176.16 |
| 114 | | 3316.7 | .04530 | 2368.8 | 20.925 | 165.72 | 164 | | 4776.6 | .16244 | 3413.1 | 8.3940 | 176.34 |
| 115 | −158.15 | 3345.9 | .04671 | 2389.7 | 20.472 | 165.97 | 165 | −108.15 | 4805.9 | .16595 | 3434.0 | 8.2666 | 176.51 |
| 116 | | 3375.0 | .04815 | 2410.6 | 20.032 | 166.22 | 166 | | 4835.1 | .16952 | 3454.9 | 8.1419 | 176.69 |
| 117 | | 3404.2 | .04962 | 2431.4 | 19.605 | 166.48 | 167 | | 4864.3 | .17313 | 3475.8 | 8.0198 | 176.87 |
| 118 | | 3433.4 | .05112 | 2452.3 | 19.191 | 166.72 | 168 | | 4893.6 | .17681 | 3496.8 | 7.9003 | 177.04 |
| 119 | | 3462.6 | .05266 | 2473.2 | 18.788 | 166.97 | 169 | | 4922.8 | .18054 | 3517.7 | 7.7832 | 177.21 |
| 120 | −153.15 | 3491.8 | .05423 | 2494.0 | 18.398 | 167.21 | 170 | −103.15 | 4952.1 | .18432 | 3538.6 | 7.6685 | 177.39 |
| 121 | | 3520.9 | .05583 | 2514.9 | 18.019 | 167.46 | 171 | | 4981.3 | .18816 | 3559.5 | 7.5561 | 177.56 |
| 122 | | 3550.1 | .05747 | 2535.8 | 17.650 | 167.70 | 172 | | 5010.5 | .19206 | 3580.4 | 7.4460 | 177.73 |
| 123 | | 3579.3 | .05914 | 2556.6 | 17.292 | 167.93 | 173 | | 5039.8 | .19601 | 3601.4 | 7.3382 | 177.90 |
| 124 | | 3608.5 | .06085 | 2577.5 | 16.944 | 168.17 | 174 | | 5069.0 | .20003 | 3622.3 | 7.2325 | 178.07 |
| 125 | −148.15 | 3637.7 | .06259 | 2598.4 | 16.606 | 168.41 | 175 | −98.15 | 5098.3 | .20410 | 3643.2 | 7.1289 | 178.23 |
| 126 | | 3666.8 | .06436 | 2619.2 | 16.277 | 168.64 | 176 | | 5127.5 | .20823 | 3664.2 | 7.0274 | 178.40 |
| 127 | | 3696.0 | .06617 | 2640.1 | 15.957 | 168.87 | 177 | | 5156.7 | .21243 | 3685.1 | 6.9278 | 178.57 |
| 128 | | 3725.2 | .06802 | 2661.0 | 15.646 | 169.10 | 178 | | 5186.0 | .21668 | 3706.0 | 6.8303 | 178.73 |
| 129 | | 3754.4 | .06990 | 2681.8 | 15.344 | 169.32 | 179 | | 5215.3 | .22099 | 3727.0 | 6.7346 | 178.89 |
| 130 | −143.15 | 3783.6 | .07182 | 2702.7 | 15.049 | 169.55 | 180 | −93.15 | 5244.5 | .22536 | 3747.9 | 6.6408 | 179.06 |
| 131 | | 3812.8 | .07378 | 2723.6 | 14.762 | 169.77 | 181 | | 5273.7 | .22980 | 3768.8 | 6.5487 | 179.22 |
| 132 | | 3842.0 | .07578 | 2744.5 | 14.483 | 170.00 | 182 | | 5303.0 | .23430 | 3789.8 | 6.4585 | 179.38 |
| 133 | | 3871.2 | .07781 | 2765.4 | 14.211 | 170.22 | 183 | | 5332.3 | .23886 | 3810.7 | 6.3700 | 179.54 |
| 134 | | 3900.4 | .07989 | 2786.2 | 13.946 | 170.43 | 184 | | 5361.5 | .24349 | 3831.7 | 6.2831 | 179.70 |
| 135 | −138.15 | 3929.6 | .08200 | 2807.1 | 13.688 | 170.65 | 185 | −88.15 | 5390.8 | .24818 | 3852.6 | 6.1979 | 179.86 |
| 136 | | 3958.8 | .08415 | 2828.0 | 13.437 | 170.87 | 186 | | 5420.0 | .25293 | 3873.6 | 6.1143 | 180.02 |
| 137 | | 3987.9 | .08635 | 2848.9 | 13.192 | 171.08 | 187 | | 5449.3 | .25775 | 3894.5 | 6.0322 | 180.17 |
| 138 | | 4017.1 | .08858 | 2869.8 | 12.953 | 171.29 | 188 | | 5478.6 | .26263 | 3915.5 | 5.9517 | 180.33 |
| 139 | | 4046.3 | .09085 | 2890.6 | 12.720 | 171.50 | 189 | | 5507.8 | .26758 | 3936.4 | 5.8727 | 180.49 |
| 140 | −133.15 | 4075.5 | .09317 | 2911.5 | 12.493 | 171.71 | 190 | −83.15 | 5537.1 | .27260 | 3957.4 | 5.7951 | 180.64 |
| 141 | | 4104.7 | .09553 | 2932.4 | 12.272 | 171.92 | 191 | | 5566.4 | .27768 | 3978.3 | 5.7189 | 180.79 |
| 142 | | 4133.9 | .09793 | 2953.3 | 12.056 | 172.13 | 192 | | 5595.7 | .28284 | 3999.3 | 5.6441 | 180.95 |
| 143 | | 4163.1 | .10037 | 2974.2 | 11.845 | 172.33 | 193 | | 5624.9 | .28806 | 4020.3 | 5.5707 | 181.10 |
| 144 | | 4192.3 | .10286 | 2995.1 | 11.640 | 172.54 | 194 | | 5654.2 | .29335 | 4041.2 | 5.4986 | 181.25 |
| 145 | −128.15 | 4221.5 | .10539 | 3016.0 | 11.439 | 172.74 | 195 | −78.15 | 5683.5 | .29871 | 4062.2 | 5.4278 | 181.40 |
| 146 | | 4250.8 | .10797 | 3036.8 | 11.243 | 172.94 | 196 | | 5712.8 | .30414 | 4083.1 | 5.3582 | 181.55 |
| 147 | | 4280.0 | .11059 | 3057.7 | 11.052 | 173.14 | 197 | | 5742.0 | .30964 | 4104.1 | 5.2899 | 181.70 |
| 148 | | 4309.2 | .11325 | 3078.6 | 10.866 | 173.34 | 198 | | 5771.3 | .31521 | 4125.1 | 5.2228 | 181.85 |
| 149 | | 4338.4 | .11596 | 3099.5 | 10.683 | 173.53 | 199 | | 5800.6 | .32085 | 4146.0 | 5.1568 | 181.99 |

## Table 3 Products—400% Theoretical Air (for One Gram-Mole)

| T | t | $\bar{h}$ | $p_r$ | $\bar{u}$ | $v_r$ | $\bar{\phi}$ | T | t | $\bar{h}$ | $p_r$ | $\bar{u}$ | $v_r$ | $\bar{\phi}$ |
|---|---|---|---|---|---|---|---|---|---|---|---|---|---|
| 200 | −73.15 | 5829.9 | .32656 | 4167.0 | 5.0920 | 182.14 | 250 | −23.15 | 7296.7 | .7176 | 5218.1 | 2.8966 | 188.69 |
| 201 | | 5859.2 | .33235 | 4188.0 | 5.0284 | 182.29 | 251 | | 7326.1 | .7278 | 5239.2 | 2.8674 | 188.80 |
| 202 | | 5888.5 | .33821 | 4209.0 | 4.9658 | 182.43 | 252 | | 7355.5 | .7381 | 5260.3 | 2.8386 | 188.92 |
| 203 | | 5917.8 | .34415 | 4229.9 | 4.9043 | 182.58 | 253 | | 7384.9 | .7485 | 5281.4 | 2.8103 | 189.04 |
| 204 | | 5947.1 | .35016 | 4250.9 | 4.8439 | 182.72 | 254 | | 7414.3 | .7590 | 5302.5 | 2.7823 | 189.15 |
| 205 | −68.15 | 5976.4 | .35625 | 4271.9 | 4.7845 | 182.86 | 255 | −18.15 | 7443.7 | .7697 | 5323.5 | 2.7547 | 189.27 |
| 206 | | 6005.7 | .36241 | 4292.9 | 4.7261 | 183.01 | 256 | | 7473.1 | .7804 | 5344.6 | 2.7275 | 189.38 |
| 207 | | 6034.9 | .36864 | 4313.9 | 4.6687 | 183.15 | 257 | | 7502.5 | .7912 | 5365.7 | 2.7007 | 189.50 |
| 208 | | 6064.2 | .37496 | 4334.9 | 4.6122 | 183.29 | 258 | | 7531.9 | .8022 | 5386.8 | 2.6742 | 189.61 |
| 209 | | 6093.6 | .38135 | 4355.8 | 4.5567 | 183.43 | 259 | | 7561.3 | .8132 | 5407.9 | 2.6481 | 189.73 |
| 210 | −63.15 | 6122.9 | .38782 | 4376.8 | 4.5021 | 183.57 | 260 | −13.15 | 7590.8 | .8244 | 5429.0 | 2.6223 | 189.84 |
| 211 | | 6152.2 | .39437 | 4397.8 | 4.4484 | 183.71 | 261 | | 7620.2 | .8356 | 5450.1 | 2.5969 | 189.95 |
| 212 | | 6181.5 | .40100 | 4418.8 | 4.3957 | 183.85 | 262 | | 7649.6 | .8470 | 5471.2 | 2.5718 | 190.07 |
| 213 | | 6210.8 | .40771 | 4439.8 | 4.3437 | 183.99 | 263 | | 7679.0 | .8585 | 5492.3 | 2.5471 | 190.18 |
| 214 | | 6240.1 | .41450 | 4460.8 | 4.2926 | 184.12 | 264 | | 7708.4 | .8701 | 5513.4 | 2.5226 | 190.29 |
| 215 | −58.15 | 6269.4 | .42136 | 4481.8 | 4.2424 | 184.26 | 265 | −8.15 | 7737.8 | .8818 | 5534.5 | 2.4985 | 190.40 |
| 216 | | 6298.7 | .42832 | 4502.8 | 4.1930 | 184.40 | 266 | | 7767.3 | .8937 | 5555.6 | 2.4748 | 190.51 |
| 217 | | 6328.1 | .43535 | 4523.8 | 4.1443 | 184.53 | 267 | | 7796.7 | .9056 | 5576.8 | 2.4513 | 190.62 |
| 218 | | 6357.4 | .44247 | 4544.8 | 4.0964 | 184.67 | 268 | | 7826.1 | .9177 | 5597.9 | 2.4281 | 190.73 |
| 219 | | 6386.7 | .44967 | 4565.8 | 4.0494 | 184.80 | 269 | | 7855.6 | .9299 | 5619.0 | 2.4053 | 190.84 |
| 220 | −53.15 | 6416.0 | .45695 | 4586.9 | 4.0030 | 184.93 | 270 | −3.15 | 7885.0 | .9422 | 5640.1 | 2.3827 | 190.95 |
| 221 | | 6445.4 | .46432 | 4607.9 | 3.9574 | 185.07 | 271 | | 7914.4 | .9546 | 5661.2 | 2.3604 | 191.06 |
| 222 | | 6474.7 | .47177 | 4628.9 | 3.9125 | 185.20 | 272 | | 7943.9 | .9671 | 5682.4 | 2.3384 | 191.17 |
| 223 | | 6504.0 | .47931 | 4649.9 | 3.8683 | 185.33 | 273 | | 7973.3 | .9798 | 5703.5 | 2.3167 | 191.28 |
| 224 | | 6533.3 | .48694 | 4670.9 | 3.8248 | 185.46 | 274 | | 8002.8 | .9925 | 5724.6 | 2.2953 | 191.38 |
| 225 | −48.15 | 6562.7 | .49465 | 4691.9 | 3.7820 | 185.59 | 275 | 1.85 | 8032.2 | 1.0054 | 5745.8 | 2.2741 | 191.49 |
| 226 | | 6592.0 | .50245 | 4713.0 | 3.7398 | 185.72 | 276 | | 8061.7 | 1.0184 | 5766.9 | 2.2532 | 191.60 |
| 227 | | 6621.4 | .51034 | 4734.0 | 3.6983 | 185.85 | 277 | | 8091.1 | 1.0316 | 5788.0 | 2.2326 | 191.70 |
| 228 | | 6650.7 | .51832 | 4755.0 | 3.6574 | 185.98 | 278 | | 8120.6 | 1.0448 | 5809.2 | 2.2122 | 191.81 |
| 229 | | 6680.0 | .52638 | 4776.0 | 3.6171 | 186.11 | 279 | | 8150.0 | 1.0582 | 5830.3 | 2.1921 | 191.92 |
| 230 | −43.15 | 6709.4 | .53454 | 4797.1 | 3.5775 | 186.24 | 280 | 6.85 | 8179.5 | 1.0717 | 5851.5 | 2.1723 | 192.02 |
| 231 | | 6738.7 | .54279 | 4818.1 | 3.5384 | 186.37 | 281 | | 8209.0 | 1.0853 | 5872.6 | 2.1527 | 192.13 |
| 232 | | 6768.1 | .55113 | 4839.1 | 3.5000 | 186.49 | 282 | | 8238.4 | 1.0991 | 5893.8 | 2.1333 | 192.23 |
| 233 | | 6797.4 | .55956 | 4860.2 | 3.4621 | 186.62 | 283 | | 8267.9 | 1.1130 | 5914.9 | 2.1142 | 192.34 |
| 234 | | 6826.8 | .56809 | 4881.2 | 3.4248 | 186.74 | 284 | | 8297.4 | 1.1270 | 5936.1 | 2.0953 | 192.44 |
| 235 | −38.15 | 6856.1 | .57671 | 4902.3 | 3.3880 | 186.87 | 285 | 11.85 | 8326.8 | 1.1411 | 5957.2 | 2.0766 | 192.54 |
| 236 | | 6885.5 | .58542 | 4923.3 | 3.3518 | 186.99 | 286 | | 8356.3 | 1.1553 | 5978.4 | 2.0582 | 192.65 |
| 237 | | 6914.9 | .59423 | 4944.3 | 3.3161 | 187.12 | 287 | | 8385.8 | 1.1697 | 5999.6 | 2.0400 | 192.75 |
| 238 | | 6944.2 | .60313 | 4965.4 | 3.2809 | 187.24 | 288 | | 8415.3 | 1.1842 | 6020.7 | 2.0220 | 192.85 |
| 239 | | 6973.6 | .61213 | 4986.4 | 3.2463 | 187.37 | 289 | | 8444.8 | 1.1989 | 6041.9 | 2.0042 | 192.95 |
| 240 | −33.15 | 7003.0 | .62122 | 5007.5 | 3.2122 | 187.49 | 290 | 16.85 | 8474.3 | 1.2137 | 6063.1 | 1.9867 | 193.06 |
| 241 | | 7032.3 | .63041 | 5028.6 | 3.1785 | 187.61 | 291 | | 8503.8 | 1.2286 | 6084.3 | 1.9693 | 193.16 |
| 242 | | 7061.7 | .63970 | 5049.6 | 3.1454 | 187.73 | 292 | | 8533.2 | 1.2436 | 6105.4 | 1.9522 | 193.26 |
| 243 | | 7091.1 | .64909 | 5070.7 | 3.1127 | 187.85 | 293 | | 8562.7 | 1.2588 | 6126.6 | 1.9353 | 193.36 |
| 244 | | 7120.5 | .65858 | 5091.7 | 3.0805 | 187.97 | 294 | | 8592.2 | 1.2741 | 6147.8 | 1.9186 | 193.46 |
| 245 | −28.15 | 7149.8 | .66816 | 5112.8 | 3.0487 | 188.09 | 295 | 21.85 | 8621.7 | 1.2895 | 6169.0 | 1.9020 | 193.56 |
| 246 | | 7179.2 | .67785 | 5133.9 | 3.0174 | 188.21 | 296 | | 8651.2 | 1.3051 | 6190.2 | 1.8857 | 193.66 |
| 247 | | 7208.6 | .68764 | 5154.9 | 2.9865 | 188.33 | 297 | | 8680.7 | 1.3208 | 6211.4 | 1.8696 | 193.76 |
| 248 | | 7238.0 | .69753 | 5176.0 | 2.9561 | 188.45 | 298 | | 8710.3 | 1.3367 | 6232.6 | 1.8536 | 193.86 |
| 249 | | 7267.4 | .70752 | 5197.1 | 2.9261 | 188.57 | 299 | | 8739.8 | 1.3527 | 6253.8 | 1.8378 | 193.96 |

# Table 3    Products—400% Theoretical Air (for One Gram-Mole)

| T | t | $\bar{h}$ | $p_r$ | $\bar{u}$ | $v_r$ | $\bar{\phi}$ | T | t | $\bar{h}$ | $p_r$ | $\bar{u}$ | $v_r$ | $\bar{\phi}$ |
|---|---|---|---|---|---|---|---|---|---|---|---|---|---|
| 300 | 26.85 | 8769.3 | 1.3688 | 6275.0 | 1.8223 | 194.06 | 350 | 76.85 | 10249.1 | 2.3693 | 7339.0 | 1.22822 | 198.62 |
| 301 | | 8798.8 | 1.3851 | 6296.2 | 1.8069 | 194.15 | 351 | | 10278.7 | 2.3936 | 7360.4 | 1.21925 | 198.70 |
| 302 | | 8828.3 | 1.4015 | 6317.4 | 1.7916 | 194.25 | 352 | | 10308.4 | 2.4180 | 7381.7 | 1.21037 | 198.79 |
| 303 | | 8857.9 | 1.4180 | 6338.6 | 1.7766 | 194.35 | 353 | | 10338.1 | 2.4426 | 7403.1 | 1.20157 | 198.87 |
| 304 | | 8887.4 | 1.4347 | 6359.8 | 1.7617 | 194.45 | 354 | | 10367.8 | 2.4674 | 7424.5 | 1.19286 | 198.96 |
| 305 | 31.85 | 8916.9 | 1.4516 | 6381.0 | 1.7470 | 194.54 | 355 | 81.85 | 10397.5 | 2.4924 | 7445.9 | 1.18424 | 199.04 |
| 306 | | 8946.4 | 1.4685 | 6402.2 | 1.7325 | 194.64 | 356 | | 10427.2 | 2.5176 | 7467.3 | 1.17570 | 199.12 |
| 307 | | 8976.0 | 1.4856 | 6423.5 | 1.7181 | 194.74 | 357 | | 10456.9 | 2.5429 | 7488.7 | 1.16725 | 199.21 |
| 308 | | 9005.5 | 1.5029 | 6444.7 | 1.7039 | 194.83 | 358 | | 10486.6 | 2.5685 | 7510.1 | 1.15888 | 199.29 |
| 309 | | 9035.1 | 1.5203 | 6465.9 | 1.6899 | 194.93 | 359 | | 10516.3 | 2.5942 | 7531.5 | 1.15059 | 199.37 |
| 310 | 36.85 | 9064.6 | 1.5379 | 6487.2 | 1.6760 | 195.02 | 360 | 86.85 | 10546.0 | 2.6201 | 7552.9 | 1.14238 | 199.45 |
| 311 | | 9094.2 | 1.5556 | 6508.4 | 1.6623 | 195.12 | 361 | | 10575.8 | 2.6462 | 7574.3 | 1.13425 | 199.54 |
| 312 | | 9123.7 | 1.5734 | 6529.6 | 1.6487 | 195.21 | 362 | | 10605.5 | 2.6725 | 7595.7 | 1.12620 | 199.62 |
| 313 | | 9153.3 | 1.5914 | 6550.9 | 1.6353 | 195.31 | 363 | | 10635.2 | 2.6990 | 7617.1 | 1.11823 | 199.70 |
| 314 | | 9182.8 | 1.6096 | 6572.1 | 1.6220 | 195.40 | 364 | | 10664.9 | 2.7257 | 7638.5 | 1.11033 | 199.78 |
| 315 | 41.85 | 9212.4 | 1.6279 | 6593.4 | 1.6089 | 195.50 | 365 | 91.85 | 10694.7 | 2.7526 | 7659.9 | 1.10251 | 199.86 |
| 316 | | 9241.9 | 1.6463 | 6614.6 | 1.5959 | 195.59 | 366 | | 10724.4 | 2.7797 | 7681.3 | 1.09476 | 199.95 |
| 317 | | 9271.5 | 1.6649 | 6635.8 | 1.5830 | 195.68 | 367 | | 10754.2 | 2.8069 | 7702.8 | 1.08709 | 200.03 |
| 318 | | 9301.1 | 1.6837 | 6657.1 | 1.5703 | 195.78 | 368 | | 10783.9 | 2.8344 | 7724.2 | 1.07949 | 200.11 |
| 319 | | 9330.6 | 1.7026 | 6678.4 | 1.5578 | 195.87 | 369 | | 10813.7 | 2.8620 | 7745.6 | 1.07197 | 200.19 |
| 320 | 46.85 | 9360.2 | 1.7217 | 6699.6 | 1.5454 | 195.96 | 370 | 96.85 | 10843.4 | 2.8899 | 7767.1 | 1.06451 | 200.27 |
| 321 | | 9389.8 | 1.7409 | 6720.9 | 1.5331 | 196.06 | 371 | | 10873.2 | 2.9180 | 7788.5 | 1.05712 | 200.35 |
| 322 | | 9419.4 | 1.7603 | 6742.2 | 1.5209 | 196.15 | 372 | | 10902.9 | 2.9462 | 7810.0 | 1.04981 | 200.43 |
| 323 | | 9449.0 | 1.7798 | 6763.4 | 1.5089 | 196.24 | 373 | | 10932.7 | 2.9747 | 7831.4 | 1.04256 | 200.51 |
| 324 | | 9478.6 | 1.7995 | 6784.7 | 1.4970 | 196.33 | 374 | | 10962.5 | 3.0033 | 7852.9 | 1.03538 | 200.59 |
| 325 | 51.85 | 9508.2 | 1.8193 | 6806.0 | 1.4853 | 196.42 | 375 | 101.85 | 10992.3 | 3.0322 | 7874.4 | 1.02827 | 200.67 |
| 326 | | 9537.8 | 1.8393 | 6827.3 | 1.4736 | 196.51 | 376 | | 11022.0 | 3.0612 | 7895.8 | 1.02123 | 200.75 |
| 327 | | 9567.4 | 1.8595 | 6848.6 | 1.4621 | 196.60 | 377 | | 11051.8 | 3.0905 | 7917.3 | 1.01425 | 200.83 |
| 328 | | 9597.0 | 1.8798 | 6869.8 | 1.4508 | 196.69 | 378 | | 11081.6 | 3.1200 | 7938.8 | 1.00733 | 200.91 |
| 329 | | 9626.6 | 1.9003 | 6891.1 | 1.4395 | 196.78 | 379 | | 11111.4 | 3.1496 | 7960.2 | 1.00048 | 200.99 |
| 330 | 56.85 | 9656.2 | 1.9209 | 6912.4 | 1.4283 | 196.87 | 380 | 106.85 | 11141.2 | 3.1795 | 7981.7 | .99369 | 201.06 |
| 331 | | 9685.8 | 1.9417 | 6933.7 | 1.4173 | 196.96 | 381 | | 11171.0 | 3.2096 | 8003.2 | .98697 | 201.14 |
| 332 | | 9715.4 | 1.9627 | 6955.0 | 1.4064 | 197.05 | 382 | | 11200.8 | 3.2399 | 8024.7 | .98030 | 201.22 |
| 333 | | 9745.0 | 1.9839 | 6976.3 | 1.3956 | 197.14 | 383 | | 11230.6 | 3.2704 | 8046.2 | .97370 | 201.30 |
| 334 | | 9774.6 | 2.0052 | 6997.6 | 1.3849 | 197.23 | 384 | | 11260.4 | 3.3012 | 8067.7 | .96715 | 201.38 |
| 335 | 61.85 | 9804.3 | 2.0266 | 7018.9 | 1.3744 | 197.32 | 385 | 111.85 | 11290.2 | 3.3321 | 8089.2 | .96067 | 201.45 |
| 336 | | 9833.9 | 2.0483 | 7040.2 | 1.3639 | 197.41 | 386 | | 11320.1 | 3.3633 | 8110.7 | .95424 | 201.53 |
| 337 | | 9863.5 | 2.0701 | 7061.6 | 1.3535 | 197.50 | 387 | | 11349.9 | 3.3946 | 8132.2 | .94787 | 201.61 |
| 338 | | 9893.2 | 2.0921 | 7082.9 | 1.3433 | 197.58 | 388 | | 11379.7 | 3.4262 | 8153.7 | .94156 | 201.68 |
| 339 | | 9922.8 | 2.1142 | 7104.2 | 1.3332 | 197.67 | 389 | | 11409.6 | 3.4580 | 8175.3 | .93531 | 201.76 |
| 340 | 66.85 | 9952.4 | 2.1365 | 7125.5 | 1.3231 | 197.76 | 390 | 116.85 | 11439.4 | 3.4900 | 8196.8 | .92912 | 201.84 |
| 341 | | 9982.1 | 2.1590 | 7146.9 | 1.3132 | 197.85 | 391 | | 11469.3 | 3.5222 | 8218.3 | .92298 | 201.91 |
| 342 | | 10011.7 | 2.1817 | 7168.2 | 1.3034 | 197.93 | 392 | | 11499.1 | 3.5547 | 8239.9 | .91689 | 201.99 |
| 343 | | 10041.4 | 2.2045 | 7189.5 | 1.2936 | 198.02 | 393 | | 11529.0 | 3.5873 | 8261.4 | .91086 | 202.07 |
| 344 | | 10071.0 | 2.2275 | 7210.9 | 1.2840 | 198.11 | 394 | | 11558.8 | 3.6202 | 8282.9 | .90488 | 202.14 |
| 345 | 71.85 | 10100.7 | 2.2507 | 7232.2 | 1.2745 | 198.19 | 395 | 121.85 | 11588.7 | 3.6533 | 8304.5 | .89895 | 202.22 |
| 346 | | 10130.4 | 2.2741 | 7253.6 | 1.2650 | 198.28 | 396 | | 11618.5 | 3.6867 | 8326.0 | .89308 | 202.29 |
| 347 | | 10160.0 | 2.2976 | 7274.9 | 1.2557 | 198.36 | 397 | | 11648.4 | 3.7202 | 8347.6 | .88726 | 202.37 |
| 348 | | 10189.7 | 2.3213 | 7296.3 | 1.2464 | 198.45 | 398 | | 11678.3 | 3.7540 | 8369.2 | .88150 | 202.44 |
| 349 | | 10219.4 | 2.3452 | 7317.7 | 1.2373 | 198.53 | 399 | | 11708.2 | 3.7880 | 8390.7 | .87577 | 202.52 |

## Table 3  Products—400% Theoretical Air (for One Gram-Mole)

| T | t | $\bar{h}$ | $p_r$ | $\bar{u}$ | $v_r$ | $\bar{\phi}$ | T | t | $\bar{h}$ | $p_r$ | $\bar{u}$ | $v_r$ | $\bar{\phi}$ |
|---|---|-----------|-------|-----------|-------|--------------|---|---|-----------|-------|-----------|-------|--------------|
| 400 | 126.85 | 11738.1 | 3.8223 | 8412.3 | .87011 | 202.59 | 450 | 176.85 | 13238.6 | 5.8471 | 9497.1 | .63988 | 206.13 |
| 401 | | 11767.9 | 3.8567 | 8433.9 | .86449 | 202.67 | 451 | | 13268.8 | 5.8944 | 9519.0 | .63616 | 206.20 |
| 402 | | 11797.8 | 3.8914 | 8455.4 | .85891 | 202.74 | 452 | | 13298.9 | 5.9419 | 9540.8 | .63248 | 206.26 |
| 403 | | 11827.7 | 3.9263 | 8477.0 | .85339 | 202.82 | 453 | | 13329.1 | 5.9898 | 9562.7 | .62881 | 206.33 |
| 404 | | 11857.6 | 3.9615 | 8498.6 | .84792 | 202.89 | 454 | | 13359.2 | 6.0378 | 9584.5 | .62518 | 206.40 |
| 405 | 131.85 | 11887.6 | 3.9969 | 8520.2 | .84249 | 202.97 | 455 | 181.85 | 13389.4 | 6.0863 | 9606.4 | .62157 | 206.46 |
| 406 | | 11917.5 | 4.0325 | 8541.8 | .83711 | 203.04 | 456 | | 13419.6 | 6.1349 | 9628.2 | .61800 | 206.53 |
| 407 | | 11947.4 | 4.0684 | 8563.4 | .83178 | 203.11 | 457 | | 13449.8 | 6.1839 | 9650.1 | .61445 | 206.59 |
| 408 | | 11977.3 | 4.1044 | 8585.0 | .82649 | 203.19 | 458 | | 13479.9 | 6.2332 | 9671.9 | .61092 | 206.66 |
| 409 | | 12007.2 | 4.1408 | 8606.6 | .82125 | 203.26 | 459 | | 13510.1 | 6.2827 | 9693.8 | .60743 | 206.73 |
| 410 | 136.85 | 12037.2 | 4.1773 | 8628.3 | .81605 | 203.33 | 460 | 186.85 | 13540.3 | 6.3326 | 9715.7 | .60396 | 206.79 |
| 411 | | 12067.1 | 4.2141 | 8649.9 | .81089 | 203.41 | 461 | | 13570.5 | 6.3827 | 9737.6 | .60052 | 206.86 |
| 412 | | 12097.0 | 4.2512 | 8671.5 | .80579 | 203.48 | 462 | | 13600.7 | 6.4332 | 9759.5 | .59710 | 206.92 |
| 413 | | 12127.0 | 4.2885 | 8693.1 | .80072 | 203.55 | 463 | | 13631.0 | 6.4839 | 9781.4 | .59371 | 206.99 |
| 414 | | 12156.9 | 4.3260 | 8714.8 | .79569 | 203.62 | 464 | | 13661.2 | 6.5350 | 9803.3 | .59035 | 207.05 |
| 415 | 141.85 | 12186.9 | 4.3638 | 8736.4 | .79071 | 203.70 | 465 | 191.85 | 13691.4 | 6.5863 | 9825.2 | .58701 | 207.12 |
| 416 | | 12216.9 | 4.4018 | 8758.1 | .78578 | 203.77 | 466 | | 13721.6 | 6.6379 | 9847.1 | .58369 | 207.18 |
| 417 | | 12246.8 | 4.4400 | 8779.7 | .78088 | 203.84 | 467 | | 13751.9 | 6.6899 | 9869.0 | .58040 | 207.25 |
| 418 | | 12276.8 | 4.4785 | 8801.4 | .77602 | 203.91 | 468 | | 13782.1 | 6.7422 | 9891.0 | .57714 | 207.31 |
| 419 | | 12306.8 | 4.5173 | 8823.0 | .77121 | 203.98 | 469 | | 13812.4 | 6.7947 | 9912.9 | .57390 | 207.38 |
| 420 | 146.85 | 12336.8 | 4.5563 | 8844.7 | .76643 | 204.05 | 470 | 196.85 | 13842.6 | 6.8476 | 9934.9 | .57068 | 207.44 |
| 421 | | 12366.8 | 4.5955 | 8866.4 | .76169 | 204.13 | 471 | | 13872.9 | 6.9008 | 9956.8 | .56749 | 207.51 |
| 422 | | 12396.7 | 4.6350 | 8888.1 | .75700 | 204.20 | 472 | | 13903.1 | 6.9542 | 9978.7 | .56432 | 207.57 |
| 423 | | 12426.7 | 4.6747 | 8909.7 | .75234 | 204.27 | 473 | | 13933.4 | 7.0080 | 10000.7 | .56117 | 207.63 |
| 424 | | 12456.7 | 4.7147 | 8931.4 | .74772 | 204.34 | 474 | | 13963.7 | 7.0621 | 10022.7 | .55805 | 207.70 |
| 425 | 151.85 | 12486.8 | 4.7550 | 8953.1 | .74314 | 204.41 | 475 | 201.85 | 13994.0 | 7.1165 | 10044.6 | .55495 | 207.76 |
| 426 | | 12516.8 | 4.7955 | 8974.8 | .73859 | 204.48 | 476 | | 14024.3 | 7.1713 | 10066.6 | .55188 | 207.83 |
| 427 | | 12546.8 | 4.8363 | 8996.5 | .73409 | 204.55 | 477 | | 14054.6 | 7.2263 | 10088.6 | .54882 | 207.89 |
| 428 | | 12576.8 | 4.8773 | 9018.2 | .72962 | 204.62 | 478 | | 14084.9 | 7.2817 | 10110.6 | .54579 | 207.95 |
| 429 | | 12606.8 | 4.9186 | 9040.0 | .72518 | 204.69 | 479 | | 14115.2 | 7.3374 | 10132.6 | .54278 | 208.02 |
| 430 | 156.85 | 12636.9 | 4.9601 | 9061.7 | .72078 | 204.76 | 480 | 206.85 | 14145.5 | 7.3934 | 10154.6 | .53980 | 208.08 |
| 431 | | 12666.9 | 5.0019 | 9083.4 | .71642 | 204.83 | 481 | | 14175.8 | 7.4497 | 10176.6 | .53683 | 208.14 |
| 432 | | 12696.9 | 5.0440 | 9105.1 | .71210 | 204.90 | 482 | | 14206.1 | 7.5063 | 10198.6 | .53389 | 208.21 |
| 433 | | 12727.0 | 5.0863 | 9126.9 | .70781 | 204.97 | 483 | | 14236.5 | 7.5633 | 10220.6 | .53096 | 208.27 |
| 434 | | 12757.1 | 5.1289 | 9148.6 | .70355 | 205.04 | 484 | | 14266.8 | 7.6206 | 10242.6 | .52806 | 208.33 |
| 435 | 161.85 | 12787.1 | 5.1718 | 9170.3 | .69933 | 205.11 | 485 | 211.85 | 14297.1 | 7.6783 | 10264.7 | .52518 | 208.39 |
| 436 | | 12817.2 | 5.2149 | 9192.1 | .69514 | 205.18 | 486 | | 14327.5 | 7.7362 | 10286.7 | .52232 | 208.46 |
| 437 | | 12847.2 | 5.2583 | 9213.8 | .69099 | 205.25 | 487 | | 14357.9 | 7.7945 | 10308.7 | .51948 | 208.52 |
| 438 | | 12877.3 | 5.3019 | 9235.6 | .68687 | 205.32 | 488 | | 14388.2 | 7.8531 | 10330.8 | .51667 | 208.58 |
| 439 | | 12907.4 | 5.3458 | 9257.4 | .68278 | 205.38 | 489 | | 14418.6 | 7.9120 | 10352.8 | .51387 | 208.64 |
| 440 | 166.85 | 12937.5 | 5.3900 | 9279.1 | .67872 | 205.45 | 490 | 216.85 | 14449.0 | 7.9713 | 10374.9 | .51109 | 208.71 |
| 441 | | 12967.6 | 5.4345 | 9300.9 | .67470 | 205.52 | 491 | | 14479.3 | 8.0309 | 10397.0 | .50833 | 208.77 |
| 442 | | 12997.7 | 5.4792 | 9322.7 | .67071 | 205.59 | 492 | | 14509.7 | 8.0908 | 10419.0 | .50559 | 208.83 |
| 443 | | 13027.8 | 5.5243 | 9344.5 | .66675 | 205.66 | 493 | | 14540.1 | 8.1511 | 10441.1 | .50287 | 208.89 |
| 444 | | 13057.9 | 5.5695 | 9366.3 | .66282 | 205.72 | 494 | | 14570.5 | 8.2118 | 10463.2 | .50018 | 208.95 |
| 445 | 171.85 | 13088.0 | 5.6151 | 9388.1 | .65892 | 205.79 | 495 | 221.85 | 14600.9 | 8.2727 | 10485.3 | .49750 | 209.01 |
| 446 | | 13118.1 | 5.6610 | 9409.9 | .65505 | 205.86 | 496 | | 14631.3 | 8.3340 | 10507.4 | .49484 | 209.08 |
| 447 | | 13148.2 | 5.7071 | 9431.7 | .65121 | 205.93 | 497 | | 14661.8 | 8.3956 | 10529.5 | .49219 | 209.14 |
| 448 | | 13178.4 | 5.7535 | 9453.5 | .64741 | 205.99 | 498 | | 14692.2 | 8.4576 | 10551.6 | .48957 | 209.20 |
| 449 | | 13208.5 | 5.8002 | 9475.3 | .64363 | 206.06 | 499 | | 14722.6 | 8.5199 | 10573.7 | .48696 | 209.26 |

## Table 3    Products—400% Theoretical Air (for One Gram-Mole)

| T | t | $\bar{h}$ | $p_r$ | $\bar{u}$ | $v_r$ | $\bar{\phi}$ | T | t | $\bar{h}$ | $p_r$ | $\bar{u}$ | $v_r$ | $\bar{\phi}$ |
|---|---|---|---|---|---|---|---|---|---|---|---|---|---|
| 500 | 226.85 | 14753.1 | 8.583 | 10595.9 | .48438 | 209.32 | 550 | 276.85 | 16283.3 | 12.189 | 11710.4 | .37517 | 212.24 |
| 501 | | 14783.5 | 8.646 | 10618.0 | .48181 | 209.38 | 551 | | 16314.1 | 12.271 | 11732.9 | .37333 | 212.29 |
| 502 | | 14813.9 | 8.709 | 10640.1 | .47926 | 209.44 | 552 | | 16344.9 | 12.354 | 11755.3 | .37150 | 212.35 |
| 503 | | 14844.4 | 8.773 | 10662.3 | .47672 | 209.50 | 553 | | 16375.7 | 12.437 | 11777.8 | .36969 | 212.40 |
| 504 | | 14874.9 | 8.837 | 10684.4 | .47421 | 209.56 | 554 | | 16406.5 | 12.521 | 11800.3 | .36789 | 212.46 |
| 505 | 231.85 | 14905.3 | 8.901 | 10706.6 | .47171 | 209.62 | 555 | 281.85 | 16437.3 | 12.605 | 11822.8 | .36610 | 212.52 |
| 506 | | 14935.8 | 8.966 | 10728.7 | .46923 | 209.68 | 556 | | 16468.1 | 12.689 | 11845.3 | .36432 | 212.57 |
| 507 | | 14966.3 | 9.031 | 10750.9 | .46676 | 209.74 | 557 | | 16498.9 | 12.774 | 11867.8 | .36255 | 212.63 |
| 508 | | 14996.8 | 9.097 | 10773.1 | .46432 | 209.80 | 558 | | 16529.7 | 12.859 | 11890.3 | .36080 | 212.68 |
| 509 | | 15027.3 | 9.162 | 10795.3 | .46189 | 209.86 | 559 | | 16560.6 | 12.945 | 11912.8 | .35905 | 212.74 |
| 510 | 236.85 | 15057.8 | 9.229 | 10817.4 | .45948 | 209.92 | 560 | 286.85 | 16591.4 | 13.031 | 11935.3 | .35732 | 212.79 |
| 511 | | 15088.3 | 9.295 | 10839.6 | .45708 | 209.98 | 561 | | 16622.2 | 13.117 | 11957.9 | .35559 | 212.85 |
| 512 | | 15118.8 | 9.362 | 10861.8 | .45470 | 210.04 | 562 | | 16653.1 | 13.204 | 11980.4 | .35388 | 212.90 |
| 513 | | 15149.3 | 9.429 | 10884.0 | .45234 | 210.10 | 563 | | 16684.0 | 13.292 | 12003.0 | .35218 | 212.96 |
| 514 | | 15179.9 | 9.497 | 10906.3 | .44999 | 210.16 | 564 | | 16714.8 | 13.379 | 12025.5 | .35049 | 213.01 |
| 515 | 241.85 | 15210.4 | 9.565 | 10928.5 | .44766 | 210.22 | 565 | 291.85 | 16745.7 | 13.468 | 12048.1 | .34881 | 213.07 |
| 516 | | 15240.9 | 9.634 | 10950.7 | .44534 | 210.28 | 566 | | 16776.6 | 13.557 | 12070.6 | .34714 | 213.12 |
| 517 | | 15271.5 | 9.702 | 10972.9 | .44304 | 210.34 | 567 | | 16807.5 | 13.646 | 12093.2 | .34548 | 213.18 |
| 518 | | 15302.0 | 9.772 | 10995.2 | .44076 | 210.40 | 568 | | 16838.4 | 13.735 | 12115.8 | .34383 | 213.23 |
| 519 | | 15332.6 | 9.841 | 11017.4 | .43849 | 210.46 | 569 | | 16869.3 | 13.825 | 12138.4 | .34219 | 213.28 |
| 520 | 246.85 | 15363.2 | 9.911 | 11039.7 | .43624 | 210.52 | 570 | 296.85 | 16900.2 | 13.916 | 12161.0 | .34056 | 213.34 |
| 521 | | 15393.7 | 9.981 | 11061.9 | .43400 | 210.58 | 571 | | 16931.1 | 14.007 | 12183.6 | .33894 | 213.39 |
| 522 | | 15424.3 | 10.052 | 11084.2 | .43177 | 210.63 | 572 | | 16962.0 | 14.098 | 12206.2 | .33733 | 213.45 |
| 523 | | 15454.9 | 10.123 | 11106.5 | .42957 | 210.69 | 573 | | 16993.0 | 14.190 | 12228.8 | .33573 | 213.50 |
| 524 | | 15485.5 | 10.194 | 11128.8 | .42737 | 210.75 | 574 | | 17023.9 | 14.283 | 12251.4 | .33414 | 213.55 |
| 525 | 251.85 | 15516.1 | 10.266 | 11151.0 | .42520 | 210.81 | 575 | 301.85 | 17054.9 | 14.376 | 12274.1 | .33256 | 213.61 |
| 526 | | 15546.7 | 10.338 | 11173.3 | .42303 | 210.87 | 576 | | 17085.8 | 14.469 | 12296.7 | .33099 | 213.66 |
| 527 | | 15577.3 | 10.411 | 11195.6 | .42088 | 210.93 | 577 | | 17116.8 | 14.563 | 12319.4 | .32943 | 213.72 |
| 528 | | 15607.9 | 10.484 | 11217.9 | .41875 | 210.98 | 578 | | 17147.7 | 14.657 | 12342.0 | .32788 | 213.77 |
| 529 | | 15638.6 | 10.557 | 11240.2 | .41663 | 211.04 | 579 | | 17178.7 | 14.752 | 12364.7 | .32634 | 213.82 |
| 530 | 256.85 | 15669.2 | 10.631 | 11262.6 | .41452 | 211.10 | 580 | 306.85 | 17209.7 | 14.847 | 12387.3 | .32481 | 213.88 |
| 531 | | 15699.8 | 10.705 | 11284.9 | .41243 | 211.16 | 581 | | 17240.7 | 14.942 | 12410.0 | .32329 | 213.93 |
| 532 | | 15730.5 | 10.779 | 11307.2 | .41035 | 211.21 | 582 | | 17271.7 | 15.039 | 12432.7 | .32177 | 213.98 |
| 533 | | 15761.1 | 10.854 | 11329.6 | .40828 | 211.27 | 583 | | 17302.7 | 15.135 | 12455.4 | .32027 | 214.04 |
| 534 | | 15791.8 | 10.929 | 11351.9 | .40623 | 211.33 | 584 | | 17333.7 | 15.232 | 12478.1 | .31877 | 214.09 |
| 535 | 261.85 | 15822.5 | 11.005 | 11374.3 | .40419 | 211.39 | 585 | 311.85 | 17364.7 | 15.330 | 12500.8 | .31729 | 214.14 |
| 536 | | 15853.1 | 11.081 | 11396.6 | .40217 | 211.44 | 586 | | 17395.8 | 15.428 | 12523.5 | .31581 | 214.20 |
| 537 | | 15883.8 | 11.158 | 11419.0 | .40016 | 211.50 | 587 | | 17426.8 | 15.526 | 12546.2 | .31434 | 214.25 |
| 538 | | 15914.5 | 11.235 | 11441.4 | .39816 | 211.56 | 588 | | 17457.8 | 15.625 | 12569.0 | .31288 | 214.30 |
| 539 | | 15945.2 | 11.312 | 11463.7 | .39617 | 211.62 | 589 | | 17488.9 | 15.725 | 12591.7 | .31143 | 214.35 |
| 540 | 266.85 | 15975.9 | 11.390 | 11486.1 | .39420 | 211.67 | 590 | 316.85 | 17519.9 | 15.825 | 12614.4 | .30999 | 214.41 |
| 541 | | 16006.6 | 11.468 | 11508.5 | .39224 | 211.73 | 591 | | 17551.0 | 15.925 | 12637.2 | .30856 | 214.46 |
| 542 | | 16037.3 | 11.546 | 11530.9 | .39029 | 211.79 | 592 | | 17582.1 | 16.026 | 12659.9 | .30713 | 214.51 |
| 543 | | 16068.1 | 11.625 | 11553.3 | .38836 | 211.84 | 593 | | 17613.1 | 16.127 | 12682.7 | .30572 | 214.56 |
| 544 | | 16098.8 | 11.704 | 11575.7 | .38644 | 211.90 | 594 | | 17644.2 | 16.229 | 12705.5 | .30431 | 214.62 |
| 545 | 271.85 | 16129.5 | 11.784 | 11598.2 | .38453 | 211.96 | 595 | 321.85 | 17675.3 | 16.332 | 12728.2 | .30291 | 214.67 |
| 546 | | 16160.3 | 11.864 | 11620.6 | .38263 | 212.01 | 596 | | 17706.4 | 16.435 | 12751.0 | .30152 | 214.72 |
| 547 | | 16191.0 | 11.945 | 11643.0 | .38075 | 212.07 | 597 | | 17737.5 | 16.538 | 12773.8 | .30014 | 214.77 |
| 548 | | 16221.8 | 12.026 | 11665.5 | .37888 | 212.12 | 598 | | 17768.6 | 16.642 | 12796.6 | .29877 | 214.83 |
| 549 | | 16252.5 | 12.107 | 11687.9 | .37702 | 212.18 | 599 | | 17799.7 | 16.746 | 12819.4 | .29740 | 214.88 |

# Table 3 Products—400% Theoretical Air (for One Gram-Mole)

| T | t | $\bar{h}$ | $p_r$ | $\bar{u}$ | $v_r$ | $\bar{\phi}$ | T | t | $\bar{h}$ | $p_r$ | $\bar{u}$ | $v_r$ | $\bar{\phi}$ |
|---|---|-----------|-------|-----------|-------|--------------|---|---|-----------|-------|-----------|-------|--------------|
| 600 | 326.85 | 17830.9 | 16.851 | 12842.2 | .29604 | 214.93 | 650 | 376.85 | 19396.7 | 22.780 | 13992.4 | .23724 | 217.44 |
| 601 | | 17862.0 | 16.957 | 12865.1 | .29469 | 214.98 | 651 | | 19428.3 | 22.913 | 14015.6 | .23622 | 217.48 |
| 602 | | 17893.2 | 17.063 | 12887.9 | .29335 | 215.03 | 652 | | 19459.8 | 23.047 | 14038.8 | .23522 | 217.53 |
| 603 | | 17924.3 | 17.169 | 12910.7 | .29201 | 215.08 | 653 | | 19491.3 | 23.181 | 14062.0 | .23421 | 217.58 |
| 604 | | 17955.5 | 17.276 | 12933.6 | .29069 | 215.14 | 654 | | 19522.8 | 23.316 | 14085.2 | .23321 | 217.63 |
| 605 | 331.85 | 17986.6 | 17.383 | 12956.4 | .28937 | 215.19 | 655 | 381.85 | 19554.4 | 23.452 | 14108.4 | .23222 | 217.68 |
| 606 | | 18017.8 | 17.491 | 12979.3 | .28806 | 215.24 | 656 | | 19585.9 | 23.588 | 14131.6 | .23123 | 217.73 |
| 607 | | 18049.0 | 17.600 | 13002.1 | .28676 | 215.29 | 657 | | 19617.5 | 23.725 | 14154.9 | .23025 | 217.77 |
| 608 | | 18080.2 | 17.709 | 13025.0 | .28546 | 215.34 | 658 | | 19649.0 | 23.862 | 14178.1 | .22927 | 217.82 |
| 609 | | 18111.4 | 17.818 | 13047.9 | .28417 | 215.39 | 659 | | 19680.6 | 24.000 | 14201.4 | .22830 | 217.87 |
| 610 | 336.85 | 18142.6 | 17.928 | 13070.8 | .28289 | 215.44 | 660 | 386.85 | 19712.2 | 24.138 | 14224.7 | .22733 | 217.92 |
| 611 | | 18173.8 | 18.039 | 13093.7 | .28162 | 215.50 | 661 | | 19743.7 | 24.278 | 14247.9 | .22637 | 217.97 |
| 612 | | 18205.0 | 18.150 | 13116.6 | .28035 | 215.55 | 662 | | 19775.3 | 24.418 | 14271.2 | .22542 | 218.01 |
| 613 | | 18236.2 | 18.262 | 13139.5 | .27909 | 215.60 | 663 | | 19806.9 | 24.558 | 14294.5 | .22447 | 218.06 |
| 614 | | 18267.4 | 18.374 | 13162.4 | .27784 | 215.65 | 664 | | 19838.5 | 24.699 | 14317.8 | .22352 | 218.11 |
| 615 | 341.85 | 18298.7 | 18.487 | 13185.3 | .27660 | 215.70 | 665 | 391.85 | 19870.1 | 24.841 | 14341.1 | .22258 | 218.16 |
| 616 | | 18329.9 | 18.600 | 13208.3 | .27536 | 215.75 | 666 | | 19901.8 | 24.983 | 14364.4 | .22164 | 218.20 |
| 617 | | 18361.2 | 18.714 | 13231.2 | .27413 | 215.80 | 667 | | 19933.4 | 25.126 | 14387.7 | .22071 | 218.25 |
| 618 | | 18392.4 | 18.828 | 13254.1 | .27291 | 215.85 | 668 | | 19965.0 | 25.270 | 14411.0 | .21979 | 218.30 |
| 619 | | 18423.7 | 18.943 | 13277.1 | .27170 | 215.90 | 669 | | 19996.7 | 25.414 | 14434.3 | .21887 | 218.35 |
| 620 | 346.85 | 18455.0 | 19.058 | 13300.1 | .27049 | 215.95 | 670 | 396.85 | 20028.3 | 25.559 | 14457.7 | .21795 | 218.39 |
| 621 | | 18486.3 | 19.174 | 13323.0 | .26928 | 216.00 | 671 | | 20060.0 | 25.705 | 14481.0 | .21704 | 218.44 |
| 622 | | 18517.6 | 19.290 | 13346.0 | .26809 | 216.05 | 672 | | 20091.7 | 25.851 | 14504.4 | .21614 | 218.49 |
| 623 | | 18548.9 | 19.407 | 13369.0 | .26690 | 216.10 | 673 | | 20123.3 | 25.998 | 14527.7 | .21523 | 218.53 |
| 624 | | 18580.2 | 19.525 | 13392.0 | .26572 | 216.15 | 674 | | 20155.0 | 26.145 | 14551.1 | .21434 | 218.58 |
| 625 | 351.85 | 18611.5 | 19.643 | 13415.0 | .26455 | 216.20 | 675 | 401.85 | 20186.7 | 26.293 | 14574.5 | .21345 | 218.63 |
| 626 | | 18642.8 | 19.762 | 13438.0 | .26338 | 216.25 | 676 | | 20218.4 | 26.442 | 14597.9 | .21256 | 218.68 |
| 627 | | 18674.1 | 19.881 | 13461.0 | .26222 | 216.30 | 677 | | 20250.1 | 26.592 | 14621.2 | .21168 | 218.72 |
| 628 | | 18705.5 | 20.001 | 13484.0 | .26106 | 216.35 | 678 | | 20281.8 | 26.742 | 14644.6 | .21080 | 218.77 |
| 629 | | 18736.8 | 20.121 | 13507.1 | .25992 | 216.40 | 679 | | 20313.5 | 26.892 | 14668.0 | .20993 | 218.82 |
| 630 | 356.85 | 18768.2 | 20.242 | 13530.1 | .25877 | 216.45 | 680 | 406.85 | 20345.3 | 27.044 | 14691.5 | .20906 | 218.86 |
| 631 | | 18799.5 | 20.363 | 13553.1 | .25764 | 216.50 | 681 | | 20377.0 | 27.196 | 14714.9 | .20820 | 218.91 |
| 632 | | 18830.9 | 20.485 | 13576.2 | .25651 | 216.55 | 682 | | 20408.7 | 27.349 | 14738.3 | .20734 | 218.96 |
| 633 | | 18862.3 | 20.608 | 13599.2 | .25539 | 216.60 | 683 | | 20440.5 | 27.502 | 14761.7 | .20648 | 219.00 |
| 634 | | 18893.6 | 20.731 | 13622.3 | .25427 | 216.65 | 684 | | 20472.2 | 27.656 | 14785.2 | .20563 | 219.05 |
| 635 | 361.85 | 18925.0 | 20.855 | 13645.4 | .25316 | 216.70 | 685 | 411.85 | 20504.0 | 27.811 | 14808.6 | .20479 | 219.10 |
| 636 | | 18956.4 | 20.979 | 13668.5 | .25206 | 216.75 | 686 | | 20535.8 | 27.967 | 14832.1 | .20395 | 219.14 |
| 637 | | 18987.8 | 21.104 | 13691.6 | .25096 | 216.80 | 687 | | 20567.6 | 28.123 | 14855.6 | .20311 | 219.19 |
| 638 | | 19019.2 | 21.229 | 13714.6 | .24987 | 216.85 | 688 | | 20599.3 | 28.280 | 14879.0 | .20228 | 219.23 |
| 639 | | 19050.7 | 21.355 | 13737.7 | .24879 | 216.90 | 689 | | 20631.1 | 28.437 | 14902.5 | .20145 | 219.28 |
| 640 | 366.85 | 19082.1 | 21.482 | 13760.9 | .24771 | 216.95 | 690 | 416.85 | 20662.9 | 28.595 | 14926.0 | .20062 | 219.33 |
| 641 | | 19113.5 | 21.609 | 13784.0 | .24663 | 217.00 | 691 | | 20694.7 | 28.754 | 14949.5 | .19981 | 219.37 |
| 642 | | 19144.9 | 21.737 | 13807.1 | .24557 | 217.05 | 692 | | 20726.6 | 28.914 | 14973.0 | .19899 | 219.42 |
| 643 | | 19176.4 | 21.865 | 13830.2 | .24450 | 217.10 | 693 | | 20758.4 | 29.074 | 14996.5 | .19818 | 219.46 |
| 644 | | 19207.9 | 21.994 | 13853.4 | .24345 | 217.14 | 694 | | 20790.2 | 29.235 | 15020.0 | .19737 | 219.51 |
| 645 | 371.85 | 19239.3 | 22.124 | 13876.5 | .24240 | 217.19 | 695 | 421.85 | 20822.1 | 29.397 | 15043.6 | .19657 | 219.56 |
| 646 | | 19270.8 | 22.254 | 13899.7 | .24136 | 217.24 | 696 | | 20853.9 | 29.559 | 15067.1 | .19577 | 219.60 |
| 647 | | 19302.3 | 22.385 | 13922.8 | .24032 | 217.29 | 697 | | 20885.8 | 29.722 | 15090.6 | .19498 | 219.65 |
| 648 | | 19333.7 | 22.516 | 13946.0 | .23929 | 217.34 | 698 | | 20917.6 | 29.886 | 15114.2 | .19419 | 219.69 |
| 649 | | 19365.2 | 22.648 | 13969.2 | .23826 | 217.39 | 699 | | 20949.5 | 30.050 | 15137.7 | .19340 | 219.74 |

# Table 3  Products—400% Theoretical Air (for One Gram-Mole)

| T | t | $\bar{h}$ | $p_r$ | $\bar{u}$ | $v_r$ | $\bar{\phi}$ | T | t | $\bar{h}$ | $p_r$ | $\bar{u}$ | $v_r$ | $\bar{\phi}$ |
|---|---|---|---|---|---|---|---|---|---|---|---|---|---|
| 700 | 426.85 | 20981.4 | 30.215 | 15161.3 | .19262 | 219.78 | 750 | 476.85 | 22584.9 | 39.427 | 16349.1 | .15816 | 222.00 |
| 701 | | 21013.3 | 30.381 | 15184.9 | .19184 | 219.83 | 751 | | 22617.1 | 39.631 | 16373.0 | .15756 | 222.04 |
| 702 | | 21045.2 | 30.548 | 15208.4 | .19107 | 219.88 | 752 | | 22649.4 | 39.836 | 16397.0 | .15695 | 222.08 |
| 703 | | 21077.0 | 30.715 | 15232.0 | .19030 | 219.92 | 753 | | 22681.7 | 40.043 | 16420.9 | .15635 | 222.13 |
| 704 | | 21109.0 | 30.883 | 15255.6 | .18953 | 219.97 | 754 | | 22713.9 | 40.249 | 16444.9 | .15576 | 222.17 |
| 705 | 431.85 | 21140.9 | 31.052 | 15279.2 | .18877 | 220.01 | 755 | 481.85 | 22746.2 | 40.457 | 16468.9 | .15516 | 222.21 |
| 706 | | 21172.8 | 31.221 | 15302.8 | .18801 | 220.06 | 756 | | 22778.5 | 40.666 | 16492.8 | .15457 | 222.25 |
| 707 | | 21204.7 | 31.392 | 15326.4 | .18726 | 220.10 | 757 | | 22810.8 | 40.875 | 16516.8 | .15398 | 222.30 |
| 708 | | 21236.7 | 31.562 | 15350.1 | .18651 | 220.15 | 758 | | 22843.2 | 41.085 | 16540.8 | .15340 | 222.34 |
| 709 | | 21268.6 | 31.734 | 15373.7 | .18576 | 220.19 | 759 | | 22875.5 | 41.296 | 16564.8 | .15281 | 222.38 |
| 710 | 436.85 | 21300.6 | 31.906 | 15397.3 | .18502 | 220.24 | 760 | 486.85 | 22907.8 | 41.508 | 16588.9 | .15223 | 222.42 |
| 711 | | 21332.5 | 32.079 | 15421.0 | .18428 | 220.28 | 761 | | 22940.1 | 41.721 | 16612.9 | .15166 | 222.47 |
| 712 | | 21364.5 | 32.253 | 15444.6 | .18354 | 220.33 | 762 | | 22972.5 | 41.935 | 16636.9 | .15108 | 222.51 |
| 713 | | 21396.5 | 32.428 | 15468.3 | .18281 | 220.37 | 763 | | 23004.8 | 42.149 | 16660.9 | .15051 | 222.55 |
| 714 | | 21428.5 | 32.603 | 15492.0 | .18208 | 220.42 | 764 | | 23037.2 | 42.365 | 16685.0 | .14994 | 222.59 |
| 715 | 441.85 | 21460.5 | 32.779 | 15515.7 | .18136 | 220.46 | 765 | 491.85 | 23069.5 | 42.581 | 16709.0 | .14937 | 222.64 |
| 716 | | 21492.5 | 32.956 | 15539.3 | .18064 | 220.51 | 766 | | 23101.9 | 42.798 | 16733.1 | .14881 | 222.68 |
| 717 | | 21524.5 | 33.134 | 15563.0 | .17992 | 220.55 | 767 | | 23134.3 | 43.016 | 16757.2 | .14825 | 222.72 |
| 718 | | 21556.5 | 33.312 | 15586.7 | .17921 | 220.60 | 768 | | 23166.7 | 43.235 | 16781.2 | .14769 | 222.76 |
| 719 | | 21588.5 | 33.491 | 15610.4 | .17850 | 220.64 | 769 | | 23199.1 | 43.455 | 16805.3 | .14714 | 222.81 |
| 720 | 446.85 | 21620.5 | 33.671 | 15634.1 | .17779 | 220.68 | 770 | 496.85 | 23231.5 | 43.675 | 16829.4 | .14658 | 222.85 |
| 721 | | 21652.6 | 33.851 | 15657.9 | .17709 | 220.73 | 771 | | 23263.9 | 43.897 | 16853.5 | .14603 | 222.89 |
| 722 | | 21684.6 | 34.033 | 15681.6 | .17639 | 220.77 | 772 | | 23296.3 | 44.119 | 16877.6 | .14549 | 222.93 |
| 723 | | 21716.6 | 34.215 | 15705.3 | .17569 | 220.82 | 773 | | 23328.7 | 44.343 | 16901.7 | .14494 | 222.97 |
| 724 | | 21748.7 | 34.397 | 15729.1 | .17500 | 220.86 | 774 | | 23361.2 | 44.567 | 16925.8 | .14440 | 223.02 |
| 725 | 451.85 | 21780.8 | 34.581 | 15752.8 | .17431 | 220.91 | 775 | 501.85 | 23393.6 | 44.792 | 16949.9 | .14386 | 223.06 |
| 726 | | 21812.8 | 34.766 | 15776.6 | .17363 | 220.95 | 776 | | 23426.0 | 45.018 | 16974.1 | .14332 | 223.10 |
| 727 | | 21844.9 | 34.951 | 15800.4 | .17295 | 221.00 | 777 | | 23458.5 | 45.245 | 16998.2 | .14279 | 223.14 |
| 728 | | 21877.0 | 35.137 | 15824.1 | .17227 | 221.04 | 778 | | 23491.0 | 45.472 | 17022.3 | .14225 | 223.18 |
| 729 | | 21909.1 | 35.323 | 15847.9 | .17159 | 221.08 | 779 | | 23523.4 | 45.701 | 17046.5 | .14172 | 223.22 |
| 730 | 456.85 | 21941.2 | 35.511 | 15871.7 | .17092 | 221.13 | 780 | 506.85 | 23555.9 | 45.931 | 17070.7 | .14120 | 223.27 |
| 731 | | 21973.3 | 35.699 | 15895.5 | .17025 | 221.17 | 781 | | 23588.4 | 46.161 | 17094.8 | .14067 | 223.31 |
| 732 | | 22005.4 | 35.888 | 15919.3 | .16959 | 221.22 | 782 | | 23620.9 | 46.392 | 17119.0 | .14015 | 223.35 |
| 733 | | 22037.6 | 36.078 | 15943.1 | .16893 | 221.26 | 783 | | 23653.4 | 46.625 | 17143.2 | .13963 | 223.39 |
| 734 | | 22069.7 | 36.268 | 15966.9 | .16827 | 221.30 | 784 | | 23685.9 | 46.858 | 17167.4 | .13911 | 223.43 |
| 735 | 461.85 | 22101.8 | 36.460 | 15990.8 | .16761 | 221.35 | 785 | 511.85 | 23718.4 | 47.092 | 17191.6 | .13860 | 223.47 |
| 736 | | 22134.0 | 36.652 | 16014.6 | .16696 | 221.39 | 786 | | 23750.9 | 47.327 | 17215.8 | .13808 | 223.52 |
| 737 | | 22166.1 | 36.845 | 16038.4 | .16631 | 221.43 | 787 | | 23783.4 | 47.563 | 17240.0 | .13757 | 223.56 |
| 738 | | 22198.3 | 37.038 | 16062.3 | .16567 | 221.48 | 788 | | 23816.0 | 47.800 | 17264.2 | .13707 | 223.60 |
| 739 | | 22230.5 | 37.233 | 16086.1 | .16502 | 221.52 | 789 | | 23848.5 | 48.038 | 17288.4 | .13656 | 223.64 |
| 740 | 466.85 | 22262.7 | 37.428 | 16110.0 | .16438 | 221.56 | 790 | 516.85 | 23881.1 | 48.277 | 17312.7 | .13606 | 223.68 |
| 741 | | 22294.9 | 37.625 | 16133.9 | .16375 | 221.61 | 791 | | 23913.6 | 48.516 | 17336.9 | .13556 | 223.72 |
| 742 | | 22327.0 | 37.822 | 16157.8 | .16312 | 221.65 | 792 | | 23946.2 | 48.757 | 17361.2 | .13506 | 223.76 |
| 743 | | 22359.2 | 38.019 | 16181.6 | .16249 | 221.69 | 793 | | 23978.7 | 48.999 | 17385.4 | .13456 | 223.80 |
| 744 | | 22391.5 | 38.218 | 16205.5 | .16186 | 221.74 | 794 | | 24011.3 | 49.241 | 17409.7 | .13407 | 223.85 |
| 745 | 471.85 | 22423.7 | 38.417 | 16229.4 | .16124 | 221.78 | 795 | 521.85 | 24043.9 | 49.485 | 17434.0 | .13358 | 223.89 |
| 746 | | 22455.9 | 38.618 | 16253.4 | .16061 | 221.82 | 796 | | 24076.5 | 49.729 | 17458.2 | .13309 | 223.93 |
| 747 | | 22488.1 | 38.819 | 16277.3 | .16000 | 221.87 | 797 | | 24109.1 | 49.975 | 17482.5 | .13260 | 223.97 |
| 748 | | 22520.4 | 39.021 | 16301.2 | .15938 | 221.91 | 798 | | 24141.7 | 50.221 | 17506.8 | .13211 | 224.01 |
| 749 | | 22552.6 | 39.223 | 16325.1 | .15877 | 221.95 | 799 | | 24174.3 | 50.468 | 17531.1 | .13163 | 224.05 |

## Table 3 Products—400% Theoretical Air (for One Gram-Mole)

| T | t | $\bar{h}$ | $p_r$ | $\bar{u}$ | $v_r$ | $\bar{\phi}$ | T | t | $\bar{h}$ | $p_r$ | $\bar{u}$ | $v_r$ | $\bar{\phi}$ |
|---|---|---|---|---|---|---|---|---|---|---|---|---|---|
| 800 | 526.85 | 24206.9 | 50.717 | 17555.4 | .13115 | 224.09 | 850 | 576.85 | 25847.1 | 64.420 | 18779.9 | .109705 | 226.08 |
| 801 | | 24239.6 | 50.966 | 17579.7 | .13067 | 224.13 | 851 | | 25880.1 | 64.721 | 18804.5 | .109323 | 226.12 |
| 802 | | 24272.2 | 51.216 | 17604.1 | .13020 | 224.17 | 852 | | 25913.1 | 65.024 | 18829.2 | .108943 | 226.16 |
| 803 | | 24304.8 | 51.467 | 17628.4 | .12972 | 224.21 | 853 | | 25946.1 | 65.327 | 18853.9 | .108565 | 226.20 |
| 804 | | 24337.5 | 51.719 | 17652.7 | .12925 | 224.25 | 854 | | 25979.1 | 65.632 | 18878.6 | .108187 | 226.23 |
| 805 | 531.85 | 24370.1 | 51.973 | 17677.1 | .12878 | 224.29 | 855 | 581.85 | 26012.1 | 65.937 | 18903.3 | .107812 | 226.27 |
| 806 | | 24402.8 | 52.227 | 17701.4 | .12831 | 224.33 | 856 | | 26045.1 | 66.244 | 18928.0 | .107438 | 226.31 |
| 807 | | 24435.5 | 52.482 | 17725.8 | .12785 | 224.38 | 857 | | 26078.1 | 66.552 | 18952.7 | .107066 | 226.35 |
| 808 | | 24468.2 | 52.738 | 17750.1 | .12739 | 224.42 | 858 | | 26111.2 | 66.861 | 18977.4 | .106695 | 226.39 |
| 809 | | 24500.8 | 52.995 | 17774.5 | .12692 | 224.46 | 859 | | 26144.2 | 67.171 | 19002.1 | .106326 | 226.43 |
| 810 | 536.85 | 24533.5 | 53.253 | 17798.9 | .12647 | 224.50 | 860 | 586.85 | 26177.3 | 67.483 | 19026.9 | .105959 | 226.47 |
| 811 | | 24566.2 | 53.512 | 17823.3 | .12601 | 224.54 | 861 | | 26210.3 | 67.795 | 19051.6 | .105593 | 226.50 |
| 812 | | 24599.0 | 53.772 | 17847.7 | .12555 | 224.58 | 862 | | 26243.4 | 68.109 | 19076.4 | .105229 | 226.54 |
| 813 | | 24631.7 | 54.033 | 17872.1 | .12510 | 224.62 | 863 | | 26276.5 | 68.424 | 19101.1 | .104866 | 226.58 |
| 814 | | 24664.4 | 54.295 | 17896.5 | .12465 | 224.66 | 864 | | 26309.5 | 68.740 | 19125.9 | .104505 | 226.62 |
| 815 | 541.85 | 24697.1 | 54.558 | 17920.9 | .12420 | 224.70 | 865 | 591.85 | 26342.6 | 69.057 | 19150.6 | .104146 | 226.66 |
| 816 | | 24729.9 | 54.822 | 17945.3 | .12376 | 224.74 | 866 | | 26375.7 | 69.375 | 19175.4 | .103788 | 226.70 |
| 817 | | 24762.6 | 55.087 | 17969.7 | .12331 | 224.78 | 867 | | 26408.8 | 69.694 | 19200.2 | .103432 | 226.73 |
| 818 | | 24795.3 | 55.353 | 17994.2 | .12287 | 224.82 | 868 | | 26441.9 | 70.015 | 19225.0 | .103077 | 226.77 |
| 819 | | 24828.1 | 55.620 | 18018.6 | .12243 | 224.86 | 869 | | 26475.0 | 70.337 | 19249.8 | .102724 | 226.81 |
| 820 | 546.85 | 24860.9 | 55.888 | 18043.1 | .12199 | 224.90 | 870 | 596.85 | 26508.1 | 70.660 | 19274.6 | .102372 | 226.85 |
| 821 | | 24893.6 | 56.157 | 18067.5 | .12155 | 224.94 | 871 | | 26541.2 | 70.984 | 19299.4 | .102021 | 226.89 |
| 822 | | 24926.4 | 56.428 | 18092.0 | .12112 | 224.98 | 872 | | 26574.4 | 71.309 | 19324.2 | .101673 | 226.92 |
| 823 | | 24959.2 | 56.699 | 18116.5 | .12069 | 225.02 | 873 | | 26607.5 | 71.635 | 19349.0 | .101325 | 226.96 |
| 824 | | 24992.0 | 56.971 | 18140.9 | .12026 | 225.06 | 874 | | 26640.6 | 71.963 | 19373.8 | .100980 | 227.00 |
| 825 | 551.85 | 25024.8 | 57.244 | 18165.4 | .11983 | 225.10 | 875 | 601.85 | 26673.8 | 72.292 | 19398.7 | .100635 | 227.04 |
| 826 | | 25057.6 | 57.519 | 18189.9 | .11940 | 225.14 | 876 | | 26706.9 | 72.622 | 19423.5 | .100293 | 227.08 |
| 827 | | 25090.4 | 57.794 | 18214.4 | .11897 | 225.18 | 877 | | 26740.1 | 72.953 | 19448.4 | .099951 | 227.11 |
| 828 | | 25123.2 | 58.070 | 18238.9 | .11855 | 225.22 | 878 | | 26773.3 | 73.285 | 19473.2 | .099611 | 227.15 |
| 829 | | 25156.1 | 58.348 | 18263.4 | .11813 | 225.26 | 879 | | 26806.5 | 73.619 | 19498.1 | .099273 | 227.19 |
| 830 | 556.85 | 25188.9 | 58.626 | 18288.0 | .11771 | 225.30 | 880 | 606.85 | 26839.6 | 73.954 | 19523.0 | .098936 | 227.23 |
| 831 | | 25221.8 | 58.906 | 18312.5 | .11729 | 225.34 | 881 | | 26872.8 | 74.290 | 19547.8 | .098600 | 227.26 |
| 832 | | 25254.6 | 59.186 | 18337.0 | .11688 | 225.37 | 882 | | 26906.0 | 74.627 | 19572.7 | .098266 | 227.30 |
| 833 | | 25287.5 | 59.468 | 18361.6 | .11646 | 225.41 | 883 | | 26939.2 | 74.966 | 19597.6 | .097933 | 227.34 |
| 834 | | 25320.3 | 59.751 | 18386.1 | .11605 | 225.45 | 884 | | 26972.4 | 75.305 | 19622.5 | .097602 | 227.38 |
| 835 | 561.85 | 25353.2 | 60.034 | 18410.7 | .11564 | 225.49 | 885 | 611.85 | 27005.7 | 75.646 | 19647.4 | .097272 | 227.41 |
| 836 | | 25386.1 | 60.319 | 18435.2 | .11523 | 225.53 | 886 | | 27038.9 | 75.989 | 19672.3 | .096943 | 227.45 |
| 837 | | 25419.0 | 60.605 | 18459.8 | .11483 | 225.57 | 887 | | 27072.1 | 76.332 | 19697.2 | .096616 | 227.49 |
| 838 | | 25451.8 | 60.892 | 18484.4 | .11442 | 225.61 | 888 | | 27105.4 | 76.677 | 19722.2 | .096290 | 227.53 |
| 839 | | 25484.8 | 61.180 | 18509.0 | .11402 | 225.65 | 889 | | 27138.6 | 77.023 | 19747.1 | .095966 | 227.56 |
| 840 | 566.85 | 25517.7 | 61.469 | 18533.6 | .11362 | 225.69 | 890 | 616.85 | 27171.9 | 77.369 | 19772.0 | .095643 | 227.60 |
| 841 | | 25550.6 | 61.759 | 18558.2 | .11322 | 225.73 | 891 | | 27205.1 | 77.718 | 19797.0 | .095321 | 227.64 |
| 842 | | 25583.5 | 62.051 | 18582.8 | .11282 | 225.77 | 892 | | 27238.4 | 78.067 | 19821.9 | .095001 | 227.68 |
| 843 | | 25616.4 | 62.343 | 18607.4 | .11243 | 225.81 | 893 | | 27271.6 | 78.418 | 19846.9 | .094682 | 227.71 |
| 844 | | 25649.4 | 62.637 | 18632.0 | .11203 | 225.85 | 894 | | 27304.9 | 78.770 | 19871.8 | .094364 | 227.75 |
| 845 | 571.85 | 25682.3 | 62.931 | 18656.6 | .11164 | 225.88 | 895 | 621.85 | 27338.2 | 79.123 | 19896.8 | .094048 | 227.79 |
| 846 | | 25715.2 | 63.227 | 18681.3 | .11125 | 225.92 | 896 | | 27371.5 | 79.478 | 19921.8 | .093733 | 227.83 |
| 847 | | 25748.2 | 63.523 | 18705.9 | .11086 | 225.96 | 897 | | 27404.8 | 79.834 | 19946.8 | .093419 | 227.86 |
| 848 | | 25781.2 | 63.821 | 18730.6 | .11047 | 226.00 | 898 | | 27438.1 | 80.191 | 19971.8 | .093107 | 227.90 |
| 849 | | 25814.1 | 64.120 | 18755.2 | .11009 | 226.04 | 899 | | 27471.4 | 80.549 | 19996.8 | .092796 | 227.94 |

# Table 3 Products—400% Theoretical Air (for One Gram-Mole)

| T | t | $\bar{h}$ | $p_r$ | $\bar{u}$ | $v_r$ | $\bar{\phi}$ | T | t | $\bar{h}$ | $p_r$ | $\bar{u}$ | $v_r$ | $\bar{\phi}$ |
|---|---|---|---|---|---|---|---|---|---|---|---|---|---|
| 900 | 626.85 | 27504.7 | 80.909 | 20021.8 | .092486 | 227.97 | 950 | 676.85 | 29179.0 | 100.59 | 21280.3 | .078523 | 229.78 |
| 901 | | 27538.1 | 81.270 | 20046.8 | .092178 | 228.01 | 951 | | 29212.7 | 101.02 | 21305.7 | .078271 | 229.82 |
| 902 | | 27571.4 | 81.632 | 20071.8 | .091871 | 228.05 | 952 | | 29246.3 | 101.45 | 21331.0 | .078021 | 229.86 |
| 903 | | 27604.7 | 81.996 | 20096.8 | .091565 | 228.08 | 953 | | 29280.0 | 101.88 | 21356.3 | .077772 | 229.89 |
| 904 | | 27638.1 | 82.361 | 20121.8 | .091260 | 228.12 | 954 | | 29313.6 | 102.32 | 21381.7 | .077523 | 229.93 |
| 905 | 631.85 | 27671.4 | 82.726 | 20146.9 | .090957 | 228.16 | 955 | 681.85 | 29347.3 | 102.75 | 21407.1 | .077276 | 229.96 |
| 906 | | 27704.8 | 83.094 | 20171.9 | .090655 | 228.20 | 956 | | 29381.0 | 103.19 | 21432.4 | .077030 | 230.00 |
| 907 | | 27738.1 | 83.462 | 20197.0 | .090354 | 228.23 | 957 | | 29414.7 | 103.63 | 21457.8 | .076784 | 230.03 |
| 908 | | 27771.5 | 83.832 | 20222.0 | .090055 | 228.27 | 958 | | 29448.4 | 104.07 | 21483.2 | .076540 | 230.07 |
| 909 | | 27804.9 | 84.204 | 20247.1 | .089756 | 228.31 | 959 | | 29482.1 | 104.51 | 21508.6 | .076296 | 230.10 |
| 910 | 636.85 | 27838.3 | 84.576 | 20272.2 | .089459 | 228.34 | 960 | 686.85 | 29515.8 | 104.95 | 21534.0 | .076054 | 230.14 |
| 911 | | 27871.7 | 84.950 | 20297.2 | .089163 | 228.38 | 961 | | 29549.5 | 105.39 | 21559.4 | .075813 | 230.17 |
| 912 | | 27905.1 | 85.325 | 20322.3 | .088869 | 228.42 | 962 | | 29583.2 | 105.84 | 21584.8 | .075572 | 230.21 |
| 913 | | 27938.5 | 85.702 | 20347.4 | .088575 | 228.45 | 963 | | 29616.9 | 106.29 | 21610.2 | .075332 | 230.24 |
| 914 | | 27971.9 | 86.080 | 20372.5 | .088283 | 228.49 | 964 | | 29650.7 | 106.73 | 21635.6 | .075094 | 230.28 |
| 915 | 641.85 | 28005.3 | 86.459 | 20397.6 | .087992 | 228.53 | 965 | 691.85 | 29684.4 | 107.18 | 21661.0 | .074856 | 230.31 |
| 916 | | 28038.7 | 86.839 | 20422.7 | .087702 | 228.56 | 966 | | 29718.2 | 107.64 | 21686.4 | .074619 | 230.35 |
| 917 | | 28072.2 | 87.221 | 20447.9 | .087413 | 228.60 | 967 | | 29751.9 | 108.09 | 21711.9 | .074383 | 230.38 |
| 918 | | 28105.6 | 87.605 | 20473.0 | .087126 | 228.64 | 968 | | 29785.7 | 108.54 | 21737.3 | .074148 | 230.42 |
| 919 | | 28139.0 | 87.989 | 20498.1 | .086840 | 228.67 | 969 | | 29819.4 | 109.00 | 21762.8 | .073914 | 230.45 |
| 920 | 646.85 | 28172.5 | 88.375 | 20523.2 | .086555 | 228.71 | 970 | 696.85 | 29853.2 | 109.46 | 21788.2 | .073681 | 230.49 |
| 921 | | 28205.9 | 88.762 | 20548.4 | .086271 | 228.74 | 971 | | 29887.0 | 109.92 | 21813.7 | .073449 | 230.52 |
| 922 | | 28239.4 | 89.150 | 20573.5 | .085988 | 228.78 | 972 | | 29920.8 | 110.38 | 21839.2 | .073218 | 230.56 |
| 923 | | 28272.9 | 89.540 | 20598.7 | .085707 | 228.82 | 973 | | 29954.5 | 110.84 | 21864.6 | .072988 | 230.59 |
| 924 | | 28306.4 | 89.932 | 20623.8 | .085426 | 228.85 | 974 | | 29988.3 | 111.30 | 21890.1 | .072758 | 230.63 |
| 925 | 651.85 | 28339.8 | 90.324 | 20649.0 | .085147 | 228.89 | 975 | 701.85 | 30022.1 | 111.77 | 21915.6 | .072530 | 230.66 |
| 926 | | 28373.3 | 90.718 | 20674.2 | .084869 | 228.93 | 976 | | 30055.9 | 112.24 | 21941.1 | .072302 | 230.70 |
| 927 | | 28406.8 | 91.113 | 20699.4 | .084592 | 228.96 | 977 | | 30089.8 | 112.70 | 21966.6 | .072076 | 230.73 |
| 928 | | 28440.3 | 91.510 | 20724.6 | .084316 | 229.00 | 978 | | 30123.6 | 113.17 | 21992.1 | .071850 | 230.76 |
| 929 | | 28473.8 | 91.908 | 20749.8 | .084041 | 229.03 | 979 | | 30157.4 | 113.65 | 22017.6 | .071625 | 230.80 |
| 930 | 656.85 | 28507.3 | 92.307 | 20775.0 | .083768 | 229.07 | 980 | 706.85 | 30191.2 | 114.12 | 22043.1 | .071401 | 230.83 |
| 931 | | 28540.9 | 92.708 | 20800.2 | .083495 | 229.11 | 981 | | 30225.1 | 114.59 | 22068.6 | .071178 | 230.87 |
| 932 | | 28574.4 | 93.110 | 20825.4 | .083224 | 229.14 | 982 | | 30258.9 | 115.07 | 22094.2 | .070955 | 230.90 |
| 933 | | 28607.9 | 93.514 | 20850.6 | .082954 | 229.18 | 983 | | 30292.8 | 115.55 | 22119.7 | .070734 | 230.94 |
| 934 | | 28641.5 | 93.919 | 20875.8 | .082685 | 229.21 | 984 | | 30326.6 | 116.03 | 22145.2 | .070513 | 230.97 |
| 935 | 661.85 | 28675.0 | 94.326 | 20901.1 | .082416 | 229.25 | 985 | 711.85 | 30360.5 | 116.51 | 22170.8 | .070294 | 231.01 |
| 936 | | 28708.6 | 94.733 | 20926.3 | .082149 | 229.29 | 986 | | 30394.3 | 116.99 | 22196.3 | .070075 | 231.04 |
| 937 | | 28742.1 | 95.142 | 20951.5 | .081884 | 229.32 | 987 | | 30428.2 | 117.47 | 22221.9 | .069857 | 231.07 |
| 938 | | 28775.7 | 95.553 | 20976.8 | .081619 | 229.36 | 988 | | 30462.1 | 117.96 | 22247.5 | .069639 | 231.11 |
| 939 | | 28809.3 | 95.965 | 21002.1 | .081355 | 229.39 | 989 | | 30496.0 | 118.45 | 22273.0 | .069423 | 231.14 |
| 940 | 666.85 | 28842.9 | 96.379 | 21027.3 | .081092 | 229.43 | 990 | 716.85 | 30529.9 | 118.94 | 22298.6 | .069208 | 231.18 |
| 941 | | 28876.4 | 96.793 | 21052.6 | .080831 | 229.46 | 991 | | 30563.8 | 119.43 | 22324.2 | .068993 | 231.21 |
| 942 | | 28910.0 | 97.210 | 21077.9 | .080570 | 229.50 | 992 | | 30597.7 | 119.92 | 22349.8 | .068779 | 231.25 |
| 943 | | 28943.6 | 97.627 | 21103.2 | .080310 | 229.54 | 993 | | 30631.6 | 120.41 | 22375.4 | .068566 | 231.28 |
| 944 | | 28977.2 | 98.047 | 21128.4 | .080052 | 229.57 | 994 | | 30665.5 | 120.91 | 22401.0 | .068354 | 231.31 |
| 945 | 671.85 | 29010.9 | 98.467 | 21153.7 | .079794 | 229.61 | 995 | 721.85 | 30699.4 | 121.40 | 22426.6 | .068143 | 231.35 |
| 946 | | 29044.5 | 98.889 | 21179.0 | .079538 | 229.64 | 996 | | 30733.3 | 121.90 | 22452.2 | .067932 | 231.38 |
| 947 | | 29078.1 | 99.312 | 21204.4 | .079283 | 229.68 | 997 | | 30767.3 | 122.40 | 22477.8 | .067722 | 231.42 |
| 948 | | 29111.7 | 99.737 | 21229.7 | .079028 | 229.71 | 998 | | 30801.2 | 122.91 | 22503.5 | .067514 | 231.45 |
| 949 | | 29145.4 | 100.163 | 21255.0 | .078775 | 229.75 | 999 | | 30835.2 | 123.41 | 22529.1 | .067305 | 231.48 |

## Table 3 Products—400% Theoretical Air (for One Gram-Mole)

| T | t | $\bar{h}$ | $p_r$ | $\bar{u}$ | $v_r$ | $\bar{\phi}$ | T | t | $\bar{h}$ | $p_r$ | $\bar{u}$ | $v_r$ | $\bar{\phi}$ |
|---|---|---|---|---|---|---|---|---|---|---|---|---|---|
| 1000 | 726.85 | 30869.1 | 123.91 | 22554.7 | .067098 | 231.52 | 1050 | 776.85 | 32574.2 | 151.36 | 23844.1 | .057676 | 233.18 |
| 1001 | | 30903.1 | 124.42 | 22580.3 | .066893 | 231.55 | 1051 | | 32608.4 | 151.96 | 23870.0 | .057506 | 233.21 |
| 1002 | | 30937.0 | 124.93 | 22606.0 | .066686 | 231.59 | 1052 | | 32642.7 | 152.56 | 23895.9 | .057334 | 233.25 |
| 1003 | | 30971.0 | 125.44 | 22631.7 | .066482 | 231.62 | 1053 | | 32677.0 | 153.15 | 23921.9 | .057165 | 233.28 |
| 1004 | | 31005.0 | 125.95 | 22657.3 | .066277 | 231.65 | 1054 | | 32711.2 | 153.75 | 23947.9 | .056996 | 233.31 |
| 1005 | 731.85 | 31039.0 | 126.46 | 22683.0 | .066074 | 231.69 | 1055 | 781.85 | 32745.5 | 154.35 | 23973.8 | .056828 | 233.34 |
| 1006 | | 31073.0 | 126.98 | 22708.7 | .065871 | 231.72 | 1056 | | 32779.8 | 154.96 | 23999.7 | .056659 | 233.38 |
| 1007 | | 31107.0 | 127.50 | 22734.4 | .065670 | 231.76 | 1057 | | 32814.0 | 155.57 | 24025.7 | .056492 | 233.41 |
| 1008 | | 31141.0 | 128.02 | 22760.0 | .065468 | 231.79 | 1058 | | 32848.3 | 156.17 | 24051.7 | .056326 | 233.44 |
| 1009 | | 31175.0 | 128.53 | 22785.7 | .065269 | 231.82 | 1059 | | 32882.7 | 156.78 | 24077.7 | .056161 | 233.47 |
| 1010 | 736.85 | 31209.0 | 129.06 | 22811.4 | .065068 | 231.86 | 1060 | 786.85 | 32916.9 | 157.40 | 24103.7 | .055994 | 233.51 |
| 1011 | | 31243.0 | 129.58 | 22837.1 | .064870 | 231.89 | 1061 | | 32951.2 | 158.01 | 24129.6 | .055829 | 233.54 |
| 1012 | | 31277.0 | 130.11 | 22862.9 | .064671 | 231.92 | 1062 | | 32985.5 | 158.62 | 24155.6 | .055666 | 233.57 |
| 1013 | | 31311.1 | 130.63 | 22888.6 | .064475 | 231.96 | 1063 | | 33019.9 | 159.24 | 24181.7 | .055502 | 233.60 |
| 1014 | | 31345.1 | 131.16 | 22914.3 | .064277 | 231.99 | 1064 | | 33054.2 | 159.86 | 24207.7 | .055338 | 233.64 |
| 1015 | 741.85 | 31379.1 | 131.69 | 22940.0 | .064083 | 232.02 | 1065 | 791.85 | 33088.5 | 160.48 | 24233.7 | .055176 | 233.67 |
| 1016 | | 31413.2 | 132.23 | 22965.8 | .063886 | 232.06 | 1066 | | 33122.8 | 161.11 | 24259.7 | .055015 | 233.70 |
| 1017 | | 31447.2 | 132.76 | 22991.5 | .063693 | 232.09 | 1067 | | 33157.1 | 161.73 | 24285.7 | .054854 | 233.73 |
| 1018 | | 31481.3 | 133.30 | 23017.2 | .063498 | 232.13 | 1068 | | 33191.5 | 162.36 | 24311.8 | .054692 | 233.77 |
| 1019 | | 31515.4 | 133.83 | 23043.0 | .063306 | 232.16 | 1069 | | 33225.9 | 162.99 | 24337.8 | .054532 | 233.80 |
| 1020 | 746.85 | 31549.4 | 134.37 | 23068.7 | .063115 | 232.19 | 1070 | 796.85 | 33260.2 | 163.62 | 24363.8 | .054373 | 233.83 |
| 1021 | | 31583.5 | 134.91 | 23094.5 | .062923 | 232.23 | 1071 | | 33294.6 | 164.25 | 24389.8 | .054214 | 233.86 |
| 1022 | | 31617.6 | 135.45 | 23120.3 | .062733 | 232.26 | 1072 | | 33329.0 | 164.88 | 24415.9 | .054056 | 233.89 |
| 1023 | | 31651.7 | 136.00 | 23146.0 | .062542 | 232.29 | 1073 | | 33363.3 | 165.53 | 24442.0 | .053897 | 233.93 |
| 1024 | | 31685.8 | 136.54 | 23171.8 | .062354 | 232.33 | 1074 | | 33397.7 | 166.16 | 24468.0 | .053740 | 233.96 |
| 1025 | 751.85 | 31719.9 | 137.09 | 23197.6 | .062166 | 232.36 | 1075 | 801.85 | 33432.1 | 166.80 | 24494.1 | .053584 | 233.99 |
| 1026 | | 31754.0 | 137.64 | 23223.4 | .061977 | 232.39 | 1076 | | 33466.5 | 167.45 | 24520.2 | .053428 | 234.02 |
| 1027 | | 31788.1 | 138.19 | 23249.2 | .061791 | 232.42 | 1077 | | 33500.9 | 168.09 | 24546.2 | .053273 | 234.05 |
| 1028 | | 31822.2 | 138.75 | 23275.0 | .061603 | 232.46 | 1078 | | 33535.2 | 168.74 | 24572.3 | .053117 | 234.09 |
| 1029 | | 31856.3 | 139.30 | 23300.8 | .061418 | 232.49 | 1079 | | 33569.7 | 169.39 | 24598.4 | .052963 | 234.12 |
| 1030 | 756.85 | 31890.4 | 139.85 | 23326.6 | .061234 | 232.52 | 1080 | 806.85 | 33604.1 | 170.04 | 24624.5 | .052809 | 234.15 |
| 1031 | | 31924.6 | 140.42 | 23352.4 | .061049 | 232.56 | 1081 | | 33638.5 | 170.69 | 24650.6 | .052656 | 234.18 |
| 1032 | | 31958.7 | 140.97 | 23378.2 | .060866 | 232.59 | 1082 | | 33672.9 | 171.34 | 24676.8 | .052504 | 234.21 |
| 1033 | | 31992.8 | 141.53 | 23404.1 | .060684 | 232.62 | 1083 | | 33707.3 | 172.00 | 24702.8 | .052352 | 234.24 |
| 1034 | | 32027.0 | 142.10 | 23429.9 | .060500 | 232.66 | 1084 | | 33741.8 | 172.66 | 24728.9 | .052201 | 234.28 |
| 1035 | 761.85 | 32061.1 | 142.66 | 23455.7 | .060320 | 232.69 | 1085 | 811.85 | 33776.2 | 173.32 | 24755.1 | .052048 | 234.31 |
| 1036 | | 32095.3 | 143.23 | 23481.6 | .060139 | 232.72 | 1086 | | 33810.7 | 173.98 | 24781.2 | .051898 | 234.34 |
| 1037 | | 32129.4 | 143.80 | 23507.4 | .059958 | 232.76 | 1087 | | 33845.1 | 174.65 | 24807.3 | .051748 | 234.37 |
| 1038 | | 32163.6 | 144.37 | 23533.3 | .059779 | 232.79 | 1088 | | 33879.6 | 175.32 | 24833.5 | .051599 | 234.40 |
| 1039 | | 32197.9 | 144.95 | 23559.2 | .059599 | 232.82 | 1089 | | 33914.0 | 175.98 | 24859.6 | .051450 | 234.43 |
| 1040 | 766.85 | 32232.0 | 145.52 | 23585.1 | .059422 | 232.85 | 1090 | 816.85 | 33948.5 | 176.65 | 24885.8 | .051302 | 234.47 |
| 1041 | | 32266.2 | 146.09 | 23610.9 | .059245 | 232.89 | 1091 | | 33983.0 | 177.33 | 24912.0 | .051155 | 234.50 |
| 1042 | | 32300.4 | 146.68 | 23636.8 | .059067 | 232.92 | 1092 | | 34017.4 | 178.00 | 24938.1 | .051008 | 234.53 |
| 1043 | | 32334.6 | 147.25 | 23662.7 | .058891 | 232.95 | 1093 | | 34051.9 | 178.68 | 24964.2 | .050861 | 234.56 |
| 1044 | | 32368.8 | 147.83 | 23688.6 | .058716 | 232.99 | 1094 | | 34086.4 | 179.36 | 24990.4 | .050713 | 234.59 |
| 1045 | 771.85 | 32403.0 | 148.42 | 23714.5 | .058542 | 233.02 | 1095 | 821.85 | 34120.9 | 180.04 | 25016.6 | .050568 | 234.62 |
| 1046 | | 32437.2 | 149.00 | 23740.4 | .058366 | 233.05 | 1096 | | 34155.3 | 180.72 | 25042.7 | .050423 | 234.66 |
| 1047 | | 32471.5 | 149.59 | 23766.3 | .058193 | 233.08 | 1097 | | 34189.9 | 181.41 | 25069.0 | .050278 | 234.69 |
| 1048 | | 32505.7 | 150.18 | 23792.2 | .058021 | 233.12 | 1098 | | 34224.4 | 182.10 | 25095.1 | .050134 | 234.72 |
| 1049 | | 32540.0 | 150.77 | 23818.2 | .057847 | 233.15 | 1099 | | 34258.9 | 182.78 | 25121.4 | .049991 | 234.75 |

# Table 3  Products—400% Theoretical Air (for One Gram-Mole)

| T | t | $\bar{h}$ | $p_r$ | $\bar{u}$ | $v_r$ | $\bar{\phi}$ | T | t | $\bar{h}$ | $p_r$ | $\bar{u}$ | $v_r$ | $\bar{\phi}$ |
|---|---|---|---|---|---|---|---|---|---|---|---|---|---|
| 1100 | 826.85 | 34293.4 | 183.48 | 25147.6 | .049848 | 234.78 | 1150 | 876.85 | 36025.9 | 220.81 | 26464.4 | .043302 | 236.32 |
| 1101 | | 34327.9 | 184.17 | 25173.7 | .049705 | 234.81 | 1151 | | 36060.7 | 221.62 | 26490.8 | .043182 | 236.35 |
| 1102 | | 34362.5 | 184.86 | 25200.0 | .049563 | 234.84 | 1152 | | 36095.5 | 222.43 | 26517.3 | .043062 | 236.38 |
| 1103 | | 34397.0 | 185.56 | 25226.2 | .049422 | 234.88 | 1153 | | 36130.3 | 223.24 | 26543.8 | .042943 | 236.41 |
| 1104 | | 34431.6 | 186.26 | 25252.5 | .049281 | 234.91 | 1154 | | 36165.1 | 224.04 | 26570.3 | .042826 | 236.44 |
| 1105 | 831.85 | 34466.1 | 186.96 | 25278.7 | .049140 | 234.94 | 1155 | 881.85 | 36199.9 | 224.86 | 26596.7 | .042708 | 236.47 |
| 1106 | | 34500.6 | 187.67 | 25304.9 | .049000 | 234.97 | 1156 | | 36234.7 | 225.67 | 26623.2 | .042590 | 236.50 |
| 1107 | | 34535.2 | 188.37 | 25331.1 | .048861 | 235.00 | 1157 | | 36269.5 | 226.49 | 26649.7 | .042473 | 236.53 |
| 1108 | | 34569.7 | 189.08 | 25357.4 | .048722 | 235.03 | 1158 | | 36304.3 | 227.32 | 26676.2 | .042356 | 236.56 |
| 1109 | | 34604.3 | 189.79 | 25383.7 | .048583 | 235.06 | 1159 | | 36339.1 | 228.13 | 26702.8 | .042240 | 236.59 |
| 1110 | 836.85 | 34638.9 | 190.50 | 25409.9 | .048445 | 235.09 | 1160 | 886.85 | 36373.9 | 228.96 | 26729.2 | .042124 | 236.62 |
| 1111 | | 34673.5 | 191.22 | 25436.2 | .048308 | 235.13 | 1161 | | 36408.8 | 229.79 | 26755.8 | .042008 | 236.65 |
| 1112 | | 34708.0 | 191.94 | 25462.4 | .048170 | 235.16 | 1162 | | 36443.6 | 230.62 | 26782.3 | .041893 | 236.68 |
| 1113 | | 34742.6 | 192.65 | 25488.6 | .048034 | 235.19 | 1163 | | 36478.4 | 231.46 | 26808.8 | .041778 | 236.71 |
| 1114 | | 34777.2 | 193.38 | 25515.0 | .047898 | 235.22 | 1164 | | 36513.3 | 232.28 | 26835.3 | .041664 | 236.74 |
| 1115 | 841.85 | 34811.8 | 194.10 | 25541.2 | .047762 | 235.25 | 1165 | 891.85 | 36548.2 | 233.12 | 26861.9 | .041550 | 236.77 |
| 1116 | | 34846.4 | 194.82 | 25567.5 | .047627 | 235.28 | 1166 | | 36583.0 | 233.97 | 26888.4 | .041436 | 236.80 |
| 1117 | | 34881.0 | 195.55 | 25593.8 | .047492 | 235.31 | 1167 | | 36617.9 | 234.81 | 26915.0 | .041322 | 236.83 |
| 1118 | | 34915.6 | 196.28 | 25620.1 | .047358 | 235.34 | 1168 | | 36652.8 | 235.65 | 26941.5 | .041211 | 236.86 |
| 1119 | | 34950.2 | 197.01 | 25646.4 | .047224 | 235.37 | 1169 | | 36687.6 | 236.50 | 26968.0 | .041098 | 236.89 |
| 1120 | 846.85 | 34984.9 | 197.75 | 25672.7 | .047091 | 235.40 | 1170 | 896.85 | 36722.5 | 237.35 | 26994.6 | .040985 | 236.92 |
| 1121 | | 35019.5 | 198.49 | 25699.0 | .046958 | 235.44 | 1171 | | 36757.4 | 238.20 | 27021.2 | .040873 | 236.95 |
| 1122 | | 35054.1 | 199.22 | 25725.4 | .046825 | 235.47 | 1172 | | 36792.3 | 239.06 | 27047.8 | .040762 | 236.98 |
| 1123 | | 35088.7 | 199.97 | 25751.7 | .046694 | 235.50 | 1173 | | 36827.1 | 239.91 | 27074.3 | .040652 | 237.01 |
| 1124 | | 35123.4 | 200.71 | 25778.0 | .046562 | 235.53 | 1174 | | 36862.0 | 240.77 | 27100.9 | .040541 | 237.04 |
| 1125 | 851.85 | 35158.0 | 201.45 | 25804.3 | .046431 | 235.56 | 1175 | 901.85 | 36896.9 | 241.64 | 27127.5 | .040430 | 237.07 |
| 1126 | | 35192.7 | 202.20 | 25830.7 | .046300 | 235.59 | 1176 | | 36931.9 | 242.49 | 27154.1 | .040321 | 237.10 |
| 1127 | | 35227.3 | 202.95 | 25857.0 | .046170 | 235.62 | 1177 | | 36966.7 | 243.36 | 27180.7 | .040212 | 237.13 |
| 1128 | | 35262.1 | 203.70 | 25883.4 | .046041 | 235.65 | 1178 | | 37001.7 | 244.24 | 27207.3 | .040102 | 237.16 |
| 1129 | | 35296.7 | 204.45 | 25909.7 | .045913 | 235.68 | 1179 | | 37036.6 | 245.10 | 27233.9 | .039994 | 237.19 |
| 1130 | 856.85 | 35331.4 | 205.21 | 25936.1 | .045784 | 235.71 | 1180 | 906.85 | 37071.6 | 245.98 | 27260.6 | .039886 | 237.22 |
| 1131 | | 35366.0 | 205.97 | 25962.4 | .045656 | 235.74 | 1181 | | 37106.4 | 246.86 | 27287.1 | .039777 | 237.25 |
| 1132 | | 35400.7 | 206.73 | 25988.8 | .045528 | 235.77 | 1182 | | 37141.4 | 247.74 | 27313.8 | .039669 | 237.28 |
| 1133 | | 35435.4 | 207.49 | 26015.2 | .045400 | 235.80 | 1183 | | 37176.3 | 248.62 | 27340.4 | .039563 | 237.31 |
| 1134 | | 35470.1 | 208.26 | 26041.6 | .045273 | 235.83 | 1184 | | 37211.3 | 249.50 | 27367.0 | .039456 | 237.34 |
| 1135 | 861.85 | 35504.8 | 209.03 | 26067.9 | .045147 | 235.87 | 1185 | 911.85 | 37246.2 | 250.39 | 27393.6 | .039349 | 237.37 |
| 1136 | | 35539.5 | 209.80 | 26094.4 | .045021 | 235.90 | 1186 | | 37281.2 | 251.27 | 27420.3 | .039243 | 237.40 |
| 1137 | | 35574.2 | 210.57 | 26120.7 | .044895 | 235.93 | 1187 | | 37316.1 | 252.17 | 27446.9 | .039137 | 237.43 |
| 1138 | | 35608.9 | 211.34 | 26147.1 | .044770 | 235.96 | 1188 | | 37351.1 | 253.07 | 27473.6 | .039031 | 237.46 |
| 1139 | | 35643.6 | 212.12 | 26173.5 | .044645 | 235.99 | 1189 | | 37386.0 | 253.97 | 27500.2 | .038926 | 237.48 |
| 1140 | 866.85 | 35678.4 | 212.89 | 26200.0 | .044522 | 236.02 | 1190 | 916.85 | 37421.0 | 254.86 | 27526.9 | .038822 | 237.51 |
| 1141 | | 35713.1 | 213.68 | 26226.4 | .044398 | 236.05 | 1191 | | 37456.0 | 255.76 | 27553.5 | .038717 | 237.54 |
| 1142 | | 35747.8 | 214.46 | 26252.8 | .044274 | 236.08 | 1192 | | 37491.0 | 256.67 | 27580.2 | .038613 | 237.57 |
| 1143 | | 35782.6 | 215.25 | 26279.2 | .044151 | 236.11 | 1193 | | 37525.9 | 257.57 | 27606.8 | .038510 | 237.60 |
| 1144 | | 35817.3 | 216.04 | 26305.6 | .044028 | 236.14 | 1194 | | 37560.9 | 258.49 | 27633.5 | .038406 | 237.63 |
| 1145 | 871.85 | 35852.1 | 216.83 | 26332.1 | .043906 | 236.17 | 1195 | 921.85 | 37595.9 | 259.40 | 27660.2 | .038302 | 237.66 |
| 1146 | | 35886.9 | 217.62 | 26358.6 | .043784 | 236.20 | 1196 | | 37631.0 | 260.31 | 27686.9 | .038201 | 237.69 |
| 1147 | | 35921.6 | 218.41 | 26385.0 | .043664 | 236.23 | 1197 | | 37666.0 | 261.23 | 27713.6 | .038098 | 237.72 |
| 1148 | | 35956.4 | 219.21 | 26411.4 | .043543 | 236.26 | 1198 | | 37700.9 | 262.15 | 27740.3 | .037995 | 237.75 |
| 1149 | | 35991.1 | 220.01 | 26437.9 | .043422 | 236.29 | 1199 | | 37736.0 | 263.07 | 27767.0 | .037895 | 237.78 |

# Table 3  Products—400% Theoretical Air (for One Gram-Mole)

| T | t | $\bar{h}$ | $p_r$ | $\bar{u}$ | $v_r$ | $\bar{\phi}$ | T | t | $\bar{h}$ | $p_r$ | $\bar{u}$ | $v_r$ | $\bar{\phi}$ |
|---|---|---|---|---|---|---|---|---|---|---|---|---|---|
| 1200 | 926.85 | 37771.0 | 264.00 | 27793.7 | .037793 | 237.81 | 1250 | 976.85 | 39527.8 | 313.72 | 29134.8 | .033128 | 239.24 |
| 1201 |  | 37806.0 | 264.93 | 27820.4 | .037691 | 237.84 | 1251 |  | 39563.0 | 314.78 | 29161.7 | .033043 | 239.27 |
| 1202 |  | 37841.1 | 265.86 | 27847.2 | .037591 | 237.87 | 1252 |  | 39598.3 | 315.85 | 29188.6 | .032957 | 239.30 |
| 1203 |  | 37876.0 | 266.79 | 27873.8 | .037491 | 237.89 | 1253 |  | 39633.5 | 316.92 | 29215.6 | .032872 | 239.33 |
| 1204 |  | 37911.1 | 267.73 | 27900.5 | .037390 | 237.92 | 1254 |  | 39668.8 | 318.00 | 29242.5 | .032787 | 239.35 |
| 1205 | 931.85 | 37946.1 | 268.66 | 27927.3 | .037291 | 237.95 | 1255 | 981.85 | 39704.1 | 319.07 | 29269.5 | .032703 | 239.38 |
| 1206 |  | 37981.2 | 269.61 | 27954.0 | .037192 | 237.98 | 1256 |  | 39739.3 | 320.16 | 29296.4 | .032618 | 239.41 |
| 1207 |  | 38016.3 | 270.56 | 27980.8 | .037092 | 238.01 | 1257 |  | 39774.6 | 321.24 | 29323.4 | .032534 | 239.44 |
| 1208 |  | 38051.3 | 271.50 | 28007.5 | .036994 | 238.04 | 1258 |  | 39809.9 | 322.33 | 29350.4 | .032450 | 239.47 |
| 1209 |  | 38086.3 | 272.45 | 28034.2 | .036895 | 238.07 | 1259 |  | 39845.2 | 323.41 | 29377.3 | .032367 | 239.49 |
| 1210 | 936.85 | 38121.4 | 273.40 | 28061.0 | .036798 | 238.10 | 1260 | 986.85 | 39880.5 | 324.50 | 29404.3 | .032284 | 239.52 |
| 1211 |  | 38156.5 | 274.35 | 28087.7 | .036700 | 238.13 | 1261 |  | 39915.7 | 325.60 | 29431.3 | .032200 | 239.55 |
| 1212 |  | 38191.6 | 275.31 | 28114.5 | .036602 | 238.16 | 1262 |  | 39951.0 | 326.69 | 29458.3 | .032118 | 239.58 |
| 1213 |  | 38226.7 | 276.27 | 28141.3 | .036506 | 238.18 | 1263 |  | 39986.3 | 327.80 | 29485.3 | .032035 | 239.61 |
| 1214 |  | 38261.8 | 277.23 | 28168.1 | .036409 | 238.21 | 1264 |  | 40021.6 | 328.90 | 29512.2 | .031953 | 239.63 |
| 1215 | 941.85 | 38296.9 | 278.20 | 28194.9 | .036312 | 238.24 | 1265 | 991.85 | 40057.0 | 330.01 | 29539.2 | .031871 | 239.66 |
| 1216 |  | 38331.9 | 279.16 | 28221.6 | .036216 | 238.27 | 1266 |  | 40092.3 | 331.12 | 29566.2 | .031789 | 239.69 |
| 1217 |  | 38367.0 | 280.14 | 28248.4 | .036120 | 238.30 | 1267 |  | 40127.6 | 332.23 | 29593.2 | .031708 | 239.72 |
| 1218 |  | 38402.1 | 281.11 | 28275.2 | .036025 | 238.33 | 1268 |  | 40163.0 | 333.35 | 29620.3 | .031627 | 239.75 |
| 1219 |  | 38437.2 | 282.09 | 28302.0 | .035930 | 238.36 | 1269 |  | 40198.3 | 334.46 | 29647.3 | .031546 | 239.77 |
| 1220 | 946.85 | 38472.3 | 283.07 | 28328.8 | .035834 | 238.39 | 1270 | 996.85 | 40233.6 | 335.59 | 29674.3 | .031465 | 239.80 |
| 1221 |  | 38507.5 | 284.04 | 28355.6 | .035740 | 238.42 | 1271 |  | 40269.0 | 336.71 | 29701.4 | .031385 | 239.83 |
| 1222 |  | 38542.6 | 285.03 | 28382.4 | .035646 | 238.44 | 1272 |  | 40304.3 | 337.83 | 29728.4 | .031305 | 239.86 |
| 1223 |  | 38577.7 | 286.01 | 28409.2 | .035552 | 238.47 | 1273 |  | 40339.6 | 338.97 | 29755.4 | .031225 | 239.89 |
| 1224 |  | 38612.8 | 287.01 | 28436.0 | .035458 | 238.50 | 1274 |  | 40375.0 | 340.10 | 29782.4 | .031145 | 239.91 |
| 1225 | 951.85 | 38647.9 | 287.99 | 28462.8 | .035366 | 238.53 | 1275 | 1001.85 | 40410.3 | 341.23 | 29809.5 | .031066 | 239.94 |
| 1226 |  | 38683.1 | 288.99 | 28489.6 | .035272 | 238.56 | 1276 |  | 40445.7 | 342.37 | 29836.5 | .030988 | 239.97 |
| 1227 |  | 38718.2 | 290.00 | 28516.4 | .035179 | 238.59 | 1277 |  | 40481.1 | 343.52 | 29863.6 | .030908 | 240.00 |
| 1228 |  | 38753.4 | 291.00 | 28543.3 | .035086 | 238.62 | 1278 |  | 40516.4 | 344.66 | 29890.6 | .030830 | 240.02 |
| 1229 |  | 38788.5 | 292.00 | 28570.1 | .034995 | 238.64 | 1279 |  | 40551.9 | 345.80 | 29917.7 | .030752 | 240.05 |
| 1230 | 956.85 | 38823.7 | 293.01 | 28597.0 | .034902 | 238.67 | 1280 | 1006.85 | 40587.2 | 346.96 | 29944.8 | .030673 | 240.08 |
| 1231 |  | 38858.9 | 294.01 | 28623.8 | .034812 | 238.70 | 1281 |  | 40622.6 | 348.12 | 29971.8 | .030596 | 240.11 |
| 1232 |  | 38894.0 | 295.03 | 28650.7 | .034720 | 238.73 | 1282 |  | 40658.0 | 349.27 | 29998.9 | .030518 | 240.13 |
| 1233 |  | 38929.2 | 296.04 | 28677.6 | .034630 | 238.76 | 1283 |  | 40693.4 | 350.44 | 30026.0 | .030440 | 240.16 |
| 1234 |  | 38964.4 | 297.06 | 28704.4 | .034538 | 238.79 | 1284 |  | 40728.7 | 351.60 | 30053.0 | .030363 | 240.19 |
| 1235 | 961.85 | 38999.5 | 298.07 | 28731.2 | .034449 | 238.82 | 1285 | 1011.85 | 40764.1 | 352.76 | 30080.1 | .030287 | 240.22 |
| 1236 |  | 39034.7 | 299.10 | 28758.1 | .034358 | 238.84 | 1286 |  | 40799.5 | 353.94 | 30107.2 | .030210 | 240.24 |
| 1237 |  | 39069.9 | 300.13 | 28785.0 | .034268 | 238.87 | 1287 |  | 40834.9 | 355.11 | 30134.3 | .030133 | 240.27 |
| 1238 |  | 39105.1 | 301.16 | 28811.8 | .034179 | 238.90 | 1288 |  | 40870.3 | 356.28 | 30161.4 | .030058 | 240.30 |
| 1239 |  | 39140.3 | 302.19 | 28838.7 | .034089 | 238.93 | 1289 |  | 40905.8 | 357.47 | 30188.5 | .029981 | 240.33 |
| 1240 | 966.85 | 39175.5 | 303.22 | 28865.6 | .034001 | 238.96 | 1290 | 1016.85 | 40941.2 | 358.65 | 30215.6 | .029906 | 240.35 |
| 1241 |  | 39210.7 | 304.26 | 28892.5 | .033912 | 238.99 | 1291 |  | 40976.6 | 359.83 | 30242.7 | .029830 | 240.38 |
| 1242 |  | 39245.9 | 305.30 | 28919.4 | .033824 | 239.02 | 1292 |  | 41012.0 | 361.03 | 30269.8 | .029755 | 240.41 |
| 1243 |  | 39281.1 | 306.35 | 28946.3 | .033736 | 239.04 | 1293 |  | 41047.5 | 362.22 | 30296.9 | .029680 | 240.44 |
| 1244 |  | 39316.3 | 307.39 | 28973.2 | .033649 | 239.07 | 1294 |  | 41082.9 | 363.41 | 30324.1 | .029605 | 240.46 |
| 1245 | 971.85 | 39351.6 | 308.44 | 29000.1 | .033561 | 239.10 | 1295 | 1021.85 | 41118.3 | 364.62 | 30351.2 | .029530 | 240.49 |
| 1246 |  | 39386.8 | 309.49 | 29027.1 | .033474 | 239.13 | 1296 |  | 41153.7 | 365.81 | 30378.3 | .029456 | 240.52 |
| 1247 |  | 39422.0 | 310.54 | 29054.0 | .033387 | 239.16 | 1297 |  | 41189.3 | 367.01 | 30405.5 | .029382 | 240.55 |
| 1248 |  | 39457.3 | 311.60 | 29080.9 | .033301 | 239.19 | 1298 |  | 41224.7 | 368.22 | 30432.6 | .029309 | 240.57 |
| 1249 |  | 39492.5 | 312.66 | 29107.8 | .033214 | 239.21 | 1299 |  | 41260.1 | 369.44 | 30459.7 | .029235 | 240.60 |

# Table 3    Products—400% Theoretical Air (for One Gram-Mole)

| T | t | $\bar{h}$ | $p_r$ | $\bar{u}$ | $v_r$ | $\bar{\phi}$ | T | t | $\bar{h}$ | $p_r$ | $\bar{u}$ | $v_r$ | $\bar{\phi}$ |
|---|---|---|---|---|---|---|---|---|---|---|---|---|---|
| **1300** | 1026.85 | 41295.6 | 370.65 | 30486.9 | .029162 | 240.63 | **1350** | 1076.85 | 43073.7 | 435.59 | 31849.3 | .025768 | 241.97 |
| **1301** | | 41331.0 | 371.86 | 30514.0 | .029089 | 240.66 | **1351** | | 43109.3 | 436.98 | 31876.6 | .025706 | 242.00 |
| **1302** | | 41366.5 | 373.09 | 30541.1 | .029015 | 240.68 | **1352** | | 43145.1 | 438.37 | 31904.0 | .025643 | 242.02 |
| **1303** | | 41402.0 | 374.31 | 30568.3 | .028943 | 240.71 | **1353** | | 43180.7 | 439.76 | 31931.3 | .025581 | 242.05 |
| **1304** | | 41437.5 | 375.54 | 30595.5 | .028871 | 240.74 | **1354** | | 43216.4 | 441.15 | 31958.7 | .025519 | 242.08 |
| **1305** | 1031.85 | 41473.0 | 376.76 | 30622.7 | .028799 | 240.76 | **1355** | 1081.85 | 43252.1 | 442.56 | 31986.1 | .025456 | 242.10 |
| **1306** | | 41508.4 | 378.01 | 30649.8 | .028726 | 240.79 | **1356** | | 43287.7 | 443.97 | 32013.4 | .025395 | 242.13 |
| **1307** | | 41543.9 | 379.24 | 30677.0 | .028654 | 240.82 | **1357** | | 43323.4 | 445.37 | 32040.8 | .025333 | 242.15 |
| **1308** | | 41579.4 | 380.48 | 30704.1 | .028583 | 240.85 | **1358** | | 43359.2 | 446.78 | 32068.2 | .025272 | 242.18 |
| **1309** | | 41614.9 | 381.72 | 30731.3 | .028512 | 240.87 | **1359** | | 43394.8 | 448.19 | 32095.5 | .025211 | 242.21 |
| **1310** | 1036.85 | 41650.4 | 382.97 | 30758.6 | .028440 | 240.90 | **1360** | 1086.85 | 43430.5 | 449.61 | 32122.9 | .025150 | 242.23 |
| **1311** | | 41685.9 | 384.22 | 30785.7 | .028370 | 240.93 | **1361** | | 43466.3 | 451.03 | 32150.3 | .025089 | 242.26 |
| **1312** | | 41721.4 | 385.47 | 30812.9 | .028299 | 240.95 | **1362** | | 43501.9 | 452.45 | 32177.7 | .025028 | 242.29 |
| **1313** | | 41756.9 | 386.74 | 30840.1 | .028228 | 240.98 | **1363** | | 43537.6 | 453.88 | 32205.1 | .024968 | 242.31 |
| **1314** | | 41792.4 | 387.99 | 30867.3 | .028158 | 241.01 | **1364** | | 43573.4 | 455.31 | 32232.5 | .024908 | 242.34 |
| **1315** | 1041.85 | 41828.0 | 389.25 | 30894.6 | .028088 | 241.04 | **1365** | 1091.85 | 43609.1 | 456.76 | 32259.9 | .024847 | 242.36 |
| **1316** | | 41863.5 | 390.52 | 30921.8 | .028019 | 241.06 | **1366** | | 43644.8 | 458.20 | 32287.3 | .024787 | 242.39 |
| **1317** | | 41899.0 | 391.80 | 30949.0 | .027948 | 241.09 | **1367** | | 43680.5 | 459.64 | 32314.7 | .024727 | 242.42 |
| **1318** | | 41934.5 | 393.07 | 30976.2 | .027879 | 241.12 | **1368** | | 43716.2 | 461.09 | 32342.1 | .024668 | 242.44 |
| **1319** | | 41970.1 | 394.34 | 31003.4 | .027810 | 241.14 | **1369** | | 43751.9 | 462.54 | 32369.5 | .024609 | 242.47 |
| **1320** | 1046.85 | 42005.7 | 395.62 | 31030.7 | .027741 | 241.17 | **1370** | 1096.85 | 43787.7 | 463.99 | 32397.0 | .024550 | 242.50 |
| **1321** | | 42041.2 | 396.90 | 31057.9 | .027673 | 241.20 | **1371** | | 43823.5 | 465.45 | 32424.4 | .024491 | 242.52 |
| **1322** | | 42076.7 | 398.19 | 31085.1 | .027604 | 241.22 | **1372** | | 43859.2 | 466.91 | 32451.8 | .024432 | 242.55 |
| **1323** | | 42112.3 | 399.48 | 31112.3 | .027536 | 241.25 | **1373** | | 43895.0 | 468.37 | 32479.3 | .024373 | 242.57 |
| **1324** | | 42147.8 | 400.77 | 31139.5 | .027468 | 241.28 | **1374** | | 43930.7 | 469.84 | 32506.7 | .024315 | 242.60 |
| **1325** | 1051.85 | 42183.4 | 402.06 | 31166.8 | .027400 | 241.30 | **1375** | 1101.85 | 43966.5 | 471.31 | 32534.2 | .024256 | 242.63 |
| **1326** | | 42219.0 | 403.38 | 31194.1 | .027332 | 241.33 | **1376** | | 44002.2 | 472.79 | 32561.6 | .024198 | 242.65 |
| **1327** | | 42254.5 | 404.68 | 31221.3 | .027264 | 241.36 | **1377** | | 44038.0 | 474.27 | 32589.0 | .024140 | 242.68 |
| **1328** | | 42290.1 | 405.98 | 31248.5 | .027197 | 241.39 | **1378** | | 44073.8 | 475.75 | 32616.5 | .024083 | 242.70 |
| **1329** | | 42325.6 | 407.29 | 31275.8 | .027130 | 241.41 | **1379** | | 44109.5 | 477.24 | 32643.9 | .024025 | 242.73 |
| **1330** | 1056.85 | 42361.3 | 408.61 | 31303.1 | .027063 | 241.44 | **1380** | 1106.85 | 44145.3 | 478.73 | 32671.5 | .023968 | 242.76 |
| **1331** | | 42396.8 | 409.92 | 31330.4 | .026996 | 241.47 | **1381** | | 44181.1 | 480.22 | 32698.9 | .023910 | 242.78 |
| **1332** | | 42432.4 | 411.24 | 31357.6 | .026930 | 241.49 | **1382** | | 44216.8 | 481.72 | 32726.3 | .023853 | 242.81 |
| **1333** | | 42468.0 | 412.56 | 31384.9 | .026864 | 241.52 | **1383** | | 44252.7 | 483.22 | 32753.8 | .023796 | 242.83 |
| **1334** | | 42503.6 | 413.89 | 31412.2 | .026798 | 241.55 | **1384** | | 44288.4 | 484.73 | 32781.3 | .023740 | 242.86 |
| **1335** | 1061.85 | 42539.2 | 415.23 | 31439.5 | .026732 | 241.57 | **1385** | 1111.85 | 44324.3 | 486.23 | 32808.8 | .023683 | 242.88 |
| **1336** | | 42574.8 | 416.56 | 31466.8 | .026666 | 241.60 | **1386** | | 44360.0 | 487.75 | 32836.3 | .023627 | 242.91 |
| **1337** | | 42610.4 | 417.90 | 31494.0 | .026601 | 241.63 | **1387** | | 44395.8 | 489.26 | 32863.7 | .023570 | 242.94 |
| **1338** | | 42646.1 | 419.23 | 31521.4 | .026536 | 241.65 | **1388** | | 44431.6 | 490.79 | 32891.3 | .023514 | 242.96 |
| **1339** | | 42681.7 | 420.57 | 31548.7 | .026471 | 241.68 | **1389** | | 44467.4 | 492.31 | 32918.7 | .023458 | 242.99 |
| **1340** | 1066.85 | 42717.3 | 421.93 | 31576.0 | .026405 | 241.71 | **1390** | 1116.85 | 44503.3 | 493.84 | 32946.3 | .023402 | 243.01 |
| **1341** | | 42752.9 | 423.28 | 31603.2 | .026341 | 241.73 | **1391** | | 44539.1 | 495.37 | 32973.7 | .023347 | 243.04 |
| **1342** | | 42788.5 | 424.64 | 31630.6 | .026277 | 241.76 | **1392** | | 44574.8 | 496.91 | 33001.2 | .023291 | 243.07 |
| **1343** | | 42824.2 | 425.99 | 31657.9 | .026212 | 241.79 | **1393** | | 44610.7 | 498.45 | 33028.8 | .023236 | 243.09 |
| **1344** | | 42859.8 | 427.35 | 31685.2 | .026149 | 241.81 | **1394** | | 44646.5 | 499.99 | 33056.2 | .023181 | 243.12 |
| **1345** | 1071.85 | 42895.5 | 428.71 | 31712.6 | .026085 | 241.84 | **1395** | 1121.85 | 44682.4 | 501.54 | 33083.8 | .023126 | 243.14 |
| **1346** | | 42931.1 | 430.08 | 31739.9 | .026021 | 241.86 | **1396** | | 44718.2 | 503.09 | 33111.3 | .023071 | 243.17 |
| **1347** | | 42966.7 | 431.46 | 31767.2 | .025957 | 241.89 | **1397** | | 44754.1 | 504.65 | 33138.8 | .023017 | 243.19 |
| **1348** | | 43002.3 | 432.84 | 31794.5 | .025894 | 241.92 | **1398** | | 44789.9 | 506.21 | 33166.3 | .022962 | 243.22 |
| **1349** | | 43038.1 | 434.21 | 31821.9 | .025831 | 241.94 | **1399** | | 44825.8 | 507.77 | 33193.9 | .022908 | 243.25 |

# Table 3 Products—400% Theoretical Air (for One Gram-Mole)

| T | t | $\bar{h}$ | $p_r$ | $\bar{u}$ | $v_r$ | $\bar{\phi}$ | T | t | $\bar{h}$ | $p_r$ | $\bar{u}$ | $v_r$ | $\bar{\phi}$ |
|---|---|---|---|---|---|---|---|---|---|---|---|---|---|
| 1400 | 1126.85 | 44861.6 | 509.34 | 33221.4 | .022854 | 243.27 | 1450 | 1176.85 | 46658.6 | 592.75 | 34602.7 | .020339 | 244.53 |
| 1401 | | 44897.4 | 510.91 | 33248.9 | .022800 | 243.30 | 1451 | | 46694.7 | 594.53 | 34630.5 | .020292 | 244.56 |
| 1402 | | 44933.3 | 512.48 | 33276.5 | .022746 | 243.32 | 1452 | | 46730.7 | 596.31 | 34658.1 | .020245 | 244.58 |
| 1403 | | 44969.1 | 514.06 | 33304.0 | .022692 | 243.35 | 1453 | | 46766.7 | 598.08 | 34685.9 | .020200 | 244.61 |
| 1404 | | 45005.0 | 515.65 | 33331.6 | .022638 | 243.37 | 1454 | | 46802.7 | 599.87 | 34713.6 | .020153 | 244.63 |
| 1405 | 1131.85 | 45040.9 | 517.23 | 33359.1 | .022585 | 243.40 | 1455 | 1181.85 | 46838.8 | 601.66 | 34741.3 | .020107 | 244.66 |
| 1406 | | 45076.8 | 518.82 | 33386.7 | .022532 | 243.42 | 1456 | | 46874.9 | 603.46 | 34769.1 | .020061 | 244.68 |
| 1407 | | 45112.6 | 520.42 | 33414.2 | .022479 | 243.45 | 1457 | | 46910.9 | 605.26 | 34796.8 | .020015 | 244.71 |
| 1408 | | 45148.5 | 522.02 | 33441.8 | .022426 | 243.48 | 1458 | | 46947.0 | 607.05 | 34824.6 | .019969 | 244.73 |
| 1409 | | 45184.4 | 523.62 | 33469.4 | .022373 | 243.50 | 1459 | | 46983.0 | 608.86 | 34852.3 | .019924 | 244.75 |
| 1410 | 1136.85 | 45220.3 | 525.21 | 33497.0 | .022321 | 243.53 | 1460 | 1186.85 | 47019.1 | 610.68 | 34880.0 | .019878 | 244.78 |
| 1411 | | 45256.1 | 526.82 | 33524.5 | .022269 | 243.55 | 1461 | | 47055.2 | 612.50 | 34907.8 | .019832 | 244.80 |
| 1412 | | 45292.1 | 528.44 | 33552.1 | .022216 | 243.58 | 1462 | | 47091.2 | 614.32 | 34935.5 | .019787 | 244.83 |
| 1413 | | 45327.9 | 530.05 | 33579.7 | .022164 | 243.60 | 1463 | | 47127.3 | 616.13 | 34963.3 | .019742 | 244.85 |
| 1414 | | 45363.9 | 531.68 | 33607.3 | .022112 | 243.63 | 1464 | | 47163.3 | 617.97 | 34991.0 | .019697 | 244.88 |
| 1415 | 1141.85 | 45399.7 | 533.30 | 33634.9 | .022060 | 243.65 | 1465 | 1191.85 | 47199.4 | 619.81 | 35018.8 | .019652 | 244.90 |
| 1416 | | 45435.7 | 534.94 | 33662.5 | .022009 | 243.68 | 1466 | | 47235.5 | 621.65 | 35046.6 | .019607 | 244.93 |
| 1417 | | 45471.6 | 536.57 | 33690.1 | .021957 | 243.70 | 1467 | | 47271.6 | 623.50 | 35074.3 | .019563 | 244.95 |
| 1418 | | 45507.5 | 538.21 | 33717.7 | .021906 | 243.73 | 1468 | | 47307.7 | 625.33 | 35102.1 | .019519 | 244.98 |
| 1419 | | 45543.4 | 539.85 | 33745.3 | .021854 | 243.75 | 1469 | | 47343.7 | 627.18 | 35129.9 | .019474 | 245.00 |
| 1420 | 1146.85 | 45579.4 | 541.50 | 33772.9 | .021803 | 243.78 | 1470 | 1196.85 | 47379.9 | 629.04 | 35157.7 | .019430 | 245.03 |
| 1421 | | 45615.2 | 543.13 | 33800.5 | .021753 | 243.80 | 1471 | | 47416.0 | 630.91 | 35185.5 | .019386 | 245.05 |
| 1422 | | 45651.2 | 544.79 | 33828.1 | .021702 | 243.83 | 1472 | | 47452.0 | 632.76 | 35213.2 | .019342 | 245.07 |
| 1423 | | 45687.1 | 546.45 | 33855.7 | .021651 | 243.86 | 1473 | | 47488.2 | 634.63 | 35241.0 | .019298 | 245.10 |
| 1424 | | 45723.1 | 548.11 | 33883.4 | .021601 | 243.88 | 1474 | | 47524.3 | 636.51 | 35268.9 | .019254 | 245.12 |
| 1425 | 1151.85 | 45759.0 | 549.78 | 33910.9 | .021551 | 243.91 | 1475 | 1201.85 | 47560.4 | 638.39 | 35296.6 | .019210 | 245.15 |
| 1426 | | 45795.0 | 551.45 | 33938.6 | .021500 | 243.93 | 1476 | | 47596.5 | 640.28 | 35324.4 | .019167 | 245.17 |
| 1427 | | 45830.9 | 553.13 | 33966.2 | .021450 | 243.96 | 1477 | | 47632.6 | 642.15 | 35352.2 | .019124 | 245.20 |
| 1428 | | 45866.8 | 554.79 | 33993.9 | .021401 | 243.98 | 1478 | | 47668.7 | 644.05 | 35380.0 | .019080 | 245.22 |
| 1429 | | 45902.8 | 556.47 | 34021.5 | .021351 | 244.01 | 1479 | | 47704.9 | 645.95 | 35407.9 | .019037 | 245.25 |
| 1430 | 1156.85 | 45938.7 | 558.16 | 34049.2 | .021301 | 244.03 | 1480 | 1206.85 | 47741.0 | 647.85 | 35435.6 | .018994 | 245.27 |
| 1431 | | 45974.7 | 559.85 | 34076.8 | .021252 | 244.06 | 1481 | | 47777.1 | 649.74 | 35463.5 | .018952 | 245.30 |
| 1432 | | 46010.7 | 561.55 | 34104.4 | .021202 | 244.08 | 1482 | | 47813.3 | 651.66 | 35491.3 | .018909 | 245.32 |
| 1433 | | 46046.6 | 563.25 | 34132.1 | .021153 | 244.11 | 1483 | | 47849.4 | 653.57 | 35519.1 | .018866 | 245.34 |
| 1434 | | 46082.6 | 564.96 | 34159.7 | .021104 | 244.13 | 1484 | | 47885.5 | 655.50 | 35547.0 | .018823 | 245.37 |
| 1435 | 1161.85 | 46118.6 | 566.65 | 34187.4 | .021056 | 244.16 | 1485 | 1211.85 | 47921.7 | 657.40 | 35574.8 | .018781 | 245.39 |
| 1436 | | 46154.5 | 568.36 | 34215.1 | .021007 | 244.18 | 1486 | | 47957.8 | 659.34 | 35602.6 | .018739 | 245.42 |
| 1437 | | 46190.6 | 570.08 | 34242.8 | .020958 | 244.21 | 1487 | | 47994.0 | 661.28 | 35630.5 | .018696 | 245.44 |
| 1438 | | 46226.5 | 571.80 | 34270.4 | .020910 | 244.23 | 1488 | | 48030.2 | 663.22 | 35658.3 | .018654 | 245.47 |
| 1439 | | 46262.5 | 573.52 | 34298.1 | .020861 | 244.26 | 1489 | | 48066.3 | 665.14 | 35686.1 | .018613 | 245.49 |
| 1440 | 1166.85 | 46298.5 | 575.26 | 34325.7 | .020813 | 244.28 | 1490 | 1216.85 | 48102.4 | 667.09 | 35714.0 | .018571 | 245.51 |
| 1441 | | 46334.5 | 576.99 | 34353.5 | .020765 | 244.31 | 1491 | | 48138.6 | 669.05 | 35741.9 | .018529 | 245.54 |
| 1442 | | 46370.5 | 578.71 | 34381.1 | .020717 | 244.33 | 1492 | | 48174.7 | 670.99 | 35769.7 | .018488 | 245.56 |
| 1443 | | 46406.5 | 580.45 | 34408.8 | .020670 | 244.36 | 1493 | | 48210.9 | 672.95 | 35797.5 | .018446 | 245.59 |
| 1444 | | 46442.5 | 582.20 | 34436.5 | .020622 | 244.38 | 1494 | | 48247.1 | 674.92 | 35825.4 | .018405 | 245.61 |
| 1445 | 1171.85 | 46478.5 | 583.95 | 34464.2 | .020574 | 244.41 | 1495 | 1221.85 | 48283.3 | 676.90 | 35853.3 | .018363 | 245.64 |
| 1446 | | 46514.6 | 585.71 | 34491.9 | .020527 | 244.43 | 1496 | | 48319.5 | 678.86 | 35881.1 | .018323 | 245.66 |
| 1447 | | 46550.5 | 587.47 | 34519.6 | .020479 | 244.46 | 1497 | | 48355.7 | 680.84 | 35909.0 | .018281 | 245.68 |
| 1448 | | 46586.6 | 589.21 | 34547.3 | .020433 | 244.48 | 1498 | | 48391.9 | 682.83 | 35936.9 | .018240 | 245.71 |
| 1449 | | 46622.6 | 590.98 | 34575.0 | .020386 | 244.51 | 1499 | | 48428.0 | 684.80 | 35964.7 | .018200 | 245.73 |

## Table 3    Products—400% Theoretical Air (for One Gram-Mole)

| T | t | $\bar{h}$ | $p_r$ | $\bar{u}$ | $v_r$ | $\bar{\phi}$ | T | t | $\bar{h}$ | $p_r$ | $\bar{u}$ | $v_r$ | $\bar{\phi}$ |
|---|---|---|---|---|---|---|---|---|---|---|---|---|---|
| 1500 | 1226.85 | 48464.2 | 686.80 | 35992.6 | .018159 | 245.76 | 1550 | 1276.85 | 50278.0 | 792.43 | 37390.7 | .016263 | 246.95 |
| 1501 |  | 48500.5 | 688.80 | 36020.5 | .018118 | 245.78 | 1551 |  | 50314.3 | 794.63 | 37418.7 | .016229 | 246.97 |
| 1502 |  | 48536.6 | 690.78 | 36048.4 | .018078 | 245.80 | 1552 |  | 50350.7 | 796.88 | 37446.8 | .016193 | 246.99 |
| 1503 |  | 48572.8 | 692.80 | 36076.3 | .018038 | 245.83 | 1553 |  | 50387.1 | 799.12 | 37474.8 | .016158 | 247.02 |
| 1504 |  | 48609.1 | 694.81 | 36104.2 | .017997 | 245.85 | 1554 |  | 50423.4 | 801.38 | 37502.8 | .016123 | 247.04 |
| 1505 | 1231.85 | 48645.3 | 696.81 | 36132.1 | .017958 | 245.88 | 1555 | 1281.85 | 50459.8 | 803.65 | 37530.9 | .016088 | 247.06 |
| 1506 |  | 48681.4 | 698.84 | 36160.0 | .017918 | 245.90 | 1556 |  | 50496.2 | 805.90 | 37558.9 | .016053 | 247.09 |
| 1507 |  | 48717.7 | 700.87 | 36187.9 | .017877 | 245.92 | 1557 |  | 50532.5 | 808.18 | 37587.0 | .016018 | 247.11 |
| 1508 |  | 48753.9 | 702.88 | 36215.8 | .017838 | 245.95 | 1558 |  | 50568.9 | 810.44 | 37615.1 | .015984 | 247.13 |
| 1509 |  | 48790.1 | 704.93 | 36243.7 | .017798 | 245.97 | 1559 |  | 50605.3 | 812.73 | 37643.2 | .015949 | 247.16 |
| 1510 | 1236.85 | 48826.3 | 706.97 | 36271.6 | .017759 | 246.00 | 1560 | 1286.85 | 50641.6 | 815.00 | 37671.2 | .015915 | 247.18 |
| 1511 |  | 48862.6 | 709.00 | 36299.5 | .017719 | 246.02 | 1561 |  | 50678.0 | 817.30 | 37699.3 | .015880 | 247.20 |
| 1512 |  | 48898.8 | 711.05 | 36327.5 | .017680 | 246.04 | 1562 |  | 50714.4 | 819.58 | 37727.3 | .015846 | 247.23 |
| 1513 |  | 48935.0 | 713.11 | 36355.3 | .017641 | 246.07 | 1563 |  | 50750.8 | 821.89 | 37755.4 | .015812 | 247.25 |
| 1514 |  | 48971.3 | 715.15 | 36383.3 | .017602 | 246.09 | 1564 |  | 50787.2 | 824.18 | 37783.5 | .015778 | 247.27 |
| 1515 | 1241.85 | 49007.5 | 717.22 | 36411.2 | .017563 | 246.12 | 1565 | 1291.85 | 50823.7 | 826.51 | 37811.6 | .015743 | 247.30 |
| 1516 |  | 49043.8 | 719.30 | 36439.2 | .017524 | 246.14 | 1566 |  | 50860.1 | 828.81 | 37839.7 | .015710 | 247.32 |
| 1517 |  | 49080.0 | 721.35 | 36467.1 | .017485 | 246.16 | 1567 |  | 50896.4 | 831.14 | 37867.7 | .015676 | 247.34 |
| 1518 |  | 49116.3 | 723.44 | 36495.0 | .017446 | 246.19 | 1568 |  | 50932.8 | 833.45 | 37895.8 | .015642 | 247.37 |
| 1519 |  | 49152.5 | 725.53 | 36523.0 | .017407 | 246.21 | 1569 |  | 50969.2 | 835.80 | 37923.9 | .015608 | 247.39 |
| 1520 | 1246.85 | 49188.8 | 727.60 | 36550.9 | .017369 | 246.24 | 1570 | 1296.85 | 51005.7 | 838.12 | 37952.0 | .015575 | 247.41 |
| 1521 |  | 49225.0 | 729.70 | 36578.8 | .017331 | 246.26 | 1571 |  | 51042.1 | 840.47 | 37980.2 | .015541 | 247.44 |
| 1522 |  | 49261.3 | 731.78 | 36606.8 | .017293 | 246.28 | 1572 |  | 51078.5 | 842.80 | 38008.3 | .015508 | 247.46 |
| 1523 |  | 49297.6 | 733.89 | 36634.8 | .017255 | 246.31 | 1573 |  | 51115.0 | 845.17 | 38036.4 | .015475 | 247.48 |
| 1524 |  | 49333.9 | 736.00 | 36662.7 | .017216 | 246.33 | 1574 |  | 51151.3 | 847.51 | 38064.4 | .015442 | 247.50 |
| 1525 | 1251.85 | 49370.1 | 738.09 | 36690.6 | .017179 | 246.36 | 1575 | 1301.85 | 51187.7 | 849.89 | 38092.6 | .015408 | 247.53 |
| 1526 |  | 49406.4 | 740.22 | 36718.6 | .017141 | 246.38 | 1576 |  | 51224.2 | 852.24 | 38120.7 | .015375 | 247.55 |
| 1527 |  | 49442.7 | 742.32 | 36746.6 | .017103 | 246.40 | 1577 |  | 51260.6 | 854.63 | 38148.8 | .015342 | 247.57 |
| 1528 |  | 49479.0 | 744.45 | 36774.6 | .017065 | 246.43 | 1578 |  | 51297.1 | 856.99 | 38176.9 | .015310 | 247.60 |
| 1529 |  | 49515.2 | 746.59 | 36802.5 | .017028 | 246.45 | 1579 |  | 51333.5 | 859.39 | 38205.1 | .015277 | 247.62 |
| 1530 | 1256.85 | 49551.5 | 748.71 | 36830.5 | .016991 | 246.47 | 1580 | 1306.85 | 51370.0 | 861.76 | 38233.2 | .015244 | 247.64 |
| 1531 |  | 49587.8 | 750.86 | 36858.5 | .016953 | 246.50 | 1581 |  | 51406.4 | 864.17 | 38261.3 | .015211 | 247.67 |
| 1532 |  | 49624.2 | 752.99 | 36886.5 | .016916 | 246.52 | 1582 |  | 51442.9 | 866.56 | 38289.5 | .015179 | 247.69 |
| 1533 |  | 49660.4 | 755.15 | 36914.4 | .016879 | 246.55 | 1583 |  | 51479.3 | 868.95 | 38317.6 | .015147 | 247.71 |
| 1534 |  | 49696.7 | 757.31 | 36942.4 | .016842 | 246.57 | 1584 |  | 51515.7 | 871.37 | 38345.7 | .015114 | 247.74 |
| 1535 | 1261.85 | 49733.0 | 759.45 | 36970.4 | .016805 | 246.59 | 1585 | 1311.85 | 51552.2 | 873.77 | 38373.9 | .015082 | 247.76 |
| 1536 |  | 49769.4 | 761.63 | 36998.4 | .016768 | 246.62 | 1586 |  | 51588.7 | 876.21 | 38402.0 | .015050 | 247.78 |
| 1537 |  | 49805.7 | 763.78 | 37026.4 | .016732 | 246.64 | 1587 |  | 51625.1 | 878.62 | 38430.2 | .015018 | 247.80 |
| 1538 |  | 49841.9 | 765.96 | 37054.4 | .016695 | 246.66 | 1588 |  | 51661.6 | 881.07 | 38458.3 | .014986 | 247.83 |
| 1539 |  | 49878.3 | 768.15 | 37082.4 | .016658 | 246.69 | 1589 |  | 51698.1 | 883.49 | 38486.5 | .014954 | 247.85 |
| 1540 | 1266.85 | 49914.6 | 770.32 | 37110.4 | .016622 | 246.71 | 1590 | 1316.85 | 51734.6 | 885.95 | 38514.6 | .014922 | 247.87 |
| 1541 |  | 49950.9 | 772.52 | 37138.5 | .016585 | 246.73 | 1591 |  | 51771.0 | 888.38 | 38542.8 | .014890 | 247.90 |
| 1542 |  | 49987.3 | 774.70 | 37166.5 | .016549 | 246.76 | 1592 |  | 51807.5 | 890.85 | 38571.0 | .014858 | 247.92 |
| 1543 |  | 50023.6 | 776.91 | 37194.4 | .016513 | 246.78 | 1593 |  | 51844.0 | 893.29 | 38599.2 | .014827 | 247.94 |
| 1544 |  | 50059.9 | 779.09 | 37222.5 | .016477 | 246.80 | 1594 |  | 51880.4 | 895.74 | 38627.3 | .014796 | 247.96 |
| 1545 | 1271.85 | 50096.3 | 781.31 | 37250.5 | .016441 | 246.83 | 1595 | 1321.85 | 51916.9 | 898.23 | 38655.4 | .014764 | 247.99 |
| 1546 |  | 50132.6 | 783.51 | 37278.5 | .016406 | 246.85 | 1596 |  | 51953.4 | 900.69 | 38683.6 | .014733 | 248.01 |
| 1547 |  | 50169.0 | 785.74 | 37306.6 | .016370 | 246.88 | 1597 |  | 51989.9 | 903.19 | 38711.8 | .014701 | 248.03 |
| 1548 |  | 50205.3 | 787.97 | 37334.6 | .016334 | 246.90 | 1598 |  | 52026.4 | 905.66 | 38740.0 | .014670 | 248.06 |
| 1549 |  | 50241.6 | 790.19 | 37362.6 | .016299 | 246.92 | 1599 |  | 52062.9 | 908.17 | 38768.2 | .014639 | 248.08 |

# Table 3 Products—400% Theoretical Air (for One Gram-Mole)

| T | t | $\bar{h}$ | $p_r$ | $\bar{u}$ | $v_r$ | $\bar{\phi}$ | T | t | $\bar{h}$ | $p_r$ | $\bar{u}$ | $v_r$ | $\bar{\phi}$ |
|---|---|---|---|---|---|---|---|---|---|---|---|---|---|
| 1600 | 1326.85 | 52099.4 | 910.6 | 38796.4 | .014608 | 248.10 | 1650 | 1376.85 | 53928.1 | 1042.6 | 40209.3 | .013158 | 249.23 |
| 1601 | | 52135.9 | 913.1 | 38824.6 | .014578 | 248.12 | 1651 | | 53964.7 | 1045.4 | 40237.6 | .013130 | 249.25 |
| 1602 | | 52172.4 | 915.7 | 38852.8 | .014546 | 248.15 | 1652 | | 54001.3 | 1048.2 | 40266.0 | .013103 | 249.27 |
| 1603 | | 52208.9 | 918.2 | 38881.0 | .014516 | 248.17 | 1653 | | 54038.0 | 1051.0 | 40294.3 | .013077 | 249.29 |
| 1604 | | 52245.5 | 920.7 | 38909.2 | .014485 | 248.19 | 1654 | | 54074.6 | 1053.8 | 40322.6 | .013050 | 249.32 |
| 1605 | 1331.85 | 52282.0 | 923.2 | 38937.4 | .014455 | 248.22 | 1655 | 1381.85 | 54111.3 | 1056.6 | 40351.0 | .013023 | 249.34 |
| 1606 | | 52318.5 | 925.8 | 38965.6 | .014424 | 248.24 | 1656 | | 54148.0 | 1059.5 | 40379.3 | .012996 | 249.36 |
| 1607 | | 52355.0 | 928.3 | 38993.8 | .014394 | 248.26 | 1657 | | 54184.6 | 1062.3 | 40407.6 | .012969 | 249.38 |
| 1608 | | 52391.6 | 930.8 | 39022.0 | .014363 | 248.28 | 1658 | | 54221.3 | 1065.1 | 40436.1 | .012942 | 249.40 |
| 1609 | | 52428.0 | 933.4 | 39050.1 | .014333 | 248.31 | 1659 | | 54258.0 | 1067.9 | 40464.4 | .012916 | 249.43 |
| 1610 | 1336.85 | 52464.5 | 935.9 | 39078.4 | .014303 | 248.33 | 1660 | 1386.85 | 54294.7 | 1070.8 | 40492.8 | .012890 | 249.45 |
| 1611 | | 52501.1 | 938.5 | 39106.6 | .014272 | 248.35 | 1661 | | 54331.3 | 1073.7 | 40521.1 | .012863 | 249.47 |
| 1612 | | 52537.6 | 941.0 | 39134.8 | .014243 | 248.37 | 1662 | | 54368.0 | 1076.5 | 40549.5 | .012837 | 249.49 |
| 1613 | | 52574.2 | 943.6 | 39163.0 | .014213 | 248.40 | 1663 | | 54404.7 | 1079.3 | 40577.8 | .012810 | 249.51 |
| 1614 | | 52610.7 | 946.2 | 39191.3 | .014183 | 248.42 | 1664 | | 54441.4 | 1082.2 | 40606.2 | .012784 | 249.54 |
| 1615 | 1341.85 | 52647.2 | 948.7 | 39219.5 | .014153 | 248.44 | 1665 | 1391.85 | 54478.0 | 1085.1 | 40634.6 | .012758 | 249.56 |
| 1616 | | 52683.8 | 951.4 | 39247.7 | .014123 | 248.47 | 1666 | | 54514.7 | 1088.0 | 40662.9 | .012732 | 249.58 |
| 1617 | | 52720.4 | 953.9 | 39276.0 | .014094 | 248.49 | 1667 | | 54551.4 | 1090.8 | 40691.3 | .012706 | 249.60 |
| 1618 | | 52756.9 | 956.5 | 39304.2 | .014064 | 248.51 | 1668 | | 54588.1 | 1093.7 | 40719.7 | .012680 | 249.62 |
| 1619 | | 52793.5 | 959.1 | 39332.4 | .014035 | 248.53 | 1669 | | 54624.8 | 1096.6 | 40748.0 | .012654 | 249.65 |
| 1620 | 1346.85 | 52830.0 | 961.7 | 39360.7 | .014005 | 248.56 | 1670 | 1396.85 | 54661.5 | 1099.5 | 40776.4 | .012628 | 249.67 |
| 1621 | | 52866.6 | 964.3 | 39388.9 | .013976 | 248.58 | 1671 | | 54698.2 | 1102.4 | 40804.8 | .012602 | 249.69 |
| 1622 | | 52903.2 | 967.0 | 39417.2 | .013947 | 248.60 | 1672 | | 54734.8 | 1105.4 | 40833.2 | .012576 | 249.71 |
| 1623 | | 52939.7 | 969.6 | 39445.4 | .013918 | 248.62 | 1673 | | 54771.5 | 1108.3 | 40861.6 | .012551 | 249.73 |
| 1624 | | 52976.3 | 972.2 | 39473.7 | .013888 | 248.65 | 1674 | | 54808.2 | 1111.2 | 40889.9 | .012525 | 249.76 |
| 1625 | 1351.85 | 53012.9 | 974.9 | 39502.0 | .013859 | 248.67 | 1675 | 1401.85 | 54845.0 | 1114.1 | 40918.4 | .012500 | 249.78 |
| 1626 | | 53049.4 | 977.5 | 39530.2 | .013831 | 248.69 | 1676 | | 54881.7 | 1117.1 | 40946.8 | .012474 | 249.80 |
| 1627 | | 53086.0 | 980.2 | 39558.5 | .013801 | 248.71 | 1677 | | 54918.4 | 1120.0 | 40975.2 | .012449 | 249.82 |
| 1628 | | 53122.6 | 982.8 | 39586.8 | .013773 | 248.74 | 1678 | | 54955.2 | 1123.0 | 41003.6 | .012424 | 249.84 |
| 1629 | | 53159.2 | 985.4 | 39615.0 | .013744 | 248.76 | 1679 | | 54991.9 | 1125.9 | 41032.0 | .012398 | 249.87 |
| 1630 | 1356.85 | 53195.8 | 988.1 | 39643.3 | .013715 | 248.78 | 1680 | 1406.85 | 55028.6 | 1128.9 | 41060.4 | .012373 | 249.89 |
| 1631 | | 53232.4 | 990.8 | 39671.6 | .013687 | 248.80 | 1681 | | 55065.3 | 1131.9 | 41088.8 | .012348 | 249.91 |
| 1632 | | 53269.0 | 993.5 | 39699.8 | .013658 | 248.83 | 1682 | | 55102.0 | 1134.9 | 41117.2 | .012323 | 249.93 |
| 1633 | | 53305.5 | 996.2 | 39728.1 | .013630 | 248.85 | 1683 | | 55138.7 | 1137.8 | 41145.6 | .012298 | 249.95 |
| 1634 | | 53342.1 | 998.8 | 39756.4 | .013602 | 248.87 | 1684 | | 55175.5 | 1140.9 | 41174.0 | .012273 | 249.98 |
| 1635 | 1361.85 | 53378.7 | 1001.5 | 39784.7 | .013573 | 248.89 | 1685 | 1411.85 | 55212.2 | 1143.8 | 41202.4 | .012248 | 250.00 |
| 1636 | | 53415.4 | 1004.2 | 39813.0 | .013545 | 248.92 | 1686 | | 55249.0 | 1146.8 | 41230.9 | .012223 | 250.02 |
| 1637 | | 53452.0 | 1006.9 | 39841.3 | .013517 | 248.94 | 1687 | | 55285.7 | 1149.8 | 41259.3 | .012199 | 250.04 |
| 1638 | | 53488.6 | 1009.6 | 39869.6 | .013489 | 248.96 | 1688 | | 55322.5 | 1152.8 | 41287.8 | .012174 | 250.06 |
| 1639 | | 53525.2 | 1012.4 | 39897.9 | .013461 | 248.98 | 1689 | | 55359.2 | 1155.9 | 41316.2 | .012149 | 250.08 |
| 1640 | 1366.85 | 53561.8 | 1015.1 | 39926.2 | .013433 | 249.00 | 1690 | 1416.85 | 55395.9 | 1158.9 | 41344.6 | .012125 | 250.11 |
| 1641 | | 53598.4 | 1017.8 | 39954.5 | .013405 | 249.03 | 1691 | | 55432.7 | 1161.9 | 41373.0 | .012100 | 250.13 |
| 1642 | | 53635.0 | 1020.6 | 39982.8 | .013377 | 249.05 | 1692 | | 55469.4 | 1165.0 | 41401.5 | .012076 | 250.15 |
| 1643 | | 53671.6 | 1023.3 | 40011.1 | .013350 | 249.07 | 1693 | | 55506.2 | 1168.0 | 41429.9 | .012051 | 250.17 |
| 1644 | | 53708.3 | 1026.0 | 40039.4 | .013322 | 249.09 | 1694 | | 55542.9 | 1171.1 | 41458.3 | .012027 | 250.19 |
| 1645 | 1371.85 | 53744.9 | 1028.8 | 40067.7 | .013294 | 249.12 | 1695 | 1421.85 | 55579.7 | 1174.1 | 41486.8 | .012003 | 250.21 |
| 1646 | | 53781.5 | 1031.5 | 40096.0 | .013267 | 249.14 | 1696 | | 55616.5 | 1177.2 | 41515.3 | .011979 | 250.24 |
| 1647 | | 53818.2 | 1034.3 | 40124.3 | .013240 | 249.16 | 1697 | | 55653.3 | 1180.3 | 41543.7 | .011955 | 250.26 |
| 1648 | | 53854.8 | 1037.1 | 40152.7 | .013212 | 249.18 | 1698 | | 55690.0 | 1183.4 | 41572.2 | .011930 | 250.28 |
| 1649 | | 53891.4 | 1039.9 | 40181.0 | .013185 | 249.21 | 1699 | | 55726.8 | 1186.5 | 41600.6 | .011906 | 250.30 |

# Table 3 Products—400% Theoretical Air (for One Gram-Mole)

| T | t | $\bar{h}$ | $p_r$ | $\bar{u}$ | $v_r$ | $\bar{\phi}$ | T | t | $\bar{h}$ | $p_r$ | $\bar{u}$ | $v_r$ | $\bar{\phi}$ |
|---|---|---|---|---|---|---|---|---|---|---|---|---|---|
| 1700 | 1426.85 | 55763.6 | 1189.5 | 41629.1 | .011882 | 250.32 | 1750 | 1476.85 | 57605.5 | 1352.6 | 43055.3 | .010758 | 251.39 |
| 1701 | | 55800.3 | 1192.6 | 41657.5 | .011859 | 250.34 | 1751 | | 57642.4 | 1356.0 | 43083.9 | .010737 | 251.41 |
| 1702 | | 55837.1 | 1195.7 | 41686.0 | .011835 | 250.37 | 1752 | | 57679.3 | 1359.4 | 43112.5 | .010715 | 251.43 |
| 1703 | | 55873.9 | 1198.9 | 41714.4 | .011811 | 250.39 | 1753 | | 57716.3 | 1362.9 | 43141.1 | .010695 | 251.45 |
| 1704 | | 55910.7 | 1202.0 | 41742.9 | .011787 | 250.41 | 1754 | | 57753.1 | 1366.3 | 43169.7 | .010674 | 251.48 |
| 1705 | 1431.85 | 55947.5 | 1205.1 | 41771.4 | .011763 | 250.43 | 1755 | 1481.85 | 57790.0 | 1369.8 | 43198.3 | .010653 | 251.50 |
| 1706 | | 55984.3 | 1208.2 | 41799.9 | .011740 | 250.45 | 1756 | | 57826.9 | 1373.2 | 43226.8 | .010632 | 251.52 |
| 1707 | | 56021.1 | 1211.3 | 41828.4 | .011716 | 250.47 | 1757 | | 57863.8 | 1376.7 | 43255.4 | .010611 | 251.54 |
| 1708 | | 56057.8 | 1214.5 | 41856.8 | .011693 | 250.50 | 1758 | | 57900.8 | 1380.2 | 43284.1 | .010590 | 251.56 |
| 1709 | | 56094.6 | 1217.7 | 41885.3 | .011669 | 250.52 | 1759 | | 57937.7 | 1383.7 | 43312.7 | .010569 | 251.58 |
| 1710 | 1436.85 | 56131.4 | 1220.8 | 41913.8 | .011646 | 250.54 | 1760 | 1486.85 | 57974.6 | 1387.2 | 43341.3 | .010549 | 251.60 |
| 1711 | | 56168.2 | 1224.0 | 41942.3 | .011623 | 250.56 | 1761 | | 58011.5 | 1390.7 | 43369.9 | .010528 | 251.62 |
| 1712 | | 56205.1 | 1227.1 | 41970.8 | .011600 | 250.58 | 1762 | | 58048.5 | 1394.2 | 43398.5 | .010508 | 251.64 |
| 1713 | | 56241.9 | 1230.3 | 41999.3 | .011576 | 250.60 | 1763 | | 58085.4 | 1397.7 | 43427.2 | .010487 | 251.66 |
| 1714 | | 56278.7 | 1233.5 | 42027.8 | .011553 | 250.63 | 1764 | | 58122.4 | 1401.3 | 43455.8 | .010467 | 251.69 |
| 1715 | 1441.85 | 56315.5 | 1236.7 | 42056.3 | .011530 | 250.65 | 1765 | 1491.85 | 58159.3 | 1404.8 | 43484.4 | .010446 | 251.71 |
| 1716 | | 56352.3 | 1239.9 | 42084.8 | .011507 | 250.67 | 1766 | | 58196.2 | 1408.3 | 43513.0 | .010426 | 251.73 |
| 1717 | | 56389.1 | 1243.1 | 42113.3 | .011484 | 250.69 | 1767 | | 58233.2 | 1411.9 | 43541.7 | .010406 | 251.75 |
| 1718 | | 56425.9 | 1246.3 | 42141.8 | .011461 | 250.71 | 1768 | | 58270.1 | 1415.4 | 43570.3 | .010386 | 251.77 |
| 1719 | | 56462.8 | 1249.5 | 42170.3 | .011439 | 250.73 | 1769 | | 58307.1 | 1419.0 | 43598.9 | .010365 | 251.79 |
| 1720 | 1446.85 | 56499.6 | 1252.8 | 42198.8 | .011415 | 250.75 | 1770 | 1496.85 | 58344.0 | 1422.5 | 43627.5 | .010345 | 251.81 |
| 1721 | | 56536.4 | 1256.0 | 42227.3 | .011393 | 250.78 | 1771 | | 58381.0 | 1426.1 | 43656.2 | .010325 | 251.83 |
| 1722 | | 56573.2 | 1259.2 | 42255.8 | .011370 | 250.80 | 1772 | | 58417.9 | 1429.7 | 43684.8 | .010305 | 251.85 |
| 1723 | | 56610.1 | 1262.4 | 42284.3 | .011348 | 250.82 | 1773 | | 58454.9 | 1433.3 | 43713.4 | .010285 | 251.87 |
| 1724 | | 56646.9 | 1265.7 | 42312.9 | .011325 | 250.84 | 1774 | | 58491.8 | 1436.9 | 43742.1 | .010265 | 251.89 |
| 1725 | 1451.85 | 56683.7 | 1268.9 | 42341.4 | .011303 | 250.86 | 1775 | 1501.85 | 58528.7 | 1440.5 | 43770.7 | .010245 | 251.91 |
| 1726 | | 56720.6 | 1272.2 | 42370.0 | .011280 | 250.88 | 1776 | | 58565.8 | 1444.1 | 43799.4 | .010225 | 251.94 |
| 1727 | | 56757.4 | 1275.5 | 42398.5 | .011257 | 250.90 | 1777 | | 58602.7 | 1447.8 | 43828.0 | .010205 | 251.96 |
| 1728 | | 56794.3 | 1278.8 | 42427.0 | .011235 | 250.92 | 1778 | | 58639.7 | 1451.4 | 43856.7 | .010185 | 251.98 |
| 1729 | | 56831.1 | 1282.1 | 42455.5 | .011213 | 250.95 | 1779 | | 58676.6 | 1455.0 | 43885.3 | .010166 | 252.00 |
| 1730 | 1456.85 | 56868.0 | 1285.3 | 42484.0 | .011191 | 250.97 | 1780 | 1506.85 | 58713.7 | 1458.7 | 43914.0 | .010146 | 252.02 |
| 1731 | | 56904.8 | 1288.6 | 42512.6 | .011169 | 250.99 | 1781 | | 58750.6 | 1462.3 | 43942.7 | .010126 | 252.04 |
| 1732 | | 56941.7 | 1291.9 | 42541.2 | .011147 | 251.01 | 1782 | | 58787.6 | 1466.0 | 43971.3 | .010107 | 252.06 |
| 1733 | | 56978.6 | 1295.2 | 42569.7 | .011125 | 251.03 | 1783 | | 58824.5 | 1469.6 | 43999.9 | .010087 | 252.08 |
| 1734 | | 57015.4 | 1298.6 | 42598.2 | .011102 | 251.05 | 1784 | | 58861.6 | 1473.3 | 44028.7 | .010068 | 252.10 |
| 1735 | 1461.85 | 57052.2 | 1301.9 | 42626.8 | .011080 | 251.07 | 1785 | 1511.85 | 58898.5 | 1477.0 | 44057.3 | .010048 | 252.12 |
| 1736 | | 57089.1 | 1305.2 | 42655.3 | .011058 | 251.09 | 1786 | | 58935.5 | 1480.6 | 44086.0 | .010029 | 252.14 |
| 1737 | | 57126.0 | 1308.6 | 42683.9 | .011037 | 251.12 | 1787 | | 58972.5 | 1484.3 | 44114.6 | .010010 | 252.16 |
| 1738 | | 57162.9 | 1311.9 | 42712.5 | .011015 | 251.14 | 1788 | | 59009.5 | 1488.0 | 44143.4 | .009990 | 252.18 |
| 1739 | | 57199.8 | 1315.2 | 42741.0 | .010993 | 251.16 | 1789 | | 59046.5 | 1491.7 | 44172.0 | .009971 | 252.21 |
| 1740 | 1466.85 | 57236.6 | 1318.6 | 42769.6 | .010972 | 251.18 | 1790 | 1516.85 | 59083.5 | 1495.5 | 44200.7 | .009952 | 252.23 |
| 1741 | | 57273.5 | 1322.0 | 42798.1 | .010950 | 251.20 | 1791 | | 59120.5 | 1499.2 | 44229.4 | .009933 | 252.25 |
| 1742 | | 57310.3 | 1325.3 | 42826.7 | .010928 | 251.22 | 1792 | | 59157.5 | 1502.9 | 44258.1 | .009914 | 252.27 |
| 1743 | | 57347.3 | 1328.7 | 42855.3 | .010907 | 251.24 | 1793 | | 59194.5 | 1506.6 | 44286.8 | .009895 | 252.29 |
| 1744 | | 57384.2 | 1332.1 | 42883.8 | .010885 | 251.26 | 1794 | | 59231.5 | 1510.4 | 44315.4 | .009876 | 252.31 |
| 1745 | 1471.85 | 57421.0 | 1335.5 | 42912.4 | .010864 | 251.29 | 1795 | 1521.85 | 59268.5 | 1514.1 | 44344.2 | .009857 | 252.33 |
| 1746 | | 57457.9 | 1338.9 | 42941.0 | .010842 | 251.31 | 1796 | | 59305.5 | 1517.9 | 44372.8 | .009838 | 252.35 |
| 1747 | | 57494.8 | 1342.3 | 42969.5 | .010821 | 251.33 | 1797 | | 59342.5 | 1521.7 | 44401.5 | .009819 | 252.37 |
| 1748 | | 57531.7 | 1345.7 | 42998.2 | .010800 | 251.35 | 1798 | | 59379.5 | 1525.4 | 44430.2 | .009800 | 252.39 |
| 1749 | | 57568.6 | 1349.1 | 43026.7 | .010779 | 251.37 | 1799 | | 59416.6 | 1529.2 | 44459.0 | .009781 | 252.41 |

# Table 3    Products—400% Theoretical Air (for One Gram-Mole)

| T | t | $\bar{h}$ | $p_r$ | $\bar{u}$ | $v_r$ | $\bar{\phi}$ | T | t | $\bar{h}$ | $p_r$ | $\bar{u}$ | $v_r$ | $\bar{\phi}$ |
|---|---|---|---|---|---|---|---|---|---|---|---|---|---|
| 1800 | 1526.85 | 59453.6 | 1533.0 | 44487.6 | .0097625 | 252.43 | 1850 | 1576.85 | 61307.5 | 1732.2 | 45925.8 | .0088797 | 253.45 |
| 1801 | | 59490.6 | 1536.8 | 44516.3 | .0097438 | 252.45 | 1851 | | 61344.6 | 1736.4 | 45954.6 | .0088630 | 253.47 |
| 1802 | | 59527.7 | 1540.6 | 44545.1 | .0097251 | 252.47 | 1852 | | 61381.8 | 1740.6 | 45983.5 | .0088464 | 253.49 |
| 1803 | | 59564.7 | 1544.4 | 44573.8 | .0097065 | 252.49 | 1853 | | 61418.9 | 1744.8 | 46012.3 | .0088298 | 253.51 |
| 1804 | | 59601.7 | 1548.2 | 44602.5 | .0096879 | 252.51 | 1854 | | 61456.0 | 1749.1 | 46041.1 | .0088133 | 253.53 |
| 1805 | 1531.85 | 59638.7 | 1552.1 | 44631.2 | .0096694 | 252.53 | 1855 | 1581.85 | 61493.2 | 1753.3 | 46070.0 | .0087968 | 253.55 |
| 1806 | | 59675.8 | 1555.9 | 44659.9 | .0096509 | 252.56 | 1856 | | 61530.3 | 1757.5 | 46098.8 | .0087803 | 253.57 |
| 1807 | | 59712.8 | 1559.7 | 44688.6 | .0096325 | 252.58 | 1857 | | 61567.5 | 1761.8 | 46127.7 | .0087639 | 253.59 |
| 1808 | | 59749.8 | 1563.6 | 44717.3 | .0096141 | 252.60 | 1858 | | 61604.6 | 1766.0 | 46156.5 | .0087475 | 253.61 |
| 1809 | | 59786.9 | 1567.4 | 44746.1 | .0095957 | 252.62 | 1859 | | 61641.7 | 1770.2 | 46185.3 | .0087315 | 253.63 |
| 1810 | 1536.85 | 59823.9 | 1571.3 | 44774.8 | .0095774 | 252.64 | 1860 | 1586.85 | 61679.0 | 1774.5 | 46214.2 | .0087152 | 253.65 |
| 1811 | | 59860.9 | 1575.2 | 44803.5 | .0095591 | 252.66 | 1861 | | 61716.1 | 1778.7 | 46243.0 | .0086989 | 253.67 |
| 1812 | | 59897.9 | 1579.1 | 44832.2 | .0095409 | 252.68 | 1862 | | 61753.2 | 1783.0 | 46271.8 | .0086826 | 253.69 |
| 1813 | | 59935.1 | 1583.0 | 44861.0 | .0095227 | 252.70 | 1863 | | 61790.4 | 1787.3 | 46300.7 | .0086664 | 253.71 |
| 1814 | | 59972.1 | 1586.9 | 44889.7 | .0095046 | 252.72 | 1864 | | 61827.6 | 1791.6 | 46329.5 | .0086503 | 253.73 |
| 1815 | 1541.85 | 60009.1 | 1590.8 | 44918.5 | .0094865 | 252.74 | 1865 | 1591.85 | 61864.8 | 1795.9 | 46358.4 | .0086342 | 253.75 |
| 1816 | | 60046.2 | 1594.7 | 44947.3 | .0094684 | 252.76 | 1866 | | 61901.9 | 1800.3 | 46387.2 | .0086181 | 253.77 |
| 1817 | | 60083.3 | 1598.6 | 44976.0 | .0094504 | 252.78 | 1867 | | 61939.0 | 1804.5 | 46416.1 | .0086023 | 253.79 |
| 1818 | | 60120.3 | 1602.5 | 45004.7 | .0094324 | 252.80 | 1868 | | 61976.3 | 1808.8 | 46445.0 | .0085863 | 253.81 |
| 1819 | | 60157.4 | 1606.5 | 45033.5 | .0094145 | 252.82 | 1869 | | 62013.4 | 1813.2 | 46473.8 | .0085703 | 253.83 |
| 1820 | 1546.85 | 60194.4 | 1610.4 | 45062.2 | .0093966 | 252.84 | 1870 | 1596.85 | 62050.6 | 1817.5 | 46502.6 | .0085544 | 253.85 |
| 1821 | | 60231.5 | 1614.3 | 45091.0 | .0093791 | 252.86 | 1871 | | 62087.8 | 1821.9 | 46531.5 | .0085385 | 253.87 |
| 1822 | | 60268.6 | 1618.3 | 45119.8 | .0093612 | 252.88 | 1872 | | 62124.9 | 1826.3 | 46560.4 | .0085226 | 253.89 |
| 1823 | | 60305.7 | 1622.2 | 45148.5 | .0093435 | 252.90 | 1873 | | 62162.2 | 1830.6 | 46589.3 | .0085068 | 253.91 |
| 1824 | | 60342.7 | 1626.2 | 45177.2 | .0093258 | 252.92 | 1874 | | 62199.3 | 1835.0 | 46618.1 | .0084913 | 253.93 |
| 1825 | 1551.85 | 60379.9 | 1630.2 | 45206.1 | .0093081 | 252.94 | 1875 | 1601.85 | 62236.5 | 1839.4 | 46647.0 | .0084755 | 253.95 |
| 1826 | | 60416.9 | 1634.2 | 45234.8 | .0092904 | 252.96 | 1876 | | 62273.7 | 1843.8 | 46675.9 | .0084598 | 253.97 |
| 1827 | | 60454.0 | 1638.2 | 45263.5 | .0092728 | 252.98 | 1877 | | 62310.9 | 1848.2 | 46704.8 | .0084441 | 253.99 |
| 1828 | | 60491.0 | 1642.2 | 45292.3 | .0092553 | 253.00 | 1878 | | 62348.1 | 1852.6 | 46733.7 | .0084284 | 254.01 |
| 1829 | | 60528.2 | 1646.2 | 45321.1 | .0092377 | 253.02 | 1879 | | 62385.3 | 1857.0 | 46762.5 | .0084128 | 254.03 |
| 1830 | 1556.85 | 60565.2 | 1650.2 | 45349.9 | .0092203 | 253.04 | 1880 | 1606.85 | 62422.5 | 1861.5 | 46791.5 | .0083972 | 254.05 |
| 1831 | | 60602.3 | 1654.2 | 45378.6 | .0092028 | 253.07 | 1881 | | 62459.7 | 1865.8 | 46820.3 | .0083820 | 254.07 |
| 1832 | | 60639.4 | 1658.3 | 45407.4 | .0091854 | 253.09 | 1882 | | 62496.9 | 1870.3 | 46849.2 | .0083665 | 254.09 |
| 1833 | | 60676.5 | 1662.3 | 45436.2 | .0091681 | 253.11 | 1883 | | 62534.1 | 1874.8 | 46878.1 | .0083510 | 254.11 |
| 1834 | | 60713.6 | 1666.4 | 45465.0 | .0091507 | 253.13 | 1884 | | 62571.3 | 1879.2 | 46907.0 | .0083356 | 254.13 |
| 1835 | 1561.85 | 60750.7 | 1670.4 | 45493.8 | .0091335 | 253.15 | 1885 | 1611.85 | 62608.6 | 1883.7 | 46935.9 | .0083201 | 254.15 |
| 1836 | | 60787.8 | 1674.5 | 45522.6 | .0091162 | 253.17 | 1886 | | 62645.7 | 1888.2 | 46964.8 | .0083048 | 254.16 |
| 1837 | | 60824.9 | 1678.6 | 45551.3 | .0090990 | 253.19 | 1887 | | 62682.9 | 1892.6 | 46993.6 | .0082897 | 254.18 |
| 1838 | | 60862.0 | 1682.6 | 45580.2 | .0090822 | 253.21 | 1888 | | 62720.2 | 1897.1 | 47022.6 | .0082744 | 254.20 |
| 1839 | | 60899.1 | 1686.7 | 45608.9 | .0090651 | 253.23 | 1889 | | 62757.4 | 1901.6 | 47051.5 | .0082591 | 254.22 |
| 1840 | 1566.85 | 60936.3 | 1690.8 | 45637.8 | .0090480 | 253.25 | 1890 | 1616.85 | 62794.6 | 1906.2 | 47080.4 | .0082439 | 254.24 |
| 1841 | | 60973.4 | 1694.9 | 45666.5 | .0090310 | 253.27 | 1891 | | 62831.8 | 1910.7 | 47109.3 | .0082287 | 254.26 |
| 1842 | | 61010.5 | 1699.0 | 45695.3 | .0090140 | 253.29 | 1892 | | 62869.1 | 1915.2 | 47138.2 | .0082138 | 254.28 |
| 1843 | | 61047.6 | 1703.2 | 45724.2 | .0089970 | 253.31 | 1893 | | 62906.3 | 1919.7 | 47167.1 | .0081987 | 254.30 |
| 1844 | | 61084.7 | 1707.3 | 45753.0 | .0089801 | 253.33 | 1894 | | 62943.5 | 1924.3 | 47196.0 | .0081836 | 254.32 |
| 1845 | 1571.85 | 61121.8 | 1711.5 | 45781.7 | .0089632 | 253.35 | 1895 | 1621.85 | 62980.8 | 1928.8 | 47225.0 | .0081685 | 254.34 |
| 1846 | | 61159.0 | 1715.6 | 45810.6 | .0089463 | 253.37 | 1896 | | 63018.0 | 1933.4 | 47253.9 | .0081535 | 254.36 |
| 1847 | | 61196.1 | 1719.8 | 45839.4 | .0089295 | 253.39 | 1897 | | 63055.2 | 1938.0 | 47282.8 | .0081385 | 254.38 |
| 1848 | | 61233.2 | 1723.9 | 45868.2 | .0089128 | 253.41 | 1898 | | 63092.4 | 1942.5 | 47311.7 | .0081238 | 254.40 |
| 1849 | | 61270.4 | 1728.1 | 45897.0 | .0088963 | 253.43 | 1899 | | 63129.7 | 1947.1 | 47340.7 | .0081089 | 254.42 |

# Table 3   Products—400% Theoretical Air (for One Gram-Mole)

| T | t | $\bar{h}$ | $p_r$ | $\bar{u}$ | $v_r$ | $\bar{\phi}$ | T | t | $\bar{h}$ | $p_r$ | $\bar{u}$ | $v_r$ | $\bar{\phi}$ |
|---|---|---|---|---|---|---|---|---|---|---|---|---|---|
| 1900 | 1626.85 | 63166.9 | 1951.7 | 47369.6 | .0080940 | 254.44 | 1950 | 1676.85 | 65031.7 | 2192.9 | 48818.6 | .0073933 | 255.41 |
| 1901 |  | 63204.1 | 1956.4 | 47398.5 | .0080791 | 254.46 | 1951 |  | 65069.0 | 2197.9 | 48847.6 | .0073803 | 255.43 |
| 1902 |  | 63241.4 | 1961.0 | 47427.4 | .0080643 | 254.48 | 1952 |  | 65106.4 | 2203.0 | 48876.7 | .0073670 | 255.45 |
| 1903 |  | 63278.6 | 1965.6 | 47456.3 | .0080498 | 254.50 | 1953 |  | 65143.7 | 2208.1 | 48905.7 | .0073538 | 255.47 |
| 1904 |  | 63315.9 | 1970.2 | 47485.3 | .0080350 | 254.52 | 1954 |  | 65181.1 | 2213.1 | 48934.8 | .0073408 | 255.49 |
| 1905 | 1631.85 | 63353.2 | 1974.9 | 47514.2 | .0080203 | 254.54 | 1955 | 1681.85 | 65218.5 | 2218.3 | 48963.9 | .0073276 | 255.50 |
| 1906 |  | 63390.5 | 1979.5 | 47543.2 | .0080056 | 254.56 | 1956 |  | 65255.8 | 2223.4 | 48992.9 | .0073145 | 255.52 |
| 1907 |  | 63427.7 | 1984.2 | 47572.1 | .0079910 | 254.58 | 1957 |  | 65293.2 | 2228.4 | 49021.9 | .0073016 | 255.54 |
| 1908 |  | 63465.0 | 1988.8 | 47601.1 | .0079766 | 254.60 | 1958 |  | 65330.5 | 2233.6 | 49050.9 | .0072885 | 255.56 |
| 1909 |  | 63502.2 | 1993.5 | 47630.0 | .0079620 | 254.62 | 1959 |  | 65367.9 | 2238.7 | 49080.0 | .0072755 | 255.58 |
| 1910 | 1636.85 | 63539.4 | 1998.2 | 47658.9 | .0079474 | 254.64 | 1960 | 1686.85 | 65405.3 | 2243.9 | 49109.0 | .0072624 | 255.60 |
| 1911 |  | 63576.7 | 2002.9 | 47687.9 | .0079329 | 254.66 | 1961 |  | 65442.7 | 2249.0 | 49138.1 | .0072497 | 255.62 |
| 1912 |  | 63614.0 | 2007.6 | 47716.8 | .0079184 | 254.67 | 1962 |  | 65480.0 | 2254.2 | 49167.1 | .0072367 | 255.64 |
| 1913 |  | 63651.3 | 2012.3 | 47745.8 | .0079042 | 254.69 | 1963 |  | 65517.4 | 2259.4 | 49196.2 | .0072237 | 255.66 |
| 1914 |  | 63688.5 | 2017.0 | 47774.8 | .0078898 | 254.71 | 1964 |  | 65554.7 | 2264.5 | 49225.3 | .0072111 | 255.68 |
| 1915 | 1641.85 | 63725.8 | 2021.8 | 47803.8 | .0078754 | 254.73 | 1965 | 1691.85 | 65592.2 | 2269.7 | 49254.4 | .0071982 | 255.70 |
| 1916 |  | 63763.1 | 2026.5 | 47832.7 | .0078610 | 254.75 | 1966 |  | 65629.5 | 2274.9 | 49283.4 | .0071853 | 255.71 |
| 1917 |  | 63800.4 | 2031.2 | 47861.7 | .0078470 | 254.77 | 1967 |  | 65666.9 | 2280.1 | 49312.5 | .0071727 | 255.73 |
| 1918 |  | 63837.6 | 2036.0 | 47890.6 | .0078326 | 254.79 | 1968 |  | 65704.3 | 2285.3 | 49341.5 | .0071599 | 255.75 |
| 1919 |  | 63875.0 | 2040.8 | 47919.6 | .0078184 | 254.81 | 1969 |  | 65741.7 | 2290.6 | 49370.6 | .0071471 | 255.77 |
| 1920 | 1646.85 | 63912.2 | 2045.5 | 47948.6 | .0078041 | 254.83 | 1970 | 1696.85 | 65779.0 | 2295.8 | 49399.6 | .0071346 | 255.79 |
| 1921 |  | 63949.5 | 2050.3 | 47977.6 | .0077902 | 254.85 | 1971 |  | 65816.4 | 2301.0 | 49428.7 | .0071219 | 255.81 |
| 1922 |  | 63986.8 | 2055.1 | 48006.5 | .0077760 | 254.87 | 1972 |  | 65853.8 | 2306.3 | 49457.8 | .0071092 | 255.83 |
| 1923 |  | 64024.1 | 2059.9 | 48035.5 | .0077619 | 254.89 | 1973 |  | 65891.2 | 2311.5 | 49486.9 | .0070967 | 255.85 |
| 1924 |  | 64061.4 | 2064.7 | 48064.5 | .0077477 | 254.91 | 1974 |  | 65928.6 | 2316.8 | 49516.0 | .0070841 | 255.87 |
| 1925 | 1651.85 | 64098.6 | 2069.6 | 48093.4 | .0077337 | 254.93 | 1975 | 1701.85 | 65966.0 | 2322.1 | 49545.0 | .0070715 | 255.88 |
| 1926 |  | 64136.0 | 2074.3 | 48122.4 | .0077199 | 254.95 | 1976 |  | 66003.4 | 2327.4 | 49574.2 | .0070591 | 255.90 |
| 1927 |  | 64173.2 | 2079.2 | 48151.4 | .0077059 | 254.97 | 1977 |  | 66040.8 | 2332.7 | 49603.2 | .0070466 | 255.92 |
| 1928 |  | 64210.6 | 2084.0 | 48180.4 | .0076919 | 254.99 | 1978 |  | 66078.2 | 2338.0 | 49632.3 | .0070340 | 255.94 |
| 1929 |  | 64247.8 | 2088.9 | 48209.4 | .0076779 | 255.00 | 1979 |  | 66115.6 | 2343.3 | 49661.4 | .0070218 | 255.96 |
| 1930 | 1656.85 | 64285.2 | 2093.7 | 48238.4 | .0076642 | 255.02 | 1980 | 1706.85 | 66153.0 | 2348.7 | 49690.5 | .0070093 | 255.98 |
| 1931 |  | 64322.5 | 2098.6 | 48267.3 | .0076503 | 255.04 | 1981 |  | 66190.4 | 2354.0 | 49719.6 | .0069968 | 256.00 |
| 1932 |  | 64359.8 | 2103.5 | 48296.4 | .0076365 | 255.06 | 1982 |  | 66227.8 | 2359.3 | 49748.7 | .0069846 | 256.02 |
| 1933 |  | 64397.1 | 2108.4 | 48325.3 | .0076226 | 255.08 | 1983 |  | 66265.2 | 2364.7 | 49777.7 | .0069722 | 256.04 |
| 1934 |  | 64434.4 | 2113.3 | 48354.4 | .0076091 | 255.10 | 1984 |  | 66302.6 | 2370.1 | 49806.9 | .0069599 | 256.05 |
| 1935 | 1661.85 | 64471.7 | 2118.2 | 48383.3 | .0075953 | 255.12 | 1985 | 1711.85 | 66340.1 | 2375.5 | 49836.0 | .0069478 | 256.07 |
| 1936 |  | 64509.1 | 2123.1 | 48412.4 | .0075816 | 255.14 | 1986 |  | 66377.5 | 2380.9 | 49865.1 | .0069354 | 256.09 |
| 1937 |  | 64546.4 | 2128.0 | 48441.4 | .0075681 | 255.16 | 1987 |  | 66414.9 | 2386.3 | 49894.2 | .0069231 | 256.11 |
| 1938 |  | 64583.7 | 2133.0 | 48470.4 | .0075544 | 255.18 | 1988 |  | 66452.3 | 2391.7 | 49923.3 | .0069111 | 256.13 |
| 1939 |  | 64621.0 | 2137.9 | 48499.4 | .0075408 | 255.20 | 1989 |  | 66489.8 | 2397.1 | 49952.4 | .0068989 | 256.15 |
| 1940 | 1666.85 | 64658.4 | 2142.9 | 48528.4 | .0075272 | 255.22 | 1990 | 1716.85 | 66527.1 | 2402.6 | 49981.5 | .0068867 | 256.17 |
| 1941 |  | 64695.7 | 2147.8 | 48557.4 | .0075138 | 255.24 | 1991 |  | 66564.6 | 2408.0 | 50010.6 | .0068747 | 256.19 |
| 1942 |  | 64733.0 | 2152.8 | 48586.4 | .0075003 | 255.26 | 1992 |  | 66602.0 | 2413.4 | 50039.7 | .0068626 | 256.21 |
| 1943 |  | 64770.3 | 2157.8 | 48615.4 | .0074867 | 255.27 | 1993 |  | 66639.5 | 2418.9 | 50068.8 | .0068504 | 256.22 |
| 1944 |  | 64807.7 | 2162.8 | 48644.5 | .0074732 | 255.29 | 1994 |  | 66676.9 | 2424.3 | 50098.0 | .0068385 | 256.24 |
| 1945 | 1671.85 | 64845.0 | 2167.8 | 48673.5 | .0074600 | 255.31 | 1995 | 1721.85 | 66714.3 | 2429.8 | 50127.1 | .0068265 | 256.26 |
| 1946 |  | 64882.4 | 2172.8 | 48702.5 | .0074466 | 255.33 | 1996 |  | 66751.8 | 2435.3 | 50156.2 | .0068146 | 256.28 |
| 1947 |  | 64919.7 | 2177.8 | 48731.5 | .0074331 | 255.35 | 1997 |  | 66789.2 | 2440.8 | 50185.3 | .0068026 | 256.30 |
| 1948 |  | 64957.0 | 2182.8 | 48760.6 | .0074200 | 255.37 | 1998 |  | 66826.6 | 2446.3 | 50214.5 | .0067906 | 256.32 |
| 1949 |  | 64994.3 | 2187.9 | 48789.6 | .0074067 | 255.39 | 1999 |  | 66864.0 | 2451.8 | 50243.6 | .0067789 | 256.34 |

# Table 4    Products—400% Theoretical Air (for One Gram-Mole)

| T | t | Fuel—$(CH_1)_n$ | | | Fuel—$(CH_2)_n$ | | | Fuel—$(CH_3)_n$ | | |
|---|---|---|---|---|---|---|---|---|---|---|
| | | $\bar{c}_p$ $\frac{J}{g\text{-mol K}}$ | $\bar{c}_v$ $\frac{J}{g\text{-mol K}}$ | $k = \frac{\bar{c}_p}{\bar{c}_v}$ | $\bar{c}_p$ $\frac{J}{g\text{-mol K}}$ | $\bar{c}_v$ $\frac{J}{g\text{-mol K}}$ | $k = \frac{\bar{c}_p}{\bar{c}_v}$ | $\bar{c}_p$ $\frac{J}{g\text{-mol K}}$ | $\bar{c}_v$ $\frac{J}{g\text{-mol K}}$ | $k = \frac{\bar{c}_p}{\bar{c}_v}$ |
| K | C | | | | | | | | | |
| 100 | −173.15 | 29.115 | 20.801 | 1.400 | 29.172 | 20.858 | 1.399 | 29.212 | 20.898 | 1.398 |
| 120 | −153.15 | 29.126 | 20.812 | 1.399 | 29.182 | 20.867 | 1.398 | 29.221 | 20.906 | 1.398 |
| 140 | −133.15 | 29.147 | 20.832 | 1.399 | 29.199 | 20.884 | 1.398 | 29.235 | 20.921 | 1.397 |
| 160 | −113.15 | 29.176 | 20.861 | 1.399 | 29.223 | 20.909 | 1.398 | 29.256 | 20.942 | 1.397 |
| 180 | −93.15 | 29.212 | 20.897 | 1.398 | 29.253 | 20.939 | 1.397 | 29.282 | 20.968 | 1.397 |
| 200 | −73.15 | 29.253 | 20.938 | 1.397 | 29.288 | 20.973 | 1.396 | 29.312 | 20.998 | 1.396 |
| 220 | −53.15 | 29.298 | 20.983 | 1.396 | 29.326 | 21.012 | 1.396 | 29.346 | 21.031 | 1.395 |
| 240 | −33.15 | 29.346 | 21.032 | 1.395 | 29.368 | 21.053 | 1.395 | 29.383 | 21.068 | 1.395 |
| 260 | −13.15 | 29.398 | 21.084 | 1.394 | 29.413 | 21.099 | 1.394 | 29.424 | 21.109 | 1.394 |
| 280 | 6.85 | 29.454 | 21.139 | 1.393 | 29.463 | 21.148 | 1.393 | 29.469 | 21.154 | 1.393 |
| 300 | 26.85 | 29.514 | 21.200 | 1.392 | 29.517 | 21.203 | 1.392 | 29.519 | 21.205 | 1.392 |
| 350 | 76.85 | 29.689 | 21.375 | 1.389 | 29.680 | 21.366 | 1.389 | 29.674 | 21.359 | 1.389 |
| 400 | 126.85 | 29.906 | 21.592 | 1.385 | 29.888 | 21.573 | 1.385 | 29.875 | 21.560 | 1.386 |
| 450 | 176.85 | 30.169 | 21.854 | 1.380 | 30.143 | 21.828 | 1.381 | 30.124 | 21.810 | 1.381 |
| 500 | 226.85 | 30.473 | 22.159 | 1.375 | 30.441 | 22.127 | 1.376 | 30.418 | 22.104 | 1.376 |
| 550 | 276.85 | 30.811 | 22.496 | 1.370 | 30.774 | 22.460 | 1.370 | 30.748 | 22.434 | 1.371 |
| 600 | 326.85 | 31.172 | 22.858 | 1.364 | 31.132 | 22.817 | 1.364 | 31.104 | 22.789 | 1.365 |
| 650 | 376.85 | 31.547 | 23.233 | 1.358 | 31.504 | 23.190 | 1.359 | 31.474 | 23.159 | 1.359 |
| 700 | 426.85 | 31.926 | 23.612 | 1.352 | 31.882 | 23.567 | 1.353 | 31.850 | 23.536 | 1.353 |
| 750 | 476.85 | 32.303 | 23.988 | 1.347 | 32.257 | 23.943 | 1.347 | 32.225 | 23.910 | 1.348 |
| 800 | 526.85 | 32.670 | 24.356 | 1.341 | 32.625 | 24.310 | 1.342 | 32.592 | 24.278 | 1.342 |
| 850 | 576.85 | 33.026 | 24.711 | 1.336 | 32.980 | 24.666 | 1.337 | 32.948 | 24.634 | 1.338 |
| 900 | 626.85 | 33.366 | 25.051 | 1.332 | 33.322 | 25.007 | 1.332 | 33.291 | 24.976 | 1.333 |
| 950 | 676.85 | 33.689 | 25.374 | 1.328 | 33.647 | 25.332 | 1.328 | 33.617 | 25.303 | 1.329 |
| 1000 | 726.85 | 33.995 | 25.680 | 1.324 | 33.955 | 25.641 | 1.324 | 33.927 | 25.613 | 1.325 |
| 1050 | 776.85 | 34.283 | 25.969 | 1.320 | 34.246 | 25.932 | 1.321 | 34.220 | 25.906 | 1.321 |
| 1100 | 826.85 | 34.554 | 26.239 | 1.317 | 34.520 | 26.206 | 1.317 | 34.496 | 26.182 | 1.318 |
| 1150 | 876.85 | 34.809 | 26.494 | 1.314 | 34.778 | 26.464 | 1.314 | 34.757 | 26.443 | 1.314 |
| 1200 | 926.85 | 35.047 | 26.733 | 1.311 | 35.020 | 26.706 | 1.311 | 35.002 | 26.687 | 1.312 |
| 1250 | 976.85 | 35.271 | 26.956 | 1.308 | 35.248 | 26.934 | 1.309 | 35.232 | 26.918 | 1.309 |
| 1300 | 1026.85 | 35.481 | 27.166 | 1.306 | 35.462 | 27.148 | 1.306 | 35.449 | 27.134 | 1.306 |
| 1350 | 1076.85 | 35.677 | 27.363 | 1.304 | 35.662 | 27.348 | 1.304 | 35.652 | 27.338 | 1.304 |
| 1400 | 1126.85 | 35.862 | 27.547 | 1.302 | 35.851 | 27.537 | 1.302 | 35.843 | 27.529 | 1.302 |
| 1450 | 1176.85 | 36.035 | 27.721 | 1.300 | 36.028 | 27.714 | 1.300 | 36.024 | 27.709 | 1.300 |
| 1500 | 1226.85 | 36.198 | 27.884 | 1.298 | 36.195 | 27.881 | 1.298 | 36.193 | 27.879 | 1.298 |
| 1550 | 1276.85 | 36.352 | 28.037 | 1.297 | 36.353 | 28.038 | 1.297 | 36.354 | 28.039 | 1.297 |
| 1600 | 1326.85 | 36.497 | 28.182 | 1.295 | 36.502 | 28.187 | 1.295 | 36.505 | 28.191 | 1.295 |
| 1650 | 1376.85 | 36.634 | 28.319 | 1.294 | 36.642 | 28.328 | 1.294 | 36.648 | 28.334 | 1.293 |
| 1700 | 1426.85 | 36.763 | 28.449 | 1.292 | 36.775 | 28.461 | 1.292 | 36.784 | 28.470 | 1.292 |
| 1750 | 1476.85 | 36.886 | 28.571 | 1.291 | 36.901 | 28.587 | 1.291 | 36.913 | 28.598 | 1.291 |
| 1800 | 1526.85 | 37.002 | 28.687 | 1.290 | 37.021 | 28.707 | 1.290 | 37.035 | 28.720 | 1.289 |
| 1900 | 1626.85 | 37.218 | 28.903 | 1.288 | 37.243 | 28.929 | 1.287 | 37.261 | 28.947 | 1.287 |
| 2000 | 1726.85 | 37.413 | 29.099 | 1.286 | 37.445 | 29.130 | 1.285 | 37.467 | 29.153 | 1.285 |
| 2100 | 1826.85 | 37.591 | 29.277 | 1.284 | 37.629 | 29.314 | 1.284 | 37.655 | 29.340 | 1.283 |
| 2200 | 1926.85 | 37.755 | 29.440 | 1.282 | 37.797 | 29.483 | 1.282 | 37.827 | 29.513 | 1.282 |

# Table 5 Products—200% Theoretical Air (for One Gram-Mole)

| T | t | $\bar{h}$ | $p_r$ | $\bar{u}$ | $v_r$ | $\bar{\phi}$ | T | t | $\bar{h}$ | $p_r$ | $\bar{u}$ | $v_r$ | $\bar{\phi}$ |
|---|---|---|---|---|---|---|---|---|---|---|---|---|---|
| 100 | −173.15 | 2921.3 | .02752 | 2089.9 | 30.216 | 161.57 | 150 | −123.15 | 4388.6 | .11509 | 3141.4 | 10.8361 | 173.47 |
| 101 | | 2950.7 | .02850 | 2110.9 | 29.466 | 161.86 | 151 | | 4418.0 | .11783 | 3162.5 | 10.6551 | 173.67 |
| 102 | | 2980.0 | .02951 | 2131.9 | 28.742 | 162.15 | 152 | | 4447.4 | .12061 | 3183.6 | 10.4783 | 173.86 |
| 103 | | 3009.3 | .03054 | 2152.9 | 28.042 | 162.44 | 153 | | 4476.8 | .12344 | 3204.7 | 10.3055 | 174.05 |
| 104 | | 3038.6 | .03160 | 2173.9 | 27.366 | 162.72 | 154 | | 4506.2 | .12632 | 3225.8 | 10.1367 | 174.24 |
| 105 | −168.15 | 3067.9 | .03268 | 2194.9 | 26.712 | 163.00 | 155 | −118.15 | 4535.6 | .12924 | 3246.9 | 9.9717 | 174.43 |
| 106 | | 3097.3 | .03379 | 2215.9 | 26.080 | 163.28 | 156 | | 4565.0 | .13221 | 3267.9 | 9.8104 | 174.62 |
| 107 | | 3126.6 | .03493 | 2236.9 | 25.468 | 163.56 | 157 | | 4594.4 | .13523 | 3289.0 | 9.6526 | 174.81 |
| 108 | | 3155.9 | .03610 | 2257.9 | 24.877 | 163.83 | 158 | | 4623.8 | .13831 | 3310.1 | 9.4983 | 175.00 |
| 109 | | 3185.2 | .03729 | 2279.0 | 24.304 | 164.10 | 159 | | 4653.2 | .14143 | 3331.2 | 9.3475 | 175.18 |
| 110 | −163.15 | 3214.5 | .03851 | 2299.9 | 23.750 | 164.37 | 160 | −113.15 | 4682.6 | .14460 | 3352.3 | 9.2000 | 175.37 |
| 111 | | 3243.9 | .03976 | 2321.0 | 23.213 | 164.63 | 161 | | 4712.0 | .14782 | 3373.4 | 9.0556 | 175.55 |
| 112 | | 3273.2 | .04104 | 2342.0 | 22.693 | 164.90 | 162 | | 4741.5 | .15110 | 3394.5 | 8.9144 | 175.73 |
| 113 | | 3302.5 | .04234 | 2363.0 | 22.189 | 165.16 | 163 | | 4770.9 | .15442 | 3415.6 | 8.7763 | 175.91 |
| 114 | | 3331.8 | .04368 | 2384.0 | 21.700 | 165.41 | 164 | | 4800.3 | .15780 | 3436.7 | 8.6411 | 176.09 |
| 115 | −158.15 | 3361.2 | .04505 | 2405.0 | 21.226 | 165.67 | 165 | −108.15 | 4829.7 | .16123 | 3457.8 | 8.5087 | 176.27 |
| 116 | | 3390.5 | .04644 | 2426.0 | 20.767 | 165.92 | 166 | | 4859.2 | .16472 | 3479.0 | 8.3791 | 176.45 |
| 117 | | 3419.8 | .04787 | 2447.0 | 20.321 | 166.18 | 167 | | 4888.6 | .16826 | 3500.1 | 8.2523 | 176.63 |
| 118 | | 3449.2 | .04933 | 2468.1 | 19.889 | 166.43 | 168 | | 4918.0 | .17185 | 3521.2 | 8.1281 | 176.80 |
| 119 | | 3478.5 | .05082 | 2489.1 | 19.469 | 166.67 | 169 | | 4947.5 | .17550 | 3542.3 | 8.0064 | 176.98 |
| 120 | −153.15 | 3507.8 | .05234 | 2510.1 | 19.061 | 166.92 | 170 | −103.15 | 4976.9 | .17920 | 3563.4 | 7.8873 | 177.15 |
| 121 | | 3537.2 | .05390 | 2531.1 | 18.666 | 167.16 | 171 | | 5006.3 | .18297 | 3584.6 | 7.7706 | 177.32 |
| 122 | | 3566.5 | .05549 | 2552.1 | 18.281 | 167.40 | 172 | | 5035.8 | .18678 | 3605.7 | 7.6563 | 177.50 |
| 123 | | 3595.8 | .05711 | 2573.2 | 17.908 | 167.64 | 173 | | 5065.2 | .19066 | 3626.8 | 7.5444 | 177.67 |
| 124 | | 3625.2 | .05876 | 2594.2 | 17.545 | 167.88 | 174 | | 5094.7 | .19459 | 3648.0 | 7.4346 | 177.84 |
| 125 | −148.15 | 3654.5 | .06045 | 2615.2 | 17.192 | 168.12 | 175 | −98.15 | 5124.1 | .19858 | 3669.1 | 7.3271 | 178.01 |
| 126 | | 3683.9 | .06218 | 2636.2 | 16.849 | 168.35 | 176 | | 5153.6 | .20263 | 3690.3 | 7.2217 | 178.17 |
| 127 | | 3713.2 | .06394 | 2657.3 | 16.515 | 168.58 | 177 | | 5183.1 | .20674 | 3711.4 | 7.1184 | 178.34 |
| 128 | | 3742.5 | .06573 | 2678.3 | 16.191 | 168.81 | 178 | | 5212.5 | .21091 | 3732.6 | 7.0171 | 178.51 |
| 129 | | 3771.9 | .06756 | 2699.3 | 15.875 | 169.04 | 179 | | 5242.0 | .21514 | 3753.7 | 6.9177 | 178.67 |
| 130 | −143.15 | 3801.2 | .06943 | 2720.4 | 15.568 | 169.27 | 180 | −93.15 | 5271.4 | .21943 | 3774.9 | 6.8204 | 178.84 |
| 131 | | 3830.6 | .07133 | 2741.4 | 15.269 | 169.49 | 181 | | 5300.9 | .22378 | 3796.0 | 6.7249 | 179.00 |
| 132 | | 3859.9 | .07327 | 2762.4 | 14.978 | 169.72 | 182 | | 5330.4 | .22819 | 3817.2 | 6.6313 | 179.16 |
| 133 | | 3889.3 | .07525 | 2783.5 | 14.695 | 169.94 | 183 | | 5359.9 | .23267 | 3838.3 | 6.5394 | 179.32 |
| 134 | | 3918.6 | .07727 | 2804.5 | 14.419 | 170.16 | 184 | | 5389.4 | .23721 | 3859.5 | 6.4493 | 179.48 |
| 135 | −138.15 | 3948.0 | .07932 | 2825.6 | 14.150 | 170.38 | 185 | −88.15 | 5418.8 | .24182 | 3880.7 | 6.3609 | 179.64 |
| 136 | | 3977.4 | .08142 | 2846.6 | 13.888 | 170.59 | 186 | | 5448.3 | .24648 | 3901.9 | 6.2742 | 179.80 |
| 137 | | 4006.7 | .08355 | 2867.6 | 13.633 | 170.81 | 187 | | 5477.8 | .25122 | 3923.0 | 6.1890 | 179.96 |
| 138 | | 4036.1 | .08573 | 2888.7 | 13.384 | 171.02 | 188 | | 5507.3 | .25602 | 3944.2 | 6.1055 | 180.12 |
| 139 | | 4065.4 | .08794 | 2909.7 | 13.142 | 171.23 | 189 | | 5536.8 | .26088 | 3965.4 | 6.0236 | 180.27 |
| 140 | −133.15 | 4094.8 | .09019 | 2930.8 | 12.906 | 171.44 | 190 | −83.15 | 5566.3 | .26581 | 3986.6 | 5.9431 | 180.43 |
| 141 | | 4124.2 | .09249 | 2951.9 | 12.675 | 171.65 | 191 | | 5595.8 | .27081 | 4007.8 | 5.8641 | 180.58 |
| 142 | | 4153.6 | .09483 | 2972.9 | 12.450 | 171.86 | 192 | | 5625.3 | .27587 | 4029.0 | 5.7866 | 180.74 |
| 143 | | 4182.9 | .09721 | 2994.0 | 12.231 | 172.07 | 193 | | 5654.9 | .28101 | 4050.2 | 5.7104 | 180.89 |
| 144 | | 4212.3 | .09963 | 3015.0 | 12.017 | 172.27 | 194 | | 5684.4 | .28621 | 4071.4 | 5.6357 | 181.04 |
| 145 | −128.15 | 4241.7 | .10210 | 3036.1 | 11.808 | 172.47 | 195 | −78.15 | 5713.9 | .29148 | 4092.6 | 5.5622 | 181.20 |
| 146 | | 4271.1 | .10461 | 3057.2 | 11.604 | 172.68 | 196 | | 5743.4 | .29683 | 4113.8 | 5.4902 | 181.35 |
| 147 | | 4300.5 | .10716 | 3078.2 | 11.405 | 172.88 | 197 | | 5772.9 | .30224 | 4135.0 | 5.4194 | 181.50 |
| 148 | | 4329.8 | .10976 | 3099.3 | 11.211 | 173.08 | 198 | | 5802.5 | .30772 | 4156.2 | 5.3498 | 181.65 |
| 149 | | 4359.2 | .11240 | 3120.4 | 11.021 | 173.27 | 199 | | 5832.0 | .31328 | 4177.4 | 5.2815 | 181.80 |

## Table 5  Products—200% Theoretical Air (for One Gram-Mole)

| T | t | $\bar{h}$ | $p_r$ | $\bar{u}$ | $v_r$ | $\bar{\phi}$ | T | t | $\bar{h}$ | $p_r$ | $\bar{u}$ | $v_r$ | $\bar{\phi}$ |
|---|---|---|---|---|---|---|---|---|---|---|---|---|---|
| 200 | −73.15 | 5861.5 | .31891 | 4198.6 | 5.2143 | 181.94 | 250 | −23.15 | 7342.9 | .7062 | 5264.3 | 2.9433 | 188.55 |
| 201 | | 5891.1 | .32461 | 4219.9 | 5.1484 | 182.09 | 251 | | 7372.6 | .7164 | 5285.7 | 2.9132 | 188.67 |
| 202 | | 5920.6 | .33038 | 4241.1 | 5.0835 | 182.24 | 252 | | 7402.3 | .7266 | 5307.1 | 2.8835 | 188.79 |
| 203 | | 5950.1 | .33623 | 4262.3 | 5.0199 | 182.38 | 253 | | 7432.0 | .7370 | 5328.5 | 2.8542 | 188.91 |
| 204 | | 5979.7 | .34216 | 4283.6 | 4.9572 | 182.53 | 254 | | 7461.8 | .7475 | 5349.9 | 2.8254 | 189.03 |
| 205 | −68.15 | 6009.3 | .34815 | 4304.8 | 4.8957 | 182.67 | 255 | −18.15 | 7491.5 | .7580 | 5371.3 | 2.7969 | 189.14 |
| 206 | | 6038.8 | .35423 | 4326.0 | 4.8352 | 182.82 | 256 | | 7521.2 | .7687 | 5392.8 | 2.7689 | 189.26 |
| 207 | | 6068.4 | .36038 | 4347.3 | 4.7758 | 182.96 | 257 | | 7551.0 | .7795 | 5414.2 | 2.7412 | 189.38 |
| 208 | | 6097.9 | .36661 | 4368.5 | 4.7173 | 183.10 | 258 | | 7580.7 | .7904 | 5435.6 | 2.7139 | 189.49 |
| 209 | | 6127.5 | .37291 | 4389.8 | 4.6598 | 183.24 | 259 | | 7610.5 | .8014 | 5457.1 | 2.6869 | 189.61 |
| 210 | −63.15 | 6157.1 | .37930 | 4411.0 | 4.6033 | 183.39 | 260 | −13.15 | 7640.2 | .8126 | 5478.5 | 2.6604 | 189.72 |
| 211 | | 6186.6 | .38576 | 4432.3 | 4.5477 | 183.53 | 261 | | 7670.0 | .8238 | 5499.9 | 2.6342 | 189.83 |
| 212 | | 6216.2 | .39230 | 4453.6 | 4.4931 | 183.67 | 262 | | 7699.8 | .8352 | 5521.4 | 2.6083 | 189.95 |
| 213 | | 6245.8 | .39893 | 4474.8 | 4.4393 | 183.81 | 263 | | 7729.5 | .8466 | 5542.8 | 2.5828 | 190.06 |
| 214 | | 6275.4 | .40563 | 4496.1 | 4.3864 | 183.94 | 264 | | 7759.3 | .8582 | 5564.3 | 2.5576 | 190.18 |
| 215 | −58.15 | 6305.0 | .41242 | 4517.4 | 4.3344 | 184.08 | 265 | −8.15 | 7789.1 | .8699 | 5585.8 | 2.5328 | 190.29 |
| 216 | | 6334.6 | .41929 | 4538.7 | 4.2832 | 184.22 | 266 | | 7818.9 | .8817 | 5607.2 | 2.5083 | 190.40 |
| 217 | | 6364.2 | .42624 | 4559.9 | 4.2329 | 184.36 | 267 | | 7848.6 | .8937 | 5628.7 | 2.4841 | 190.51 |
| 218 | | 6393.8 | .43327 | 4581.2 | 4.1834 | 184.49 | 268 | | 7878.4 | .9057 | 5650.2 | 2.4602 | 190.62 |
| 219 | | 6423.4 | .44039 | 4602.5 | 4.1347 | 184.63 | 269 | | 7908.2 | .9179 | 5671.6 | 2.4366 | 190.73 |
| 220 | −53.15 | 6453.0 | .44759 | 4623.8 | 4.0867 | 184.76 | 270 | −3.15 | 7938.0 | .9302 | 5693.1 | 2.4134 | 190.84 |
| 221 | | 6482.6 | .45488 | 4645.1 | 4.0395 | 184.90 | 271 | | 7967.8 | .9426 | 5714.6 | 2.3905 | 190.95 |
| 222 | | 6512.2 | .46225 | 4666.4 | 3.9931 | 185.03 | 272 | | 7997.6 | .9551 | 5736.1 | 2.3678 | 191.06 |
| 223 | | 6541.8 | .46971 | 4687.7 | 3.9474 | 185.16 | 273 | | 8027.4 | .9678 | 5757.6 | 2.3454 | 191.17 |
| 224 | | 6571.4 | .47726 | 4709.0 | 3.9023 | 185.30 | 274 | | 8057.2 | .9805 | 5779.1 | 2.3234 | 191.28 |
| 225 | −48.15 | 6601.0 | .48489 | 4730.3 | 3.8580 | 185.43 | 275 | 1.85 | 8087.1 | .9934 | 5800.6 | 2.3016 | 191.39 |
| 226 | | 6630.7 | .49262 | 4751.6 | 3.8144 | 185.56 | 276 | | 8116.9 | 1.0065 | 5822.1 | 2.2801 | 191.50 |
| 227 | | 6660.3 | .50043 | 4772.9 | 3.7715 | 185.69 | 277 | | 8146.7 | 1.0196 | 5843.6 | 2.2588 | 191.61 |
| 228 | | 6689.9 | .50833 | 4794.2 | 3.7293 | 185.82 | 278 | | 8176.6 | 1.0329 | 5865.1 | 2.2379 | 191.72 |
| 229 | | 6719.6 | .51632 | 4815.6 | 3.6876 | 185.95 | 279 | | 8206.4 | 1.0463 | 5886.7 | 2.2172 | 191.82 |
| 230 | −43.15 | 6749.2 | .52441 | 4836.9 | 3.6466 | 186.08 | 280 | 6.85 | 8236.2 | 1.0598 | 5908.2 | 2.1967 | 191.93 |
| 231 | | 6778.9 | .53258 | 4858.2 | 3.6063 | 186.21 | 281 | | 8266.1 | 1.0734 | 5929.7 | 2.1765 | 192.04 |
| 232 | | 6808.5 | .54085 | 4879.6 | 3.5665 | 186.34 | 282 | | 8295.9 | 1.0872 | 5951.2 | 2.1566 | 192.14 |
| 233 | | 6838.2 | .54921 | 4900.9 | 3.5274 | 186.46 | 283 | | 8325.8 | 1.1011 | 5972.8 | 2.1369 | 192.25 |
| 234 | | 6867.8 | .55767 | 4922.3 | 3.4888 | 186.59 | 284 | | 8355.6 | 1.1151 | 5994.3 | 2.1175 | 192.35 |
| 235 | −38.15 | 6897.5 | .56621 | 4943.6 | 3.4508 | 186.72 | 285 | 11.85 | 8385.5 | 1.1293 | 6015.8 | 2.0983 | 192.46 |
| 236 | | 6927.1 | .57486 | 4965.0 | 3.4134 | 186.84 | 286 | | 8415.3 | 1.1436 | 6037.4 | 2.0793 | 192.56 |
| 237 | | 6956.8 | .58360 | 4986.3 | 3.3765 | 186.97 | 287 | | 8445.2 | 1.1580 | 6058.9 | 2.0606 | 192.67 |
| 238 | | 6986.5 | .59243 | 5007.7 | 3.3402 | 187.09 | 288 | | 8475.1 | 1.1726 | 6080.5 | 2.0421 | 192.77 |
| 239 | | 7016.2 | .60137 | 5029.0 | 3.3044 | 187.22 | 289 | | 8504.9 | 1.1873 | 6102.1 | 2.0238 | 192.87 |
| 240 | −33.15 | 7045.8 | .61040 | 5050.4 | 3.2691 | 187.34 | 290 | 16.85 | 8534.8 | 1.2021 | 6123.6 | 2.0058 | 192.98 |
| 241 | | 7075.5 | .61952 | 5071.8 | 3.2344 | 187.47 | 291 | | 8564.7 | 1.2171 | 6145.2 | 1.9880 | 193.08 |
| 242 | | 7105.2 | .62875 | 5093.1 | 3.2001 | 187.59 | 292 | | 8594.6 | 1.2322 | 6166.8 | 1.9704 | 193.18 |
| 243 | | 7134.9 | .63808 | 5114.5 | 3.1664 | 187.71 | 293 | | 8624.5 | 1.2474 | 6188.3 | 1.9530 | 193.28 |
| 244 | | 7164.6 | .64751 | 5135.9 | 3.1331 | 187.83 | 294 | | 8654.4 | 1.2628 | 6209.9 | 1.9358 | 193.39 |
| 245 | −28.15 | 7194.3 | .65704 | 5157.3 | 3.1003 | 187.95 | 295 | 21.85 | 8684.3 | 1.2783 | 6231.5 | 1.9188 | 193.49 |
| 246 | | 7224.0 | .66667 | 5178.7 | 3.0680 | 188.08 | 296 | | 8714.1 | 1.2939 | 6253.1 | 1.9020 | 193.59 |
| 247 | | 7253.7 | .67640 | 5200.1 | 3.0361 | 188.20 | 297 | | 8744.1 | 1.3097 | 6274.7 | 1.8854 | 193.69 |
| 248 | | 7283.4 | .68624 | 5221.5 | 3.0047 | 188.32 | 298 | | 8774.0 | 1.3257 | 6296.3 | 1.8690 | 193.79 |
| 249 | | 7313.2 | .69618 | 5242.9 | 2.9738 | 188.44 | 299 | | 8803.9 | 1.3417 | 6317.9 | 1.8528 | 193.89 |

# Table 5  Products—200% Theoretical Air (for One Gram-Mole)

| T | t | $\bar{h}$ | $p_r$ | $\bar{u}$ | $v_r$ | $\bar{\phi}$ | T | t | $\bar{h}$ | $p_r$ | $\bar{u}$ | $v_r$ | $\bar{\phi}$ |
|---|---|---|---|---|---|---|---|---|---|---|---|---|---|
| 300 | 26.85 | 8833.8 | 1.3580 | 6339.5 | 1.8368 | 193.99 | 350 | 76.85 | 10335.5 | 2.3697 | 7425.4 | 1.22804 | 198.62 |
| 301 | | 8863.7 | 1.3743 | 6361.1 | 1.8210 | 194.09 | 351 | | 10365.6 | 2.3943 | 7447.3 | 1.21887 | 198.71 |
| 302 | | 8893.7 | 1.3908 | 6382.7 | 1.8054 | 194.19 | 352 | | 10395.8 | 2.4191 | 7469.1 | 1.20980 | 198.79 |
| 303 | | 8923.6 | 1.4075 | 6404.3 | 1.7899 | 194.29 | 353 | | 10425.9 | 2.4442 | 7491.0 | 1.20081 | 198.88 |
| 304 | | 8953.5 | 1.4243 | 6426.0 | 1.7746 | 194.39 | 354 | | 10456.1 | 2.4694 | 7512.8 | 1.19192 | 198.96 |
| 305 | 31.85 | 8983.5 | 1.4412 | 6447.6 | 1.7595 | 194.49 | 355 | 81.85 | 10486.3 | 2.4948 | 7534.7 | 1.18312 | 199.05 |
| 306 | | 9013.4 | 1.4583 | 6469.2 | 1.7446 | 194.58 | 356 | | 10516.5 | 2.5204 | 7556.5 | 1.17440 | 199.13 |
| 307 | | 9043.4 | 1.4756 | 6490.8 | 1.7299 | 194.68 | 357 | | 10546.6 | 2.5462 | 7578.4 | 1.16577 | 199.22 |
| 308 | | 9073.3 | 1.4929 | 6512.5 | 1.7153 | 194.78 | 358 | | 10576.8 | 2.5722 | 7600.3 | 1.15721 | 199.30 |
| 309 | | 9103.3 | 1.5105 | 6534.1 | 1.7009 | 194.88 | 359 | | 10607.0 | 2.5984 | 7622.1 | 1.14876 | 199.39 |
| 310 | 36.85 | 9133.3 | 1.5282 | 6555.8 | 1.6866 | 194.97 | 360 | 86.85 | 10637.2 | 2.6247 | 7644.0 | 1.14037 | 199.47 |
| 311 | | 9163.2 | 1.5460 | 6577.4 | 1.6725 | 195.07 | 361 | | 10667.4 | 2.6513 | 7665.9 | 1.13208 | 199.55 |
| 312 | | 9193.2 | 1.5640 | 6599.1 | 1.6586 | 195.16 | 362 | | 10697.6 | 2.6781 | 7687.8 | 1.12386 | 199.64 |
| 313 | | 9223.2 | 1.5822 | 6620.8 | 1.6448 | 195.26 | 363 | | 10727.8 | 2.7051 | 7709.7 | 1.11572 | 199.72 |
| 314 | | 9253.1 | 1.6005 | 6642.4 | 1.6312 | 195.36 | 364 | | 10758.0 | 2.7323 | 7731.6 | 1.10767 | 199.80 |
| 315 | 41.85 | 9283.1 | 1.6189 | 6664.1 | 1.6178 | 195.45 | 365 | 91.85 | 10788.3 | 2.7597 | 7753.5 | 1.09969 | 199.89 |
| 316 | | 9313.1 | 1.6375 | 6685.8 | 1.6044 | 195.55 | 366 | | 10818.5 | 2.7873 | 7775.4 | 1.09178 | 199.97 |
| 317 | | 9343.1 | 1.6563 | 6707.4 | 1.5913 | 195.64 | 367 | | 10848.7 | 2.8150 | 7797.3 | 1.08396 | 200.05 |
| 318 | | 9373.1 | 1.6752 | 6729.1 | 1.5783 | 195.74 | 368 | | 10879.0 | 2.8430 | 7819.3 | 1.07621 | 200.13 |
| 319 | | 9403.1 | 1.6943 | 6750.8 | 1.5654 | 195.83 | 369 | | 10909.2 | 2.8712 | 7841.2 | 1.06854 | 200.22 |
| 320 | 46.85 | 9433.1 | 1.7136 | 6772.5 | 1.5527 | 195.92 | 370 | 96.85 | 10939.4 | 2.8997 | 7863.1 | 1.06093 | 200.30 |
| 321 | | 9463.1 | 1.7330 | 6794.2 | 1.5401 | 196.02 | 371 | | 10969.7 | 2.9283 | 7885.1 | 1.05340 | 200.38 |
| 322 | | 9493.1 | 1.7526 | 6815.9 | 1.5276 | 196.11 | 372 | | 10999.9 | 2.9571 | 7907.0 | 1.04594 | 200.46 |
| 323 | | 9523.2 | 1.7723 | 6837.6 | 1.5153 | 196.20 | 373 | | 11030.2 | 2.9861 | 7928.9 | 1.03856 | 200.54 |
| 324 | | 9553.2 | 1.7922 | 6859.3 | 1.5031 | 196.30 | 374 | | 11060.5 | 3.0154 | 7950.9 | 1.03124 | 200.62 |
| 325 | 51.85 | 9583.2 | 1.8123 | 6881.0 | 1.4911 | 196.39 | 375 | 101.85 | 11090.8 | 3.0448 | 7972.9 | 1.02400 | 200.70 |
| 326 | | 9613.3 | 1.8325 | 6902.8 | 1.4792 | 196.48 | 376 | | 11121.0 | 3.0745 | 7994.8 | 1.01682 | 200.78 |
| 327 | | 9643.3 | 1.8529 | 6924.5 | 1.4674 | 196.57 | 377 | | 11151.3 | 3.1044 | 8016.8 | 1.00971 | 200.86 |
| 328 | | 9673.3 | 1.8734 | 6946.2 | 1.4557 | 196.67 | 378 | | 11181.6 | 3.1345 | 8038.8 | 1.00266 | 200.95 |
| 329 | | 9703.4 | 1.8941 | 6968.0 | 1.4442 | 196.76 | 379 | | 11211.9 | 3.1648 | 8060.7 | .99569 | 201.03 |
| 330 | 56.85 | 9733.4 | 1.9150 | 6989.7 | 1.4328 | 196.85 | 380 | 106.85 | 11242.2 | 3.1953 | 8082.7 | .98878 | 201.10 |
| 331 | | 9763.5 | 1.9361 | 7011.4 | 1.4215 | 196.94 | 381 | | 11272.5 | 3.2261 | 8104.7 | .98193 | 201.18 |
| 332 | | 9793.6 | 1.9573 | 7033.2 | 1.4103 | 197.03 | 382 | | 11302.8 | 3.2571 | 8126.7 | .97515 | 201.26 |
| 333 | | 9823.6 | 1.9787 | 7054.9 | 1.3992 | 197.12 | 383 | | 11333.1 | 3.2883 | 8148.7 | .96842 | 201.34 |
| 334 | | 9853.7 | 2.0003 | 7076.7 | 1.3883 | 197.21 | 384 | | 11363.4 | 3.3197 | 8170.7 | .96176 | 201.42 |
| 335 | 61.85 | 9883.8 | 2.0220 | 7098.5 | 1.3775 | 197.30 | 385 | 111.85 | 11393.8 | 3.3514 | 8192.7 | .95515 | 201.50 |
| 336 | | 9913.8 | 2.0440 | 7120.2 | 1.3668 | 197.39 | 386 | | 11424.1 | 3.3832 | 8214.7 | .94862 | 201.58 |
| 337 | | 9943.9 | 2.0661 | 7142.0 | 1.3562 | 197.48 | 387 | | 11454.5 | 3.4153 | 8236.8 | .94213 | 201.66 |
| 338 | | 9974.0 | 2.0883 | 7163.8 | 1.3457 | 197.57 | 388 | | 11484.8 | 3.4476 | 8258.8 | .93571 | 201.74 |
| 339 | | 10004.1 | 2.1108 | 7185.5 | 1.3353 | 197.66 | 389 | | 11515.1 | 3.4802 | 8280.8 | .92936 | 201.81 |
| 340 | 66.85 | 10034.2 | 2.1334 | 7207.3 | 1.3251 | 197.75 | 390 | 116.85 | 11545.5 | 3.5129 | 8302.9 | .92305 | 201.89 |
| 341 | | 10064.3 | 2.1562 | 7229.1 | 1.3149 | 197.83 | 391 | | 11575.8 | 3.5459 | 8324.9 | .91681 | 201.97 |
| 342 | | 10094.4 | 2.1792 | 7250.9 | 1.3048 | 197.92 | 392 | | 11606.2 | 3.5792 | 8347.0 | .91062 | 202.05 |
| 343 | | 10124.5 | 2.2024 | 7272.7 | 1.2949 | 198.01 | 393 | | 11636.6 | 3.6126 | 8369.0 | .90448 | 202.13 |
| 344 | | 10154.7 | 2.2257 | 7294.5 | 1.2851 | 198.10 | 394 | | 11667.0 | 3.6463 | 8391.1 | .89841 | 202.20 |
| 345 | 71.85 | 10184.8 | 2.2492 | 7316.3 | 1.2753 | 198.19 | 395 | 121.85 | 11697.3 | 3.6803 | 8413.2 | .89238 | 202.28 |
| 346 | | 10214.9 | 2.2730 | 7338.1 | 1.2657 | 198.27 | 396 | | 11727.7 | 3.7144 | 8435.2 | .88641 | 202.36 |
| 347 | | 10245.0 | 2.2969 | 7359.9 | 1.2561 | 198.36 | 397 | | 11758.1 | 3.7488 | 8457.3 | .88050 | 202.43 |
| 348 | | 10275.2 | 2.3210 | 7381.8 | 1.2466 | 198.45 | 398 | | 11788.5 | 3.7834 | 8479.4 | .87464 | 202.51 |
| 349 | | 10305.3 | 2.3452 | 7403.6 | 1.2373 | 198.53 | 399 | | 11818.9 | 3.8183 | 8501.5 | .86882 | 202.59 |

# Table 5 Products—200% Theoretical Air (for One Gram-Mole)

| T | t | $\bar{h}$ | $p_r$ | $\bar{u}$ | $v_r$ | $\bar{\phi}$ | T | t | $\bar{h}$ | $p_r$ | $\bar{u}$ | $v_r$ | $\bar{\phi}$ |
|---|---|---|---|---|---|---|---|---|---|---|---|---|---|
| 400 | 126.85 | 11849.3 | 3.8534 | 8523.6 | .86307 | 202.66 | 450 | 176.85 | 13377.2 | 5.9406 | 9635.8 | .62981 | 206.26 |
| 401 | | 11879.7 | 3.8888 | 8545.7 | .85736 | 202.74 | 451 | | 13408.0 | 5.9896 | 9658.2 | .62606 | 206.33 |
| 402 | | 11910.2 | 3.9244 | 8567.8 | .85170 | 202.81 | 452 | | 13438.7 | 6.0388 | 9680.6 | .62233 | 206.40 |
| 403 | | 11940.6 | 3.9602 | 8589.9 | .84609 | 202.89 | 453 | | 13469.4 | 6.0883 | 9703.0 | .61863 | 206.47 |
| 404 | | 11971.0 | 3.9963 | 8612.0 | .84053 | 202.96 | 454 | | 13500.1 | 6.1381 | 9725.4 | .61497 | 206.53 |
| 405 | 131.85 | 12001.5 | 4.0326 | 8634.1 | .83502 | 203.04 | 455 | 181.85 | 13530.9 | 6.1883 | 9747.8 | .61132 | 206.60 |
| 406 | | 12031.9 | 4.0692 | 8656.2 | .82956 | 203.12 | 456 | | 13561.6 | 6.2387 | 9770.3 | .60772 | 206.67 |
| 407 | | 12062.3 | 4.1061 | 8678.4 | .82414 | 203.19 | 457 | | 13592.4 | 6.2895 | 9792.7 | .60413 | 206.74 |
| 408 | | 12092.8 | 4.1431 | 8700.5 | .81877 | 203.26 | 458 | | 13623.1 | 6.3406 | 9815.1 | .60058 | 206.80 |
| 409 | | 12123.2 | 4.1805 | 8722.7 | .81345 | 203.34 | 459 | | 13653.9 | 6.3919 | 9837.6 | .59705 | 206.87 |
| 410 | 136.85 | 12153.7 | 4.2180 | 8744.8 | .80818 | 203.41 | 460 | 186.85 | 13684.7 | 6.4436 | 9860.1 | .59355 | 206.94 |
| 411 | | 12184.2 | 4.2559 | 8767.0 | .80294 | 203.49 | 461 | | 13715.5 | 6.4956 | 9882.5 | .59008 | 207.00 |
| 412 | | 12214.6 | 4.2939 | 8789.1 | .79776 | 203.56 | 462 | | 13746.2 | 6.5480 | 9905.0 | .58663 | 207.07 |
| 413 | | 12245.1 | 4.3323 | 8811.3 | .79262 | 203.64 | 463 | | 13777.0 | 6.6006 | 9927.5 | .58322 | 207.14 |
| 414 | | 12275.6 | 4.3709 | 8833.5 | .78753 | 203.71 | 464 | | 13807.9 | 6.6535 | 9950.0 | .57982 | 207.20 |
| 415 | 141.85 | 12306.1 | 4.4097 | 8855.6 | .78247 | 203.78 | 465 | 191.85 | 13838.6 | 6.7068 | 9972.4 | .57646 | 207.27 |
| 416 | | 12336.6 | 4.4488 | 8877.8 | .77747 | 203.86 | 466 | | 13869.5 | 6.7604 | 9995.0 | .57312 | 207.34 |
| 417 | | 12367.1 | 4.4881 | 8900.0 | .77251 | 203.93 | 467 | | 13900.3 | 6.8144 | 10017.4 | .56980 | 207.40 |
| 418 | | 12397.7 | 4.5278 | 8922.2 | .76758 | 204.00 | 468 | | 13931.1 | 6.8686 | 10040.0 | .56651 | 207.47 |
| 419 | | 12428.2 | 4.5677 | 8944.4 | .76270 | 204.08 | 469 | | 13961.9 | 6.9232 | 10062.5 | .56324 | 207.53 |
| 420 | 146.85 | 12458.7 | 4.6078 | 8966.6 | .75786 | 204.15 | 470 | 196.85 | 13992.8 | 6.9781 | 10085.0 | .56000 | 207.60 |
| 421 | | 12489.2 | 4.6482 | 8988.9 | .75306 | 204.22 | 471 | | 14023.6 | 7.0334 | 10107.5 | .55679 | 207.66 |
| 422 | | 12519.8 | 4.6889 | 9011.1 | .74830 | 204.29 | 472 | | 14054.5 | 7.0889 | 10130.1 | .55360 | 207.73 |
| 423 | | 12550.3 | 4.7298 | 9033.3 | .74358 | 204.37 | 473 | | 14085.3 | 7.1448 | 10152.6 | .55043 | 207.80 |
| 424 | | 12580.9 | 4.7710 | 9055.5 | .73890 | 204.44 | 474 | | 14116.2 | 7.2011 | 10175.2 | .54728 | 207.86 |
| 425 | 151.85 | 12611.4 | 4.8126 | 9077.8 | .73425 | 204.51 | 475 | 201.85 | 14147.1 | 7.2576 | 10197.7 | .54417 | 207.93 |
| 426 | | 12642.0 | 4.8543 | 9100.0 | .72965 | 204.58 | 476 | | 14177.9 | 7.3145 | 10220.3 | .54107 | 207.99 |
| 427 | | 12672.5 | 4.8963 | 9122.3 | .72509 | 204.65 | 477 | | 14208.8 | 7.3718 | 10242.9 | .53799 | 208.06 |
| 428 | | 12703.1 | 4.9386 | 9144.5 | .72056 | 204.72 | 478 | | 14239.7 | 7.4294 | 10265.4 | .53494 | 208.12 |
| 429 | | 12733.7 | 4.9812 | 9166.8 | .71607 | 204.80 | 479 | | 14270.6 | 7.4873 | 10288.0 | .53191 | 208.18 |
| 430 | 156.85 | 12764.3 | 5.0241 | 9189.1 | .71162 | 204.87 | 480 | 206.85 | 14301.5 | 7.5456 | 10310.6 | .52891 | 208.25 |
| 431 | | 12794.9 | 5.0672 | 9211.4 | .70720 | 204.94 | 481 | | 14332.5 | 7.6042 | 10333.2 | .52592 | 208.31 |
| 432 | | 12825.5 | 5.1106 | 9233.6 | .70282 | 205.01 | 482 | | 14363.4 | 7.6631 | 10355.8 | .52296 | 208.38 |
| 433 | | 12856.1 | 5.1543 | 9255.9 | .69848 | 205.08 | 483 | | 14394.3 | 7.7225 | 10378.4 | .52002 | 208.44 |
| 434 | | 12886.7 | 5.1982 | 9278.2 | .69418 | 205.15 | 484 | | 14425.2 | 7.7821 | 10401.1 | .51710 | 208.51 |
| 435 | 161.85 | 12917.3 | 5.2424 | 9300.5 | .68990 | 205.22 | 485 | 211.85 | 14456.2 | 7.8421 | 10423.7 | .51421 | 208.57 |
| 436 | | 12947.9 | 5.2869 | 9322.8 | .68567 | 205.29 | 486 | | 14487.1 | 7.9025 | 10446.3 | .51133 | 208.63 |
| 437 | | 12978.5 | 5.3317 | 9345.1 | .68147 | 205.36 | 487 | | 14518.1 | 7.9632 | 10469.0 | .50848 | 208.70 |
| 438 | | 13009.2 | 5.3768 | 9367.4 | .67730 | 205.43 | 488 | | 14549.0 | 8.0243 | 10491.6 | .50565 | 208.76 |
| 439 | | 13039.8 | 5.4222 | 9389.8 | .67316 | 205.50 | 489 | | 14580.0 | 8.0857 | 10514.3 | .50283 | 208.82 |
| 440 | 166.85 | 13070.4 | 5.4679 | 9412.1 | .66906 | 205.57 | 490 | 216.85 | 14611.0 | 8.1475 | 10536.9 | .50004 | 208.89 |
| 441 | | 13101.1 | 5.5139 | 9434.4 | .66499 | 205.64 | 491 | | 14642.0 | 8.2096 | 10559.6 | .49727 | 208.95 |
| 442 | | 13131.7 | 5.5601 | 9456.8 | .66096 | 205.71 | 492 | | 14673.0 | 8.2721 | 10582.3 | .49452 | 209.01 |
| 443 | | 13162.4 | 5.6066 | 9479.1 | .65695 | 205.78 | 493 | | 14703.9 | 8.3350 | 10604.9 | .49178 | 209.08 |
| 444 | | 13193.1 | 5.6534 | 9501.5 | .65298 | 205.85 | 494 | | 14734.9 | 8.3982 | 10627.6 | .48907 | 209.14 |
| 445 | 171.85 | 13223.8 | 5.7006 | 9523.9 | .64904 | 205.92 | 495 | 221.85 | 14765.9 | 8.4617 | 10650.3 | .48638 | 209.20 |
| 446 | | 13254.4 | 5.7480 | 9546.2 | .64514 | 205.99 | 496 | | 14797.0 | 8.5257 | 10673.0 | .48371 | 209.26 |
| 447 | | 13285.1 | 5.7957 | 9568.6 | .64126 | 206.06 | 497 | | 14828.0 | 8.5900 | 10695.7 | .48105 | 209.33 |
| 448 | | 13315.8 | 5.8437 | 9591.0 | .63741 | 206.12 | 498 | | 14859.0 | 8.6547 | 10718.5 | .47842 | 209.39 |
| 449 | | 13346.5 | 5.8920 | 9613.4 | .63360 | 206.19 | 499 | | 14890.1 | 8.7198 | 10741.2 | .47580 | 209.45 |

## Table 5 Products—200% Theoretical Air (for One Gram-Mole)

| T | t | $\bar{h}$ | $p_r$ | $\bar{u}$ | $v_r$ | $\bar{\phi}$ | T | t | $\bar{h}$ | $p_r$ | $\bar{u}$ | $v_r$ | $\bar{\phi}$ |
|---|---|---|---|---|---|---|---|---|---|---|---|---|---|
| 500 | 226.85 | 14921.1 | 8.785 | 10763.9 | .47321 | 209.51 | 550 | 276.85 | 16482.7 | 12.566 | 11909.7 | .36390 | 212.49 |
| 501 |  | 14952.2 | 8.851 | 10786.7 | .47063 | 209.58 | 551 |  | 16514.1 | 12.653 | 11932.8 | .36207 | 212.55 |
| 502 |  | 14983.2 | 8.917 | 10809.4 | .46807 | 209.64 | 552 |  | 16545.5 | 12.740 | 11956.0 | .36025 | 212.60 |
| 503 |  | 15014.3 | 8.984 | 10832.2 | .46553 | 209.70 | 553 |  | 16576.9 | 12.828 | 11979.1 | .35844 | 212.66 |
| 504 |  | 15045.4 | 9.051 | 10854.9 | .46300 | 209.76 | 554 |  | 16608.4 | 12.916 | 12002.2 | .35664 | 212.72 |
| 505 | 231.85 | 15076.4 | 9.118 | 10877.7 | .46050 | 209.82 | 555 | 281.85 | 16639.8 | 13.004 | 12025.3 | .35485 | 212.77 |
| 506 |  | 15107.6 | 9.186 | 10900.5 | .45801 | 209.88 | 556 |  | 16671.3 | 13.093 | 12048.5 | .35308 | 212.83 |
| 507 |  | 15138.6 | 9.254 | 10923.2 | .45554 | 209.95 | 557 |  | 16702.8 | 13.182 | 12071.6 | .35132 | 212.89 |
| 508 |  | 15169.8 | 9.322 | 10946.0 | .45309 | 210.01 | 558 |  | 16734.2 | 13.272 | 12094.8 | .34957 | 212.94 |
| 509 |  | 15200.8 | 9.391 | 10968.8 | .45065 | 210.07 | 559 |  | 16765.7 | 13.362 | 12118.0 | .34783 | 213.00 |
| 510 | 236.85 | 15232.0 | 9.460 | 10991.6 | .44823 | 210.13 | 560 | 286.85 | 16797.2 | 13.453 | 12141.1 | .34610 | 213.06 |
| 511 |  | 15263.1 | 9.530 | 11014.4 | .44583 | 210.19 | 561 |  | 16828.7 | 13.544 | 12164.3 | .34438 | 213.11 |
| 512 |  | 15294.2 | 9.600 | 11037.3 | .44344 | 210.25 | 562 |  | 16860.2 | 13.636 | 12187.5 | .34267 | 213.17 |
| 513 |  | 15325.4 | 9.670 | 11060.1 | .44107 | 210.31 | 563 |  | 16891.7 | 13.728 | 12210.7 | .34098 | 213.23 |
| 514 |  | 15356.5 | 9.741 | 11082.9 | .43872 | 210.37 | 564 |  | 16923.3 | 13.821 | 12233.9 | .33929 | 213.28 |
| 515 | 241.85 | 15387.7 | 9.812 | 11105.7 | .43638 | 210.43 | 565 | 291.85 | 16954.8 | 13.914 | 12257.1 | .33762 | 213.34 |
| 516 |  | 15418.8 | 9.884 | 11128.6 | .43406 | 210.49 | 566 |  | 16986.3 | 14.008 | 12280.4 | .33596 | 213.39 |
| 517 |  | 15450.0 | 9.956 | 11151.5 | .43176 | 210.55 | 567 |  | 17017.8 | 14.102 | 12303.6 | .33431 | 213.45 |
| 518 |  | 15481.2 | 10.028 | 11174.3 | .42947 | 210.61 | 568 |  | 17049.4 | 14.196 | 12326.8 | .33266 | 213.50 |
| 519 |  | 15512.4 | 10.101 | 11197.2 | .42720 | 210.67 | 569 |  | 17081.0 | 14.291 | 12350.1 | .33103 | 213.56 |
| 520 | 246.85 | 15543.5 | 10.174 | 11220.1 | .42494 | 210.73 | 570 | 296.85 | 17112.5 | 14.387 | 12373.3 | .32941 | 213.62 |
| 521 |  | 15574.7 | 10.248 | 11242.9 | .42270 | 210.79 | 571 |  | 17144.1 | 14.483 | 12396.6 | .32780 | 213.67 |
| 522 |  | 15605.9 | 10.322 | 11265.8 | .42048 | 210.85 | 572 |  | 17175.7 | 14.580 | 12419.8 | .32620 | 213.73 |
| 523 |  | 15637.1 | 10.396 | 11288.7 | .41827 | 210.91 | 573 |  | 17207.3 | 14.677 | 12443.1 | .32461 | 213.78 |
| 524 |  | 15668.4 | 10.471 | 11311.6 | .41607 | 210.97 | 574 |  | 17238.9 | 14.774 | 12466.4 | .32303 | 213.84 |
| 525 | 251.85 | 15699.6 | 10.546 | 11334.5 | .41389 | 211.03 | 575 | 301.85 | 17270.5 | 14.872 | 12489.7 | .32145 | 213.89 |
| 526 |  | 15730.8 | 10.622 | 11357.4 | .41173 | 211.09 | 576 |  | 17302.1 | 14.971 | 12513.0 | .31989 | 213.95 |
| 527 |  | 15762.1 | 10.698 | 11380.4 | .40958 | 211.15 | 577 |  | 17333.7 | 15.070 | 12536.3 | .31834 | 214.00 |
| 528 |  | 15793.3 | 10.775 | 11403.3 | .40744 | 211.21 | 578 |  | 17365.4 | 15.170 | 12559.6 | .31680 | 214.06 |
| 529 |  | 15824.6 | 10.852 | 11426.2 | .40532 | 211.27 | 579 |  | 17397.0 | 15.270 | 12583.0 | .31527 | 214.11 |
| 530 | 256.85 | 15855.8 | 10.929 | 11449.2 | .40321 | 211.33 | 580 | 306.85 | 17428.6 | 15.370 | 12606.3 | .31374 | 214.16 |
| 531 |  | 15887.1 | 11.007 | 11472.1 | .40112 | 211.39 | 581 |  | 17460.3 | 15.472 | 12629.6 | .31223 | 214.22 |
| 532 |  | 15918.4 | 11.085 | 11495.1 | .39904 | 211.45 | 582 |  | 17491.9 | 15.573 | 12653.0 | .31073 | 214.27 |
| 533 |  | 15949.7 | 11.163 | 11518.1 | .39698 | 211.51 | 583 |  | 17523.6 | 15.675 | 12676.3 | .30923 | 214.33 |
| 534 |  | 15981.0 | 11.242 | 11541.1 | .39492 | 211.56 | 584 |  | 17555.3 | 15.778 | 12699.7 | .30774 | 214.38 |
| 535 | 261.85 | 16012.3 | 11.322 | 11564.1 | .39289 | 211.62 | 585 | 311.85 | 17587.0 | 15.881 | 12723.0 | .30627 | 214.44 |
| 536 |  | 16043.6 | 11.402 | 11587.0 | .39086 | 211.68 | 586 |  | 17618.7 | 15.985 | 12746.4 | .30480 | 214.49 |
| 537 |  | 16074.9 | 11.482 | 11610.0 | .38885 | 211.74 | 587 |  | 17650.4 | 16.089 | 12769.8 | .30334 | 214.54 |
| 538 |  | 16106.2 | 11.563 | 11633.0 | .38686 | 211.80 | 588 |  | 17682.1 | 16.194 | 12793.2 | .30189 | 214.60 |
| 539 |  | 16137.5 | 11.644 | 11656.1 | .38487 | 211.86 | 589 |  | 17713.8 | 16.299 | 12816.6 | .30045 | 214.65 |
| 540 | 266.85 | 16168.9 | 11.726 | 11679.1 | .38290 | 211.91 | 590 | 316.85 | 17745.5 | 16.405 | 12840.0 | .29902 | 214.71 |
| 541 |  | 16200.2 | 11.808 | 11702.1 | .38094 | 211.97 | 591 |  | 17777.2 | 16.512 | 12863.4 | .29760 | 214.76 |
| 542 |  | 16231.6 | 11.890 | 11725.2 | .37900 | 212.03 | 592 |  | 17809.0 | 16.619 | 12886.8 | .29618 | 214.81 |
| 543 |  | 16262.9 | 11.973 | 11748.2 | .37707 | 212.09 | 593 |  | 17840.7 | 16.726 | 12910.3 | .29478 | 214.87 |
| 544 |  | 16294.3 | 12.057 | 11771.3 | .37515 | 212.15 | 594 |  | 17872.5 | 16.834 | 12933.7 | .29338 | 214.92 |
| 545 | 271.85 | 16325.7 | 12.140 | 11794.3 | .37324 | 212.20 | 595 | 321.85 | 17904.2 | 16.942 | 12957.2 | .29199 | 214.97 |
| 546 |  | 16357.1 | 12.225 | 11817.4 | .37135 | 212.26 | 596 |  | 17936.0 | 17.051 | 12980.6 | .29061 | 215.03 |
| 547 |  | 16388.4 | 12.309 | 11840.5 | .36947 | 212.32 | 597 |  | 17967.8 | 17.161 | 13004.1 | .28924 | 215.08 |
| 548 |  | 16419.8 | 12.395 | 11863.5 | .36760 | 212.38 | 598 |  | 17999.6 | 17.271 | 13027.6 | .28788 | 215.13 |
| 549 |  | 16451.2 | 12.480 | 11886.6 | .36574 | 212.43 | 599 |  | 18031.4 | 17.382 | 13051.0 | .28653 | 215.19 |

## Table 5    Products—200% Theoretical Air (for One Gram-Mole)

| T | t | $\bar{h}$ | $p_r$ | $\bar{u}$ | $v_r$ | $\bar{\phi}$ | T | t | $\bar{h}$ | $p_r$ | $\bar{u}$ | $v_r$ | $\bar{\phi}$ |
|---|---|---|---|---|---|---|---|---|---|---|---|---|---|
| 600 | 326.85 | 18063.2 | 17.493 | 13074.5 | .28518 | 215.24 | 650 | 376.85 | 19663.5 | 23.805 | 14259.1 | .22702 | 217.80 |
| 601 | | 18095.0 | 17.605 | 13098.0 | .28384 | 215.29 | 651 | | 19695.7 | 23.948 | 14283.0 | .22602 | 217.85 |
| 602 | | 18126.8 | 17.717 | 13121.5 | .28251 | 215.35 | 652 | | 19727.9 | 24.090 | 14306.9 | .22503 | 217.90 |
| 603 | | 18158.6 | 17.830 | 13145.0 | .28119 | 215.40 | 653 | | 19760.1 | 24.234 | 14330.8 | .22404 | 217.95 |
| 604 | | 18190.4 | 17.944 | 13168.5 | .27987 | 215.45 | 654 | | 19792.4 | 24.378 | 14354.8 | .22305 | 218.00 |
| 605 | 331.85 | 18222.3 | 18.058 | 13192.1 | .27856 | 215.50 | 655 | 381.85 | 19824.6 | 24.523 | 14378.7 | .22208 | 218.05 |
| 606 | | 18254.1 | 18.172 | 13215.6 | .27726 | 215.56 | 656 | | 19856.9 | 24.668 | 14402.6 | .22110 | 218.10 |
| 607 | | 18286.0 | 18.288 | 13239.2 | .27597 | 215.61 | 657 | | 19889.1 | 24.815 | 14426.6 | .22013 | 218.15 |
| 608 | | 18317.9 | 18.403 | 13262.7 | .27469 | 215.66 | 658 | | 19921.4 | 24.962 | 14450.5 | .21917 | 218.20 |
| 609 | | 18349.8 | 18.520 | 13286.3 | .27341 | 215.71 | 659 | | 19953.7 | 25.109 | 14474.5 | .21822 | 218.25 |
| 610 | 336.85 | 18381.6 | 18.636 | 13309.8 | .27214 | 215.77 | 660 | 386.85 | 19986.0 | 25.257 | 14498.5 | .21726 | 218.29 |
| 611 | | 18413.5 | 18.754 | 13333.4 | .27088 | 215.82 | 661 | | 20018.3 | 25.406 | 14522.4 | .21632 | 218.34 |
| 612 | | 18445.4 | 18.872 | 13357.0 | .26963 | 215.87 | 662 | | 20050.5 | 25.556 | 14546.4 | .21538 | 218.39 |
| 613 | | 18477.3 | 18.991 | 13380.6 | .26838 | 215.92 | 663 | | 20082.9 | 25.706 | 14570.4 | .21444 | 218.44 |
| 614 | | 18509.2 | 19.110 | 13404.2 | .26714 | 215.98 | 664 | | 20115.2 | 25.857 | 14594.4 | .21351 | 218.49 |
| 615 | 341.85 | 18541.2 | 19.230 | 13427.8 | .26591 | 216.03 | 665 | 391.85 | 20147.5 | 26.009 | 14618.4 | .21258 | 218.54 |
| 616 | | 18573.1 | 19.350 | 13451.4 | .26469 | 216.08 | 666 | | 20179.8 | 26.162 | 14642.5 | .21166 | 218.59 |
| 617 | | 18605.0 | 19.471 | 13475.0 | .26347 | 216.13 | 667 | | 20212.2 | 26.315 | 14666.5 | .21075 | 218.64 |
| 618 | | 18637.0 | 19.592 | 13498.7 | .26226 | 216.18 | 668 | | 20244.5 | 26.469 | 14690.5 | .20984 | 218.68 |
| 619 | | 18668.9 | 19.714 | 13522.3 | .26106 | 216.23 | 669 | | 20276.9 | 26.623 | 14714.6 | .20893 | 218.73 |
| 620 | 346.85 | 18700.9 | 19.837 | 13546.0 | .25986 | 216.29 | 670 | 396.85 | 20309.2 | 26.778 | 14738.6 | .20803 | 218.78 |
| 621 | | 18732.9 | 19.960 | 13569.6 | .25867 | 216.34 | 671 | | 20341.6 | 26.934 | 14762.7 | .20713 | 218.83 |
| 622 | | 18764.8 | 20.084 | 13593.3 | .25749 | 216.39 | 672 | | 20374.0 | 27.091 | 14786.7 | .20624 | 218.88 |
| 623 | | 18796.8 | 20.209 | 13616.9 | .25632 | 216.44 | 673 | | 20406.4 | 27.248 | 14810.8 | .20536 | 218.93 |
| 624 | | 18828.8 | 20.334 | 13640.6 | .25515 | 216.49 | 674 | | 20438.8 | 27.406 | 14834.9 | .20448 | 218.97 |
| 625 | 351.85 | 18860.8 | 20.460 | 13664.3 | .25399 | 216.54 | 675 | 401.85 | 20471.2 | 27.565 | 14859.0 | .20360 | 219.02 |
| 626 | | 18892.8 | 20.586 | 13688.0 | .25283 | 216.59 | 676 | | 20503.6 | 27.725 | 14883.1 | .20273 | 219.07 |
| 627 | | 18924.8 | 20.713 | 13711.7 | .25169 | 216.65 | 677 | | 20536.0 | 27.885 | 14907.2 | .20186 | 219.12 |
| 628 | | 18956.9 | 20.840 | 13735.4 | .25055 | 216.70 | 678 | | 20568.5 | 28.046 | 14931.3 | .20100 | 219.17 |
| 629 | | 18988.9 | 20.969 | 13759.2 | .24941 | 216.75 | 679 | | 20600.9 | 28.208 | 14955.4 | .20014 | 219.21 |
| 630 | 356.85 | 19020.9 | 21.097 | 13782.9 | .24828 | 216.80 | 680 | 406.85 | 20633.4 | 28.370 | 14979.6 | .19929 | 219.26 |
| 631 | | 19053.0 | 21.227 | 13806.6 | .24716 | 216.85 | 681 | | 20665.8 | 28.533 | 15003.7 | .19844 | 219.31 |
| 632 | | 19085.0 | 21.357 | 13830.3 | .24605 | 216.90 | 682 | | 20698.3 | 28.697 | 15027.8 | .19759 | 219.36 |
| 633 | | 19117.1 | 21.487 | 13854.1 | .24494 | 216.95 | 683 | | 20730.7 | 28.862 | 15052.0 | .19675 | 219.40 |
| 634 | | 19149.2 | 21.619 | 13877.9 | .24383 | 217.00 | 684 | | 20763.2 | 29.027 | 15076.2 | .19592 | 219.45 |
| 635 | 361.85 | 19181.3 | 21.751 | 13901.6 | .24274 | 217.05 | 685 | 411.85 | 20795.7 | 29.194 | 15100.3 | .19509 | 219.50 |
| 636 | | 19213.4 | 21.883 | 13925.4 | .24165 | 217.10 | 686 | | 20828.2 | 29.361 | 15124.5 | .19426 | 219.55 |
| 637 | | 19245.5 | 22.016 | 13949.2 | .24056 | 217.15 | 687 | | 20860.7 | 29.528 | 15148.7 | .19344 | 219.59 |
| 638 | | 19277.6 | 22.150 | 13973.0 | .23949 | 217.20 | 688 | | 20893.2 | 29.697 | 15172.9 | .19262 | 219.64 |
| 639 | | 19309.7 | 22.284 | 13996.8 | .23841 | 217.25 | 689 | | 20925.8 | 29.866 | 15197.1 | .19181 | 219.69 |
| 640 | 366.85 | 19341.8 | 22.419 | 14020.6 | .23735 | 217.30 | 690 | 416.85 | 20958.3 | 30.036 | 15221.3 | .19100 | 219.74 |
| 641 | | 19373.9 | 22.555 | 14044.4 | .23629 | 217.35 | 691 | | 20990.8 | 30.207 | 15245.6 | .19020 | 219.78 |
| 642 | | 19406.1 | 22.691 | 14068.2 | .23524 | 217.40 | 692 | | 21023.4 | 30.378 | 15269.8 | .18940 | 219.83 |
| 643 | | 19438.2 | 22.828 | 14092.1 | .23419 | 217.45 | 693 | | 21055.9 | 30.550 | 15294.0 | .18860 | 219.88 |
| 644 | | 19470.4 | 22.966 | 14115.9 | .23315 | 217.50 | 694 | | 21088.5 | 30.724 | 15318.3 | .18781 | 219.92 |
| 645 | 371.85 | 19502.5 | 23.104 | 14139.8 | .23211 | 217.55 | 695 | 421.85 | 21121.0 | 30.897 | 15342.5 | .18702 | 219.97 |
| 646 | | 19534.7 | 23.243 | 14163.6 | .23108 | 217.60 | 696 | | 21153.6 | 31.072 | 15366.8 | .18624 | 220.02 |
| 647 | | 19566.9 | 23.383 | 14187.5 | .23006 | 217.65 | 697 | | 21186.2 | 31.247 | 15391.0 | .18546 | 220.06 |
| 648 | | 19599.1 | 23.523 | 14211.3 | .22904 | 217.70 | 698 | | 21218.8 | 31.423 | 15415.3 | .18469 | 220.11 |
| 649 | | 19631.3 | 23.664 | 14235.2 | .22803 | 217.75 | 699 | | 21251.4 | 31.600 | 15439.6 | .18392 | 220.16 |

# Table 5  Products—200% Theoretical Air (for One Gram-Mole)

| T | t | $\bar{h}$ | $p_r$ | $\bar{u}$ | $v_r$ | $\bar{\phi}$ | T | t | $\bar{h}$ | $p_r$ | $\bar{u}$ | $v_r$ | $\bar{\phi}$ |
|---|---|---|---|---|---|---|---|---|---|---|---|---|---|
| 700 | 426.85 | 21284.0 | 31.778 | 15463.9 | .18315 | 220.20 | 750 | 476.85 | 22924.8 | 41.723 | 16689.0 | .14946 | 222.47 |
| 701 | | 21316.6 | 31.956 | 15488.2 | .18239 | 220.25 | 751 | | 22957.8 | 41.944 | 16713.7 | .14887 | 222.51 |
| 702 | | 21349.2 | 32.135 | 15512.5 | .18163 | 220.30 | 752 | | 22990.8 | 42.166 | 16738.4 | .14828 | 222.56 |
| 703 | | 21381.9 | 32.316 | 15536.8 | .18087 | 220.34 | 753 | | 23023.8 | 42.390 | 16763.1 | .14770 | 222.60 |
| 704 | | 21414.5 | 32.496 | 15561.2 | .18012 | 220.39 | 754 | | 23056.9 | 42.614 | 16787.8 | .14711 | 222.64 |
| 705 | 431.85 | 21447.2 | 32.678 | 15585.5 | .17938 | 220.44 | 755 | 481.85 | 23089.9 | 42.839 | 16812.5 | .14654 | 222.69 |
| 706 | | 21479.8 | 32.860 | 15609.8 | .17863 | 220.48 | 756 | | 23123.0 | 43.065 | 16837.3 | .14596 | 222.73 |
| 707 | | 21512.5 | 33.044 | 15634.2 | .17789 | 220.53 | 757 | | 23156.0 | 43.292 | 16862.0 | .14539 | 222.77 |
| 708 | | 21545.2 | 33.228 | 15658.6 | .17716 | 220.57 | 758 | | 23189.1 | 43.520 | 16886.8 | .14482 | 222.82 |
| 709 | | 21577.8 | 33.413 | 15682.9 | .17643 | 220.62 | 759 | | 23222.2 | 43.749 | 16911.6 | .14425 | 222.86 |
| 710 | 436.85 | 21610.5 | 33.598 | 15707.3 | .17570 | 220.67 | 760 | 486.85 | 23255.3 | 43.978 | 16936.4 | .14368 | 222.91 |
| 711 | | 21643.2 | 33.785 | 15731.7 | .17498 | 220.71 | 761 | | 23288.4 | 44.209 | 16961.1 | .14312 | 222.95 |
| 712 | | 21675.9 | 33.972 | 15756.1 | .17426 | 220.76 | 762 | | 23321.5 | 44.441 | 16985.9 | .14256 | 222.99 |
| 713 | | 21708.7 | 34.160 | 15780.5 | .17354 | 220.80 | 763 | | 23354.6 | 44.674 | 17010.7 | .14200 | 223.04 |
| 714 | | 21741.4 | 34.349 | 15804.9 | .17283 | 220.85 | 764 | | 23387.8 | 44.908 | 17035.5 | .14145 | 223.08 |
| 715 | 441.85 | 21774.1 | 34.539 | 15829.3 | .17212 | 220.90 | 765 | 491.85 | 23420.9 | 45.142 | 17060.4 | .14090 | 223.12 |
| 716 | | 21806.9 | 34.729 | 15853.7 | .17141 | 220.94 | 766 | | 23454.0 | 45.378 | 17085.2 | .14035 | 223.17 |
| 717 | | 21839.6 | 34.921 | 15878.2 | .17071 | 220.99 | 767 | | 23487.2 | 45.615 | 17110.0 | .13980 | 223.21 |
| 718 | | 21872.4 | 35.113 | 15902.6 | .17001 | 221.03 | 768 | | 23520.3 | 45.852 | 17134.9 | .13926 | 223.25 |
| 719 | | 21905.1 | 35.306 | 15927.1 | .16932 | 221.08 | 769 | | 23553.5 | 46.091 | 17159.7 | .13872 | 223.30 |
| 720 | 446.85 | 21937.9 | 35.500 | 15951.5 | .16863 | 221.12 | 770 | 496.85 | 23586.7 | 46.331 | 17184.6 | .13818 | 223.34 |
| 721 | | 21970.7 | 35.695 | 15976.0 | .16794 | 221.17 | 771 | | 23619.8 | 46.571 | 17209.4 | .13765 | 223.38 |
| 722 | | 22003.4 | 35.891 | 16000.4 | .16726 | 221.22 | 772 | | 23653.0 | 46.813 | 17234.3 | .13711 | 223.42 |
| 723 | | 22036.2 | 36.087 | 16024.9 | .16658 | 221.26 | 773 | | 23686.2 | 47.055 | 17259.2 | .13659 | 223.47 |
| 724 | | 22069.1 | 36.284 | 16049.4 | .16590 | 221.31 | 774 | | 23719.4 | 47.299 | 17284.1 | .13606 | 223.51 |
| 725 | 451.85 | 22101.9 | 36.483 | 16073.9 | .16523 | 221.35 | 775 | 501.85 | 23752.6 | 47.544 | 17309.0 | .13553 | 223.55 |
| 726 | | 22134.7 | 36.682 | 16098.4 | .16456 | 221.40 | 776 | | 23785.9 | 47.789 | 17333.9 | .13501 | 223.60 |
| 727 | | 22167.5 | 36.882 | 16122.9 | .16389 | 221.44 | 777 | | 23819.1 | 48.035 | 17358.8 | .13449 | 223.64 |
| 728 | | 22200.3 | 37.083 | 16147.4 | .16323 | 221.49 | 778 | | 23852.3 | 48.283 | 17383.7 | .13397 | 223.68 |
| 729 | | 22233.2 | 37.284 | 16172.0 | .16257 | 221.53 | 779 | | 23885.6 | 48.532 | 17408.6 | .13346 | 223.72 |
| 730 | 456.85 | 22266.0 | 37.487 | 16196.5 | .16191 | 221.58 | 780 | 506.85 | 23918.8 | 48.781 | 17433.6 | .13295 | 223.77 |
| 731 | | 22298.9 | 37.690 | 16221.1 | .16126 | 221.62 | 781 | | 23952.1 | 49.032 | 17458.5 | .13244 | 223.81 |
| 732 | | 22331.8 | 37.894 | 16245.6 | .16061 | 221.67 | 782 | | 23985.3 | 49.283 | 17483.5 | .13193 | 223.85 |
| 733 | | 22364.6 | 38.099 | 16270.2 | .15996 | 221.71 | 783 | | 24018.6 | 49.536 | 17508.4 | .13142 | 223.89 |
| 734 | | 22397.5 | 38.305 | 16294.7 | .15932 | 221.76 | 784 | | 24051.9 | 49.790 | 17533.4 | .13092 | 223.94 |
| 735 | 461.85 | 22430.4 | 38.512 | 16319.3 | .15868 | 221.80 | 785 | 511.85 | 24085.2 | 50.045 | 17558.4 | .13042 | 223.98 |
| 736 | | 22463.3 | 38.719 | 16343.9 | .15804 | 221.85 | 786 | | 24118.5 | 50.300 | 17583.3 | .12992 | 224.02 |
| 737 | | 22496.2 | 38.928 | 16368.5 | .15741 | 221.89 | 787 | | 24151.8 | 50.557 | 17608.3 | .12943 | 224.06 |
| 738 | | 22529.1 | 39.138 | 16393.1 | .15678 | 221.94 | 788 | | 24185.1 | 50.815 | 17633.3 | .12893 | 224.11 |
| 739 | | 22562.1 | 39.348 | 16417.7 | .15615 | 221.98 | 789 | | 24218.4 | 51.074 | 17658.3 | .12844 | 224.15 |
| 740 | 466.85 | 22595.0 | 39.559 | 16442.3 | .15553 | 222.02 | 790 | 516.85 | 24251.7 | 51.334 | 17683.4 | .12795 | 224.19 |
| 741 | | 22627.9 | 39.772 | 16467.0 | .15491 | 222.07 | 791 | | 24285.1 | 51.595 | 17708.4 | .12747 | 224.23 |
| 742 | | 22660.9 | 39.985 | 16491.6 | .15429 | 222.11 | 792 | | 24318.4 | 51.857 | 17733.4 | .12698 | 224.28 |
| 743 | | 22693.8 | 40.199 | 16516.2 | .15368 | 222.16 | 793 | | 24351.8 | 52.120 | 17758.4 | .12650 | 224.32 |
| 744 | | 22726.8 | 40.414 | 16540.9 | .15306 | 222.20 | 794 | | 24385.1 | 52.385 | 17783.5 | .12602 | 224.36 |
| 745 | 471.85 | 22759.8 | 40.629 | 16565.5 | .15246 | 222.25 | 795 | 521.85 | 24418.5 | 52.650 | 17808.5 | .12555 | 224.40 |
| 746 | | 22792.8 | 40.846 | 16590.2 | .15185 | 222.29 | 796 | | 24451.9 | 52.916 | 17833.6 | .12507 | 224.44 |
| 747 | | 22825.7 | 41.064 | 16614.9 | .15125 | 222.34 | 797 | | 24485.3 | 53.184 | 17858.7 | .12460 | 224.49 |
| 748 | | 22858.7 | 41.283 | 16639.6 | .15065 | 222.38 | 798 | | 24518.6 | 53.452 | 17883.7 | .12413 | 224.53 |
| 749 | | 22891.7 | 41.502 | 16664.3 | .15005 | 222.42 | 799 | | 24552.0 | 53.722 | 17908.8 | .12366 | 224.57 |

# Table 5    Products—200% Theoretical Air (for One Gram-Mole)

| T | t | $\bar{h}$ | $p_r$ | $\bar{u}$ | $v_r$ | $\bar{\phi}$ | T | t | $\bar{h}$ | $p_r$ | $\bar{u}$ | $v_r$ | $\bar{\phi}$ |
|---|---|---|---|---|---|---|---|---|---|---|---|---|---|
| 800 | 526.85 | 24585.4 | 53.992 | 17933.9 | .12319 | 224.61 | 850 | 576.85 | 26265.6 | 68.982 | 19198.3 | .102451 | 226.65 |
| 801 | | 24618.8 | 54.264 | 17959.0 | .12273 | 224.65 | 851 | | 26299.4 | 69.312 | 19223.8 | .102083 | 226.69 |
| 802 | | 24652.3 | 54.537 | 17984.1 | .12227 | 224.69 | 852 | | 26333.1 | 69.644 | 19249.3 | .101716 | 226.73 |
| 803 | | 24685.7 | 54.811 | 18009.3 | .12181 | 224.74 | 853 | | 26367.0 | 69.977 | 19274.8 | .101351 | 226.77 |
| 804 | | 24719.1 | 55.086 | 18034.4 | .12135 | 224.78 | 854 | | 26400.8 | 70.311 | 19300.3 | .100987 | 226.81 |
| 805 | 531.85 | 24752.6 | 55.362 | 18059.5 | .12090 | 224.82 | 855 | 581.85 | 26434.6 | 70.647 | 19325.8 | .100625 | 226.85 |
| 806 | | 24786.0 | 55.639 | 18084.6 | .12044 | 224.86 | 856 | | 26468.4 | 70.983 | 19351.3 | .100265 | 226.89 |
| 807 | | 24819.5 | 55.918 | 18109.8 | .11999 | 224.90 | 857 | | 26502.3 | 71.322 | 19376.8 | .099906 | 226.93 |
| 808 | | 24853.0 | 56.197 | 18134.9 | .11954 | 224.94 | 858 | | 26536.1 | 71.661 | 19402.4 | .099549 | 226.96 |
| 809 | | 24886.4 | 56.478 | 18160.1 | .11910 | 224.99 | 859 | | 26570.0 | 72.002 | 19427.9 | .099193 | 227.00 |
| 810 | 536.85 | 24919.9 | 56.759 | 18185.3 | .11865 | 225.03 | 860 | 586.85 | 26603.9 | 72.344 | 19453.5 | .098839 | 227.04 |
| 811 | | 24953.4 | 57.042 | 18210.4 | .11821 | 225.07 | 861 | | 26637.7 | 72.687 | 19479.0 | .098487 | 227.08 |
| 812 | | 24986.9 | 57.326 | 18235.6 | .11777 | 225.11 | 862 | | 26671.6 | 73.031 | 19504.6 | .098136 | 227.12 |
| 813 | | 25020.4 | 57.611 | 18260.8 | .11733 | 225.15 | 863 | | 26705.5 | 73.378 | 19530.2 | .097786 | 227.16 |
| 814 | | 25053.9 | 57.897 | 18286.0 | .11690 | 225.19 | 864 | | 26739.4 | 73.725 | 19555.7 | .097439 | 227.20 |
| 815 | 541.85 | 25087.5 | 58.185 | 18311.2 | .11646 | 225.23 | 865 | 591.85 | 26773.3 | 74.073 | 19581.3 | .097092 | 227.24 |
| 816 | | 25121.0 | 58.473 | 18336.4 | .11603 | 225.27 | 866 | | 26807.2 | 74.423 | 19606.9 | .096748 | 227.28 |
| 817 | | 25154.5 | 58.762 | 18361.7 | .11560 | 225.31 | 867 | | 26841.1 | 74.775 | 19632.5 | .096404 | 227.32 |
| 818 | | 25188.1 | 59.053 | 18386.9 | .11517 | 225.36 | 868 | | 26875.0 | 75.127 | 19658.1 | .096063 | 227.36 |
| 819 | | 25221.6 | 59.345 | 18412.1 | .11474 | 225.40 | 869 | | 26909.0 | 75.481 | 19683.8 | .095723 | 227.40 |
| 820 | 546.85 | 25255.2 | 59.638 | 18437.4 | .11432 | 225.44 | 870 | 596.85 | 26942.9 | 75.836 | 19709.4 | .095384 | 227.44 |
| 821 | | 25288.8 | 59.932 | 18462.6 | .11390 | 225.48 | 871 | | 26976.8 | 76.193 | 19735.0 | .095047 | 227.47 |
| 822 | | 25322.3 | 60.228 | 18487.9 | .11348 | 225.52 | 872 | | 27010.8 | 76.550 | 19760.6 | .094711 | 227.51 |
| 823 | | 25355.9 | 60.524 | 18513.2 | .11306 | 225.56 | 873 | | 27044.8 | 76.910 | 19786.3 | .094377 | 227.55 |
| 824 | | 25389.5 | 60.822 | 18538.4 | .11264 | 225.60 | 874 | | 27078.7 | 77.270 | 19811.9 | .094044 | 227.59 |
| 825 | 551.85 | 25423.1 | 61.121 | 18563.7 | .11223 | 225.64 | 875 | 601.85 | 27112.7 | 77.632 | 19837.6 | .093713 | 227.63 |
| 826 | | 25456.7 | 61.421 | 18589.0 | .11181 | 225.68 | 876 | | 27146.7 | 77.995 | 19863.3 | .093383 | 227.67 |
| 827 | | 25490.3 | 61.722 | 18614.3 | .11140 | 225.72 | 877 | | 27180.7 | 78.360 | 19888.9 | .093055 | 227.71 |
| 828 | | 25524.0 | 62.024 | 18639.6 | .11099 | 225.76 | 878 | | 27214.7 | 78.726 | 19914.6 | .092728 | 227.75 |
| 829 | | 25557.6 | 62.328 | 18664.9 | .11059 | 225.80 | 879 | | 27248.7 | 79.093 | 19940.3 | .092402 | 227.79 |
| 830 | 556.85 | 25591.2 | 62.633 | 18690.3 | .11018 | 225.85 | 880 | 606.85 | 27282.7 | 79.462 | 19966.0 | .092078 | 227.82 |
| 831 | | 25624.9 | 62.939 | 18715.6 | .10978 | 225.89 | 881 | | 27316.7 | 79.832 | 19991.7 | .091755 | 227.86 |
| 832 | | 25658.5 | 63.246 | 18740.9 | .10938 | 225.93 | 882 | | 27350.7 | 80.203 | 20017.4 | .091434 | 227.90 |
| 833 | | 25692.2 | 63.554 | 18766.3 | .10898 | 225.97 | 883 | | 27384.8 | 80.576 | 20043.1 | .091114 | 227.94 |
| 834 | | 25725.8 | 63.864 | 18791.6 | .10858 | 226.01 | 884 | | 27418.8 | 80.950 | 20068.9 | .090796 | 227.98 |
| 835 | 561.85 | 25759.5 | 64.174 | 18817.0 | .10818 | 226.05 | 885 | 611.85 | 27452.8 | 81.326 | 20094.6 | .090478 | 228.02 |
| 836 | | 25793.2 | 64.486 | 18842.4 | .10779 | 226.09 | 886 | | 27486.9 | 81.703 | 20120.3 | .090162 | 228.06 |
| 837 | | 25826.9 | 64.799 | 18867.7 | .10740 | 226.13 | 887 | | 27521.0 | 82.082 | 20146.1 | .089848 | 228.09 |
| 838 | | 25860.6 | 65.114 | 18893.1 | .10700 | 226.17 | 888 | | 27555.0 | 82.462 | 20171.9 | .089535 | 228.13 |
| 839 | | 25894.3 | 65.429 | 18918.5 | .10662 | 226.21 | 889 | | 27589.1 | 82.843 | 20197.6 | .089223 | 228.17 |
| 840 | 566.85 | 25928.0 | 65.746 | 18943.9 | .10623 | 226.25 | 890 | 616.85 | 27623.2 | 83.226 | 20223.4 | .088913 | 228.21 |
| 841 | | 25961.7 | 66.064 | 18969.3 | .10584 | 226.29 | 891 | | 27657.3 | 83.609 | 20249.2 | .088604 | 228.25 |
| 842 | | 25995.5 | 66.383 | 18994.7 | .10546 | 226.33 | 892 | | 27691.4 | 83.995 | 20274.9 | .088296 | 228.29 |
| 843 | | 26029.2 | 66.704 | 19020.1 | .10508 | 226.37 | 893 | | 27725.5 | 84.382 | 20300.7 | .087990 | 228.32 |
| 844 | | 26062.9 | 67.026 | 19045.6 | .10470 | 226.41 | 894 | | 27759.6 | 84.770 | 20326.5 | .087685 | 228.36 |
| 845 | 571.85 | 26096.7 | 67.348 | 19071.0 | .10432 | 226.45 | 895 | 621.85 | 27793.7 | 85.160 | 20352.3 | .087381 | 228.40 |
| 846 | | 26130.4 | 67.672 | 19096.5 | .10394 | 226.49 | 896 | | 27827.8 | 85.551 | 20378.1 | .087079 | 228.44 |
| 847 | | 26164.2 | 67.998 | 19121.9 | .10357 | 226.53 | 897 | | 27862.0 | 85.944 | 20404.0 | .086778 | 228.48 |
| 848 | | 26198.0 | 68.324 | 19147.4 | .10319 | 226.57 | 898 | | 27896.1 | 86.338 | 20429.8 | .086478 | 228.51 |
| 849 | | 26231.8 | 68.652 | 19172.8 | .10282 | 226.61 | 899 | | 27930.3 | 86.734 | 20455.6 | .086179 | 228.55 |

# Table 5 Products—200% Theoretical Air (for One Gram-Mole)

| T | t | $\bar{h}$ | $p_r$ | $\bar{u}$ | $v_r$ | $\bar{\phi}$ | T | t | $\bar{h}$ | $p_r$ | $\bar{u}$ | $v_r$ | $\bar{\phi}$ |
|---|---|---|---|---|---|---|---|---|---|---|---|---|---|
| 900 | 626.85 | 27964.4 | 87.13 | 20481.5 | .085882 | 228.59 | 950 | 676.85 | 29681.2 | 108.93 | 21782.6 | .072514 | 230.45 |
| 901 | | 27998.6 | 87.53 | 20507.3 | .085586 | 228.63 | 951 | | 29715.8 | 109.40 | 21808.8 | .072274 | 230.48 |
| 902 | | 28032.8 | 87.93 | 20533.2 | .085292 | 228.67 | 952 | | 29750.3 | 109.88 | 21835.0 | .072034 | 230.52 |
| 903 | | 28066.9 | 88.33 | 20559.0 | .084998 | 228.70 | 953 | | 29784.8 | 110.36 | 21861.2 | .071796 | 230.56 |
| 904 | | 28101.1 | 88.73 | 20584.9 | .084706 | 228.74 | 954 | | 29819.3 | 110.84 | 21887.4 | .071560 | 230.59 |
| 905 | 631.85 | 28135.3 | 89.14 | 20610.8 | .084415 | 228.78 | 955 | 681.85 | 29853.9 | 111.33 | 21913.6 | .071323 | 230.63 |
| 906 | | 28169.5 | 89.54 | 20636.7 | .084125 | 228.82 | 956 | | 29888.4 | 111.81 | 21939.9 | .071088 | 230.66 |
| 907 | | 28203.7 | 89.95 | 20662.6 | .083837 | 228.85 | 957 | | 29923.0 | 112.30 | 21966.1 | .070854 | 230.70 |
| 908 | | 28237.9 | 90.36 | 20688.5 | .083550 | 228.89 | 958 | | 29957.5 | 112.79 | 21992.3 | .070621 | 230.74 |
| 909 | | 28272.2 | 90.77 | 20714.4 | .083263 | 228.93 | 959 | | 29992.1 | 113.28 | 22018.6 | .070388 | 230.77 |
| 910 | 636.85 | 28306.4 | 91.18 | 20740.3 | .082978 | 228.97 | 960 | 686.85 | 30026.7 | 113.77 | 22044.9 | .070157 | 230.81 |
| 911 | | 28340.6 | 91.59 | 20766.2 | .082695 | 229.01 | 961 | | 30061.3 | 114.26 | 22071.1 | .069927 | 230.84 |
| 912 | | 28374.9 | 92.01 | 20792.1 | .082413 | 229.04 | 962 | | 30095.9 | 114.76 | 22097.4 | .069697 | 230.88 |
| 913 | | 28409.1 | 92.43 | 20818.1 | .082131 | 229.08 | 963 | | 30130.4 | 115.26 | 22123.7 | .069469 | 230.92 |
| 914 | | 28443.4 | 92.84 | 20844.0 | .081851 | 229.12 | 964 | | 30165.1 | 115.76 | 22150.0 | .069241 | 230.95 |
| 915 | 641.85 | 28477.6 | 93.26 | 20869.9 | .081572 | 229.16 | 965 | 691.85 | 30199.7 | 116.26 | 22176.3 | .069014 | 230.99 |
| 916 | | 28511.9 | 93.68 | 20895.9 | .081294 | 229.19 | 966 | | 30234.3 | 116.76 | 22202.6 | .068789 | 231.02 |
| 917 | | 28546.2 | 94.11 | 20921.9 | .081018 | 229.23 | 967 | | 30268.9 | 117.26 | 22228.9 | .068564 | 231.06 |
| 918 | | 28580.5 | 94.53 | 20947.8 | .080742 | 229.27 | 968 | | 30303.5 | 117.77 | 22255.2 | .068340 | 231.10 |
| 919 | | 28614.7 | 94.96 | 20973.8 | .080468 | 229.31 | 969 | | 30338.2 | 118.28 | 22281.5 | .068117 | 231.13 |
| 920 | 646.85 | 28649.0 | 95.38 | 20999.8 | .080195 | 229.34 | 970 | 696.85 | 30372.8 | 118.79 | 22307.8 | .067895 | 231.17 |
| 921 | | 28683.3 | 95.81 | 21025.8 | .079923 | 229.38 | 971 | | 30407.5 | 119.30 | 22334.2 | .067673 | 231.20 |
| 922 | | 28717.7 | 96.24 | 21051.8 | .079653 | 229.42 | 972 | | 30442.1 | 119.81 | 22360.5 | .067453 | 231.24 |
| 923 | | 28752.0 | 96.67 | 21077.8 | .079383 | 229.45 | 973 | | 30476.8 | 120.33 | 22386.9 | .067234 | 231.27 |
| 924 | | 28786.3 | 97.11 | 21103.8 | .079115 | 229.49 | 974 | | 30511.4 | 120.84 | 22413.2 | .067015 | 231.31 |
| 925 | 651.85 | 28820.6 | 97.54 | 21129.8 | .078847 | 229.53 | 975 | 701.85 | 30546.1 | 121.36 | 22439.6 | .066797 | 231.35 |
| 926 | | 28855.0 | 97.98 | 21155.8 | .078581 | 229.57 | 976 | | 30580.8 | 121.88 | 22465.9 | .066581 | 231.38 |
| 927 | | 28889.3 | 98.41 | 21181.9 | .078316 | 229.60 | 977 | | 30615.5 | 122.40 | 22492.3 | .066365 | 231.42 |
| 928 | | 28923.7 | 98.85 | 21207.9 | .078052 | 229.64 | 978 | | 30650.2 | 122.93 | 22518.7 | .066150 | 231.45 |
| 929 | | 28958.0 | 99.30 | 21234.0 | .077789 | 229.68 | 979 | | 30684.9 | 123.45 | 22545.1 | .065935 | 231.49 |
| 930 | 656.85 | 28992.4 | 99.74 | 21260.0 | .077527 | 229.71 | 980 | 706.85 | 30719.6 | 123.98 | 22571.5 | .065722 | 231.52 |
| 931 | | 29026.8 | 100.18 | 21286.1 | .077267 | 229.75 | 981 | | 30754.3 | 124.51 | 22597.9 | .065510 | 231.56 |
| 932 | | 29061.2 | 100.63 | 21312.1 | .077007 | 229.79 | 982 | | 30789.0 | 125.04 | 22624.3 | .065298 | 231.59 |
| 933 | | 29095.6 | 101.07 | 21338.2 | .076749 | 229.82 | 983 | | 30823.8 | 125.57 | 22650.7 | .065087 | 231.63 |
| 934 | | 29129.9 | 101.52 | 21364.3 | .076491 | 229.86 | 984 | | 30858.5 | 126.11 | 22677.1 | .064877 | 231.66 |
| 935 | 661.85 | 29164.4 | 101.97 | 21390.4 | .076234 | 229.90 | 985 | 711.85 | 30893.2 | 126.64 | 22703.6 | .064668 | 231.70 |
| 936 | | 29198.8 | 102.43 | 21416.5 | .075979 | 229.93 | 986 | | 30928.0 | 127.18 | 22730.0 | .064460 | 231.73 |
| 937 | | 29233.2 | 102.88 | 21442.6 | .075725 | 229.97 | 987 | | 30962.7 | 127.72 | 22756.4 | .064252 | 231.77 |
| 938 | | 29267.6 | 103.34 | 21468.7 | .075472 | 230.01 | 988 | | 30997.5 | 128.26 | 22782.9 | .064046 | 231.80 |
| 939 | | 29302.0 | 103.79 | 21494.8 | .075220 | 230.04 | 989 | | 31032.3 | 128.81 | 22809.3 | .063840 | 231.84 |
| 940 | 666.85 | 29336.5 | 104.25 | 21520.9 | .074968 | 230.08 | 990 | 716.85 | 31067.1 | 129.35 | 22835.8 | .063635 | 231.88 |
| 941 | | 29370.9 | 104.71 | 21547.1 | .074719 | 230.12 | 991 | | 31101.8 | 129.90 | 22862.3 | .063431 | 231.91 |
| 942 | | 29405.4 | 105.17 | 21573.2 | .074469 | 230.15 | 992 | | 31136.6 | 130.45 | 22888.7 | .063228 | 231.95 |
| 943 | | 29439.8 | 105.64 | 21599.4 | .074221 | 230.19 | 993 | | 31171.4 | 131.00 | 22915.2 | .063025 | 231.98 |
| 944 | | 29474.3 | 106.10 | 21625.5 | .073974 | 230.23 | 994 | | 31206.2 | 131.55 | 22941.7 | .062823 | 232.02 |
| 945 | 671.85 | 29508.8 | 106.57 | 21651.7 | .073728 | 230.26 | 995 | 721.85 | 31241.0 | 132.11 | 22968.2 | .062622 | 232.05 |
| 946 | | 29543.3 | 107.04 | 21677.8 | .073483 | 230.30 | 996 | | 31275.8 | 132.66 | 22994.7 | .062422 | 232.09 |
| 947 | | 29577.7 | 107.51 | 21704.0 | .073239 | 230.34 | 997 | | 31310.7 | 133.22 | 23021.2 | .062223 | 232.12 |
| 948 | | 29612.2 | 107.98 | 21730.2 | .072996 | 230.37 | 998 | | 31345.5 | 133.78 | 23047.7 | .062024 | 232.16 |
| 949 | | 29646.7 | 108.45 | 21756.4 | .072755 | 230.41 | 999 | | 31380.3 | 134.35 | 23074.2 | .061826 | 232.19 |

## Table 5   Products—200% Theoretical Air (for One Gram-Mole)

| T | t | $\bar{h}$ | $p_r$ | $\bar{u}$ | $v_r$ | $\bar{\phi}$ | T | t | $\bar{h}$ | $p_r$ | $\bar{u}$ | $v_r$ | $\bar{\phi}$ |
|---|---|---|---|---|---|---|---|---|---|---|---|---|---|
| 1000 | 726.85 | 31415.2 | 134.91 | 23100.8 | .061629 | 232.23 | 1050 | 776.85 | 33165.3 | 165.67 | 24435.1 | .052697 | 233.93 |
| 1001 | | 31450.0 | 135.47 | 23127.3 | .061435 | 232.26 | 1051 | | 33200.4 | 166.33 | 24462.0 | .052536 | 233.97 |
| 1002 | | 31484.8 | 136.05 | 23153.8 | .061237 | 232.29 | 1052 | | 33235.5 | 167.01 | 24488.8 | .052372 | 234.00 |
| 1003 | | 31519.7 | 136.61 | 23180.4 | .061045 | 232.33 | 1053 | | 33270.8 | 167.68 | 24515.8 | .052213 | 234.03 |
| 1004 | | 31554.6 | 137.19 | 23206.9 | .060849 | 232.36 | 1054 | | 33306.0 | 168.35 | 24542.6 | .052054 | 234.07 |
| 1005 | 731.85 | 31589.5 | 137.76 | 23233.5 | .060658 | 232.40 | 1055 | 781.85 | 33341.2 | 169.03 | 24569.5 | .051895 | 234.10 |
| 1006 | | 31624.4 | 138.34 | 23260.1 | .060463 | 232.43 | 1056 | | 33376.3 | 169.71 | 24596.3 | .051734 | 234.13 |
| 1007 | | 31659.2 | 138.91 | 23286.6 | .060274 | 232.47 | 1057 | | 33411.5 | 170.39 | 24623.2 | .051577 | 234.17 |
| 1008 | | 31694.1 | 139.49 | 23313.2 | .060081 | 232.50 | 1058 | | 33446.7 | 171.07 | 24650.0 | .051420 | 234.20 |
| 1009 | | 31729.1 | 140.07 | 23339.8 | .059892 | 232.54 | 1059 | | 33482.0 | 171.76 | 24677.0 | .051264 | 234.23 |
| 1010 | 736.85 | 31764.0 | 140.66 | 23366.4 | .059701 | 232.57 | 1060 | 786.85 | 33517.2 | 172.45 | 24703.9 | .051105 | 234.27 |
| 1011 | | 31798.9 | 141.24 | 23393.0 | .059514 | 232.61 | 1061 | | 33552.4 | 173.14 | 24730.8 | .050950 | 234.30 |
| 1012 | | 31833.8 | 141.83 | 23419.6 | .059324 | 232.64 | 1062 | | 33587.6 | 173.83 | 24757.7 | .050796 | 234.33 |
| 1013 | | 31868.7 | 142.42 | 23446.2 | .059139 | 232.68 | 1063 | | 33622.9 | 174.52 | 24784.7 | .050642 | 234.37 |
| 1014 | | 31903.7 | 143.02 | 23472.9 | .058950 | 232.71 | 1064 | | 33658.1 | 175.23 | 24811.6 | .050486 | 234.40 |
| 1015 | 741.85 | 31938.6 | 143.61 | 23499.5 | .058766 | 232.74 | 1065 | 791.85 | 33693.3 | 175.93 | 24838.5 | .050333 | 234.43 |
| 1016 | | 31973.5 | 144.21 | 23526.1 | .058579 | 232.78 | 1066 | | 33728.6 | 176.62 | 24865.4 | .050181 | 234.47 |
| 1017 | | 32008.5 | 144.80 | 23552.8 | .058396 | 232.81 | 1067 | | 33763.8 | 177.33 | 24892.3 | .050029 | 234.50 |
| 1018 | | 32043.4 | 145.40 | 23579.4 | .058211 | 232.85 | 1068 | | 33799.2 | 178.04 | 24919.4 | .049875 | 234.53 |
| 1019 | | 32078.4 | 146.00 | 23606.0 | .058029 | 232.88 | 1069 | | 33834.4 | 178.75 | 24946.3 | .049724 | 234.56 |
| 1020 | 746.85 | 32113.4 | 146.60 | 23632.7 | .057849 | 232.92 | 1070 | 796.85 | 33869.6 | 179.46 | 24973.2 | .049575 | 234.60 |
| 1021 | | 32148.4 | 147.21 | 23659.4 | .057665 | 232.95 | 1071 | | 33904.9 | 180.17 | 25000.2 | .049425 | 234.63 |
| 1022 | | 32183.3 | 147.82 | 23686.0 | .057486 | 232.98 | 1072 | | 33940.3 | 180.88 | 25027.2 | .049276 | 234.66 |
| 1023 | | 32218.3 | 148.43 | 23712.7 | .057304 | 233.02 | 1073 | | 33975.5 | 181.61 | 25054.2 | .049125 | 234.70 |
| 1024 | | 32253.3 | 149.04 | 23739.3 | .057126 | 233.05 | 1074 | | 34010.8 | 182.32 | 25081.1 | .048977 | 234.73 |
| 1025 | 751.85 | 32288.3 | 149.65 | 23766.0 | .056949 | 233.09 | 1075 | 801.85 | 34046.2 | 183.04 | 25108.2 | .048830 | 234.76 |
| 1026 | | 32323.3 | 150.27 | 23792.7 | .056769 | 233.12 | 1076 | | 34081.5 | 183.77 | 25135.2 | .048683 | 234.79 |
| 1027 | | 32358.3 | 150.88 | 23819.4 | .056593 | 233.16 | 1077 | | 34116.8 | 184.49 | 25162.1 | .048537 | 234.83 |
| 1028 | | 32393.3 | 151.51 | 23846.1 | .056414 | 233.19 | 1078 | | 34152.0 | 185.23 | 25189.1 | .048388 | 234.86 |
| 1029 | | 32428.3 | 152.13 | 23872.8 | .056239 | 233.22 | 1079 | | 34187.5 | 185.96 | 25216.2 | .048243 | 234.89 |
| 1030 | 756.85 | 32463.4 | 152.75 | 23899.5 | .056066 | 233.26 | 1080 | 806.85 | 34222.8 | 186.69 | 25243.2 | .048099 | 234.93 |
| 1031 | | 32498.4 | 153.38 | 23926.2 | .055889 | 233.29 | 1081 | | 34258.1 | 187.42 | 25270.2 | .047955 | 234.96 |
| 1032 | | 32533.4 | 154.00 | 23953.0 | .055716 | 233.33 | 1082 | | 34293.5 | 188.16 | 25297.3 | .047811 | 234.99 |
| 1033 | | 32568.5 | 154.63 | 23979.7 | .055544 | 233.36 | 1083 | | 34328.8 | 188.90 | 25324.3 | .047668 | 235.02 |
| 1034 | | 32603.5 | 155.27 | 24006.4 | .055369 | 233.39 | 1084 | | 34364.1 | 189.64 | 25351.3 | .047526 | 235.06 |
| 1035 | 761.85 | 32638.6 | 155.90 | 24033.2 | .055199 | 233.43 | 1085 | 811.85 | 34399.6 | 190.40 | 25378.5 | .047380 | 235.09 |
| 1036 | | 32673.6 | 156.53 | 24059.9 | .055029 | 233.46 | 1086 | | 34434.9 | 191.14 | 25405.5 | .047239 | 235.12 |
| 1037 | | 32708.7 | 157.18 | 24086.6 | .054856 | 233.50 | 1087 | | 34470.3 | 191.89 | 25432.5 | .047098 | 235.15 |
| 1038 | | 32743.7 | 157.81 | 24113.4 | .054687 | 233.53 | 1088 | | 34505.7 | 192.64 | 25459.7 | .046958 | 235.19 |
| 1039 | | 32779.0 | 158.46 | 24140.3 | .054515 | 233.56 | 1089 | | 34541.1 | 193.40 | 25486.7 | .046818 | 235.22 |
| 1040 | 766.85 | 32814.0 | 159.10 | 24167.1 | .054348 | 233.60 | 1090 | 816.85 | 34576.4 | 194.15 | 25513.7 | .046678 | 235.25 |
| 1041 | | 32849.1 | 159.75 | 24193.8 | .054181 | 233.63 | 1091 | | 34611.9 | 194.91 | 25540.9 | .046539 | 235.28 |
| 1042 | | 32884.2 | 160.40 | 24220.6 | .054011 | 233.66 | 1092 | | 34647.3 | 195.67 | 25568.0 | .046401 | 235.32 |
| 1043 | | 32919.3 | 161.05 | 24247.4 | .053846 | 233.70 | 1093 | | 34682.6 | 196.43 | 25595.0 | .046263 | 235.35 |
| 1044 | | 32954.4 | 161.70 | 24274.2 | .053681 | 233.73 | 1094 | | 34718.1 | 197.21 | 25622.2 | .046122 | 235.38 |
| 1045 | 771.85 | 32989.5 | 162.35 | 24301.0 | .053516 | 233.76 | 1095 | 821.85 | 34753.5 | 197.98 | 25649.2 | .045985 | 235.41 |
| 1046 | | 33024.6 | 163.02 | 24327.8 | .053349 | 233.80 | 1096 | | 34788.9 | 198.75 | 25676.3 | .045849 | 235.45 |
| 1047 | | 33059.9 | 163.67 | 24354.7 | .053186 | 233.83 | 1097 | | 34824.4 | 199.53 | 25703.5 | .045713 | 235.48 |
| 1048 | | 33095.0 | 164.33 | 24381.5 | .053023 | 233.87 | 1098 | | 34859.8 | 200.30 | 25730.6 | .045578 | 235.51 |
| 1049 | | 33130.1 | 165.00 | 24408.3 | .052858 | 233.90 | 1099 | | 34895.4 | 201.08 | 25757.8 | .045443 | 235.54 |

# Table 5 Products—200% Theoretical Air (for One Gram-Mole)

| T | t | $\bar{h}$ | $p_r$ | $\bar{u}$ | $v_r$ | $\bar{\phi}$ | T | t | $\bar{h}$ | $p_r$ | $\bar{u}$ | $v_r$ | $\bar{\phi}$ |
|---|---|---|---|---|---|---|---|---|---|---|---|---|---|
| 1100 | 826.85 | 34930.7 | 201.86 | 25784.9 | .045308 | 235.58 | 1150 | 876.85 | 36710.7 | 244.17 | 27149.1 | .039160 | 237.16 |
| 1101 |  | 34966.1 | 202.64 | 25812.0 | .045174 | 235.61 | 1151 |  | 36746.5 | 245.09 | 27176.6 | .039047 | 237.19 |
| 1102 |  | 35001.7 | 203.43 | 25839.2 | .045040 | 235.64 | 1152 |  | 36782.2 | 246.00 | 27204.0 | .038935 | 237.22 |
| 1103 |  | 35037.1 | 204.22 | 25866.3 | .044907 | 235.67 | 1153 |  | 36818.0 | 246.93 | 27231.5 | .038823 | 237.25 |
| 1104 |  | 35072.7 | 205.01 | 25893.6 | .044775 | 235.70 | 1154 |  | 36853.8 | 247.84 | 27259.0 | .038714 | 237.28 |
| 1105 | 831.85 | 35108.1 | 205.80 | 25920.7 | .044642 | 235.74 | 1155 | 881.85 | 36889.5 | 248.76 | 27286.3 | .038603 | 237.31 |
| 1106 |  | 35143.5 | 206.60 | 25947.8 | .044511 | 235.77 | 1156 |  | 36925.3 | 249.69 | 27313.8 | .038493 | 237.34 |
| 1107 |  | 35179.1 | 207.39 | 25975.1 | .044380 | 235.80 | 1157 |  | 36960.9 | 250.63 | 27341.2 | .038383 | 237.37 |
| 1108 |  | 35214.5 | 208.19 | 26002.2 | .044249 | 235.83 | 1158 |  | 36996.8 | 251.57 | 27368.7 | .038273 | 237.41 |
| 1109 |  | 35250.1 | 209.00 | 26029.5 | .044119 | 235.86 | 1159 |  | 37032.6 | 252.49 | 27396.2 | .038166 | 237.44 |
| 1110 | 836.85 | 35285.6 | 209.80 | 26056.6 | .043989 | 235.90 | 1160 | 886.85 | 37068.3 | 253.43 | 27423.6 | .038057 | 237.47 |
| 1111 |  | 35321.2 | 210.61 | 26083.9 | .043859 | 235.93 | 1161 |  | 37104.2 | 254.38 | 27451.1 | .037948 | 237.50 |
| 1112 |  | 35356.6 | 211.42 | 26111.0 | .043730 | 235.96 | 1162 |  | 37140.0 | 255.32 | 27478.7 | .037840 | 237.53 |
| 1113 |  | 35392.1 | 212.24 | 26138.2 | .043602 | 235.99 | 1163 |  | 37175.7 | 256.27 | 27506.1 | .037732 | 237.56 |
| 1114 |  | 35427.7 | 213.05 | 26165.5 | .043474 | 236.02 | 1164 |  | 37211.6 | 257.21 | 27533.6 | .037627 | 237.59 |
| 1115 | 841.85 | 35463.2 | 213.87 | 26192.6 | .043346 | 236.06 | 1165 | 891.85 | 37247.5 | 258.17 | 27561.2 | .037519 | 237.62 |
| 1116 |  | 35498.8 | 214.69 | 26219.9 | .043219 | 236.09 | 1166 |  | 37283.2 | 259.13 | 27588.6 | .037412 | 237.65 |
| 1117 |  | 35534.3 | 215.52 | 26247.1 | .043093 | 236.12 | 1167 |  | 37319.1 | 260.09 | 27616.1 | .037306 | 237.68 |
| 1118 |  | 35569.9 | 216.34 | 26274.4 | .042966 | 236.15 | 1168 |  | 37354.9 | 261.04 | 27643.7 | .037202 | 237.71 |
| 1119 |  | 35605.4 | 217.17 | 26301.6 | .042841 | 236.18 | 1169 |  | 37390.7 | 262.01 | 27671.1 | .037096 | 237.74 |
| 1120 | 846.85 | 35641.1 | 218.01 | 26328.9 | .042715 | 236.22 | 1170 | 896.85 | 37426.6 | 262.98 | 27698.7 | .036991 | 237.77 |
| 1121 |  | 35676.6 | 218.84 | 26356.1 | .042590 | 236.25 | 1171 |  | 37462.5 | 263.95 | 27726.3 | .036886 | 237.81 |
| 1122 |  | 35712.2 | 219.68 | 26383.5 | .042466 | 236.28 | 1172 |  | 37498.4 | 264.93 | 27753.9 | .036781 | 237.84 |
| 1123 |  | 35747.8 | 220.52 | 26410.7 | .042342 | 236.31 | 1173 |  | 37534.1 | 265.90 | 27781.3 | .036679 | 237.87 |
| 1124 |  | 35783.4 | 221.36 | 26438.0 | .042218 | 236.34 | 1174 |  | 37570.0 | 266.88 | 27808.9 | .036575 | 237.90 |
| 1125 | 851.85 | 35818.9 | 222.20 | 26465.2 | .042095 | 236.37 | 1175 | 901.85 | 37605.9 | 267.87 | 27836.5 | .036471 | 237.93 |
| 1126 |  | 35854.6 | 223.05 | 26492.6 | .041972 | 236.41 | 1176 |  | 37641.9 | 268.84 | 27864.1 | .036371 | 237.96 |
| 1127 |  | 35890.2 | 223.90 | 26519.8 | .041850 | 236.44 | 1177 |  | 37677.6 | 269.83 | 27891.6 | .036268 | 237.99 |
| 1128 |  | 35925.8 | 224.76 | 26547.2 | .041728 | 236.47 | 1178 |  | 37713.6 | 270.83 | 27919.2 | .036165 | 238.02 |
| 1129 |  | 35961.4 | 225.60 | 26574.4 | .041610 | 236.50 | 1179 |  | 37749.5 | 271.81 | 27946.8 | .036065 | 238.05 |
| 1130 | 856.85 | 35997.1 | 226.45 | 26601.8 | .041489 | 236.53 | 1180 | 906.85 | 37785.4 | 272.81 | 27974.4 | .035963 | 238.08 |
| 1131 |  | 36032.6 | 227.32 | 26629.0 | .041368 | 236.56 | 1181 |  | 37821.2 | 273.81 | 28001.9 | .035861 | 238.11 |
| 1132 |  | 36068.3 | 228.18 | 26656.4 | .041248 | 236.59 | 1182 |  | 37857.2 | 274.82 | 28029.6 | .035760 | 238.14 |
| 1133 |  | 36103.9 | 229.05 | 26683.7 | .041128 | 236.63 | 1183 |  | 37893.1 | 275.81 | 28057.2 | .035662 | 238.17 |
| 1134 |  | 36139.6 | 229.92 | 26711.1 | .041009 | 236.66 | 1184 |  | 37929.1 | 276.83 | 28084.8 | .035561 | 238.20 |
| 1135 | 861.85 | 36175.2 | 230.79 | 26738.3 | .040890 | 236.69 | 1185 | 911.85 | 37964.9 | 277.84 | 28112.3 | .035461 | 238.23 |
| 1136 |  | 36210.9 | 231.66 | 26765.8 | .040771 | 236.72 | 1186 |  | 38000.9 | 278.85 | 28140.0 | .035363 | 238.26 |
| 1137 |  | 36246.5 | 232.54 | 26793.0 | .040653 | 236.75 | 1187 |  | 38036.8 | 279.87 | 28167.6 | .035264 | 238.29 |
| 1138 |  | 36282.2 | 233.42 | 26820.4 | .040535 | 236.78 | 1188 |  | 38072.8 | 280.90 | 28195.3 | .035164 | 238.32 |
| 1139 |  | 36317.8 | 234.30 | 26847.7 | .040418 | 236.81 | 1189 |  | 38108.6 | 281.92 | 28222.8 | .035065 | 238.35 |
| 1140 | 866.85 | 36353.6 | 235.17 | 26875.1 | .040304 | 236.85 | 1190 | 916.85 | 38144.6 | 282.94 | 28250.5 | .034969 | 238.38 |
| 1141 |  | 36389.3 | 236.06 | 26902.6 | .040187 | 236.88 | 1191 |  | 38180.6 | 283.97 | 28278.2 | .034871 | 238.41 |
| 1142 |  | 36424.9 | 236.95 | 26929.9 | .040071 | 236.91 | 1192 |  | 38216.6 | 285.01 | 28305.8 | .034773 | 238.44 |
| 1143 |  | 36460.7 | 237.85 | 26957.3 | .039956 | 236.94 | 1193 |  | 38252.5 | 286.04 | 28333.4 | .034678 | 238.47 |
| 1144 |  | 36496.3 | 238.74 | 26984.6 | .039840 | 236.97 | 1194 |  | 38288.5 | 287.08 | 28361.1 | .034581 | 238.50 |
| 1145 | 871.85 | 36532.1 | 239.64 | 27012.1 | .039726 | 237.00 | 1195 | 921.85 | 38324.5 | 288.13 | 28388.8 | .034484 | 238.53 |
| 1146 |  | 36567.8 | 240.55 | 27039.5 | .039611 | 237.03 | 1196 |  | 38360.5 | 289.16 | 28416.5 | .034389 | 238.56 |
| 1147 |  | 36603.5 | 241.44 | 27066.9 | .039500 | 237.06 | 1197 |  | 38396.5 | 290.22 | 28444.2 | .034293 | 238.59 |
| 1148 |  | 36639.3 | 242.34 | 27094.3 | .039386 | 237.10 | 1198 |  | 38432.4 | 291.27 | 28471.7 | .034197 | 238.62 |
| 1149 |  | 36674.9 | 243.25 | 27121.7 | .039273 | 237.13 | 1199 |  | 38468.4 | 292.31 | 28499.4 | .034104 | 238.65 |

## Table 5   Products—200% Theoretical Air (for One Gram-Mole)

| T | t | $\bar{h}$ | $p_r$ | $\bar{u}$ | $v_r$ | $\bar{\phi}$ | T | t | $\bar{h}$ | $p_r$ | $\bar{u}$ | $v_r$ | $\bar{\phi}$ |
|---|---|---|---|---|---|---|---|---|---|---|---|---|---|
| 1200 | 926.85 | 38504.4 | 293.38 | 28527.2 | .034008 | 238.68 | 1250 | 976.85 | 40310.9 | 350.35 | 29917.9 | .029664 | 240.16 |
| 1201 |  | 38540.5 | 294.45 | 28554.9 | .033913 | 238.71 | 1251 |  | 40347.2 | 351.57 | 29945.9 | .029586 | 240.19 |
| 1202 |  | 38576.5 | 295.50 | 28582.6 | .033821 | 238.74 | 1252 |  | 40383.5 | 352.81 | 29973.8 | .029505 | 240.22 |
| 1203 |  | 38612.4 | 296.57 | 28610.2 | .033726 | 238.77 | 1253 |  | 40419.7 | 354.03 | 30001.8 | .029427 | 240.25 |
| 1204 |  | 38648.4 | 297.65 | 28637.9 | .033632 | 238.80 | 1254 |  | 40456.0 | 355.27 | 30029.7 | .029347 | 240.28 |
| 1205 | 931.85 | 38684.5 | 298.71 | 28665.6 | .033541 | 238.83 | 1255 | 981.85 | 40492.3 | 356.50 | 30057.7 | .029269 | 240.30 |
| 1206 |  | 38720.6 | 299.79 | 28693.4 | .033447 | 238.86 | 1256 |  | 40528.6 | 357.76 | 30085.7 | .029190 | 240.33 |
| 1207 |  | 38756.6 | 300.88 | 28721.1 | .033354 | 238.89 | 1257 |  | 40564.8 | 358.99 | 30113.6 | .029113 | 240.36 |
| 1208 |  | 38792.7 | 301.95 | 28748.9 | .033264 | 238.92 | 1258 |  | 40601.1 | 360.25 | 30141.6 | .029034 | 240.39 |
| 1209 |  | 38828.6 | 303.04 | 28776.5 | .033171 | 238.95 | 1259 |  | 40637.4 | 361.49 | 30169.6 | .028957 | 240.42 |
| 1210 | 936.85 | 38864.7 | 304.12 | 28804.3 | .033081 | 238.98 | 1260 | 986.85 | 40673.7 | 362.74 | 30197.6 | .028881 | 240.45 |
| 1211 |  | 38900.8 | 305.21 | 28832.0 | .032989 | 239.01 | 1261 |  | 40710.0 | 364.01 | 30225.5 | .028803 | 240.48 |
| 1212 |  | 38936.9 | 306.32 | 28859.8 | .032897 | 239.04 | 1262 |  | 40746.3 | 365.26 | 30253.5 | .028727 | 240.51 |
| 1213 |  | 38972.9 | 307.40 | 28887.6 | .032808 | 239.07 | 1263 |  | 40782.6 | 366.54 | 30281.5 | .028649 | 240.54 |
| 1214 |  | 39009.0 | 308.51 | 28915.4 | .032717 | 239.10 | 1264 |  | 40818.9 | 367.80 | 30309.5 | .028574 | 240.56 |
| 1215 | 941.85 | 39045.1 | 309.63 | 28943.1 | .032627 | 239.13 | 1265 | 991.85 | 40855.3 | 369.09 | 30337.5 | .028496 | 240.59 |
| 1216 |  | 39081.1 | 310.72 | 28970.8 | .032538 | 239.16 | 1266 |  | 40891.6 | 370.36 | 30365.5 | .028421 | 240.62 |
| 1217 |  | 39117.2 | 311.84 | 28998.6 | .032448 | 239.19 | 1267 |  | 40927.9 | 371.63 | 30393.6 | .028347 | 240.65 |
| 1218 |  | 39153.3 | 312.94 | 29026.4 | .032361 | 239.22 | 1268 |  | 40964.4 | 372.92 | 30421.7 | .028270 | 240.68 |
| 1219 |  | 39189.4 | 314.07 | 29054.2 | .032271 | 239.25 | 1269 |  | 41000.7 | 374.20 | 30449.7 | .028196 | 240.71 |
| 1220 | 946.85 | 39225.5 | 315.20 | 29082.0 | .032182 | 239.28 | 1270 | 996.85 | 41037.1 | 375.51 | 30477.8 | .028120 | 240.74 |
| 1221 |  | 39261.7 | 316.31 | 29109.8 | .032095 | 239.31 | 1271 |  | 41073.4 | 376.79 | 30505.8 | .028046 | 240.76 |
| 1222 |  | 39297.8 | 317.44 | 29137.6 | .032006 | 239.34 | 1272 |  | 41109.8 | 378.08 | 30533.8 | .027973 | 240.79 |
| 1223 |  | 39333.9 | 318.56 | 29165.4 | .031920 | 239.37 | 1273 |  | 41146.1 | 379.40 | 30561.9 | .027897 | 240.82 |
| 1224 |  | 39369.9 | 319.71 | 29193.1 | .031832 | 239.40 | 1274 |  | 41182.5 | 380.69 | 30589.9 | .027824 | 240.85 |
| 1225 | 951.85 | 39406.1 | 320.83 | 29220.9 | .031746 | 239.43 | 1275 | 1001.85 | 41218.8 | 381.99 | 30618.0 | .027751 | 240.88 |
| 1226 |  | 39442.2 | 321.98 | 29248.8 | .031659 | 239.46 | 1276 |  | 41255.2 | 383.30 | 30646.0 | .027679 | 240.91 |
| 1227 |  | 39478.4 | 323.13 | 29276.6 | .031572 | 239.49 | 1277 |  | 41291.5 | 384.63 | 30674.1 | .027605 | 240.94 |
| 1228 |  | 39514.5 | 324.29 | 29304.4 | .031485 | 239.52 | 1278 |  | 41327.9 | 385.94 | 30702.1 | .027532 | 240.96 |
| 1229 |  | 39550.7 | 325.43 | 29332.3 | .031400 | 239.55 | 1279 |  | 41364.4 | 387.25 | 30730.3 | .027460 | 240.99 |
| 1230 | 956.85 | 39586.9 | 326.59 | 29360.1 | .031314 | 239.58 | 1280 | 1006.85 | 41400.8 | 388.60 | 30758.4 | .027387 | 241.02 |
| 1231 |  | 39623.0 | 327.73 | 29388.0 | .031230 | 239.60 | 1281 |  | 41437.2 | 389.92 | 30786.5 | .027315 | 241.05 |
| 1232 |  | 39659.2 | 328.90 | 29415.9 | .031144 | 239.63 | 1282 |  | 41473.6 | 391.24 | 30814.5 | .027244 | 241.08 |
| 1233 |  | 39695.4 | 330.06 | 29443.7 | .031060 | 239.66 | 1283 |  | 41510.0 | 392.60 | 30842.6 | .027171 | 241.11 |
| 1234 |  | 39731.6 | 331.23 | 29471.6 | .030975 | 239.69 | 1284 |  | 41546.4 | 393.93 | 30870.7 | .027100 | 241.13 |
| 1235 | 961.85 | 39767.6 | 332.39 | 29499.3 | .030892 | 239.72 | 1285 | 1011.85 | 41582.8 | 395.27 | 30898.8 | .027030 | 241.16 |
| 1236 |  | 39803.8 | 333.57 | 29527.2 | .030808 | 239.75 | 1286 |  | 41619.2 | 396.64 | 30926.9 | .026958 | 241.19 |
| 1237 |  | 39840.0 | 334.76 | 29555.1 | .030723 | 239.78 | 1287 |  | 41655.6 | 397.98 | 30955.0 | .026887 | 241.22 |
| 1238 |  | 39876.2 | 335.93 | 29582.9 | .030641 | 239.81 | 1288 |  | 41692.0 | 399.33 | 30983.1 | .026817 | 241.25 |
| 1239 |  | 39912.4 | 337.12 | 29610.8 | .030557 | 239.84 | 1289 |  | 41728.6 | 400.71 | 31011.3 | .026746 | 241.28 |
| 1240 | 966.85 | 39948.6 | 338.30 | 29638.7 | .030476 | 239.87 | 1290 | 1016.85 | 41765.0 | 402.06 | 31039.4 | .026676 | 241.30 |
| 1241 |  | 39984.8 | 339.50 | 29666.6 | .030392 | 239.90 | 1291 |  | 41801.4 | 403.42 | 31067.5 | .026607 | 241.33 |
| 1242 |  | 40021.0 | 340.68 | 29694.6 | .030311 | 239.93 | 1292 |  | 41837.9 | 404.81 | 31095.7 | .026536 | 241.36 |
| 1243 |  | 40057.3 | 341.89 | 29722.5 | .030228 | 239.96 | 1293 |  | 41874.3 | 406.18 | 31123.8 | .026467 | 241.39 |
| 1244 |  | 40093.5 | 343.08 | 29750.4 | .030148 | 239.99 | 1294 |  | 41910.7 | 407.55 | 31151.9 | .026399 | 241.42 |
| 1245 | 971.85 | 40129.7 | 344.29 | 29778.3 | .030066 | 240.01 | 1295 | 1021.85 | 41947.2 | 408.95 | 31180.0 | .026329 | 241.45 |
| 1246 |  | 40166.0 | 345.49 | 29806.2 | .029986 | 240.04 | 1296 |  | 41983.6 | 410.33 | 31208.2 | .026260 | 241.47 |
| 1247 |  | 40202.2 | 346.71 | 29834.1 | .029904 | 240.07 | 1297 |  | 42020.2 | 411.71 | 31236.5 | .026193 | 241.50 |
| 1248 |  | 40238.5 | 347.91 | 29862.1 | .029825 | 240.10 | 1298 |  | 42056.7 | 413.10 | 31264.6 | .026125 | 241.53 |
| 1249 |  | 40274.7 | 349.14 | 29890.0 | .029743 | 240.13 | 1299 |  | 42093.1 | 414.52 | 31292.7 | .026055 | 241.56 |

# Table 5 Products—200% Theoretical Air (for One Gram-Mole)

| T | t | $\bar{h}$ | $p_r$ | $\bar{u}$ | $v_r$ | $\bar{\phi}$ | T | t | $\bar{h}$ | $p_r$ | $\bar{u}$ | $v_r$ | $\bar{\phi}$ |
|---|---|---|---|---|---|---|---|---|---|---|---|---|---|
| 1300 | 1026.85 | 42129.6 | 415.91 | 31320.9 | .025988 | 241.59 | 1350 | 1076.85 | 43959.5 | 491.11 | 32735.0 | .022855 | 242.97 |
| 1301 | | 42166.1 | 417.31 | 31349.0 | .025921 | 241.61 | 1351 | | 43996.1 | 492.71 | 32763.4 | .022798 | 242.99 |
| 1302 | | 42202.5 | 418.74 | 31377.2 | .025852 | 241.64 | 1352 | | 44033.0 | 494.32 | 32791.9 | .022741 | 243.02 |
| 1303 | | 42239.0 | 420.14 | 31405.3 | .025786 | 241.67 | 1353 | | 44069.6 | 495.93 | 32820.2 | .022683 | 243.05 |
| 1304 | | 42275.6 | 421.55 | 31433.7 | .025719 | 241.70 | 1354 | | 44106.3 | 497.55 | 32848.6 | .022626 | 243.08 |
| 1305 | 1031.85 | 42312.1 | 422.97 | 31461.8 | .025653 | 241.73 | 1355 | 1081.85 | 44143.1 | 499.20 | 32877.1 | .022568 | 243.10 |
| 1306 | | 42348.6 | 424.41 | 31490.0 | .025585 | 241.75 | 1356 | | 44179.8 | 500.83 | 32905.5 | .022511 | 243.13 |
| 1307 | | 42385.1 | 425.84 | 31518.2 | .025519 | 241.78 | 1357 | | 44216.5 | 502.46 | 32933.9 | .022455 | 243.16 |
| 1308 | | 42421.6 | 427.26 | 31546.4 | .025453 | 241.81 | 1358 | | 44253.4 | 504.10 | 32962.4 | .022398 | 243.18 |
| 1309 | | 42458.1 | 428.69 | 31574.5 | .025388 | 241.84 | 1359 | | 44290.1 | 505.74 | 32990.8 | .022342 | 243.21 |
| 1310 | 1036.85 | 42494.7 | 430.15 | 31602.9 | .025321 | 241.87 | 1360 | 1086.85 | 44326.8 | 507.38 | 33019.2 | .022286 | 243.24 |
| 1311 | | 42531.3 | 431.59 | 31631.1 | .025256 | 241.89 | 1361 | | 44363.6 | 509.03 | 33047.7 | .022230 | 243.27 |
| 1312 | | 42567.8 | 433.03 | 31659.3 | .025191 | 241.92 | 1362 | | 44400.3 | 510.68 | 33076.1 | .022175 | 243.29 |
| 1313 | | 42604.3 | 434.51 | 31687.5 | .025125 | 241.95 | 1363 | | 44437.0 | 512.34 | 33104.5 | .022119 | 243.32 |
| 1314 | | 42640.8 | 435.96 | 31715.7 | .025060 | 241.98 | 1364 | | 44473.9 | 514.00 | 33133.0 | .022064 | 243.35 |
| 1315 | 1041.85 | 42677.5 | 437.41 | 31744.0 | .024996 | 242.01 | 1365 | 1091.85 | 44510.6 | 515.70 | 33161.4 | .022007 | 243.37 |
| 1316 | | 42714.0 | 438.87 | 31772.3 | .024932 | 242.03 | 1366 | | 44547.3 | 517.37 | 33189.8 | .021952 | 243.40 |
| 1317 | | 42750.5 | 440.36 | 31800.5 | .024866 | 242.06 | 1367 | | 44584.2 | 519.05 | 33218.4 | .021898 | 243.43 |
| 1318 | | 42787.1 | 441.82 | 31828.7 | .024803 | 242.09 | 1368 | | 44620.9 | 520.72 | 33246.8 | .021843 | 243.45 |
| 1319 | | 42823.6 | 443.29 | 31856.9 | .024739 | 242.12 | 1369 | | 44657.7 | 522.41 | 33275.2 | .021788 | 243.48 |
| 1320 | 1046.85 | 42860.3 | 444.77 | 31885.3 | .024676 | 242.14 | 1370 | 1096.85 | 44694.5 | 524.10 | 33303.8 | .021734 | 243.51 |
| 1321 | | 42896.9 | 446.25 | 31913.5 | .024613 | 242.17 | 1371 | | 44731.3 | 525.79 | 33332.2 | .021680 | 243.54 |
| 1322 | | 42933.4 | 447.76 | 31941.8 | .024548 | 242.20 | 1372 | | 44768.0 | 527.49 | 33360.7 | .021626 | 243.56 |
| 1323 | | 42970.0 | 449.24 | 31970.0 | .024485 | 242.23 | 1373 | | 44804.9 | 529.19 | 33389.2 | .021572 | 243.59 |
| 1324 | | 43006.5 | 450.73 | 31998.3 | .024423 | 242.25 | 1374 | | 44841.7 | 530.90 | 33417.7 | .021518 | 243.62 |
| 1325 | 1051.85 | 43043.2 | 452.23 | 32026.7 | .024361 | 242.28 | 1375 | 1101.85 | 44878.6 | 532.61 | 33446.3 | .021465 | 243.64 |
| 1326 | | 43079.8 | 453.76 | 32054.9 | .024297 | 242.31 | 1376 | | 44915.3 | 534.32 | 33474.7 | .021411 | 243.67 |
| 1327 | | 43116.4 | 455.26 | 32083.2 | .024235 | 242.34 | 1377 | | 44952.1 | 536.04 | 33503.2 | .021358 | 243.70 |
| 1328 | | 43152.9 | 456.77 | 32111.4 | .024173 | 242.37 | 1378 | | 44989.0 | 537.77 | 33531.8 | .021305 | 243.72 |
| 1329 | | 43189.5 | 458.28 | 32139.7 | .024112 | 242.39 | 1379 | | 45025.8 | 539.50 | 33560.2 | .021252 | 243.75 |
| 1330 | 1056.85 | 43226.3 | 459.82 | 32168.1 | .024049 | 242.42 | 1380 | 1106.85 | 45062.7 | 541.23 | 33588.8 | .021200 | 243.78 |
| 1331 | | 43262.8 | 461.34 | 32196.4 | .023988 | 242.45 | 1381 | | 45099.5 | 542.97 | 33617.3 | .021147 | 243.80 |
| 1332 | | 43299.4 | 462.87 | 32224.6 | .023927 | 242.48 | 1382 | | 45136.3 | 544.72 | 33645.8 | .021095 | 243.83 |
| 1333 | | 43336.0 | 464.39 | 32252.9 | .023866 | 242.50 | 1383 | | 45173.2 | 546.46 | 33674.4 | .021042 | 243.86 |
| 1334 | | 43372.8 | 465.92 | 32281.4 | .023805 | 242.53 | 1384 | | 45210.0 | 548.22 | 33702.9 | .020990 | 243.88 |
| 1335 | 1061.85 | 43409.4 | 467.49 | 32309.6 | .023743 | 242.56 | 1385 | 1111.85 | 45246.9 | 549.97 | 33731.5 | .020938 | 243.91 |
| 1336 | | 43446.0 | 469.03 | 32337.9 | .023683 | 242.59 | 1386 | | 45283.7 | 551.74 | 33760.0 | .020886 | 243.94 |
| 1337 | | 43482.6 | 470.58 | 32366.2 | .023623 | 242.61 | 1387 | | 45320.5 | 553.50 | 33788.5 | .020835 | 243.96 |
| 1338 | | 43519.4 | 472.13 | 32394.7 | .023563 | 242.64 | 1388 | | 45357.5 | 555.27 | 33817.1 | .020783 | 243.99 |
| 1339 | | 43556.0 | 473.68 | 32423.0 | .023503 | 242.67 | 1389 | | 45394.3 | 557.05 | 33845.6 | .020732 | 244.02 |
| 1340 | 1066.85 | 43592.6 | 475.27 | 32451.3 | .023442 | 242.70 | 1390 | 1116.85 | 45431.3 | 558.83 | 33874.2 | .020681 | 244.04 |
| 1341 | | 43629.2 | 476.83 | 32479.6 | .023383 | 242.72 | 1391 | | 45468.1 | 560.62 | 33902.7 | .020630 | 244.07 |
| 1342 | | 43666.0 | 478.39 | 32508.1 | .023324 | 242.75 | 1392 | | 45504.9 | 562.41 | 33931.2 | .020579 | 244.09 |
| 1343 | | 43702.6 | 479.96 | 32536.4 | .023265 | 242.78 | 1393 | | 45541.9 | 564.20 | 33959.9 | .020528 | 244.12 |
| 1344 | | 43739.3 | 481.54 | 32564.7 | .023206 | 242.80 | 1394 | | 45578.7 | 566.00 | 33988.4 | .020477 | 244.15 |
| 1345 | 1071.85 | 43776.1 | 483.12 | 32593.2 | .023147 | 242.83 | 1395 | 1121.85 | 45615.7 | 567.81 | 34017.1 | .020427 | 244.17 |
| 1346 | | 43812.7 | 484.70 | 32621.5 | .023089 | 242.86 | 1396 | | 45652.5 | 569.62 | 34045.6 | .020377 | 244.20 |
| 1347 | | 43849.4 | 486.32 | 32649.9 | .023029 | 242.89 | 1397 | | 45689.5 | 571.43 | 34074.3 | .020327 | 244.23 |
| 1348 | | 43886.0 | 487.91 | 32678.2 | .022971 | 242.91 | 1398 | | 45726.3 | 573.25 | 34102.8 | .020277 | 244.25 |
| 1349 | | 43922.8 | 489.51 | 32706.7 | .022913 | 242.94 | 1399 | | 45763.3 | 575.07 | 34131.5 | .020227 | 244.28 |

# Table 5 Products—200% Theoretical Air (for One Gram-Mole)

| T | t | $\bar{h}$ | $p_r$ | $\bar{u}$ | $v_r$ | $\bar{\phi}$ | T | t | $\bar{h}$ | $p_r$ | $\bar{u}$ | $v_r$ | $\bar{\phi}$ |
|---|---|---|---|---|---|---|---|---|---|---|---|---|---|
| 1400 | 1126.85 | 45800.2 | 576.90 | 34160.0 | .020177 | 244.31 | 1450 | 1176.85 | 47650.9 | 674.42 | 35595.1 | .017876 | 245.61 |
| 1401 | | 45837.0 | 578.74 | 34188.5 | .020127 | 244.33 | 1451 | | 47688.1 | 676.51 | 35623.9 | .017833 | 245.63 |
| 1402 | | 45874.0 | 580.57 | 34217.2 | .020078 | 244.36 | 1452 | | 47725.1 | 678.60 | 35652.6 | .017790 | 245.66 |
| 1403 | | 45910.9 | 582.42 | 34245.8 | .020029 | 244.39 | 1453 | | 47762.3 | 680.65 | 35681.5 | .017749 | 245.68 |
| 1404 | | 45947.9 | 584.26 | 34274.5 | .019980 | 244.41 | 1454 | | 47799.4 | 682.76 | 35710.2 | .017706 | 245.71 |
| 1405 | 1131.85 | 45984.7 | 586.12 | 34303.0 | .019931 | 244.44 | 1455 | 1181.85 | 47836.6 | 684.86 | 35739.1 | .017664 | 245.73 |
| 1406 | | 46021.8 | 587.98 | 34331.7 | .019882 | 244.46 | 1456 | | 47873.8 | 686.98 | 35768.0 | .017622 | 245.76 |
| 1407 | | 46058.7 | 589.84 | 34360.3 | .019833 | 244.49 | 1457 | | 47910.8 | 689.10 | 35796.7 | .017580 | 245.78 |
| 1408 | | 46095.7 | 591.71 | 34389.0 | .019785 | 244.52 | 1458 | | 47948.0 | 691.17 | 35825.6 | .017539 | 245.81 |
| 1409 | | 46132.6 | 593.58 | 34417.6 | .019736 | 244.54 | 1459 | | 47985.1 | 693.30 | 35854.3 | .017497 | 245.83 |
| 1410 | 1136.85 | 46169.6 | 595.41 | 34446.3 | .019689 | 244.57 | 1460 | 1186.85 | 48022.3 | 695.44 | 35883.2 | .017455 | 245.86 |
| 1411 | | 46206.5 | 597.30 | 34474.8 | .019641 | 244.60 | 1461 | | 48059.5 | 697.58 | 35912.1 | .017414 | 245.89 |
| 1412 | | 46243.5 | 599.18 | 34503.6 | .019593 | 244.62 | 1462 | | 48096.5 | 699.73 | 35940.9 | .017372 | 245.91 |
| 1413 | | 46280.4 | 601.08 | 34532.2 | .019545 | 244.65 | 1463 | | 48133.8 | 701.83 | 35969.8 | .017332 | 245.94 |
| 1414 | | 46317.5 | 602.97 | 34560.9 | .019498 | 244.67 | 1464 | | 48170.8 | 703.99 | 35998.5 | .017290 | 245.96 |
| 1415 | 1141.85 | 46354.4 | 604.87 | 34589.5 | .019450 | 244.70 | 1465 | 1191.85 | 48208.1 | 706.15 | 36027.5 | .017249 | 245.99 |
| 1416 | | 46391.4 | 606.78 | 34618.2 | .019403 | 244.73 | 1466 | | 48245.3 | 708.32 | 36056.4 | .017208 | 246.01 |
| 1417 | | 46428.3 | 608.69 | 34646.8 | .019355 | 244.75 | 1467 | | 48282.4 | 710.49 | 36085.1 | .017167 | 246.04 |
| 1418 | | 46465.4 | 610.61 | 34675.6 | .019308 | 244.78 | 1468 | | 48319.6 | 712.62 | 36114.0 | .017128 | 246.06 |
| 1419 | | 46502.3 | 612.53 | 34704.2 | .019261 | 244.80 | 1469 | | 48356.7 | 714.81 | 36142.8 | .017087 | 246.09 |
| 1420 | 1146.85 | 46539.4 | 614.46 | 34732.9 | .019214 | 244.83 | 1470 | 1196.85 | 48393.9 | 717.00 | 36171.7 | .017046 | 246.11 |
| 1421 | | 46576.3 | 616.35 | 34761.5 | .019169 | 244.86 | 1471 | | 48431.2 | 719.20 | 36200.7 | .017006 | 246.14 |
| 1422 | | 46613.4 | 618.28 | 34790.3 | .019122 | 244.88 | 1472 | | 48468.3 | 721.35 | 36229.5 | .016967 | 246.16 |
| 1423 | | 46650.3 | 620.23 | 34818.9 | .019076 | 244.91 | 1473 | | 48505.5 | 723.55 | 36258.4 | .016926 | 246.19 |
| 1424 | | 46687.4 | 622.17 | 34847.7 | .019030 | 244.93 | 1474 | | 48542.8 | 725.77 | 36287.3 | .016886 | 246.22 |
| 1425 | 1151.85 | 46724.3 | 624.13 | 34876.3 | .018983 | 244.96 | 1475 | 1201.85 | 48579.9 | 727.98 | 36316.1 | .016846 | 246.24 |
| 1426 | | 46761.4 | 626.08 | 34905.1 | .018937 | 244.99 | 1476 | | 48617.1 | 730.21 | 36345.1 | .016806 | 246.27 |
| 1427 | | 46798.3 | 628.04 | 34933.7 | .018891 | 245.01 | 1477 | | 48654.2 | 732.39 | 36373.9 | .016768 | 246.29 |
| 1428 | | 46835.4 | 629.97 | 34962.5 | .018847 | 245.04 | 1478 | | 48691.5 | 734.62 | 36402.8 | .016728 | 246.32 |
| 1429 | | 46872.4 | 631.94 | 34991.1 | .018801 | 245.06 | 1479 | | 48728.8 | 736.86 | 36431.8 | .016688 | 246.34 |
| 1430 | 1156.85 | 46909.5 | 633.92 | 35019.9 | .018756 | 245.09 | 1480 | 1206.85 | 48765.9 | 739.11 | 36460.6 | .016649 | 246.37 |
| 1431 | | 46946.4 | 635.90 | 35048.5 | .018710 | 245.12 | 1481 | | 48803.2 | 741.31 | 36489.6 | .016611 | 246.39 |
| 1432 | | 46983.6 | 637.89 | 35077.3 | .018665 | 245.14 | 1482 | | 48840.5 | 743.57 | 36518.5 | .016571 | 246.42 |
| 1433 | | 47020.5 | 639.88 | 35106.0 | .018620 | 245.17 | 1483 | | 48877.6 | 745.83 | 36547.3 | .016532 | 246.44 |
| 1434 | | 47057.6 | 641.88 | 35134.8 | .018575 | 245.19 | 1484 | | 48914.9 | 748.10 | 36576.3 | .016493 | 246.47 |
| 1435 | 1161.85 | 47094.8 | 643.84 | 35163.6 | .018531 | 245.22 | 1485 | 1211.85 | 48952.2 | 750.33 | 36605.3 | .016455 | 246.49 |
| 1436 | | 47131.7 | 645.85 | 35192.3 | .018487 | 245.25 | 1486 | | 48989.3 | 752.61 | 36634.1 | .016417 | 246.52 |
| 1437 | | 47168.9 | 647.86 | 35221.1 | .018442 | 245.27 | 1487 | | 49026.6 | 754.89 | 36663.1 | .016378 | 246.54 |
| 1438 | | 47205.8 | 649.88 | 35249.7 | .018397 | 245.30 | 1488 | | 49063.9 | 757.18 | 36692.1 | .016339 | 246.57 |
| 1439 | | 47243.0 | 651.90 | 35278.6 | .018353 | 245.32 | 1489 | | 49101.1 | 759.43 | 36720.9 | .016302 | 246.59 |
| 1440 | 1166.85 | 47280.0 | 653.93 | 35307.2 | .018309 | 245.35 | 1490 | 1216.85 | 49138.4 | 761.73 | 36749.9 | .016264 | 246.62 |
| 1441 | | 47317.1 | 655.97 | 35336.1 | .018265 | 245.37 | 1491 | | 49175.7 | 764.04 | 36778.9 | .016225 | 246.64 |
| 1442 | | 47354.1 | 657.96 | 35364.7 | .018222 | 245.40 | 1492 | | 49212.8 | 766.30 | 36807.8 | .016188 | 246.67 |
| 1443 | | 47391.3 | 660.01 | 35393.6 | .018178 | 245.43 | 1493 | | 49250.2 | 768.62 | 36836.8 | .016150 | 246.69 |
| 1444 | | 47428.4 | 662.06 | 35422.4 | .018134 | 245.45 | 1494 | | 49287.5 | 770.95 | 36865.8 | .016112 | 246.72 |
| 1445 | 1171.85 | 47465.4 | 664.11 | 35451.1 | .018091 | 245.48 | 1495 | 1221.85 | 49324.8 | 773.28 | 36894.8 | .016074 | 246.74 |
| 1446 | | 47502.6 | 666.17 | 35479.9 | .018047 | 245.50 | 1496 | | 49362.0 | 775.57 | 36923.6 | .016038 | 246.77 |
| 1447 | | 47539.6 | 668.24 | 35508.6 | .018004 | 245.53 | 1497 | | 49399.3 | 777.91 | 36952.7 | .016000 | 246.79 |
| 1448 | | 47576.8 | 670.26 | 35537.5 | .017962 | 245.55 | 1498 | | 49436.6 | 780.26 | 36981.7 | .015963 | 246.82 |
| 1449 | | 47613.8 | 672.34 | 35566.2 | .017919 | 245.58 | 1499 | | 49473.8 | 782.56 | 37010.5 | .015926 | 246.84 |

# Table 5 Products—200% Theoretical Air (for One Gram-Mole)

| T | t | $\bar{h}$ | $p_r$ | $\bar{u}$ | $v_r$ | $\bar{\phi}$ | T | t | $\bar{h}$ | $p_r$ | $\bar{u}$ | $v_r$ | $\bar{\phi}$ |
|---|---|---|---|---|---|---|---|---|---|---|---|---|---|
| 1500 | 1226.85 | 49511.2 | 784.93 | 37039.6 | .015889 | 246.87 | 1550 | 1276.85 | 51380.4 | 909.6 | 38493.0 | .014168 | 248.09 |
| 1501 | | 49548.5 | 787.29 | 37068.6 | .015852 | 246.89 | 1551 | | 51417.9 | 912.2 | 38522.2 | .014137 | 248.12 |
| 1502 | | 49585.7 | 789.61 | 37097.5 | .015816 | 246.92 | 1552 | | 51455.4 | 914.9 | 38551.4 | .014105 | 248.14 |
| 1503 | | 49623.1 | 791.99 | 37126.5 | .015779 | 246.94 | 1553 | | 51492.9 | 917.5 | 38580.6 | .014073 | 248.16 |
| 1504 | | 49660.4 | 794.38 | 37155.5 | .015742 | 246.97 | 1554 | | 51530.2 | 920.2 | 38609.6 | .014041 | 248.19 |
| 1505 | 1231.85 | 49697.8 | 796.71 | 37184.6 | .015706 | 246.99 | 1555 | 1281.85 | 51567.7 | 922.9 | 38638.8 | .014009 | 248.21 |
| 1506 | | 49735.0 | 799.11 | 37213.5 | .015669 | 247.02 | 1556 | | 51605.2 | 925.5 | 38668.0 | .013978 | 248.24 |
| 1507 | | 49772.3 | 801.51 | 37242.5 | .015633 | 247.04 | 1557 | | 51642.8 | 928.2 | 38697.2 | .013946 | 248.26 |
| 1508 | | 49809.7 | 803.87 | 37271.6 | .015597 | 247.06 | 1558 | | 51680.3 | 930.9 | 38726.4 | .013915 | 248.28 |
| 1509 | | 49846.9 | 806.28 | 37300.5 | .015561 | 247.09 | 1559 | | 51717.8 | 933.6 | 38755.7 | .013884 | 248.31 |
| 1510 | 1236.85 | 49884.3 | 808.70 | 37329.5 | .015525 | 247.11 | 1560 | 1286.85 | 51755.2 | 936.3 | 38784.7 | .013853 | 248.33 |
| 1511 | | 49921.7 | 811.08 | 37358.6 | .015489 | 247.14 | 1561 | | 51792.7 | 939.0 | 38813.9 | .013821 | 248.36 |
| 1512 | | 49959.0 | 813.51 | 37387.7 | .015453 | 247.16 | 1562 | | 51830.2 | 941.7 | 38843.1 | .013791 | 248.38 |
| 1513 | | 49996.3 | 815.95 | 37416.6 | .015417 | 247.19 | 1563 | | 51867.8 | 944.5 | 38872.4 | .013759 | 248.41 |
| 1514 | | 50033.7 | 818.34 | 37445.7 | .015382 | 247.21 | 1564 | | 51905.3 | 947.2 | 38901.6 | .013729 | 248.43 |
| 1515 | 1241.85 | 50071.0 | 820.79 | 37474.7 | .015347 | 247.24 | 1565 | 1291.85 | 51942.9 | 949.9 | 38930.8 | .013698 | 248.45 |
| 1516 | | 50108.4 | 823.24 | 37503.8 | .015311 | 247.26 | 1566 | | 51980.4 | 952.7 | 38960.0 | .013667 | 248.48 |
| 1517 | | 50145.7 | 825.65 | 37532.7 | .015276 | 247.29 | 1567 | | 52017.8 | 955.4 | 38989.1 | .013636 | 248.50 |
| 1518 | | 50183.1 | 828.12 | 37561.8 | .015241 | 247.31 | 1568 | | 52055.4 | 958.2 | 39018.4 | .013606 | 248.52 |
| 1519 | | 50220.5 | 830.59 | 37590.9 | .015206 | 247.34 | 1569 | | 52092.9 | 961.0 | 39047.6 | .013575 | 248.55 |
| 1520 | 1246.85 | 50257.9 | 833.02 | 37620.0 | .015171 | 247.36 | 1570 | 1296.85 | 52130.5 | 963.7 | 39076.8 | .013546 | 248.57 |
| 1521 | | 50295.1 | 835.51 | 37648.9 | .015136 | 247.39 | 1571 | | 52168.0 | 966.5 | 39106.1 | .013515 | 248.60 |
| 1522 | | 50332.5 | 837.94 | 37678.0 | .015102 | 247.41 | 1572 | | 52205.6 | 969.2 | 39135.3 | .013485 | 248.62 |
| 1523 | | 50370.0 | 840.44 | 37707.1 | .015067 | 247.43 | 1573 | | 52243.2 | 972.1 | 39164.6 | .013454 | 248.64 |
| 1524 | | 50407.4 | 842.95 | 37736.2 | .015032 | 247.46 | 1574 | | 52280.6 | 974.8 | 39193.7 | .013425 | 248.67 |
| 1525 | 1251.85 | 50444.6 | 845.40 | 37765.2 | .014998 | 247.48 | 1575 | 1301.85 | 52318.1 | 977.7 | 39222.9 | .013394 | 248.69 |
| 1526 | | 50482.1 | 847.92 | 37794.3 | .014963 | 247.51 | 1576 | | 52355.7 | 980.4 | 39252.2 | .013365 | 248.72 |
| 1527 | | 50519.5 | 850.39 | 37823.4 | .014930 | 247.53 | 1577 | | 52393.3 | 983.3 | 39281.5 | .013335 | 248.74 |
| 1528 | | 50556.9 | 852.92 | 37852.5 | .014895 | 247.56 | 1578 | | 52430.9 | 986.1 | 39310.7 | .013305 | 248.76 |
| 1529 | | 50594.2 | 855.45 | 37881.5 | .014861 | 247.58 | 1579 | | 52468.5 | 988.9 | 39340.0 | .013275 | 248.79 |
| 1530 | 1256.85 | 50631.6 | 857.94 | 37910.6 | .014828 | 247.61 | 1580 | 1306.85 | 52506.0 | 991.7 | 39369.3 | .013246 | 248.81 |
| 1531 | | 50669.1 | 860.48 | 37939.7 | .014793 | 247.63 | 1581 | | 52543.6 | 994.6 | 39398.5 | .013216 | 248.84 |
| 1532 | | 50706.5 | 862.98 | 37968.8 | .014760 | 247.65 | 1582 | | 52581.2 | 997.4 | 39427.8 | .013187 | 248.86 |
| 1533 | | 50743.8 | 865.54 | 37997.8 | .014726 | 247.68 | 1583 | | 52618.7 | 1000.2 | 39457.0 | .013159 | 248.88 |
| 1534 | | 50781.2 | 868.11 | 38026.9 | .014692 | 247.70 | 1584 | | 52656.3 | 1003.1 | 39486.2 | .013129 | 248.91 |
| 1535 | 1261.85 | 50818.7 | 870.63 | 38056.1 | .014659 | 247.73 | 1585 | 1311.85 | 52693.9 | 1006.0 | 39515.5 | .013100 | 248.93 |
| 1536 | | 50856.1 | 873.20 | 38085.2 | .014625 | 247.75 | 1586 | | 52731.5 | 1008.9 | 39544.8 | .013071 | 248.95 |
| 1537 | | 50893.6 | 875.73 | 38114.4 | .014593 | 247.78 | 1587 | | 52769.1 | 1011.7 | 39574.1 | .013042 | 248.98 |
| 1538 | | 50930.9 | 878.32 | 38143.4 | .014559 | 247.80 | 1588 | | 52806.7 | 1014.7 | 39603.4 | .013013 | 249.00 |
| 1539 | | 50968.4 | 880.92 | 38172.5 | .014526 | 247.83 | 1589 | | 52844.3 | 1017.5 | 39632.7 | .012984 | 249.02 |
| 1540 | 1266.85 | 51005.8 | 883.47 | 38201.7 | .014493 | 247.85 | 1590 | 1316.85 | 52881.9 | 1020.4 | 39662.0 | .012955 | 249.05 |
| 1541 | | 51043.3 | 886.08 | 38230.8 | .014460 | 247.87 | 1591 | | 52919.5 | 1023.3 | 39691.3 | .012927 | 249.07 |
| 1542 | | 51080.8 | 888.64 | 38260.0 | .014427 | 247.90 | 1592 | | 52957.1 | 1026.3 | 39720.6 | .012898 | 249.10 |
| 1543 | | 51118.1 | 891.26 | 38289.0 | .014394 | 247.92 | 1593 | | 52994.8 | 1029.2 | 39749.9 | .012870 | 249.12 |
| 1544 | | 51155.6 | 893.83 | 38318.1 | .014362 | 247.95 | 1594 | | 53032.2 | 1032.1 | 39779.1 | .012841 | 249.14 |
| 1545 | 1271.85 | 51193.1 | 896.47 | 38347.3 | .014329 | 247.97 | 1595 | 1321.85 | 53069.9 | 1035.0 | 39808.4 | .012813 | 249.17 |
| 1546 | | 51230.5 | 899.05 | 38376.5 | .014297 | 248.00 | 1596 | | 53107.5 | 1037.9 | 39837.7 | .012785 | 249.19 |
| 1547 | | 51268.0 | 901.70 | 38405.7 | .014265 | 248.02 | 1597 | | 53145.1 | 1040.9 | 39867.0 | .012756 | 249.21 |
| 1548 | | 51305.4 | 904.36 | 38434.7 | .014232 | 248.04 | 1598 | | 53182.8 | 1043.8 | 39896.4 | .012728 | 249.24 |
| 1549 | | 51342.9 | 906.96 | 38463.9 | .014200 | 248.07 | 1599 | | 53220.4 | 1046.8 | 39925.7 | .012700 | 249.26 |

## Table 5 Products—200% Theoretical Air (for One Gram-Mole)

| T | t | $\bar{h}$ | $p_r$ | $\bar{u}$ | $v_r$ | $\bar{\phi}$ | T | t | $\bar{h}$ | $p_r$ | $\bar{u}$ | $v_r$ | $\bar{\phi}$ |
|---|---|---|---|---|---|---|---|---|---|---|---|---|---|
| 1600 | 1326.85 | 53258.1 | 1049.8 | 39955.0 | .012672 | 249.28 | 1650 | 1376.85 | 55143.7 | 1207.0 | 41424.9 | .011366 | 250.44 |
| 1601 |  | 53295.7 | 1052.7 | 39984.3 | .012645 | 249.31 | 1651 |  | 55181.5 | 1210.4 | 41454.4 | .011341 | 250.47 |
| 1602 |  | 53333.4 | 1055.7 | 40013.7 | .012616 | 249.33 | 1652 |  | 55219.3 | 1213.7 | 41483.9 | .011317 | 250.49 |
| 1603 |  | 53371.0 | 1058.7 | 40043.0 | .012589 | 249.35 | 1653 |  | 55257.1 | 1217.0 | 41513.4 | .011293 | 250.51 |
| 1604 |  | 53408.7 | 1061.7 | 40072.4 | .012561 | 249.38 | 1654 |  | 55294.9 | 1220.3 | 41542.9 | .011269 | 250.54 |
| 1605 | 1331.85 | 53446.3 | 1064.7 | 40101.7 | .012534 | 249.40 | 1655 | 1381.85 | 55332.7 | 1223.8 | 41572.3 | .011244 | 250.56 |
| 1606 |  | 53484.0 | 1067.8 | 40131.0 | .012506 | 249.43 | 1656 |  | 55370.5 | 1227.1 | 41601.8 | .011221 | 250.58 |
| 1607 |  | 53521.6 | 1070.7 | 40160.4 | .012478 | 249.45 | 1657 |  | 55408.3 | 1230.4 | 41631.3 | .011197 | 250.60 |
| 1608 |  | 53559.3 | 1073.7 | 40189.7 | .012451 | 249.47 | 1658 |  | 55446.2 | 1233.9 | 41661.0 | .011172 | 250.63 |
| 1609 |  | 53596.8 | 1076.8 | 40219.0 | .012424 | 249.50 | 1659 |  | 55484.1 | 1237.2 | 41690.5 | .011149 | 250.65 |
| 1610 | 1336.85 | 53634.5 | 1079.8 | 40248.3 | .012397 | 249.52 | 1660 | 1386.85 | 55521.9 | 1240.6 | 41720.0 | .011125 | 250.67 |
| 1611 |  | 53672.2 | 1082.9 | 40277.7 | .012369 | 249.54 | 1661 |  | 55559.7 | 1244.1 | 41749.5 | .011101 | 250.70 |
| 1612 |  | 53709.9 | 1085.9 | 40307.0 | .012342 | 249.57 | 1662 |  | 55597.5 | 1247.5 | 41778.9 | .011077 | 250.72 |
| 1613 |  | 53747.5 | 1088.9 | 40336.4 | .012316 | 249.59 | 1663 |  | 55635.3 | 1250.9 | 41808.4 | .011054 | 250.74 |
| 1614 |  | 53785.2 | 1092.1 | 40365.8 | .012288 | 249.61 | 1664 |  | 55673.1 | 1254.3 | 41838.0 | .011031 | 250.76 |
| 1615 | 1341.85 | 53822.9 | 1095.1 | 40395.1 | .012262 | 249.64 | 1665 | 1391.85 | 55711.0 | 1257.8 | 41867.5 | .011006 | 250.79 |
| 1616 |  | 53860.6 | 1098.2 | 40424.5 | .012234 | 249.66 | 1666 |  | 55748.8 | 1261.2 | 41897.0 | .010983 | 250.81 |
| 1617 |  | 53898.3 | 1101.3 | 40453.9 | .012208 | 249.68 | 1667 |  | 55786.6 | 1264.6 | 41926.5 | .010960 | 250.83 |
| 1618 |  | 53936.0 | 1104.3 | 40483.3 | .012182 | 249.71 | 1668 |  | 55824.4 | 1268.0 | 41956.0 | .010937 | 250.85 |
| 1619 |  | 53973.7 | 1107.5 | 40512.7 | .012155 | 249.73 | 1669 |  | 55862.3 | 1271.6 | 41985.5 | .010913 | 250.88 |
| 1620 | 1346.85 | 54011.4 | 1110.6 | 40542.1 | .012129 | 249.75 | 1670 | 1396.85 | 55900.1 | 1275.0 | 42015.0 | .010890 | 250.90 |
| 1621 |  | 54049.1 | 1113.6 | 40571.4 | .012102 | 249.77 | 1671 |  | 55938.0 | 1278.5 | 42044.6 | .010867 | 250.92 |
| 1622 |  | 54086.8 | 1116.8 | 40600.8 | .012076 | 249.80 | 1672 |  | 55975.8 | 1282.0 | 42074.1 | .010844 | 250.95 |
| 1623 |  | 54124.5 | 1119.9 | 40630.2 | .012050 | 249.82 | 1673 |  | 56013.6 | 1285.5 | 42103.6 | .010821 | 250.97 |
| 1624 |  | 54162.2 | 1123.1 | 40659.6 | .012023 | 249.85 | 1674 |  | 56051.5 | 1289.0 | 42133.2 | .010798 | 250.99 |
| 1625 | 1351.85 | 54199.9 | 1126.2 | 40689.0 | .011997 | 249.87 | 1675 | 1401.85 | 56089.5 | 1292.5 | 42162.9 | .010775 | 251.01 |
| 1626 |  | 54237.7 | 1129.3 | 40718.4 | .011971 | 249.89 | 1676 |  | 56127.4 | 1296.0 | 42192.4 | .010752 | 251.04 |
| 1627 |  | 54275.4 | 1132.5 | 40747.8 | .011945 | 249.91 | 1677 |  | 56165.2 | 1299.5 | 42222.0 | .010729 | 251.06 |
| 1628 |  | 54313.1 | 1135.6 | 40777.2 | .011919 | 249.94 | 1678 |  | 56203.1 | 1303.1 | 42251.5 | .010707 | 251.08 |
| 1629 |  | 54350.8 | 1138.8 | 40806.7 | .011893 | 249.96 | 1679 |  | 56240.9 | 1306.6 | 42281.0 | .010684 | 251.10 |
| 1630 | 1356.85 | 54388.6 | 1142.0 | 40836.1 | .011867 | 249.98 | 1680 | 1406.85 | 56278.8 | 1310.2 | 42310.6 | .010661 | 251.13 |
| 1631 |  | 54426.3 | 1145.2 | 40865.5 | .011842 | 250.01 | 1681 |  | 56316.7 | 1313.7 | 42340.2 | .010639 | 251.15 |
| 1632 |  | 54464.0 | 1148.3 | 40894.9 | .011816 | 250.03 | 1682 |  | 56354.5 | 1317.3 | 42369.7 | .010616 | 251.17 |
| 1633 |  | 54501.8 | 1151.6 | 40924.3 | .011790 | 250.05 | 1683 |  | 56392.4 | 1320.8 | 42399.3 | .010594 | 251.19 |
| 1634 |  | 54539.5 | 1154.8 | 40953.8 | .011765 | 250.08 | 1684 |  | 56430.3 | 1324.5 | 42428.8 | .010571 | 251.22 |
| 1635 | 1361.85 | 54577.2 | 1157.9 | 40983.2 | .011740 | 250.10 | 1685 | 1411.85 | 56468.2 | 1328.0 | 42458.4 | .010549 | 251.24 |
| 1636 |  | 54615.0 | 1161.2 | 41012.6 | .011714 | 250.12 | 1686 |  | 56506.2 | 1331.6 | 42488.1 | .010527 | 251.26 |
| 1637 |  | 54652.7 | 1164.4 | 41042.1 | .011689 | 250.15 | 1687 |  | 56544.1 | 1335.2 | 42517.7 | .010505 | 251.28 |
| 1638 |  | 54690.5 | 1167.6 | 41071.5 | .011664 | 250.17 | 1688 |  | 56582.0 | 1338.8 | 42547.2 | .010483 | 251.31 |
| 1639 |  | 54728.2 | 1170.9 | 41100.9 | .011638 | 250.19 | 1689 |  | 56619.8 | 1342.5 | 42576.8 | .010460 | 251.33 |
| 1640 | 1366.85 | 54766.0 | 1174.1 | 41130.4 | .011613 | 250.21 | 1690 | 1416.85 | 56657.7 | 1346.1 | 42606.4 | .010439 | 251.35 |
| 1641 |  | 54803.8 | 1177.4 | 41159.8 | .011589 | 250.24 | 1691 |  | 56695.6 | 1349.7 | 42636.0 | .010417 | 251.37 |
| 1642 |  | 54841.5 | 1180.7 | 41189.3 | .011563 | 250.26 | 1692 |  | 56733.5 | 1353.3 | 42665.5 | .010395 | 251.40 |
| 1643 |  | 54879.3 | 1183.9 | 41218.7 | .011538 | 250.28 | 1693 |  | 56771.4 | 1357.1 | 42695.1 | .010373 | 251.42 |
| 1644 |  | 54917.0 | 1187.2 | 41248.2 | .011514 | 250.31 | 1694 |  | 56809.3 | 1360.7 | 42724.7 | .010351 | 251.44 |
| 1645 | 1371.85 | 54954.8 | 1190.5 | 41277.6 | .011489 | 250.33 | 1695 | 1421.85 | 56847.2 | 1364.3 | 42754.3 | .010329 | 251.46 |
| 1646 |  | 54992.6 | 1193.8 | 41307.1 | .011464 | 250.35 | 1696 |  | 56885.3 | 1368.0 | 42784.1 | .010308 | 251.49 |
| 1647 |  | 55030.4 | 1197.0 | 41336.6 | .011440 | 250.38 | 1697 |  | 56923.2 | 1371.7 | 42813.6 | .010286 | 251.51 |
| 1648 |  | 55068.1 | 1200.4 | 41366.0 | .011415 | 250.40 | 1698 |  | 56961.1 | 1375.4 | 42843.2 | .010264 | 251.53 |
| 1649 |  | 55105.9 | 1203.7 | 41395.5 | .011390 | 250.42 | 1699 |  | 56999.0 | 1379.1 | 42872.8 | .010243 | 251.55 |

# Table 5    Products—200% Theoretical Air (for One Gram-Mole)

| T | t | $\bar{h}$ | $p_r$ | $\bar{u}$ | $v_r$ | $\bar{\phi}$ | T | t | $\bar{h}$ | $p_r$ | $\bar{u}$ | $v_r$ | $\bar{\phi}$ |
|---|---|---|---|---|---|---|---|---|---|---|---|---|---|
| 1700 | 1426.85 | 57036.9 | 1382.8 | 42902.5 | .0102217 | 251.57 | 1750 | 1476.85 | 58937.2 | 1578.7 | 44387.0 | .0092166 | 252.68 |
| 1701 | | 57074.8 | 1386.5 | 42932.0 | .0102004 | 251.60 | 1751 | | 58975.2 | 1582.8 | 44416.7 | .0091978 | 252.70 |
| 1702 | | 57112.8 | 1390.2 | 42961.7 | .0101792 | 251.62 | 1752 | | 59013.3 | 1587.0 | 44446.5 | .0091791 | 252.72 |
| 1703 | | 57150.7 | 1394.0 | 42991.3 | .0101574 | 251.64 | 1753 | | 59051.5 | 1591.1 | 44476.4 | .0091604 | 252.74 |
| 1704 | | 57188.6 | 1397.7 | 43020.9 | .0101363 | 251.66 | 1754 | | 59089.5 | 1595.3 | 44506.1 | .0091418 | 252.76 |
| 1705 | 1431.85 | 57226.7 | 1401.5 | 43050.6 | .0101152 | 251.69 | 1755 | 1481.85 | 59127.6 | 1599.4 | 44535.8 | .0091232 | 252.78 |
| 1706 | | 57264.6 | 1405.2 | 43080.3 | .0100942 | 251.71 | 1756 | | 59165.7 | 1603.6 | 44565.6 | .0091047 | 252.81 |
| 1707 | | 57302.6 | 1408.9 | 43109.9 | .0100733 | 251.73 | 1757 | | 59203.7 | 1607.8 | 44595.3 | .0090862 | 252.83 |
| 1708 | | 57340.5 | 1412.8 | 43139.5 | .0100517 | 251.75 | 1758 | | 59241.9 | 1612.1 | 44625.2 | .0090671 | 252.85 |
| 1709 | | 57378.4 | 1416.6 | 43169.1 | .0100309 | 251.78 | 1759 | | 59280.0 | 1616.3 | 44655.0 | .0090487 | 252.87 |
| 1710 | 1436.85 | 57416.4 | 1420.3 | 43198.8 | .0100101 | 251.80 | 1760 | 1486.85 | 59318.1 | 1620.5 | 44684.7 | .0090303 | 252.89 |
| 1711 | | 57454.3 | 1424.1 | 43228.4 | .0099894 | 251.82 | 1761 | | 59356.1 | 1624.7 | 44714.5 | .0090120 | 252.92 |
| 1712 | | 57492.4 | 1427.9 | 43258.2 | .0099687 | 251.84 | 1762 | | 59394.2 | 1628.9 | 44744.2 | .0089937 | 252.94 |
| 1713 | | 57530.4 | 1431.7 | 43287.8 | .0099481 | 251.86 | 1763 | | 59432.4 | 1633.1 | 44774.1 | .0089755 | 252.96 |
| 1714 | | 57568.3 | 1435.6 | 43317.4 | .0099268 | 251.89 | 1764 | | 59470.5 | 1637.4 | 44803.9 | .0089573 | 252.98 |
| 1715 | 1441.85 | 57606.3 | 1439.4 | 43347.1 | .0099063 | 251.91 | 1765 | 1491.85 | 59508.6 | 1641.6 | 44833.7 | .0089392 | 253.00 |
| 1716 | | 57644.3 | 1443.2 | 43376.7 | .0098858 | 251.93 | 1766 | | 59546.7 | 1645.9 | 44863.4 | .0089211 | 253.02 |
| 1717 | | 57682.2 | 1447.1 | 43406.4 | .0098654 | 251.95 | 1767 | | 59584.9 | 1650.2 | 44893.4 | .0089030 | 253.04 |
| 1718 | | 57720.2 | 1450.9 | 43436.0 | .0098450 | 251.97 | 1768 | | 59623.0 | 1654.5 | 44923.1 | .0088850 | 253.07 |
| 1719 | | 57758.3 | 1454.7 | 43465.8 | .0098247 | 252.00 | 1769 | | 59661.1 | 1658.7 | 44952.9 | .0088671 | 253.09 |
| 1720 | 1446.85 | 57796.3 | 1458.7 | 43495.5 | .0098038 | 252.02 | 1770 | 1496.85 | 59699.2 | 1663.0 | 44982.7 | .0088492 | 253.11 |
| 1721 | | 57834.2 | 1462.6 | 43525.1 | .0097836 | 252.04 | 1771 | | 59737.4 | 1667.3 | 45012.6 | .0088313 | 253.13 |
| 1722 | | 57872.2 | 1466.4 | 43554.8 | .0097634 | 252.06 | 1772 | | 59775.5 | 1671.7 | 45042.4 | .0088134 | 253.15 |
| 1723 | | 57910.2 | 1470.3 | 43584.5 | .0097433 | 252.09 | 1773 | | 59813.6 | 1676.0 | 45072.2 | .0087957 | 253.17 |
| 1724 | | 57948.1 | 1474.2 | 43614.1 | .0097232 | 252.11 | 1774 | | 59851.7 | 1680.3 | 45102.0 | .0087779 | 253.20 |
| 1725 | 1451.85 | 57986.1 | 1478.1 | 43643.8 | .0097032 | 252.13 | 1775 | 1501.85 | 59889.8 | 1684.8 | 45131.7 | .0087596 | 253.22 |
| 1726 | | 58024.3 | 1482.0 | 43673.6 | .0096832 | 252.15 | 1776 | | 59928.1 | 1689.1 | 45161.7 | .0087419 | 253.24 |
| 1727 | | 58062.2 | 1486.0 | 43703.3 | .0096626 | 252.17 | 1777 | | 59966.2 | 1693.5 | 45191.5 | .0087243 | 253.26 |
| 1728 | | 58100.2 | 1490.0 | 43732.9 | .0096428 | 252.20 | 1778 | | 60004.3 | 1697.9 | 45221.3 | .0087068 | 253.28 |
| 1729 | | 58138.2 | 1493.9 | 43762.6 | .0096230 | 252.22 | 1779 | | 60042.4 | 1702.3 | 45251.1 | .0086892 | 253.30 |
| 1730 | 1456.85 | 58176.2 | 1497.8 | 43792.3 | .0096032 | 252.24 | 1780 | 1506.85 | 60080.7 | 1706.7 | 45281.0 | .0086717 | 253.32 |
| 1731 | | 58214.2 | 1501.8 | 43822.0 | .0095835 | 252.26 | 1781 | | 60118.8 | 1711.1 | 45310.8 | .0086543 | 253.35 |
| 1732 | | 58252.4 | 1505.7 | 43851.8 | .0095638 | 252.28 | 1782 | | 60156.9 | 1715.5 | 45340.6 | .0086369 | 253.37 |
| 1733 | | 58290.4 | 1509.7 | 43881.5 | .0095442 | 252.30 | 1783 | | 60195.0 | 1719.9 | 45370.4 | .0086195 | 253.39 |
| 1734 | | 58328.4 | 1513.8 | 43911.2 | .0095240 | 252.33 | 1784 | | 60233.3 | 1724.3 | 45400.4 | .0086022 | 253.41 |
| 1735 | 1461.85 | 58366.4 | 1517.8 | 43940.9 | .0095044 | 252.35 | 1785 | 1511.85 | 60271.4 | 1728.8 | 45430.2 | .0085849 | 253.43 |
| 1736 | | 58404.4 | 1521.8 | 43970.6 | .0094849 | 252.37 | 1786 | | 60309.5 | 1733.2 | 45460.0 | .0085677 | 253.45 |
| 1737 | | 58442.5 | 1525.8 | 44000.4 | .0094655 | 252.39 | 1787 | | 60347.7 | 1737.7 | 45489.8 | .0085505 | 253.47 |
| 1738 | | 58480.6 | 1529.8 | 44030.1 | .0094461 | 252.41 | 1788 | | 60386.0 | 1742.1 | 45519.8 | .0085333 | 253.50 |
| 1739 | | 58518.6 | 1533.8 | 44059.8 | .0094268 | 252.44 | 1789 | | 60424.1 | 1746.6 | 45549.6 | .0085162 | 253.52 |
| 1740 | 1466.85 | 58556.6 | 1537.8 | 44089.5 | .0094075 | 252.46 | 1790 | 1516.85 | 60462.2 | 1751.1 | 45579.5 | .0084992 | 253.54 |
| 1741 | | 58594.6 | 1541.9 | 44119.2 | .0093883 | 252.48 | 1791 | | 60500.5 | 1755.6 | 45609.4 | .0084821 | 253.56 |
| 1742 | | 58632.6 | 1545.9 | 44148.9 | .0093691 | 252.50 | 1792 | | 60538.7 | 1760.1 | 45639.3 | .0084651 | 253.58 |
| 1743 | | 58670.8 | 1550.1 | 44178.8 | .0093493 | 252.52 | 1793 | | 60576.8 | 1764.6 | 45669.1 | .0084482 | 253.60 |
| 1744 | | 58708.8 | 1554.1 | 44208.5 | .0093302 | 252.55 | 1794 | | 60615.0 | 1769.1 | 45698.9 | .0084313 | 253.62 |
| 1745 | 1471.85 | 58746.9 | 1558.2 | 44238.2 | .0093112 | 252.57 | 1795 | 1521.85 | 60653.3 | 1773.7 | 45728.9 | .0084144 | 253.64 |
| 1746 | | 58784.9 | 1562.3 | 44268.0 | .0092922 | 252.59 | 1796 | | 60691.4 | 1778.2 | 45758.8 | .0083976 | 253.67 |
| 1747 | | 58822.9 | 1566.4 | 44297.7 | .0092732 | 252.61 | 1797 | | 60729.6 | 1782.8 | 45788.6 | .0083808 | 253.69 |
| 1748 | | 58861.1 | 1570.5 | 44327.5 | .0092543 | 252.63 | 1798 | | 60767.7 | 1787.3 | 45818.4 | .0083641 | 253.71 |
| 1749 | | 58899.2 | 1574.6 | 44357.3 | .0092354 | 252.65 | 1799 | | 60806.0 | 1791.9 | 45848.4 | .0083474 | 253.73 |

## Table 5    Products—200% Theoretical Air (for One Gram-Mole)

| T | t | $\bar{h}$ | $p_r$ | $\bar{u}$ | $v_r$ | $\bar{\phi}$ | T | t | $\bar{h}$ | $p_r$ | $\bar{u}$ | $v_r$ | $\bar{\phi}$ |
|---|---|---|---|---|---|---|---|---|---|---|---|---|---|
| 1800 | 1526.85 | 60844.2 | 1796.5 | 45878.3 | .0083307 | 253.75 | 1850 | 1576.85 | 62757.6 | 2037.9 | 47376.0 | .0075477 | 254.80 |
| 1801 | | 60882.4 | 1801.1 | 45908.1 | .0083141 | 253.77 | 1851 | | 62795.9 | 2043.0 | 47405.9 | .0075329 | 254.82 |
| 1802 | | 60920.7 | 1805.7 | 45938.1 | .0082975 | 253.79 | 1852 | | 62834.3 | 2048.1 | 47436.0 | .0075182 | 254.84 |
| 1803 | | 60958.9 | 1810.3 | 45968.0 | .0082809 | 253.81 | 1853 | | 62872.6 | 2053.3 | 47466.0 | .0075035 | 254.86 |
| 1804 | | 60997.0 | 1814.9 | 45997.8 | .0082644 | 253.84 | 1854 | | 62910.9 | 2058.4 | 47496.0 | .0074888 | 254.88 |
| 1805 | 1531.85 | 61035.2 | 1819.5 | 46027.7 | .0082480 | 253.86 | 1855 | 1581.85 | 62949.3 | 2063.5 | 47526.1 | .0074742 | 254.90 |
| 1806 | | 61073.5 | 1824.2 | 46057.7 | .0082315 | 253.88 | 1856 | | 62987.6 | 2068.7 | 47556.1 | .0074596 | 254.92 |
| 1807 | | 61111.7 | 1828.8 | 46087.6 | .0082152 | 253.90 | 1857 | | 63026.0 | 2073.8 | 47586.2 | .0074450 | 254.94 |
| 1808 | | 61149.9 | 1833.5 | 46117.4 | .0081988 | 253.92 | 1858 | | 63064.3 | 2079.0 | 47616.2 | .0074305 | 254.97 |
| 1809 | | 61188.2 | 1838.2 | 46147.5 | .0081825 | 253.94 | 1859 | | 63102.6 | 2084.1 | 47646.2 | .0074165 | 254.99 |
| 1810 | 1536.85 | 61226.4 | 1842.8 | 46177.3 | .0081662 | 253.96 | 1860 | 1586.85 | 63141.1 | 2089.3 | 47676.3 | .0074020 | 255.01 |
| 1811 | | 61264.6 | 1847.5 | 46207.2 | .0081500 | 253.98 | 1861 | | 63179.4 | 2094.5 | 47706.3 | .0073876 | 255.03 |
| 1812 | | 61302.8 | 1852.2 | 46237.1 | .0081338 | 254.01 | 1862 | | 63217.7 | 2099.7 | 47736.3 | .0073732 | 255.05 |
| 1813 | | 61341.1 | 1856.9 | 46267.1 | .0081176 | 254.03 | 1863 | | 63256.1 | 2104.9 | 47766.4 | .0073589 | 255.07 |
| 1814 | | 61379.3 | 1861.7 | 46297.0 | .0081015 | 254.05 | 1864 | | 63294.5 | 2110.1 | 47796.4 | .0073445 | 255.09 |
| 1815 | 1541.85 | 61417.5 | 1866.4 | 46326.9 | .0080855 | 254.07 | 1865 | 1591.85 | 63332.9 | 2115.4 | 47826.5 | .0073303 | 255.11 |
| 1816 | | 61455.9 | 1871.1 | 46356.9 | .0080694 | 254.09 | 1866 | | 63371.2 | 2120.7 | 47856.5 | .0073160 | 255.13 |
| 1817 | | 61494.1 | 1875.9 | 46386.8 | .0080534 | 254.11 | 1867 | | 63409.5 | 2125.8 | 47886.5 | .0073023 | 255.15 |
| 1818 | | 61532.3 | 1880.6 | 46416.7 | .0080374 | 254.13 | 1868 | | 63448.0 | 2131.1 | 47916.7 | .0072881 | 255.17 |
| 1819 | | 61570.6 | 1885.4 | 46446.7 | .0080215 | 254.15 | 1869 | | 63486.3 | 2136.3 | 47946.7 | .0072739 | 255.19 |
| 1820 | 1546.85 | 61608.8 | 1890.2 | 46476.6 | .0080056 | 254.17 | 1870 | 1596.85 | 63524.6 | 2141.6 | 47976.7 | .0072598 | 255.21 |
| 1821 | | 61647.0 | 1894.9 | 46506.5 | .0079903 | 254.19 | 1871 | | 63563.1 | 2147.0 | 48006.9 | .0072457 | 255.23 |
| 1822 | | 61685.4 | 1899.7 | 46536.6 | .0079745 | 254.22 | 1872 | | 63601.4 | 2152.3 | 48036.9 | .0072316 | 255.25 |
| 1823 | | 61723.6 | 1904.5 | 46566.5 | .0079587 | 254.24 | 1873 | | 63639.9 | 2157.6 | 48067.0 | .0072176 | 255.27 |
| 1824 | | 61761.8 | 1909.3 | 46596.4 | .0079430 | 254.26 | 1874 | | 63678.2 | 2162.8 | 48097.0 | .0072041 | 255.29 |
| 1825 | 1551.85 | 61800.2 | 1914.1 | 46626.4 | .0079273 | 254.28 | 1875 | 1601.85 | 63716.6 | 2168.2 | 48127.1 | .0071902 | 255.31 |
| 1826 | | 61838.4 | 1919.0 | 46656.3 | .0079116 | 254.30 | 1876 | | 63755.0 | 2173.5 | 48157.2 | .0071762 | 255.34 |
| 1827 | | 61876.6 | 1923.8 | 46686.2 | .0078960 | 254.32 | 1877 | | 63793.4 | 2178.9 | 48187.2 | .0071623 | 255.36 |
| 1828 | | 61914.9 | 1928.7 | 46716.1 | .0078804 | 254.34 | 1878 | | 63831.9 | 2184.3 | 48217.4 | .0071485 | 255.38 |
| 1829 | | 61953.2 | 1933.5 | 46746.2 | .0078649 | 254.36 | 1879 | | 63870.2 | 2189.7 | 48247.4 | .0071346 | 255.40 |
| 1830 | 1556.85 | 61991.5 | 1938.4 | 46776.1 | .0078494 | 254.38 | 1880 | 1606.85 | 63908.7 | 2195.1 | 48277.6 | .0071208 | 255.42 |
| 1831 | | 62029.7 | 1943.3 | 46806.0 | .0078339 | 254.40 | 1881 | | 63947.0 | 2200.4 | 48307.6 | .0071075 | 255.44 |
| 1832 | | 62068.1 | 1948.2 | 46836.1 | .0078184 | 254.43 | 1882 | | 63985.4 | 2205.8 | 48337.7 | .0070938 | 255.46 |
| 1833 | | 62106.3 | 1953.1 | 46866.0 | .0078030 | 254.45 | 1883 | | 64023.9 | 2211.3 | 48367.9 | .0070801 | 255.48 |
| 1834 | | 62144.6 | 1958.0 | 46896.0 | .0077877 | 254.47 | 1884 | | 64062.2 | 2216.7 | 48397.9 | .0070664 | 255.50 |
| 1835 | 1561.85 | 62183.0 | 1963.0 | 46926.0 | .0077723 | 254.49 | 1885 | 1611.85 | 64100.7 | 2222.2 | 48428.1 | .0070528 | 255.52 |
| 1836 | | 62221.2 | 1967.9 | 46956.0 | .0077570 | 254.51 | 1886 | | 64139.1 | 2227.7 | 48458.1 | .0070392 | 255.54 |
| 1837 | | 62259.5 | 1972.9 | 46985.9 | .0077418 | 254.53 | 1887 | | 64177.4 | 2233.0 | 48488.1 | .0070261 | 255.56 |
| 1838 | | 62297.9 | 1977.7 | 47016.0 | .0077271 | 254.55 | 1888 | | 64215.9 | 2238.5 | 48518.3 | .0070125 | 255.58 |
| 1839 | | 62336.1 | 1982.7 | 47045.9 | .0077119 | 254.57 | 1889 | | 64254.3 | 2244.0 | 48548.4 | .0069990 | 255.60 |
| 1840 | 1566.85 | 62374.5 | 1987.7 | 47076.0 | .0076967 | 254.59 | 1890 | 1616.85 | 64292.8 | 2249.6 | 48578.6 | .0069855 | 255.62 |
| 1841 | | 62412.8 | 1992.7 | 47105.9 | .0076816 | 254.61 | 1891 | | 64331.2 | 2255.1 | 48608.6 | .0069720 | 255.64 |
| 1842 | | 62451.0 | 1997.7 | 47135.9 | .0076665 | 254.63 | 1892 | | 64369.7 | 2260.5 | 48638.8 | .0069591 | 255.66 |
| 1843 | | 62489.4 | 2002.7 | 47166.0 | .0076515 | 254.65 | 1893 | | 64408.0 | 2266.0 | 48668.9 | .0069457 | 255.68 |
| 1844 | | 62527.7 | 2007.7 | 47195.9 | .0076365 | 254.68 | 1894 | | 64446.4 | 2271.6 | 48698.9 | .0069323 | 255.70 |
| 1845 | 1571.85 | 62566.0 | 2012.7 | 47225.9 | .0076215 | 254.70 | 1895 | 1621.85 | 64484.9 | 2277.2 | 48729.1 | .0069190 | 255.72 |
| 1846 | | 62604.4 | 2017.8 | 47256.0 | .0076066 | 254.72 | 1896 | | 64523.3 | 2282.8 | 48759.2 | .0069057 | 255.74 |
| 1847 | | 62642.6 | 2022.8 | 47285.9 | .0075917 | 254.74 | 1897 | | 64561.8 | 2288.4 | 48789.4 | .0068924 | 255.76 |
| 1848 | | 62680.9 | 2027.9 | 47315.9 | .0075768 | 254.76 | 1898 | | 64600.2 | 2293.8 | 48819.5 | .0068796 | 255.78 |
| 1849 | | 62719.3 | 2032.8 | 47346.0 | .0075625 | 254.78 | 1899 | | 64638.8 | 2299.5 | 48849.7 | .0068664 | 255.80 |

# Table 5    Products—200% Theoretical Air (for One Gram-Mole)

| T | t | $\bar{h}$ | $p_r$ | $\bar{u}$ | $v_r$ | $\bar{\phi}$ | T | t | $\bar{h}$ | $p_r$ | $\bar{u}$ | $v_r$ | $\bar{\phi}$ |
|---|---|---|---|---|---|---|---|---|---|---|---|---|---|
| 1900 | 1626.85 | 64677.1 | 2305.1 | 48879.8 | .0068533 | 255.82 | 1950 | 1676.85 | 66602.5 | 2599.8 | 50389.4 | .0062363 | 256.82 |
| 1901 | | 64715.5 | 2310.7 | 48909.8 | .0068401 | 255.84 | 1951 | | 66641.0 | 2605.9 | 50419.6 | .0062250 | 256.84 |
| 1902 | | 64754.1 | 2316.4 | 48940.1 | .0068270 | 255.86 | 1952 | | 66679.7 | 2612.1 | 50449.9 | .0062133 | 256.86 |
| 1903 | | 64792.5 | 2321.9 | 48970.1 | .0068144 | 255.88 | 1953 | | 66718.1 | 2618.4 | 50480.1 | .0062016 | 256.88 |
| 1904 | | 64831.0 | 2327.6 | 49000.4 | .0068013 | 255.90 | 1954 | | 66756.8 | 2624.5 | 50510.4 | .0061903 | 256.90 |
| 1905 | 1631.85 | 64869.4 | 2333.3 | 49030.4 | .0067883 | 255.92 | 1955 | 1681.85 | 66795.4 | 2630.8 | 50540.8 | .0061787 | 256.92 |
| 1906 | | 64907.9 | 2339.0 | 49060.7 | .0067753 | 255.94 | 1956 | | 66833.9 | 2637.1 | 50571.0 | .0061671 | 256.94 |
| 1907 | | 64946.3 | 2344.7 | 49090.8 | .0067623 | 255.97 | 1957 | | 66872.6 | 2643.2 | 50601.3 | .0061559 | 256.96 |
| 1908 | | 64984.9 | 2350.3 | 49121.0 | .0067498 | 255.98 | 1958 | | 66911.1 | 2649.5 | 50631.5 | .0061444 | 256.98 |
| 1909 | | 65023.3 | 2356.0 | 49151.1 | .0067369 | 256.01 | 1959 | | 66949.7 | 2655.9 | 50661.8 | .0061328 | 257.00 |
| 1910 | 1636.85 | 65061.7 | 2361.8 | 49181.2 | .0067240 | 256.03 | 1960 | 1686.85 | 66988.2 | 2662.2 | 50692.0 | .0061213 | 257.02 |
| 1911 | | 65100.3 | 2367.5 | 49211.4 | .0067112 | 256.05 | 1961 | | 67026.9 | 2668.4 | 50722.4 | .0061102 | 257.04 |
| 1912 | | 65138.6 | 2373.3 | 49241.5 | .0066983 | 256.07 | 1962 | | 67065.4 | 2674.8 | 50752.6 | .0060988 | 257.06 |
| 1913 | | 65177.2 | 2378.9 | 49271.8 | .0066860 | 256.09 | 1963 | | 67104.1 | 2681.2 | 50782.9 | .0060873 | 257.08 |
| 1914 | | 65215.6 | 2384.7 | 49301.9 | .0066732 | 256.11 | 1964 | | 67142.6 | 2687.4 | 50813.1 | .0060764 | 257.10 |
| 1915 | 1641.85 | 65254.2 | 2390.5 | 49332.1 | .0066605 | 256.13 | 1965 | 1691.85 | 67181.3 | 2693.8 | 50843.5 | .0060650 | 257.12 |
| 1916 | | 65292.6 | 2396.3 | 49362.2 | .0066478 | 256.15 | 1966 | | 67219.8 | 2700.2 | 50873.6 | .0060536 | 257.14 |
| 1917 | | 65331.2 | 2402.0 | 49392.5 | .0066356 | 256.17 | 1967 | | 67258.4 | 2706.5 | 50904.0 | .0060427 | 257.16 |
| 1918 | | 65369.6 | 2407.9 | 49422.6 | .0066229 | 256.19 | 1968 | | 67297.0 | 2712.9 | 50934.2 | .0060314 | 257.18 |
| 1919 | | 65408.2 | 2413.7 | 49452.8 | .0066103 | 256.21 | 1969 | | 67335.6 | 2719.4 | 50964.6 | .0060201 | 257.20 |
| 1920 | 1646.85 | 65446.6 | 2419.6 | 49482.9 | .0065977 | 256.23 | 1970 | 1696.85 | 67374.2 | 2725.7 | 50994.8 | .0060093 | 257.22 |
| 1921 | | 65485.2 | 2425.3 | 49513.2 | .0065856 | 256.25 | 1971 | | 67412.8 | 2732.2 | 51025.1 | .0059980 | 257.24 |
| 1922 | | 65523.6 | 2431.2 | 49543.3 | .0065730 | 256.27 | 1972 | | 67451.4 | 2738.7 | 51055.4 | .0059868 | 257.26 |
| 1923 | | 65562.2 | 2437.1 | 49573.6 | .0065605 | 256.29 | 1973 | | 67490.0 | 2745.0 | 51085.7 | .0059760 | 257.28 |
| 1924 | | 65600.6 | 2443.0 | 49603.7 | .0065480 | 256.31 | 1974 | | 67528.7 | 2751.5 | 51116.1 | .0059649 | 257.30 |
| 1925 | 1651.85 | 65639.0 | 2448.9 | 49633.8 | .0065356 | 256.33 | 1975 | 1701.85 | 67567.3 | 2758.1 | 51146.3 | .0059538 | 257.32 |
| 1926 | | 65677.6 | 2454.7 | 49664.1 | .0065236 | 256.35 | 1976 | | 67605.9 | 2764.5 | 51176.7 | .0059430 | 257.33 |
| 1927 | | 65716.1 | 2460.7 | 49694.2 | .0065112 | 256.37 | 1977 | | 67644.5 | 2771.0 | 51206.9 | .0059320 | 257.35 |
| 1928 | | 65754.7 | 2466.6 | 49724.5 | .0064989 | 256.39 | 1978 | | 67683.2 | 2777.6 | 51237.3 | .0059209 | 257.37 |
| 1929 | | 65793.1 | 2472.6 | 49754.6 | .0064865 | 256.41 | 1979 | | 67721.7 | 2784.0 | 51267.5 | .0059103 | 257.39 |
| 1930 | 1656.85 | 65831.7 | 2478.4 | 49784.9 | .0064746 | 256.43 | 1980 | 1706.85 | 67760.4 | 2790.6 | 51297.9 | .0058993 | 257.41 |
| 1931 | | 65870.1 | 2484.4 | 49815.0 | .0064624 | 256.45 | 1981 | | 67799.0 | 2797.2 | 51328.1 | .0058883 | 257.43 |
| 1932 | | 65908.8 | 2490.4 | 49845.3 | .0064501 | 256.47 | 1982 | | 67837.7 | 2803.7 | 51358.5 | .0058777 | 257.45 |
| 1933 | | 65947.2 | 2496.4 | 49875.5 | .0064379 | 256.49 | 1983 | | 67876.2 | 2810.3 | 51388.7 | .0058668 | 257.47 |
| 1934 | | 65985.8 | 2502.3 | 49905.7 | .0064261 | 256.51 | 1984 | | 67914.9 | 2817.0 | 51419.1 | .0058559 | 257.49 |
| 1935 | 1661.85 | 66024.3 | 2508.3 | 49935.9 | .0064139 | 256.53 | 1985 | 1711.85 | 67953.6 | 2823.4 | 51449.5 | .0058454 | 257.51 |
| 1936 | | 66062.9 | 2514.4 | 49966.2 | .0064018 | 256.55 | 1986 | | 67992.2 | 2830.1 | 51479.8 | .0058345 | 257.53 |
| 1937 | | 66101.3 | 2520.3 | 49996.3 | .0063901 | 256.57 | 1987 | | 68030.9 | 2836.8 | 51510.2 | .0058237 | 257.55 |
| 1938 | | 66139.9 | 2526.4 | 50026.6 | .0063780 | 256.59 | 1988 | | 68069.4 | 2843.3 | 51540.4 | .0058132 | 257.57 |
| 1939 | | 66178.4 | 2532.5 | 50056.8 | .0063660 | 256.61 | 1989 | | 68108.2 | 2850.1 | 51570.8 | .0058025 | 257.59 |
| 1940 | 1666.85 | 66217.0 | 2538.6 | 50087.1 | .0063539 | 256.63 | 1990 | 1716.85 | 68146.7 | 2856.8 | 51601.1 | .0057917 | 257.61 |
| 1941 | | 66255.5 | 2544.5 | 50117.2 | .0063423 | 256.65 | 1991 | | 68185.4 | 2863.4 | 51631.5 | .0057813 | 257.63 |
| 1942 | | 66294.1 | 2550.7 | 50147.5 | .0063304 | 256.67 | 1992 | | 68224.0 | 2870.1 | 51661.7 | .0057706 | 257.65 |
| 1943 | | 66332.6 | 2556.8 | 50177.7 | .0063184 | 256.69 | 1993 | | 68262.7 | 2876.9 | 51692.1 | .0057599 | 257.67 |
| 1944 | | 66371.2 | 2563.0 | 50208.0 | .0063065 | 256.71 | 1994 | | 68301.5 | 2883.5 | 51722.5 | .0057496 | 257.69 |
| 1945 | 1671.85 | 66409.7 | 2568.9 | 50238.2 | .0062950 | 256.72 | 1995 | 1721.85 | 68340.0 | 2890.3 | 51752.8 | .0057390 | 257.70 |
| 1946 | | 66448.3 | 2575.1 | 50268.5 | .0062831 | 256.74 | 1996 | | 68378.7 | 2896.9 | 51783.2 | .0057287 | 257.72 |
| 1947 | | 66486.8 | 2581.3 | 50298.6 | .0062713 | 256.76 | 1997 | | 68417.3 | 2903.7 | 51813.5 | .0057181 | 257.74 |
| 1948 | | 66525.4 | 2587.3 | 50329.0 | .0062599 | 256.78 | 1998 | | 68456.1 | 2910.6 | 51843.9 | .0057075 | 257.76 |
| 1949 | | 66563.9 | 2593.6 | 50359.1 | .0062481 | 256.80 | 1999 | | 68494.6 | 2917.2 | 51874.1 | .0056974 | 257.78 |

# Table 6  Products—200% Theoretical Air (for One Gram-Mole)

| T | t | Fuel—$(CH_1)_n$ | | | Fuel—$(CH_2)_n$ | | | Fuel—$(CH_3)_n$ | | |
|---|---|---|---|---|---|---|---|---|---|---|
| | | $\bar{c}_p$ $\dfrac{J}{\text{g-mol K}}$ | $\bar{c}_v$ $\dfrac{J}{\text{g-mol K}}$ | $k = \dfrac{\bar{c}_p}{\bar{c}_v}$ | $\bar{c}_p$ $\dfrac{J}{\text{g-mol K}}$ | $\bar{c}_v$ $\dfrac{J}{\text{g-mol K}}$ | $k = \dfrac{\bar{c}_p}{\bar{c}_v}$ | $\bar{c}_p$ $\dfrac{J}{\text{g-mol K}}$ | $\bar{c}_v$ $\dfrac{J}{\text{g-mol K}}$ | $k = \dfrac{\bar{c}_p}{\bar{c}_v}$ |
| K | C | | | | | | | | | |
| 100 | −173.15 | 29.205 | 20.891 | 1.398 | 29.316 | 21.002 | 1.396 | 29.393 | 21.079 | 1.394 |
| 120 | −153.15 | 29.227 | 20.913 | 1.398 | 29.335 | 21.020 | 1.396 | 29.409 | 21.095 | 1.394 |
| 140 | −133.15 | 29.267 | 20.952 | 1.397 | 29.367 | 21.053 | 1.395 | 29.438 | 21.123 | 1.394 |
| 160 | −113.15 | 29.323 | 21.009 | 1.396 | 29.414 | 21.100 | 1.394 | 29.477 | 21.163 | 1.393 |
| 180 | −93.15 | 29.392 | 21.078 | 1.394 | 29.472 | 21.157 | 1.393 | 29.527 | 21.212 | 1.392 |
| 200 | −73.15 | 29.471 | 21.157 | 1.393 | 29.537 | 21.222 | 1.392 | 29.583 | 21.268 | 1.391 |
| 220 | −53.15 | 29.555 | 21.241 | 1.391 | 29.607 | 21.293 | 1.390 | 29.644 | 21.329 | 1.390 |
| 240 | −33.15 | 29.643 | 21.329 | 1.390 | 29.682 | 21.367 | 1.389 | 29.708 | 21.394 | 1.389 |
| 260 | −13.15 | 29.734 | 21.419 | 1.388 | 29.759 | 21.444 | 1.388 | 29.776 | 21.462 | 1.387 |
| 280 | 6.85 | 29.826 | 21.512 | 1.387 | 29.839 | 21.524 | 1.386 | 29.847 | 21.533 | 1.386 |
| 300 | 26.85 | 29.921 | 21.607 | 1.385 | 29.922 | 21.608 | 1.385 | 29.922 | 21.608 | 1.385 |
| 350 | 76.85 | 30.174 | 21.859 | 1.380 | 30.149 | 21.835 | 1.381 | 30.132 | 21.818 | 1.381 |
| 400 | 126.85 | 30.455 | 22.141 | 1.376 | 30.411 | 22.097 | 1.376 | 30.380 | 22.066 | 1.377 |
| 450 | 176.85 | 30.771 | 22.457 | 1.370 | 30.712 | 22.397 | 1.371 | 30.670 | 22.356 | 1.372 |
| 500 | 226.85 | 31.121 | 22.806 | 1.365 | 31.049 | 22.735 | 1.366 | 30.999 | 22.685 | 1.367 |
| 550 | 276.85 | 31.498 | 23.184 | 1.359 | 31.417 | 23.103 | 1.360 | 31.360 | 23.046 | 1.361 |
| 600 | 326.85 | 31.895 | 23.580 | 1.353 | 31.806 | 23.492 | 1.354 | 31.744 | 23.430 | 1.355 |
| 650 | 376.85 | 32.302 | 23.988 | 1.347 | 32.208 | 23.893 | 1.348 | 32.142 | 23.827 | 1.349 |
| 700 | 426.85 | 32.712 | 24.397 | 1.341 | 32.613 | 24.299 | 1.342 | 32.545 | 24.230 | 1.343 |
| 750 | 476.85 | 33.117 | 24.802 | 1.335 | 33.016 | 24.701 | 1.337 | 32.946 | 24.631 | 1.338 |
| 800 | 526.85 | 33.512 | 25.197 | 1.330 | 33.410 | 25.096 | 1.331 | 33.339 | 25.025 | 1.332 |
| 850 | 576.85 | 33.893 | 25.579 | 1.325 | 33.792 | 25.478 | 1.326 | 33.722 | 25.408 | 1.327 |
| 900 | 626.85 | 34.258 | 25.944 | 1.320 | 34.160 | 25.845 | 1.322 | 34.091 | 25.776 | 1.323 |
| 950 | 676.85 | 34.605 | 26.291 | 1.316 | 34.510 | 26.196 | 1.317 | 34.444 | 26.130 | 1.318 |
| 1000 | 726.85 | 34.934 | 26.620 | 1.312 | 34.844 | 26.529 | 1.313 | 34.780 | 26.466 | 1.314 |
| 1050 | 776.85 | 35.245 | 26.931 | 1.309 | 35.160 | 26.845 | 1.310 | 35.100 | 26.786 | 1.310 |
| 1100 | 826.85 | 35.537 | 27.223 | 1.305 | 35.458 | 27.143 | 1.306 | 35.402 | 27.088 | 1.307 |
| 1150 | 876.85 | 35.812 | 27.498 | 1.302 | 35.739 | 27.425 | 1.303 | 35.688 | 27.374 | 1.304 |
| 1200 | 926.85 | 36.070 | 27.755 | 1.300 | 36.004 | 27.690 | 1.300 | 35.958 | 27.644 | 1.301 |
| 1250 | 976.85 | 36.312 | 27.998 | 1.297 | 36.254 | 27.939 | 1.298 | 36.213 | 27.898 | 1.298 |
| 1300 | 1026.85 | 36.540 | 28.225 | 1.295 | 36.489 | 28.174 | 1.295 | 36.453 | 28.138 | 1.295 |
| 1350 | 1076.85 | 36.752 | 28.438 | 1.292 | 36.709 | 28.394 | 1.293 | 36.678 | 28.364 | 1.293 |
| 1400 | 1126.85 | 36.952 | 28.638 | 1.290 | 36.916 | 28.602 | 1.291 | 36.891 | 28.577 | 1.291 |
| 1450 | 1176.85 | 37.140 | 28.826 | 1.288 | 37.112 | 28.798 | 1.289 | 37.092 | 28.778 | 1.289 |
| 1500 | 1226.85 | 37.316 | 29.002 | 1.287 | 37.296 | 28.982 | 1.287 | 37.282 | 28.967 | 1.287 |
| 1550 | 1276.85 | 37.483 | 29.168 | 1.285 | 37.470 | 29.155 | 1.285 | 37.461 | 29.146 | 1.285 |
| 1600 | 1326.85 | 37.640 | 29.326 | 1.284 | 37.634 | 29.320 | 1.284 | 37.630 | 29.316 | 1.284 |
| 1650 | 1376.85 | 37.788 | 29.473 | 1.282 | 37.789 | 29.475 | 1.282 | 37.790 | 29.476 | 1.282 |
| 1700 | 1426.85 | 37.927 | 29.613 | 1.281 | 37.936 | 29.621 | 1.281 | 37.941 | 29.627 | 1.281 |
| 1750 | 1476.85 | 38.059 | 29.745 | 1.280 | 38.074 | 29.760 | 1.279 | 38.085 | 29.771 | 1.279 |
| 1800 | 1526.85 | 38.184 | 29.870 | 1.278 | 38.206 | 29.891 | 1.278 | 38.221 | 29.907 | 1.278 |
| 1900 | 1626.85 | 38.416 | 30.101 | 1.276 | 38.450 | 30.135 | 1.276 | 38.474 | 30.159 | 1.276 |
| 2000 | 1726.85 | 38.624 | 30.310 | 1.274 | 38.670 | 30.355 | 1.274 | 38.702 | 30.387 | 1.274 |
| 2100 | 1826.85 | 38.813 | 30.499 | 1.273 | 38.870 | 30.556 | 1.272 | 38.909 | 30.595 | 1.272 |
| 2200 | 1926.85 | 38.986 | 30.671 | 1.271 | 39.052 | 30.738 | 1.270 | 39.099 | 30.784 | 1.270 |

# Table 7     Products—100% Theoretical Air (for One Gram-Mole)

| T | t | $\bar{h}$ | $p_r$ | $\bar{u}$ | $v_r$ | $\bar{\phi}$ | T | t | $\bar{h}$ | $p_r$ | $\bar{u}$ | $v_r$ | $\bar{\phi}$ |
|---|---|---|---|---|---|---|---|---|---|---|---|---|---|
| 100 | −173.15 | 2946.3 | .02557 | 2114.9 | 32.521 | 160.96 | 150 | −123.15 | 4428.6 | .10850 | 3181.4 | 11.4945 | 172.98 |
| 101 | | 2975.9 | .02649 | 2136.1 | 31.703 | 161.26 | 151 | | 4458.3 | .11111 | 3202.9 | 11.2995 | 173.18 |
| 102 | | 3005.5 | .02743 | 2157.4 | 30.914 | 161.55 | 152 | | 4488.1 | .11376 | 3224.3 | 11.1090 | 173.37 |
| 103 | | 3035.1 | .02840 | 2178.7 | 30.152 | 161.84 | 153 | | 4517.8 | .11646 | 3245.7 | 10.9228 | 173.57 |
| 104 | | 3064.7 | .02940 | 2200.0 | 29.415 | 162.12 | 154 | | 4547.6 | .11921 | 3267.1 | 10.7409 | 173.76 |
| 105 | −168.15 | 3094.2 | .03041 | 2221.2 | 28.704 | 162.41 | 155 | −118.15 | 4577.3 | .12200 | 3288.6 | 10.5632 | 173.96 |
| 106 | | 3123.9 | .03146 | 2242.5 | 28.016 | 162.69 | 156 | | 4607.0 | .12484 | 3310.0 | 10.3895 | 174.15 |
| 107 | | 3153.4 | .03253 | 2263.8 | 27.350 | 162.96 | 157 | | 4636.8 | .12773 | 3331.5 | 10.2195 | 174.34 |
| 108 | | 3183.1 | .03362 | 2285.1 | 26.707 | 163.24 | 158 | | 4666.6 | .13067 | 3352.9 | 10.0534 | 174.53 |
| 109 | | 3212.7 | .03474 | 2306.4 | 26.084 | 163.51 | 159 | | 4696.3 | .13365 | 3374.3 | 9.8912 | 174.71 |
| 110 | −163.15 | 3242.3 | .03589 | 2327.7 | 25.481 | 163.78 | 160 | −113.15 | 4726.1 | .13669 | 3395.8 | 9.7324 | 174.90 |
| 111 | | 3271.9 | .03707 | 2349.0 | 24.897 | 164.05 | 161 | | 4755.9 | .13977 | 3417.3 | 9.5770 | 175.09 |
| 112 | | 3301.5 | .03827 | 2370.3 | 24.332 | 164.32 | 162 | | 4785.7 | .14291 | 3438.8 | 9.4251 | 175.27 |
| 113 | | 3331.1 | .03950 | 2391.6 | 23.784 | 164.58 | 163 | | 4815.5 | .14609 | 3460.2 | 9.2766 | 175.45 |
| 114 | | 3360.7 | .04076 | 2412.9 | 23.253 | 164.84 | 164 | | 4845.2 | .14933 | 3481.7 | 9.1312 | 175.64 |
| 115 | −158.15 | 3390.3 | .04205 | 2434.2 | 22.739 | 165.10 | 165 | −108.15 | 4875.1 | .15262 | 3503.2 | 8.9889 | 175.82 |
| 116 | | 3419.9 | .04337 | 2455.5 | 22.240 | 165.35 | 166 | | 4904.9 | .15596 | 3524.7 | 8.8495 | 176.00 |
| 117 | | 3449.6 | .04471 | 2476.8 | 21.756 | 165.61 | 167 | | 4934.7 | .15936 | 3546.1 | 8.7133 | 176.18 |
| 118 | | 3479.2 | .04609 | 2498.1 | 21.287 | 165.86 | 168 | | 4964.5 | .16281 | 3567.7 | 8.5797 | 176.35 |
| 119 | | 3508.8 | .04750 | 2519.4 | 20.831 | 166.11 | 169 | | 4994.3 | .16631 | 3589.2 | 8.4490 | 176.53 |
| 120 | −153.15 | 3538.4 | .04893 | 2540.7 | 20.390 | 166.36 | 170 | −103.15 | 5024.2 | .16987 | 3610.7 | 8.3210 | 176.71 |
| 121 | | 3568.1 | .05040 | 2562.0 | 19.961 | 166.60 | 171 | | 5054.0 | .17348 | 3632.2 | 8.1956 | 176.88 |
| 122 | | 3597.7 | .05190 | 2583.3 | 19.544 | 166.85 | 172 | | 5083.8 | .17715 | 3653.7 | 8.0728 | 177.06 |
| 123 | | 3627.3 | .05343 | 2604.6 | 19.139 | 167.09 | 173 | | 5113.6 | .18087 | 3675.3 | 7.9527 | 177.23 |
| 124 | | 3656.9 | .05500 | 2625.9 | 18.746 | 167.33 | 174 | | 5143.5 | .18465 | 3696.8 | 7.8348 | 177.40 |
| 125 | −148.15 | 3686.5 | .05660 | 2647.2 | 18.363 | 167.57 | 175 | −98.15 | 5173.4 | .18849 | 3718.3 | 7.7193 | 177.57 |
| 126 | | 3716.2 | .05823 | 2668.6 | 17.992 | 167.80 | 176 | | 5203.2 | .19239 | 3739.9 | 7.6061 | 177.74 |
| 127 | | 3745.9 | .05989 | 2689.9 | 17.631 | 168.04 | 177 | | 5233.1 | .19634 | 3761.4 | 7.4953 | 177.91 |
| 128 | | 3775.5 | .06159 | 2711.2 | 17.280 | 168.27 | 178 | | 5263.0 | .20036 | 3783.0 | 7.3866 | 178.08 |
| 129 | | 3805.1 | .06332 | 2732.6 | 16.938 | 168.50 | 179 | | 5292.9 | .20444 | 3804.6 | 7.2799 | 178.25 |
| 130 | −143.15 | 3834.8 | .06509 | 2753.9 | 16.606 | 168.73 | 180 | −93.15 | 5322.7 | .20857 | 3826.1 | 7.1755 | 178.41 |
| 131 | | 3864.4 | .06689 | 2775.3 | 16.282 | 168.96 | 181 | | 5352.6 | .21277 | 3847.7 | 7.0731 | 178.58 |
| 132 | | 3894.1 | .06873 | 2796.6 | 15.967 | 169.18 | 182 | | 5382.5 | .21702 | 3869.3 | 6.9727 | 178.74 |
| 133 | | 3923.8 | .07061 | 2817.9 | 15.661 | 169.41 | 183 | | 5412.4 | .22134 | 3890.9 | 6.8742 | 178.91 |
| 134 | | 3953.4 | .07252 | 2839.3 | 15.363 | 169.63 | 184 | | 5442.3 | .22572 | 3912.5 | 6.7776 | 179.07 |
| 135 | −138.15 | 3983.1 | .07447 | 2860.6 | 15.072 | 169.85 | 185 | −88.15 | 5472.2 | .23017 | 3934.0 | 6.6827 | 179.23 |
| 136 | | 4012.8 | .07646 | 2882.0 | 14.789 | 170.07 | 186 | | 5502.1 | .23468 | 3955.7 | 6.5898 | 179.39 |
| 137 | | 4042.4 | .07849 | 2903.3 | 14.513 | 170.29 | 187 | | 5532.1 | .23925 | 3977.3 | 6.4986 | 179.55 |
| 138 | | 4072.1 | .08055 | 2924.7 | 14.244 | 170.50 | 188 | | 5562.0 | .24389 | 3998.9 | 6.4090 | 179.71 |
| 139 | | 4101.8 | .08265 | 2946.1 | 13.982 | 170.72 | 189 | | 5592.0 | .24859 | 4020.6 | 6.3213 | 179.87 |
| 140 | −133.15 | 4131.5 | .08480 | 2967.5 | 13.727 | 170.93 | 190 | −83.15 | 5621.9 | .25336 | 4042.2 | 6.2351 | 180.03 |
| 141 | | 4161.2 | .08698 | 2988.8 | 13.478 | 171.14 | 191 | | 5651.8 | .25820 | 4063.8 | 6.1506 | 180.19 |
| 142 | | 4190.9 | .08920 | 3010.2 | 13.236 | 171.35 | 192 | | 5681.8 | .26310 | 4085.4 | 6.0675 | 180.34 |
| 143 | | 4220.6 | .09147 | 3031.6 | 12.999 | 171.56 | 193 | | 5711.8 | .26807 | 4107.1 | 5.9860 | 180.50 |
| 144 | | 4250.3 | .09377 | 3053.0 | 12.768 | 171.77 | 194 | | 5741.7 | .27311 | 4128.8 | 5.9060 | 180.66 |
| 145 | −128.15 | 4280.0 | .09612 | 3074.4 | 12.543 | 171.97 | 195 | −78.15 | 5771.7 | .27823 | 4150.4 | 5.8273 | 180.81 |
| 146 | | 4309.7 | .09851 | 3095.8 | 12.323 | 172.18 | 196 | | 5801.7 | .28340 | 4172.1 | 5.7502 | 180.96 |
| 147 | | 4339.4 | .10094 | 3117.2 | 12.108 | 172.38 | 197 | | 5831.7 | .28865 | 4193.8 | 5.6744 | 181.12 |
| 148 | | 4369.1 | .10342 | 3138.6 | 11.899 | 172.58 | 198 | | 5861.7 | .29398 | 4215.4 | 5.5999 | 181.27 |
| 149 | | 4398.9 | .10594 | 3160.0 | 11.694 | 172.78 | 199 | | 5891.7 | .29937 | 4237.1 | 5.5269 | 181.42 |

# Table 7 Products—100% Theoretical Air (for One Gram-Mole)

| T | t | $\bar{h}$ | $p_r$ | $\bar{u}$ | $v_r$ | $\bar{\phi}$ | T | t | $\bar{h}$ | $p_r$ | $\bar{u}$ | $v_r$ | $\bar{\phi}$ |
|---|---|-----|-----|-----|-----|------|---|---|-----|-----|-----|-----|------|
| 200 | −73.15 | 5921.7 | .30483 | 4258.8 | 5.4551 | 181.57 | 250 | −23.15 | 7430.6 | .6851 | 5352.0 | 3.0342 | 188.30 |
| 201 | | 5951.7 | .31037 | 4280.5 | 5.3845 | 181.72 | 251 | | 7460.9 | .6951 | 5374.0 | 3.0023 | 188.42 |
| 202 | | 5981.7 | .31598 | 4302.2 | 5.3152 | 181.87 | 252 | | 7491.3 | .7053 | 5396.0 | 2.9708 | 188.54 |
| 203 | | 6011.7 | .32166 | 4323.9 | 5.2472 | 182.02 | 253 | | 7521.7 | .7156 | 5418.1 | 2.9397 | 188.66 |
| 204 | | 6041.8 | .32743 | 4345.6 | 5.1801 | 182.16 | 254 | | 7552.0 | .7259 | 5440.2 | 2.9091 | 188.78 |
| 205 | −68.15 | 6071.8 | .33326 | 4367.4 | 5.1144 | 182.31 | 255 | −18.15 | 7582.4 | .7364 | 5462.2 | 2.8790 | 188.90 |
| 206 | | 6101.9 | .33918 | 4389.1 | 5.0498 | 182.46 | 256 | | 7612.8 | .7470 | 5484.3 | 2.8492 | 189.02 |
| 207 | | 6131.9 | .34516 | 4410.8 | 4.9863 | 182.60 | 257 | | 7643.1 | .7578 | 5506.3 | 2.8199 | 189.14 |
| 208 | | 6161.9 | .35123 | 4432.6 | 4.9238 | 182.75 | 258 | | 7673.5 | .7686 | 5528.4 | 2.7910 | 189.26 |
| 209 | | 6192.0 | .35738 | 4454.3 | 4.8624 | 182.89 | 259 | | 7703.9 | .7795 | 5550.5 | 2.7624 | 189.38 |
| 210 | −63.15 | 6222.1 | .36360 | 4476.1 | 4.8020 | 183.03 | 260 | −13.15 | 7734.4 | .7906 | 5572.6 | 2.7343 | 189.49 |
| 211 | | 6252.2 | .36991 | 4497.9 | 4.7427 | 183.18 | 261 | | 7764.8 | .8018 | 5594.7 | 2.7065 | 189.61 |
| 212 | | 6282.3 | .37629 | 4519.6 | 4.6844 | 183.32 | 262 | | 7795.2 | .8131 | 5616.8 | 2.6791 | 189.73 |
| 213 | | 6312.4 | .38275 | 4541.4 | 4.6269 | 183.46 | 263 | | 7825.6 | .8245 | 5638.9 | 2.6522 | 189.84 |
| 214 | | 6342.5 | .38930 | 4563.2 | 4.5705 | 183.60 | 264 | | 7856.1 | .8360 | 5661.1 | 2.6255 | 189.96 |
| 215 | −58.15 | 6372.6 | .39592 | 4585.0 | 4.5150 | 183.74 | 265 | −8.15 | 7886.5 | .8477 | 5683.2 | 2.5992 | 190.07 |
| 216 | | 6402.7 | .40264 | 4606.8 | 4.4604 | 183.88 | 266 | | 7917.0 | .8595 | 5705.4 | 2.5732 | 190.19 |
| 217 | | 6432.8 | .40943 | 4628.6 | 4.4066 | 184.02 | 267 | | 7947.4 | .8714 | 5727.5 | 2.5476 | 190.30 |
| 218 | | 6463.0 | .41631 | 4650.4 | 4.3538 | 184.16 | 268 | | 7977.9 | .8834 | 5749.6 | 2.5224 | 190.42 |
| 219 | | 6493.1 | .42327 | 4672.2 | 4.3019 | 184.30 | 269 | | 8008.4 | .8955 | 5771.8 | 2.4975 | 190.53 |
| 220 | −53.15 | 6523.2 | .43032 | 4694.1 | 4.2507 | 184.44 | 270 | −3.15 | 8038.9 | .9078 | 5794.0 | 2.4729 | 190.64 |
| 221 | | 6553.4 | .43746 | 4715.9 | 4.2004 | 184.57 | 271 | | 8069.4 | .9202 | 5816.2 | 2.4486 | 190.75 |
| 222 | | 6583.5 | .44467 | 4737.7 | 4.1509 | 184.71 | 272 | | 8099.8 | .9327 | 5838.3 | 2.4247 | 190.87 |
| 223 | | 6613.6 | .45198 | 4759.5 | 4.1022 | 184.84 | 273 | | 8130.4 | .9454 | 5860.5 | 2.4010 | 190.98 |
| 224 | | 6643.8 | .45938 | 4781.4 | 4.0542 | 184.98 | 274 | | 8160.8 | .9581 | 5882.7 | 2.3777 | 191.09 |
| 225 | −48.15 | 6674.0 | .46687 | 4803.3 | 4.0070 | 185.11 | 275 | 1.85 | 8191.4 | .9710 | 5904.9 | 2.3547 | 191.20 |
| 226 | | 6704.2 | .47444 | 4825.1 | 3.9606 | 185.25 | 276 | | 8221.9 | .9841 | 5927.1 | 2.3320 | 191.31 |
| 227 | | 6734.4 | .48211 | 4847.0 | 3.9148 | 185.38 | 277 | | 8252.4 | .9972 | 5949.4 | 2.3095 | 191.42 |
| 228 | | 6764.6 | .48986 | 4868.9 | 3.8699 | 185.51 | 278 | | 8283.0 | 1.0105 | 5971.6 | 2.2874 | 191.53 |
| 229 | | 6794.8 | .49771 | 4890.8 | 3.8255 | 185.64 | 279 | | 8313.5 | 1.0239 | 5993.8 | 2.2655 | 191.64 |
| 230 | −43.15 | 6825.0 | .50565 | 4912.7 | 3.7819 | 185.78 | 280 | 6.85 | 8344.1 | 1.0375 | 6016.0 | 2.2440 | 191.75 |
| 231 | | 6855.2 | .51369 | 4934.6 | 3.7389 | 185.91 | 281 | | 8374.6 | 1.0512 | 6038.3 | 2.2226 | 191.86 |
| 232 | | 6885.4 | .52182 | 4956.5 | 3.6966 | 186.04 | 282 | | 8405.2 | 1.0650 | 6060.5 | 2.2016 | 191.97 |
| 233 | | 6915.6 | .53004 | 4978.4 | 3.6549 | 186.17 | 283 | | 8435.8 | 1.0789 | 6082.8 | 2.1809 | 192.08 |
| 234 | | 6945.9 | .53837 | 5000.3 | 3.6138 | 186.30 | 284 | | 8466.3 | 1.0930 | 6105.1 | 2.1604 | 192.19 |
| 235 | −38.15 | 6976.1 | .54677 | 5022.2 | 3.5735 | 186.43 | 285 | 11.85 | 8496.9 | 1.1072 | 6127.3 | 2.1401 | 192.29 |
| 236 | | 7006.4 | .55529 | 5044.2 | 3.5337 | 186.56 | 286 | | 8527.5 | 1.1216 | 6149.6 | 2.1201 | 192.40 |
| 237 | | 7036.6 | .56390 | 5066.1 | 3.4944 | 186.68 | 287 | | 8558.1 | 1.1361 | 6171.9 | 2.1004 | 192.51 |
| 238 | | 7066.9 | .57261 | 5088.0 | 3.4558 | 186.81 | 288 | | 8588.7 | 1.1507 | 6194.2 | 2.0809 | 192.61 |
| 239 | | 7097.1 | .58142 | 5110.0 | 3.4177 | 186.94 | 289 | | 8619.3 | 1.1655 | 6216.5 | 2.0617 | 192.72 |
| 240 | −33.15 | 7127.4 | .59033 | 5132.0 | 3.3803 | 187.06 | 290 | 16.85 | 8649.9 | 1.1804 | 6238.8 | 2.0426 | 192.83 |
| 241 | | 7157.7 | .59934 | 5153.9 | 3.3433 | 187.19 | 291 | | 8680.6 | 1.1955 | 6261.1 | 2.0239 | 192.93 |
| 242 | | 7188.0 | .60845 | 5175.9 | 3.3069 | 187.32 | 292 | | 8711.2 | 1.2107 | 6283.4 | 2.0053 | 193.04 |
| 243 | | 7218.3 | .61765 | 5197.9 | 3.2711 | 187.44 | 293 | | 8741.8 | 1.2260 | 6305.7 | 1.9870 | 193.14 |
| 244 | | 7248.6 | .62698 | 5219.9 | 3.2357 | 187.56 | 294 | | 8772.5 | 1.2415 | 6328.1 | 1.9689 | 193.24 |
| 245 | −28.15 | 7278.9 | .63640 | 5241.9 | 3.2009 | 187.69 | 295 | 21.85 | 8803.2 | 1.2572 | 6350.4 | 1.9510 | 193.35 |
| 246 | | 7309.2 | .64592 | 5263.9 | 3.1666 | 187.81 | 296 | | 8833.8 | 1.2730 | 6372.7 | 1.9334 | 193.45 |
| 247 | | 7339.6 | .65554 | 5285.9 | 3.1328 | 187.94 | 297 | | 8864.5 | 1.2889 | 6395.1 | 1.9159 | 193.56 |
| 248 | | 7369.9 | .66527 | 5307.9 | 3.0994 | 188.06 | 298 | | 8895.2 | 1.3050 | 6417.5 | 1.8987 | 193.66 |
| 249 | | 7400.2 | .67512 | 5330.0 | 3.0666 | 188.18 | 299 | | 8925.8 | 1.3212 | 6439.8 | 1.8816 | 193.76 |

# Table 7 Products—100% Theoretical Air (for One Gram-Mole)

| T | t | $\bar{h}$ | $p_r$ | $\bar{u}$ | $v_r$ | $\bar{\phi}$ | T | t | $\bar{h}$ | $p_r$ | $\bar{u}$ | $v_r$ | $\bar{\phi}$ |
|---|---|---|---|---|---|---|---|---|---|---|---|---|---|
| 300 | 26.85 | 8956.5 | 1.3376 | 6462.2 | 1.8648 | 193.86 | 350 | 76.85 | 10499.8 | 2.3703 | 7589.8 | 1.22769 | 198.62 |
| 301 | | 8987.2 | 1.3541 | 6484.6 | 1.8482 | 193.97 | 351 | | 10530.9 | 2.3957 | 7612.5 | 1.21816 | 198.71 |
| 302 | | 9017.9 | 1.3708 | 6507.0 | 1.8318 | 194.07 | 352 | | 10561.9 | 2.4213 | 7635.2 | 1.20872 | 198.80 |
| 303 | | 9048.6 | 1.3876 | 6529.4 | 1.8155 | 194.17 | 353 | | 10593.0 | 2.4471 | 7658.0 | 1.19936 | 198.89 |
| 304 | | 9079.3 | 1.4046 | 6551.8 | 1.7995 | 194.27 | 354 | | 10624.1 | 2.4731 | 7680.8 | 1.19013 | 198.97 |
| 305 | 31.85 | 9110.1 | 1.4218 | 6574.2 | 1.7836 | 194.37 | 355 | 81.85 | 10655.1 | 2.4993 | 7703.5 | 1.18098 | 199.06 |
| 306 | | 9140.8 | 1.4391 | 6596.6 | 1.7679 | 194.47 | 356 | | 10686.2 | 2.5257 | 7726.3 | 1.17191 | 199.15 |
| 307 | | 9171.5 | 1.4565 | 6619.0 | 1.7525 | 194.57 | 357 | | 10717.3 | 2.5523 | 7749.0 | 1.16295 | 199.24 |
| 308 | | 9202.3 | 1.4742 | 6641.5 | 1.7371 | 194.67 | 358 | | 10748.4 | 2.5792 | 7771.9 | 1.15405 | 199.32 |
| 309 | | 9233.1 | 1.4920 | 6663.9 | 1.7220 | 194.77 | 359 | | 10779.5 | 2.6063 | 7794.6 | 1.14527 | 199.41 |
| 310 | 36.85 | 9263.8 | 1.5099 | 6686.3 | 1.7070 | 194.87 | 360 | 86.85 | 10810.6 | 2.6335 | 7817.4 | 1.13657 | 199.50 |
| 311 | | 9294.6 | 1.5280 | 6708.8 | 1.6923 | 194.97 | 361 | | 10841.7 | 2.6610 | 7840.2 | 1.12796 | 199.58 |
| 312 | | 9325.3 | 1.5463 | 6731.2 | 1.6777 | 195.07 | 362 | | 10872.8 | 2.6887 | 7863.0 | 1.11942 | 199.67 |
| 313 | | 9356.1 | 1.5647 | 6753.7 | 1.6632 | 195.17 | 363 | | 10904.0 | 2.7166 | 7885.8 | 1.11098 | 199.76 |
| 314 | | 9386.9 | 1.5833 | 6776.2 | 1.6489 | 195.27 | 364 | | 10935.1 | 2.7448 | 7908.6 | 1.10263 | 199.84 |
| 315 | 41.85 | 9417.7 | 1.6020 | 6798.7 | 1.6348 | 195.36 | 365 | 91.85 | 10966.2 | 2.7731 | 7931.5 | 1.09434 | 199.93 |
| 316 | | 9448.5 | 1.6210 | 6821.1 | 1.6209 | 195.46 | 366 | | 10997.4 | 2.8017 | 7954.3 | 1.08614 | 200.01 |
| 317 | | 9479.3 | 1.6400 | 6843.6 | 1.6071 | 195.56 | 367 | | 11028.6 | 2.8305 | 7977.2 | 1.07803 | 200.10 |
| 318 | | 9510.1 | 1.6593 | 6866.1 | 1.5934 | 195.66 | 368 | | 11059.7 | 2.8595 | 8000.0 | 1.07000 | 200.18 |
| 319 | | 9540.9 | 1.6787 | 6888.6 | 1.5799 | 195.75 | 369 | | 11090.9 | 2.8888 | 8022.9 | 1.06205 | 200.27 |
| 320 | 46.85 | 9571.7 | 1.6983 | 6911.1 | 1.5666 | 195.85 | 370 | 96.85 | 11122.1 | 2.9183 | 8045.8 | 1.05416 | 200.35 |
| 321 | | 9602.6 | 1.7181 | 6933.6 | 1.5534 | 195.95 | 371 | | 11153.3 | 2.9480 | 8068.6 | 1.04636 | 200.44 |
| 322 | | 9633.4 | 1.7380 | 6956.2 | 1.5404 | 196.04 | 372 | | 11184.4 | 2.9779 | 8091.5 | 1.03863 | 200.52 |
| 323 | | 9664.3 | 1.7581 | 6978.7 | 1.5275 | 196.14 | 373 | | 11215.7 | 3.0080 | 8114.4 | 1.03100 | 200.60 |
| 324 | | 9695.1 | 1.7784 | 7001.3 | 1.5147 | 196.23 | 374 | | 11246.9 | 3.0384 | 8137.3 | 1.02342 | 200.69 |
| 325 | 51.85 | 9726.0 | 1.7989 | 7023.8 | 1.5021 | 196.33 | 375 | 101.85 | 11278.1 | 3.0690 | 8160.2 | 1.01593 | 200.77 |
| 326 | | 9756.9 | 1.8195 | 7046.4 | 1.4897 | 196.42 | 376 | | 11309.3 | 3.0999 | 8183.1 | 1.00850 | 200.85 |
| 327 | | 9787.8 | 1.8403 | 7069.0 | 1.4774 | 196.52 | 377 | | 11340.5 | 3.1310 | 8206.0 | 1.00114 | 200.94 |
| 328 | | 9818.6 | 1.8613 | 7091.5 | 1.4652 | 196.61 | 378 | | 11371.8 | 3.1623 | 8228.9 | .99385 | 201.02 |
| 329 | | 9849.5 | 1.8825 | 7114.1 | 1.4531 | 196.71 | 379 | | 11403.0 | 3.1938 | 8251.9 | .98665 | 201.10 |
| 330 | 56.85 | 9880.4 | 1.9038 | 7136.7 | 1.4412 | 196.80 | 380 | 106.85 | 11434.3 | 3.2256 | 8274.8 | .97950 | 201.18 |
| 331 | | 9911.3 | 1.9254 | 7159.3 | 1.4294 | 196.89 | 381 | | 11465.6 | 3.2576 | 8297.8 | .97242 | 201.27 |
| 332 | | 9942.3 | 1.9471 | 7181.9 | 1.4177 | 196.99 | 382 | | 11496.8 | 3.2899 | 8320.7 | .96541 | 201.35 |
| 333 | | 9973.2 | 1.9690 | 7204.5 | 1.4062 | 197.08 | 383 | | 11528.1 | 3.3224 | 8343.7 | .95846 | 201.43 |
| 334 | | 10004.1 | 1.9911 | 7227.1 | 1.3947 | 197.17 | 384 | | 11559.4 | 3.3552 | 8366.6 | .95158 | 201.51 |
| 335 | 61.85 | 10035.0 | 2.0133 | 7249.7 | 1.3834 | 197.26 | 385 | 111.85 | 11590.7 | 3.3882 | 8389.6 | .94475 | 201.59 |
| 336 | | 10065.9 | 2.0358 | 7272.3 | 1.3723 | 197.36 | 386 | | 11622.0 | 3.4215 | 8412.6 | .93801 | 201.67 |
| 337 | | 10096.9 | 2.0584 | 7294.9 | 1.3612 | 197.45 | 387 | | 11653.3 | 3.4550 | 8435.6 | .93131 | 201.75 |
| 338 | | 10127.8 | 2.0812 | 7317.6 | 1.3503 | 197.54 | 388 | | 11684.6 | 3.4887 | 8458.6 | .92469 | 201.84 |
| 339 | | 10158.8 | 2.1042 | 7340.2 | 1.3395 | 197.63 | 389 | | 11715.9 | 3.5227 | 8481.6 | .91813 | 201.92 |
| 340 | 66.85 | 10189.7 | 2.1275 | 7362.8 | 1.3288 | 197.72 | 390 | 116.85 | 11747.3 | 3.5570 | 8504.6 | .91163 | 202.00 |
| 341 | | 10220.7 | 2.1509 | 7385.5 | 1.3182 | 197.81 | 391 | | 11778.6 | 3.5914 | 8527.7 | .90519 | 202.08 |
| 342 | | 10251.7 | 2.1745 | 7408.2 | 1.3077 | 197.90 | 392 | | 11809.9 | 3.6262 | 8550.7 | .89881 | 202.16 |
| 343 | | 10282.7 | 2.1982 | 7430.8 | 1.2973 | 198.00 | 393 | | 11841.3 | 3.6612 | 8573.7 | .89248 | 202.24 |
| 344 | | 10313.7 | 2.2222 | 7453.5 | 1.2871 | 198.09 | 394 | | 11872.6 | 3.6965 | 8596.8 | .88622 | 202.32 |
| 345 | 71.85 | 10344.7 | 2.2464 | 7476.2 | 1.2769 | 198.18 | 395 | 121.85 | 11904.0 | 3.7320 | 8619.8 | .88001 | 202.40 |
| 346 | | 10375.7 | 2.2708 | 7499.0 | 1.2668 | 198.27 | 396 | | 11935.4 | 3.7678 | 8642.9 | .87387 | 202.48 |
| 347 | | 10406.7 | 2.2954 | 7521.6 | 1.2569 | 198.35 | 397 | | 11966.8 | 3.8037 | 8666.0 | .86778 | 202.55 |
| 348 | | 10437.8 | 2.3202 | 7544.3 | 1.2471 | 198.44 | 398 | | 11998.2 | 3.8400 | 8689.0 | .86175 | 202.63 |
| 349 | | 10468.8 | 2.3452 | 7567.1 | 1.2373 | 198.53 | 399 | | 12029.5 | 3.8766 | 8712.1 | .85575 | 202.71 |

## Table 7 Products—100% Theoretical Air (for One Gram-Mole)

| T | t | $\bar{h}$ | $p_r$ | $\bar{u}$ | $v_r$ | $\bar{\phi}$ | T | t | $\bar{h}$ | $p_r$ | $\bar{u}$ | $v_r$ | $\bar{\phi}$ |
|---|---|-----------|-------|-----------|-------|--------------|---|---|-----------|-------|-----------|-------|--------------|
| 400 | 126.85 | 12060.9 | 3.9134 | 8735.2 | .84983 | 202.79 | 450 | 176.85 | 13640.9 | 6.1226 | 9899.4 | .61109 | 206.51 |
| 401 | | 12092.4 | 3.9505 | 8758.3 | .84396 | 202.87 | 451 | | 13672.7 | 6.1748 | 9922.9 | .60727 | 206.58 |
| 402 | | 12123.8 | 3.9879 | 8781.4 | .83813 | 202.95 | 452 | | 13704.5 | 6.2274 | 9946.3 | .60348 | 206.65 |
| 403 | | 12155.2 | 4.0255 | 8804.5 | .83237 | 203.03 | 453 | | 13736.3 | 6.2803 | 9969.9 | .59972 | 206.72 |
| 404 | | 12186.6 | 4.0634 | 8827.6 | .82666 | 203.10 | 454 | | 13768.1 | 6.3335 | 9993.4 | .59600 | 206.79 |
| 405 | 131.85 | 12218.1 | 4.1016 | 8850.7 | .82099 | 203.18 | 455 | 181.85 | 13799.9 | 6.3871 | 10016.9 | .59229 | 206.86 |
| 406 | | 12249.5 | 4.1400 | 8873.9 | .81537 | 203.26 | 456 | | 13831.8 | 6.4410 | 10040.4 | .58864 | 206.93 |
| 407 | | 12281.0 | 4.1787 | 8897.0 | .80980 | 203.34 | 457 | | 13863.6 | 6.4953 | 10063.9 | .58499 | 207.00 |
| 408 | | 12312.4 | 4.2177 | 8920.1 | .80429 | 203.41 | 458 | | 13895.4 | 6.5499 | 10087.5 | .58138 | 207.07 |
| 409 | | 12343.9 | 4.2570 | 8943.3 | .79882 | 203.49 | 459 | | 13927.3 | 6.6048 | 10111.0 | .57781 | 207.14 |
| 410 | 136.85 | 12375.4 | 4.2965 | 8966.5 | .79341 | 203.57 | 460 | 186.85 | 13959.2 | 6.6602 | 10134.6 | .57425 | 207.21 |
| 411 | | 12406.9 | 4.3364 | 8989.7 | .78804 | 203.64 | 461 | | 13991.1 | 6.7158 | 10158.1 | .57073 | 207.28 |
| 412 | | 12438.3 | 4.3764 | 9012.8 | .78272 | 203.72 | 462 | | 14023.0 | 6.7720 | 10181.7 | .56723 | 207.35 |
| 413 | | 12469.9 | 4.4168 | 9036.0 | .77745 | 203.80 | 463 | | 14054.9 | 6.8283 | 10205.3 | .56377 | 207.42 |
| 414 | | 12501.4 | 4.4575 | 9059.2 | .77222 | 203.87 | 464 | | 14086.8 | 6.8850 | 10228.9 | .56033 | 207.49 |
| 415 | 141.85 | 12532.9 | 4.4984 | 9082.4 | .76704 | 203.95 | 465 | 191.85 | 14118.7 | 6.9422 | 10252.5 | .55691 | 207.56 |
| 416 | | 12564.4 | 4.5396 | 9105.6 | .76191 | 204.02 | 466 | | 14150.6 | 6.9996 | 10276.1 | .55353 | 207.62 |
| 417 | | 12595.9 | 4.5811 | 9128.8 | .75682 | 204.10 | 467 | | 14182.5 | 7.0575 | 10299.7 | .55017 | 207.69 |
| 418 | | 12627.5 | 4.6229 | 9152.1 | .75178 | 204.18 | 468 | | 14214.5 | 7.1157 | 10323.3 | .54684 | 207.76 |
| 419 | | 12659.0 | 4.6651 | 9175.3 | .74677 | 204.25 | 469 | | 14246.4 | 7.1744 | 10347.0 | .54353 | 207.83 |
| 420 | 146.85 | 12690.6 | 4.7074 | 9198.5 | .74181 | 204.33 | 470 | 196.85 | 14278.4 | 7.2333 | 10370.6 | .54025 | 207.90 |
| 421 | | 12722.2 | 4.7501 | 9221.8 | .73690 | 204.40 | 471 | | 14310.3 | 7.2926 | 10394.2 | .53699 | 207.97 |
| 422 | | 12753.7 | 4.7931 | 9245.1 | .73203 | 204.48 | 472 | | 14342.3 | 7.3523 | 10417.9 | .53377 | 208.03 |
| 423 | | 12785.3 | 4.8364 | 9268.3 | .72719 | 204.55 | 473 | | 14374.3 | 7.4124 | 10441.6 | .53056 | 208.10 |
| 424 | | 12816.9 | 4.8800 | 9291.6 | .72240 | 204.63 | 474 | | 14406.2 | 7.4729 | 10465.2 | .52738 | 208.17 |
| 425 | 151.85 | 12848.5 | 4.9239 | 9314.9 | .71765 | 204.70 | 475 | 201.85 | 14438.2 | 7.5336 | 10488.9 | .52423 | 208.24 |
| 426 | | 12880.1 | 4.9681 | 9338.2 | .71294 | 204.77 | 476 | | 14470.2 | 7.5949 | 10512.6 | .52109 | 208.30 |
| 427 | | i2911.7 | 5.0126 | 9361.5 | .70827 | 204.85 | 477 | | 14502.2 | 7.6565 | 10536.3 | .51799 | 208.37 |
| 428 | | 12943.3 | 5.0573 | 9384.8 | .70365 | 204.92 | 478 | | 14534.3 | 7.7185 | 10560.0 | .51490 | 208.44 |
| 429 | | 12974.9 | 5.1024 | 9408.1 | .69905 | 205.00 | 479 | | 14566.3 | 7.7810 | 10583.7 | .51184 | 208.50 |
| 430 | 156.85 | 13006.6 | 5.1479 | 9431.4 | .69450 | 205.07 | 480 | 206.85 | 14598.3 | 7.8437 | 10607.4 | .50880 | 208.57 |
| 431 | | 13038.2 | 5.1936 | 9454.7 | .68999 | 205.14 | 481 | | 14630.4 | 7.9069 | 10631.1 | .50579 | 208.64 |
| 432 | | 13069.9 | 5.2396 | 9478.1 | .68551 | 205.22 | 482 | | 14662.4 | 7.9704 | 10654.9 | .50280 | 208.70 |
| 433 | | 13101.5 | 5.2860 | 9501.4 | .68108 | 205.29 | 483 | | 14694.5 | 8.0344 | 10678.6 | .49983 | 208.77 |
| 434 | | 13133.2 | 5.3325 | 9524.8 | .67669 | 205.36 | 484 | | 14726.5 | 8.0988 | 10702.4 | .49689 | 208.84 |
| 435 | 161.85 | 13164.9 | 5.3795 | 9548.1 | .67232 | 205.44 | 485 | 211.85 | 14758.6 | 8.1636 | 10726.1 | .49396 | 208.90 |
| 436 | | 13196.6 | 5.4267 | 9571.5 | .66801 | 205.51 | 486 | | 14790.7 | 8.2288 | 10749.9 | .49106 | 208.97 |
| 437 | | 13228.2 | 5.4743 | 9594.8 | .66372 | 205.58 | 487 | | 14822.8 | 8.2943 | 10773.7 | .48818 | 209.04 |
| 438 | | 13259.9 | 5.5222 | 9618.2 | .65947 | 205.65 | 488 | | 14854.9 | 8.3602 | 10797.5 | .48533 | 209.10 |
| 439 | | 13291.6 | 5.5704 | 9641.6 | .65525 | 205.73 | 489 | | 14887.0 | 8.4266 | 10821.3 | .48249 | 209.17 |
| 440 | 166.85 | 13323.3 | 5.6190 | 9665.0 | .65106 | 205.80 | 490 | 216.85 | 14919.1 | 8.4933 | 10845.0 | .47968 | 209.23 |
| 441 | | 13355.0 | 5.6679 | 9688.4 | .64691 | 205.87 | 491 | | 14951.2 | 8.5605 | 10868.9 | .47688 | 209.30 |
| 442 | | 13386.8 | 5.7171 | 9711.8 | .64281 | 205.94 | 492 | | 14983.4 | 8.6280 | 10892.7 | .47412 | 209.36 |
| 443 | | 13418.5 | 5.7666 | 9735.2 | .63872 | 206.01 | 493 | | 15015.5 | 8.6961 | 10916.5 | .47136 | 209.43 |
| 444 | | 13450.3 | 5.8165 | 9758.7 | .63468 | 206.09 | 494 | | 15047.7 | 8.7645 | 10940.4 | .46863 | 209.49 |
| 445 | 171.85 | 13482.0 | 5.8667 | 9782.1 | .63067 | 206.16 | 495 | 221.85 | 15079.8 | 8.8333 | 10964.2 | .46592 | 209.56 |
| 446 | | 13513.7 | 5.9172 | 9805.5 | .62669 | 206.23 | 496 | | 15112.0 | 8.9025 | 10988.0 | .46324 | 209.62 |
| 447 | | 13545.5 | 5.9681 | 9829.0 | .62274 | 206.30 | 497 | | 15144.2 | 8.9722 | 11011.9 | .46056 | 209.69 |
| 448 | | 13577.3 | 6.0192 | 9852.4 | .61883 | 206.37 | 498 | | 15176.3 | 9.0422 | 11035.8 | .45792 | 209.75 |
| 449 | | 13609.1 | 6.0708 | 9875.9 | .61494 | 206.44 | 499 | | 15208.6 | 9.1128 | 11059.7 | .45528 | 209.82 |

## Table 7 Products—100% Theoretical Air (for One Gram-Mole)

| T | t | $\bar{h}$ | $p_r$ | $\bar{u}$ | $v_r$ | $\bar{\phi}$ | T | t | $\bar{h}$ | $p_r$ | $\bar{u}$ | $v_r$ | $\bar{\phi}$ |
|---|---|-----------|-------|-----------|-------|--------------|---|---|-----------|-------|-----------|-------|--------------|
| 500 | 226.85 | 15240.7 | 9.184 | 11083.5 | .45268 | 209.88 | 550 | 276.85 | 16861.8 | 13.317 | 12288.9 | .34339 | 212.97 |
| 501 | | 15273.0 | 9.255 | 11107.4 | .45009 | 209.95 | 551 | | 16894.5 | 13.412 | 12313.2 | .34157 | 213.03 |
| 502 | | 15305.2 | 9.327 | 11131.4 | .44751 | 210.01 | 552 | | 16927.1 | 13.508 | 12337.5 | .33977 | 213.09 |
| 503 | | 15337.4 | 9.399 | 11155.3 | .44496 | 210.08 | 553 | | 16959.8 | 13.604 | 12361.9 | .33797 | 213.15 |
| 504 | | 15369.6 | 9.472 | 11179.2 | .44242 | 210.14 | 554 | | 16992.5 | 13.701 | 12386.3 | .33618 | 213.21 |
| 505 | 231.85 | 15401.9 | 9.545 | 11203.1 | .43990 | 210.20 | 555 | 281.85 | 17025.1 | 13.799 | 12410.6 | .33441 | 213.27 |
| 506 | | 15434.2 | 9.618 | 11227.1 | .43741 | 210.27 | 556 | | 17057.8 | 13.897 | 12435.0 | .33265 | 213.33 |
| 507 | | 15466.4 | 9.692 | 11251.0 | .43493 | 210.33 | 557 | | 17090.5 | 13.995 | 12459.4 | .33090 | 213.39 |
| 508 | | 15498.7 | 9.767 | 11275.0 | .43246 | 210.39 | 558 | | 17123.2 | 14.095 | 12483.8 | .32917 | 213.44 |
| 509 | | 15531.0 | 9.841 | 11298.9 | .43002 | 210.46 | 559 | | 17155.9 | 14.194 | 12508.2 | .32744 | 213.50 |
| 510 | 236.85 | 15563.3 | 9.917 | 11322.9 | .42759 | 210.52 | 560 | 286.85 | 17188.6 | 14.294 | 12532.6 | .32573 | 213.56 |
| 511 | | 15595.5 | 9.993 | 11346.9 | .42518 | 210.58 | 561 | | 17221.4 | 14.395 | 12557.0 | .32403 | 213.62 |
| 512 | | 15627.9 | 10.069 | 11370.9 | .42279 | 210.65 | 562 | | 17254.1 | 14.496 | 12581.4 | .32233 | 213.68 |
| 513 | | 15660.2 | 10.145 | 11394.9 | .42042 | 210.71 | 563 | | 17286.9 | 14.598 | 12605.9 | .32065 | 213.74 |
| 514 | | 15692.5 | 10.222 | 11418.9 | .41806 | 210.77 | 564 | | 17319.7 | 14.701 | 12630.3 | .31899 | 213.79 |
| 515 | 241.85 | 15724.8 | 10.300 | 11442.9 | .41572 | 210.84 | 565 | 291.85 | 17352.4 | 14.804 | 12654.8 | .31733 | 213.85 |
| 516 | | 15757.2 | 10.378 | 11466.9 | .41339 | 210.90 | 566 | | 17385.2 | 14.907 | 12679.2 | .31568 | 213.91 |
| 517 | | 15789.5 | 10.457 | 11491.0 | .41109 | 210.96 | 567 | | 17417.9 | 15.011 | 12703.7 | .31405 | 213.97 |
| 518 | | 15821.9 | 10.536 | 11515.0 | .40880 | 211.02 | 568 | | 17450.8 | 15.116 | 12728.2 | .31242 | 214.03 |
| 519 | | 15854.2 | 10.615 | 11539.1 | .40653 | 211.09 | 569 | | 17483.6 | 15.221 | 12752.7 | .31081 | 214.08 |
| 520 | 246.85 | 15886.6 | 10.695 | 11563.1 | .40427 | 211.15 | 570 | 296.85 | 17516.4 | 15.327 | 12777.2 | .30920 | 214.14 |
| 521 | | 15919.0 | 10.775 | 11587.2 | .40203 | 211.21 | 571 | | 17549.2 | 15.433 | 12801.7 | .30761 | 214.20 |
| 522 | | 15951.4 | 10.856 | 11611.2 | .39980 | 211.27 | 572 | | 17582.0 | 15.540 | 12826.2 | .30603 | 214.26 |
| 523 | | 15983.7 | 10.937 | 11635.3 | .39759 | 211.34 | 573 | | 17614.9 | 15.648 | 12850.7 | .30446 | 214.31 |
| 524 | | 16016.2 | 11.019 | 11659.4 | .39539 | 211.40 | 574 | | 17647.7 | 15.756 | 12875.3 | .30289 | 214.37 |
| 525 | 251.85 | 16048.5 | 11.101 | 11683.5 | .39321 | 211.46 | 575 | 301.85 | 17680.6 | 15.865 | 12899.8 | .30134 | 214.43 |
| 526 | | 16081.0 | 11.184 | 11707.6 | .39105 | 211.52 | 576 | | 17713.4 | 15.974 | 12924.3 | .29980 | 214.49 |
| 527 | | 16113.4 | 11.267 | 11731.7 | .38890 | 211.58 | 577 | | 17746.3 | 16.084 | 12948.9 | .29827 | 214.54 |
| 528 | | 16145.9 | 11.350 | 11755.9 | .38677 | 211.64 | 578 | | 17779.2 | 16.195 | 12973.5 | .29674 | 214.60 |
| 529 | | 16178.3 | 11.435 | 11780.0 | .38465 | 211.71 | 579 | | 17812.1 | 16.306 | 12998.1 | .29523 | 214.66 |
| 530 | 256.85 | 16210.8 | 11.519 | 11804.1 | .38255 | 211.77 | 580 | 306.85 | 17845.0 | 16.418 | 13022.6 | .29373 | 214.71 |
| 531 | | 16243.2 | 11.604 | 11828.3 | .38046 | 211.83 | 581 | | 17877.9 | 16.530 | 13047.2 | .29223 | 214.77 |
| 532 | | 16275.7 | 11.690 | 11852.5 | .37839 | 211.89 | 582 | | 17910.8 | 16.643 | 13071.9 | .29075 | 214.83 |
| 533 | | 16308.2 | 11.776 | 11876.6 | .37633 | 211.95 | 583 | | 17943.8 | 16.757 | 13096.5 | .28928 | 214.88 |
| 534 | | 16340.7 | 11.863 | 11900.8 | .37428 | 212.01 | 584 | | 17976.7 | 16.871 | 13121.1 | .28781 | 214.94 |
| 535 | 261.85 | 16373.2 | 11.950 | 11925.0 | .37225 | 212.07 | 585 | 311.85 | 18009.7 | 16.986 | 13145.7 | .28635 | 215.00 |
| 536 | | 16405.7 | 12.037 | 11949.2 | .37023 | 212.13 | 586 | | 18042.6 | 17.101 | 13170.4 | .28491 | 215.05 |
| 537 | | 16438.2 | 12.125 | 11973.4 | .36822 | 212.19 | 587 | | 18075.6 | 17.217 | 13195.0 | .28347 | 215.11 |
| 538 | | 16470.8 | 12.214 | 11997.6 | .36624 | 212.25 | 588 | | 18108.6 | 17.334 | 13219.7 | .28204 | 215.16 |
| 539 | | 16503.3 | 12.303 | 12021.9 | .36426 | 212.31 | 589 | | 18141.5 | 17.451 | 13244.4 | .28063 | 215.22 |
| 540 | 266.85 | 16535.9 | 12.393 | 12046.1 | .36230 | 212.37 | 590 | 316.85 | 18174.5 | 17.569 | 13269.0 | .27922 | 215.28 |
| 541 | | 16568.4 | 12.483 | 12070.3 | .36035 | 212.43 | 591 | | 18207.5 | 17.687 | 13293.7 | .27782 | 215.33 |
| 542 | | 16601.0 | 12.573 | 12094.6 | .35841 | 212.49 | 592 | | 18240.5 | 17.807 | 13318.4 | .27642 | 215.39 |
| 543 | | 16633.5 | 12.664 | 12118.8 | .35649 | 212.55 | 593 | | 18273.6 | 17.926 | 13343.1 | .27504 | 215.44 |
| 544 | | 16666.1 | 12.756 | 12143.1 | .35458 | 212.61 | 594 | | 18306.6 | 18.047 | 13367.8 | .27367 | 215.50 |
| 545 | 271.85 | 16698.7 | 12.848 | 12167.4 | .35269 | 212.67 | 595 | 321.85 | 18339.6 | 18.168 | 13392.6 | .27230 | 215.55 |
| 546 | | 16731.3 | 12.941 | 12191.7 | .35080 | 212.73 | 596 | | 18372.7 | 18.289 | 13417.3 | .27095 | 215.61 |
| 547 | | 16763.9 | 13.034 | 12215.9 | .34893 | 212.79 | 597 | | 18405.7 | 18.411 | 13442.0 | .26960 | 215.67 |
| 548 | | 16796.5 | 13.128 | 12240.2 | .34707 | 212.85 | 598 | | 18438.8 | 18.534 | 13466.8 | .26826 | 215.72 |
| 549 | | 16829.2 | 13.222 | 12264.6 | .34523 | 212.91 | 599 | | 18471.9 | 18.658 | 13491.5 | .26693 | 215.78 |

## Table 7    Products—100% Theoretical Air (for One Gram-Mole)

| T | t | $\bar{h}$ | $p_r$ | $\bar{u}$ | $v_r$ | $\bar{\phi}$ | T | t | $\bar{h}$ | $p_r$ | $\bar{u}$ | $v_r$ | $\bar{\phi}$ |
|---|---|-----------|-------|-----------|-------|--------------|---|---|-----------|-------|-----------|-------|--------------|
| 600 | 326.85 | 18504.9 | 18.782 | 13516.3 | .26561 | 215.83 | 650 | 376.85 | 20170.8 | 25.884 | 14766.4 | .20879 | 218.50 |
| 601 |        | 18538.0 | 18.907 | 13541.1 | .26429 | 215.89 | 651 |        | 20204.4 | 26.045 | 14791.7 | .20782 | 218.55 |
| 602 |        | 18571.1 | 19.033 | 13565.9 | .26298 | 215.94 | 652 |        | 20237.9 | 26.207 | 14816.9 | .20685 | 218.60 |
| 603 |        | 18604.3 | 19.159 | 13590.7 | .26169 | 216.00 | 653 |        | 20271.5 | 26.370 | 14842.2 | .20589 | 218.65 |
| 604 |        | 18637.4 | 19.286 | 13615.5 | .26039 | 216.05 | 654 |        | 20305.1 | 26.533 | 14867.4 | .20494 | 218.70 |
| 605 | 331.85 | 18670.5 | 19.413 | 13640.3 | .25911 | 216.11 | 655 | 381.85 | 20338.6 | 26.697 | 14892.7 | .20399 | 218.76 |
| 606 |        | 18703.6 | 19.542 | 13665.1 | .25784 | 216.16 | 656 |        | 20372.2 | 26.862 | 14918.0 | .20305 | 218.81 |
| 607 |        | 18736.8 | 19.670 | 13689.9 | .25657 | 216.22 | 657 |        | 20405.8 | 27.028 | 14943.3 | .20211 | 218.86 |
| 608 |        | 18769.9 | 19.800 | 13714.8 | .25531 | 216.27 | 658 |        | 20439.4 | 27.195 | 14968.6 | .20117 | 218.91 |
| 609 |        | 18803.1 | 19.930 | 13739.7 | .25406 | 216.32 | 659 |        | 20473.1 | 27.362 | 14993.9 | .20025 | 218.96 |
| 610 | 336.85 | 18836.3 | 20.061 | 13764.5 | .25282 | 216.38 | 660 | 386.85 | 20506.7 | 27.531 | 15019.2 | .19932 | 219.01 |
| 611 |        | 18869.5 | 20.193 | 13789.4 | .25158 | 216.43 | 661 |        | 20540.4 | 27.700 | 15044.5 | .19840 | 219.06 |
| 612 |        | 18902.7 | 20.325 | 13814.2 | .25035 | 216.49 | 662 |        | 20574.0 | 27.870 | 15069.9 | .19749 | 219.11 |
| 613 |        | 18935.9 | 20.458 | 13839.2 | .24913 | 216.54 | 663 |        | 20607.7 | 28.041 | 15095.2 | .19659 | 219.16 |
| 614 |        | 18969.1 | 20.592 | 13864.0 | .24792 | 216.60 | 664 |        | 20641.3 | 28.212 | 15120.6 | .19569 | 219.21 |
| 615 | 341.85 | 19002.3 | 20.726 | 13889.0 | .24671 | 216.65 | 665 | 391.85 | 20675.0 | 28.385 | 15145.9 | .19479 | 219.27 |
| 616 |        | 19035.6 | 20.861 | 13913.9 | .24551 | 216.70 | 666 |        | 20708.7 | 28.558 | 15171.3 | .19390 | 219.32 |
| 617 |        | 19068.8 | 20.997 | 13938.8 | .24432 | 216.76 | 667 |        | 20742.4 | 28.732 | 15196.7 | .19301 | 219.37 |
| 618 |        | 19102.0 | 21.133 | 13963.7 | .24314 | 216.81 | 668 |        | 20776.1 | 28.907 | 15222.1 | .19213 | 219.42 |
| 619 |        | 19135.3 | 21.270 | 13988.7 | .24197 | 216.87 | 669 |        | 20809.8 | 29.083 | 15247.5 | .19126 | 219.47 |
| 620 | 346.85 | 19168.6 | 21.408 | 14013.6 | .24080 | 216.92 | 670 | 396.85 | 20843.5 | 29.260 | 15272.9 | .19038 | 219.52 |
| 621 |        | 19201.8 | 21.547 | 14038.6 | .23963 | 216.97 | 671 |        | 20877.3 | 29.438 | 15298.3 | .18952 | 219.57 |
| 622 |        | 19235.1 | 21.686 | 14063.5 | .23848 | 217.03 | 672 |        | 20911.0 | 29.616 | 15323.8 | .18866 | 219.62 |
| 623 |        | 19268.4 | 21.826 | 14088.5 | .23733 | 217.08 | 673 |        | 20944.8 | 29.795 | 15349.2 | .18780 | 219.67 |
| 624 |        | 19301.7 | 21.966 | 14113.5 | .23619 | 217.13 | 674 |        | 20978.5 | 29.975 | 15374.6 | .18695 | 219.72 |
| 625 | 351.85 | 19335.0 | 22.108 | 14138.5 | .23506 | 217.19 | 675 | 401.85 | 21012.3 | 30.157 | 15400.1 | .18610 | 219.77 |
| 626 |        | 19368.4 | 22.250 | 14163.5 | .23393 | 217.24 | 676 |        | 21046.1 | 30.338 | 15425.5 | .18526 | 219.82 |
| 627 |        | 19401.7 | 22.392 | 14188.5 | .23281 | 217.29 | 677 |        | 21079.9 | 30.521 | 15451.0 | .18442 | 219.87 |
| 628 |        | 19435.0 | 22.536 | 14213.6 | .23169 | 217.35 | 678 |        | 21113.7 | 30.705 | 15476.5 | .18359 | 219.92 |
| 629 |        | 19468.4 | 22.680 | 14238.6 | .23059 | 217.40 | 679 |        | 21147.5 | 30.889 | 15502.0 | .18276 | 219.97 |
| 630 | 356.85 | 19501.7 | 22.825 | 14263.6 | .22949 | 217.45 | 680 | 406.85 | 21181.3 | 31.075 | 15527.5 | .18194 | 220.02 |
| 631 |        | 19535.1 | 22.971 | 14288.7 | .22839 | 217.51 | 681 |        | 21215.1 | 31.261 | 15553.0 | .18112 | 220.07 |
| 632 |        | 19568.5 | 23.117 | 14313.8 | .22730 | 217.56 | 682 |        | 21248.9 | 31.448 | 15578.5 | .18031 | 220.12 |
| 633 |        | 19601.9 | 23.265 | 14338.8 | .22622 | 217.61 | 683 |        | 21282.8 | 31.637 | 15604.1 | .17950 | 220.17 |
| 634 |        | 19635.2 | 23.413 | 14363.9 | .22515 | 217.66 | 684 |        | 21316.7 | 31.826 | 15629.6 | .17869 | 220.22 |
| 635 | 361.85 | 19668.7 | 23.562 | 14389.0 | .22408 | 217.72 | 685 | 411.85 | 21350.5 | 32.016 | 15655.1 | .17789 | 220.27 |
| 636 |        | 19702.1 | 23.711 | 14414.1 | .22302 | 217.77 | 686 |        | 21384.4 | 32.207 | 15680.7 | .17710 | 220.32 |
| 637 |        | 19735.5 | 23.861 | 14439.2 | .22196 | 217.82 | 687 |        | 21418.3 | 32.398 | 15706.3 | .17631 | 220.36 |
| 638 |        | 19768.9 | 24.012 | 14464.3 | .22091 | 217.87 | 688 |        | 21452.2 | 32.591 | 15731.9 | .17552 | 220.41 |
| 639 |        | 19802.4 | 24.164 | 14489.4 | .21987 | 217.93 | 689 |        | 21486.1 | 32.785 | 15757.5 | .17473 | 220.46 |
| 640 | 366.85 | 19835.8 | 24.317 | 14514.6 | .21883 | 217.98 | 690 | 416.85 | 21520.0 | 32.979 | 15783.0 | .17396 | 220.51 |
| 641 |        | 19869.3 | 24.470 | 14539.7 | .21780 | 218.03 | 691 |        | 21553.9 | 33.175 | 15808.6 | .17318 | 220.56 |
| 642 |        | 19902.7 | 24.624 | 14564.9 | .21678 | 218.08 | 692 |        | 21587.8 | 33.371 | 15834.2 | .17241 | 220.61 |
| 643 |        | 19936.2 | 24.779 | 14590.0 | .21576 | 218.14 | 693 |        | 21621.7 | 33.568 | 15859.9 | .17165 | 220.66 |
| 644 |        | 19969.7 | 24.934 | 14615.2 | .21474 | 218.19 | 694 |        | 21655.7 | 33.767 | 15885.5 | .17088 | 220.71 |
| 645 | 371.85 | 20003.2 | 25.091 | 14640.4 | .21374 | 218.24 | 695 | 421.85 | 21689.6 | 33.966 | 15911.1 | .17013 | 220.76 |
| 646 |        | 20036.7 | 25.248 | 14665.6 | .21274 | 218.29 | 696 |        | 21723.6 | 34.166 | 15936.8 | .16937 | 220.81 |
| 647 |        | 20070.2 | 25.406 | 14690.8 | .21174 | 218.34 | 697 |        | 21757.6 | 34.367 | 15962.4 | .16863 | 220.85 |
| 648 |        | 20103.7 | 25.564 | 14716.0 | .21075 | 218.39 | 698 |        | 21791.6 | 34.569 | 15988.1 | .16788 | 220.90 |
| 649 |        | 20137.3 | 25.724 | 14741.2 | .20977 | 218.45 | 699 |        | 21825.5 | 34.772 | 16013.8 | .16714 | 220.95 |

## Table 7    Products—100% Theoretical Air (for One Gram-Mole)

| T | t | $\bar{h}$ | $p_r$ | $\bar{u}$ | $v_r$ | $\bar{\phi}$ | T | t | $\bar{h}$ | $p_r$ | $\bar{u}$ | $v_r$ | $\bar{\phi}$ |
|---|---|---|---|---|---|---|---|---|---|---|---|---|---|
| 700 | 426.85 | 21859.5 | 34.976 | 16039.5 | .16640 | 221.00 | 750 | 476.85 | 23571.2 | 46.464 | 17335.4 | .13421 | 223.36 |
| 701 | | 21893.6 | 35.180 | 16065.2 | .16567 | 221.05 | 751 | | 23605.7 | 46.722 | 17361.5 | .13365 | 223.41 |
| 702 | | 21927.6 | 35.386 | 16090.9 | .16494 | 221.10 | 752 | | 23640.1 | 46.980 | 17387.7 | .13309 | 223.45 |
| 703 | | 21961.6 | 35.593 | 16116.6 | .16422 | 221.15 | 753 | | 23674.6 | 47.240 | 17413.9 | .13253 | 223.50 |
| 704 | | 21995.6 | 35.801 | 16142.3 | .16350 | 221.19 | 754 | | 23709.1 | 47.500 | 17440.0 | .13198 | 223.55 |
| 705 | 431.85 | 22029.7 | 36.009 | 16168.0 | .16278 | 221.24 | 755 | 481.85 | 23743.6 | 47.762 | 17466.2 | .13143 | 223.59 |
| 706 | | 22063.7 | 36.219 | 16193.8 | .16207 | 221.29 | 756 | | 23778.1 | 48.025 | 17492.4 | .13088 | 223.64 |
| 707 | | 22097.8 | 36.430 | 16219.5 | .16136 | 221.34 | 757 | | 23812.6 | 48.290 | 17518.6 | .13034 | 223.68 |
| 708 | | 22131.9 | 36.641 | 16245.3 | .16066 | 221.39 | 758 | | 23847.1 | 48.555 | 17544.8 | .12980 | 223.73 |
| 709 | | 22166.0 | 36.854 | 16271.0 | .15995 | 221.44 | 759 | | 23881.7 | 48.822 | 17571.1 | .12926 | 223.77 |
| 710 | 436.85 | 22200.1 | 37.067 | 16296.8 | .15926 | 221.48 | 760 | 486.85 | 23916.2 | 49.089 | 17597.3 | .12872 | 223.82 |
| 711 | | 22234.2 | 37.282 | 16322.6 | .15856 | 221.53 | 761 | | 23950.8 | 49.358 | 17623.5 | .12819 | 223.86 |
| 712 | | 22268.3 | 37.497 | 16348.4 | .15787 | 221.58 | 762 | | 23985.4 | 49.628 | 17649.8 | .12766 | 223.91 |
| 713 | | 22302.4 | 37.714 | 16374.2 | .15719 | 221.63 | 763 | | 24019.9 | 49.900 | 17676.0 | .12713 | 223.96 |
| 714 | | 22336.5 | 37.932 | 16400.0 | .15650 | 221.68 | 764 | | 24054.5 | 50.173 | 17702.3 | .12661 | 224.00 |
| 715 | 441.85 | 22370.7 | 38.150 | 16425.9 | .15583 | 221.72 | 765 | 491.85 | 24089.1 | 50.447 | 17728.6 | .12608 | 224.05 |
| 716 | | 22404.8 | 38.370 | 16451.7 | .15515 | 221.77 | 766 | | 24123.7 | 50.722 | 17754.9 | .12556 | 224.09 |
| 717 | | 22439.0 | 38.590 | 16477.5 | .15448 | 221.82 | 767 | | 24158.3 | 50.998 | 17781.2 | .12505 | 224.14 |
| 718 | | 22473.1 | 38.812 | 16503.4 | .15381 | 221.87 | 768 | | 24192.9 | 51.276 | 17807.5 | .12453 | 224.18 |
| 719 | | 22507.3 | 39.035 | 16529.3 | .15315 | 221.91 | 769 | | 24227.5 | 51.554 | 17833.8 | .12402 | 224.23 |
| 720 | 446.85 | 22541.5 | 39.259 | 16555.1 | .15249 | 221.96 | 770 | 496.85 | 24262.2 | 51.834 | 17860.1 | .12351 | 224.27 |
| 721 | | 22575.7 | 39.483 | 16581.0 | .15183 | 222.01 | 771 | | 24296.8 | 52.115 | 17886.4 | .12300 | 224.32 |
| 722 | | 22609.9 | 39.709 | 16606.9 | .15117 | 222.06 | 772 | | 24331.5 | 52.397 | 17912.8 | .12250 | 224.36 |
| 723 | | 22644.1 | 39.936 | 16632.8 | .15052 | 222.10 | 773 | | 24366.2 | 52.681 | 17939.1 | .12200 | 224.41 |
| 724 | | 22678.3 | 40.164 | 16658.7 | .14988 | 222.15 | 774 | | 24400.8 | 52.965 | 17965.5 | .12150 | 224.45 |
| 725 | 451.85 | 22712.5 | 40.392 | 16684.6 | .14923 | 222.20 | 775 | 501.85 | 24435.5 | 53.252 | 17991.8 | .12100 | 224.50 |
| 726 | | 22746.8 | 40.623 | 16710.5 | .14859 | 222.25 | 776 | | 24470.2 | 53.538 | 18018.2 | .12051 | 224.54 |
| 727 | | 22781.0 | 40.854 | 16736.4 | .14796 | 222.29 | 777 | | 24504.9 | 53.827 | 18044.6 | .12002 | 224.59 |
| 728 | | 22815.3 | 41.086 | 16762.4 | .14732 | 222.34 | 778 | | 24539.5 | 54.117 | 18070.9 | .11953 | 224.63 |
| 729 | | 22849.5 | 41.319 | 16788.3 | .14669 | 222.39 | 779 | | 24574.3 | 54.408 | 18097.4 | .11904 | 224.67 |
| 730 | 456.85 | 22883.8 | 41.553 | 16814.3 | .14607 | 222.43 | 780 | 506.85 | 24609.0 | 54.700 | 18123.8 | .11856 | 224.72 |
| 731 | | 22918.1 | 41.788 | 16840.3 | .14544 | 222.48 | 781 | | 24643.7 | 54.994 | 18150.2 | .11808 | 224.76 |
| 732 | | 22952.4 | 42.024 | 16866.3 | .14482 | 222.53 | 782 | | 24678.5 | 55.288 | 18176.6 | .11760 | 224.81 |
| 733 | | 22986.7 | 42.261 | 16892.2 | .14421 | 222.57 | 783 | | 24713.2 | 55.585 | 18203.1 | .11712 | 224.85 |
| 734 | | 23021.0 | 42.500 | 16918.2 | .14359 | 222.62 | 784 | | 24748.0 | 55.882 | 18229.5 | .11665 | 224.90 |
| 735 | 461.85 | 23055.3 | 42.740 | 16944.2 | .14298 | 222.67 | 785 | 511.85 | 24782.7 | 56.181 | 18255.9 | .11618 | 224.94 |
| 736 | | 23089.6 | 42.980 | 16970.2 | .14238 | 222.71 | 786 | | 24817.5 | 56.480 | 18282.4 | .11571 | 224.99 |
| 737 | | 23124.0 | 43.221 | 16996.3 | .14178 | 222.76 | 787 | | 24852.3 | 56.782 | 18308.9 | .11524 | 225.03 |
| 738 | | 23158.3 | 43.464 | 17022.3 | .14118 | 222.81 | 788 | | 24887.1 | 57.084 | 18335.4 | .11477 | 225.07 |
| 739 | | 23192.7 | 43.708 | 17048.4 | .14058 | 222.85 | 789 | | 24921.9 | 57.388 | 18361.8 | .11431 | 225.12 |
| 740 | 466.85 | 23227.1 | 43.953 | 17074.4 | .13998 | 222.90 | 790 | 516.85 | 24956.7 | 57.693 | 18388.3 | .11385 | 225.16 |
| 741 | | 23261.4 | 44.199 | 17100.5 | .13939 | 222.95 | 791 | | 24991.6 | 58.000 | 18414.9 | .11339 | 225.21 |
| 742 | | 23295.8 | 44.446 | 17126.5 | .13880 | 222.99 | 792 | | 25026.4 | 58.308 | 18441.4 | .11294 | 225.25 |
| 743 | | 23330.2 | 44.694 | 17152.6 | .13822 | 223.04 | 793 | | 25061.2 | 58.617 | 18467.9 | .11248 | 225.29 |
| 744 | | 23364.6 | 44.944 | 17178.7 | .13764 | 223.09 | 794 | | 25096.0 | 58.927 | 18494.4 | .11203 | 225.34 |
| 745 | 471.85 | 23399.0 | 45.194 | 17204.8 | .13706 | 223.13 | 795 | 521.85 | 25130.9 | 59.239 | 18521.0 | .11158 | 225.38 |
| 746 | | 23433.4 | 45.446 | 17230.9 | .13648 | 223.18 | 796 | | 25165.8 | 59.552 | 18547.5 | .11113 | 225.43 |
| 747 | | 23467.9 | 45.699 | 17257.0 | .13591 | 223.22 | 797 | | 25200.7 | 59.867 | 18574.1 | .11069 | 225.47 |
| 748 | | 23502.3 | 45.953 | 17283.1 | .13534 | 223.27 | 798 | | 25235.5 | 60.182 | 18600.6 | .11025 | 225.51 |
| 749 | | 23536.8 | 46.208 | 17309.3 | .13477 | 223.32 | 799 | | 25270.4 | 60.500 | 18627.2 | .10981 | 225.56 |

# Table 7  Products—100% Theoretical Air (for One Gram-Mole)

| T | t | $\bar{h}$ | $p_r$ | $\bar{u}$ | $v_r$ | $\bar{\phi}$ | T | t | $\bar{h}$ | $p_r$ | $\bar{u}$ | $v_r$ | $\bar{\phi}$ |
|---|---|---|---|---|---|---|---|---|---|---|---|---|---|
| 800 | 526.85 | 25305.3 | 60.818 | 18653.8 | .109368 | 225.60 | 850 | 576.85 | 27061.4 | 78.567 | 19994.2 | .089952 | 227.73 |
| 801 | | 25340.2 | 61.137 | 18680.4 | .108932 | 225.64 | 851 | | 27096.7 | 78.961 | 20021.2 | .089608 | 227.77 |
| 802 | | 25375.1 | 61.459 | 18707.0 | .108497 | 225.69 | 852 | | 27132.1 | 79.356 | 20048.2 | .089267 | 227.81 |
| 803 | | 25410.1 | 61.782 | 18733.6 | .108066 | 225.73 | 853 | | 27167.4 | 79.752 | 20075.3 | .088928 | 227.85 |
| 804 | | 25445.0 | 62.106 | 18760.2 | .107636 | 225.78 | 854 | | 27202.8 | 80.151 | 20102.3 | .088589 | 227.90 |
| 805 | 531.85 | 25479.9 | 62.431 | 18786.9 | .107208 | 225.82 | 855 | 581.85 | 27238.2 | 80.552 | 20129.4 | .088252 | 227.94 |
| 806 | | 25514.9 | 62.758 | 18813.5 | .106783 | 225.86 | 856 | | 27273.6 | 80.953 | 20156.4 | .087917 | 227.98 |
| 807 | | 25549.9 | 63.086 | 18840.2 | .106359 | 225.91 | 857 | | 27309.0 | 81.357 | 20183.5 | .087583 | 228.02 |
| 808 | | 25584.8 | 63.415 | 18866.8 | .105937 | 225.95 | 858 | | 27344.4 | 81.761 | 20210.6 | .087251 | 228.06 |
| 809 | | 25619.8 | 63.746 | 18893.4 | .105519 | 225.99 | 859 | | 27379.8 | 82.168 | 20237.7 | .086920 | 228.10 |
| 810 | 536.85 | 25654.8 | 64.078 | 18920.1 | .105101 | 226.03 | 860 | 586.85 | 27415.2 | 82.576 | 20264.8 | .086591 | 228.14 |
| 811 | | 25689.8 | 64.412 | 18946.8 | .104685 | 226.08 | 861 | | 27450.6 | 82.986 | 20291.9 | .086264 | 228.18 |
| 812 | | 25724.8 | 64.747 | 18973.5 | .104272 | 226.12 | 862 | | 27486.1 | 83.397 | 20319.1 | .085938 | 228.23 |
| 813 | | 25759.8 | 65.083 | 19000.2 | .103862 | 226.16 | 863 | | 27521.5 | 83.811 | 20346.2 | .085613 | 228.27 |
| 814 | | 25794.8 | 65.421 | 19026.9 | .103452 | 226.21 | 864 | | 27557.0 | 84.226 | 20373.3 | .085290 | 228.31 |
| 815 | 541.85 | 25829.9 | 65.760 | 19053.6 | .103045 | 226.25 | 865 | 591.85 | 27592.4 | 84.643 | 20400.5 | .084968 | 228.35 |
| 816 | | 25864.9 | 66.101 | 19080.4 | .102640 | 226.29 | 866 | | 27627.9 | 85.061 | 20427.6 | .084648 | 228.39 |
| 817 | | 25899.9 | 66.443 | 19107.1 | .102236 | 226.34 | 867 | | 27663.3 | 85.481 | 20454.8 | .084330 | 228.43 |
| 818 | | 25935.0 | 66.786 | 19133.8 | .101835 | 226.38 | 868 | | 27698.8 | 85.903 | 20481.9 | .084012 | 228.47 |
| 819 | | 25970.1 | 67.131 | 19160.6 | .101436 | 226.42 | 869 | | 27734.3 | 86.325 | 20509.1 | .083698 | 228.51 |
| 820 | 546.85 | 26005.2 | 67.478 | 19187.4 | .101038 | 226.46 | 870 | 596.85 | 27769.8 | 86.751 | 20536.3 | .083383 | 228.55 |
| 821 | | 26040.2 | 67.826 | 19214.1 | .100642 | 226.51 | 871 | | 27805.3 | 87.178 | 20563.5 | .083070 | 228.59 |
| 822 | | 26075.3 | 68.175 | 19240.9 | .100248 | 226.55 | 872 | | 27840.9 | 87.605 | 20590.7 | .082759 | 228.64 |
| 823 | | 26110.4 | 68.526 | 19267.7 | .099856 | 226.59 | 873 | | 27876.4 | 88.036 | 20617.9 | .082449 | 228.68 |
| 824 | | 26145.5 | 68.878 | 19294.5 | .099467 | 226.64 | 874 | | 27911.9 | 88.467 | 20645.1 | .082141 | 228.72 |
| 825 | 551.85 | 26180.7 | 69.232 | 19321.3 | .099078 | 226.68 | 875 | 601.85 | 27947.4 | 88.900 | 20672.3 | .081834 | 228.76 |
| 826 | | 26215.8 | 69.588 | 19348.1 | .098691 | 226.72 | 876 | | 27983.0 | 89.335 | 20699.6 | .081529 | 228.80 |
| 827 | | 26250.9 | 69.945 | 19374.9 | .098306 | 226.76 | 877 | | 28018.6 | 89.773 | 20726.9 | .081224 | 228.84 |
| 828 | | 26286.1 | 70.303 | 19401.7 | .097924 | 226.81 | 878 | | 28054.2 | 90.212 | 20754.1 | .080922 | 228.88 |
| 829 | | 26321.2 | 70.663 | 19428.6 | .097543 | 226.85 | 879 | | 28089.7 | 90.652 | 20781.4 | .080620 | 228.92 |
| 830 | 556.85 | 26356.4 | 71.024 | 19455.4 | .097164 | 226.89 | 880 | 606.85 | 28125.3 | 91.094 | 20808.6 | .080320 | 228.96 |
| 831 | | 26391.5 | 71.386 | 19482.3 | .096787 | 226.93 | 881 | | 28160.9 | 91.538 | 20835.9 | .080021 | 229.00 |
| 832 | | 26426.7 | 71.750 | 19509.1 | .096412 | 226.98 | 882 | | 28196.5 | 91.984 | 20863.2 | .079724 | 229.04 |
| 833 | | 26461.9 | 72.117 | 19536.0 | .096037 | 227.02 | 883 | | 28232.1 | 92.431 | 20890.5 | .079428 | 229.08 |
| 834 | | 26497.1 | 72.484 | 19562.9 | .095666 | 227.06 | 884 | | 28267.7 | 92.881 | 20917.8 | .079133 | 229.12 |
| 835 | 561.85 | 26532.3 | 72.852 | 19589.8 | .095297 | 227.10 | 885 | 611.85 | 28303.4 | 93.332 | 20945.1 | .078839 | 229.16 |
| 836 | | 26567.5 | 73.222 | 19616.7 | .094928 | 227.14 | 886 | | 28339.0 | 93.785 | 20972.4 | .078547 | 229.20 |
| 837 | | 26602.7 | 73.594 | 19643.6 | .094561 | 227.19 | 887 | | 28374.6 | 94.239 | 20999.7 | .078257 | 229.24 |
| 838 | | 26638.0 | 73.968 | 19670.5 | .094196 | 227.23 | 888 | | 28410.3 | 94.696 | 21027.1 | .077967 | 229.28 |
| 839 | | 26673.2 | 74.342 | 19697.4 | .093834 | 227.27 | 889 | | 28445.9 | 95.155 | 21054.4 | .077679 | 229.32 |
| 840 | 566.85 | 26708.5 | 74.718 | 19724.4 | .093472 | 227.31 | 890 | 616.85 | 28481.6 | 95.614 | 21081.8 | .077392 | 229.36 |
| 841 | | 26743.7 | 75.097 | 19751.3 | .093112 | 227.35 | 891 | | 28517.3 | 96.076 | 21109.2 | .077107 | 229.40 |
| 842 | | 26779.0 | 75.476 | 19778.3 | .092755 | 227.40 | 892 | | 28553.0 | 96.540 | 21136.5 | .076823 | 229.44 |
| 843 | | 26814.2 | 75.856 | 19805.2 | .092399 | 227.44 | 893 | | 28588.6 | 97.005 | 21163.9 | .076540 | 229.48 |
| 844 | | 26849.5 | 76.240 | 19832.2 | .092044 | 227.48 | 894 | | 28624.4 | 97.472 | 21191.3 | .076258 | 229.52 |
| 845 | 571.85 | 26884.8 | 76.623 | 19859.2 | .091691 | 227.52 | 895 | 621.85 | 28660.1 | 97.941 | 21218.7 | .075978 | 229.56 |
| 846 | | 26920.1 | 77.009 | 19886.1 | .091340 | 227.56 | 896 | | 28695.8 | 98.412 | 21246.1 | .075699 | 229.60 |
| 847 | | 26955.4 | 77.396 | 19913.1 | .090991 | 227.60 | 897 | | 28731.5 | 98.885 | 21273.5 | .075421 | 229.64 |
| 848 | | 26990.7 | 77.785 | 19940.1 | .090643 | 227.65 | 898 | | 28767.2 | 99.359 | 21300.9 | .075145 | 229.68 |
| 849 | | 27026.1 | 78.175 | 19967.1 | .090296 | 227.69 | 899 | | 28803.0 | 99.836 | 21328.3 | .074869 | 229.72 |

## Table 7  Products—100% Theoretical Air (for One Gram-Mole)

| T | t | $\bar{h}$ | $p_r$ | $\bar{u}$ | $v_r$ | $\bar{\phi}$ | T | t | $\bar{h}$ | $p_r$ | $\bar{u}$ | $v_r$ | $\bar{\phi}$ |
|---|---|---|---|---|---|---|---|---|---|---|---|---|---|
| 900 | 626.85 | 28838.7 | 100.31 | 21355.8 | .074595 | 229.76 | 950 | 676.85 | 30636.5 | 126.73 | 22737.8 | .062325 | 231.71 |
| 901 | | 28874.5 | 100.80 | 21383.2 | .074322 | 229.80 | 951 | | 30672.6 | 127.32 | 22765.6 | .062105 | 231.74 |
| 902 | | 28910.3 | 101.28 | 21410.7 | .074051 | 229.84 | 952 | | 30708.8 | 127.90 | 22793.5 | .061887 | 231.78 |
| 903 | | 28946.0 | 101.76 | 21438.1 | .073780 | 229.88 | 953 | | 30744.9 | 128.49 | 22821.3 | .061670 | 231.82 |
| 904 | | 28981.8 | 102.25 | 21465.6 | .073510 | 229.92 | 954 | | 30781.1 | 129.07 | 22849.2 | .061454 | 231.86 |
| 905 | 631.85 | 29017.6 | 102.73 | 21493.1 | .073243 | 229.96 | 955 | 681.85 | 30817.3 | 129.66 | 22877.0 | .061238 | 231.90 |
| 906 | | 29053.4 | 103.22 | 21520.6 | .072976 | 230.00 | 956 | | 30853.5 | 130.26 | 22904.9 | .061023 | 231.93 |
| 907 | | 29089.2 | 103.72 | 21548.0 | .072710 | 230.04 | 957 | | 30889.7 | 130.85 | 22932.8 | .060810 | 231.97 |
| 908 | | 29125.0 | 104.21 | 21575.5 | .072446 | 230.08 | 958 | | 30925.9 | 131.45 | 22960.7 | .060597 | 232.01 |
| 909 | | 29160.8 | 104.70 | 21603.0 | .072182 | 230.12 | 959 | | 30962.1 | 132.05 | 22988.6 | .060385 | 232.05 |
| 910 | 636.85 | 29196.7 | 105.20 | 21630.6 | .071920 | 230.16 | 960 | 686.85 | 30998.4 | 132.65 | 23016.5 | .060174 | 232.08 |
| 911 | | 29232.5 | 105.70 | 21658.1 | .071659 | 230.20 | 961 | | 31034.6 | 133.25 | 23044.5 | .059965 | 232.12 |
| 912 | | 29268.4 | 106.20 | 21685.6 | .071400 | 230.24 | 962 | | 31070.8 | 133.85 | 23072.4 | .059755 | 232.16 |
| 913 | | 29304.2 | 106.71 | 21713.2 | .071140 | 230.28 | 963 | | 31107.1 | 134.46 | 23100.3 | .059547 | 232.20 |
| 914 | | 29340.1 | 107.21 | 21740.7 | .070882 | 230.31 | 964 | | 31143.4 | 135.07 | 23128.3 | .059340 | 232.24 |
| 915 | 641.85 | 29375.9 | 107.72 | 21768.3 | .070626 | 230.35 | 965 | 691.85 | 31179.6 | 135.68 | 23156.2 | .059133 | 232.27 |
| 916 | | 29411.8 | 108.23 | 21795.8 | .070371 | 230.39 | 966 | | 31215.9 | 136.30 | 23184.2 | .058927 | 232.31 |
| 917 | | 29447.7 | 108.74 | 21823.4 | .070116 | 230.43 | 967 | | 31252.2 | 136.92 | 23212.2 | .058722 | 232.35 |
| 918 | | 29483.6 | 109.25 | 21851.0 | .069863 | 230.47 | 968 | | 31288.5 | 137.53 | 23240.1 | .058519 | 232.39 |
| 919 | | 29519.5 | 109.77 | 21878.5 | .069612 | 230.51 | 969 | | 31324.8 | 138.16 | 23268.1 | .058316 | 232.42 |
| 920 | 646.85 | 29555.4 | 110.28 | 21906.2 | .069361 | 230.55 | 970 | 696.85 | 31361.1 | 138.78 | 23296.1 | .058113 | 232.46 |
| 921 | | 29591.3 | 110.80 | 21933.8 | .069111 | 230.59 | 971 | | 31397.4 | 139.41 | 23324.1 | .057912 | 232.50 |
| 922 | | 29627.3 | 111.32 | 21961.4 | .068863 | 230.63 | 972 | | 31433.7 | 140.03 | 23352.1 | .057711 | 232.54 |
| 923 | | 29663.2 | 111.85 | 21989.0 | .068615 | 230.67 | 973 | | 31470.0 | 140.66 | 23380.1 | .057512 | 232.57 |
| 924 | | 29699.1 | 112.37 | 22016.6 | .068369 | 230.71 | 974 | | 31506.3 | 141.30 | 23408.1 | .057313 | 232.61 |
| 925 | 651.85 | 29735.1 | 112.90 | 22044.3 | .068124 | 230.74 | 975 | 701.85 | 31542.7 | 141.93 | 23436.2 | .057115 | 232.65 |
| 926 | | 29771.0 | 113.42 | 22071.9 | .067880 | 230.78 | 976 | | 31579.0 | 142.57 | 23464.2 | .056919 | 232.68 |
| 927 | | 29807.0 | 113.96 | 22099.6 | .067636 | 230.82 | 977 | | 31615.4 | 143.21 | 23492.2 | .056722 | 232.72 |
| 928 | | 29843.0 | 114.49 | 22127.2 | .067394 | 230.86 | 978 | | 31651.8 | 143.85 | 23520.3 | .056527 | 232.76 |
| 929 | | 29879.0 | 115.02 | 22154.9 | .067153 | 230.90 | 979 | | 31688.1 | 144.50 | 23548.3 | .056332 | 232.80 |
| 930 | 656.85 | 29915.0 | 115.56 | 22182.6 | .066913 | 230.94 | 980 | 706.85 | 31724.5 | 145.14 | 23576.4 | .056138 | 232.83 |
| 931 | | 29950.9 | 116.10 | 22210.2 | .066674 | 230.98 | 981 | | 31760.9 | 145.79 | 23604.5 | .055946 | 232.87 |
| 932 | | 29987.0 | 116.64 | 22237.9 | .066436 | 231.02 | 982 | | 31797.3 | 146.45 | 23632.5 | .055753 | 232.91 |
| 933 | | 30023.0 | 117.18 | 22265.6 | .066199 | 231.05 | 983 | | 31833.7 | 147.10 | 23660.6 | .055562 | 232.94 |
| 934 | | 30059.0 | 117.73 | 22293.3 | .065963 | 231.09 | 984 | | 31870.1 | 147.76 | 23688.7 | .055371 | 232.98 |
| 935 | 661.85 | 30095.0 | 118.28 | 22321.1 | .065728 | 231.13 | 985 | 711.85 | 31906.5 | 148.41 | 23716.8 | .055182 | 233.02 |
| 936 | | 30131.1 | 118.82 | 22348.8 | .065494 | 231.17 | 986 | | 31942.9 | 149.07 | 23744.9 | .054993 | 233.06 |
| 937 | | 30167.1 | 119.38 | 22376.5 | .065261 | 231.21 | 987 | | 31979.4 | 149.74 | 23773.0 | .054804 | 233.09 |
| 938 | | 30203.2 | 119.93 | 22404.3 | .065029 | 231.25 | 988 | | 32015.8 | 150.40 | 23801.2 | .054617 | 233.13 |
| 939 | | 30239.2 | 120.48 | 22432.0 | .064799 | 231.28 | 989 | | 32052.2 | 151.07 | 23829.3 | .054430 | 233.17 |
| 940 | 666.85 | 30275.3 | 121.04 | 22459.8 | .064569 | 231.32 | 990 | 716.85 | 32088.7 | 151.74 | 23857.5 | .054244 | 233.20 |
| 941 | | 30311.4 | 121.60 | 22487.5 | .064340 | 231.36 | 991 | | 32125.2 | 152.42 | 23885.6 | .054059 | 233.24 |
| 942 | | 30347.5 | 122.16 | 22515.3 | .064112 | 231.40 | 992 | | 32161.7 | 153.09 | 23913.8 | .053875 | 233.28 |
| 943 | | 30383.6 | 122.73 | 22543.1 | .063885 | 231.44 | 993 | | 32198.1 | 153.77 | 23941.9 | .053691 | 233.31 |
| 944 | | 30419.7 | 123.29 | 22570.9 | .063659 | 231.48 | 994 | | 32234.6 | 154.45 | 23970.1 | .053508 | 233.35 |
| 945 | 671.85 | 30455.8 | 123.86 | 22598.7 | .063434 | 231.51 | 995 | 721.85 | 32271.1 | 155.13 | 23998.3 | .053327 | 233.39 |
| 946 | | 30491.9 | 124.43 | 22626.5 | .063211 | 231.55 | 996 | | 32307.6 | 155.82 | 24026.5 | .053145 | 233.42 |
| 947 | | 30528.0 | 125.00 | 22654.3 | .062988 | 231.59 | 997 | | 32344.1 | 156.51 | 24054.7 | .052965 | 233.46 |
| 948 | | 30564.1 | 125.58 | 22682.1 | .062766 | 231.63 | 998 | | 32380.7 | 157.20 | 24082.9 | .052786 | 233.50 |
| 949 | | 30600.3 | 126.16 | 22709.9 | .062545 | 231.67 | 999 | | 32417.2 | 157.89 | 24111.1 | .052606 | 233.53 |

## Table 7 Products—100% Theoretical Air (for One Gram-Mole)

| T | t | $\bar{h}$ | $p_r$ | $\bar{u}$ | $v_r$ | $\bar{\phi}$ | T | t | $\bar{h}$ | $p_r$ | $\bar{u}$ | $v_r$ | $\bar{\phi}$ |
|---|---|-----------|-------|-----------|-------|--------------|---|---|-----------|-------|-----------|-------|--------------|
| 1000 | 726.85 | 32453.7 | 158.59 | 24139.3 | .052428 | 233.57 | 1050 | 776.85 | 34289.4 | 196.71 | 25559.3 | .044381 | 235.36 |
| 1001 | | 32490.2 | 159.28 | 24167.5 | .052253 | 233.61 | 1051 | | 34326.3 | 197.53 | 25587.9 | .044238 | 235.40 |
| 1002 | | 32526.7 | 159.99 | 24195.7 | .052073 | 233.64 | 1052 | | 34363.1 | 198.39 | 25616.4 | .044089 | 235.43 |
| 1003 | | 32563.3 | 160.68 | 24223.9 | .051901 | 233.68 | 1053 | | 34400.3 | 199.22 | 25645.2 | .043947 | 235.47 |
| 1004 | | 32599.9 | 161.40 | 24252.2 | .051722 | 233.72 | 1054 | | 34437.2 | 200.05 | 25673.8 | .043805 | 235.50 |
| 1005 | 731.85 | 32636.5 | 162.09 | 24280.5 | .051551 | 233.75 | 1055 | 781.85 | 34474.0 | 200.89 | 25702.3 | .043664 | 235.54 |
| 1006 | | 32673.1 | 162.81 | 24308.8 | .051373 | 233.79 | 1056 | | 34510.9 | 201.76 | 25730.9 | .043518 | 235.57 |
| 1007 | | 32709.7 | 163.52 | 24337.1 | .051204 | 233.82 | 1057 | | 34547.8 | 202.60 | 25759.5 | .043377 | 235.61 |
| 1008 | | 32746.3 | 164.24 | 24365.3 | .051028 | 233.86 | 1058 | | 34584.7 | 203.45 | 25788.1 | .043238 | 235.64 |
| 1009 | | 32782.9 | 164.95 | 24393.6 | .050859 | 233.90 | 1059 | | 34621.9 | 204.30 | 25816.9 | .043099 | 235.68 |
| 1010 | 736.85 | 32819.5 | 165.68 | 24422.0 | .050684 | 233.93 | 1060 | 786.85 | 34658.8 | 205.18 | 25845.5 | .042955 | 235.71 |
| 1011 | | 32856.1 | 166.40 | 24450.2 | .050517 | 233.97 | 1061 | | 34695.7 | 206.03 | 25874.1 | .042816 | 235.75 |
| 1012 | | 32892.7 | 167.13 | 24478.6 | .050344 | 234.01 | 1062 | | 34732.6 | 206.89 | 25902.7 | .042679 | 235.78 |
| 1013 | | 32929.4 | 167.85 | 24506.9 | .050178 | 234.04 | 1063 | | 34769.8 | 207.75 | 25931.6 | .042542 | 235.81 |
| 1014 | | 32966.0 | 168.60 | 24535.2 | .050006 | 234.08 | 1064 | | 34806.7 | 208.65 | 25960.2 | .042400 | 235.85 |
| 1015 | 741.85 | 33002.7 | 169.32 | 24563.5 | .049842 | 234.11 | 1065 | 791.85 | 34843.7 | 209.51 | 25988.8 | .042264 | 235.88 |
| 1016 | | 33039.3 | 170.07 | 24591.9 | .049671 | 234.15 | 1066 | | 34880.6 | 210.38 | 26017.4 | .042128 | 235.92 |
| 1017 | | 33076.0 | 170.80 | 24620.2 | .049508 | 234.19 | 1067 | | 34917.6 | 211.26 | 26046.1 | .041993 | 235.95 |
| 1018 | | 33112.6 | 171.55 | 24648.5 | .049338 | 234.22 | 1068 | | 34954.8 | 212.16 | 26075.0 | .041853 | 235.99 |
| 1019 | | 33149.3 | 172.29 | 24676.9 | .049176 | 234.26 | 1069 | | 34991.8 | 213.04 | 26103.7 | .041720 | 236.02 |
| 1020 | 746.85 | 33186.0 | 173.02 | 24705.3 | .049015 | 234.29 | 1070 | 796.85 | 35028.7 | 213.93 | 26132.3 | .041586 | 236.06 |
| 1021 | | 33222.6 | 173.79 | 24733.6 | .048848 | 234.33 | 1071 | | 35065.7 | 214.81 | 26160.9 | .041453 | 236.09 |
| 1022 | | 33259.3 | 174.53 | 24762.0 | .048687 | 234.37 | 1072 | | 35102.9 | 215.70 | 26189.9 | .041321 | 236.13 |
| 1023 | | 33296.0 | 175.30 | 24790.4 | .048521 | 234.40 | 1073 | | 35139.9 | 216.63 | 26218.6 | .041183 | 236.16 |
| 1024 | | 33332.7 | 176.05 | 24818.7 | .048362 | 234.44 | 1074 | | 35176.9 | 217.52 | 26247.2 | .041052 | 236.20 |
| 1025 | 751.85 | 33369.4 | 176.80 | 24847.2 | .048204 | 234.47 | 1075 | 801.85 | 35214.2 | 218.42 | 26276.2 | .040921 | 236.23 |
| 1026 | | 33406.1 | 177.57 | 24875.5 | .048040 | 234.51 | 1076 | | 35251.1 | 219.32 | 26304.9 | .040790 | 236.27 |
| 1027 | | 33442.8 | 178.33 | 24903.9 | .047883 | 234.55 | 1077 | | 35288.1 | 220.23 | 26333.5 | .040660 | 236.30 |
| 1028 | | 33479.5 | 179.11 | 24932.3 | .047720 | 234.58 | 1078 | | 35325.1 | 221.17 | 26362.2 | .040525 | 236.34 |
| 1029 | | 33516.3 | 179.87 | 24960.7 | .047564 | 234.62 | 1079 | | 35362.4 | 222.08 | 26391.2 | .040396 | 236.37 |
| 1030 | 756.85 | 33553.0 | 180.64 | 24989.2 | .047409 | 234.65 | 1080 | 806.85 | 35399.5 | 223.00 | 26419.9 | .040268 | 236.40 |
| 1031 | | 33589.7 | 181.43 | 25017.6 | .047248 | 234.69 | 1081 | | 35436.5 | 223.91 | 26448.6 | .040140 | 236.44 |
| 1032 | | 33626.5 | 182.20 | 25046.0 | .047093 | 234.72 | 1082 | | 35473.8 | 224.84 | 26477.6 | .040012 | 236.47 |
| 1033 | | 33663.2 | 182.97 | 25074.5 | .046940 | 234.76 | 1083 | | 35510.8 | 225.76 | 26506.3 | .039885 | 236.51 |
| 1034 | | 33700.0 | 183.77 | 25102.9 | .046781 | 234.80 | 1084 | | 35547.9 | 226.69 | 26535.1 | .039758 | 236.54 |
| 1035 | 761.85 | 33736.8 | 184.55 | 25131.4 | .046628 | 234.83 | 1085 | 811.85 | 35585.2 | 227.65 | 26564.1 | .039627 | 236.58 |
| 1036 | | 33773.5 | 185.33 | 25159.8 | .046477 | 234.87 | 1086 | | 35622.3 | 228.59 | 26592.8 | .039501 | 236.61 |
| 1037 | | 33810.3 | 186.14 | 25188.3 | .046319 | 234.90 | 1087 | | 35659.3 | 229.53 | 26621.5 | .039376 | 236.64 |
| 1038 | | 33847.1 | 186.93 | 25216.7 | .046169 | 234.94 | 1088 | | 35696.7 | 230.47 | 26650.6 | .039251 | 236.68 |
| 1039 | | 33884.2 | 187.75 | 25245.5 | .046012 | 234.97 | 1089 | | 35733.7 | 231.41 | 26679.3 | .039127 | 236.71 |
| 1040 | 766.85 | 33921.0 | 188.54 | 25274.0 | .045863 | 235.01 | 1090 | 816.85 | 35770.8 | 232.36 | 26708.1 | .039003 | 236.75 |
| 1041 | | 33957.7 | 189.34 | 25302.5 | .045714 | 235.04 | 1091 | | 35808.1 | 233.31 | 26737.1 | .038879 | 236.78 |
| 1042 | | 33994.5 | 190.16 | 25330.9 | .045560 | 235.08 | 1092 | | 35845.2 | 234.27 | 26765.9 | .038756 | 236.81 |
| 1043 | | 34031.3 | 190.96 | 25359.4 | .045412 | 235.11 | 1093 | | 35882.3 | 235.23 | 26794.6 | .038634 | 236.85 |
| 1044 | | 34068.2 | 191.77 | 25387.9 | .045265 | 235.15 | 1094 | | 35919.7 | 236.22 | 26823.7 | .038507 | 236.88 |
| 1045 | 771.85 | 34105.0 | 192.57 | 25416.4 | .045118 | 235.18 | 1095 | 821.85 | 35956.8 | 237.18 | 26852.5 | .038385 | 236.92 |
| 1046 | | 34141.8 | 193.41 | 25444.9 | .044966 | 235.22 | 1096 | | 35993.9 | 238.15 | 26881.3 | .038264 | 236.95 |
| 1047 | | 34178.9 | 194.22 | 25473.7 | .044820 | 235.25 | 1097 | | 36031.3 | 239.12 | 26910.4 | .038143 | 236.98 |
| 1048 | | 34215.7 | 195.04 | 25502.3 | .044675 | 235.29 | 1098 | | 36068.4 | 240.10 | 26939.2 | .038023 | 237.02 |
| 1049 | | 34252.6 | 195.89 | 25530.8 | .044525 | 235.33 | 1099 | | 36105.8 | 241.08 | 26968.3 | .037903 | 237.05 |

## Table 7 Products—100% Theoretical Air (for One Gram-Mole)

| T | t | $\bar{h}$ | $p_r$ | $\bar{u}$ | $v_r$ | $\bar{\phi}$ | T | t | $\bar{h}$ | $p_r$ | $\bar{u}$ | $v_r$ | $\bar{\phi}$ |
|---|---|---|---|---|---|---|---|---|---|---|---|---|---|
| 1100 | 826.85 | 36142.9 | 242.06 | 26997.1 | .037784 | 237.09 | 1150 | 876.85 | 38013.1 | 295.62 | 28451.5 | .032344 | 238.75 |
| 1101 |  | 36180.0 | 243.04 | 27025.9 | .037665 | 237.12 | 1151 |  | 38050.8 | 296.79 | 28480.9 | .032244 | 238.78 |
| 1102 |  | 36217.5 | 244.03 | 27055.0 | .037546 | 237.15 | 1152 |  | 38088.2 | 297.97 | 28510.0 | .032145 | 238.81 |
| 1103 |  | 36254.6 | 245.03 | 27083.8 | .037428 | 237.19 | 1153 |  | 38125.9 | 299.14 | 28539.4 | .032046 | 238.85 |
| 1104 |  | 36292.0 | 246.02 | 27112.9 | .037310 | 237.22 | 1154 |  | 38163.6 | 300.29 | 28568.7 | .031952 | 238.88 |
| 1105 | 831.85 | 36329.2 | 247.02 | 27141.8 | .037193 | 237.25 | 1155 | 881.85 | 38201.0 | 301.47 | 28597.9 | .031854 | 238.91 |
| 1106 |  | 36366.4 | 248.02 | 27170.6 | .037076 | 237.29 | 1156 |  | 38238.7 | 302.66 | 28627.3 | .031757 | 238.94 |
| 1107 |  | 36403.8 | 249.03 | 27199.7 | .036960 | 237.32 | 1157 |  | 38276.1 | 303.85 | 28656.3 | .031659 | 238.98 |
| 1108 |  | 36441.0 | 250.04 | 27228.6 | .036844 | 237.36 | 1158 |  | 38313.8 | 305.05 | 28685.8 | .031562 | 239.01 |
| 1109 |  | 36478.4 | 251.05 | 27257.8 | .036728 | 237.39 | 1159 |  | 38351.6 | 306.21 | 28715.2 | .031470 | 239.04 |
| 1110 | 836.85 | 36515.6 | 252.07 | 27286.6 | .036613 | 237.42 | 1160 | 886.85 | 38389.0 | 307.42 | 28744.3 | .031373 | 239.07 |
| 1111 |  | 36553.1 | 253.09 | 27315.8 | .036498 | 237.46 | 1161 |  | 38426.7 | 308.63 | 28773.7 | .031277 | 239.11 |
| 1112 |  | 36590.3 | 254.11 | 27344.6 | .036384 | 237.49 | 1162 |  | 38464.5 | 309.84 | 28803.1 | .031182 | 239.14 |
| 1113 |  | 36627.4 | 255.14 | 27373.5 | .036270 | 237.52 | 1163 |  | 38501.9 | 311.06 | 28832.3 | .031086 | 239.17 |
| 1114 |  | 36664.9 | 256.17 | 27402.7 | .036156 | 237.56 | 1164 |  | 38539.7 | 312.24 | 28861.7 | .030995 | 239.20 |
| 1115 | 841.85 | 36702.1 | 257.21 | 27431.6 | .036043 | 237.59 | 1165 | 891.85 | 38577.4 | 313.46 | 28891.1 | .030901 | 239.23 |
| 1116 |  | 36739.6 | 258.24 | 27460.8 | .035931 | 237.62 | 1166 |  | 38614.9 | 314.69 | 28920.3 | .030806 | 239.27 |
| 1117 |  | 36776.9 | 259.29 | 27489.7 | .035818 | 237.66 | 1167 |  | 38652.7 | 315.93 | 28949.8 | .030712 | 239.30 |
| 1118 |  | 36814.4 | 260.33 | 27518.9 | .035706 | 237.69 | 1168 |  | 38690.4 | 317.12 | 28979.2 | .030623 | 239.33 |
| 1119 |  | 36851.6 | 261.38 | 27547.8 | .035595 | 237.72 | 1169 |  | 38727.9 | 318.37 | 29008.4 | .030529 | 239.36 |
| 1120 | 846.85 | 36889.1 | 262.43 | 27577.0 | .035484 | 237.76 | 1170 | 896.85 | 38765.7 | 319.61 | 29037.8 | .030436 | 239.40 |
| 1121 |  | 36926.3 | 263.49 | 27605.9 | .035373 | 237.79 | 1171 |  | 38803.5 | 320.86 | 29067.3 | .030344 | 239.43 |
| 1122 |  | 36963.9 | 264.55 | 27635.1 | .035263 | 237.82 | 1172 |  | 38841.3 | 322.12 | 29096.8 | .030251 | 239.46 |
| 1123 |  | 37001.1 | 265.61 | 27664.1 | .035153 | 237.86 | 1173 |  | 38878.8 | 323.33 | 29126.0 | .030163 | 239.49 |
| 1124 |  | 37038.7 | 266.68 | 27693.3 | .035044 | 237.89 | 1174 |  | 38916.6 | 324.60 | 29155.5 | .030072 | 239.52 |
| 1125 | 851.85 | 37075.9 | 267.75 | 27722.2 | .034935 | 237.92 | 1175 | 901.85 | 38954.4 | 325.86 | 29184.9 | .029980 | 239.56 |
| 1126 |  | 37113.5 | 268.82 | 27751.5 | .034826 | 237.96 | 1176 |  | 38992.2 | 327.09 | 29214.5 | .029893 | 239.59 |
| 1127 |  | 37150.8 | 269.90 | 27780.4 | .034718 | 237.99 | 1177 |  | 39029.7 | 328.37 | 29243.7 | .029802 | 239.62 |
| 1128 |  | 37188.3 | 270.98 | 27809.7 | .034610 | 238.02 | 1178 |  | 39067.5 | 329.65 | 29273.2 | .029712 | 239.65 |
| 1129 |  | 37225.6 | 272.03 | 27838.6 | .034507 | 238.06 | 1179 |  | 39105.4 | 330.89 | 29302.7 | .029625 | 239.68 |
| 1130 | 856.85 | 37263.2 | 273.12 | 27867.9 | .034400 | 238.09 | 1180 | 906.85 | 39143.2 | 332.18 | 29332.2 | .029536 | 239.72 |
| 1131 |  | 37300.4 | 274.22 | 27896.8 | .034293 | 238.12 | 1181 |  | 39180.7 | 333.47 | 29361.4 | .029446 | 239.75 |
| 1132 |  | 37338.0 | 275.31 | 27926.1 | .034186 | 238.16 | 1182 |  | 39218.6 | 334.76 | 29391.0 | .029357 | 239.78 |
| 1133 |  | 37375.3 | 276.41 | 27955.1 | .034080 | 238.19 | 1183 |  | 39256.4 | 336.02 | 29420.5 | .029272 | 239.81 |
| 1134 |  | 37412.9 | 277.52 | 27984.4 | .033975 | 238.22 | 1184 |  | 39294.3 | 337.33 | 29450.0 | .029183 | 239.84 |
| 1135 | 861.85 | 37450.2 | 278.62 | 28013.4 | .033870 | 238.26 | 1185 | 911.85 | 39331.8 | 338.64 | 29479.3 | .029095 | 239.88 |
| 1136 |  | 37487.8 | 279.74 | 28042.7 | .033765 | 238.29 | 1186 |  | 39369.7 | 339.91 | 29508.8 | .029011 | 239.91 |
| 1137 |  | 37525.2 | 280.85 | 28071.7 | .033660 | 238.32 | 1187 |  | 39407.6 | 341.22 | 29538.4 | .028923 | 239.94 |
| 1138 |  | 37562.8 | 281.97 | 28101.0 | .033556 | 238.35 | 1188 |  | 39445.4 | 342.55 | 29567.9 | .028836 | 239.97 |
| 1139 |  | 37600.1 | 283.09 | 28130.0 | .033452 | 238.39 | 1189 |  | 39483.0 | 343.87 | 29597.2 | .028748 | 240.00 |
| 1140 | 866.85 | 37637.7 | 284.18 | 28159.3 | .033353 | 238.42 | 1190 | 916.85 | 39520.9 | 345.16 | 29626.7 | .028665 | 240.04 |
| 1141 |  | 37675.4 | 285.31 | 28188.6 | .033250 | 238.45 | 1191 |  | 39558.8 | 346.50 | 29656.3 | .028579 | 240.07 |
| 1142 |  | 37712.7 | 286.45 | 28217.6 | .033148 | 238.49 | 1192 |  | 39596.7 | 347.84 | 29685.9 | .028493 | 240.10 |
| 1143 |  | 37750.3 | 287.59 | 28247.0 | .033045 | 238.52 | 1193 |  | 39634.3 | 349.13 | 29715.2 | .028411 | 240.13 |
| 1144 |  | 37787.7 | 288.73 | 28276.0 | .032943 | 238.55 | 1194 |  | 39672.2 | 350.48 | 29744.8 | .028325 | 240.16 |
| 1145 | 871.85 | 37825.3 | 289.87 | 28305.3 | .032842 | 238.58 | 1195 | 921.85 | 39710.1 | 351.84 | 29774.4 | .028240 | 240.19 |
| 1146 |  | 37863.0 | 291.02 | 28334.7 | .032741 | 238.62 | 1196 |  | 39748.0 | 353.15 | 29804.0 | .028158 | 240.23 |
| 1147 |  | 37900.4 | 292.14 | 28363.8 | .032644 | 238.65 | 1197 |  | 39785.9 | 354.51 | 29833.6 | .028074 | 240.26 |
| 1148 |  | 37938.1 | 293.30 | 28393.1 | .032544 | 238.68 | 1198 |  | 39823.5 | 355.88 | 29862.9 | .027989 | 240.29 |
| 1149 |  | 37975.4 | 294.46 | 28422.2 | .032443 | 238.71 | 1199 |  | 39861.4 | 357.20 | 29892.5 | .027909 | 240.32 |

## Table 7 Products—100% Theoretical Air (for One Gram-Mole)

| T | t | $\bar{h}$ | $p_r$ | $\bar{u}$ | $v_r$ | $\bar{\phi}$ | T | t | $\bar{h}$ | $p_r$ | $\bar{u}$ | $v_r$ | $\bar{\phi}$ |
|---|---|---|---|---|---|---|---|---|---|---|---|---|---|
| 1200 | 926.85 | 39899.4 | 358.58 | 29922.1 | .027825 | 240.35 | 1250 | 976.85 | 41800.5 | 432.25 | 31407.5 | .024044 | 241.91 |
| 1201 | | 39937.3 | 359.96 | 29951.7 | .027741 | 240.38 | 1251 | | 41838.6 | 433.81 | 31437.3 | .023977 | 241.94 |
| 1202 | | 39975.2 | 361.29 | 29981.3 | .027661 | 240.42 | 1252 | | 41876.8 | 435.43 | 31467.2 | .023906 | 241.97 |
| 1203 | | 40012.9 | 362.68 | 30010.7 | .027578 | 240.45 | 1253 | | 41915.0 | 437.01 | 31497.1 | .023839 | 242.00 |
| 1204 | | 40050.9 | 364.08 | 30040.3 | .027496 | 240.48 | 1254 | | 41953.2 | 438.64 | 31526.9 | .023769 | 242.03 |
| 1205 | 931.85 | 40088.8 | 365.43 | 30070.0 | .027417 | 240.51 | 1255 | 981.85 | 41991.4 | 440.23 | 31556.8 | .023703 | 242.06 |
| 1206 | | 40126.8 | 366.83 | 30099.6 | .027335 | 240.54 | 1256 | | 42029.6 | 441.87 | 31586.7 | .023633 | 242.09 |
| 1207 | | 40164.7 | 368.24 | 30129.2 | .027253 | 240.57 | 1257 | | 42067.8 | 443.47 | 31616.6 | .023567 | 242.12 |
| 1208 | | 40202.7 | 369.60 | 30158.9 | .027175 | 240.60 | 1258 | | 42106.0 | 445.12 | 31646.5 | .023498 | 242.15 |
| 1209 | | 40240.4 | 371.02 | 30188.3 | .027093 | 240.64 | 1259 | | 42144.2 | 446.73 | 31676.4 | .023432 | 242.18 |
| 1210 | 936.85 | 40278.3 | 372.39 | 30217.9 | .027016 | 240.67 | 1260 | 986.85 | 42182.4 | 448.34 | 31706.3 | .023367 | 242.21 |
| 1211 | | 40316.3 | 373.81 | 30247.6 | .026935 | 240.70 | 1261 | | 42220.6 | 450.01 | 31736.1 | .023298 | 242.24 |
| 1212 | | 40354.3 | 375.24 | 30277.3 | .026855 | 240.73 | 1262 | | 42258.8 | 451.63 | 31766.1 | .023233 | 242.27 |
| 1213 | | 40392.3 | 376.63 | 30306.9 | .026778 | 240.76 | 1263 | | 42297.1 | 453.31 | 31796.0 | .023165 | 242.30 |
| 1214 | | 40430.3 | 378.07 | 30336.6 | .026698 | 240.79 | 1264 | | 42335.3 | 454.94 | 31825.9 | .023101 | 242.33 |
| 1215 | 941.85 | 40468.3 | 379.51 | 30366.3 | .026618 | 240.82 | 1265 | 991.85 | 42373.5 | 456.63 | 31855.8 | .023033 | 242.36 |
| 1216 | | 40506.0 | 380.91 | 30395.7 | .026542 | 240.86 | 1266 | | 42411.8 | 458.27 | 31885.8 | .022969 | 242.39 |
| 1217 | | 40544.0 | 382.37 | 30425.4 | .026463 | 240.89 | 1267 | | 42450.0 | 459.91 | 31915.7 | .022905 | 242.42 |
| 1218 | | 40582.1 | 383.77 | 30455.1 | .026388 | 240.92 | 1268 | | 42488.6 | 461.62 | 31945.9 | .022838 | 242.45 |
| 1219 | | 40620.1 | 385.24 | 30484.8 | .026309 | 240.95 | 1269 | | 42526.8 | 463.28 | 31975.9 | .022775 | 242.48 |
| 1220 | 946.85 | 40658.1 | 386.71 | 30514.5 | .026231 | 240.98 | 1270 | 996.85 | 42565.1 | 465.00 | 32005.8 | .022708 | 242.51 |
| 1221 | | 40696.1 | 388.13 | 30544.2 | .026156 | 241.01 | 1271 | | 42603.4 | 466.66 | 32035.8 | .022645 | 242.54 |
| 1222 | | 40734.2 | 389.60 | 30574.0 | .026078 | 241.04 | 1272 | | 42641.6 | 468.33 | 32065.7 | .022582 | 242.57 |
| 1223 | | 40772.2 | 391.04 | 30603.7 | .026004 | 241.07 | 1273 | | 42679.9 | 470.07 | 32095.6 | .022516 | 242.60 |
| 1224 | | 40810.0 | 392.52 | 30633.1 | .025927 | 241.10 | 1274 | | 42718.2 | 471.75 | 32125.6 | .022454 | 242.63 |
| 1225 | 951.85 | 40848.0 | 393.96 | 30662.9 | .025853 | 241.14 | 1275 | 1001.85 | 42756.5 | 473.43 | 32155.6 | .022392 | 242.66 |
| 1226 | | 40886.1 | 395.46 | 30692.6 | .025776 | 241.17 | 1276 | | 42794.7 | 475.12 | 32185.6 | .022329 | 242.69 |
| 1227 | | 40924.1 | 396.96 | 30722.3 | .025700 | 241.20 | 1277 | | 42833.0 | 476.88 | 32215.5 | .022264 | 242.72 |
| 1228 | | 40962.2 | 398.47 | 30752.1 | .025624 | 241.23 | 1278 | | 42871.3 | 478.58 | 32245.5 | .022203 | 242.75 |
| 1229 | | 41000.3 | 399.92 | 30781.9 | .025551 | 241.26 | 1279 | | 42909.9 | 480.29 | 32275.8 | .022141 | 242.78 |
| 1230 | 956.85 | 41038.3 | 401.44 | 30811.6 | .025475 | 241.29 | 1280 | 1006.85 | 42948.2 | 482.06 | 32305.8 | .022077 | 242.81 |
| 1231 | | 41076.4 | 402.91 | 30841.4 | .025403 | 241.32 | 1281 | | 42986.5 | 483.78 | 32335.8 | .022016 | 242.84 |
| 1232 | | 41114.5 | 404.43 | 30871.1 | .025328 | 241.35 | 1282 | | 43024.8 | 485.50 | 32365.8 | .021955 | 242.87 |
| 1233 | | 41152.6 | 405.91 | 30900.9 | .025256 | 241.38 | 1283 | | 43063.2 | 487.29 | 32395.8 | .021891 | 242.90 |
| 1234 | | 41190.7 | 407.45 | 30930.7 | .025181 | 241.41 | 1284 | | 43101.5 | 489.02 | 32425.8 | .021831 | 242.93 |
| 1235 | 961.85 | 41228.5 | 408.93 | 30960.2 | .025110 | 241.45 | 1285 | 1011.85 | 43139.8 | 490.76 | 32455.8 | .021770 | 242.96 |
| 1236 | | 41266.6 | 410.48 | 30990.0 | .025036 | 241.48 | 1286 | | 43178.1 | 492.57 | 32485.8 | .021707 | 242.99 |
| 1237 | | 41304.7 | 412.03 | 31019.8 | .024962 | 241.51 | 1287 | | 43216.5 | 494.32 | 32515.8 | .021647 | 243.02 |
| 1238 | | 41342.8 | 413.53 | 31049.5 | .024891 | 241.54 | 1288 | | 43254.8 | 496.07 | 32545.9 | .021588 | 243.05 |
| 1239 | | 41380.9 | 415.09 | 31079.3 | .024818 | 241.57 | 1289 | | 43293.4 | 497.89 | 32576.2 | .021525 | 243.08 |
| 1240 | 966.85 | 41419.0 | 416.60 | 31109.2 | .024748 | 241.60 | 1290 | 1016.85 | 43331.8 | 499.66 | 32606.2 | .021466 | 243.11 |
| 1241 | | 41457.1 | 418.17 | 31139.0 | .024675 | 241.63 | 1291 | | 43370.2 | 501.43 | 32636.3 | .021407 | 243.14 |
| 1242 | | 41495.3 | 419.69 | 31168.8 | .024605 | 241.66 | 1292 | | 43408.5 | 503.27 | 32666.3 | .021345 | 243.17 |
| 1243 | | 41533.4 | 421.27 | 31198.6 | .024533 | 241.69 | 1293 | | 43446.9 | 505.05 | 32696.3 | .021286 | 243.20 |
| 1244 | | 41571.6 | 422.80 | 31228.4 | .024464 | 241.72 | 1294 | | 43485.2 | 506.84 | 32726.4 | .021227 | 243.23 |
| 1245 | 971.85 | 41609.7 | 424.39 | 31258.3 | .024391 | 241.75 | 1295 | 1021.85 | 43523.6 | 508.70 | 32756.4 | .021166 | 243.26 |
| 1246 | | 41647.8 | 425.93 | 31288.1 | .024323 | 241.78 | 1296 | | 43562.0 | 510.49 | 32786.5 | .021108 | 243.29 |
| 1247 | | 41686.0 | 427.53 | 31317.9 | .024251 | 241.82 | 1297 | | 43600.7 | 512.30 | 32816.9 | .021050 | 243.32 |
| 1248 | | 41724.2 | 429.08 | 31347.8 | .024183 | 241.85 | 1298 | | 43639.0 | 514.10 | 32846.9 | .020992 | 243.35 |
| 1249 | | 41762.3 | 430.69 | 31377.6 | .024112 | 241.88 | 1299 | | 43677.4 | 515.99 | 32877.0 | .020932 | 243.38 |

# Table 7  Products—100% Theoretical Air (for One Gram-Mole)

| T | t | $\bar{h}$ | $p_r$ | $\bar{u}$ | $v_r$ | $\bar{\phi}$ | T | t | $\bar{h}$ | $p_r$ | $\bar{u}$ | $v_r$ | $\bar{\phi}$ |
|---|---|---|---|---|---|---|---|---|---|---|---|---|---|
| 1300 | 1026.85 | 43715.8 | 517.81 | 32907.1 | .020874 | 243.41 | 1350 | 1076.85 | 45644.2 | 616.95 | 34419.7 | .018193 | 244.86 |
| 1301 | | 43754.2 | 519.63 | 32937.2 | .020817 | 243.44 | 1351 | | 45682.8 | 619.07 | 34450.0 | .018145 | 244.89 |
| 1302 | | 43792.6 | 521.53 | 32967.3 | .020757 | 243.47 | 1352 | | 45721.7 | 621.20 | 34480.6 | .018096 | 244.92 |
| 1303 | | 43831.0 | 523.36 | 32997.4 | .020700 | 243.50 | 1353 | | 45760.3 | 623.33 | 34510.9 | .018047 | 244.95 |
| 1304 | | 43869.7 | 525.21 | 33027.7 | .020643 | 243.53 | 1354 | | 45798.9 | 625.47 | 34541.2 | .017999 | 244.98 |
| 1305 | 1031.85 | 43908.1 | 527.05 | 33057.8 | .020587 | 243.56 | 1355 | 1081.85 | 45837.9 | 627.70 | 34571.8 | .017948 | 245.01 |
| 1306 | | 43946.6 | 528.98 | 33087.9 | .020528 | 243.59 | 1356 | | 45876.5 | 629.85 | 34602.2 | .017900 | 245.04 |
| 1307 | | 43985.0 | 530.83 | 33118.0 | .020471 | 243.61 | 1357 | | 45915.1 | 632.01 | 34632.5 | .017852 | 245.06 |
| 1308 | | 44023.4 | 532.70 | 33148.2 | .020415 | 243.64 | 1358 | | 45954.1 | 634.17 | 34663.1 | .017804 | 245.09 |
| 1309 | | 44061.8 | 534.57 | 33178.3 | .020360 | 243.67 | 1359 | | 45992.7 | 636.34 | 34693.4 | .017757 | 245.12 |
| 1310 | 1036.85 | 44100.6 | 536.51 | 33208.7 | .020301 | 243.70 | 1360 | 1086.85 | 46031.3 | 638.52 | 34723.8 | .017709 | 245.15 |
| 1311 | | 44139.0 | 538.40 | 33238.8 | .020246 | 243.73 | 1361 | | 46070.3 | 640.71 | 34754.4 | .017662 | 245.18 |
| 1312 | | 44177.5 | 540.28 | 33269.0 | .020190 | 243.76 | 1362 | | 46109.0 | 642.90 | 34784.7 | .017614 | 245.21 |
| 1313 | | 44215.9 | 542.25 | 33299.1 | .020133 | 243.79 | 1363 | | 46147.6 | 645.09 | 34815.1 | .017567 | 245.24 |
| 1314 | | 44254.3 | 544.14 | 33329.2 | .020078 | 243.82 | 1364 | | 46186.6 | 647.30 | 34845.7 | .017520 | 245.26 |
| 1315 | 1041.85 | 44293.1 | 546.05 | 33359.7 | .020023 | 243.85 | 1365 | 1091.85 | 46225.2 | 649.59 | 34876.1 | .017471 | 245.29 |
| 1316 | | 44331.6 | 547.96 | 33389.8 | .019968 | 243.88 | 1366 | | 46263.9 | 651.81 | 34906.4 | .017425 | 245.32 |
| 1317 | | 44370.0 | 549.95 | 33420.0 | .019911 | 243.91 | 1367 | | 46302.9 | 654.03 | 34937.1 | .017378 | 245.35 |
| 1318 | | 44408.5 | 551.87 | 33450.1 | .019857 | 243.94 | 1368 | | 46341.6 | 656.26 | 34967.5 | .017332 | 245.38 |
| 1319 | | 44447.0 | 553.79 | 33480.3 | .019803 | 243.97 | 1369 | | 46380.3 | 658.49 | 34997.8 | .017286 | 245.41 |
| 1320 | 1046.85 | 44485.8 | 555.73 | 33510.8 | .019749 | 244.00 | 1370 | 1096.85 | 46419.2 | 660.74 | 35028.5 | .017239 | 245.43 |
| 1321 | | 44524.3 | 557.67 | 33540.9 | .019695 | 244.02 | 1371 | | 46457.9 | 662.99 | 35058.9 | .017194 | 245.46 |
| 1322 | | 44562.7 | 559.68 | 33571.1 | .019639 | 244.05 | 1372 | | 46496.6 | 665.24 | 35089.3 | .017148 | 245.49 |
| 1323 | | 44601.2 | 561.64 | 33601.3 | .019586 | 244.08 | 1373 | | 46535.6 | 667.50 | 35119.9 | .017102 | 245.52 |
| 1324 | | 44639.7 | 563.59 | 33631.5 | .019532 | 244.11 | 1374 | | 46574.3 | 669.77 | 35150.3 | .017057 | 245.55 |
| 1325 | 1051.85 | 44678.5 | 565.55 | 33661.9 | .019479 | 244.14 | 1375 | 1101.85 | 46613.3 | 672.05 | 35181.0 | .017011 | 245.58 |
| 1326 | | 44717.0 | 567.60 | 33692.1 | .019424 | 244.17 | 1376 | | 46652.0 | 674.33 | 35211.4 | .016966 | 245.60 |
| 1327 | | 44755.5 | 569.57 | 33722.3 | .019371 | 244.20 | 1377 | | 46690.8 | 676.62 | 35241.8 | .016921 | 245.63 |
| 1328 | | 44794.0 | 571.55 | 33752.5 | .019319 | 244.23 | 1378 | | 46729.8 | 678.91 | 35272.5 | .016876 | 245.66 |
| 1329 | | 44832.6 | 573.54 | 33782.7 | .019266 | 244.26 | 1379 | | 46768.5 | 681.21 | 35302.9 | .016831 | 245.69 |
| 1330 | 1056.85 | 44871.4 | 575.60 | 33813.2 | .019211 | 244.29 | 1380 | 1106.85 | 46807.5 | 683.52 | 35333.6 | .016786 | 245.72 |
| 1331 | | 44909.9 | 577.60 | 33843.4 | .019159 | 244.32 | 1381 | | 46846.3 | 685.84 | 35364.1 | .016742 | 245.74 |
| 1332 | | 44948.4 | 579.61 | 33873.6 | .019107 | 244.35 | 1382 | | 46885.0 | 688.16 | 35394.5 | .016697 | 245.77 |
| 1333 | | 44987.0 | 581.62 | 33903.9 | .019056 | 244.37 | 1383 | | 46924.0 | 690.49 | 35425.2 | .016653 | 245.80 |
| 1334 | | 45025.8 | 583.63 | 33934.4 | .019004 | 244.40 | 1384 | | 46962.8 | 692.82 | 35455.6 | .016609 | 245.83 |
| 1335 | 1061.85 | 45064.3 | 585.73 | 33964.6 | .018950 | 244.43 | 1385 | 1111.85 | 47001.8 | 695.17 | 35486.4 | .016565 | 245.86 |
| 1336 | | 45102.9 | 587.76 | 33994.8 | .018899 | 244.46 | 1386 | | 47040.6 | 697.51 | 35516.8 | .016521 | 245.88 |
| 1337 | | 45141.4 | 589.79 | 34025.1 | .018848 | 244.49 | 1387 | | 47079.3 | 699.87 | 35547.2 | .016477 | 245.91 |
| 1338 | | 45180.3 | 591.83 | 34055.6 | .018797 | 244.52 | 1388 | | 47118.4 | 702.23 | 35578.0 | .016434 | 245.94 |
| 1339 | | 45218.9 | 593.88 | 34085.9 | .018746 | 244.55 | 1389 | | 47157.1 | 704.60 | 35608.4 | .016390 | 245.97 |
| 1340 | 1066.85 | 45257.4 | 596.01 | 34116.1 | .018693 | 244.58 | 1390 | 1116.85 | 47196.2 | 706.98 | 35639.2 | .016347 | 246.00 |
| 1341 | | 45296.0 | 598.07 | 34146.4 | .018643 | 244.61 | 1391 | | 47235.0 | 709.36 | 35669.6 | .016304 | 246.02 |
| 1342 | | 45334.8 | 600.13 | 34176.9 | .018592 | 244.63 | 1392 | | 47273.7 | 711.75 | 35700.1 | .016261 | 246.05 |
| 1343 | | 45373.4 | 602.20 | 34207.2 | .018542 | 244.66 | 1393 | | 47312.8 | 714.15 | 35730.9 | .016218 | 246.08 |
| 1344 | | 45412.0 | 604.28 | 34237.4 | .018492 | 244.69 | 1394 | | 47351.6 | 716.55 | 35761.3 | .016175 | 246.11 |
| 1345 | 1071.85 | 45450.9 | 606.36 | 34268.0 | .018443 | 244.72 | 1395 | 1121.85 | 47390.7 | 718.96 | 35792.1 | .016132 | 246.14 |
| 1346 | | 45489.5 | 608.45 | 34298.3 | .018393 | 244.75 | 1396 | | 47429.5 | 721.38 | 35822.6 | .016090 | 246.16 |
| 1347 | | 45528.1 | 610.63 | 34328.6 | .018341 | 244.78 | 1397 | | 47468.6 | 723.81 | 35853.4 | .016047 | 246.19 |
| 1348 | | 45566.7 | 612.73 | 34358.9 | .018292 | 244.81 | 1398 | | 47507.4 | 726.24 | 35883.8 | .016005 | 246.22 |
| 1349 | | 45605.6 | 614.84 | 34389.4 | .018242 | 244.84 | 1399 | | 47546.5 | 728.68 | 35914.6 | .015963 | 246.25 |

## Table 7    Products—100% Theoretical Air (for One Gram-Mole)

| T | t | $\bar{h}$ | $p_r$ | $\bar{u}$ | $v_r$ | $\bar{\phi}$ | T | t | $\bar{h}$ | $p_r$ | $\bar{u}$ | $v_r$ | $\bar{\phi}$ |
|---|---|---|---|---|---|---|---|---|---|---|---|---|---|
| 1400 | 1126.85 | 47585.3 | 731.12 | 35945.1 | .015921 | 246.28 | 1450 | 1176.85 | 49538.3 | 862.08 | 37482.4 | .013985 | 247.65 |
| 1401 | | 47624.1 | 733.58 | 35975.6 | .015879 | 246.30 | 1451 | | 49577.6 | 864.91 | 37513.4 | .013949 | 247.67 |
| 1402 | | 47663.2 | 736.04 | 36006.4 | .015837 | 246.33 | 1452 | | 49616.6 | 867.74 | 37544.1 | .013913 | 247.70 |
| 1403 | | 47702.0 | 738.50 | 36036.9 | .015796 | 246.36 | 1453 | | 49655.9 | 870.47 | 37575.1 | .013879 | 247.73 |
| 1404 | | 47741.1 | 740.98 | 36067.7 | .015754 | 246.39 | 1454 | | 49694.9 | 873.32 | 37605.8 | .013843 | 247.75 |
| 1405 | 1131.85 | 47780.0 | 743.46 | 36098.2 | .015713 | 246.42 | 1455 | 1181.85 | 49734.2 | 876.18 | 37636.8 | .013807 | 247.78 |
| 1406 | | 47819.1 | 745.95 | 36129.0 | .015671 | 246.44 | 1456 | | 49773.5 | 879.04 | 37667.8 | .013772 | 247.81 |
| 1407 | | 47857.9 | 748.44 | 36159.6 | .015630 | 246.47 | 1457 | | 49812.6 | 881.91 | 37698.5 | .013736 | 247.84 |
| 1408 | | 47897.1 | 750.94 | 36190.4 | .015589 | 246.50 | 1458 | | 49851.9 | 884.68 | 37729.5 | .013703 | 247.86 |
| 1409 | | 47935.9 | 753.45 | 36220.9 | .015548 | 246.53 | 1459 | | 49890.9 | 887.57 | 37760.2 | .013667 | 247.89 |
| 1410 | 1136.85 | 47975.1 | 755.87 | 36251.7 | .015510 | 246.55 | 1460 | 1186.85 | 49930.3 | 890.47 | 37791.2 | .013632 | 247.92 |
| 1411 | | 48013.9 | 758.40 | 36282.3 | .015469 | 246.58 | 1461 | | 49969.6 | 893.37 | 37822.2 | .013597 | 247.94 |
| 1412 | | 48053.1 | 760.93 | 36313.1 | .015428 | 246.61 | 1462 | | 50008.6 | 896.28 | 37853.0 | .013562 | 247.97 |
| 1413 | | 48091.9 | 763.46 | 36343.7 | .015388 | 246.64 | 1463 | | 50048.0 | 899.09 | 37884.0 | .013529 | 248.00 |
| 1414 | | 48131.1 | 766.01 | 36374.5 | .015348 | 246.66 | 1464 | | 50087.0 | 902.02 | 37914.7 | .013495 | 248.02 |
| 1415 | 1141.85 | 48169.9 | 768.56 | 36405.1 | .015308 | 246.69 | 1465 | 1191.85 | 50126.4 | 904.96 | 37945.8 | .013460 | 248.05 |
| 1416 | | 48209.1 | 771.12 | 36435.9 | .015268 | 246.72 | 1466 | | 50165.7 | 907.90 | 37976.8 | .013425 | 248.08 |
| 1417 | | 48248.0 | 773.69 | 36466.5 | .015228 | 246.75 | 1467 | | 50204.8 | 910.86 | 38007.5 | .013391 | 248.10 |
| 1418 | | 48287.2 | 776.27 | 36497.3 | .015188 | 246.77 | 1468 | | 50244.1 | 913.70 | 38038.6 | .013358 | 248.13 |
| 1419 | | 48326.0 | 778.85 | 36527.9 | .015148 | 246.80 | 1469 | | 50283.2 | 916.67 | 38069.4 | .013324 | 248.16 |
| 1420 | 1146.85 | 48365.2 | 781.44 | 36558.8 | .015109 | 246.83 | 1470 | 1196.85 | 50322.6 | 919.65 | 38100.4 | .013290 | 248.18 |
| 1421 | | 48404.1 | 783.93 | 36589.4 | .015071 | 246.86 | 1471 | | 50362.0 | 922.63 | 38131.5 | .013256 | 248.21 |
| 1422 | | 48443.3 | 786.56 | 36620.2 | .015032 | 246.88 | 1472 | | 50401.0 | 925.51 | 38162.2 | .013224 | 248.24 |
| 1423 | | 48482.2 | 789.15 | 36650.8 | .014993 | 246.91 | 1473 | | 50440.4 | 928.51 | 38193.3 | .013190 | 248.26 |
| 1424 | | 48521.4 | 791.77 | 36681.7 | .014954 | 246.94 | 1474 | | 50479.8 | 931.52 | 38224.4 | .013156 | 248.29 |
| 1425 | 1151.85 | 48560.3 | 794.39 | 36712.3 | .014915 | 246.97 | 1475 | 1201.85 | 50518.9 | 934.54 | 38255.2 | .013123 | 248.32 |
| 1426 | | 48599.5 | 797.03 | 36743.2 | .014876 | 246.99 | 1476 | | 50558.3 | 937.57 | 38286.2 | .013089 | 248.34 |
| 1427 | | 48638.4 | 799.67 | 36773.8 | .014837 | 247.02 | 1477 | | 50597.4 | 940.48 | 38317.0 | .013058 | 248.37 |
| 1428 | | 48677.6 | 802.21 | 36804.7 | .014800 | 247.05 | 1478 | | 50636.8 | 943.53 | 38348.1 | .013024 | 248.40 |
| 1429 | | 48716.5 | 804.87 | 36835.3 | .014762 | 247.08 | 1479 | | 50676.2 | 946.58 | 38379.2 | .012991 | 248.42 |
| 1430 | 1156.85 | 48755.8 | 807.53 | 36866.2 | .014723 | 247.10 | 1480 | 1206.85 | 50715.3 | 949.64 | 38410.0 | .012958 | 248.45 |
| 1431 | | 48794.7 | 810.20 | 36896.8 | .014685 | 247.13 | 1481 | | 50754.7 | 952.59 | 38441.1 | .012926 | 248.48 |
| 1432 | | 48833.9 | 812.88 | 36927.7 | .014647 | 247.16 | 1482 | | 50794.1 | 955.67 | 38472.2 | .012894 | 248.50 |
| 1433 | | 48872.8 | 815.57 | 36958.3 | .014609 | 247.19 | 1483 | | 50833.2 | 958.76 | 38503.0 | .012861 | 248.53 |
| 1434 | | 48912.1 | 818.26 | 36989.2 | .014571 | 247.21 | 1484 | | 50872.7 | 961.85 | 38534.1 | .012828 | 248.56 |
| 1435 | 1161.85 | 48951.3 | 820.86 | 37020.2 | .014535 | 247.24 | 1485 | 1211.85 | 50912.1 | 964.83 | 38565.2 | .012797 | 248.58 |
| 1436 | | 48990.3 | 823.57 | 37050.8 | .014497 | 247.27 | 1486 | | 50951.2 | 967.94 | 38596.0 | .012764 | 248.61 |
| 1437 | | 49029.5 | 826.28 | 37081.7 | .014460 | 247.29 | 1487 | | 50990.6 | 971.06 | 38627.1 | .012732 | 248.64 |
| 1438 | | 49068.5 | 829.01 | 37112.4 | .014422 | 247.32 | 1488 | | 51030.1 | 974.19 | 38658.2 | .012700 | 248.66 |
| 1439 | | 49107.7 | 831.74 | 37143.3 | .014385 | 247.35 | 1489 | | 51069.2 | 977.20 | 38689.1 | .012669 | 248.69 |
| 1440 | 1166.85 | 49146.7 | 834.48 | 37173.9 | .014348 | 247.38 | 1490 | 1216.85 | 51108.6 | 980.35 | 38720.2 | .012637 | 248.72 |
| 1441 | | 49185.9 | 837.23 | 37204.9 | .014310 | 247.40 | 1491 | | 51148.1 | 983.51 | 38751.3 | .012605 | 248.74 |
| 1442 | | 49224.9 | 839.87 | 37235.6 | .014275 | 247.43 | 1492 | | 51187.2 | 986.54 | 38782.1 | .012574 | 248.77 |
| 1443 | | 49264.2 | 842.64 | 37266.5 | .014238 | 247.46 | 1493 | | 51226.7 | 989.71 | 38813.3 | .012542 | 248.79 |
| 1444 | | 49303.4 | 845.41 | 37297.4 | .014201 | 247.48 | 1494 | | 51266.1 | 992.90 | 38844.4 | .012511 | 248.82 |
| 1445 | 1171.85 | 49342.4 | 848.19 | 37328.1 | .014165 | 247.51 | 1495 | 1221.85 | 51305.6 | 996.09 | 38875.6 | .012479 | 248.85 |
| 1446 | | 49381.7 | 850.97 | 37359.1 | .014128 | 247.54 | 1496 | | 51344.8 | 999.15 | 38906.4 | .012449 | 248.87 |
| 1447 | | 49420.7 | 853.77 | 37389.8 | .014092 | 247.57 | 1497 | | 51384.2 | 1002.36 | 38937.6 | .012417 | 248.90 |
| 1448 | | 49460.0 | 856.46 | 37420.7 | .014057 | 247.59 | 1498 | | 51423.7 | 1005.58 | 38968.7 | .012386 | 248.93 |
| 1449 | | 49499.0 | 859.27 | 37451.4 | .014021 | 247.62 | 1499 | | 51462.9 | 1008.67 | 38999.6 | .012356 | 248.95 |

## Table 7    Products—100% Theoretical Air (for One Gram-Mole)

| T | t | $\bar{h}$ | $p_r$ | $\bar{u}$ | $v_r$ | $\bar{\phi}$ | T | t | $\bar{h}$ | $p_r$ | $\bar{u}$ | $v_r$ | $\bar{\phi}$ |
|---|---|---|---|---|---|---|---|---|---|---|---|---|---|
| 1500 | 1226.85 | 51502.4 | 1011.9 | 39030.8 | .012325 | 248.98 | 1550 | 1276.85 | 53477.0 | 1182.5 | 40589.7 | .010899 | 250.27 |
| 1501 |  | 51541.8 | 1015.1 | 39061.9 | .012294 | 249.01 | 1551 |  | 53516.6 | 1185.9 | 40621.0 | .010874 | 250.30 |
| 1502 |  | 51581.0 | 1018.3 | 39092.8 | .012264 | 249.03 | 1552 |  | 53556.3 | 1189.6 | 40652.3 | .010847 | 250.32 |
| 1503 |  | 51620.5 | 1021.5 | 39124.0 | .012233 | 249.06 | 1553 |  | 53596.0 | 1193.2 | 40683.7 | .010822 | 250.35 |
| 1504 |  | 51660.0 | 1024.8 | 39155.1 | .012202 | 249.08 | 1554 |  | 53635.3 | 1196.9 | 40714.7 | .010795 | 250.37 |
| 1505 | 1231.85 | 51699.5 | 1027.9 | 39186.3 | .012173 | 249.11 | 1555 | 1281.85 | 53675.0 | 1200.7 | 40746.1 | .010768 | 250.40 |
| 1506 |  | 51738.7 | 1031.2 | 39217.2 | .012142 | 249.14 | 1556 |  | 53714.6 | 1204.3 | 40777.4 | .010743 | 250.43 |
| 1507 |  | 51778.2 | 1034.5 | 39248.4 | .012112 | 249.16 | 1557 |  | 53754.3 | 1208.0 | 40808.8 | .010716 | 250.45 |
| 1508 |  | 51817.7 | 1037.7 | 39279.6 | .012083 | 249.19 | 1558 |  | 53794.0 | 1211.6 | 40840.1 | .010691 | 250.48 |
| 1509 |  | 51856.9 | 1041.0 | 39310.5 | .012052 | 249.21 | 1559 |  | 53833.7 | 1215.4 | 40871.5 | .010665 | 250.50 |
| 1510 | 1236.85 | 51896.4 | 1044.3 | 39341.7 | .012022 | 249.24 | 1560 | 1286.85 | 53873.0 | 1219.1 | 40902.6 | .010640 | 250.53 |
| 1511 |  | 51935.9 | 1047.5 | 39372.9 | .011993 | 249.27 | 1561 |  | 53912.7 | 1222.9 | 40933.9 | .010613 | 250.55 |
| 1512 |  | 51975.0 | 1050.9 | 39404.1 | .011963 | 249.29 | 1562 |  | 53952.4 | 1226.5 | 40965.3 | .010589 | 250.58 |
| 1513 |  | 52014.7 | 1054.2 | 39435.0 | .011933 | 249.32 | 1563 |  | 53992.1 | 1230.3 | 40996.7 | .010562 | 250.60 |
| 1514 |  | 52054.2 | 1057.4 | 39466.2 | .011904 | 249.34 | 1564 |  | 54031.8 | 1234.0 | 41028.1 | .010538 | 250.63 |
| 1515 | 1241.85 | 52093.7 | 1060.8 | 39497.4 | .011874 | 249.37 | 1565 | 1291.85 | 54071.5 | 1237.9 | 41059.5 | .010512 | 250.65 |
| 1516 |  | 52133.3 | 1064.2 | 39528.6 | .011844 | 249.40 | 1566 |  | 54111.2 | 1241.6 | 41090.8 | .010487 | 250.68 |
| 1517 |  | 52172.5 | 1067.4 | 39559.6 | .011816 | 249.42 | 1567 |  | 54150.6 | 1245.4 | 41121.9 | .010461 | 250.70 |
| 1518 |  | 52212.0 | 1070.8 | 39590.8 | .011787 | 249.45 | 1568 |  | 54190.3 | 1249.1 | 41153.3 | .010437 | 250.73 |
| 1519 |  | 52251.6 | 1074.2 | 39622.0 | .011757 | 249.48 | 1569 |  | 54230.0 | 1253.0 | 41184.7 | .010411 | 250.76 |
| 1520 | 1246.85 | 52291.1 | 1077.5 | 39653.2 | .011729 | 249.50 | 1570 | 1296.85 | 54269.7 | 1256.8 | 41216.1 | .010387 | 250.78 |
| 1521 |  | 52330.4 | 1080.9 | 39684.2 | .011700 | 249.53 | 1571 |  | 54309.5 | 1260.7 | 41247.5 | .010361 | 250.81 |
| 1522 |  | 52369.9 | 1084.2 | 39715.4 | .011672 | 249.55 | 1572 |  | 54349.2 | 1264.4 | 41278.9 | .010337 | 250.83 |
| 1523 |  | 52409.5 | 1087.7 | 39746.6 | .011642 | 249.58 | 1573 |  | 54388.9 | 1268.3 | 41310.3 | .010312 | 250.86 |
| 1524 |  | 52449.0 | 1091.1 | 39777.9 | .011613 | 249.60 | 1574 |  | 54428.3 | 1272.1 | 41341.4 | .010287 | 250.88 |
| 1525 | 1251.85 | 52488.3 | 1094.4 | 39808.8 | .011586 | 249.63 | 1575 | 1301.85 | 54468.0 | 1276.1 | 41372.8 | .010262 | 250.91 |
| 1526 |  | 52527.9 | 1097.9 | 39840.1 | .011557 | 249.66 | 1576 |  | 54507.8 | 1279.9 | 41404.3 | .010238 | 250.93 |
| 1527 |  | 52567.4 | 1101.2 | 39871.4 | .011529 | 249.68 | 1577 |  | 54547.5 | 1283.8 | 41435.7 | .010213 | 250.96 |
| 1528 |  | 52607.0 | 1104.7 | 39902.6 | .011500 | 249.71 | 1578 |  | 54587.2 | 1287.6 | 41467.1 | .010189 | 250.98 |
| 1529 |  | 52646.3 | 1108.2 | 39933.6 | .011471 | 249.73 | 1579 |  | 54627.0 | 1291.6 | 41498.6 | .010164 | 251.01 |
| 1530 | 1256.85 | 52685.9 | 1111.6 | 39964.8 | .011444 | 249.76 | 1580 | 1306.85 | 54666.7 | 1295.5 | 41530.0 | .010141 | 251.03 |
| 1531 |  | 52725.5 | 1115.1 | 39996.1 | .011416 | 249.79 | 1581 |  | 54706.5 | 1299.5 | 41561.4 | .010116 | 251.06 |
| 1532 |  | 52765.0 | 1118.5 | 40027.4 | .011389 | 249.81 | 1582 |  | 54746.2 | 1303.3 | 41592.8 | .010092 | 251.08 |
| 1533 |  | 52804.3 | 1122.0 | 40058.4 | .011360 | 249.84 | 1583 |  | 54785.7 | 1307.2 | 41624.0 | .010069 | 251.11 |
| 1534 |  | 52843.9 | 1125.5 | 40089.6 | .011332 | 249.86 | 1584 |  | 54825.5 | 1311.2 | 41655.4 | .010044 | 251.13 |
| 1535 | 1261.85 | 52883.5 | 1128.9 | 40120.9 | .011305 | 249.89 | 1585 | 1311.85 | 54865.2 | 1315.1 | 41686.9 | .010021 | 251.16 |
| 1536 |  | 52923.1 | 1132.5 | 40152.2 | .011277 | 249.91 | 1586 |  | 54905.0 | 1319.2 | 41718.3 | .009996 | 251.18 |
| 1537 |  | 52962.7 | 1135.9 | 40183.5 | .011250 | 249.94 | 1587 |  | 54944.7 | 1323.1 | 41749.8 | .009973 | 251.21 |
| 1538 |  | 53002.0 | 1139.5 | 40214.5 | .011222 | 249.97 | 1588 |  | 54984.5 | 1327.1 | 41781.2 | .009949 | 251.23 |
| 1539 |  | 53041.6 | 1143.1 | 40245.8 | .011194 | 249.99 | 1589 |  | 55024.3 | 1331.1 | 41812.7 | .009926 | 251.26 |
| 1540 | 1266.85 | 53081.3 | 1146.6 | 40277.1 | .011168 | 250.02 | 1590 | 1316.85 | 55064.1 | 1335.2 | 41844.2 | .009901 | 251.28 |
| 1541 |  | 53120.9 | 1150.2 | 40308.4 | .011140 | 250.04 | 1591 |  | 55103.8 | 1339.1 | 41875.6 | .009878 | 251.31 |
| 1542 |  | 53160.5 | 1153.6 | 40339.7 | .011113 | 250.07 | 1592 |  | 55143.6 | 1343.2 | 41907.1 | .009854 | 251.33 |
| 1543 |  | 53199.8 | 1157.3 | 40370.7 | .011086 | 250.09 | 1593 |  | 55183.4 | 1347.2 | 41938.6 | .009831 | 251.36 |
| 1544 |  | 53239.5 | 1160.8 | 40402.0 | .011060 | 250.12 | 1594 |  | 55222.9 | 1351.2 | 41969.7 | .009809 | 251.38 |
| 1545 | 1271.85 | 53279.1 | 1164.4 | 40433.3 | .011032 | 250.15 | 1595 | 1321.85 | 55262.7 | 1355.3 | 42001.2 | .009785 | 251.41 |
| 1546 |  | 53318.7 | 1167.9 | 40464.6 | .011006 | 250.17 | 1596 |  | 55302.5 | 1359.3 | 42032.7 | .009762 | 251.43 |
| 1547 |  | 53358.4 | 1171.6 | 40496.0 | .010979 | 250.20 | 1597 |  | 55342.3 | 1363.5 | 42064.2 | .009738 | 251.46 |
| 1548 |  | 53397.7 | 1175.3 | 40527.0 | .010951 | 250.22 | 1598 |  | 55382.1 | 1367.5 | 42095.7 | .009716 | 251.48 |
| 1549 |  | 53437.4 | 1178.8 | 40558.3 | .010926 | 250.25 | 1599 |  | 55421.9 | 1371.7 | 42127.2 | .009692 | 251.51 |

# Table 7  Products—100% Theoretical Air (for One Gram-Mole)

| T | t | $\bar{h}$ | $p_r$ | $\bar{u}$ | $v_r$ | $\bar{\phi}$ | T | t | $\bar{h}$ | $p_r$ | $\bar{u}$ | $v_r$ | $\bar{\phi}$ |
|---|---|---|---|---|---|---|---|---|---|---|---|---|---|
| 1600 | 1326.85 | 55461.7 | 1375.8 | 42158.6 | .0096697 | 251.53 | 1650 | 1376.85 | 57455.8 | 1594.5 | 43737.0 | .0086039 | 252.76 |
| 1601 | | 55501.5 | 1379.8 | 42190.2 | .0096474 | 251.56 | 1651 | | 57495.8 | 1599.3 | 43768.7 | .0085833 | 252.78 |
| 1602 | | 55541.3 | 1384.0 | 42221.7 | .0096238 | 251.58 | 1652 | | 57535.7 | 1603.9 | 43800.3 | .0085639 | 252.81 |
| 1603 | | 55581.1 | 1388.1 | 42253.1 | .0096017 | 251.61 | 1653 | | 57575.7 | 1608.5 | 43832.0 | .0085445 | 252.83 |
| 1604 | | 55621.0 | 1392.3 | 42284.7 | .0095783 | 251.63 | 1654 | | 57615.7 | 1613.1 | 43863.6 | .0085252 | 252.86 |
| 1605 | 1331.85 | 55660.8 | 1396.4 | 42316.2 | .0095562 | 251.66 | 1655 | 1381.85 | 57655.6 | 1618.0 | 43895.3 | .0085048 | 252.88 |
| 1606 | | 55700.6 | 1400.7 | 42347.7 | .0095330 | 251.68 | 1656 | | 57695.6 | 1622.6 | 43927.0 | .0084856 | 252.90 |
| 1607 | | 55740.4 | 1404.8 | 42379.2 | .0095110 | 251.71 | 1657 | | 57735.6 | 1627.3 | 43958.6 | .0084664 | 252.93 |
| 1608 | | 55780.3 | 1408.9 | 42410.7 | .0094891 | 251.73 | 1658 | | 57775.9 | 1632.1 | 43990.6 | .0084462 | 252.95 |
| 1609 | | 55819.8 | 1413.2 | 42441.9 | .0094661 | 251.76 | 1659 | | 57815.9 | 1636.8 | 44022.3 | .0084271 | 252.98 |
| 1610 | 1336.85 | 55859.7 | 1417.4 | 42473.5 | .0094443 | 251.78 | 1660 | 1386.85 | 57855.9 | 1641.5 | 44054.0 | .0084081 | 253.00 |
| 1611 | | 55899.5 | 1421.7 | 42505.0 | .0094214 | 251.81 | 1661 | | 57895.9 | 1646.4 | 44085.6 | .0083880 | 253.03 |
| 1612 | | 55939.3 | 1425.9 | 42536.5 | .0093997 | 251.83 | 1662 | | 57935.8 | 1651.1 | 44117.3 | .0083691 | 253.05 |
| 1613 | | 55979.2 | 1430.0 | 42568.1 | .0093781 | 251.85 | 1663 | | 57975.8 | 1655.9 | 44149.0 | .0083502 | 253.07 |
| 1614 | | 56019.0 | 1434.4 | 42599.6 | .0093554 | 251.88 | 1664 | | 58015.8 | 1660.6 | 44180.7 | .0083314 | 253.10 |
| 1615 | 1341.85 | 56058.9 | 1438.6 | 42631.1 | .0093339 | 251.90 | 1665 | 1391.85 | 58055.9 | 1665.6 | 44212.4 | .0083116 | 253.12 |
| 1616 | | 56098.8 | 1443.0 | 42662.7 | .0093113 | 251.93 | 1666 | | 58095.9 | 1670.3 | 44244.1 | .0082929 | 253.15 |
| 1617 | | 56138.6 | 1447.2 | 42694.2 | .0092899 | 251.95 | 1667 | | 58135.9 | 1675.1 | 44275.8 | .0082742 | 253.17 |
| 1618 | | 56178.5 | 1451.4 | 42725.8 | .0092686 | 251.98 | 1668 | | 58175.9 | 1679.9 | 44307.5 | .0082556 | 253.19 |
| 1619 | | 56218.3 | 1455.8 | 42757.3 | .0092462 | 252.00 | 1669 | | 58215.9 | 1684.9 | 44339.2 | .0082359 | 253.22 |
| 1620 | 1346.85 | 56258.2 | 1460.1 | 42788.9 | .0092250 | 252.03 | 1670 | 1396.85 | 58255.9 | 1689.7 | 44370.8 | .0082174 | 253.24 |
| 1621 | | 56298.1 | 1464.3 | 42820.5 | .0092039 | 252.05 | 1671 | | 58295.9 | 1694.5 | 44402.6 | .0081989 | 253.27 |
| 1622 | | 56338.0 | 1468.8 | 42852.0 | .0091816 | 252.08 | 1672 | | 58336.0 | 1699.6 | 44434.3 | .0081794 | 253.29 |
| 1623 | | 56377.9 | 1473.1 | 42883.6 | .0091606 | 252.10 | 1673 | | 58376.0 | 1704.4 | 44466.0 | .0081611 | 253.31 |
| 1624 | | 56417.8 | 1477.6 | 42915.2 | .0091385 | 252.13 | 1674 | | 58416.0 | 1709.3 | 44497.7 | .0081427 | 253.34 |
| 1625 | 1351.85 | 56457.6 | 1481.9 | 42946.7 | .0091176 | 252.15 | 1675 | 1401.85 | 58456.4 | 1714.2 | 44529.7 | .0081245 | 253.36 |
| 1626 | | 56497.5 | 1486.2 | 42978.3 | .0090967 | 252.17 | 1676 | | 58496.4 | 1719.3 | 44561.5 | .0081052 | 253.39 |
| 1627 | | 56537.4 | 1490.7 | 43009.9 | .0090748 | 252.20 | 1677 | | 58536.4 | 1724.2 | 44593.2 | .0080870 | 253.41 |
| 1628 | | 56577.3 | 1495.0 | 43041.5 | .0090540 | 252.22 | 1678 | | 58576.5 | 1729.1 | 44624.9 | .0080689 | 253.43 |
| 1629 | | 56617.2 | 1499.3 | 43073.1 | .0090334 | 252.25 | 1679 | | 58616.5 | 1734.0 | 44656.6 | .0080508 | 253.46 |
| 1630 | 1356.85 | 56657.1 | 1503.9 | 43104.6 | .0090116 | 252.27 | 1680 | 1406.85 | 58656.6 | 1739.1 | 44688.4 | .0080317 | 253.48 |
| 1631 | | 56697.0 | 1508.3 | 43136.2 | .0089910 | 252.30 | 1681 | | 58696.6 | 1744.1 | 44720.1 | .0080137 | 253.50 |
| 1632 | | 56736.9 | 1512.6 | 43167.8 | .0089705 | 252.32 | 1682 | | 58736.7 | 1749.0 | 44751.9 | .0079958 | 253.53 |
| 1633 | | 56776.8 | 1517.2 | 43199.4 | .0089489 | 252.35 | 1683 | | 58776.7 | 1754.0 | 44783.6 | .0079779 | 253.55 |
| 1634 | | 56816.7 | 1521.6 | 43231.0 | .0089285 | 252.37 | 1684 | | 58816.8 | 1759.2 | 44815.4 | .0079590 | 253.58 |
| 1635 | 1361.85 | 56856.7 | 1526.0 | 43262.6 | .0089082 | 252.39 | 1685 | 1411.85 | 58856.9 | 1764.2 | 44847.1 | .0079412 | 253.60 |
| 1636 | | 56896.6 | 1530.6 | 43294.2 | .0088867 | 252.42 | 1686 | | 58897.2 | 1769.2 | 44879.2 | .0079235 | 253.62 |
| 1637 | | 56936.5 | 1535.1 | 43325.8 | .0088665 | 252.44 | 1687 | | 58937.3 | 1774.2 | 44910.9 | .0079057 | 253.65 |
| 1638 | | 56976.4 | 1539.5 | 43357.4 | .0088463 | 252.47 | 1688 | | 58977.4 | 1779.2 | 44942.7 | .0078881 | 253.67 |
| 1639 | | 57016.4 | 1544.2 | 43389.0 | .0088251 | 252.49 | 1689 | | 59017.5 | 1784.5 | 44974.4 | .0078694 | 253.70 |
| 1640 | 1366.85 | 57056.3 | 1548.6 | 43420.7 | .0088050 | 252.52 | 1690 | 1416.85 | 59057.5 | 1789.6 | 45006.2 | .0078519 | 253.72 |
| 1641 | | 57096.2 | 1553.1 | 43452.3 | .0087850 | 252.54 | 1691 | | 59097.6 | 1794.6 | 45038.0 | .0078344 | 253.74 |
| 1642 | | 57136.2 | 1557.8 | 43483.9 | .0087639 | 252.57 | 1692 | | 59137.7 | 1799.7 | 45069.7 | .0078169 | 253.77 |
| 1643 | | 57176.1 | 1562.3 | 43515.5 | .0087440 | 252.59 | 1693 | | 59177.8 | 1805.0 | 45101.5 | .0077984 | 253.79 |
| 1644 | | 57216.0 | 1566.8 | 43547.2 | .0087241 | 252.61 | 1694 | | 59217.9 | 1810.1 | 45133.3 | .0077811 | 253.81 |
| 1645 | 1371.85 | 57256.0 | 1571.5 | 43578.8 | .0087032 | 252.64 | 1695 | 1421.85 | 59258.0 | 1815.2 | 45165.1 | .0077637 | 253.84 |
| 1646 | | 57295.9 | 1576.0 | 43610.4 | .0086835 | 252.66 | 1696 | | 59298.4 | 1820.4 | 45197.1 | .0077464 | 253.86 |
| 1647 | | 57335.9 | 1580.6 | 43642.1 | .0086638 | 252.69 | 1697 | | 59338.4 | 1825.5 | 45228.9 | .0077292 | 253.88 |
| 1648 | | 57375.9 | 1585.3 | 43673.7 | .0086430 | 252.71 | 1698 | | 59378.6 | 1830.9 | 45260.7 | .0077110 | 253.91 |
| 1649 | | 57415.8 | 1589.9 | 43705.4 | .0086234 | 252.74 | 1699 | | 59418.7 | 1836.0 | 45292.5 | .0076938 | 253.93 |

# Table 7  Products—100% Theoretical Air (for One Gram-Mole)

| T | t | $\overline{h}$ | $p_r$ | $\overline{u}$ | $v_r$ | $\overline{\phi}$ | T | t | $\overline{h}$ | $p_r$ | $\overline{u}$ | $v_r$ | $\overline{\phi}$ |
|---|---|---|---|---|---|---|---|---|---|---|---|---|---|
| 1700 | 1426.85 | 59458.8 | 1841.2 | 45324.3 | .0076767 | 253.96 | 1750 | 1476.85 | 61470.0 | 2118.4 | 46919.8 | .0068686 | 255.12 |
| 1701 | | 59498.9 | 1846.4 | 45356.1 | .0076596 | 253.98 | 1751 | | 61510.2 | 2124.2 | 46951.7 | .0068536 | 255.14 |
| 1702 | | 59539.0 | 1851.6 | 45387.9 | .0076426 | 254.00 | 1752 | | 61550.5 | 2130.1 | 46983.7 | .0068386 | 255.17 |
| 1703 | | 59579.1 | 1857.1 | 45419.7 | .0076246 | 254.03 | 1753 | | 61591.0 | 2136.0 | 47015.9 | .0068237 | 255.19 |
| 1704 | | 59619.2 | 1862.3 | 45451.5 | .0076077 | 254.05 | 1754 | | 61631.3 | 2141.9 | 47047.8 | .0068088 | 255.21 |
| 1705 | 1431.85 | 59659.6 | 1867.5 | 45483.6 | .0075908 | 254.07 | 1755 | 1481.85 | 61671.5 | 2147.8 | 47079.8 | .0067940 | 255.24 |
| 1706 | | 59699.8 | 1872.8 | 45515.4 | .0075740 | 254.10 | 1756 | | 61711.8 | 2153.7 | 47111.7 | .0067791 | 255.26 |
| 1707 | | 59739.9 | 1878.0 | 45547.2 | .0075572 | 254.12 | 1757 | | 61752.1 | 2159.6 | 47143.7 | .0067644 | 255.28 |
| 1708 | | 59780.0 | 1883.6 | 45579.0 | .0075394 | 254.14 | 1758 | | 61792.6 | 2165.8 | 47175.9 | .0067488 | 255.31 |
| 1709 | | 59820.1 | 1888.9 | 45610.8 | .0075227 | 254.17 | 1759 | | 61832.9 | 2171.8 | 47207.9 | .0067341 | 255.33 |
| 1710 | 1436.85 | 59860.3 | 1894.2 | 45642.6 | .0075060 | 254.19 | 1760 | 1486.85 | 61873.2 | 2177.8 | 47239.8 | .0067194 | 255.35 |
| 1711 | | 59900.4 | 1899.5 | 45674.5 | .0074894 | 254.21 | 1761 | | 61913.4 | 2183.8 | 47271.8 | .0067048 | 255.37 |
| 1712 | | 59940.9 | 1904.8 | 45706.6 | .0074728 | 254.24 | 1762 | | 61953.7 | 2189.8 | 47303.7 | .0066902 | 255.40 |
| 1713 | | 59981.0 | 1910.1 | 45738.4 | .0074563 | 254.26 | 1763 | | 61994.3 | 2195.8 | 47336.0 | .0066756 | 255.42 |
| 1714 | | 60021.1 | 1915.7 | 45770.2 | .0074388 | 254.29 | 1764 | | 62034.6 | 2201.8 | 47367.9 | .0066611 | 255.44 |
| 1715 | 1441.85 | 60061.3 | 1921.1 | 45802.1 | .0074224 | 254.31 | 1765 | 1491.85 | 62074.8 | 2207.9 | 47399.9 | .0066466 | 255.47 |
| 1716 | | 60101.4 | 1926.5 | 45833.9 | .0074060 | 254.33 | 1766 | | 62115.1 | 2213.9 | 47431.9 | .0066322 | 255.49 |
| 1717 | | 60141.6 | 1931.9 | 45865.8 | .0073896 | 254.36 | 1767 | | 62155.7 | 2220.0 | 47464.2 | .0066178 | 255.51 |
| 1718 | | 60181.8 | 1937.3 | 45897.6 | .0073732 | 254.38 | 1768 | | 62196.0 | 2226.1 | 47496.1 | .0066034 | 255.53 |
| 1719 | | 60222.2 | 1942.7 | 45929.7 | .0073570 | 254.40 | 1769 | | 62236.3 | 2232.2 | 47528.1 | .0065891 | 255.56 |
| 1720 | 1446.85 | 60262.4 | 1948.4 | 45961.6 | .0073398 | 254.43 | 1770 | 1496.85 | 62276.6 | 2238.3 | 47560.1 | .0065748 | 255.58 |
| 1721 | | 60302.5 | 1953.8 | 45993.4 | .0073236 | 254.45 | 1771 | | 62317.2 | 2244.5 | 47592.4 | .0065605 | 255.60 |
| 1722 | | 60342.7 | 1959.3 | 46025.3 | .0073074 | 254.47 | 1772 | | 62357.5 | 2250.6 | 47624.4 | .0065463 | 255.62 |
| 1723 | | 60382.9 | 1964.8 | 46057.2 | .0072913 | 254.50 | 1773 | | 62397.8 | 2256.8 | 47656.4 | .0065321 | 255.65 |
| 1724 | | 60423.0 | 1970.3 | 46089.0 | .0072752 | 254.52 | 1774 | | 62438.1 | 2262.9 | 47688.4 | .0065180 | 255.67 |
| 1725 | 1451.85 | 60463.2 | 1975.8 | 46120.9 | .0072592 | 254.54 | 1775 | 1501.85 | 62478.4 | 2269.4 | 47720.3 | .0065030 | 255.69 |
| 1726 | | 60503.7 | 1981.3 | 46153.0 | .0072432 | 254.56 | 1776 | | 62519.0 | 2275.6 | 47752.6 | .0064889 | 255.72 |
| 1727 | | 60543.9 | 1987.1 | 46184.9 | .0072263 | 254.59 | 1777 | | 62559.3 | 2281.9 | 47784.7 | .0064749 | 255.74 |
| 1728 | | 60584.1 | 1992.6 | 46216.8 | .0072104 | 254.61 | 1778 | | 62599.7 | 2288.1 | 47816.6 | .0064609 | 255.76 |
| 1729 | | 60624.2 | 1998.1 | 46248.6 | .0071945 | 254.64 | 1779 | | 62640.0 | 2294.3 | 47848.6 | .0064469 | 255.78 |
| 1730 | 1456.85 | 60664.4 | 2003.7 | 46280.5 | .0071787 | 254.66 | 1780 | 1506.85 | 62680.6 | 2300.6 | 47881.0 | .0064329 | 255.81 |
| 1731 | | 60704.6 | 2009.3 | 46312.4 | .0071629 | 254.68 | 1781 | | 62720.9 | 2306.9 | 47913.0 | .0064190 | 255.83 |
| 1732 | | 60745.1 | 2014.9 | 46344.6 | .0071471 | 254.70 | 1782 | | 62761.3 | 2313.2 | 47945.0 | .0064052 | 255.85 |
| 1733 | | 60785.3 | 2020.5 | 46376.5 | .0071314 | 254.73 | 1783 | | 62801.6 | 2319.5 | 47977.0 | .0063913 | 255.88 |
| 1734 | | 60825.5 | 2026.4 | 46408.4 | .0071148 | 254.75 | 1784 | | 62842.2 | 2325.8 | 48009.3 | .0063775 | 255.90 |
| 1735 | 1461.85 | 60865.7 | 2032.0 | 46440.2 | .0070992 | 254.78 | 1785 | 1511.85 | 62882.5 | 2332.1 | 48041.3 | .0063638 | 255.92 |
| 1736 | | 60905.9 | 2037.6 | 46472.1 | .0070836 | 254.80 | 1786 | | 62922.9 | 2338.5 | 48073.3 | .0063500 | 255.94 |
| 1737 | | 60946.4 | 2043.3 | 46504.3 | .0070680 | 254.82 | 1787 | | 62963.2 | 2344.9 | 48105.4 | .0063363 | 255.97 |
| 1738 | | 60986.6 | 2049.0 | 46536.2 | .0070525 | 254.84 | 1788 | | 63003.9 | 2351.2 | 48137.7 | .0063227 | 255.99 |
| 1739 | | 61026.9 | 2054.7 | 46568.1 | .0070371 | 254.87 | 1789 | | 63044.2 | 2357.6 | 48169.7 | .0063091 | 256.01 |
| 1740 | 1466.85 | 61067.1 | 2060.4 | 46600.0 | .0070216 | 254.89 | 1790 | 1516.85 | 63084.6 | 2364.1 | 48201.8 | .0062955 | 256.03 |
| 1741 | | 61107.3 | 2066.1 | 46631.9 | .0070062 | 254.91 | 1791 | | 63125.2 | 2370.5 | 48234.1 | .0062819 | 256.06 |
| 1742 | | 61147.5 | 2071.8 | 46663.8 | .0069909 | 254.94 | 1792 | | 63165.5 | 2376.9 | 48266.1 | .0062684 | 256.08 |
| 1743 | | 61188.0 | 2077.8 | 46696.0 | .0069747 | 254.96 | 1793 | | 63205.9 | 2383.4 | 48298.2 | .0062549 | 256.10 |
| 1744 | | 61228.3 | 2083.6 | 46728.0 | .0069594 | 254.98 | 1794 | | 63246.3 | 2389.8 | 48330.2 | .0062414 | 256.12 |
| 1745 | 1471.85 | 61268.5 | 2089.3 | 46759.9 | .0069442 | 255.01 | 1795 | 1521.85 | 63286.9 | 2396.3 | 48362.6 | .0062280 | 256.15 |
| 1746 | | 61308.7 | 2095.1 | 46791.8 | .0069290 | 255.03 | 1796 | | 63327.3 | 2402.8 | 48394.6 | .0062146 | 256.17 |
| 1747 | | 61349.0 | 2100.9 | 46823.7 | .0069138 | 255.05 | 1797 | | 63367.7 | 2409.4 | 48426.7 | .0062012 | 256.19 |
| 1748 | | 61389.5 | 2106.7 | 46855.9 | .0068987 | 255.08 | 1798 | | 63408.0 | 2415.9 | 48458.7 | .0061879 | 256.21 |
| 1749 | | 61429.8 | 2112.5 | 46887.9 | .0068836 | 255.10 | 1799 | | 63448.7 | 2422.4 | 48491.1 | .0061746 | 256.24 |

## Table 7  Products—100% Theoretical Air (for One Gram-Mole)

| T | t | $\bar{h}$ | $p_r$ | $\bar{u}$ | $v_r$ | $\bar{\phi}$ | T | t | $\bar{h}$ | $p_r$ | $\bar{u}$ | $v_r$ | $\bar{\phi}$ |
|---|---|---|---|---|---|---|---|---|---|---|---|---|---|
| 1800 | 1526.85 | 63489.1 | 2429.0 | 48523.2 | .0061614 | 256.26 | 1850 | 1576.85 | 65515.6 | 2776.1 | 50134.0 | .0055408 | 257.37 |
| 1801 | | 63529.4 | 2435.6 | 48555.2 | .0061481 | 256.28 | 1851 | | 65556.2 | 2783.5 | 50166.2 | .0055291 | 257.39 |
| 1802 | | 63570.1 | 2442.2 | 48587.6 | .0061350 | 256.30 | 1852 | | 65597.0 | 2790.9 | 50198.7 | .0055174 | 257.41 |
| 1803 | | 63610.5 | 2448.8 | 48619.6 | .0061218 | 256.33 | 1853 | | 65637.5 | 2798.3 | 50230.9 | .0055058 | 257.44 |
| 1804 | | 63650.9 | 2455.4 | 48651.7 | .0061087 | 256.35 | 1854 | | 65678.0 | 2805.7 | 50263.1 | .0054941 | 257.46 |
| 1805 | 1531.85 | 63691.3 | 2462.0 | 48683.8 | .0060956 | 256.37 | 1855 | 1581.85 | 65718.8 | 2813.1 | 50295.5 | .0054826 | 257.48 |
| 1806 | | 63731.9 | 2468.7 | 48716.1 | .0060825 | 256.39 | 1856 | | 65759.3 | 2820.6 | 50327.7 | .0054710 | 257.50 |
| 1807 | | 63772.3 | 2475.4 | 48748.2 | .0060695 | 256.42 | 1857 | | 65800.1 | 2828.1 | 50360.2 | .0054595 | 257.52 |
| 1808 | | 63812.7 | 2482.0 | 48780.3 | .0060565 | 256.44 | 1858 | | 65840.6 | 2835.6 | 50392.4 | .0054480 | 257.55 |
| 1809 | | 63853.4 | 2488.7 | 48812.7 | .0060435 | 256.46 | 1859 | | 65881.1 | 2842.7 | 50424.6 | .0054372 | 257.57 |
| 1810 | 1536.85 | 63893.8 | 2495.5 | 48844.8 | .0060306 | 256.48 | 1860 | 1586.85 | 65921.9 | 2850.2 | 50457.1 | .0054258 | 257.59 |
| 1811 | | 63934.2 | 2502.2 | 48876.8 | .0060177 | 256.51 | 1861 | | 65962.5 | 2857.8 | 50489.4 | .0054144 | 257.61 |
| 1812 | | 63974.6 | 2508.9 | 48908.9 | .0060048 | 256.53 | 1862 | | 66003.0 | 2865.4 | 50521.6 | .0054030 | 257.63 |
| 1813 | | 64015.3 | 2515.7 | 48941.3 | .0059920 | 256.55 | 1863 | | 66043.8 | 2872.9 | 50554.1 | .0053916 | 257.65 |
| 1814 | | 64055.7 | 2522.5 | 48973.4 | .0059792 | 256.57 | 1864 | | 66084.4 | 2880.5 | 50586.3 | .0053803 | 257.68 |
| 1815 | 1541.85 | 64096.2 | 2529.3 | 49005.5 | .0059664 | 256.60 | 1865 | 1591.85 | 66125.2 | 2888.1 | 50618.8 | .0053690 | 257.70 |
| 1816 | | 64136.9 | 2536.1 | 49037.9 | .0059536 | 256.62 | 1866 | | 66165.7 | 2895.8 | 50651.0 | .0053577 | 257.72 |
| 1817 | | 64177.3 | 2542.9 | 49070.0 | .0059409 | 256.64 | 1867 | | 66206.2 | 2903.0 | 50683.2 | .0053472 | 257.74 |
| 1818 | | 64217.7 | 2549.8 | 49102.1 | .0059283 | 256.66 | 1868 | | 66247.1 | 2910.7 | 50715.8 | .0053359 | 257.76 |
| 1819 | | 64258.4 | 2556.6 | 49134.5 | .0059156 | 256.68 | 1869 | | 66287.6 | 2918.4 | 50748.0 | .0053247 | 257.79 |
| 1820 | 1546.85 | 64298.8 | 2563.5 | 49166.6 | .0059030 | 256.71 | 1870 | 1596.85 | 66328.2 | 2926.1 | 50780.2 | .0053136 | 257.81 |
| 1821 | | 64339.3 | 2570.0 | 49198.7 | .0058912 | 256.73 | 1871 | | 66369.0 | 2933.8 | 50812.8 | .0053024 | 257.83 |
| 1822 | | 64380.0 | 2576.9 | 49231.2 | .0058786 | 256.75 | 1872 | | 66409.6 | 2941.5 | 50845.0 | .0052913 | 257.85 |
| 1823 | | 64420.4 | 2583.9 | 49263.3 | .0058661 | 256.77 | 1873 | | 66450.4 | 2949.3 | 50877.5 | .0052802 | 257.87 |
| 1824 | | 64460.9 | 2590.8 | 49295.4 | .0058536 | 256.80 | 1874 | | 66491.0 | 2956.7 | 50909.8 | .0052699 | 257.89 |
| 1825 | 1551.85 | 64501.6 | 2597.7 | 49327.8 | .0058411 | 256.82 | 1875 | 1601.85 | 66531.5 | 2964.5 | 50942.0 | .0052588 | 257.92 |
| 1826 | | 64542.0 | 2604.7 | 49359.9 | .0058287 | 256.84 | 1876 | | 66572.4 | 2972.3 | 50974.5 | .0052478 | 257.94 |
| 1827 | | 64582.5 | 2611.7 | 49392.0 | .0058163 | 256.86 | 1877 | | 66612.9 | 2980.1 | 51006.8 | .0052368 | 257.96 |
| 1828 | | 64622.9 | 2618.7 | 49424.2 | .0058039 | 256.88 | 1878 | | 66653.8 | 2987.9 | 51039.3 | .0052259 | 257.98 |
| 1829 | | 64663.6 | 2625.7 | 49456.6 | .0057916 | 256.91 | 1879 | | 66694.4 | 2995.8 | 51071.6 | .0052149 | 258.00 |
| 1830 | 1556.85 | 64704.1 | 2632.7 | 49488.7 | .0057793 | 256.93 | 1880 | 1606.85 | 66735.2 | 3003.6 | 51104.1 | .0052040 | 258.02 |
| 1831 | | 64744.6 | 2639.8 | 49520.9 | .0057670 | 256.95 | 1881 | | 66775.8 | 3011.1 | 51136.4 | .0051938 | 258.05 |
| 1832 | | 64785.3 | 2646.9 | 49553.3 | .0057547 | 256.97 | 1882 | | 66816.4 | 3019.1 | 51168.6 | .0051830 | 258.07 |
| 1833 | | 64825.8 | 2653.9 | 49585.5 | .0057425 | 257.00 | 1883 | | 66857.2 | 3027.0 | 51201.2 | .0051722 | 258.09 |
| 1834 | | 64866.2 | 2661.0 | 49617.6 | .0057303 | 257.02 | 1884 | | 66897.8 | 3034.9 | 51233.5 | .0051614 | 258.11 |
| 1835 | 1561.85 | 64907.0 | 2668.2 | 49650.0 | .0057182 | 257.04 | 1885 | 1611.85 | 66938.7 | 3042.9 | 51266.0 | .0051506 | 258.13 |
| 1836 | | 64947.5 | 2675.3 | 49682.2 | .0057060 | 257.06 | 1886 | | 66979.3 | 3050.9 | 51298.3 | .0051399 | 258.15 |
| 1837 | | 64987.9 | 2682.4 | 49714.4 | .0056939 | 257.08 | 1887 | | 67019.8 | 3058.5 | 51330.6 | .0051298 | 258.17 |
| 1838 | | 65028.7 | 2689.2 | 49746.8 | .0056826 | 257.11 | 1888 | | 67060.7 | 3066.5 | 51363.1 | .0051191 | 258.20 |
| 1839 | | 65069.2 | 2696.4 | 49779.0 | .0056706 | 257.13 | 1889 | | 67101.3 | 3074.5 | 51395.4 | .0051084 | 258.22 |
| 1840 | 1566.85 | 65109.9 | 2703.6 | 49811.4 | .0056586 | 257.15 | 1890 | 1616.85 | 67142.2 | 3082.6 | 51428.0 | .0050978 | 258.24 |
| 1841 | | 65150.4 | 2710.8 | 49843.6 | .0056466 | 257.17 | 1891 | | 67182.8 | 3090.6 | 51460.3 | .0050872 | 258.26 |
| 1842 | | 65190.9 | 2718.0 | 49875.7 | .0056346 | 257.19 | 1892 | | 67223.7 | 3098.3 | 51492.8 | .0050772 | 258.28 |
| 1843 | | 65231.6 | 2725.3 | 49908.2 | .0056227 | 257.22 | 1893 | | 67264.3 | 3106.4 | 51525.1 | .0050667 | 258.30 |
| 1844 | | 65272.1 | 2732.5 | 49940.4 | .0056108 | 257.24 | 1894 | | 67304.9 | 3114.5 | 51557.4 | .0050561 | 258.33 |
| 1845 | 1571.85 | 65312.6 | 2739.8 | 49972.5 | .0055989 | 257.26 | 1895 | 1621.85 | 67345.8 | 3122.7 | 51590.0 | .0050456 | 258.35 |
| 1846 | | 65353.4 | 2747.1 | 50005.0 | .0055871 | 257.28 | 1896 | | 67386.4 | 3130.8 | 51622.3 | .0050351 | 258.37 |
| 1847 | | 65393.9 | 2754.4 | 50037.2 | .0055753 | 257.30 | 1897 | | 67427.3 | 3139.0 | 51654.9 | .0050246 | 258.39 |
| 1848 | | 65434.4 | 2761.8 | 50069.4 | .0055635 | 257.33 | 1898 | | 67467.9 | 3146.8 | 51687.2 | .0050149 | 258.41 |
| 1849 | | 65475.2 | 2768.7 | 50101.8 | .0055525 | 257.35 | 1899 | | 67508.8 | 3155.0 | 51719.8 | .0050044 | 258.43 |

# Table 7    Products—100% Theoretical Air (for One Gram-Mole)

| T | t | $\bar{h}$ | $p_r$ | $\bar{u}$ | $v_r$ | $\bar{\phi}$ | T | t | $\bar{h}$ | $p_r$ | $\bar{u}$ | $v_r$ | $\bar{\phi}$ |
|---|---|---|---|---|---|---|---|---|---|---|---|---|---|
| 1900 | 1626.85 | 67549.4 | 3163.2 | 51752.1 | .0049940 | 258.45 | 1950 | 1676.85 | 69590.1 | 3593.5 | 53377.0 | .0045118 | 259.52 |
| 1901 | | 67590.0 | 3171.5 | 51784.3 | .0049837 | 258.48 | 1951 | | 69630.8 | 3602.2 | 53409.4 | .0045032 | 259.54 |
| 1902 | | 67631.0 | 3179.8 | 51817.0 | .0049733 | 258.50 | 1952 | | 69671.8 | 3611.4 | 53442.1 | .0044940 | 259.56 |
| 1903 | | 67671.6 | 3187.6 | 51849.3 | .0049637 | 258.52 | 1953 | | 69712.5 | 3620.7 | 53474.5 | .0044848 | 259.58 |
| 1904 | | 67712.5 | 3195.9 | 51881.9 | .0049534 | 258.54 | 1954 | | 69753.6 | 3629.5 | 53507.2 | .0044762 | 259.60 |
| 1905 | 1631.85 | 67753.1 | 3204.3 | 51914.2 | .0049431 | 258.56 | 1955 | 1681.85 | 69794.6 | 3638.7 | 53539.9 | .0044671 | 259.62 |
| 1906 | | 67794.1 | 3212.6 | 51946.8 | .0049328 | 258.58 | 1956 | | 69835.3 | 3648.0 | 53572.4 | .0044580 | 259.64 |
| 1907 | | 67834.7 | 3221.0 | 51979.1 | .0049226 | 258.61 | 1957 | | 69876.4 | 3656.9 | 53605.1 | .0044495 | 259.66 |
| 1908 | | 67875.6 | 3228.9 | 52011.7 | .0049131 | 258.63 | 1958 | | 69917.1 | 3666.2 | 53637.5 | .0044404 | 259.68 |
| 1909 | | 67916.2 | 3237.3 | 52044.0 | .0049029 | 258.65 | 1959 | | 69958.1 | 3675.6 | 53670.2 | .0044314 | 259.70 |
| 1910 | 1636.85 | 67956.9 | 3245.7 | 52076.4 | .0048927 | 258.67 | 1960 | 1686.85 | 69998.9 | 3685.0 | 53702.7 | .0044223 | 259.72 |
| 1911 | | 67997.8 | 3254.2 | 52109.0 | .0048826 | 258.69 | 1961 | | 70040.0 | 3693.9 | 53735.4 | .0044139 | 259.74 |
| 1912 | | 68038.4 | 3262.6 | 52141.3 | .0048725 | 258.71 | 1962 | | 70080.7 | 3703.3 | 53767.8 | .0044049 | 259.77 |
| 1913 | | 68079.4 | 3270.7 | 52173.9 | .0048631 | 258.73 | 1963 | | 70121.7 | 3712.8 | 53800.6 | .0043959 | 259.79 |
| 1914 | | 68120.0 | 3279.2 | 52206.3 | .0048530 | 258.75 | 1964 | | 70162.5 | 3721.8 | 53833.0 | .0043876 | 259.81 |
| 1915 | 1641.85 | 68161.0 | 3287.7 | 52238.9 | .0048430 | 258.78 | 1965 | 1691.85 | 70203.6 | 3731.2 | 53865.8 | .0043787 | 259.83 |
| 1916 | | 68201.6 | 3296.2 | 52271.2 | .0048330 | 258.80 | 1966 | | 70244.3 | 3740.8 | 53898.2 | .0043697 | 259.85 |
| 1917 | | 68242.6 | 3304.3 | 52303.9 | .0048236 | 258.82 | 1967 | | 70285.4 | 3749.8 | 53930.9 | .0043614 | 259.87 |
| 1918 | | 68283.2 | 3312.9 | 52336.2 | .0048137 | 258.84 | 1968 | | 70326.1 | 3759.3 | 53963.4 | .0043526 | 259.89 |
| 1919 | | 68324.2 | 3321.5 | 52368.8 | .0048037 | 258.86 | 1969 | | 70367.2 | 3768.9 | 53996.1 | .0043437 | 259.91 |
| 1920 | 1646.85 | 68364.8 | 3330.1 | 52401.2 | .0047938 | 258.88 | 1970 | 1696.85 | 70408.0 | 3778.0 | 54028.6 | .0043355 | 259.93 |
| 1921 | | 68405.8 | 3338.2 | 52433.8 | .0047845 | 258.90 | 1971 | | 70449.0 | 3787.6 | 54061.3 | .0043267 | 259.95 |
| 1922 | | 68446.5 | 3346.9 | 52466.2 | .0047747 | 258.92 | 1972 | | 70489.8 | 3797.2 | 54093.8 | .0043179 | 259.97 |
| 1923 | | 68487.4 | 3355.5 | 52498.8 | .0047648 | 258.95 | 1973 | | 70530.9 | 3806.4 | 54126.5 | .0043097 | 259.99 |
| 1924 | | 68528.1 | 3364.2 | 52531.2 | .0047550 | 258.97 | 1974 | | 70571.9 | 3816.1 | 54159.3 | .0043009 | 260.01 |
| 1925 | 1651.85 | 68568.8 | 3372.9 | 52563.5 | .0047452 | 258.99 | 1975 | 1701.85 | 70612.7 | 3825.8 | 54191.8 | .0042922 | 260.04 |
| 1926 | | 68609.7 | 3381.2 | 52596.2 | .0047361 | 259.01 | 1976 | | 70653.8 | 3835.0 | 54224.5 | .0042841 | 260.06 |
| 1927 | | 68650.4 | 3389.9 | 52628.6 | .0047263 | 259.03 | 1977 | | 70694.6 | 3844.7 | 54257.0 | .0042754 | 260.08 |
| 1928 | | 68691.4 | 3398.7 | 52661.2 | .0047166 | 259.05 | 1978 | | 70735.6 | 3854.5 | 54289.8 | .0042667 | 260.10 |
| 1929 | | 68732.1 | 3407.5 | 52693.6 | .0047069 | 259.07 | 1979 | | 70776.4 | 3863.7 | 54322.2 | .0042586 | 260.12 |
| 1930 | 1656.85 | 68773.0 | 3415.8 | 52726.2 | .0046978 | 259.09 | 1980 | 1706.85 | 70817.5 | 3873.5 | 54355.0 | .0042500 | 260.14 |
| 1931 | | 68813.7 | 3424.6 | 52758.6 | .0046881 | 259.12 | 1981 | | 70858.3 | 3883.3 | 54387.5 | .0042414 | 260.16 |
| 1932 | | 68854.7 | 3433.5 | 52791.3 | .0046785 | 259.14 | 1982 | | 70899.4 | 3892.7 | 54420.2 | .0042334 | 260.18 |
| 1933 | | 68895.4 | 3442.3 | 52823.6 | .0046689 | 259.16 | 1983 | | 70940.2 | 3902.5 | 54452.7 | .0042248 | 260.20 |
| 1934 | | 68936.4 | 3450.7 | 52856.3 | .0046599 | 259.18 | 1984 | | 70981.3 | 3912.4 | 54485.5 | .0042163 | 260.22 |
| 1935 | 1661.85 | 68977.1 | 3459.6 | 52888.7 | .0046503 | 259.20 | 1985 | 1711.85 | 71022.4 | 3921.8 | 54518.3 | .0042083 | 260.24 |
| 1936 | | 69018.0 | 3468.5 | 52921.4 | .0046408 | 259.22 | 1986 | | 71063.2 | 3931.7 | 54550.8 | .0041998 | 260.26 |
| 1937 | | 69058.8 | 3477.0 | 52953.8 | .0046319 | 259.24 | 1987 | | 71104.3 | 3941.7 | 54583.6 | .0041913 | 260.28 |
| 1938 | | 69099.7 | 3486.0 | 52986.4 | .0046224 | 259.26 | 1988 | | 71145.1 | 3951.1 | 54616.1 | .0041834 | 260.30 |
| 1939 | | 69140.4 | 3494.9 | 53018.8 | .0046129 | 259.28 | 1989 | | 71186.2 | 3961.1 | 54648.9 | .0041749 | 260.33 |
| 1940 | 1666.85 | 69181.4 | 3503.9 | 53051.5 | .0046034 | 259.31 | 1990 | 1716.85 | 71227.0 | 3971.1 | 54681.3 | .0041665 | 260.35 |
| 1941 | | 69222.1 | 3512.5 | 53083.9 | .0045946 | 259.33 | 1991 | | 71268.1 | 3980.6 | 54714.1 | .0041586 | 260.37 |
| 1942 | | 69263.1 | 3521.5 | 53116.6 | .0045852 | 259.35 | 1992 | | 71309.0 | 3990.7 | 54746.7 | .0041503 | 260.39 |
| 1943 | | 69303.9 | 3530.5 | 53149.0 | .0045758 | 259.37 | 1993 | | 71350.1 | 4000.7 | 54779.5 | .0041419 | 260.41 |
| 1944 | | 69344.9 | 3539.6 | 53181.7 | .0045664 | 259.39 | 1994 | | 71391.2 | 4010.3 | 54812.3 | .0041341 | 260.43 |
| 1945 | 1671.85 | 69385.6 | 3548.2 | 53214.1 | .0045577 | 259.41 | 1995 | 1721.85 | 71432.0 | 4020.4 | 54844.8 | .0041257 | 260.45 |
| 1946 | | 69426.6 | 3557.3 | 53246.8 | .0045483 | 259.43 | 1996 | | 71473.1 | 4030.0 | 54877.5 | .0041180 | 260.47 |
| 1947 | | 69467.3 | 3566.4 | 53279.2 | .0045390 | 259.45 | 1997 | | 71513.9 | 4040.2 | 54910.0 | .0041097 | 260.49 |
| 1948 | | 69508.3 | 3575.1 | 53311.9 | .0045303 | 259.47 | 1998 | | 71555.1 | 4050.4 | 54942.9 | .0041014 | 260.51 |
| 1949 | | 69549.1 | 3584.3 | 53344.3 | .0045211 | 259.49 | 1999 | | 71595.9 | 4060.0 | 54975.4 | .0040937 | 260.53 |

# Table 8    Products—100% Theoretical Air (for One Gram-Mole)

| T | t | Fuel—$(CH_1)_n$ | | | Fuel—$(CH_2)_n$ | | | Fuel—$(CH_3)_n$ | | |
|---|---|---|---|---|---|---|---|---|---|---|
| | | $\bar{c}_p$ $\dfrac{J}{\text{g-mol K}}$ | $\bar{c}_v$ $\dfrac{J}{\text{g-mol K}}$ | $k = \dfrac{\bar{c}_p}{\bar{c}_v}$ | $\bar{c}_p$ $\dfrac{J}{\text{g-mol K}}$ | $\bar{c}_v$ $\dfrac{J}{\text{g-mol K}}$ | $k = \dfrac{\bar{c}_p}{\bar{c}_v}$ | $\bar{c}_p$ $\dfrac{J}{\text{g-mol K}}$ | $\bar{c}_v$ $\dfrac{J}{\text{g-mol K}}$ | $k = \dfrac{\bar{c}_p}{\bar{c}_v}$ |
| K | C | | | | | | | | | |
| 100 | −173.15 | 29.380 | 21.066 | 1.395 | 29.590 | 21.276 | 1.391 | 29.733 | 21.419 | 1.388 |
| 120 | −153.15 | 29.423 | 21.108 | 1.394 | 29.625 | 21.311 | 1.390 | 29.764 | 21.449 | 1.388 |
| 140 | −133.15 | 29.499 | 21.185 | 1.392 | 29.688 | 21.373 | 1.389 | 29.817 | 21.502 | 1.387 |
| 160 | −113.15 | 29.609 | 21.294 | 1.390 | 29.777 | 21.463 | 1.387 | 29.892 | 21.578 | 1.385 |
| 180 | −93.15 | 29.743 | 21.429 | 1.388 | 29.887 | 21.573 | 1.385 | 29.985 | 21.671 | 1.384 |
| 200 | −73.15 | 29.894 | 21.580 | 1.385 | 30.011 | 21.696 | 1.383 | 30.090 | 21.776 | 1.382 |
| 220 | −53.15 | 30.055 | 21.741 | 1.382 | 30.143 | 21.828 | 1.381 | 30.203 | 21.888 | 1.380 |
| 240 | −33.15 | 30.220 | 21.905 | 1.380 | 30.279 | 21.964 | 1.379 | 30.319 | 22.004 | 1.378 |
| 260 | −13.15 | 30.385 | 22.070 | 1.377 | 30.416 | 22.102 | 1.376 | 30.437 | 22.123 | 1.376 |
| 280 | 6.85 | 30.549 | 22.234 | 1.374 | 30.554 | 22.239 | 1.374 | 30.557 | 22.243 | 1.374 |
| 300 | 26.85 | 30.711 | 22.397 | 1.371 | 30.692 | 22.378 | 1.372 | 30.679 | 22.364 | 1.372 |
| 350 | 76.85 | 31.114 | 22.799 | 1.365 | 31.042 | 22.727 | 1.366 | 30.993 | 22.678 | 1.367 |
| 400 | 126.85 | 31.520 | 23.205 | 1.358 | 31.407 | 23.092 | 1.360 | 31.329 | 23.015 | 1.361 |
| 450 | 176.85 | 31.939 | 23.625 | 1.352 | 31.794 | 23.480 | 1.354 | 31.695 | 23.380 | 1.356 |
| 500 | 226.85 | 32.377 | 24.063 | 1.346 | 32.206 | 23.892 | 1.348 | 32.089 | 23.775 | 1.350 |
| 550 | 276.85 | 32.831 | 24.517 | 1.339 | 32.639 | 24.325 | 1.342 | 32.508 | 24.194 | 1.344 |
| 600 | 326.85 | 33.297 | 24.982 | 1.333 | 33.088 | 24.774 | 1.336 | 32.946 | 24.632 | 1.338 |
| 650 | 376.85 | 33.767 | 25.453 | 1.327 | 33.546 | 25.231 | 1.330 | 33.395 | 25.080 | 1.332 |
| 700 | 426.85 | 34.235 | 25.921 | 1.321 | 34.005 | 25.690 | 1.324 | 33.847 | 25.533 | 1.326 |
| 750 | 476.85 | 34.695 | 26.381 | 1.315 | 34.459 | 26.145 | 1.318 | 34.298 | 25.984 | 1.320 |
| 800 | 526.85 | 35.143 | 26.829 | 1.310 | 34.904 | 26.590 | 1.313 | 34.741 | 26.427 | 1.315 |
| 850 | 576.85 | 35.575 | 27.261 | 1.305 | 35.337 | 27.022 | 1.308 | 35.174 | 26.859 | 1.310 |
| 900 | 626.85 | 35.989 | 27.675 | 1.300 | 35.753 | 27.439 | 1.303 | 35.592 | 27.278 | 1.305 |
| 950 | 676.85 | 36.383 | 28.069 | 1.296 | 36.153 | 27.838 | 1.299 | 35.995 | 27.681 | 1.300 |
| 1000 | 726.85 | 36.757 | 28.443 | 1.292 | 36.534 | 28.220 | 1.295 | 36.382 | 28.067 | 1.296 |
| 1050 | 776.85 | 37.111 | 28.797 | 1.289 | 36.897 | 28.583 | 1.291 | 36.751 | 28.436 | 1.292 |
| 1100 | 826.85 | 37.444 | 29.130 | 1.285 | 37.241 | 28.926 | 1.287 | 37.102 | 28.787 | 1.289 |
| 1150 | 876.85 | 37.759 | 29.445 | 1.282 | 37.567 | 29.253 | 1.284 | 37.436 | 29.122 | 1.286 |
| 1200 | 926.85 | 38.054 | 29.740 | 1.280 | 37.875 | 29.560 | 1.281 | 37.752 | 29.438 | 1.282 |
| 1250 | 976.85 | 38.333 | 30.018 | 1.277 | 38.166 | 29.852 | 1.279 | 38.053 | 29.738 | 1.280 |
| 1300 | 1026.85 | 38.594 | 30.280 | 1.275 | 38.441 | 30.127 | 1.276 | 38.337 | 30.022 | 1.277 |
| 1350 | 1076.85 | 38.837 | 30.523 | 1.272 | 38.699 | 30.384 | 1.274 | 38.604 | 30.290 | 1.274 |
| 1400 | 1126.85 | 39.067 | 30.753 | 1.270 | 38.943 | 30.629 | 1.271 | 38.858 | 30.544 | 1.272 |
| 1450 | 1176.85 | 39.283 | 30.969 | 1.268 | 39.173 | 30.859 | 1.269 | 39.098 | 30.783 | 1.270 |
| 1500 | 1226.85 | 39.486 | 31.171 | 1.267 | 39.389 | 31.075 | 1.268 | 39.324 | 31.009 | 1.268 |
| 1550 | 1276.85 | 39.676 | 31.362 | 1.265 | 39.594 | 31.280 | 1.266 | 39.538 | 31.223 | 1.266 |
| 1600 | 1326.85 | 39.857 | 31.543 | 1.264 | 39.788 | 31.474 | 1.264 | 39.741 | 31.427 | 1.265 |
| 1650 | 1376.85 | 40.026 | 31.712 | 1.262 | 39.970 | 31.656 | 1.263 | 39.932 | 31.618 | 1.263 |
| 1700 | 1426.85 | 40.185 | 31.871 | 1.261 | 40.142 | 31.828 | 1.261 | 40.113 | 31.799 | 1.261 |
| 1750 | 1476.85 | 40.336 | 32.021 | 1.260 | 40.306 | 31.991 | 1.260 | 40.285 | 31.971 | 1.260 |
| 1800 | 1526.85 | 40.477 | 32.163 | 1.259 | 40.459 | 32.145 | 1.259 | 40.447 | 32.133 | 1.259 |
| 1900 | 1626.85 | 40.739 | 32.425 | 1.256 | 40.745 | 32.430 | 1.256 | 40.748 | 32.434 | 1.256 |
| 2000 | 1726.85 | 40.973 | 32.658 | 1.255 | 41.000 | 32.685 | 1.254 | 41.018 | 32.704 | 1.254 |
| 2100 | 1826.85 | 41.184 | 32.869 | 1.253 | 41.231 | 32.917 | 1.253 | 41.263 | 32.949 | 1.252 |
| 2200 | 1926.85 | 41.373 | 33.059 | 1.252 | 41.440 | 33.125 | 1.251 | 41.485 | 33.171 | 1.251 |

# Table 9 Molecular Weight and Gas Constant for Products of Combustion of Hydrocarbon Fuels of Composition $(CH_x)_n$ with Air

**Hydrogen-Carbon Ratio**       **Percent of Theoretical Fuel**

## Molecular Weight

| $\dfrac{\text{kg of H}}{\text{kg of C}}$ | $\dfrac{\text{atoms of H}}{\text{atoms of C}}$ | 0 | 10 | 20 | 30 | 40 | 50 | 60 | 70 | 80 | 90 | 100 |
|---|---|---|---|---|---|---|---|---|---|---|---|---|
| 0.06 | 0.715 | 28.967 | 29.101 | 29.234 | 29.367 | 29.498 | 29.629 | 29.759 | 29.888 | 30.016 | 30.144 | 30.270 |
| 0.08 | 0.953 | 28.967 | 29.069 | 29.171 | 29.272 | 29.372 | 29.471 | 29.569 | 29.667 | 29.764 | 29.860 | 29.955 |
| 0.10 | 1.192 | 28.967 | 29.041 | 29.114 | 29.186 | 29.258 | 29.329 | 29.399 | 29.469 | 29.538 | 29.606 | 29.674 |
| 0.12 | 1.430 | 28.967 | 29.014 | 29.062 | 29.108 | 29.154 | 29.200 | 29.245 | 29.290 | 29.334 | 29.377 | 29.421 |
| 0.14 | 1.668 | 28.967 | 28.991 | 29.014 | 29.037 | 29.060 | 29.083 | 29.105 | 29.127 | 29.149 | 29.170 | 29.192 |
| 0.16 | 1.907 | 28.967 | 28.969 | 28.970 | 28.972 | 28.974 | 28.975 | 28.977 | 28.979 | 28.980 | 28.982 | 28.984 |
| 0.18 | 2.145 | 28.967 | 28.948 | 28.930 | 28.912 | 28.895 | 28.877 | 28.860 | 28.843 | 28.826 | 28.810 | 28.794 |
| 0.20 | 2.383 | 28.967 | 28.930 | 28.893 | 28.857 | 28.822 | 28.787 | 28.752 | 28.718 | 28.685 | 28.652 | 28.620 |
| 0.22 | 2.622 | 28.967 | 28.912 | 28.859 | 28.806 | 28.754 | 28.703 | 28.653 | 28.604 | 28.555 | 28.507 | 28.460 |
| 0.24 | 2.860 | 28.967 | 28.896 | 28.827 | 28.759 | 28.692 | 28.626 | 28.561 | 28.497 | 28.435 | 28.373 | 28.312 |
| 0.26 | 3.098 | 28.697 | 28.881 | 28.797 | 28.715 | 28.634 | 28.554 | 28.476 | 28.399 | 28.323 | 28.249 | 28.175 |
| 0.28 | 3.337 | 28.967 | 28.867 | 28.769 | 28.674 | 28.579 | 28.487 | 28.396 | 28.307 | 28.219 | 28.133 | 28.049 |
| 0.30 | 3.575 | 28.967 | 28.854 | 28.744 | 28.635 | 28.529 | 28.424 | 28.322 | 28.221 | 28.123 | 28.026 | 27.931 |
| 0.32 | 3.813 | 28.967 | 28.842 | 28.719 | 28.599 | 28.481 | 28.366 | 28.253 | 28.141 | 28.032 | 27.925 | 27.821 |
| 0.34 | 4.052 | 28.967 | 28.830 | 28.696 | 28.565 | 28.437 | 28.311 | 28.187 | 28.067 | 27.948 | 27.832 | 27.718 |

## Gas Constant (Newton-meter $K^{-1} kg^{-1}$)

| $\dfrac{\text{kg of H}}{\text{kg of C}}$ | $\dfrac{\text{atoms of H}}{\text{atoms of C}}$ | 0 | 10 | 20 | 30 | 40 | 50 | 60 | 70 | 80 | 90 | 100 |
|---|---|---|---|---|---|---|---|---|---|---|---|---|
| 0.06 | 0.715 | 287.03 | 285.71 | 284.41 | 283.12 | 281.86 | 280.62 | 279.39 | 278.19 | 277.00 | 275.82 | 274.67 |
| 0.08 | 0.953 | 287.03 | 286.02 | 285.02 | 284.04 | 283.07 | 282.12 | 281.19 | 280.26 | 279.34 | 278.45 | 277.56 |
| 0.10 | 1.192 | 287.03 | 286.30 | 285.58 | 284.88 | 284.18 | 283.49 | 282.81 | 282.14 | 281.48 | 280.84 | 280.19 |
| 0.12 | 1.430 | 287.03 | 286.57 | 286.09 | 285.64 | 285.19 | 284.74 | 284.30 | 283.87 | 283.44 | 283.02 | 282.60 |
| 0.14 | 1.668 | 287.03 | 286.79 | 286.57 | 286.34 | 286.11 | 285.89 | 285.67 | 285.45 | 285.24 | 285.03 | 284.82 |
| 0.16 | 1.907 | 287.03 | 287.01 | 287.00 | 286.98 | 286.96 | 286.95 | 286.93 | 286.91 | 286.90 | 286.88 | 286.86 |
| 0.18 | 2.145 | 287.03 | 287.22 | 287.40 | 287.58 | 287.75 | 287.92 | 288.09 | 288.26 | 288.43 | 288.59 | 288.75 |
| 0.20 | 2.383 | 287.03 | 287.40 | 287.77 | 288.12 | 288.47 | 288.83 | 289.18 | 289.52 | 289.85 | 290.19 | 290.51 |
| 0.22 | 2.622 | 287.03 | 287.58 | 288.10 | 288.63 | 289.16 | 289.67 | 290.18 | 290.67 | 291.17 | 291.66 | 292.14 |
| 0.24 | 2.860 | 287.03 | 287.74 | 288.42 | 289.11 | 289.78 | 290.45 | 291.11 | 291.76 | 292.40 | 293.04 | 293.67 |
| 0.26 | 3.098 | 287.03 | 287.89 | 288.72 | 289.55 | 290.37 | 291.18 | 291.98 | 292.77 | 293.56 | 294.33 | 295.10 |
| 0.28 | 3.337 | 287.03 | 288.02 | 289.01 | 289.96 | 290.93 | 291.87 | 292.80 | 293.72 | 294.64 | 295.54 | 296.42 |
| 0.30 | 3.575 | 287.03 | 288.15 | 289.26 | 290.36 | 291.44 | 292.51 | 293.57 | 294.62 | 295.64 | 296.67 | 297.68 |
| 0.32 | 3.813 | 287.03 | 288.27 | 289.51 | 290.72 | 291.93 | 293.11 | 294.28 | 295.46 | 296.60 | 297.74 | 298.85 |
| 0.34 | 4.052 | 287.03 | 288.39 | 289.74 | 291.07 | 292.38 | 293.68 | 294.97 | 296.23 | 297.50 | 298.74 | 299.96 |

# Table 10    Enthalpy of Combustion of Hydrocarbons at 298.15 K and Constant Pressure

| Compound | Formula | Enthalpy of Combustion of Liquid $-\Delta H_c^0$, kJ kg$^{-1}$ | | Enthalpy of Combustion of Gas $-\Delta H_c^0$, kJ kg$^{-1}$ | |
|---|---|---|---|---|---|
| | | Gross[a] | Net[b] | Gross[a] | Net[b] |
| **Paraffins** | | | | | |
| Methane | $CH_4$ | | | 55522 | 50032 |
| Ethane | $C_2H_6$ | 51588 | 47195 | 51902 | 47509 |
| Propane | $C_3H_8$ | 49949 | 45957 | 50325 | 46334 |
| n-Butane | $C_4H_{10}$ | 49132 | 45345 | 49509 | 45722 |
| n-Pentane | $C_5H_{12}$ | 48634 | 44973 | 49002 | 45345 |
| n-Hexane | $C_6H_{14}$ | 48313 | 44736 | 48676 | 45103 |
| n-Heptane | $C_7H_{16}$ | 48074 | 44559 | 48439 | 44924 |
| n-Octane | $C_8H_{18}$ | 47885 | 44422 | 48251 | 44782 |
| n-Nonane | $C_9H_{20}$ | 47753 | 44322 | 48116 | 44681 |
| n-Decane | $C_{10}H_{22}$ | 47639 | 44238 | 47999 | 44596 |
| n-Undecane | $C_{11}H_{24}$ | 47548 | 44166 | 47906 | 44527 |
| n-Dodecane | $C_{12}H_{26}$ | 47467 | 44108 | 47827 | 44468 |
| n-Tridecane | $C_{13}H_{28}$ | 47404 | 44061 | 47764 | 44422 |
| n-Tetradecane | $C_{14}H_{30}$ | 47346 | 44020 | 47706 | 44380 |
| n-Pentadecane | $C_{15}H_{32}$ | 47297 | 43982 | 47655 | 44343 |
| n-Hexadecane | $C_{16}H_{34}$ | 47255 | 43950 | 47613 | 44308 |
| | | | | | |
| **Olefins** | | | | | |
| Ethene | $C_2H_4$ | | | 50283 | 47146 |
| Propene | $C_3H_6$ | | | 48918 | 45780 |
| 1-Butene | $C_4H_8$ | 48053 | 44915 | 48425 | 45287 |
| 1-Pentene | $C_5H_{10}$ | 47757 | 44620 | 48125 | 44987 |
| 1-Hexene | $C_6H_{12}$ | 47571 | 44434 | 47937 | 44799 |
| 1-Heptene | $C_7H_{14}$ | 47434 | 44296 | 47797 | 44659 |
| 1-Octene | $C_8H_{16}$ | 47332 | 44194 | 47692 | 44557 |
| 1-Nonene | $C_9H_{18}$ | 47250 | 44115 | 47613 | 44475 |
| 1-Decene | $C_{10}H_{20}$ | 47188 | 44050 | 47548 | 44410 |
| 1-Undecene | $C_{11}H_{22}$ | 47134 | 43996 | 47495 | 44357 |
| 1-Dodecene | $C_{12}H_{24}$ | 47092 | 43954 | 47450 | 44313 |
| 1-Tridecene | $C_{13}H_{26}$ | 47055 | 43917 | 47413 | 44275 |
| 1-Tetradecene | $C_{14}H_{28}$ | 47022 | 43885 | 47381 | 44243 |
| 1-Pentadecene | $C_{15}H_{30}$ | 46995 | 43857 | 47353 | 44215 |
| 1-Hexadecene | $C_{16}H_{32}$ | 46971 | 43833 | 47327 | 44189 |
| | | | | | |
| **Alkylbenzenes** | | | | | |
| Benzene | $C_6H_6$ | 41835 | 40147 | 42268 | 40579 |
| Methylbenzene | $C_7H_8$ | 42438 | 40526 | 42850 | 40940 |
| Ethylbenzene | $C_8H_{10}$ | 42998 | 40928 | 43399 | 41326 |
| n-Propylbenzene | $C_9H_{12}$ | 43417 | 41221 | 43803 | 41605 |
| n-Butylbenzene | $C_{10}H_{14}$ | 43752 | 41456 | 44127 | 41831 |
| n-Pentylbenzene | $C_{11}H_{16}$ | 44024 | 41647 | 44394 | 42019 |
| n-Hexylbenzene | $C_{12}H_{18}$ | 44247 | 41808 | 44617 | 42175 |

[a] Combustion products are $H_2O$ (liq) and $CO_2$ (g).
[b] Combustion products are $H_2O$ (g) and $CO_2$ (g).

## SYMBOLS AND UNITS USED IN TABLES 11 TO 23

$a$    velocity of sound $= \sqrt{k\bar{R}T/\bar{m}}$, m s$^{-1}$

$\bar{c}_p$    specific heat at constant pressure, J K$^{-1}$ g-mol$^{-1}$

$\bar{c}_v$    specific heat at constant volume, J K$^{-1}$ g-mol$^{-1}$

$\bar{h}$    enthalpy per mole, J g-mol$^{-1}$

$k$    $\bar{c}_p/\bar{c}_v$

$\bar{m}$    molecular weight, g g-mol$^{-1}$

$p_r$    relative pressure*

$\bar{R}$    universal gas constant $= 8.31441$ J K$^{-1}$ g-mol$^{-1}$

$T$    temperature, K

$t$    temperature, °C

$\tilde{u}$    internal energy per mole, J g-mol$^{-1}$

$v_r$    relative volume*

$\bar{\phi}$    $\int_{T_0}^{T} \frac{\bar{c}_p}{T} dT$, J K$^{-1}$ g-mol$^{-1}$

*The ratio of the pressures $p_a$ and $p_b$ corresponding to the temperatures $T_a$ and $T_b$, respectively, along a given isentropic is equal to the ratio of the relative pressures $p_{ra}$ and $p_{rb}$ as tabulated for $T_a$ and $T_b$, respectively. Thus

$$\left(\frac{p_a}{p_b}\right)_{s=\text{constant}} = \frac{p_{ra}}{p_{rb}}$$

Similarly

$$\left(\frac{v_a}{v_b}\right)_{s=\text{constant}} = \frac{v_{ra}}{v_{rb}}$$

# Table 11 Nitrogen at Low Pressures (for One Gram-Mole)

$\bar{m} = 28.0134$

| T | t | $\bar{h}$ | $p_r$ | $\bar{u}$ | $v_r$ | $\bar{\phi}$ | T | t | $\bar{h}$ | $p_r$ | $\bar{u}$ | $v_r$ | $\bar{\phi}$ |
|---|---|---|---|---|---|---|---|---|---|---|---|---|---|
| 100 | −173.15 | 2902.2 | .0219 | 2070.8 | 37.998 | 159.667 | 600 | 326.85 | 17564.1 | 11.894 | 12575.5 | .419 | 212.033 |
| 110 | −163.15 | 3193.3 | .0305 | 2278.7 | 29.941 | 162.441 | 610 | 336.85 | 17865.5 | 12.629 | 12793.8 | .402 | 212.531 |
| 120 | −153.15 | 3484.3 | .0414 | 2486.6 | 24.087 | 164.974 | 620 | 346.85 | 18167.6 | 13.397 | 13012.6 | .385 | 213.022 |
| 130 | −143.15 | 3775.4 | .0548 | 2694.5 | 19.717 | 167.303 | 630 | 356.85 | 18470.2 | 14.201 | 13232.1 | .369 | 213.507 |
| 140 | −133.15 | 4066.4 | .0711 | 2902.4 | 16.382 | 169.460 | 640 | 366.85 | 18773.5 | 15.040 | 13452.3 | .354 | 213.984 |
| 150 | −123.15 | 4357.5 | .0905 | 3110.3 | 13.786 | 171.468 | 650 | 376.85 | 19077.4 | 15.917 | 13673.0 | .340 | 214.455 |
| 160 | −113.15 | 4648.5 | .1134 | 3318.2 | 11.732 | 173.347 | 660 | 386.85 | 19381.9 | 16.833 | 13894.4 | .326 | 214.920 |
| 170 | −103.15 | 4939.6 | .1402 | 3526.1 | 10.081 | 175.111 | 670 | 396.85 | 19687.2 | 17.788 | 14116.5 | .313 | 215.379 |
| 180 | −93.15 | 5230.6 | .1713 | 3734.1 | 8.739 | 176.775 | 680 | 406.85 | 19993.0 | 18.784 | 14339.2 | .301 | 215.833 |
| 190 | −83.15 | 5521.7 | .2070 | 3942.0 | 7.633 | 178.349 | 690 | 416.85 | 20299.6 | 19.823 | 14562.6 | .289 | 216.280 |
| 200 | −73.15 | 5812.8 | .2477 | 4149.9 | 6.714 | 179.842 | 700 | 426.85 | 20606.8 | 20.906 | 14786.7 | .278 | 216.722 |
| 210 | −63.15 | 6103.9 | .2938 | 4357.8 | 5.943 | 181.262 | 710 | 436.85 | 20914.7 | 22.033 | 15011.5 | .268 | 217.159 |
| 220 | −53.15 | 6394.9 | .3458 | 4565.8 | 5.290 | 182.616 | 720 | 446.85 | 21223.2 | 23.207 | 15236.9 | .258 | 217.590 |
| 230 | −43.15 | 6686.0 | .4040 | 4773.7 | 4.734 | 183.910 | 730 | 456.85 | 21532.5 | 24.428 | 15463.0 | .248 | 218.017 |
| 240 | −33.15 | 6977.1 | .4689 | 4981.7 | 4.256 | 185.149 | 740 | 466.85 | 21842.4 | 25.699 | 15689.7 | .239 | 218.439 |
| 250 | −23.15 | 7268.2 | .5409 | 5189.6 | 3.843 | 186.337 | 750 | 476.85 | 22153.0 | 27.021 | 15917.2 | .231 | 218.856 |
| 260 | −13.15 | 7559.3 | .6205 | 5397.6 | 3.484 | 187.479 | 760 | 486.85 | 22464.3 | 28.394 | 16145.3 | .223 | 219.268 |
| 270 | −3.15 | 7850.5 | .7082 | 5605.6 | 3.170 | 188.578 | 770 | 496.85 | 22776.2 | 29.822 | 16374.1 | .215 | 219.676 |
| 280 | 6.85 | 8141.6 | .8044 | 5813.6 | 2.894 | 189.637 | 780 | 506.85 | 23088.9 | 31.304 | 16603.6 | .207 | 220.079 |
| 290 | 16.85 | 8432.8 | .9096 | 6021.7 | 2.651 | 190.658 | 790 | 516.85 | 23402.2 | 32.844 | 16833.8 | .200 | 220.478 |
| 300 | 26.85 | 8724.1 | 1.0243 | 6229.7 | 2.435 | 191.646 | 800 | 526.85 | 23716.2 | 34.442 | 17064.7 | .193 | 220.873 |
| 310 | 36.85 | 9015.3 | 1.1490 | 6437.9 | 2.243 | 192.601 | 810 | 536.85 | 24030.8 | 36.100 | 17296.2 | .187 | 221.264 |
| 320 | 46.85 | 9306.7 | 1.2842 | 6646.1 | 2.072 | 193.526 | 820 | 546.85 | 24346.2 | 37.819 | 17528.4 | .180 | 221.651 |
| 330 | 56.85 | 9598.1 | 1.4304 | 6854.3 | 1.918 | 194.423 | 830 | 556.85 | 24662.2 | 39.602 | 17761.2 | .174 | 222.034 |
| 340 | 66.85 | 9889.6 | 1.5882 | 7062.7 | 1.780 | 195.293 | 840 | 566.85 | 24978.9 | 41.451 | 17994.8 | .168 | 222.413 |
| 350 | 76.85 | 10181.2 | 1.7582 | 7271.1 | 1.655 | 196.138 | 850 | 576.85 | 25296.2 | 43.366 | 18229.0 | .163 | 222.789 |
| 360 | 86.85 | 10472.9 | 1.9409 | 7479.7 | 1.542 | 196.960 | 860 | 586.85 | 25614.2 | 45.350 | 18463.8 | .158 | 223.161 |
| 370 | 96.85 | 10764.8 | 2.1368 | 7688.4 | 1.440 | 197.759 | 870 | 596.85 | 25932.8 | 47.404 | 18699.3 | .153 | 223.529 |
| 380 | 106.85 | 11056.8 | 2.3466 | 7897.3 | 1.346 | 198.538 | 880 | 606.85 | 26252.1 | 49.531 | 18935.5 | .148 | 223.894 |
| 390 | 116.85 | 11349.0 | 2.5709 | 8106.3 | 1.261 | 199.297 | 890 | 616.85 | 26572.1 | 51.732 | 19172.3 | .143 | 224.255 |
| 400 | 126.85 | 11641.3 | 2.8103 | 8315.6 | 1.183 | 200.037 | 900 | 626.85 | 26892.6 | 54.009 | 19409.7 | .139 | 224.614 |
| 410 | 136.85 | 11933.9 | 3.0654 | 8525.0 | 1.112 | 200.760 | 910 | 636.85 | 27213.9 | 56.365 | 19647.8 | .134 | 224.969 |
| 420 | 146.85 | 12226.8 | 3.3370 | 8734.7 | 1.046 | 201.466 | 920 | 646.85 | 27535.7 | 58.801 | 19886.5 | .130 | 225.320 |
| 430 | 156.85 | 12519.9 | 3.6256 | 8944.7 | .986 | 202.155 | 930 | 656.85 | 27858.2 | 61.319 | 20125.8 | .126 | 225.669 |
| 440 | 166.85 | 12813.3 | 3.9320 | 9155.0 | .930 | 202.830 | 940 | 666.85 | 28181.3 | 63.921 | 20365.7 | .122 | 226.015 |
| 450 | 176.85 | 13107.0 | 4.2568 | 9365.5 | .879 | 203.490 | 950 | 676.85 | 28504.9 | 66.609 | 20606.3 | .119 | 226.357 |
| 460 | 186.85 | 13401.0 | 4.6009 | 9576.4 | .831 | 204.136 | 960 | 686.85 | 28829.2 | 69.386 | 20847.4 | .115 | 226.697 |
| 470 | 196.85 | 13695.4 | 4.9649 | 9787.7 | .787 | 204.769 | 970 | 696.85 | 29154.2 | 72.253 | 21089.2 | .112 | 227.033 |
| 480 | 206.85 | 13990.0 | 5.3497 | 9999.3 | .746 | 205.390 | 980 | 706.85 | 29479.6 | 75.214 | 21331.5 | .108 | 227.367 |
| 490 | 216.85 | 14285.4 | 5.7560 | 10211.3 | .708 | 205.998 | 990 | 716.85 | 29805.7 | 78.269 | 21574.5 | .105 | 227.698 |
| 500 | 226.85 | 14581.0 | 6.1846 | 10423.8 | .672 | 206.596 | 1000 | 726.85 | 30132.4 | 81.421 | 21818.0 | .102 | 228.027 |
| 510 | 236.85 | 14877.0 | 6.6364 | 10636.6 | .639 | 207.182 | 1010 | 736.85 | 30459.6 | 84.673 | 22062.1 | .099 | 228.352 |
| 520 | 246.85 | 15173.5 | 7.1122 | 10850.0 | .608 | 207.758 | 1020 | 746.85 | 30787.5 | 88.027 | 22306.8 | .096 | 228.675 |
| 530 | 256.85 | 15470.5 | 7.6130 | 11063.8 | .579 | 208.323 | 1030 | 756.85 | 31115.8 | 91.485 | 22552.0 | .094 | 228.995 |
| 540 | 266.85 | 15767.9 | 8.1395 | 11278.1 | .552 | 208.879 | 1040 | 766.85 | 31444.8 | 95.050 | 22797.8 | .091 | 229.313 |
| 550 | 276.85 | 16065.9 | 8.6928 | 11493.0 | .526 | 209.426 | 1050 | 776.85 | 31774.2 | 98.723 | 23044.1 | .088 | 229.629 |
| 560 | 286.85 | 16364.4 | 9.2737 | 11708.4 | .502 | 209.964 | 1060 | 786.85 | 32104.3 | 102.508 | 23291.0 | .086 | 229.941 |
| 570 | 296.85 | 16663.5 | 9.8833 | 11924.3 | .480 | 210.493 | 1070 | 796.85 | 32434.8 | 106.408 | 23538.4 | .084 | 230.252 |
| 580 | 306.85 | 16963.1 | 10.5226 | 12140.8 | .458 | 211.014 | 1080 | 806.85 | 32765.9 | 110.423 | 23786.4 | .081 | 230.560 |
| 590 | 316.85 | 17263.3 | 11.1926 | 12357.8 | .438 | 211.528 | 1090 | 816.85 | 33097.6 | 114.558 | 24034.9 | .079 | 230.865 |

# Table 11  Nitrogen at Low Pressures (for One Gram-Mole)

$\bar{m} = 28.0134$

| T | t | $\bar{h}$ | $p_r$ | $\bar{u}$ | $v_r$ | $\bar{\phi}$ | T | t | $\bar{h}$ | $p_r$ | $\bar{u}$ | $v_r$ | $\bar{\phi}$ |
|---|---|---|---|---|---|---|---|---|---|---|---|---|---|
| 1100 | 826.85 | 33429.7 | 118.81 | 24283.9 | .0770 | 231.169 | 1600 | 1326.85 | 50571.4 | 555.53 | 37268.4 | .0239 | 243.993 |
| 1110 | 836.85 | 33762.4 | 123.20 | 24533.4 | .0749 | 231.470 | 1610 | 1336.85 | 50922.8 | 570.35 | 37536.6 | .0235 | 244.212 |
| 1120 | 846.85 | 34095.5 | 127.70 | 24783.4 | .0729 | 231.769 | 1620 | 1346.85 | 51274.4 | 585.48 | 37805.1 | .0230 | 244.429 |
| 1130 | 856.85 | 34429.2 | 132.34 | 25033.9 | .0710 | 232.065 | 1630 | 1356.85 | 51626.2 | 600.93 | 38073.8 | .0226 | 244.646 |
| 1140 | 866.85 | 34763.3 | 137.11 | 25284.9 | .0691 | 232.360 | 1640 | 1366.85 | 51978.4 | 616.70 | 38342.7 | .0221 | 244.861 |
| 1150 | 876.85 | 35098.0 | 142.02 | 25536.4 | .0673 | 232.652 | 1650 | 1376.85 | 52330.7 | 632.79 | 38612.0 | .0217 | 245.075 |
| 1160 | 886.85 | 35433.1 | 147.06 | 25788.4 | .0656 | 232.942 | 1660 | 1386.85 | 52683.4 | 649.22 | 38881.4 | .0213 | 245.288 |
| 1170 | 896.85 | 35768.7 | 152.24 | 26040.8 | .0639 | 233.230 | 1670 | 1396.85 | 53036.2 | 665.98 | 39151.2 | .0208 | 245.500 |
| 1180 | 906.85 | 36104.8 | 157.57 | 26293.8 | .0623 | 233.516 | 1680 | 1406.85 | 53389.3 | 683.08 | 39421.1 | .0204 | 245.711 |
| 1190 | 916.85 | 36441.3 | 163.05 | 26547.1 | .0607 | 233.800 | 1690 | 1416.85 | 53742.7 | 700.53 | 39691.3 | .0201 | 245.921 |
| 1200 | 926.85 | 36778.3 | 168.67 | 26801.0 | .0592 | 234.082 | 1700 | 1426.85 | 54096.3 | 718.33 | 39961.8 | .0197 | 246.129 |
| 1210 | 936.85 | 37115.7 | 174.45 | 27055.3 | .0577 | 234.362 | 1710 | 1436.85 | 54450.1 | 736.48 | 40232.5 | .0193 | 246.337 |
| 1220 | 946.85 | 37453.6 | 180.38 | 27310.0 | .0562 | 234.640 | 1720 | 1446.85 | 54804.1 | 755.00 | 40503.4 | .0189 | 246.543 |
| 1230 | 956.85 | 37791.9 | 186.48 | 27565.2 | .0548 | 234.916 | 1730 | 1456.85 | 55158.4 | 773.88 | 40774.5 | .0186 | 246.749 |
| 1240 | 966.85 | 38130.7 | 192.73 | 27820.8 | .0535 | 235.191 | 1740 | 1466.85 | 55512.9 | 793.14 | 41045.9 | .0182 | 246.953 |
| 1250 | 976.85 | 38469.9 | 199.15 | 28076.9 | .0522 | 235.463 | 1750 | 1476.85 | 55867.7 | 812.77 | 41317.4 | .0179 | 247.156 |
| 1260 | 986.85 | 38809.5 | 205.74 | 28333.3 | .0509 | 235.734 | 1760 | 1486.85 | 56222.6 | 832.78 | 41589.3 | .0176 | 247.359 |
| 1270 | 996.85 | 39149.5 | 212.50 | 28590.2 | .0497 | 236.003 | 1770 | 1496.85 | 56577.8 | 853.18 | 41861.3 | .0172 | 247.560 |
| 1280 | 1006.85 | 39489.9 | 219.43 | 28847.5 | .0485 | 236.270 | 1780 | 1506.85 | 56933.1 | 873.97 | 42133.5 | .0169 | 247.760 |
| 1290 | 1016.85 | 39830.8 | 226.55 | 29105.2 | .0473 | 236.535 | 1790 | 1516.85 | 57288.7 | 895.17 | 42406.0 | .0166 | 247.959 |
| 1300 | 1026.85 | 40172.0 | 233.84 | 29363.3 | .0462 | 236.798 | 1800 | 1526.85 | 57644.5 | 916.77 | 42678.6 | .0163 | 248.158 |
| 1310 | 1036.85 | 40513.6 | 241.32 | 29621.8 | .0451 | 237.060 | 1810 | 1536.85 | 58000.5 | 938.77 | 42951.5 | .0160 | 248.355 |
| 1320 | 1046.85 | 40855.7 | 248.99 | 29880.6 | .0441 | 237.320 | 1820 | 1546.85 | 58356.8 | 961.20 | 43224.5 | .0157 | 248.551 |
| 1330 | 1056.85 | 41198.1 | 256.85 | 30139.9 | .0431 | 237.579 | 1830 | 1556.85 | 58713.2 | 984.04 | 43497.8 | .0155 | 248.746 |
| 1340 | 1066.85 | 41540.9 | 264.91 | 30399.6 | .0421 | 237.835 | 1840 | 1566.85 | 59069.8 | 1007.31 | 43771.3 | .0152 | 248.941 |
| 1350 | 1076.85 | 41884.0 | 273.16 | 30659.6 | .0411 | 238.091 | 1850 | 1576.85 | 59426.6 | 1031.02 | 44044.9 | .0149 | 249.134 |
| 1360 | 1086.85 | 42227.6 | 281.62 | 30920.0 | .0402 | 238.344 | 1860 | 1586.85 | 59783.6 | 1055.16 | 44318.8 | .0147 | 249.326 |
| 1370 | 1096.85 | 42571.5 | 290.28 | 31180.7 | .0392 | 238.596 | 1870 | 1596.85 | 60140.8 | 1079.75 | 44592.9 | .0144 | 249.518 |
| 1380 | 1106.85 | 42915.7 | 299.16 | 31441.8 | .0384 | 238.846 | 1880 | 1606.85 | 60498.2 | 1104.79 | 44867.1 | .0141 | 249.709 |
| 1390 | 1116.85 | 43260.3 | 308.25 | 31703.3 | .0375 | 239.095 | 1890 | 1616.85 | 60855.8 | 1130.28 | 45141.5 | .0139 | 249.898 |
| 1400 | 1126.85 | 43605.3 | 317.55 | 31965.1 | .0367 | 239.343 | 1900 | 1626.85 | 61213.5 | 1156.24 | 45416.1 | .0137 | 250.087 |
| 1410 | 1136.85 | 43950.6 | 327.08 | 32227.3 | .0358 | 239.588 | 1910 | 1636.85 | 61571.5 | 1182.67 | 45690.9 | .0134 | 250.275 |
| 1420 | 1146.85 | 44296.3 | 336.83 | 32489.8 | .0351 | 239.833 | 1920 | 1646.85 | 61929.6 | 1209.57 | 45965.9 | .0132 | 250.462 |
| 1430 | 1156.85 | 44642.2 | 346.81 | 32752.6 | .0343 | 240.075 | 1930 | 1656.85 | 62287.9 | 1236.96 | 46241.1 | .0130 | 250.648 |
| 1440 | 1166.85 | 44988.6 | 357.03 | 33015.8 | .0335 | 240.317 | 1940 | 1666.85 | 62646.4 | 1264.83 | 46516.4 | .0128 | 250.833 |
| 1450 | 1176.85 | 45335.2 | 367.48 | 33279.3 | .0328 | 240.557 | 1950 | 1676.85 | 63005.0 | 1293.19 | 46791.9 | .0125 | 251.018 |
| 1460 | 1186.85 | 45682.2 | 378.17 | 33543.1 | .0321 | 240.795 | 1960 | 1686.85 | 63363.9 | 1322.06 | 47067.6 | .0123 | 251.201 |
| 1470 | 1196.85 | 46029.5 | 389.11 | 33807.3 | .0314 | 241.032 | 1970 | 1696.85 | 63722.9 | 1351.43 | 47343.5 | .0121 | 251.384 |
| 1480 | 1206.85 | 46377.1 | 400.29 | 34071.8 | .0307 | 241.268 | 1980 | 1706.85 | 64082.0 | 1381.32 | 47619.5 | .0119 | 251.566 |
| 1490 | 1216.85 | 46725.0 | 411.73 | 34336.5 | .0301 | 241.502 | 1990 | 1716.85 | 64441.4 | 1411.72 | 47895.7 | .0117 | 251.747 |
| 1500 | 1226.85 | 47073.2 | 423.43 | 34601.6 | .0295 | 241.735 | 2000 | 1726.85 | 64800.9 | 1442.65 | 48172.1 | .0115 | 251.927 |
| 1510 | 1236.85 | 47421.7 | 435.39 | 34867.0 | .0288 | 241.967 | 2010 | 1736.85 | 65160.5 | 1474.12 | 48448.6 | .0113 | 252.107 |
| 1520 | 1246.85 | 47770.6 | 447.62 | 35132.7 | .0282 | 242.197 | 2020 | 1746.85 | 65520.4 | 1506.12 | 48725.3 | .0112 | 252.285 |
| 1530 | 1256.85 | 48119.7 | 460.11 | 35398.6 | .0276 | 242.426 | 2030 | 1756.85 | 65880.4 | 1538.67 | 49002.1 | .0110 | 252.463 |
| 1540 | 1266.85 | 48469.1 | 472.88 | 35664.9 | .0271 | 242.653 | 2040 | 1766.85 | 66240.5 | 1571.77 | 49279.1 | .0108 | 252.640 |
| 1550 | 1276.85 | 48818.8 | 485.93 | 35931.5 | .0265 | 242.880 | 2050 | 1776.85 | 66600.8 | 1605.43 | 49556.3 | .0106 | 252.816 |
| 1560 | 1286.85 | 49168.8 | 499.27 | 36198.3 | .0260 | 243.105 | 2060 | 1786.85 | 66961.3 | 1639.66 | 49833.6 | .0104 | 252.991 |
| 1570 | 1296.85 | 49519.0 | 512.89 | 36465.4 | .0255 | 243.329 | 2070 | 1796.85 | 67321.9 | 1674.47 | 50111.1 | .0103 | 253.166 |
| 1580 | 1306.85 | 49869.6 | 526.80 | 36732.8 | .0249 | 243.551 | 2080 | 1806.85 | 67682.7 | 1709.85 | 50388.7 | .0101 | 253.340 |
| 1590 | 1316.85 | 50220.4 | 541.02 | 37000.5 | .0244 | 243.772 | 2090 | 1816.85 | 68043.6 | 1745.82 | 50666.5 | .0100 | 253.513 |

## Table 11 Nitrogen at Low Pressures (for One Gram-Mole)
$\bar{m} = 28.0134$

| T | t | $\bar{h}$ | $p_r$ | $\bar{u}$ | $v_r$ | $\bar{\phi}$ | T | t | $\bar{h}$ | $p_r$ | $\bar{u}$ | $v_r$ | $\bar{\phi}$ |
|---|---|---|---|---|---|---|---|---|---|---|---|---|---|
| 2100 | 1826.85 | 68404.6 | 1782.4 | 50944.4 | .00980 | 253.685 | 2600 | 2326.85 | 86617.0 | 4541.9 | 64999.5 | .00476 | 261.463 |
| 2110 | 1836.85 | 68765.8 | 1819.6 | 51222.5 | .00964 | 253.857 | 2610 | 2336.85 | 86983.9 | 4619.5 | 65283.3 | .00470 | 261.603 |
| 2120 | 1846.85 | 69127.2 | 1857.3 | 51500.7 | .00949 | 254.028 | 2620 | 2346.85 | 87350.9 | 4698.1 | 65567.2 | .00464 | 261.744 |
| 2130 | 1856.85 | 69488.7 | 1895.7 | 51779.0 | .00934 | 254.198 | 2630 | 2356.85 | 87718.0 | 4777.8 | 65851.2 | .00458 | 261.884 |
| 2140 | 1866.85 | 69850.3 | 1934.7 | 52057.5 | .00920 | 254.367 | 2640 | 2366.85 | 88085.2 | 4858.6 | 66135.2 | .00452 | 262.023 |
| 2150 | 1876.85 | 70212.1 | 1974.4 | 52336.2 | .00905 | 254.536 | 2650 | 2376.85 | 88452.5 | 4940.4 | 66419.4 | .00446 | 262.162 |
| 2160 | 1886.85 | 70574.0 | 2014.7 | 52614.9 | .00891 | 254.704 | 2660 | 2386.85 | 88819.9 | 5023.3 | 66703.6 | .00440 | 262.300 |
| 2170 | 1896.85 | 70936.1 | 2055.6 | 52893.8 | .00878 | 254.871 | 2670 | 2396.85 | 89187.4 | 5107.3 | 66987.9 | .00435 | 262.438 |
| 2180 | 1906.85 | 71298.3 | 2097.2 | 53172.9 | .00864 | 255.038 | 2680 | 2406.85 | 89554.9 | 5192.4 | 67272.3 | .00429 | 262.576 |
| 2190 | 1916.85 | 71660.6 | 2139.4 | 53452.1 | .00851 | 255.204 | 2690 | 2416.85 | 89922.5 | 5278.6 | 67556.8 | .00424 | 262.712 |
| 2200 | 1926.85 | 72023.1 | 2182.4 | 53731.4 | .00838 | 255.369 | 2700 | 2426.85 | 90290.3 | 5366.0 | 67841.4 | .00418 | 262.849 |
| 2210 | 1936.85 | 72385.7 | 2226.0 | 54010.8 | .00825 | 255.533 | 2710 | 2436.85 | 90658.1 | 5454.4 | 68126.0 | .00413 | 262.985 |
| 2220 | 1946.85 | 72748.4 | 2270.2 | 54290.4 | .00813 | 255.697 | 2720 | 2446.85 | 91025.9 | 5544.1 | 68410.8 | .00408 | 263.120 |
| 2230 | 1956.85 | 73111.2 | 2315.2 | 54570.1 | .00801 | 255.860 | 2730 | 2456.85 | 91393.9 | 5634.8 | 68695.6 | .00403 | 263.255 |
| 2240 | 1966.85 | 73474.2 | 2360.9 | 54850.0 | .00789 | 256.022 | 2740 | 2466.85 | 91761.9 | 5726.8 | 68980.5 | .00398 | 263.390 |
| 2250 | 1976.85 | 73837.3 | 2407.2 | 55129.9 | .00777 | 256.184 | 2750 | 2476.85 | 92130.1 | 5819.9 | 69265.4 | .00393 | 263.524 |
| 2260 | 1986.85 | 74200.6 | 2454.3 | 55410.0 | .00766 | 256.345 | 2760 | 2486.85 | 92498.3 | 5914.2 | 69550.5 | .00388 | 263.658 |
| 2270 | 1996.85 | 74563.9 | 2502.1 | 55690.2 | .00754 | 256.506 | 2770 | 2496.85 | 92866.5 | 6009.7 | 69835.6 | .00383 | 263.791 |
| 2280 | 2006.85 | 74927.4 | 2550.7 | 55970.6 | .00743 | 256.665 | 2780 | 2506.85 | 93234.9 | 6106.4 | 70120.8 | .00379 | 263.924 |
| 2290 | 2016.85 | 75291.0 | 2600.0 | 56251.0 | .00732 | 256.825 | 2790 | 2516.85 | 93603.3 | 6204.3 | 70406.1 | .00374 | 264.056 |
| 2300 | 2026.85 | 75654.7 | 2650.0 | 56531.6 | .00722 | 256.983 | 2800 | 2526.85 | 93971.8 | 6303.5 | 70691.5 | .00369 | 264.188 |
| 2310 | 2036.85 | 76018.6 | 2700.8 | 56812.3 | .00711 | 257.141 | 2810 | 2536.85 | 94340.4 | 6403.9 | 70976.9 | .00365 | 264.319 |
| 2320 | 2046.85 | 76382.5 | 2752.4 | 57093.1 | .00701 | 257.298 | 2820 | 2546.85 | 94709.1 | 6505.6 | 71262.5 | .00360 | 264.450 |
| 2330 | 2056.85 | 76746.6 | 2804.7 | 57374.0 | .00691 | 257.455 | 2830 | 2556.85 | 95077.8 | 6608.5 | 71548.1 | .00356 | 264.581 |
| 2340 | 2066.85 | 77110.8 | 2857.8 | 57655.1 | .00681 | 257.611 | 2840 | 2566.85 | 95446.6 | 6712.8 | 71833.7 | .00352 | 264.711 |
| 2350 | 2076.85 | 77475.1 | 2911.7 | 57936.3 | .00671 | 257.766 | 2850 | 2576.85 | 95815.5 | 6818.3 | 72119.5 | .00348 | 264.840 |
| 2360 | 2086.85 | 77839.5 | 2966.4 | 58217.5 | .00661 | 257.921 | 2860 | 2586.85 | 96184.5 | 6925.1 | 72405.3 | .00343 | 264.970 |
| 2370 | 2096.85 | 78204.0 | 3021.9 | 58498.9 | .00652 | 258.075 | 2870 | 2596.85 | 96553.5 | 7033.2 | 72691.2 | .00339 | 265.098 |
| 2380 | 2106.85 | 78568.7 | 3078.2 | 58780.4 | .00643 | 258.228 | 2880 | 2606.85 | 96922.6 | 7142.6 | 72977.1 | .00335 | 265.227 |
| 2390 | 2116.85 | 78933.4 | 3135.4 | 59062.0 | .00634 | 258.381 | 2890 | 2616.85 | 97291.8 | 7253.4 | 73263.2 | .00331 | 265.355 |
| 2400 | 2126.85 | 79298.3 | 3193.3 | 59343.7 | .00625 | 258.534 | 2900 | 2626.85 | 97661.0 | 7365.5 | 73549.3 | .00327 | 265.482 |
| 2410 | 2136.85 | 79663.3 | 3252.2 | 59625.6 | .00616 | 258.685 | 2910 | 2636.85 | 98030.4 | 7479.0 | 73835.5 | .00324 | 265.609 |
| 2420 | 2146.85 | 80028.4 | 3311.8 | 59907.5 | .00608 | 258.837 | 2920 | 2646.85 | 98399.8 | 7593.9 | 74121.7 | .00320 | 265.736 |
| 2430 | 2156.85 | 80393.5 | 3372.4 | 60189.5 | .00599 | 258.987 | 2930 | 2656.85 | 98769.2 | 7710.1 | 74408.0 | .00316 | 265.863 |
| 2440 | 2166.85 | 80758.8 | 3433.8 | 60471.7 | .00591 | 259.137 | 2940 | 2666.85 | 99138.7 | 7827.8 | 74694.4 | .00312 | 265.988 |
| 2450 | 2176.85 | 81124.2 | 3496.0 | 60753.9 | .00583 | 259.287 | 2950 | 2676.85 | 99508.3 | 7946.8 | 74980.8 | .00309 | 266.114 |
| 2460 | 2186.85 | 81489.7 | 3559.2 | 61036.3 | .00575 | 259.436 | 2960 | 2686.85 | 99878.0 | 8067.3 | 75267.4 | .00305 | 266.239 |
| 2470 | 2196.85 | 81855.3 | 3623.3 | 61318.7 | .00567 | 259.584 | 2970 | 2696.85 | 100247.7 | 8189.2 | 75554.0 | .00302 | 266.364 |
| 2480 | 2206.85 | 82221.0 | 3688.2 | 61601.3 | .00559 | 259.732 | 2980 | 2706.85 | 100617.5 | 8312.6 | 75840.6 | .00298 | 266.488 |
| 2490 | 2216.85 | 82586.8 | 3754.1 | 61883.9 | .00551 | 259.879 | 2990 | 2716.85 | 100987.4 | 8437.4 | 76127.3 | .00295 | 266.612 |
| 2500 | 2226.85 | 82952.7 | 3820.9 | 62166.7 | .00544 | 260.025 | 3000 | 2726.85 | 101357.3 | 8563.6 | 76414.1 | .00291 | 266.735 |
| 2510 | 2236.85 | 83318.7 | 3888.7 | 62449.6 | .00537 | 260.172 | 3010 | 2736.85 | 101727.3 | 8691.4 | 76701.0 | .00288 | 266.859 |
| 2520 | 2246.85 | 83684.8 | 3957.3 | 62732.5 | .00529 | 260.317 | 3020 | 2746.85 | 102097.4 | 8820.7 | 76987.9 | .00285 | 266.981 |
| 2530 | 2256.85 | 84051.0 | 4027.0 | 63015.6 | .00522 | 260.462 | 3030 | 2756.85 | 102467.5 | 8951.4 | 77274.9 | .00281 | 267.104 |
| 2540 | 2266.85 | 84417.3 | 4097.6 | 63298.7 | .00515 | 260.607 | 3040 | 2766.85 | 102837.7 | 9083.7 | 77561.9 | .00278 | 267.226 |
| 2550 | 2276.85 | 84783.7 | 4169.1 | 63581.9 | .00509 | 260.751 | 3050 | 2776.85 | 103208.0 | 9217.5 | 77849.0 | .00275 | 267.347 |
| 2560 | 2286.85 | 85150.1 | 4241.7 | 63865.3 | .00502 | 260.894 | 3060 | 2786.85 | 103578.3 | 9352.9 | 78136.2 | .00272 | 267.468 |
| 2570 | 2296.85 | 85516.7 | 4315.2 | 64148.7 | .00495 | 261.037 | 3070 | 2796.85 | 103948.7 | 9489.8 | 78423.4 | .00269 | 267.589 |
| 2580 | 2306.85 | 85883.4 | 4389.8 | 64432.2 | .00489 | 261.179 | 3080 | 2806.85 | 104319.1 | 9628.3 | 78710.7 | .00266 | 267.710 |
| 2590 | 2316.85 | 86250.1 | 4465.3 | 64715.8 | .00482 | 261.321 | 3090 | 2816.85 | 104689.6 | 9768.4 | 78998.1 | .00263 | 267.830 |

# Table 12 Nitrogen

$\bar{m} = 28.0134$

| T K | t C | $\bar{c}_p$ J / g-mol K | $\bar{c}_v$ J / g-mol K | $k = \dfrac{\bar{c}_p}{\bar{c}_v}$ | a m/s |
|---|---|---|---|---|---|
| 100 | −173.15 | 29.104 | 20.790 | 1.400 | 203.8 |
| 120 | −153.15· | 29.105 | 20.790 | 1.400 | 223.3 |
| 140 | −133.15 | 29.105 | 20.791 | 1.400 | 241.2 |
| 160 | −113.15 | 29.106 | 20.791 | 1.400 | 257.8 |
| 180 | −93.15 | 29.107 | 20.792 | 1.400 | 273.5 |
| 200 | −73.15 | 29.107 | 20.793 | 1.400 | 288.3 |
| 250 | −23.15 | 29.111 | 20.797 | 1.400 | 322.3 |
| 300 | 26.85 | 29.125 | 20.811 | 1.400 | 353.0 |
| 350 | 76.85 | 29.166 | 20.851 | 1.399 | 381.2 |
| 400 | 126.85 | 29.249 | 20.935 | 1.397 | 407.3 |
| 450 | 176.85 | 29.387 | 21.072 | 1.395 | 431.6 |
| 500 | 226.85 | 29.581 | 21.266 | 1.391 | 454.3 |
| 550 | 276.85 | 29.825 | 21.511 | 1.387 | 475.7 |
| 600 | 326.85 | 30.110 | 21.796 | 1.381 | 496.0 |
| 650 | 376.85 | 30.424 | 22.109 | 1.376 | 515.2 |
| 700 | 426.85 | 30.755 | 22.440 | 1.371 | 533.6 |
| 750 | 476.85 | 31.094 | 22.779 | 1.365 | 551.2 |
| 800 | 526.85 | 31.433 | 23.119 | 1.360 | 568.2 |
| 850 | 576.85 | 31.766 | 23.452 | 1.355 | 584.6 |
| 900 | 626.85 | 32.090 | 23.775 | 1.350 | 600.4 |
| 950 | 676.85 | 32.400 | 24.086 | 1.345 | 615.9 |
| 1000 | 726.85 | 32.696 | 24.381 | 1.341 | 630.9 |
| 1050 | 776.85 | 32.976 | 24.661 | 1.337 | 645.5 |
| 1100 | 826.85 | 33.240 | 24.925 | 1.334 | 659.8 |
| 1150 | 876.85 | 33.488 | 25.174 | 1.330 | 673.8 |
| 1200 | 926.85 | 33.721 | 25.407 | 1.327 | 687.5 |
| 1300 | 1026.85 | 34.144 | 25.830 | 1.322 | 714.2 |
| 1400 | 1126.85 | 34.514 | 26.199 | 1.317 | 739.9 |
| 1500 | 1226.85 | 34.837 | 26.523 | 1.313 | 764.7 |
| 1600 | 1326.85 | 35.121 | 26.807 | 1.310 | 788.8 |
| 1700 | 1426.85 | 35.370 | 27.056 | 1.307 | 812.2 |
| 1800 | 1526.85 | 35.590 | 27.276 | 1.305 | 834.9 |
| 1900 | 1626.85 | 35.785 | 27.471 | 1.303 | 857.1 |
| 2000 | 1726.85 | 35.959 | 27.644 | 1.301 | 878.7 |
| 2100 | 1826.85 | 36.114 | 27.799 | 1.299 | 899.8 |
| 2200 | 1926.85 | 36.253 | 27.938 | 1.298 | 920.5 |
| 2300 | 2026.85 | 36.378 | 28.064 | 1.296 | 940.7 |
| 2400 | 2126.85 | 36.492 | 28.177 | 1.295 | 960.5 |
| 2500 | 2226.85 | 36.595 | 28.280 | 1.294 | 979.9 |
| 2600 | 2326.85 | 36.689 | 28.375 | 1.293 | 998.9 |
| 2700 | 2426.85 | 36.775 | 28.461 | 1.292 | 1017.6 |
| 2800 | 2526.85 | 36.855 | 28.541 | 1.291 | 1035.9 |
| 2900 | 2626.85 | 36.928 | 28.614 | 1.291 | 1054.0 |
| 3000 | 2726.85 | 36.996 | 28.682 | 1.290 | 1071.7 |
| 3100 | 2826.85 | 37.060 | 28.745 | 1.289 | 1089.1 |
| 3200 | 2926.85 | 37.119 | 28.804 | 1.289 | 1106.3 |
| 3300 | 3026.85 | 37.174 | 28.860 | 1.288 | 1123.2 |
| 3400 | 3126.85 | 37.226 | 28.912 | 1.288 | 1139.9 |
| 3500 | 3226.85 | 37.275 | 28.960 | 1.287 | 1156.3 |
| 3600 | 3326.85 | 37.321 | 29.007 | 1.287 | 1172.5 |

# Table 13  Oxygen at Low Pressures (for One Gram-Mole)

$\bar{m} = 31.9988$

| T | t | $\bar{h}$ | $p_r$ | $\bar{u}$ | $v_r$ | $\bar{\phi}$ | T | t | $\bar{h}$ | $p_r$ | $\bar{u}$ | $v_r$ | $\bar{\phi}$ |
|---|---|---|---|---|---|---|---|---|---|---|---|---|---|
| 100 | −173.15 | 2904.5 | .1108 | 2073.1 | 7.506 | 173.152 | 600 | 326.85 | 17927.3 | 66.124 | 12938.7 | .0754 | 226.296 |
| 110 | −163.15 | 3195.6 | .1546 | 2281.0 | 5.914 | 175.926 | 610 | 336.85 | 18248.7 | 70.486 | 13176.9 | .0720 | 226.827 |
| 120 | −153.15 | 3486.7 | .2097 | 2488.9 | 4.758 | 178.459 | 620 | 346.85 | 18571.0 | 75.072 | 13416.1 | .0687 | 227.352 |
| 130 | −143.15 | 3777.7 | .2775 | 2696.9 | 3.895 | 180.789 | 630 | 356.85 | 18894.3 | 79.890 | 13656.2 | .0656 | 227.869 |
| 140 | −133.15 | 4068.8 | .3597 | 2904.8 | 3.236 | 182.946 | 640 | 366.85 | 19218.4 | 84.950 | 13897.2 | .0626 | 228.379 |
| 150 | −123.15 | 4359.9 | .4580 | 3112.7 | 2.723 | 184.954 | 650 | 376.85 | 19543.5 | 90.258 | 14139.1 | .0599 | 228.883 |
| 160 | −113.15 | 4651.0 | .5742 | 3320.7 | 2.317 | 186.833 | 660 | 386.85 | 19869.5 | 95.826 | 14382.0 | .0573 | 229.381 |
| 170 | −103.15 | 4942.1 | .7099 | 3528.7 | 1.991 | 188.598 | 670 | 396.85 | 20196.3 | 101.661 | 14625.6 | .0548 | 229.872 |
| 180 | −93.15 | 5233.3 | .8672 | 3736.7 | 1.726 | 190.262 | 680 | 406.85 | 20524.0 | 107.774 | 14870.2 | .0525 | 230.358 |
| 190 | −83.15 | 5524.4 | 1.0480 | 3944.7 | 1.507 | 191.836 | 690 | 416.85 | 20852.6 | 114.174 | 15115.6 | .0502 | 230.838 |
| 200 | −73.15 | 5815.7 | 1.2543 | 4152.8 | 1.326 | 193.330 | 700 | 426.85 | 21182.0 | 120.872 | 15361.9 | .0482 | 231.312 |
| 210 | −63.15 | 6107.0 | 1.4881 | 4360.9 | 1.173 | 194.751 | 710 | 436.85 | 21512.2 | 127.877 | 15608.9 | .0462 | 231.780 |
| 220 | −53.15 | 6398.4 | 1.7516 | 4569.2 | 1.044 | 196.107 | 720 | 446.85 | 21843.2 | 135.199 | 15856.8 | .0443 | 232.243 |
| 230 | −43.15 | 6689.9 | 2.0471 | 4777.6 | .934 | 197.403 | 730 | 456.85 | 22175.0 | 142.850 | 16105.5 | .0425 | 232.701 |
| 240 | −33.15 | 6981.6 | 2.3767 | 4986.1 | .840 | 198.644 | 740 | 466.85 | 22507.6 | 150.841 | 16355.0 | .0408 | 233.153 |
| 250 | −23.15 | 7273.5 | 2.7429 | 5194.9 | .758 | 199.836 | 750 | 476.85 | 22841.0 | 159.181 | 16605.2 | .0392 | 233.601 |
| 260 | −13.15 | 7565.6 | 3.1482 | 5403.9 | .687 | 200.981 | 760 | 486.85 | 23175.1 | 167.883 | 16856.1 | .0376 | 234.043 |
| 270 | −3.15 | 7858.0 | 3.5951 | 5613.1 | .624 | 202.085 | 770 | 496.85 | 23509.9 | 176.958 | 17107.9 | .0362 | 234.481 |
| 280 | 6.85 | 8150.8 | 4.0863 | 5822.8 | .570 | 203.150 | 780 | 506.85 | 23845.5 | 186.418 | 17360.3 | .0348 | 234.914 |
| 290 | 16.85 | 8444.0 | 4.6245 | 6032.8 | .521 | 204.179 | 790 | 516.85 | 24181.8 | 196.275 | 17613.4 | .0335 | 235.342 |
| 300 | 26.85 | 8737.6 | 5.2127 | 6243.3 | .479 | 205.174 | 800 | 526.85 | 24518.8 | 206.541 | 17867.3 | .0322 | 235.766 |
| 310 | 36.85 | 9031.7 | 5.8538 | 6454.3 | .440 | 206.138 | 810 | 536.85 | 24856.5 | 217.229 | 18121.8 | .0310 | 236.186 |
| 320 | 46.85 | 9326.4 | 6.5509 | 6665.8 | .406 | 207.074 | 820 | 546.85 | 25194.8 | 228.350 | 18377.0 | .0299 | 236.601 |
| 330 | 56.85 | 9621.6 | 7.3073 | 6877.9 | .375 | 207.983 | 830 | 556.85 | 25533.8 | 239.919 | 18632.8 | .0288 | 237.012 |
| 340 | 66.85 | 9917.5 | 8.1264 | 7090.6 | .348 | 208.866 | 840 | 566.85 | 25873.4 | 251.949 | 18889.3 | .0277 | 237.418 |
| 350 | 76.85 | 10214.1 | 9.0117 | 7304.1 | .323 | 209.726 | 850 | 576.85 | 26213.7 | 264.452 | 19146.5 | .0267 | 237.821 |
| 360 | 86.85 | 10511.4 | 9.9668 | 7518.3 | .300 | 210.563 | 860 | 586.85 | 26554.6 | 277.443 | 19404.2 | .0258 | 238.220 |
| 370 | 96.85 | 10809.5 | 10.9955 | 7733.2 | .280 | 211.380 | 870 | 596.85 | 26896.1 | 290.935 | 19662.6 | .0249 | 238.615 |
| 380 | 106.85 | 11108.4 | 12.1019 | 7949.0 | .261 | 212.177 | 880 | 606.85 | 27238.2 | 304.942 | 19921.5 | .0240 | 239.006 |
| 390 | 116.85 | 11408.2 | 13.2899 | 8165.6 | .244 | 212.956 | 890 | 616.85 | 27580.9 | 319.480 | 20181.1 | .0232 | 239.393 |
| 400 | 126.85 | 11708.8 | 14.5639 | 8383.0 | .228 | 213.717 | 900 | 626.85 | 27924.2 | 334.563 | 20441.2 | .0224 | 239.776 |
| 410 | 136.85 | 12010.3 | 15.9282 | 8601.4 | .214 | 214.461 | 910 | 636.85 | 28268.0 | 350.205 | 20701.9 | .0216 | 240.156 |
| 420 | 146.85 | 12312.8 | 17.3875 | 8820.7 | .201 | 215.190 | 920 | 646.85 | 28612.4 | 366.423 | 20963.1 | .0209 | 240.533 |
| 430 | 156.85 | 12616.2 | 18.9464 | 9041.0 | .189 | 215.904 | 930 | 656.85 | 28957.3 | 383.231 | 21224.9 | .0202 | 240.906 |
| 440 | 166.85 | 12920.5 | 20.6099 | 9262.2 | .178 | 216.604 | 940 | 666.85 | 29302.8 | 400.645 | 21487.2 | .0195 | 241.275 |
| 450 | 176.85 | 13225.9 | 22.3830 | 9484.4 | .167 | 217.290 | 950 | 676.85 | 29648.8 | 418.681 | 21750.1 | .0189 | 241.641 |
| 460 | 186.85 | 13532.2 | 24.2710 | 9707.6 | .158 | 217.963 | 960 | 686.85 | 29995.2 | 437.356 | 22013.4 | .0183 | 242.004 |
| 470 | 196.85 | 13839.6 | 26.2793 | 9931.8 | .149 | 218.624 | 970 | 696.85 | 30342.2 | 456.686 | 22277.3 | .0177 | 242.364 |
| 480 | 206.85 | 14147.9 | 28.4134 | 10157.0 | .140 | 219.273 | 980 | 706.85 | 30689.7 | 476.688 | 22541.6 | .0171 | 242.720 |
| 490 | 216.85 | 14457.3 | 30.6792 | 10383.2 | .133 | 219.911 | 990 | 716.85 | 31037.7 | 497.379 | 22806.5 | .0165 | 243.073 |
| 500 | 226.85 | 14767.7 | 33.0826 | 10610.5 | .126 | 220.538 | 1000 | 726.85 | 31386.2 | 518.777 | 23071.8 | .0160 | 243.423 |
| 510 | 236.85 | 15079.1 | 35.6297 | 10838.8 | .119 | 221.155 | 1010 | 736.85 | 31735.1 | 540.898 | 23337.5 | .0155 | 243.771 |
| 520 | 246.85 | 15391.6 | 38.3267 | 11068.1 | .113 | 221.762 | 1020 | 746.85 | 32084.5 | 563.762 | 23603.8 | .0150 | 244.115 |
| 530 | 256.85 | 15705.0 | 41.1804 | 11298.4 | .107 | 222.359 | 1030 | 756.85 | 32434.3 | 587.386 | 23870.5 | .0146 | 244.456 |
| 540 | 266.85 | 16019.5 | 44.1972 | 11529.7 | .102 | 222.947 | 1040 | 766.85 | 32784.6 | 611.788 | 24137.6 | .0141 | 244.795 |
| 550 | 276.85 | 16335.0 | 47.3840 | 11762.1 | .097 | 223.526 | 1050 | 776.85 | 33135.3 | 636.989 | 24405.2 | .0137 | 245.130 |
| 560 | 286.85 | 16651.5 | 50.7481 | 11995.4 | .092 | 224.096 | 1060 | 786.85 | 33486.4 | 663.006 | 24673.2 | .0133 | 245.463 |
| 570 | 296.85 | 16969.0 | 54.2965 | 12229.8 | .087 | 224.658 | 1070 | 796.85 | 33838.0 | 689.859 | 24941.6 | .0129 | 245.793 |
| 580 | 306.85 | 17287.4 | 58.0367 | 12465.1 | .083 | 225.212 | 1080 | 806.85 | 34190.0 | 717.568 | 25210.4 | .0125 | 246.121 |
| 590 | 316.85 | 17606.9 | 61.9765 | 12701.4 | .079 | 225.758 | 1090 | 816.85 | 34542.4 | 746.153 | 25479.7 | .0121 | 246.445 |

# Table 13 Oxygen at Low Pressures (for One Gram-Mole)

$\bar{m} = 31.9988$

| T | t | $\bar{h}$ | $p_r$ | $\bar{u}$ | $v_r$ | $\bar{\phi}$ | T | t | $\bar{h}$ | $p_r$ | $\bar{u}$ | $v_r$ | $\bar{\phi}$ |
|---|---|---|---|---|---|---|---|---|---|---|---|---|---|
| 1100 | 826.85 | 34895.2 | 775.63 | 25749.3 | .01179 | 246.768 | 1600 | 1326.85 | 52946.2 | 3938.6 | 39643.2 | .00338 | 260.278 |
| 1110 | 836.85 | 35248.4 | 806.03 | 26019.4 | .01145 | 247.087 | 1610 | 1336.85 | 53314.2 | 4048.7 | 39928.0 | .00331 | 260.507 |
| 1120 | 846.85 | 35601.9 | 837.37 | 26289.8 | .01112 | 247.404 | 1620 | 1346.85 | 53682.4 | 4161.2 | 40213.1 | .00324 | 260.735 |
| 1130 | 856.85 | 35955.9 | 869.66 | 26560.6 | .01080 | 247.719 | 1630 | 1356.85 | 54050.9 | 4276.3 | 40498.4 | .00317 | 260.962 |
| 1140 | 866.85 | 36310.2 | 902.94 | 26831.8 | .01050 | 248.031 | 1640 | 1366.85 | 54419.5 | 4393.9 | 40783.9 | .00310 | 261.187 |
| 1150 | 876.85 | 36664.9 | 937.21 | 27103.3 | .01020 | 248.341 | 1650 | 1376.85 | 54788.5 | 4514.0 | 41069.7 | .00304 | 261.411 |
| 1160 | 886.85 | 37020.0 | 972.51 | 27375.3 | .00992 | 248.648 | 1660 | 1386.85 | 55157.6 | 4636.7 | 41355.7 | .00298 | 261.634 |
| 1170 | 896.85 | 37375.4 | 1008.86 | 27647.6 | .00964 | 248.953 | 1670 | 1396.85 | 55527.0 | 4762.1 | 41642.0 | .00292 | 261.856 |
| 1180 | 906.85 | 37731.2 | 1046.28 | 27920.2 | .00938 | 249.256 | 1680 | 1406.85 | 55896.7 | 4890.2 | 41928.5 | .00286 | 262.077 |
| 1190 | 916.85 | 38087.3 | 1084.79 | 28193.2 | .00912 | 249.557 | 1690 | 1416.85 | 56266.6 | 5021.0 | 42215.2 | .00280 | 262.296 |
| 1200 | 926.85 | 38443.8 | 1124.42 | 28466.5 | .00887 | 249.855 | 1700 | 1426.85 | 56636.7 | 5154.7 | 42502.2 | .00274 | 262.515 |
| 1210 | 936.85 | 38800.6 | 1165.19 | 28740.2 | .00863 | 250.151 | 1710 | 1436.85 | 57007.0 | 5291.1 | 42789.4 | .00269 | 262.732 |
| 1220 | 946.85 | 39157.8 | 1207.12 | 29014.2 | .00840 | 250.445 | 1720 | 1446.85 | 57377.6 | 5430.4 | 43076.8 | .00263 | 262.948 |
| 1230 | 956.85 | 39515.2 | 1250.24 | 29288.5 | .00818 | 250.737 | 1730 | 1456.85 | 57748.4 | 5572.6 | 43364.5 | .00258 | 263.163 |
| 1240 | 966.85 | 39873.1 | 1294.57 | 29563.2 | .00796 | 251.027 | 1740 | 1466.85 | 58119.5 | 5717.8 | 43652.4 | .00253 | 263.377 |
| 1250 | 976.85 | 40231.2 | 1340.15 | 29838.2 | .00776 | 251.314 | 1750 | 1476.85 | 58490.7 | 5866.0 | 43940.5 | .00248 | 263.590 |
| 1260 | 986.85 | 40589.6 | 1386.98 | 30113.5 | .00755 | 251.600 | 1760 | 1486.85 | 58862.3 | 6017.3 | 44228.9 | .00243 | 263.801 |
| 1270 | 996.85 | 40948.4 | 1435.11 | 30389.1 | .00736 | 251.884 | 1770 | 1496.85 | 59234.0 | 6171.7 | 44517.5 | .00238 | 264.012 |
| 1280 | 1006.85 | 41307.5 | 1484.56 | 30665.1 | .00717 | 252.165 | 1780 | 1506.85 | 59606.0 | 6329.2 | 44806.3 | .00234 | 264.222 |
| 1290 | 1016.85 | 41666.9 | 1535.34 | 30941.3 | .00699 | 252.445 | 1790 | 1516.85 | 59978.2 | 6490.0 | 45095.4 | .00229 | 264.430 |
| 1300 | 1026.85 | 42026.6 | 1587.50 | 31217.9 | .00681 | 252.723 | 1800 | 1526.85 | 60350.6 | 6654.0 | 45384.7 | .00225 | 264.638 |
| 1310 | 1036.85 | 42386.6 | 1641.05 | 31494.7 | .00664 | 252.999 | 1810 | 1536.85 | 60723.3 | 6821.3 | 45674.2 | .00221 | 264.844 |
| 1320 | 1046.85 | 42746.9 | 1696.03 | 31771.9 | .00647 | 253.272 | 1820 | 1546.85 | 61096.2 | 6991.9 | 45963.9 | .00216 | 265.050 |
| 1330 | 1056.85 | 43107.5 | 1752.47 | 32049.3 | .00631 | 253.545 | 1830 | 1556.85 | 61469.3 | 7166.0 | 46253.9 | .00212 | 265.254 |
| 1340 | 1066.85 | 43468.3 | 1810.38 | 32327.0 | .00615 | 253.815 | 1840 | 1566.85 | 61842.6 | 7343.5 | 46544.1 | .00208 | 265.457 |
| 1350 | 1076.85 | 43829.5 | 1869.80 | 32605.1 | .00600 | 254.083 | 1850 | 1576.85 | 62216.2 | 7524.5 | 46834.5 | .00204 | 265.660 |
| 1360 | 1086.85 | 44191.0 | 1930.77 | 32883.4 | .00586 | 254.350 | 1860 | 1586.85 | 62590.0 | 7709.1 | 47125.2 | .00201 | 265.861 |
| 1370 | 1096.85 | 44552.7 | 1993.30 | 33162.0 | .00571 | 254.615 | 1870 | 1596.85 | 62964.0 | 7897.3 | 47416.1 | .00197 | 266.062 |
| 1380 | 1106.85 | 44914.7 | 2057.43 | 33440.9 | .00558 | 254.879 | 1880 | 1606.85 | 63338.3 | 8089.2 | 47707.2 | .00193 | 266.262 |
| 1390 | 1116.85 | 45277.0 | 2123.19 | 33720.0 | .00544 | 255.140 | 1890 | 1616.85 | 63712.8 | 8284.8 | 47998.5 | .00190 | 266.460 |
| 1400 | 1126.85 | 45639.6 | 2190.61 | 33999.4 | .00531 | 255.400 | 1900 | 1626.85 | 64087.5 | 8484.2 | 48290.1 | .00186 | 266.658 |
| 1410 | 1136.85 | 46002.5 | 2259.72 | 34279.2 | .00519 | 255.658 | 1910 | 1636.85 | 64462.4 | 8687.4 | 48581.9 | .00183 | 266.855 |
| 1420 | 1146.85 | 46365.6 | 2330.56 | 34559.1 | .00507 | 255.915 | 1920 | 1646.85 | 64837.5 | 8894.6 | 48873.9 | .00179 | 267.051 |
| 1430 | 1156.85 | 46729.0 | 2403.15 | 34839.4 | .00495 | 256.170 | 1930 | 1656.85 | 65212.9 | 9105.6 | 49166.1 | .00176 | 267.246 |
| 1440 | 1166.85 | 47092.6 | 2477.52 | 35119.9 | .00483 | 256.423 | 1940 | 1666.85 | 65588.5 | 9320.7 | 49458.6 | .00173 | 267.440 |
| 1450 | 1176.85 | 47456.6 | 2553.72 | 35400.7 | .00472 | 256.675 | 1950 | 1676.85 | 65964.4 | 9539.9 | 49751.3 | .00170 | 267.633 |
| 1460 | 1186.85 | 47820.8 | 2631.77 | 35681.7 | .00461 | 256.926 | 1960 | 1686.85 | 66340.4 | 9763.2 | 50044.2 | .00167 | 267.825 |
| 1470 | 1196.85 | 48185.2 | 2711.70 | 35963.0 | .00451 | 257.174 | 1970 | 1696.85 | 66716.7 | 9990.6 | 50337.3 | .00164 | 268.017 |
| 1480 | 1206.85 | 48549.9 | 2793.55 | 36244.6 | .00440 | 257.422 | 1980 | 1706.85 | 67093.2 | 10222.3 | 50630.7 | .00161 | 268.208 |
| 1490 | 1216.85 | 48914.9 | 2877.36 | 36526.4 | .00431 | 257.667 | 1990 | 1716.85 | 67469.9 | 10458.3 | 50924.2 | .00158 | 268.397 |
| 1500 | 1226.85 | 49280.1 | 2963.16 | 36808.5 | .00421 | 257.912 | 2000 | 1726.85 | 67846.8 | 10698.7 | 51218.0 | .00155 | 268.586 |
| 1510 | 1236.85 | 49645.6 | 3050.99 | 37090.9 | .00411 | 258.154 | 2010 | 1736.85 | 68224.0 | 10943.5 | 51512.0 | .00153 | 268.774 |
| 1520 | 1246.85 | 50011.3 | 3140.87 | 37373.5 | .00402 | 258.396 | 2020 | 1746.85 | 68601.4 | 11192.8 | 51806.3 | .00150 | 268.962 |
| 1530 | 1256.85 | 50377.3 | 3232.86 | 37656.3 | .00393 | 258.636 | 2030 | 1756.85 | 68979.0 | 11446.7 | 52100.7 | .00147 | 269.148 |
| 1540 | 1266.85 | 50743.6 | 3326.97 | 37939.4 | .00385 | 258.874 | 2040 | 1766.85 | 69356.8 | 11705.2 | 52395.4 | .00145 | 269.334 |
| 1550 | 1276.85 | 51110.1 | 3423.26 | 38222.7 | .00376 | 259.112 | 2050 | 1776.85 | 69734.9 | 11968.4 | 52690.3 | .00142 | 269.519 |
| 1560 | 1286.85 | 51476.8 | 3521.75 | 38506.3 | .00368 | 259.348 | 2060 | 1786.85 | 70113.1 | 12236.3 | 52985.5 | .00140 | 269.703 |
| 1570 | 1296.85 | 51843.8 | 3622.49 | 38790.2 | .00360 | 259.582 | 2070 | 1796.85 | 70491.6 | 12509.0 | 53280.8 | .00138 | 269.886 |
| 1580 | 1306.85 | 52211.0 | 3725.52 | 39074.3 | .00353 | 259.815 | 2080 | 1806.85 | 70870.3 | 12786.7 | 53576.4 | .00135 | 270.069 |
| 1590 | 1316.85 | 52578.5 | 3830.87 | 39358.6 | .00345 | 260.047 | 2090 | 1816.85 | 71249.2 | 13069.2 | 53872.1 | .00133 | 270.250 |

(93)

# Table 13    Oxygen at Low Pressures (for One Gram-Mole)

$\bar{m} = 31.9988$

| T | t | $\bar{h}$ | $p_r$ | $\bar{u}$ | $v_r$ | $\bar{\phi}$ | T | t | $\bar{h}$ | $p_r$ | $\bar{u}$ | $v_r$ | $\bar{\phi}$ |
|---|---|---|---|---|---|---|---|---|---|---|---|---|---|
| 2100 | 1826.85 | 71628.4 | 13357 | 54168.1 | .001307 | 270.431 | 2600 | 2326.85 | 90857.6 | 35853 | 69240.2 | .000603 | 278.641 |
| 2110 | 1836.85 | 72007.8 | 13649 | 54464.4 | .001285 | 270.611 | 2610 | 2336.85 | 91247.5 | 36505 | 69546.9 | .000594 | 278.791 |
| 2120 | 1846.85 | 72387.3 | 13947 | 54760.8 | .001264 | 270.791 | 2620 | 2346.85 | 91637.5 | 37165 | 69853.8 | .000586 | 278.940 |
| 2130 | 1856.85 | 72767.1 | 14250 | 55057.5 | .001243 | 270.970 | 2630 | 2356.85 | 92027.8 | 37836 | 70160.9 | .000578 | 279.088 |
| 2140 | 1866.85 | 73147.2 | 14559 | 55354.3 | .001222 | 271.148 | 2640 | 2366.85 | 92418.2 | 38516 | 70468.2 | .000570 | 279.237 |
| 2150 | 1876.85 | 73527.4 | 14872 | 55651.4 | .001202 | 271.325 | 2650 | 2376.85 | 92808.8 | 39206 | 70775.7 | .000562 | 279.384 |
| 2160 | 1886.85 | 73907.9 | 15192 | 55948.7 | .001182 | 271.501 | 2660 | 2386.85 | 93199.7 | 39907 | 71083.4 | .000554 | 279.532 |
| 2170 | 1896.85 | 74288.5 | 15516 | 56246.3 | .001163 | 271.677 | 2670 | 2396.85 | 93590.7 | 40617 | 71391.2 | .000547 | 279.678 |
| 2180 | 1906.85 | 74669.4 | 15847 | 56544.0 | .001144 | 271.852 | 2680 | 2406.85 | 93981.9 | 41338 | 71699.3 | .000539 | 279.824 |
| 2190 | 1916.85 | 75050.5 | 16183 | 56842.0 | .001125 | 272.027 | 2690 | 2416.85 | 94373.3 | 42069 | 72007.6 | .000532 | 279.970 |
| 2200 | 1926.85 | 75431.8 | 16524 | 57140.2 | .001107 | 272.201 | 2700 | 2426.85 | 94764.9 | 42811 | 72316.1 | .000524 | 280.116 |
| 2210 | 1936.85 | 75813.4 | 16872 | 57438.6 | .001089 | 272.374 | 2710 | 2436.85 | 95156.8 | 43563 | 72624.7 | .000517 | 280.260 |
| 2220 | 1946.85 | 76195.1 | 17225 | 57737.2 | .001072 | 272.546 | 2720 | 2446.85 | 95548.7 | 44326 | 72933.6 | .000510 | 280.405 |
| 2230 | 1956.85 | 76577.1 | 17584 | 58036.0 | .001054 | 272.718 | 2730 | 2456.85 | 95940.9 | 45100 | 73242.6 | .000503 | 280.549 |
| 2240 | 1966.85 | 76959.3 | 17950 | 58335.0 | .001038 | 272.889 | 2740 | 2466.85 | 96333.3 | 45885 | 73551.9 | .000496 | 280.692 |
| 2250 | 1976.85 | 77341.7 | 18321 | 58634.3 | .001021 | 273.059 | 2750 | 2476.85 | 96725.9 | 46681 | 73861.3 | .000490 | 280.835 |
| 2260 | 1986.85 | 77724.3 | 18699 | 58933.7 | .001005 | 273.229 | 2760 | 2486.85 | 97118.7 | 47489 | 74170.9 | .000483 | 280.978 |
| 2270 | 1996.85 | 78107.1 | 19083 | 59233.4 | .000989 | 273.398 | 2770 | 2496.85 | 97511.6 | 48307 | 74480.7 | .000477 | 281.120 |
| 2280 | 2006.85 | 78490.2 | 19474 | 59533.3 | .000973 | 273.566 | 2780 | 2506.85 | 97904.7 | 49138 | 74790.7 | .000470 | 281.262 |
| 2290 | 2016.85 | 78873.4 | 19870 | 59833.4 | .000958 | 273.734 | 2790 | 2516.85 | 98298.1 | 49979 | 75100.9 | .000464 | 281.403 |
| 2300 | 2026.85 | 79256.9 | 20274 | 60133.7 | .000943 | 273.901 | 2800 | 2526.85 | 98691.6 | 50833 | 75411.3 | .000458 | 281.544 |
| 2310 | 2036.85 | 79640.5 | 20684 | 60434.3 | .000929 | 274.067 | 2810 | 2536.85 | 99085.3 | 51698 | 75721.8 | .000452 | 281.684 |
| 2320 | 2046.85 | 80024.4 | 21100 | 60735.0 | .000914 | 274.233 | 2820 | 2546.85 | 99479.2 | 52576 | 76032.6 | .000446 | 281.824 |
| 2330 | 2056.85 | 80408.5 | 21524 | 61036.0 | .000900 | 274.398 | 2830 | 2556.85 | 99873.3 | 53465 | 76343.5 | .000440 | 281.963 |
| 2340 | 2066.85 | 80792.8 | 21954 | 61337.1 | .000886 | 274.563 | 2840 | 2566.85 | 100267.5 | 54367 | 76654.6 | .000434 | 282.102 |
| 2350 | 2076.85 | 81177.4 | 22391 | 61638.5 | .000873 | 274.727 | 2850 | 2576.85 | 100662.0 | 55281 | 76965.9 | .000429 | 282.241 |
| 2360 | 2086.85 | 81562.1 | 22836 | 61940.1 | .000859 | 274.890 | 2860 | 2586.85 | 101056.6 | 56208 | 77277.4 | .000423 | 282.379 |
| 2370 | 2096.85 | 81947.0 | 23287 | 62241.9 | .000846 | 275.053 | 2870 | 2596.85 | 101451.4 | 57147 | 77589.1 | .000418 | 282.517 |
| 2380 | 2106.85 | 82332.2 | 23746 | 62543.9 | .000833 | 275.215 | 2880 | 2606.85 | 101846.4 | 58100 | 77900.9 | .000412 | 282.654 |
| 2390 | 2116.85 | 82717.5 | 24212 | 62846.1 | .000821 | 275.377 | 2890 | 2616.85 | 102241.6 | 59065 | 78213.0 | .000407 | 282.791 |
| 2400 | 2126.85 | 83103.1 | 24685 | 63148.5 | .000808 | 275.538 | 2900 | 2626.85 | 102637.0 | 60043 | 78525.2 | .000402 | 282.928 |
| 2410 | 2136.85 | 83488.9 | 25166 | 63451.2 | .000796 | 275.698 | 2910 | 2636.85 | 103032.5 | 61034 | 78837.6 | .000396 | 283.064 |
| 2420 | 2146.85 | 83874.9 | 25654 | 63754.0 | .000784 | 275.858 | 2920 | 2646.85 | 103428.2 | 62039 | 79150.2 | .000391 | 283.200 |
| 2430 | 2156.85 | 84261.1 | 26151 | 64057.1 | .000773 | 276.017 | 2930 | 2656.85 | 103824.1 | 63057 | 79462.9 | .000386 | 283.335 |
| 2440 | 2166.85 | 84647.5 | 26654 | 64360.3 | .000761 | 276.176 | 2940 | 2666.85 | 104220.2 | 64089 | 79775.9 | .000381 | 283.470 |
| 2450 | 2176.85 | 85034.1 | 27166 | 64663.8 | .000750 | 276.334 | 2950 | 2676.85 | 104616.5 | 65135 | 80089.0 | .000377 | 283.605 |
| 2460 | 2186.85 | 85420.9 | 27686 | 64967.4 | .000739 | 276.492 | 2960 | 2686.85 | 105012.9 | 66194 | 80402.3 | .000372 | 283.739 |
| 2470 | 2196.85 | 85807.9 | 28214 | 65271.3 | .000728 | 276.649 | 2970 | 2696.85 | 105409.5 | 67268 | 80715.7 | .000367 | 283.873 |
| 2480 | 2206.85 | 86195.1 | 28750 | 65575.4 | .000717 | 276.805 | 2980 | 2706.85 | 105806.3 | 68356 | 81029.4 | .000362 | 284.006 |
| 2490 | 2216.85 | 86582.5 | 29294 | 65879.7 | .000707 | 276.961 | 2990 | 2716.85 | 106203.3 | 69458 | 81343.2 | .000358 | 284.139 |
| 2500 | 2226.85 | 86970.2 | 29846 | 66184.2 | .000696 | 277.116 | 3000 | 2726.85 | 106600.4 | 70574 | 81657.2 | .000353 | 284.272 |
| 2510 | 2236.85 | 87358.0 | 30407 | 66488.9 | .000686 | 277.271 | 3010 | 2736.85 | 106997.7 | 71706 | 81971.4 | .000349 | 284.404 |
| 2520 | 2246.85 | 87746.1 | 30977 | 66793.8 | .000676 | 277.425 | 3020 | 2746.85 | 107395.2 | 72852 | 82285.7 | .000345 | 284.536 |
| 2530 | 2256.85 | 88134.3 | 31555 | 67098.9 | .000667 | 277.579 | 3030 | 2756.85 | 107792.9 | 74013 | 82600.2 | .000340 | 284.667 |
| 2540 | 2266.85 | 88522.7 | 32142 | 67404.2 | .000657 | 277.732 | 3040 | 2766.85 | 108190.7 | 75189 | 82914.9 | .000336 | 284.798 |
| 2550 | 2276.85 | 88911.4 | 32738 | 67709.7 | .000648 | 277.885 | 3050 | 2776.85 | 108588.7 | 76380 | 83229.8 | .000332 | 284.929 |
| 2560 | 2286.85 | 89300.2 | 33343 | 68015.4 | .000638 | 278.037 | 3060 | 2786.85 | 108986.9 | 77587 | 83544.8 | .000328 | 285.059 |
| 2570 | 2296.85 | 89689.3 | 33956 | 68321.3 | .000629 | 278.189 | 3070 | 2796.85 | 109385.2 | 78809 | 83860.0 | .000324 | 285.189 |
| 2580 | 2306.85 | 90078.5 | 34579 | 68627.4 | .000620 | 278.340 | 3080 | 2806.85 | 109783.8 | 80047 | 84175.4 | .000320 | 285.319 |
| 2590 | 2316.85 | 90468.0 | 35212 | 68933.7 | .000612 | 278.491 | 3090 | 2816.85 | 110182.4 | 81301 | 84490.9 | .000316 | 285.448 |

# Table 14   Oxygen

$\bar{m}$ = 31.9988

| T<br>K | t<br>C | $\bar{c}_p$<br>J<br>----<br>g-mol K | $\bar{c}_v$<br>J<br>----<br>g-mol K | $k = \dfrac{\bar{c}_p}{\bar{c}_v}$ | a<br>m<br>---<br>s |
|---|---|---|---|---|---|
| 100 | −173.15 | 29.106 | 20.792 | 1.400 | 190.7 |
| 120 | −153.15 | 29.107 | 20.793 | 1.400 | 208.9 |
| 140 | −133.15 | 29.108 | 20.794 | 1.400 | 225.7 |
| 160 | −113.15 | 29.111 | 20.796 | 1.400 | 241.2 |
| 180 | −93.15 | 29.115 | 20.801 | 1.400 | 255.9 |
| 200 | −73.15 | 29.126 | 20.811 | 1.400 | 269.7 |
| 250 | −23.15 | 29.201 | 20.887 | 1.398 | 301.4 |
| 300 | 26.85 | 29.386 | 21.071 | 1.395 | 329.7 |
| 350 | 76.85 | 29.695 | 21.380 | 1.389 | 355.4 |
| 400 | 126.85 | 30.107 | 21.792 | 1.382 | 378.9 |
| 450 | 176.85 | 30.584 | 22.270 | 1.373 | 400.7 |
| 500 | 226.85 | 31.091 | 22.777 | 1.365 | 421.1 |
| 550 | 276.85 | 31.599 | 23.285 | 1.357 | 440.4 |
| 600 | 326.85 | 32.090 | 23.775 | 1.350 | 458.7 |
| 650 | 376.85 | 32.552 | 24.238 | 1.343 | 476.3 |
| 700 | 426.85 | 32.981 | 24.666 | 1.337 | 493.1 |
| 750 | 476.85 | 33.374 | 25.060 | 1.332 | 509.4 |
| 800 | 526.85 | 33.733 | 25.418 | 1.327 | 525.2 |
| 850 | 576.85 | 34.059 | 25.744 | 1.323 | 540.5 |
| 900 | 626.85 | 34.355 | 26.041 | 1.319 | 555.4 |
| 950 | 676.85 | 34.624 | 26.310 | 1.316 | 570.0 |
| 1000 | 726.85 | 34.869 | 26.555 | 1.313 | 584.1 |
| 1050 | 776.85 | 35.093 | 26.779 | 1.310 | 597.9 |
| 1100 | 826.85 | 35.299 | 26.984 | 1.308 | 611.5 |
| 1150 | 876.85 | 35.488 | 27.174 | 1.306 | 624.7 |
| 1200 | 926.85 | 35.665 | 27.350 | 1.304 | 637.6 |
| 1300 | 1026.85 | 35.985 | 27.670 | 1.300 | 662.8 |
| 1400 | 1126.85 | 36.271 | 27.957 | 1.297 | 687.0 |
| 1500 | 1226.85 | 36.536 | 28.221 | 1.295 | 710.3 |
| 1600 | 1326.85 | 36.784 | 28.470 | 1.292 | 732.9 |
| 1700 | 1426.85 | 37.023 | 28.709 | 1.290 | 754.7 |
| 1800 | 1526.85 | 37.255 | 28.940 | 1.287 | 775.9 |
| 1900 | 1626.85 | 37.482 | 29.167 | 1.285 | 796.5 |
| 2000 | 1726.85 | 37.705 | 29.391 | 1.283 | 816.5 |
| 2100 | 1826.85 | 37.926 | 29.611 | 1.281 | 836.0 |
| 2200 | 1926.85 | 38.143 | 29.828 | 1.279 | 855.0 |
| 2300 | 2026.85 | 38.357 | 30.043 | 1.277 | 873.5 |
| 2400 | 2126.85 | 38.567 | 30.253 | 1.275 | 891.6 |
| 2500 | 2226.85 | 38.773 | 30.459 | 1.273 | 909.3 |
| 2600 | 2326.85 | 38.975 | 30.660 | 1.271 | 926.7 |
| 2700 | 2426.85 | 39.171 | 30.856 | 1.269 | 943.7 |
| 2800 | 2526.85 | 39.361 | 31.047 | 1.268 | 960.4 |
| 2900 | 2626.85 | 39.545 | 31.231 | 1.266 | 976.8 |
| 3000 | 2726.85 | 39.723 | 31.408 | 1.265 | 992.9 |
| 3100 | 2826.85 | 39.894 | 31.579 | 1.263 | 1008.7 |
| 3200 | 2926.85 | 40.057 | 31.743 | 1.262 | 1024.3 |
| 3300 | 3026.85 | 40.214 | 31.900 | 1.261 | 1039.7 |
| 3400 | 3126.85 | 40.364 | 32.049 | 1.259 | 1054.8 |
| 3500 | 3226.85 | 40.506 | 32.192 | 1.258 | 1069.7 |
| 3600 | 3326.85 | 40.641 | 32.327 | 1.257 | 1084.4 |

(95)

# Table 15   Water Vapor at Low Pressures (for One Gram-Mole)

$\bar{m} = 18.0152$

| T | t | $\bar{h}$ | $p_r$ | $\bar{u}$ | $v_r$ | $\bar{\phi}$ | T | t | $\bar{h}$ | $p_r$ | $\bar{u}$ | $v_r$ | $\bar{\phi}$ |
|---|---|---|---|---|---|---|---|---|---|---|---|---|---|
| 100 | −173.15 | 3289.6 | .0090 | 2458.1 | 92.477 | 152.272 | 600 | 326.85 | 20405.3 | 13.257 | 15416.6 | .3763 | 212.935 |
| 110 | −163.15 | 3622.5 | .0132 | 2707.9 | 69.440 | 155.447 | 610 | 336.85 | 20769.1 | 14.252 | 15697.3 | .3559 | 213.537 |
| 120 | −153.15 | 3955.7 | .0187 | 2958.0 | 53.457 | 158.345 | 620 | 346.85 | 21134.1 | 15.306 | 15979.1 | .3368 | 214.130 |
| 130 | −143.15 | 4288.9 | .0257 | 3208.0 | 42.023 | 161.012 | 630 | 356.85 | 21500.2 | 16.424 | 16262.1 | .3189 | 214.716 |
| 140 | −133.15 | 4622.0 | .0346 | 3458.0 | 33.627 | 163.481 | 640 | 366.85 | 21867.4 | 17.607 | 16546.2 | .3022 | 215.294 |
| 150 | −123.15 | 4955.2 | .0456 | 3708.1 | 27.326 | 165.780 | 650 | 376.85 | 22235.8 | 18.859 | 16831.5 | .2866 | 215.866 |
| 160 | −113.15 | 5288.4 | .0591 | 3958.1 | 22.503 | 167.931 | 660 | 386.85 | 22605.6 | 20.183 | 17118.1 | .2719 | 216.430 |
| 170 | −103.15 | 5621.8 | .0754 | 4208.4 | 18.754 | 169.951 | 670 | 396.85 | 22976.2 | 21.583 | 17405.6 | .2581 | 216.987 |
| 180 | −93.15 | 5955.0 | .0948 | 4458.4 | 15.789 | 171.856 | 680 | 406.85 | 23348.2 | 23.062 | 17694.4 | .2452 | 217.538 |
| 190 | −83.15 | 6288.4 | .1177 | 4708.7 | 13.418 | 173.659 | 690 | 416.85 | 23721.4 | 24.624 | 17984.4 | .2330 | 218.083 |
| 200 | −73.15 | 6621.8 | .1446 | 4958.9 | 11.498 | 175.369 | 700 | 426.85 | 24095.7 | 26.272 | 18275.6 | .2215 | 218.622 |
| 210 | −63.15 | 6955.4 | .1759 | 5209.4 | 9.926 | 176.997 | 710 | 436.85 | 24471.3 | 28.010 | 18568.1 | .2108 | 219.155 |
| 220 | −53.15 | 7289.3 | .2120 | 5460.1 | 8.628 | 178.549 | 720 | 446.85 | 24848.1 | 29.843 | 18861.7 | .2006 | 219.681 |
| 230 | −43.15 | 7622.9 | .2534 | 5710.6 | 7.546 | 180.033 | 730 | 456.85 | 25226.0 | 31.774 | 19156.5 | .1910 | 220.203 |
| 240 | −33.15 | 7956.8 | .3007 | 5961.3 | 6.637 | 181.454 | 740 | 466.85 | 25605.3 | 33.808 | 19452.7 | .1820 | 220.719 |
| 250 | −23.15 | 8291.1 | .3543 | 6212.5 | 5.867 | 182.819 | 750 | 476.85 | 25985.7 | 35.950 | 19749.9 | .1735 | 221.229 |
| 260 | −13.15 | 8625.4 | .4148 | 6463.7 | 5.212 | 184.129 | 760 | 486.85 | 26367.3 | 38.203 | 20048.3 | .1654 | 221.735 |
| 270 | −3.15 | 8960.0 | .4828 | 6715.1 | 4.650 | 185.392 | 770 | 496.85 | 26750.2 | 40.573 | 20348.1 | .1578 | 222.235 |
| 280 | 6.85 | 9295.2 | .5590 | 6967.2 | 4.164 | 186.611 | 780 | 506.85 | 27134.3 | 43.065 | 20649.0 | .1506 | 222.731 |
| 290 | 16.85 | 9630.4 | .6440 | 7219.2 | 3.744 | 187.787 | 790 | 516.85 | 27519.7 | 45.685 | 20951.3 | .1438 | 223.222 |
| 300 | 26.85 | 9966.1 | .7384 | 7471.8 | 3.378 | 188.925 | 800 | 526.85 | 27906.2 | 48.436 | 21254.7 | .1373 | 223.708 |
| 310 | 36.85 | 10302.4 | .8431 | 7725.0 | 3.057 | 190.027 | 810 | 536.85 | 28293.9 | 51.326 | 21559.2 | .1312 | 224.190 |
| 320 | 46.85 | 10639.0 | .9588 | 7978.4 | 2.775 | 191.097 | 820 | 546.85 | 28683.2 | 54.359 | 21865.4 | .1254 | 224.667 |
| 330 | 56.85 | 10976.3 | 1.0863 | 8232.5 | 2.526 | 192.134 | 830 | 556.85 | 29073.4 | 57.542 | 22172.4 | .1199 | 225.140 |
| 340 | 66.85 | 11314.0 | 1.2264 | 8487.1 | 2.305 | 193.143 | 840 | 566.85 | 29465.2 | 60.881 | 22481.1 | .1147 | 225.609 |
| 350 | 76.85 | 11652.7 | 1.3800 | 8742.6 | 2.109 | 194.124 | 850 | 576.85 | 29858.1 | 64.383 | 22790.8 | .1098 | 226.074 |
| 360 | 86.85 | 11991.5 | 1.5480 | 8998.3 | 1.934 | 195.079 | 860 | 586.85 | 30252.1 | 68.054 | 23101.7 | .1051 | 226.535 |
| 370 | 96.85 | 12331.5 | 1.7314 | 9255.2 | 1.777 | 196.010 | 870 | 596.85 | 30647.5 | 71.900 | 23414.0 | .1006 | 226.993 |
| 380 | 106.85 | 12672.0 | 1.9312 | 9512.5 | 1.636 | 196.918 | 880 | 606.85 | 31044.1 | 75.929 | 23727.4 | .0964 | 227.446 |
| 390 | 116.85 | 13013.5 | 2.1487 | 9770.9 | 1.509 | 197.805 | 890 | 616.85 | 31442.2 | 80.148 | 24042.4 | .0923 | 227.896 |
| 400 | 126.85 | 13355.6 | 2.3846 | 10029.8 | 1.395 | 198.672 | 900 | 626.85 | 31841.2 | 84.566 | 24358.3 | .0885 | 228.342 |
| 410 | 136.85 | 13698.7 | 2.6404 | 10289.8 | 1.291 | 199.519 | 910 | 636.85 | 32241.9 | 89.189 | 24675.8 | .0848 | 228.784 |
| 420 | 146.85 | 14042.6 | 2.9170 | 10550.5 | 1.197 | 200.347 | 920 | 646.85 | 32643.6 | 94.027 | 24994.4 | .0814 | 229.223 |
| 430 | 156.85 | 14387.6 | 3.2164 | 10812.4 | 1.112 | 201.159 | 930 | 656.85 | 33046.8 | 99.085 | 25314.4 | .0780 | 229.659 |
| 440 | 166.85 | 14733.3 | 3.5387 | 11074.9 | 1.034 | 201.954 | 940 | 666.85 | 33451.0 | 104.376 | 25635.5 | .0749 | 230.092 |
| 450 | 176.85 | 15080.1 | 3.8867 | 11338.6 | .963 | 202.733 | 950 | 676.85 | 33856.7 | 109.905 | 25958.0 | .0719 | 230.521 |
| 460 | 186.85 | 15427.8 | 4.2608 | 11603.2 | .898 | 203.498 | 960 | 686.85 | 34263.7 | 115.685 | 26281.8 | .0690 | 230.947 |
| 470 | 196.85 | 15776.4 | 4.6630 | 11868.6 | .838 | 204.247 | 970 | 696.85 | 34671.8 | 121.722 | 26606.8 | .0663 | 231.370 |
| 480 | 206.85 | 16126.2 | 5.0944 | 12135.2 | .783 | 204.983 | 980 | 706.85 | 35081.3 | 128.029 | 26933.2 | .0636 | 231.790 |
| 490 | 216.85 | 16476.6 | 5.5575 | 12402.5 | .733 | 205.707 | 990 | 716.85 | 35492.1 | 134.614 | 27260.9 | .0611 | 232.207 |
| 500 | 226.85 | 16828.4 | 6.0530 | 12671.2 | .687 | 206.417 | 1000 | 726.85 | 35904.1 | 141.489 | 27589.7 | .0588 | 232.621 |
| 510 | 236.85 | 17181.3 | 6.5838 | 12941.0 | .644 | 207.116 | 1010 | 736.85 | 36317.5 | 148.663 | 27919.9 | .0565 | 233.032 |
| 520 | 246.85 | 17535.2 | 7.1510 | 13211.7 | .605 | 207.803 | 1020 | 746.85 | 36732.0 | 156.149 | 28251.3 | .0543 | 233.441 |
| 530 | 256.85 | 17889.9 | 7.7568 | 13483.3 | .568 | 208.479 | 1030 | 756.85 | 37147.8 | 163.958 | 28584.0 | .0522 | 233.846 |
| 540 | 266.85 | 18246.1 | 8.4031 | 13756.3 | .534 | 209.144 | 1040 | 766.85 | 37565.0 | 172.101 | 28918.0 | .0502 | 234.249 |
| 550 | 276.85 | 18603.1 | 9.0921 | 14030.2 | .503 | 209.799 | 1050 | 776.85 | 37983.4 | 180.592 | 29253.2 | .0483 | 234.650 |
| 560 | 286.85 | 18961.3 | 9.8261 | 14305.2 | .474 | 210.445 | 1060 | 786.85 | 38403.1 | 189.441 | 29589.8 | .0465 | 235.048 |
| 570 | 296.85 | 19320.6 | 10.6071 | 14581.3 | .447 | 211.081 | 1070 | 796.85 | 38823.9 | 198.665 | 29927.5 | .0448 | 235.443 |
| 580 | 306.85 | 19681.0 | 11.4377 | 14858.6 | .422 | 211.708 | 1080 | 806.85 | 39246.1 | 208.275 | 30266.6 | .0431 | 235.836 |
| 590 | 316.85 | 20042.6 | 12.3203 | 15137.1 | .398 | 212.326 | 1090 | 816.85 | 39669.7 | 218.284 | 30607.0 | .0415 | 236.226 |

# Table 15 Water Vapor at Low Pressures (for One Gram-Mole)
$\bar{m} = 18.0152$

| T | t | $\bar{h}$ | $p_r$ | $\bar{u}$ | $v_r$ | $\bar{\phi}$ | T | t | $\bar{h}$ | $p_r$ | $\bar{u}$ | $v_r$ | $\bar{\phi}$ |
|---|---|---|---|---|---|---|---|---|---|---|---|---|---|
| 1100 | 826.85 | 40094.4 | 228.71 | 30948.6 | .0400 | 236.614 | 1600 | 1326.85 | 62811.9 | 1758.5 | 49508.9 | .00756 | 253.573 |
| 1110 | 836.85 | 40520.5 | 239.56 | 31291.5 | .0385 | 236.999 | 1610 | 1336.85 | 63293.0 | 1823.1 | 49906.8 | .00734 | 253.873 |
| 1120 | 846.85 | 40947.5 | 250.86 | 31635.3 | .0371 | 237.383 | 1620 | 1346.85 | 63774.9 | 1889.7 | 50305.6 | .00713 | 254.171 |
| 1130 | 856.85 | 41376.0 | 262.62 | 31980.8 | .0358 | 237.763 | 1630 | 1356.85 | 64257.8 | 1958.4 | 50705.3 | .00692 | 254.468 |
| 1140 | 866.85 | 41805.7 | 274.86 | 32327.3 | .0345 | 238.142 | 1640 | 1366.85 | 64741.3 | 2029.3 | 51105.7 | .00672 | 254.764 |
| 1150 | 876.85 | 42236.8 | 287.59 | 32675.2 | .0332 | 238.518 | 1650 | 1376.85 | 65226.0 | 2102.5 | 51507.2 | .00652 | 255.059 |
| 1160 | 886.85 | 42669.0 | 300.83 | 33024.3 | .0321 | 238.893 | 1660 | 1386.85 | 65711.5 | 2178.0 | 51909.6 | .00634 | 255.352 |
| 1170 | 896.85 | 43102.3 | 314.59 | 33374.5 | .0309 | 239.265 | 1670 | 1396.85 | 66197.6 | 2255.9 | 52312.5 | .00615 | 255.644 |
| 1180 | 906.85 | 43537.0 | 328.90 | 33726.1 | .0298 | 239.634 | 1680 | 1406.85 | 66685.0 | 2336.2 | 52716.8 | .00598 | 255.935 |
| 1190 | 916.85 | 43972.9 | 343.78 | 34078.7 | .0288 | 240.002 | 1690 | 1416.85 | 67173.0 | 2419.1 | 53121.7 | .00581 | 256.225 |
| 1200 | 926.85 | 44410.1 | 359.24 | 34432.8 | .0278 | 240.368 | 1700 | 1426.85 | 67662.0 | 2504.5 | 53527.5 | .00564 | 256.513 |
| 1210 | 936.85 | 44848.2 | 375.31 | 34787.8 | .0268 | 240.732 | 1710 | 1436.85 | 68151.7 | 2592.5 | 53934.0 | .00548 | 256.801 |
| 1220 | 946.85 | 45287.7 | 391.99 | 35144.1 | .0259 | 241.093 | 1720 | 1446.85 | 68642.2 | 2683.2 | 54341.5 | .00533 | 257.087 |
| 1230 | 956.85 | 45728.5 | 409.32 | 35501.8 | .0250 | 241.453 | 1730 | 1456.85 | 69133.7 | 2776.8 | 54749.8 | .00518 | 257.372 |
| 1240 | 966.85 | 46170.2 | 427.32 | 35860.4 | .0241 | 241.811 | 1740 | 1466.85 | 69626.1 | 2873.2 | 55159.1 | .00504 | 257.655 |
| 1250 | 976.85 | 46613.3 | 446.01 | 36220.3 | .0233 | 242.167 | 1750 | 1476.85 | 70119.2 | 2972.5 | 55569.0 | .00489 | 257.938 |
| 1260 | 986.85 | 47057.6 | 465.41 | 36581.4 | .0225 | 242.521 | 1760 | 1486.85 | 70613.2 | 3074.8 | 55979.8 | .00476 | 258.219 |
| 1270 | 996.85 | 47502.9 | 485.54 | 36943.7 | .0217 | 242.873 | 1770 | 1496.85 | 71107.8 | 3180.3 | 56391.3 | .00463 | 258.500 |
| 1280 | 1006.85 | 47949.5 | 506.43 | 37307.0 | .0210 | 243.223 | 1780 | 1506.85 | 71603.4 | 3288.9 | 56803.8 | .00450 | 258.779 |
| 1290 | 1016.85 | 48397.3 | 528.10 | 37671.8 | .0203 | 243.572 | 1790 | 1516.85 | 72099.6 | 3400.7 | 57216.9 | .00438 | 259.057 |
| 1300 | 1026.85 | 48846.1 | 550.58 | 38037.4 | .0196 | 243.918 | 1800 | 1526.85 | 72596.8 | 3515.9 | 57630.9 | .00426 | 259.334 |
| 1310 | 1036.85 | 49296.0 | 573.90 | 38404.2 | .0190 | 244.263 | 1810 | 1536.85 | 73094.6 | 3634.5 | 58045.6 | .00414 | 259.610 |
| 1320 | 1046.85 | 49747.3 | 598.08 | 38772.3 | .0184 | 244.606 | 1820 | 1546.85 | 73593.2 | 3756.6 | 58460.9 | .00403 | 259.884 |
| 1330 | 1056.85 | 50199.5 | 623.14 | 39141.3 | .0177 | 244.947 | 1830 | 1556.85 | 74092.6 | 3882.3 | 58877.2 | .00392 | 260.158 |
| 1340 | 1066.85 | 50652.8 | 649.12 | 39511.5 | .0172 | 245.287 | 1840 | 1566.85 | 74592.9 | 4011.7 | 59294.4 | .00381 | 260.430 |
| 1350 | 1076.85 | 51107.5 | 676.05 | 39883.1 | .0166 | 245.625 | 1850 | 1576.85 | 75093.7 | 4144.8 | 59712.1 | .00371 | 260.702 |
| 1360 | 1086.85 | 51563.1 | 703.95 | 40255.5 | .0161 | 245.961 | 1860 | 1586.85 | 75595.4 | 4281.9 | 60130.6 | .00361 | 260.972 |
| 1370 | 1096.85 | 52019.8 | 732.86 | 40629.1 | .0155 | 246.296 | 1870 | 1596.85 | 76097.8 | 4422.9 | 60549.9 | .00352 | 261.242 |
| 1380 | 1106.85 | 52477.7 | 762.80 | 41003.8 | .0150 | 246.629 | 1880 | 1606.85 | 76601.1 | 4567.9 | 60970.0 | .00342 | 261.510 |
| 1390 | 1116.85 | 52936.5 | 793.82 | 41379.5 | .0146 | 246.960 | 1890 | 1616.85 | 77104.8 | 4717.2 | 61390.6 | .00333 | 261.777 |
| 1400 | 1126.85 | 53396.6 | 825.94 | 41756.5 | .0141 | 247.290 | 1900 | 1626.85 | 77609.5 | 4870.7 | 61812.1 | .00324 | 262.044 |
| 1410 | 1136.85 | 53857.7 | 859.19 | 42134.4 | .0136 | 247.618 | 1910 | 1636.85 | 78114.8 | 5028.6 | 62234.3 | .00316 | 262.309 |
| 1420 | 1146.85 | 54319.8 | 893.61 | 42513.4 | .0132 | 247.945 | 1920 | 1646.85 | 78620.8 | 5191.0 | 62657.2 | .00308 | 262.573 |
| 1430 | 1156.85 | 54782.9 | 929.23 | 42893.3 | .0128 | 248.270 | 1930 | 1656.85 | 79127.5 | 5358.0 | 63080.7 | .00299 | 262.836 |
| 1440 | 1166.85 | 55247.2 | 966.10 | 43274.4 | .0124 | 248.593 | 1940 | 1666.85 | 79634.9 | 5529.6 | 63505.0 | .00292 | 263.099 |
| 1450 | 1176.85 | 55712.5 | 1004.25 | 43656.6 | .0120 | 248.915 | 1950 | 1676.85 | 80143.0 | 5706.1 | 63929.9 | .00284 | 263.360 |
| 1460 | 1186.85 | 56179.0 | 1043.72 | 44040.0 | .0116 | 249.236 | 1960 | 1686.85 | 80651.9 | 5887.6 | 64355.7 | .00277 | 263.620 |
| 1470 | 1196.85 | 56646.2 | 1084.55 | 44424.0 | .0113 | 249.555 | 1970 | 1696.85 | 81161.3 | 6074.0 | 64782.0 | .00270 | 263.879 |
| 1480 | 1206.85 | 57114.7 | 1126.77 | 44809.4 | .0109 | 249.872 | 1980 | 1706.85 | 81671.5 | 6265.7 | 65208.9 | .00263 | 264.138 |
| 1490 | 1216.85 | 57584.0 | 1170.43 | 45195.5 | .0106 | 250.189 | 1990 | 1716.85 | 82182.2 | 6462.7 | 65636.6 | .00256 | 264.395 |
| 1500 | 1226.85 | 58054.5 | 1215.57 | 45582.9 | .0103 | 250.503 | 2000 | 1726.85 | 82693.7 | 6665.1 | 66064.9 | .00249 | 264.652 |
| 1510 | 1236.85 | 58525.8 | 1262.25 | 45971.1 | .0099 | 250.816 | 2010 | 1736.85 | 83205.8 | 6873.0 | 66493.9 | .00243 | 264.907 |
| 1520 | 1246.85 | 58998.2 | 1310.48 | 46360.3 | .0096 | 251.128 | 2020 | 1746.85 | 83718.4 | 7086.7 | 66923.4 | .00237 | 265.161 |
| 1530 | 1256.85 | 59471.5 | 1360.33 | 46750.5 | .0094 | 251.439 | 2030 | 1756.85 | 84232.0 | 7306.1 | 67353.7 | .00231 | 265.415 |
| 1540 | 1266.85 | 59946.0 | 1411.84 | 47141.8 | .0091 | 251.748 | 2040 | 1766.85 | 84746.2 | 7531.4 | 67784.8 | .00225 | 265.668 |
| 1550 | 1276.85 | 60421.3 | 1465.06 | 47534.0 | .0088 | 252.055 | 2050 | 1776.85 | 85260.8 | 7762.9 | 68216.3 | .00220 | 265.919 |
| 1560 | 1286.85 | 60897.6 | 1520.04 | 47927.1 | .0085 | 252.362 | 2060 | 1786.85 | 85776.1 | 8000.6 | 68648.5 | .00214 | 266.170 |
| 1570 | 1296.85 | 61374.8 | 1576.81 | 48321.1 | .0083 | 252.666 | 2070 | 1796.85 | 86292.1 | 8244.6 | 69081.3 | .00209 | 266.420 |
| 1580 | 1306.85 | 61853.1 | 1635.46 | 48716.3 | .0080 | 252.970 | 2080 | 1806.85 | 86808.6 | 8495.3 | 69514.6 | .00204 | 266.669 |
| 1590 | 1316.85 | 62332.0 | 1696.00 | 49112.1 | .0078 | 253.272 | 2090 | 1816.85 | 87325.7 | 8752.5 | 69948.6 | .00199 | 266.917 |

**Table 15 Water Vapor at Low Pressures (for One Gram-Mole)**

$\bar{m} = 18.0152$

| T | t | $\bar{h}$ | $p_r$ | $\bar{u}$ | $v_r$ | $\bar{\phi}$ | T | t | $\bar{h}$ | $p_r$ | $\bar{u}$ | $v_r$ | $\bar{\phi}$ |
|---|---|---|---|---|---|---|---|---|---|---|---|---|---|
| 2100 | 1826.85 | 87843.6 | 9017 | 70383.3 | .001936 | 267.164 | 2600 | 2326.85 | 114410.3 | 35261 | 92792.8 | .000613 | 278.502 |
| 2110 | 1836.85 | 88361.8 | 9288 | 70818.5 | .001889 | 267.410 | 2610 | 2336.85 | 114953.5 | 36156 | 93252.9 | .000600 | 278.711 |
| 2120 | 1846.85 | 88881.0 | 9566 | 71254.5 | .001843 | 267.656 | 2620 | 2346.85 | 115497.2 | 37072 | 93713.5 | .000588 | 278.919 |
| 2130 | 1856.85 | 89400.4 | 9851 | 71690.8 | .001798 | 267.900 | 2630 | 2356.85 | 116041.4 | 38008 | 94174.5 | .000575 | 279.126 |
| 2140 | 1866.85 | 89920.8 | 10144 | 72127.9 | .001754 | 268.144 | 2640 | 2366.85 | 116585.8 | 38964 | 94635.8 | .000563 | 279.333 |
| 2150 | 1876.85 | 90441.5 | 10445 | 72565.6 | .001711 | 268.387 | 2650 | 2376.85 | 117130.7 | 39941 | 95097.5 | .000552 | 279.539 |
| 2160 | 1886.85 | 90962.8 | 10753 | 73003.7 | .001670 | 268.629 | 2660 | 2386.85 | 117676.0 | 40940 | 95559.6 | .000540 | 279.744 |
| 2170 | 1896.85 | 91484.7 | 11070 | 73442.4 | .001630 | 268.870 | 2670 | 2396.85 | 118221.7 | 41961 | 96022.3 | .000529 | 279.949 |
| 2180 | 1906.85 | 92007.3 | 11394 | 73881.9 | .001591 | 269.110 | 2680 | 2406.85 | 118767.7 | 43004 | 96485.1 | .000518 | 280.153 |
| 2190 | 1916.85 | 92530.3 | 11727 | 74321.8 | .001553 | 269.349 | 2690 | 2416.85 | 119314.1 | 44069 | 96948.4 | .000508 | 280.356 |
| 2200 | 1926.85 | 93053.8 | 12068 | 74762.1 | .001516 | 269.588 | 2700 | 2426.85 | 119860.8 | 45158 | 97411.9 | .000497 | 280.559 |
| 2210 | 1936.85 | 93578.0 | 12418 | 75203.2 | .001480 | 269.825 | 2710 | 2436.85 | 120408.2 | 46270 | 97876.1 | .000487 | 280.762 |
| 2220 | 1946.85 | 94102.7 | 12777 | 75644.7 | .001445 | 270.062 | 2720 | 2446.85 | 120955.7 | 47406 | 98340.6 | .000477 | 280.963 |
| 2230 | 1956.85 | 94628.0 | 13145 | 76086.9 | .001411 | 270.298 | 2730 | 2456.85 | 121503.5 | 48567 | 98805.2 | .000467 | 281.164 |
| 2240 | 1966.85 | 95153.8 | 13522 | 76529.5 | .001377 | 270.534 | 2740 | 2466.85 | 122052.0 | 49752 | 99270.6 | .000458 | 281.365 |
| 2250 | 1976.85 | 95680.0 | 13909 | 76972.6 | .001345 | 270.768 | 2750 | 2476.85 | 122600.5 | 50963 | 99735.9 | .000449 | 281.565 |
| 2260 | 1986.85 | 96206.9 | 14305 | 77416.3 | .001314 | 271.002 | 2760 | 2486.85 | 123149.7 | 52200 | 100201.9 | .000440 | 281.764 |
| 2270 | 1996.85 | 96734.3 | 14712 | 77860.6 | .001283 | 271.235 | 2770 | 2496.85 | 123699.1 | 53462 | 100668.2 | .000431 | 281.963 |
| 2280 | 2006.85 | 97262.1 | 15128 | 78305.2 | .001253 | 271.467 | 2780 | 2506.85 | 124248.9 | 54751 | 101134.9 | .000422 | 282.161 |
| 2290 | 2016.85 | 97790.6 | 15555 | 78750.6 | .001224 | 271.698 | 2790 | 2516.85 | 124799.0 | 56068 | 101601.8 | .000414 | 282.359 |
| 2300 | 2026.85 | 98319.5 | 15992 | 79196.4 | .001196 | 271.928 | 2800 | 2526.85 | 125349.8 | 57413 | 102069.4 | .000405 | 282.556 |
| 2310 | 2036.85 | 98848.9 | 16440 | 79642.7 | .001168 | 272.158 | 2810 | 2536.85 | 125900.5 | 58785 | 102537.0 | .000397 | 282.752 |
| 2320 | 2046.85 | 99378.8 | 16899 | 80089.4 | .001141 | 272.387 | 2820 | 2546.85 | 126452.0 | 60186 | 103005.3 | .000390 | 282.948 |
| 2330 | 2056.85 | 99909.3 | 17369 | 80536.8 | .001115 | 272.615 | 2830 | 2556.85 | 127003.6 | 61617 | 103473.9 | .000382 | 283.143 |
| 2340 | 2066.85 | 100440.3 | 17851 | 80984.6 | .001090 | 272.842 | 2840 | 2566.85 | 127555.7 | 63077 | 103942.8 | .000374 | 283.338 |
| 2350 | 2076.85 | 100971.8 | 18344 | 81432.9 | .001065 | 273.069 | 2850 | 2576.85 | 128108.1 | 64567 | 104412.0 | .000367 | 283.532 |
| 2360 | 2086.85 | 101503.9 | 18849 | 81881.9 | .001041 | 273.295 | 2860 | 2586.85 | 128660.9 | 66088 | 104881.7 | .000360 | 283.726 |
| 2370 | 2096.85 | 102036.3 | 19366 | 82331.1 | .001017 | 273.520 | 2870 | 2596.85 | 129213.9 | 67641 | 105351.6 | .000353 | 283.919 |
| 2380 | 2106.85 | 102569.3 | 19896 | 82781.0 | .000995 | 273.745 | 2880 | 2606.85 | 129767.4 | 69226 | 105821.9 | .000346 | 284.111 |
| 2390 | 2116.85 | 103102.8 | 20439 | 83231.4 | .000972 | 273.968 | 2890 | 2616.85 | 130321.3 | 70843 | 106292.7 | .000339 | 284.303 |
| 2400 | 2126.85 | 103636.5 | 20994 | 83682.0 | .000950 | 274.191 | 2900 | 2626.85 | 130875.5 | 72493 | 106763.7 | .000333 | 284.495 |
| 2410 | 2136.85 | 104170.9 | 21563 | 84133.2 | .000929 | 274.413 | 2910 | 2636.85 | 131430.1 | 74176 | 107235.2 | .000326 | 284.686 |
| 2420 | 2146.85 | 104705.8 | 22145 | 84585.0 | .000909 | 274.635 | 2920 | 2646.85 | 131985.0 | 75894 | 107706.9 | .000320 | 284.876 |
| 2430 | 2156.85 | 105241.3 | 22741 | 85037.3 | .000888 | 274.856 | 2930 | 2656.85 | 132540.3 | 77647 | 108179.1 | .000314 | 285.066 |
| 2440 | 2166.85 | 105777.1 | 23351 | 85490.0 | .000869 | 275.076 | 2940 | 2666.85 | 133095.8 | 79435 | 108651.4 | .000308 | 285.255 |
| 2450 | 2176.85 | 106313.3 | 23975 | 85943.0 | .000850 | 275.295 | 2950 | 2676.85 | 133651.8 | 81259 | 109124.3 | .000302 | 285.444 |
| 2460 | 2186.85 | 106850.0 | 24613 | 86396.6 | .000831 | 275.514 | 2960 | 2686.85 | 134208.0 | 83120 | 109597.3 | .000296 | 285.632 |
| 2470 | 2196.85 | 107387.1 | 25267 | 86850.6 | .000813 | 275.731 | 2970 | 2696.85 | 134764.6 | 85018 | 110070.9 | .000290 | 285.820 |
| 2480 | 2206.85 | 107924.7 | 25936 | 87305.0 | .000795 | 275.949 | 2980 | 2706.85 | 135321.8 | 86954 | 110544.8 | .000285 | 286.007 |
| 2490 | 2216.85 | 108462.8 | 26620 | 87759.9 | .000778 | 276.165 | 2990 | 2716.85 | 135878.9 | 88929 | 111018.8 | .000280 | 286.194 |
| 2500 | 2226.85 | 109001.3 | 27320 | 88215.2 | .000761 | 276.381 | 3000 | 2726.85 | 136436.7 | 90942 | 111493.5 | .000274 | 286.380 |
| 2510 | 2236.85 | 109540.2 | 28037 | 88671.1 | .000744 | 276.596 | 3010 | 2736.85 | 136994.5 | 92996 | 111968.1 | .000269 | 286.566 |
| 2520 | 2246.85 | 110079.6 | 28769 | 89127.3 | .000728 | 276.811 | 3020 | 2746.85 | 137552.9 | 95090 | 112443.4 | .000264 | 286.751 |
| 2530 | 2256.85 | 110619.2 | 29519 | 89583.8 | .000713 | 277.025 | 3030 | 2756.85 | 138111.4 | 97226 | 112918.8 | .000259 | 286.935 |
| 2540 | 2266.85 | 111159.5 | 30285 | 90041.0 | .000697 | 277.238 | 3040 | 2766.85 | 138670.4 | 99403 | 113394.6 | .000254 | 287.120 |
| 2550 | 2276.85 | 111700.3 | 31069 | 90498.6 | .000682 | 277.450 | 3050 | 2776.85 | 139229.7 | 101623 | 113870.8 | .000250 | 287.303 |
| 2560 | 2286.85 | 112241.5 | 31870 | 90956.7 | .000668 | 277.662 | 3060 | 2786.85 | 139789.4 | 103888 | 114347.3 | .000245 | 287.486 |
| 2570 | 2296.85 | 112783.2 | 32690 | 91415.2 | .000654 | 277.873 | 3070 | 2796.85 | 140349.2 | 106196 | 114824.0 | .000240 | 287.669 |
| 2580 | 2306.85 | 113325.1 | 33528 | 91873.9 | .000640 | 278.083 | 3080 | 2806.85 | 140909.5 | 108548 | 115301.1 | .000236 | 287.851 |
| 2590 | 2316.85 | 113867.4 | 34385 | 92333.1 | .000626 | 278.293 | 3090 | 2816.85 | 141470.0 | 110946 | 115778.5 | .000232 | 288.033 |

# Table 16    Water Vapor

$\bar{m} = 18.0152$

| T<br>K | t<br>C | $\bar{c}_p$<br>J<br>g-mol K | $\bar{c}_v$<br>J<br>g-mol K | $k = \dfrac{\bar{c}_p}{\bar{c}_v}$ | a<br>$\dfrac{m}{s}$ |
|---|---|---|---|---|---|
| 100 | –173.15 | 33.301 | 24.986 | 1.333 | 248.0 |
| 120 | –153.15 | 33.313 | 24.999 | 1.333 | 271.7 |
| 140 | –133.15 | 33.320 | 25.006 | 1.333 | 293.4 |
| 160 | –113.15 | 33.327 | 25.013 | 1.332 | 313.7 |
| 180 | –93.15 | 33.337 | 25.022 | 1.332 | 332.7 |
| 200 | –73.15 | 33.350 | 25.035 | 1.332 | 350.7 |
| 250 | –23.15 | 33.426 | 25.111 | 1.331 | 391.9 |
| 300 | 26.85 | 33.596 | 25.282 | 1.329 | 428.9 |
| 350 | 76.85 | 33.879 | 25.564 | 1.325 | 462.7 |
| 400 | 126.85 | 34.262 | 25.948 | 1.320 | 493.7 |
| 450 | 176.85 | 34.720 | 26.406 | 1.315 | 522.6 |
| 500 | 226.85 | 35.226 | 26.912 | 1.309 | 549.6 |
| 550 | 276.85 | 35.764 | 27.450 | 1.303 | 575.1 |
| 600 | 326.85 | 36.325 | 28.010 | 1.297 | 599.3 |
| 650 | 376.85 | 36.902 | 28.588 | 1.291 | 622.3 |
| 700 | 426.85 | 37.495 | 29.181 | 1.285 | 644.3 |
| 750 | 476.85 | 38.102 | 29.788 | 1.279 | 665.4 |
| 800 | 526.85 | 38.721 | 30.407 | 1.273 | 685.7 |
| 850 | 576.85 | 39.350 | 31.036 | 1.268 | 705.3 |
| 900 | 626.85 | 39.987 | 31.672 | 1.263 | 724.2 |
| 950 | 676.85 | 40.627 | 32.312 | 1.257 | 742.5 |
| 1000 | 726.85 | 41.267 | 32.953 | 1.252 | 760.2 |
| 1050 | 776.85 | 41.905 | 33.590 | 1.248 | 777.5 |
| 1100 | 826.85 | 42.536 | 34.221 | 1.243 | 794.4 |
| 1150 | 876.85 | 43.158 | 34.843 | 1.239 | 810.8 |
| 1200 | 926.85 | 43.768 | 35.453 | 1.235 | 826.9 |
| 1300 | 1026.85 | 44.945 | 36.630 | 1.227 | 858.0 |
| 1400 | 1126.85 | 46.054 | 37.740 | 1.220 | 888.0 |
| 1500 | 1226.85 | 47.090 | 38.776 | 1.214 | 916.9 |
| 1600 | 1326.85 | 48.050 | 39.736 | 1.209 | 945.0 |
| 1700 | 1426.85 | 48.935 | 40.621 | 1.205 | 972.2 |
| 1800 | 1526.85 | 49.749 | 41.434 | 1.201 | 998.7 |
| 1900 | 1626.85 | 50.495 | 42.181 | 1.197 | 1024.6 |
| 2000 | 1726.85 | 51.180 | 42.866 | 1.194 | 1049.8 |
| 2100 | 1826.85 | 51.809 | 43.494 | 1.191 | 1074.5 |
| 2200 | 1926.85 | 52.387 | 44.073 | 1.189 | 1098.6 |
| 2300 | 2026.85 | 52.921 | 44.607 | 1.186 | 1122.2 |
| 2400 | 2126.85 | 53.414 | 45.100 | 1.184 | 1145.4 |
| 2500 | 2226.85 | 53.873 | 45.559 | 1.182 | 1168.1 |
| 2600 | 2326.85 | 54.301 | 45.986 | 1.181 | 1190.3 |
| 2700 | 2426.85 | 54.702 | 46.388 | 1.179 | 1212.2 |
| 2800 | 2526.85 | 55.080 | 46.765 | 1.178 | 1233.7 |
| 2900 | 2626.85 | 55.437 | 47.123 | 1.176 | 1254.8 |
| 3000 | 2726.85 | 55.778 | 47.464 | 1.175 | 1275.6 |
| 3100 | 2826.85 | 56.103 | 47.789 | 1.174 | 1296.0 |
| 3200 | 2926.85 | 56.416 | 48.101 | 1.173 | 1316.1 |
| 3300 | 3026.85 | 56.718 | 48.403 | 1.172 | 1335.9 |
| 3400 | 3126.85 | 57.010 | 48.696 | 1.171 | 1355.4 |
| 3500 | 3226.85 | 57.295 | 48.981 | 1.170 | 1374.6 |
| 3600 | 3326.85 | 57.573 | 49.259 | 1.169 | 1393.5 |

# Table 17    Carbon Dioxide at Low Pressures (for One Gram-Mole)

$\bar{m}$ = 44.0098

| T | t | $\bar{h}$ | $p_r$ | $\bar{u}$ | $v_r$ | $\bar{\phi}$ | T | t | $\bar{h}$ | $p_r$ | $\bar{u}$ | $v_r$ | $\bar{\phi}$ |
|---|---|---|---|---|---|---|---|---|---|---|---|---|---|
| 100 | −173.15 | 2909.6 | .0022 | 2078.1 | 376.315 | 178.893 | 600 | 326.85 | 22273.3 | 5.032 | 17284.6 | .9914 | 243.170 |
| 110 | −163.15 | 3202.1 | .0031 | 2287.5 | 295.993 | 181.681 | 610 | 336.85 | 22747.7 | 5.530 | 17675.9 | .9171 | 243.955 |
| 120 | −153.15 | 3496.0 | .0042 | 2498.3 | 237.448 | 184.237 | 620 | 346.85 | 23224.7 | 6.070 | 18069.7 | .8492 | 244.729 |
| 130 | −143.15 | 3791.5 | .0056 | 2710.6 | 193.522 | 186.603 | 630 | 356.85 | 23703.9 | 6.656 | 18465.8 | .7869 | 245.496 |
| 140 | −133.15 | 4089.5 | .0073 | 2925.4 | 159.796 | 188.812 | 640 | 366.85 | 24185.4 | 7.293 | 18864.2 | .7297 | 246.255 |
| 150 | −123.15 | 4390.4 | .0093 | 3143.2 | 133.405 | 190.886 | 650 | 376.85 | 24669.1 | 7.981 | 19264.8 | .6771 | 247.005 |
| 160 | −113.15 | 4694.5 | .0118 | 3364.2 | 112.374 | 192.849 | 660 | 386.85 | 25155.2 | 8.726 | 19667.7 | .6289 | 247.747 |
| 170 | −103.15 | 5002.5 | .0148 | 3589.0 | 95.379 | 194.717 | 670 | 396.85 | 25643.5 | 9.531 | 20072.8 | .5845 | 248.481 |
| 180 | −93.15 | 5314.5 | .0184 | 3817.9 | 81.493 | 196.500 | 680 | 406.85 | 26133.8 | 10.401 | 20480.0 | .5436 | 249.207 |
| 190 | −83.15 | 5631.1 | .0226 | 4051.4 | 70.013 | 198.212 | 690 | 416.85 | 26626.4 | 11.342 | 20889.5 | .5058 | 249.927 |
| 200 | −73.15 | 5952.3 | .0275 | 4289.4 | 60.453 | 199.859 | 700 | 426.85 | 27120.9 | 12.355 | 21300.8 | .4711 | 250.639 |
| 210 | −63.15 | 6278.2 | .0333 | 4532.1 | 52.429 | 201.449 | 710 | 436.85 | 27617.6 | 13.447 | 21714.3 | .4390 | 251.343 |
| 220 | −53.15 | 6609.1 | .0401 | 4779.9 | 45.640 | 202.989 | 720 | 446.85 | 28116.3 | 14.624 | 22129.9 | .4093 | 252.040 |
| 230 | −43.15 | 6945.0 | .0480 | 5032.7 | 39.874 | 204.481 | 730 | 456.85 | 28616.9 | 15.891 | 22547.4 | .3819 | 252.731 |
| 240 | −33.15 | 7285.9 | .0571 | 5290.4 | 34.946 | 205.932 | 740 | 466.85 | 29119.5 | 17.251 | 22966.8 | .3567 | 253.414 |
| 250 | −23.15 | 7631.8 | .0677 | 5553.2 | 30.717 | 207.344 | 750 | 476.85 | 29623.9 | 18.716 | 23388.1 | .3332 | 254.091 |
| 260 | −13.15 | 7982.5 | .0798 | 5820.7 | 27.076 | 208.719 | 760 | 486.85 | 30130.1 | 20.287 | 23811.2 | .3115 | 254.762 |
| 270 | −3.15 | 8338.1 | .0938 | 6093.2 | 23.924 | 210.062 | 770 | 496.85 | 30638.4 | 21.977 | 24236.3 | .2913 | 255.427 |
| 280 | 6.85 | 8698.6 | .1099 | 6370.5 | 21.190 | 211.373 | 780 | 506.85 | 31148.3 | 23.786 | 24663.1 | .2727 | 256.084 |
| 290 | 16.85 | 9063.8 | .1282 | 6652.6 | 18.813 | 212.654 | 790 | 516.85 | 31660.0 | 25.725 | 25091.6 | .2553 | 256.736 |
| 300 | 26.85 | 9433.7 | .1490 | 6939.4 | 16.737 | 213.908 | 800 | 526.85 | 32173.5 | 27.804 | 25522.0 | .2392 | 257.382 |
| 310 | 36.85 | 9808.2 | .1727 | 7230.8 | 14.920 | 215.136 | 810 | 536.85 | 32688.6 | 30.029 | 25953.9 | .2243 | 258.023 |
| 320 | 46.85 | 10187.1 | .1996 | 7526.5 | 13.327 | 216.339 | 820 | 546.85 | 33205.5 | 32.407 | 26387.7 | .2104 | 258.656 |
| 330 | 56.85 | 10570.4 | .2301 | 7826.7 | 11.926 | 217.518 | 830 | 556.85 | 33724.0 | 34.956 | 26823.1 | .1974 | 259.285 |
| 340 | 66.85 | 10957.9 | .2644 | 8131.0 | 10.692 | 218.675 | 840 | 566.85 | 34244.1 | 37.674 | 27260.0 | .1854 | 259.908 |
| 350 | 76.85 | 11349.7 | .3031 | 8439.6 | 9.601 | 219.810 | 850 | 576.85 | 34765.8 | 40.576 | 27698.6 | .1742 | 260.525 |
| 360 | 86.85 | 11745.8 | .3466 | 8752.6 | 8.636 | 220.925 | 860 | 586.85 | 35289.1 | 43.675 | 28138.7 | .1637 | 261.137 |
| 370 | 96.85 | 12145.5 | .3955 | 9069.2 | 7.779 | 222.022 | 870 | 596.85 | 35814.0 | 46.983 | 28580.4 | .1540 | 261.744 |
| 380 | 106.85 | 12549.3 | .4501 | 9389.8 | 7.019 | 223.099 | 880 | 606.85 | 36340.4 | 50.505 | 29023.7 | .1449 | 262.345 |
| 390 | 116.85 | 12957.0 | .5113 | 9714.4 | 6.342 | 224.158 | 890 | 616.85 | 36868.2 | 54.265 | 29468.4 | .1364 | 262.942 |
| 400 | 126.85 | 13368.5 | .5795 | 10042.8 | 5.739 | 225.199 | 900 | 626.85 | 37397.4 | 58.264 | 29914.5 | .1284 | 263.533 |
| 410 | 136.85 | 13783.5 | .6555 | 10374.6 | 5.200 | 226.224 | 910 | 636.85 | 37928.2 | 62.520 | 30362.1 | .1210 | 264.119 |
| 420 | 146.85 | 14202.3 | .7401 | 10710.2 | 4.719 | 227.233 | 920 | 646.85 | 38460.1 | 67.053 | 30810.8 | .1141 | 264.701 |
| 430 | 156.85 | 14624.5 | .8340 | 11049.3 | 4.287 | 228.226 | 930 | 656.85 | 38993.6 | 71.864 | 31261.2 | .1076 | 265.278 |
| 440 | 166.85 | 15050.1 | .9382 | 11391.7 | 3.899 | 229.205 | 940 | 666.85 | 39528.4 | 76.983 | 31712.9 | .1015 | 265.850 |
| 450 | 176.85 | 15478.9 | 1.0535 | 11737.4 | 3.552 | 230.168 | 950 | 676.85 | 40064.7 | 82.416 | 32166.0 | .0958 | 266.417 |
| 460 | 186.85 | 15911.1 | 1.1810 | 12086.5 | 3.238 | 231.119 | 960 | 686.85 | 40602.0 | 88.189 | 32620.2 | .0905 | 266.980 |
| 470 | 196.85 | 16346.7 | 1.3218 | 12438.9 | 2.956 | 232.055 | 970 | 696.85 | 41141.0 | 94.319 | 33076.0 | .0855 | 267.538 |
| 480 | 206.85 | 16785.3 | 1.4771 | 12794.3 | 2.702 | 232.979 | 980 | 706.85 | 41680.8 | 100.814 | 33532.7 | .0808 | 268.092 |
| 490 | 216.85 | 17227.0 | 1.6482 | 13152.9 | 2.472 | 233.890 | 990 | 716.85 | 42222.0 | 107.703 | 33990.8 | .0764 | 268.642 |
| 500 | 226.85 | 17671.7 | 1.8362 | 13514.5 | 2.264 | 234.788 | 1000 | 726.85 | 42764.6 | 114.994 | 34450.2 | .0723 | 269.186 |
| 510 | 236.85 | 18119.6 | 2.0427 | 13879.2 | 2.076 | 235.674 | 1010 | 736.85 | 43308.8 | 122.729 | 34911.2 | .0684 | 269.728 |
| 520 | 246.85 | 18570.2 | 2.2695 | 14246.7 | 1.905 | 236.550 | 1020 | 746.85 | 43853.9 | 130.840 | 35373.2 | .0648 | 270.260 |
| 530 | 256.85 | 19023.5 | 2.5178 | 14616.8 | 1.750 | 237.413 | 1030 | 756.85 | 44399.0 | 139.488 | 35835.1 | .0614 | 270.792 |
| 540 | 266.85 | 19479.8 | 2.7898 | 14990.0 | 1.609 | 238.266 | 1040 | 766.85 | 44946.3 | 148.707 | 36299.3 | .0581 | 271.324 |
| 550 | 276.85 | 19938.8 | 3.0872 | 15365.8 | 1.481 | 239.108 | 1050 | 776.85 | 45493.7 | 158.377 | 36763.5 | .0551 | 271.848 |
| 560 | 286.85 | 20400.5 | 3.4123 | 15744.4 | 1.365 | 239.940 | 1060 | 786.85 | 46043.3 | 168.676 | 37230.0 | .0522 | 272.372 |
| 570 | 296.85 | 20864.9 | 3.7666 | 16125.7 | 1.258 | 240.762 | 1070 | 796.85 | 46592.9 | 179.465 | 37696.5 | .0496 | 272.887 |
| 580 | 306.85 | 21331.9 | 4.1532 | 16509.5 | 1.161 | 241.574 | 1080 | 806.85 | 47144.8 | 190.944 | 38165.2 | .0470 | 273.403 |
| 590 | 316.85 | 21801.3 | 4.5744 | 16895.8 | 1.072 | 242.377 | 1090 | 816.85 | 47696.7 | 202.954 | 38634.0 | .0447 | 273.910 |

# Table 17  Carbon Dioxide at Low Pressures (for One Gram-Mole)

$\bar{m} = 44.0098$

| T | t | $\bar{h}$ | $p_r$ | $\bar{u}$ | $v_r$ | $\bar{\phi}$ | T | t | $\bar{h}$ | $p_r$ | $\bar{u}$ | $v_r$ | $\bar{\phi}$ |
|---|---|---|---|---|---|---|---|---|---|---|---|---|---|
| 1100 | 826.85 | 48250.8 | 215.72 | 39105.0 | .0424 | 274.417 | 1600 | 1326.85 | 76935.6 | 2846.9 | 63632.6 | .00467 | 295.868 |
| 1110 | 836.85 | 48805.0 | 229.06 | 39576.0 | .0403 | 274.916 | 1610 | 1336.85 | 77523.9 | 2974.9 | 64137.7 | .00450 | 296.234 |
| 1120 | 846.85 | 49361.4 | 243.22 | 40049.3 | .0383 | 275.415 | 1620 | 1346.85 | 78114.4 | 3108.8 | 64645.0 | .00433 | 296.600 |
| 1130 | 856.85 | 49917.9 | 258.00 | 40522.6 | .0364 | 275.905 | 1630 | 1356.85 | 78704.9 | 3248.6 | 65152.4 | .00417 | 296.966 |
| 1140 | 866.85 | 50474.3 | 273.69 | 40995.9 | .0346 | 276.396 | 1640 | 1366.85 | 79295.4 | 3391.3 | 65659.7 | .00402 | 297.323 |
| 1150 | 876.85 | 51033.0 | 290.32 | 41471.5 | .0329 | 276.886 | 1650 | 1376.85 | 79885.9 | 3540.3 | 66167.1 | .00387 | 297.681 |
| 1160 | 886.85 | 51591.7 | 307.65 | 41947.0 | .0313 | 277.368 | 1660 | 1386.85 | 80478.6 | 3695.9 | 66676.7 | .00373 | 298.038 |
| 1170 | 896.85 | 52152.7 | 326.03 | 42424.8 | .0298 | 277.851 | 1670 | 1396.85 | 81069.1 | 3858.3 | 67184.1 | .00360 | 298.396 |
| 1180 | 906.85 | 52715.9 | 345.15 | 42905.0 | .0284 | 278.325 | 1680 | 1406.85 | 81661.9 | 4027.8 | 67693.7 | .00347 | 298.753 |
| 1190 | 916.85 | 53276.9 | 365.39 | 43382.8 | .0271 | 278.799 | 1690 | 1416.85 | 82254.7 | 4200.6 | 68203.3 | .00335 | 299.102 |
| 1200 | 926.85 | 53840.2 | 386.83 | 43862.9 | .0258 | 279.272 | 1700 | 1426.85 | 82847.5 | 4380.8 | 68713.0 | .00323 | 299.452 |
| 1210 | 936.85 | 54403.4 | 409.11 | 44343.0 | .0246 | 279.738 | 1710 | 1436.85 | 83440.2 | 4568.7 | 69222.6 | .00311 | 299.801 |
| 1220 | 946.85 | 54968.9 | 433.10 | 44825.4 | .0234 | 280.212 | 1720 | 1446.85 | 84035.3 | 4764.6 | 69734.5 | .00300 | 300.150 |
| 1230 | 956.85 | 55534.5 | 458.05 | 45307.7 | .0223 | 280.678 | 1730 | 1456.85 | 84628.1 | 4964.1 | 70244.1 | .00290 | 300.491 |
| 1240 | 966.85 | 56100.0 | 483.95 | 45790.1 | .0213 | 281.135 | 1740 | 1466.85 | 85223.1 | 5171.8 | 70756.0 | .00280 | 300.832 |
| 1250 | 976.85 | 56667.8 | 511.31 | 46274.8 | .0203 | 281.592 | 1750 | 1476.85 | 85818.1 | 5388.3 | 71267.9 | .00270 | 301.173 |
| 1260 | 986.85 | 57235.6 | 539.68 | 46759.4 | .0194 | 282.041 | 1760 | 1486.85 | 86413.2 | 5613.8 | 71779.8 | .00261 | 301.514 |
| 1270 | 996.85 | 57805.6 | 570.20 | 47246.3 | .0185 | 282.498 | 1770 | 1496.85 | 87008.2 | 5842.9 | 72291.7 | .00252 | 301.846 |
| 1280 | 1006.85 | 58375.7 | 601.23 | 47733.2 | .0177 | 282.939 | 1780 | 1506.85 | 87605.5 | 6087.4 | 72805.9 | .00243 | 302.187 |
| 1290 | 1016.85 | 58945.7 | 633.96 | 48220.2 | .0169 | 283.380 | 1790 | 1516.85 | 88200.6 | 6335.8 | 73317.8 | .00235 | 302.520 |
| 1300 | 1026.85 | 59515.8 | 668.46 | 48707.1 | .0162 | 283.820 | 1800 | 1526.85 | 88797.9 | 6594.4 | 73832.0 | .00227 | 302.852 |
| 1310 | 1036.85 | 60088.1 | 704.85 | 49196.3 | .0155 | 284.261 | 1810 | 1536.85 | 89395.2 | 6863.5 | 74346.2 | .00219 | 303.185 |
| 1320 | 1046.85 | 60660.5 | 742.47 | 49685.5 | .0148 | 284.693 | 1820 | 1546.85 | 89992.5 | 7143.6 | 74860.3 | .00212 | 303.517 |
| 1330 | 1056.85 | 61232.8 | 782.88 | 50174.7 | .0141 | 285.134 | 1830 | 1556.85 | 90589.7 | 7427.8 | 75374.5 | .00205 | 303.842 |
| 1340 | 1066.85 | 61805.1 | 824.67 | 50663.8 | .0135 | 285.566 | 1840 | 1566.85 | 91189.4 | 7723.2 | 75890.9 | .00198 | 304.166 |
| 1350 | 1076.85 | 62379.7 | 867.82 | 51155.3 | .0129 | 285.991 | 1850 | 1576.85 | 91786.8 | 8030.3 | 76405.1 | .00192 | 304.490 |
| 1360 | 1086.85 | 62954.3 | 913.22 | 51646.8 | .0124 | 286.415 | 1860 | 1586.85 | 92386.3 | 8349.7 | 76921.6 | .00185 | 304.814 |
| 1370 | 1096.85 | 63531.2 | 961.00 | 52140.5 | .0119 | 286.839 | 1870 | 1596.85 | 92983.7 | 8681.8 | 77435.7 | .00179 | 305.139 |
| 1380 | 1106.85 | 64108.1 | 1010.28 | 52634.2 | .0114 | 287.254 | 1880 | 1606.85 | 93585.5 | 9027.0 | 77954.4 | .00173 | 305.463 |
| 1390 | 1116.85 | 64685.0 | 1062.07 | 53128.0 | .0109 | 287.670 | 1890 | 1616.85 | 94185.1 | 9376.7 | 78470.9 | .00168 | 305.779 |
| 1400 | 1126.85 | 65261.8 | 1116.53 | 53621.7 | .0104 | 288.086 | 1900 | 1626.85 | 94784.7 | 9739.8 | 78987.3 | .00162 | 306.095 |
| 1410 | 1136.85 | 65841.0 | 1172.60 | 54117.7 | .0100 | 288.493 | 1910 | 1636.85 | 95384.3 | 10117.1 | 79503.8 | .00157 | 306.411 |
| 1420 | 1146.85 | 66420.1 | 1232.72 | 54613.7 | .0096 | 288.909 | 1920 | 1646.85 | 95986.1 | 10508.9 | 80022.5 | .00152 | 306.727 |
| 1430 | 1156.85 | 66999.3 | 1293.33 | 55109.7 | .0092 | 289.308 | 1930 | 1656.85 | 96588.0 | 10905.0 | 80541.2 | .00147 | 307.034 |
| 1440 | 1166.85 | 67578.4 | 1358.29 | 55605.7 | .0088 | 289.715 | 1940 | 1666.85 | 97189.8 | 11327.4 | 81059.9 | .00142 | 307.350 |
| 1450 | 1176.85 | 68159.9 | 1425.07 | 56104.0 | .0085 | 290.114 | 1950 | 1676.85 | 97791.7 | 11754.4 | 81578.6 | .00138 | 307.658 |
| 1460 | 1186.85 | 68741.3 | 1495.15 | 56602.2 | .0081 | 290.514 | 1960 | 1686.85 | 98393.6 | 12197.4 | 82097.3 | .00134 | 307.965 |
| 1470 | 1196.85 | 69322.7 | 1568.66 | 57100.5 | .0078 | 290.913 | 1970 | 1696.85 | 98995.4 | 12644.5 | 82616.0 | .00130 | 308.265 |
| 1480 | 1206.85 | 69904.1 | 1645.80 | 57598.8 | .0075 | 291.312 | 1980 | 1706.85 | 99599.6 | 13121.1 | 83137.0 | .00125 | 308.572 |
| 1490 | 1216.85 | 70487.8 | 1724.99 | 58099.3 | .0072 | 291.703 | 1990 | 1716.85 | 100201.4 | 13615.7 | 83655.7 | .00122 | 308.880 |
| 1500 | 1226.85 | 71071.5 | 1808.00 | 58599.9 | .0069 | 292.093 | 2000 | 1726.85 | 100805.5 | 14114.8 | 84176.7 | .00118 | 309.179 |
| 1510 | 1236.85 | 71655.2 | 1895.01 | 59100.4 | .0066 | 292.484 | 2010 | 1736.85 | 101409.7 | 14632.2 | 84697.7 | .00114 | 309.479 |
| 1520 | 1246.85 | 72241.1 | 1984.22 | 59603.2 | .0064 | 292.867 | 2020 | 1746.85 | 102013.8 | 15168.5 | 85218.7 | .00111 | 309.778 |
| 1530 | 1256.85 | 72824.8 | 2077.62 | 60103.8 | .0061 | 293.249 | 2030 | 1756.85 | 102617.9 | 15724.6 | 85739.7 | .00107 | 310.077 |
| 1540 | 1266.85 | 73410.8 | 2175.42 | 60606.6 | .0059 | 293.631 | 2040 | 1766.85 | 103222.1 | 16301.0 | 86260.7 | .00104 | 310.377 |
| 1550 | 1276.85 | 73996.7 | 2277.83 | 61109.4 | .0057 | 294.014 | 2050 | 1776.85 | 103826.2 | 16881.6 | 86781.7 | .00101 | 310.668 |
| 1560 | 1286.85 | 74582.7 | 2380.29 | 61612.2 | .0054 | 294.380 | 2060 | 1786.85 | 104430.3 | 17500.4 | 87302.6 | .00098 | 310.967 |
| 1570 | 1296.85 | 75170.9 | 2489.85 | 62117.3 | .0052 | 294.754 | 2070 | 1796.85 | 105036.7 | 18123.8 | 87825.9 | .00095 | 311.258 |
| 1580 | 1306.85 | 75759.2 | 2604.46 | 62622.4 | .0050 | 295.128 | 2080 | 1806.85 | 105640.8 | 18769.3 | 88346.9 | .00092 | 311.549 |
| 1590 | 1316.85 | 76347.4 | 2724.33 | 63127.5 | .0049 | 295.502 | 2090 | 1816.85 | 106247.2 | 19437.9 | 88870.1 | .00089 | 311.840 |

# Table 17  Carbon Dioxide at Low Pressures (for One Gram-Mole)

$\bar{m} = 44.0098$

| T | t | $\bar{h}$ | $p_r$ | $\bar{u}$ | $v_r$ | $\bar{\phi}$ | T | t | $\bar{h}$ | $p_r$ | $\bar{u}$ | $v_r$ | $\bar{\phi}$ |
|---|---|---|---|---|---|---|---|---|---|---|---|---|---|
| 2100 | 1826.85 | 106853.6 | 20130 | 89393.4 | .000867 | 312.131 | 2600 | 2326.85 | 137437.1 | 96856 | 115819.7 | .000223 | 325.193 |
| 2110 | 1836.85 | 107460.0 | 20847 | 89916.7 | .000842 | 312.422 | 2610 | 2336.85 | 138052.6 | 99606 | 116352.0 | .000218 | 325.426 |
| 2120 | 1846.85 | 108066.4 | 21568 | 90439.9 | .000817 | 312.705 | 2620 | 2346.85 | 138670.4 | 102537 | 116886.6 | .000212 | 325.667 |
| 2130 | 1856.85 | 108672.8 | 22337 | 90963.2 | .000793 | 312.996 | 2630 | 2356.85 | 139288.1 | 105449 | 117421.2 | .000207 | 325.900 |
| 2140 | 1866.85 | 109281.5 | 23109 | 91488.7 | .000770 | 313.278 | 2640 | 2366.85 | 139903.6 | 108443 | 117953.6 | .000202 | 326.132 |
| 2150 | 1876.85 | 109887.9 | 23908 | 92012.0 | .000748 | 313.561 | 2650 | 2376.85 | 140521.4 | 111522 | 118488.2 | .000198 | 326.365 |
| 2160 | 1886.85 | 110496.6 | 24735 | 92537.5 | .000726 | 313.844 | 2660 | 2386.85 | 141139.1 | 114689 | 119022.8 | .000193 | 326.598 |
| 2170 | 1896.85 | 111105.3 | 25565 | 93063.0 | .000706 | 314.118 | 2670 | 2396.85 | 141756.9 | 117946 | 119557.4 | .000188 | 326.831 |
| 2180 | 1906.85 | 111711.7 | 26449 | 93586.3 | .000685 | 314.401 | 2680 | 2406.85 | 142374.6 | 121295 | 120092.0 | .000184 | 327.064 |
| 2190 | 1916.85 | 112320.3 | 27364 | 94111.8 | .000665 | 314.684 | 2690 | 2416.85 | 142992.4 | 124614 | 120626.6 | .000179 | 327.288 |
| 2200 | 1926.85 | 112929.0 | 28282 | 94637.3 | .000647 | 314.958 | 2700 | 2426.85 | 143610.1 | 128153 | 121161.2 | .000175 | 327.521 |
| 2210 | 1936.85 | 113537.7 | 29231 | 95162.8 | .000629 | 315.232 | 2710 | 2436.85 | 144227.9 | 131792 | 121695.9 | .000171 | 327.754 |
| 2220 | 1946.85 | 114146.3 | 30212 | 95688.4 | .000611 | 315.507 | 2720 | 2446.85 | 144845.6 | 135399 | 122230.5 | .000167 | 327.978 |
| 2230 | 1956.85 | 114755.0 | 31225 | 96213.9 | .000594 | 315.781 | 2730 | 2456.85 | 145465.7 | 139104 | 122767.4 | .000163 | 328.203 |
| 2240 | 1966.85 | 115363.7 | 32273 | 96739.4 | .000577 | 316.055 | 2740 | 2466.85 | 146083.4 | 142911 | 123302.0 | .000159 | 328.427 |
| 2250 | 1976.85 | 115974.6 | 33356 | 97267.2 | .000561 | 316.330 | 2750 | 2476.85 | 146701.2 | 146969 | 123836.6 | .000156 | 328.660 |
| 2260 | 1986.85 | 116583.3 | 34440 | 97792.8 | .000546 | 316.596 | 2760 | 2486.85 | 147321.2 | 150992 | 124373.5 | .000152 | 328.885 |
| 2270 | 1996.85 | 117194.2 | 35596 | 98320.6 | .000530 | 316.870 | 2770 | 2496.85 | 147939.0 | 155124 | 124908.1 | .000148 | 329.109 |
| 2280 | 2006.85 | 117802.9 | 36753 | 98846.1 | .000516 | 317.136 | 2780 | 2506.85 | 148559.0 | 159369 | 125445.0 | .000145 | 329.334 |
| 2290 | 2016.85 | 118413.9 | 37948 | 99373.9 | .000502 | 317.402 | 2790 | 2516.85 | 149176.8 | 163567 | 125979.6 | .000142 | 329.550 |
| 2300 | 2026.85 | 119024.8 | 39182 | 99901.7 | .000488 | 317.668 | 2800 | 2526.85 | 149796.8 | 168044 | 126516.5 | .000139 | 329.774 |
| 2310 | 2036.85 | 119635.7 | 40456 | 100429.5 | .000475 | 317.934 | 2810 | 2536.85 | 150416.8 | 172643 | 127053.3 | .000135 | 329.999 |
| 2320 | 2046.85 | 120246.7 | 41772 | 100957.3 | .000462 | 318.201 | 2820 | 2546.85 | 151036.8 | 177190 | 127590.2 | .000132 | 330.215 |
| 2330 | 2056.85 | 120857.6 | 43087 | 101485.1 | .000450 | 318.458 | 2830 | 2556.85 | 151656.9 | 182039 | 128127.1 | .000129 | 330.439 |
| 2340 | 2066.85 | 121468.6 | 44488 | 102012.9 | .000437 | 318.724 | 2840 | 2566.85 | 152276.9 | 186835 | 128664.0 | .000126 | 330.655 |
| 2350 | 2076.85 | 122081.8 | 45889 | 102542.9 | .000426 | 318.982 | 2850 | 2576.85 | 152896.9 | 191756 | 129200.9 | .000124 | 330.872 |
| 2360 | 2086.85 | 122692.7 | 47334 | 103070.7 | .000415 | 319.240 | 2860 | 2586.85 | 153516.9 | 196807 | 129737.8 | .000121 | 331.088 |
| 2370 | 2096.85 | 123305.9 | 48824 | 103600.8 | .000404 | 319.498 | 2870 | 2596.85 | 154137.0 | 201991 | 130274.6 | .000118 | 331.304 |
| 2380 | 2106.85 | 123916.9 | 50361 | 104128.6 | .000393 | 319.755 | 2880 | 2606.85 | 154757.0 | 207312 | 130811.5 | .000116 | 331.520 |
| 2390 | 2116.85 | 124530.1 | 51947 | 104658.7 | .000383 | 320.013 | 2890 | 2616.85 | 155379.3 | 212773 | 131350.7 | .000113 | 331.736 |
| 2400 | 2126.85 | 125143.3 | 53583 | 105188.7 | .000372 | 320.271 | 2900 | 2626.85 | 155999.3 | 218377 | 131887.6 | .000110 | 331.953 |
| 2410 | 2136.85 | 125756.5 | 55270 | 105718.8 | .000363 | 320.529 | 2910 | 2636.85 | 156619.4 | 224129 | 132424.4 | .000108 | 332.169 |
| 2420 | 2146.85 | 126369.7 | 56953 | 106248.9 | .000353 | 320.778 | 2920 | 2646.85 | 157241.7 | 229803 | 132963.6 | .000106 | 332.377 |
| 2430 | 2156.85 | 126983.0 | 58746 | 106779.0 | .000344 | 321.036 | 2930 | 2656.85 | 157861.7 | 235857 | 133500.5 | .000103 | 332.593 |
| 2440 | 2166.85 | 127596.2 | 60535 | 107309.0 | .000335 | 321.285 | 2940 | 2666.85 | 158484.0 | 241827 | 134039.6 | .000101 | 332.801 |
| 2450 | 2176.85 | 128211.7 | 62379 | 107841.4 | .000327 | 321.535 | 2950 | 2676.85 | 159106.3 | 248197 | 134578.8 | .000099 | 333.017 |
| 2460 | 2186.85 | 128824.9 | 64343 | 108371.4 | .000318 | 321.792 | 2960 | 2686.85 | 159726.3 | 254480 | 135115.7 | .000097 | 333.225 |
| 2470 | 2196.85 | 129438.1 | 66302 | 108901.5 | .000310 | 322.042 | 2970 | 2696.85 | 160348.6 | 260923 | 135654.8 | .000095 | 333.433 |
| 2480 | 2206.85 | 130053.6 | 68321 | 109433.8 | .000302 | 322.291 | 2980 | 2706.85 | 160970.9 | 267528 | 136194.0 | .000093 | 333.640 |
| 2490 | 2216.85 | 130666.8 | 70332 | 109963.9 | .000294 | 322.532 | 2990 | 2716.85 | 161593.2 | 274300 | 136733.1 | .000091 | 333.848 |
| 2500 | 2226.85 | 131282.3 | 72474 | 110496.3 | .000287 | 322.782 | 3000 | 2726.85 | 162215.5 | 281244 | 137272.3 | .000089 | 334.056 |
| 2510 | 2236.85 | 131897.8 | 74681 | 111028.6 | .000279 | 323.031 | 3010 | 2736.85 | 162837.8 | 288364 | 137811.4 | .000087 | 334.264 |
| 2520 | 2246.85 | 132511.0 | 76878 | 111558.7 | .000273 | 323.272 | 3020 | 2746.85 | 163460.1 | 295664 | 138350.6 | .000085 | 334.472 |
| 2530 | 2256.85 | 133126.5 | 79140 | 112091.0 | .000266 | 323.513 | 3030 | 2756.85 | 164082.4 | 302846 | 138889.8 | .000083 | 334.671 |
| 2540 | 2266.85 | 133741.9 | 81469 | 112623.4 | .000259 | 323.755 | 3040 | 2766.85 | 164707.0 | 310512 | 139431.2 | .000081 | 334.879 |
| 2550 | 2276.85 | 134357.4 | 83866 | 113155.7 | .000253 | 323.996 | 3050 | 2776.85 | 165329.3 | 318055 | 139970.3 | .000080 | 335.079 |
| 2560 | 2286.85 | 134972.9 | 86334 | 113688.0 | .000247 | 324.237 | 3060 | 2786.85 | 165951.6 | 326106 | 140509.5 | .000078 | 335.287 |
| 2570 | 2296.85 | 135588.4 | 88874 | 114220.4 | .000240 | 324.478 | 3070 | 2796.85 | 166576.1 | 334028 | 141050.9 | .000076 | 335.486 |
| 2580 | 2306.85 | 136203.9 | 91489 | 114752.7 | .000234 | 324.719 | 3080 | 2806.85 | 167198.4 | 342141 | 141590.1 | .000075 | 335.686 |
| 2590 | 2316.85 | 136821.6 | 94087 | 115287.3 | .000229 | 324.952 | 3090 | 2816.85 | 167820.7 | 350803 | 142129.2 | .000073 | 335.894 |

# Table 18    Carbon Dioxide

$\bar{m} = 44.0098$

| T K | t C | $\bar{c}_p$ J / g-mol K | $\bar{c}_v$ J / g-mol K | $k = \dfrac{\bar{c}_p}{\bar{c}_v}$ | a m / s |
|---|---|---|---|---|---|
| 100 | −173.15 | 29.207 | 20.892 | 1.398 | 162.5 |
| 120 | −153.15 | 29.460 | 21.145 | 1.393 | 177.7 |
| 140 | −133.15 | 29.928 | 21.613 | 1.385 | 191.4 |
| 160 | −113.15 | 30.600 | 22.286 | 1.373 | 203.7 |
| 180 | −93.15 | 31.428 | 23.114 | 1.360 | 215.0 |
| 200 | −73.15 | 32.356 | 24.042 | 1.346 | 225.5 |
| 250 | −23.15 | 34.831 | 26.516 | 1.314 | 249.1 |
| 300 | 26.85 | 37.218 | 28.903 | 1.288 | 270.1 |
| 350 | 76.85 | 39.386 | 31.072 | 1.268 | 289.5 |
| 400 | 126.85 | 41.326 | 33.012 | 1.252 | 307.6 |
| 450 | 176.85 | 43.062 | 34.748 | 1.239 | 324.6 |
| 500 | 226.85 | 44.624 | 36.310 | 1.229 | 340.7 |
| 550 | 276.85 | 46.038 | 37.723 | 1.220 | 356.1 |
| 600 | 326.85 | 47.321 | 39.007 | 1.213 | 370.8 |
| 650 | 376.85 | 48.493 | 40.179 | 1.207 | 385.0 |
| 700 | 426.85 | 49.563 | 41.249 | 1.202 | 398.6 |
| 750 | 476.85 | 50.540 | 42.226 | 1.197 | 411.8 |
| 800 | 526.85 | 51.433 | 43.118 | 1.193 | 424.6 |
| 850 | 576.85 | 52.250 | 43.936 | 1.189 | 437.0 |
| 900 | 626.85 | 52.998 | 44.683 | 1.186 | 449.1 |
| 950 | 676.85 | 53.682 | 45.368 | 1.183 | 460.8 |
| 1000 | 726.85 | 54.308 | 45.994 | 1.181 | 472.3 |
| 1050 | 776.85 | 54.883 | 46.569 | 1.179 | 483.5 |
| 1100 | 826.85 | 55.407 | 47.093 | 1.177 | 494.5 |
| 1150 | 876.85 | 55.898 | 47.583 | 1.175 | 505.2 |
| 1200 | 926.85 | 56.338 | 48.024 | 1.173 | 515.7 |
| 1300 | 1026.85 | 57.137 | 48.822 | 1.170 | 536.1 |
| 1400 | 1126.85 | 57.802 | 49.487 | 1.168 | 555.8 |
| 1500 | 1226.85 | 58.375 | 50.061 | 1.166 | 574.8 |
| 1600 | 1326.85 | 58.883 | 50.568 | 1.164 | 593.3 |
| 1700 | 1426.85 | 59.315 | 51.001 | 1.163 | 611.2 |
| 1800 | 1526.85 | 59.697 | 51.383 | 1.162 | 628.6 |
| 1900 | 1626.85 | 60.047 | 51.732 | 1.161 | 645.5 |
| 2000 | 1726.85 | 60.346 | 52.032 | 1.160 | 662.0 |
| 2100 | 1826.85 | 60.620 | 52.306 | 1.159 | 678.1 |
| 2200 | 1926.85 | 60.861 | 52.547 | 1.158 | 693.8 |
| 2300 | 2026.85 | 61.086 | 52.772 | 1.158 | 709.2 |
| 2400 | 2126.85 | 61.285 | 52.971 | 1.157 | 724.3 |
| 2500 | 2226.85 | 61.468 | 53.154 | 1.156 | 739.0 |
| 2600 | 2326.85 | 61.643 | 53.329 | 1.156 | 753.5 |
| 2700 | 2426.85 | 61.801 | 53.487 | 1.155 | 767.7 |
| 2800 | 2526.85 | 61.951 | 53.636 | 1.155 | 781.7 |
| 2900 | 2626.85 | 62.092 | 53.778 | 1.155 | 795.3 |
| 3000 | 2726.85 | 62.225 | 53.911 | 1.154 | 808.8 |
| 3100 | 2826.85 | 62.350 | 54.035 | 1.154 | 822.1 |
| 3200 | 2926.85 | 62.466 | 54.152 | 1.154 | 835.1 |
| 3300 | 3026.85 | 62.574 | 54.260 | 1.153 | 847.9 |
| 3400 | 3126.85 | 62.682 | 54.368 | 1.153 | 860.6 |
| 3500 | 3226.85 | 62.782 | 54.468 | 1.153 | 873.0 |
| 3600 | 3326.85 | 931.851 | 923.536 | 1.000 | 824.8 |

# Table 19    Hydrogen at Low Pressures (for One Gram-Mole)

$\bar{m} = 2.0158$

| T | t | $\bar{h}$ | $p_r$ | $\bar{u}$ | $v_r$ | $\bar{\phi}$ | T | t | $\bar{h}$ | $p_r$ | $\bar{u}$ | $v_r$ | $\bar{\phi}$ |
|---|---|---|---|---|---|---|---|---|---|---|---|---|---|
| 100 | −173.15 | 3175.3 | .2136 | 2343.9 | 3.892 | 102.034 | 600 | 326.85 | 17278.3 | 76.899 | 12289.7 | .0649 | 150.973 |
| 110 | −163.15 | 3404.0 | .2779 | 2489.4 | 3.291 | 104.221 | 610 | 336.85 | 17571.5 | 81.572 | 12499.7 | .0622 | 151.463 |
| 120 | −153.15 | 3638.6 | .3550 | 2640.9 | 2.810 | 106.258 | 620 | 346.85 | 17865.0 | 86.357 | 12710.0 | .0597 | 151.937 |
| 130 | −143.15 | 3878.9 | .4473 | 2798.0 | 2.416 | 108.179 | 630 | 356.85 | 18158.4 | 91.331 | 12920.3 | .0574 | 152.403 |
| 140 | −133.15 | 4125.1 | .5574 | 2961.1 | 2.088 | 110.008 | 640 | 366.85 | 18451.8 | 96.591 | 13130.6 | .0551 | 152.869 |
| 150 | −123.15 | 4376.5 | .6863 | 3129.4 | 1.817 | 111.737 | 650 | 376.85 | 18745.5 | 102.053 | 13341.1 | .0530 | 153.326 |
| 160 | −113.15 | 4632.7 | .8374 | 3302.4 | 1.589 | 113.392 | 660 | 386.85 | 19039.4 | 107.607 | 13551.9 | .0510 | 153.767 |
| 170 | −103.15 | 4893.2 | 1.0126 | 3479.8 | 1.396 | 114.972 | 670 | 396.85 | 19333.3 | 113.578 | 13762.6 | .0490 | 154.216 |
| 180 | −93.15 | 5157.8 | 1.2147 | 3661.2 | 1.232 | 116.485 | 680 | 406.85 | 19627.4 | 119.640 | 13973.6 | .0473 | 154.648 |
| 190 | −83.15 | 5425.8 | 1.4470 | 3846.1 | 1.092 | 117.940 | 690 | 416.85 | 19921.7 | 126.026 | 14184.8 | .0455 | 155.080 |
| 200 | −73.15 | 5697.2 | 1.7100 | 4034.3 | .972 | 119.328 | 700 | 426.85 | 20216.1 | 132.620 | 14396.0 | .0439 | 155.504 |
| 210 | −63.15 | 5971.1 | 2.0087 | 4225.1 | .869 | 120.667 | 710 | 436.85 | 20510.6 | 139.420 | 14607.4 | .0423 | 155.920 |
| 220 | −53.15 | 6247.7 | 2.3431 | 4418.6 | .781 | 121.947 | 720 | 446.85 | 20805.2 | 146.421 | 14818.8 | .0409 | 156.327 |
| 230 | −43.15 | 6526.6 | 2.7196 | 4614.3 | .703 | 123.186 | 730 | 456.85 | 21100.0 | 153.775 | 15030.5 | .0395 | 156.735 |
| 240 | −33.15 | 6807.4 | 3.1408 | 4811.9 | .635 | 124.383 | 740 | 466.85 | 21395.0 | 161.497 | 15242.4 | .0381 | 157.142 |
| 250 | −23.15 | 7089.9 | 3.6092 | 5011.3 | .576 | 125.539 | 750 | 476.85 | 21690.3 | 169.269 | 15454.5 | .0368 | 157.533 |
| 260 | −13.15 | 7373.8 | 4.1226 | 5212.0 | .524 | 126.645 | 760 | 486.85 | 21985.5 | 177.415 | 15666.6 | .0356 | 157.924 |
| 270 | −3.15 | 7659.0 | 4.6950 | 5414.2 | .478 | 127.726 | 770 | 496.85 | 22281.0 | 185.952 | 15878.9 | .0344 | 158.315 |
| 280 | 6.85 | 7945.2 | 5.3254 | 5617.2 | .437 | 128.773 | 780 | 506.85 | 22576.9 | 194.706 | 16091.7 | .0333 | 158.697 |
| 290 | 16.85 | 8232.5 | 6.0104 | 5821.3 | .401 | 129.780 | 790 | 516.85 | 22872.9 | 203.668 | 16304.5 | .0323 | 159.071 |
| 300 | 26.85 | 8520.7 | 6.7564 | 6026.4 | .369 | 130.752 | 800 | 526.85 | 23168.1 | 213.042 | 16516.6 | .0312 | 159.445 |
| 310 | 36.85 | 8809.4 | 7.5722 | 6231.9 | .340 | 131.700 | 810 | 536.85 | 23463.4 | 222.625 | 16728.7 | .0303 | 159.811 |
| 320 | 46.85 | 9099.0 | 8.4612 | 6438.3 | .314 | 132.623 | 820 | 546.85 | 23760.9 | 232.639 | 16943.1 | .0293 | 160.177 |
| 330 | 56.85 | 9388.8 | 9.4167 | 6645.0 | .291 | 133.513 | 830 | 556.85 | 24058.4 | 242.861 | 17157.5 | .0284 | 160.534 |
| 340 | 66.85 | 9679.0 | 10.4488 | 6852.1 | .271 | 134.377 | 840 | 566.85 | 24355.9 | 253.532 | 17371.8 | .0275 | 160.892 |
| 350 | 76.85 | 9969.7 | 11.5709 | 7059.7 | .251 | 135.225 | 850 | 576.85 | 24653.5 | 264.407 | 17586.2 | .0267 | 161.241 |
| 360 | 86.85 | 10260.7 | 12.7622 | 7267.5 | .235 | 136.040 | 860 | 586.85 | 24951.0 | 275.749 | 17800.6 | .0259 | 161.590 |
| 370 | 96.85 | 10551.8 | 14.0481 | 7475.5 | .219 | 136.838 | 870 | 596.85 | 25248.5 | 287.289 | 18015.0 | .0252 | 161.931 |
| 380 | 106.85 | 10843.2 | 15.4327 | 7683.7 | .205 | 137.620 | 880 | 606.85 | 25546.0 | 299.313 | 18229.4 | .0244 | 162.272 |
| 390 | 116.85 | 11134.8 | 16.9030 | 7892.2 | .192 | 138.377 | 890 | 616.85 | 25843.6 | 311.840 | 18443.7 | .0237 | 162.613 |
| 400 | 126.85 | 11426.7 | 18.4763 | 8100.9 | .180 | 139.117 | 900 | 626.85 | 26143.3 | 324.566 | 18660.4 | .0231 | 162.946 |
| 410 | 136.85 | 11718.5 | 20.1557 | 8309.6 | .169 | 139.840 | 910 | 636.85 | 26443.1 | 337.812 | 18877.0 | .0224 | 163.278 |
| 420 | 146.85 | 12010.6 | 21.9220 | 8518.5 | .159 | 140.538 | 920 | 646.85 | 26742.9 | 351.247 | 19093.7 | .0218 | 163.603 |
| 430 | 156.85 | 12302.7 | 23.8191 | 8727.5 | .150 | 141.228 | 930 | 656.85 | 27042.7 | 365.216 | 19310.3 | .0212 | 163.927 |
| 440 | 166.85 | 12594.7 | 25.8030 | 8936.4 | .142 | 141.894 | 940 | 666.85 | 27342.5 | 379.741 | 19527.0 | .0206 | 164.251 |
| 450 | 176.85 | 12887.0 | 27.9241 | 9145.5 | .134 | 142.550 | 950 | 676.85 | 27642.3 | 394.449 | 19743.6 | .0200 | 164.567 |
| 460 | 186.85 | 13179.3 | 30.1894 | 9354.7 | .127 | 143.199 | 960 | 686.85 | 27942.1 | 409.726 | 19960.3 | .0195 | 164.883 |
| 470 | 196.85 | 13471.6 | 32.5406 | 9563.9 | .120 | 143.823 | 970 | 696.85 | 28244.2 | 425.170 | 20179.2 | .0190 | 165.191 |
| 480 | 206.85 | 13764.2 | 35.0400 | 9773.2 | .114 | 144.438 | 980 | 706.85 | 28544.0 | 441.196 | 20395.9 | .0185 | 165.498 |
| 490 | 216.85 | 14056.7 | 37.6936 | 9982.6 | .108 | 145.045 | 990 | 716.85 | 28846.0 | 457.826 | 20614.8 | .0180 | 165.806 |
| 500 | 226.85 | 14349.2 | 40.4671 | 10192.0 | .103 | 145.635 | 1000 | 726.85 | 29148.1 | 474.608 | 20833.7 | .0175 | 166.105 |
| 510 | 236.85 | 14641.7 | 43.4013 | 10401.4 | .098 | 146.217 | 1010 | 736.85 | 29450.2 | 492.005 | 21052.6 | .0171 | 166.404 |
| 520 | 246.85 | 14934.5 | 46.5017 | 10611.0 | .093 | 146.791 | 1020 | 746.85 | 29752.2 | 510.040 | 21271.5 | .0166 | 166.704 |
| 530 | 256.85 | 15227.2 | 49.6744 | 10820.6 | .089 | 147.340 | 1030 | 756.85 | 30056.6 | 528.736 | 21492.7 | .0162 | 167.003 |
| 540 | 266.85 | 15520.0 | 53.0635 | 11030.2 | .085 | 147.888 | 1040 | 766.85 | 30358.6 | 547.570 | 21711.6 | .0158 | 167.294 |
| 550 | 276.85 | 15813.0 | 56.6272 | 11240.0 | .081 | 148.429 | 1050 | 776.85 | 30663.0 | 567.074 | 21932.8 | .0154 | 167.585 |
| 560 | 286.85 | 16105.9 | 60.3095 | 11449.9 | .077 | 148.953 | 1060 | 786.85 | 30967.3 | 587.273 | 22154.0 | .0150 | 167.876 |
| 570 | 296.85 | 16398.9 | 64.2312 | 11659.7 | .074 | 149.476 | 1070 | 796.85 | 31271.6 | 607.583 | 22375.2 | .0146 | 168.159 |
| 580 | 306.85 | 16691.9 | 68.2031 | 11869.6 | .071 | 149.975 | 1080 | 806.85 | 31576.0 | 628.596 | 22596.4 | .0143 | 168.441 |
| 590 | 316.85 | 16985.1 | 72.4930 | 12079.6 | .068 | 150.482 | 1090 | 816.85 | 31880.3 | 650.336 | 22817.6 | .0139 | 168.724 |

# Table 19   Hydrogen at Low Pressures (for One Gram-Mole)

$\bar{m} = 2.0158$

| T | t | $\bar{h}$ | $p_r$ | $\bar{u}$ | $v_r$ | $\bar{\phi}$ | T | t | $\bar{h}$ | $p_r$ | $\bar{u}$ | $v_r$ | $\bar{\phi}$ |
|---|---|---|---|---|---|---|---|---|---|---|---|---|---|
| 1100 | 826.85 | 32186.9 | 672.83 | 23041.1 | .01359 | 169.007 | 1600 | 1326.85 | 48010.1 | 2791.9 | 34707.1 | .00476 | 180.838 |
| 1110 | 836.85 | 32493.5 | 695.40 | 23264.5 | .01327 | 169.281 | 1610 | 1336.85 | 48337.1 | 2859.7 | 34951.0 | .00468 | 181.038 |
| 1120 | 846.85 | 32800.1 | 718.73 | 23488.0 | .01296 | 169.556 | 1620 | 1346.85 | 48666.5 | 2932.1 | 35197.1 | .00459 | 181.246 |
| 1130 | 856.85 | 33104.5 | 742.85 | 23709.2 | .01265 | 169.830 | 1630 | 1356.85 | 48993.5 | 3003.4 | 35441.0 | .00451 | 181.445 |
| 1140 | 866.85 | 33413.3 | 767.77 | 23934.9 | .01235 | 170.104 | 1640 | 1366.85 | 49322.8 | 3076.3 | 35687.2 | .00443 | 181.645 |
| 1150 | 876.85 | 33719.9 | 792.74 | 24158.4 | .01206 | 170.370 | 1650 | 1376.85 | 49652.2 | 3151.0 | 35933.4 | .00435 | 181.844 |
| 1160 | 886.85 | 34028.8 | 818.51 | 24384.1 | .01178 | 170.636 | 1660 | 1386.85 | 49981.5 | 3227.6 | 36179.6 | .00428 | 182.044 |
| 1170 | 896.85 | 34335.4 | 845.13 | 24607.6 | .01151 | 170.903 | 1670 | 1396.85 | 50310.8 | 3306.0 | 36425.7 | .00420 | 182.243 |
| 1180 | 906.85 | 34646.6 | 872.61 | 24835.6 | .01124 | 171.169 | 1680 | 1406.85 | 50642.4 | 3386.3 | 36674.2 | .00412 | 182.443 |
| 1190 | 916.85 | 34955.5 | 900.09 | 25061.3 | .01099 | 171.426 | 1690 | 1416.85 | 50971.7 | 3465.1 | 36920.4 | .00406 | 182.634 |
| 1200 | 926.85 | 35264.3 | 928.42 | 25287.1 | .01075 | 171.684 | 1700 | 1426.85 | 51303.3 | 3549.2 | 37168.8 | .00398 | 182.834 |
| 1210 | 936.85 | 35573.2 | 957.66 | 25512.8 | .01051 | 171.942 | 1710 | 1436.85 | 51634.9 | 3631.8 | 37417.2 | .00391 | 183.025 |
| 1220 | 946.85 | 35884.4 | 986.82 | 25740.8 | .01028 | 172.191 | 1720 | 1446.85 | 51966.5 | 3720.0 | 37665.7 | .00384 | 183.224 |
| 1230 | 956.85 | 36195.5 | 1017.89 | 25968.8 | .01005 | 172.449 | 1730 | 1456.85 | 52298.1 | 3806.6 | 37914.1 | .00378 | 183.416 |
| 1240 | 966.85 | 36506.7 | 1048.89 | 26196.8 | .00983 | 172.698 | 1740 | 1466.85 | 52629.6 | 3895.2 | 38162.6 | .00371 | 183.607 |
| 1250 | 976.85 | 36817.8 | 1080.84 | 26424.8 | .00962 | 172.948 | 1750 | 1476.85 | 52963.5 | 3985.8 | 38413.3 | .00365 | 183.798 |
| 1260 | 986.85 | 37131.2 | 1113.75 | 26655.1 | .00941 | 173.197 | 1760 | 1486.85 | 53297.4 | 4078.5 | 38664.0 | .00359 | 183.989 |
| 1270 | 996.85 | 37442.4 | 1147.67 | 26883.1 | .00920 | 173.447 | 1770 | 1496.85 | 53631.2 | 4173.4 | 38914.7 | .00353 | 184.181 |
| 1280 | 1006.85 | 37758.1 | 1181.44 | 27115.6 | .00901 | 173.688 | 1780 | 1506.85 | 53965.1 | 4266.2 | 39165.5 | .00347 | 184.364 |
| 1290 | 1016.85 | 38071.5 | 1217.42 | 27345.9 | .00881 | 173.937 | 1790 | 1516.85 | 54298.9 | 4365.5 | 39416.2 | .00341 | 184.555 |
| 1300 | 1026.85 | 38384.9 | 1253.24 | 27576.2 | .00862 | 174.178 | 1800 | 1526.85 | 54635.1 | 4462.6 | 39669.2 | .00335 | 184.738 |
| 1310 | 1036.85 | 38698.3 | 1290.12 | 27806.5 | .00844 | 174.420 | 1810 | 1536.85 | 54971.2 | 4561.9 | 39922.1 | .00330 | 184.921 |
| 1320 | 1046.85 | 39014.0 | 1328.08 | 28039.0 | .00826 | 174.661 | 1820 | 1546.85 | 55307.3 | 4668.0 | 40175.1 | .00324 | 185.112 |
| 1330 | 1056.85 | 39329.7 | 1367.16 | 28271.6 | .00809 | 174.902 | 1830 | 1556.85 | 55643.5 | 4771.9 | 40428.1 | .00319 | 185.295 |
| 1340 | 1066.85 | 39645.4 | 1407.39 | 28504.1 | .00792 | 175.143 | 1840 | 1566.85 | 55981.9 | 4878.0 | 40683.4 | .00314 | 185.478 |
| 1350 | 1076.85 | 39961.1 | 1447.35 | 28736.7 | .00776 | 175.376 | 1850 | 1576.85 | 56318.0 | 4986.5 | 40936.4 | .00308 | 185.661 |
| 1360 | 1086.85 | 40276.8 | 1488.45 | 28969.2 | .00760 | 175.609 | 1860 | 1586.85 | 56656.4 | 5097.4 | 41191.6 | .00303 | 185.844 |
| 1370 | 1096.85 | 40594.8 | 1530.71 | 29204.0 | .00744 | 175.841 | 1870 | 1596.85 | 56992.5 | 5205.6 | 41444.6 | .00299 | 186.018 |
| 1380 | 1106.85 | 40912.7 | 1574.18 | 29438.8 | .00729 | 176.074 | 1880 | 1606.85 | 57331.0 | 5321.4 | 41699.9 | .00294 | 186.201 |
| 1390 | 1116.85 | 41230.7 | 1618.88 | 29673.7 | .00714 | 176.307 | 1890 | 1616.85 | 57669.4 | 5434.3 | 41955.1 | .00289 | 186.376 |
| 1400 | 1126.85 | 41548.6 | 1663.18 | 29908.5 | .00700 | 176.531 | 1900 | 1626.85 | 58007.8 | 5555.2 | 42210.4 | .00284 | 186.559 |
| 1410 | 1136.85 | 41866.6 | 1708.70 | 30143.3 | .00686 | 176.756 | 1910 | 1636.85 | 58346.2 | 5678.8 | 42465.7 | .00280 | 186.741 |
| 1420 | 1146.85 | 42186.8 | 1755.47 | 30380.4 | .00673 | 176.980 | 1920 | 1646.85 | 58686.8 | 5799.3 | 42723.2 | .00275 | 186.916 |
| 1430 | 1156.85 | 42507.1 | 1803.51 | 30617.5 | .00659 | 177.205 | 1930 | 1656.85 | 59027.5 | 5928.3 | 42980.7 | .00271 | 187.099 |
| 1440 | 1166.85 | 42827.3 | 1851.02 | 30854.6 | .00647 | 177.421 | 1940 | 1666.85 | 59368.2 | 6054.1 | 43238.2 | .00266 | 187.274 |
| 1450 | 1176.85 | 43147.5 | 1901.67 | 31091.7 | .00634 | 177.646 | 1950 | 1676.85 | 59708.9 | 6182.6 | 43495.8 | .00262 | 187.448 |
| 1460 | 1186.85 | 43467.8 | 1953.72 | 31328.7 | .00621 | 177.870 | 1960 | 1686.85 | 60049.5 | 6313.8 | 43753.3 | .00258 | 187.623 |
| 1470 | 1196.85 | 43790.3 | 2005.18 | 31568.1 | .00610 | 178.086 | 1970 | 1696.85 | 60392.5 | 6447.8 | 44013.1 | .00254 | 187.797 |
| 1480 | 1206.85 | 44112.8 | 2060.06 | 31807.5 | .00597 | 178.311 | 1980 | 1706.85 | 60733.2 | 6584.6 | 44270.6 | .00250 | 187.972 |
| 1490 | 1216.85 | 44435.3 | 2114.32 | 32046.8 | .00586 | 178.527 | 1990 | 1716.85 | 61076.1 | 6717.6 | 44530.4 | .00246 | 188.138 |
| 1500 | 1226.85 | 44757.8 | 2170.02 | 32286.2 | .00575 | 178.743 | 2000 | 1726.85 | 61419.0 | 6860.2 | 44790.2 | .00242 | 188.313 |
| 1510 | 1236.85 | 45080.3 | 2227.18 | 32525.6 | .00564 | 178.959 | 2010 | 1736.85 | 61762.0 | 7005.8 | 45050.0 | .00239 | 188.488 |
| 1520 | 1246.85 | 45405.1 | 2285.84 | 32767.2 | .00553 | 179.175 | 2020 | 1746.85 | 62104.9 | 7147.3 | 45309.8 | .00235 | 188.654 |
| 1530 | 1256.85 | 45729.9 | 2343.71 | 33008.8 | .00543 | 179.383 | 2030 | 1756.85 | 62450.2 | 7291.7 | 45571.9 | .00231 | 188.820 |
| 1540 | 1266.85 | 46054.6 | 2405.44 | 33250.4 | .00532 | 179.599 | 2040 | 1766.85 | 62793.1 | 7446.4 | 45831.7 | .00228 | 188.995 |
| 1550 | 1276.85 | 46379.4 | 2466.34 | 33492.1 | .00523 | 179.807 | 2050 | 1776.85 | 63138.3 | 7596.9 | 46093.8 | .00224 | 189.161 |
| 1560 | 1286.85 | 46704.2 | 2528.77 | 33733.7 | .00513 | 180.015 | 2060 | 1786.85 | 63483.5 | 7750.3 | 46355.9 | .00221 | 189.327 |
| 1570 | 1296.85 | 47031.2 | 2592.79 | 33977.6 | .00503 | 180.223 | 2070 | 1796.85 | 63828.7 | 7906.9 | 46617.9 | .00218 | 189.494 |
| 1580 | 1306.85 | 47356.0 | 2658.43 | 34219.2 | .00494 | 180.431 | 2080 | 1806.85 | 64171.7 | 8066.6 | 46877.7 | .00214 | 189.660 |
| 1590 | 1316.85 | 47683.1 | 2723.00 | 34463.2 | .00485 | 180.630 | 2090 | 1816.85 | 64519.2 | 8229.6 | 47142.1 | .00211 | 189.826 |

## Table 19   Hydrogen at Low Pressures (for One Gram-Mole)
$\bar{m} = 2.0158$

| T | t | $\bar{h}$ | $p_r$ | $\bar{u}$ | $v_r$ | $\bar{\phi}$ | T | t | $\bar{h}$ | $p_r$ | $\bar{u}$ | $v_r$ | $\bar{\phi}$ |
|---|---|---|---|---|---|---|---|---|---|---|---|---|---|
| 2100 | 1826.85 | 64864.4 | 8396 | 47404.1 | .002080 | 189.992 | 2600 | 2326.85 | 82563.6 | 20837 | 60946.1 | .001037 | 197.550 |
| 2110 | 1836.85 | 65211.9 | 8565 | 47668.5 | .002048 | 190.159 | 2610 | 2336.85 | 82924.7 | 21194 | 61224.1 | .001024 | 197.692 |
| 2120 | 1846.85 | 65557.1 | 8730 | 47930.6 | .002019 | 190.317 | 2620 | 2346.85 | 83285.8 | 21536 | 61502.1 | .001011 | 197.825 |
| 2130 | 1856.85 | 65906.9 | 8906 | 48197.2 | .001988 | 190.483 | 2630 | 2356.85 | 83646.9 | 21905 | 61780.0 | .000998 | 197.966 |
| 2140 | 1866.85 | 66254.3 | 9077 | 48461.5 | .001960 | 190.641 | 2640 | 2366.85 | 84010.3 | 22281 | 62060.3 | .000985 | 198.107 |
| 2150 | 1876.85 | 66601.8 | 9260 | 48725.9 | .001930 | 190.807 | 2650 | 2376.85 | 84371.4 | 22640 | 62338.2 | .000973 | 198.240 |
| 2160 | 1886.85 | 66949.3 | 9447 | 48990.2 | .001901 | 190.974 | 2660 | 2386.85 | 84734.8 | 23006 | 62618.5 | .000961 | 198.373 |
| 2170 | 1896.85 | 67299.1 | 9629 | 49256.8 | .001874 | 191.131 | 2670 | 2396.85 | 85095.9 | 23400 | 62896.5 | .000949 | 198.515 |
| 2180 | 1906.85 | 67646.6 | 9823 | 49521.2 | .001845 | 191.298 | 2680 | 2406.85 | 85459.3 | 23754 | 63176.7 | .000938 | 198.639 |
| 2190 | 1916.85 | 67994.1 | 10011 | 49785.5 | .001819 | 191.456 | 2690 | 2416.85 | 85822.7 | 24161 | 63456.9 | .000926 | 198.781 |
| 2200 | 1926.85 | 68343.8 | 10204 | 50052.1 | .001793 | 191.614 | 2700 | 2426.85 | 86186.1 | 24551 | 63737.2 | .000914 | 198.914 |
| 2210 | 1936.85 | 68693.6 | 10399 | 50318.7 | .001767 | 191.772 | 2710 | 2436.85 | 86549.5 | 24947 | 64017.4 | .000903 | 199.047 |
| 2220 | 1946.85 | 69043.3 | 10599 | 50585.4 | .001742 | 191.930 | 2720 | 2446.85 | 86912.8 | 25374 | 64297.7 | .000891 | 199.188 |
| 2230 | 1956.85 | 69395.4 | 10802 | 50854.2 | .001716 | 192.088 | 2730 | 2456.85 | 87278.5 | 25783 | 64580.2 | .000880 | 199.321 |
| 2240 | 1966.85 | 69745.1 | 10998 | 51120.9 | .001693 | 192.237 | 2740 | 2466.85 | 87641.9 | 26199 | 64860.4 | .000870 | 199.454 |
| 2250 | 1976.85 | 70097.2 | 11209 | 51389.7 | .001669 | 192.395 | 2750 | 2476.85 | 88007.5 | 26622 | 65142.9 | .000859 | 199.587 |
| 2260 | 1986.85 | 70449.2 | 11424 | 51658.6 | .001645 | 192.553 | 2760 | 2486.85 | 88373.2 | 27051 | 65425.4 | .000848 | 199.720 |
| 2270 | 1996.85 | 70798.9 | 11632 | 51925.2 | .001623 | 192.703 | 2770 | 2496.85 | 88736.6 | 27488 | 65705.7 | .000838 | 199.853 |
| 2280 | 2006.85 | 71151.0 | 11855 | 52194.1 | .001599 | 192.861 | 2780 | 2506.85 | 89102.2 | 27931 | 65988.2 | .000828 | 199.986 |
| 2290 | 2016.85 | 71503.0 | 12070 | 52463.0 | .001577 | 193.011 | 2790 | 2516.85 | 89467.9 | 28353 | 66270.7 | .000818 | 200.111 |
| 2300 | 2026.85 | 71855.0 | 12302 | 52731.9 | .001555 | 193.169 | 2800 | 2526.85 | 89833.6 | 28810 | 66553.2 | .000808 | 200.244 |
| 2310 | 2036.85 | 72207.1 | 12538 | 53000.8 | .001532 | 193.326 | 2810 | 2536.85 | 90199.2 | 29275 | 66835.7 | .000798 | 200.377 |
| 2320 | 2046.85 | 72561.4 | 12765 | 53271.9 | .001511 | 193.476 | 2820 | 2546.85 | 90564.9 | 29717 | 67118.3 | .000789 | 200.502 |
| 2330 | 2056.85 | 72913.4 | 12997 | 53540.8 | .001491 | 193.626 | 2830 | 2556.85 | 90932.8 | 30197 | 67403.0 | .000779 | 200.635 |
| 2340 | 2066.85 | 73267.7 | 13247 | 53812.0 | .001469 | 193.784 | 2840 | 2566.85 | 91298.5 | 30653 | 67685.5 | .000770 | 200.760 |
| 2350 | 2076.85 | 73622.0 | 13487 | 54083.2 | .001449 | 193.933 | 2850 | 2576.85 | 91666.4 | 31148 | 67970.3 | .000761 | 200.893 |
| 2360 | 2086.85 | 73976.3 | 13732 | 54354.3 | .001429 | 194.083 | 2860 | 2586.85 | 92034.3 | 31618 | 68255.1 | .000752 | 201.017 |
| 2370 | 2096.85 | 74330.6 | 13981 | 54625.5 | .001409 | 194.233 | 2870 | 2596.85 | 92402.2 | 32128 | 68539.9 | .000743 | 201.150 |
| 2380 | 2106.85 | 74684.9 | 14235 | 54896.6 | .001390 | 194.382 | 2880 | 2606.85 | 92770.2 | 32614 | 68824.7 | .000734 | 201.275 |
| 2390 | 2116.85 | 75039.2 | 14494 | 55167.8 | .001371 | 194.532 | 2890 | 2616.85 | 93138.1 | 33140 | 69109.5 | .000725 | 201.408 |
| 2400 | 2126.85 | 75395.8 | 14757 | 55441.2 | .001352 | 194.682 | 2900 | 2626.85 | 93506.0 | 33641 | 69394.3 | .000717 | 201.533 |
| 2410 | 2136.85 | 75752.3 | 15025 | 55714.6 | .001334 | 194.831 | 2910 | 2636.85 | 93874.0 | 34149 | 69679.0 | .000709 | 201.658 |
| 2420 | 2146.85 | 76108.9 | 15283 | 55988.1 | .001317 | 194.973 | 2920 | 2646.85 | 94244.2 | 34700 | 69966.1 | .000700 | 201.791 |
| 2430 | 2156.85 | 76465.5 | 15560 | 56261.5 | .001298 | 195.122 | 2930 | 2656.85 | 94612.1 | 35224 | 70250.9 | .000692 | 201.915 |
| 2440 | 2166.85 | 76822.1 | 15827 | 56534.9 | .001282 | 195.264 | 2940 | 2666.85 | 94980.0 | 35757 | 70535.7 | .000684 | 202.040 |
| 2450 | 2176.85 | 77178.6 | 16115 | 56808.4 | .001264 | 195.413 | 2950 | 2676.85 | 95350.2 | 36297 | 70822.7 | .000676 | 202.165 |
| 2460 | 2186.85 | 77535.2 | 16391 | 57081.8 | .001248 | 195.555 | 2960 | 2686.85 | 95720.4 | 36846 | 71109.8 | .000668 | 202.289 |
| 2470 | 2196.85 | 77894.1 | 16689 | 57357.5 | .001231 | 195.704 | 2970 | 2696.85 | 96090.6 | 37402 | 71396.8 | .000660 | 202.414 |
| 2480 | 2206.85 | 78250.6 | 16975 | 57630.9 | .001215 | 195.846 | 2980 | 2706.85 | 96460.8 | 37930 | 71683.9 | .000653 | 202.531 |
| 2490 | 2216.85 | 78607.2 | 17283 | 57904.3 | .001198 | 195.995 | 2990 | 2716.85 | 96831.0 | 38503 | 71970.9 | .000646 | 202.655 |
| 2500 | 2226.85 | 78966.0 | 17580 | 58180.0 | .001182 | 196.137 | 3000 | 2726.85 | 97201.2 | 39085 | 72258.0 | .000638 | 202.780 |
| 2510 | 2236.85 | 79324.9 | 17881 | 58455.7 | .001167 | 196.278 | 3010 | 2736.85 | 97571.4 | 39676 | 72545.1 | .000631 | 202.905 |
| 2520 | 2246.85 | 79683.7 | 18206 | 58731.4 | .001151 | 196.428 | 3020 | 2746.85 | 97943.9 | 40275 | 72834.4 | .000623 | 203.029 |
| 2530 | 2256.85 | 80042.6 | 18518 | 59007.1 | .001136 | 196.569 | 3030 | 2756.85 | 98314.1 | 40884 | 73121.4 | .000616 | 203.154 |
| 2540 | 2266.85 | 80401.4 | 18835 | 59282.8 | .001121 | 196.710 | 3040 | 2766.85 | 98684.3 | 41460 | 73408.5 | .000610 | 203.271 |
| 2550 | 2276.85 | 80760.3 | 19158 | 59558.5 | .001107 | 196.852 | 3050 | 2776.85 | 99056.7 | 42087 | 73697.8 | .000603 | 203.395 |
| 2560 | 2286.85 | 81119.1 | 19487 | 59834.2 | .001092 | 196.993 | 3060 | 2786.85 | 99429.2 | 42723 | 73987.1 | .000596 | 203.520 |
| 2570 | 2296.85 | 81480.2 | 19821 | 60112.2 | .001078 | 197.134 | 3070 | 2796.85 | 99801.7 | 43325 | 74276.5 | .000589 | 203.636 |
| 2580 | 2306.85 | 81841.3 | 20141 | 60390.2 | .001065 | 197.268 | 3080 | 2806.85 | 100176.4 | 43980 | 74568.1 | .000582 | 203.761 |
| 2590 | 2316.85 | 82202.4 | 20486 | 60668.1 | .001051 | 197.409 | 3090 | 2816.85 | 100548.9 | 44600 | 74857.4 | .000576 | 203.877 |

# Table 20    Hydrogen

$\bar{m} = 2.0158$

| T K | t C | $\bar{c}_p$ $\dfrac{J}{\text{g-mol K}}$ | $\bar{c}_v$ $\dfrac{J}{\text{g-mol K}}$ | $k = \dfrac{\bar{c}_p}{\bar{c}_v}$ | a $\dfrac{m}{s}$ |
|---|---|---|---|---|---|
| 100 | −173.15 | 22.565 | 14.251 | 1.583 | 808.1 |
| 120 | −153.15 | 23.754 | 15.440 | 1.539 | 872.6 |
| 140 | −133.15 | 24.885 | 16.571 | 1.502 | 931.2 |
| 160 | −113.15 | 25.841 | 17.527 | 1.474 | 986.4 |
| 180 | −93.15 | 26.639 | 18.325 | 1.454 | 1038.9 |
| 200 | −73.15 | 27.271 | 18.957 | 1.439 | 1089.4 |
| 250 | −23.15 | 28.327 | 20.013 | 1.415 | 1208.1 |
| 300 | 26.85 | 28.843 | 20.528 | 1.405 | 1318.5 |
| 350 | 76.85 | 29.084 | 20.769 | 1.400 | 1421.8 |
| 400 | 126.85 | 29.184 | 20.869 | 1.398 | 1518.9 |
| 450 | 176.85 | 29.233 | 20.919 | 1.397 | 1610.5 |
| 500 | 226.85 | 29.258 | 20.944 | 1.397 | 1697.4 |
| 550 | 276.85 | 29.292 | 20.977 | 1.396 | 1779.8 |
| 600 | 326.85 | 29.325 | 21.010 | 1.396 | 1858.5 |
| 650 | 376.85 | 29.375 | 21.060 | 1.395 | 1933.8 |
| 700 | 426.85 | 29.441 | 21.127 | 1.394 | 2005.9 |
| 750 | 476.85 | 29.524 | 21.210 | 1.392 | 2075.1 |
| 800 | 526.85 | 29.624 | 21.310 | 1.390 | 2141.8 |
| 850 | 576.85 | 29.749 | 21.435 | 1.388 | 2205.9 |
| 900 | 626.85 | 29.882 | 21.568 | 1.386 | 2267.9 |
| 950 | 676.85 | 30.040 | 21.726 | 1.383 | 2327.7 |
| 1000 | 726.85 | 30.206 | 21.892 | 1.380 | 2385.6 |
| 1050 | 776.85 | 30.381 | 22.066 | 1.377 | 2441.9 |
| 1100 | 826.85 | 30.580 | 22.266 | 1.373 | 2496.3 |
| 1150 | 876.85 | 30.780 | 22.466 | 1.370 | 2549.3 |
| 1200 | 926.85 | 30.988 | 22.673 | 1.367 | 2600.9 |
| 1300 | 1026.85 | 31.428 | 23.114 | 1.360 | 2700.1 |
| 1400 | 1126.85 | 31.869 | 23.555 | 1.353 | 2795.1 |
| 1500 | 1226.85 | 32.301 | 23.987 | 1.347 | 2886.4 |
| 1600 | 1326.85 | 32.725 | 24.411 | 1.341 | 2974.4 |
| 1700 | 1426.85 | 33.133 | 24.818 | 1.335 | 3059.6 |
| 1800 | 1526.85 | 33.532 | 25.218 | 1.330 | 3142.0 |
| 1900 | 1626.85 | 33.923 | 25.608 | 1.325 | 3222.0 |
| 2000 | 1726.85 | 34.289 | 25.974 | 1.320 | 3300.0 |
| 2100 | 1826.85 | 34.629 | 26.315 | 1.316 | 3376.2 |
| 2200 | 1926.85 | 34.954 | 26.639 | 1.312 | 3450.5 |
| 2300 | 2026.85 | 35.261 | 26.947 | 1.309 | 3523.3 |
| 2400 | 2126.85 | 35.552 | 27.238 | 1.305 | 3594.5 |
| 2500 | 2226.85 | 35.835 | 27.521 | 1.302 | 3664.3 |
| 2600 | 2326.85 | 36.101 | 27.787 | 1.299 | 3732.7 |
| 2700 | 2426.85 | 36.359 | 28.044 | 1.296 | 3799.8 |
| 2800 | 2526.85 | 36.600 | 28.286 | 1.294 | 3865.7 |
| 2900 | 2626.85 | 36.833 | 28.518 | 1.292 | 3930.5 |
| 3000 | 2726.85 | 37.066 | 28.751 | 1.289 | 3994.0 |
| 3100 | 2826.85 | 37.298 | 28.984 | 1.287 | 4056.4 |
| 3200 | 2926.85 | 37.523 | 29.208 | 1.285 | 4117.8 |
| 3300 | 3026.85 | 37.747 | 29.433 | 1.282 | 4178.1 |
| 3400 | 3126.85 | 37.964 | 29.649 | 1.280 | 4237.5 |
| 3500 | 3226.85 | 100.410 | 92.096 | 1.000 | 3800.0 |
| 3600 | 3326.85 | 931.851 | 923.536 | 1.000 | 3853.9 |

# Table 21    Carbon monoxide at Low Pressures (for One Gram-Mole)

$\bar{m} = 28.0104$

| T | t | $\bar{h}$ | $p_r$ | $\bar{u}$ | $v_r$ | $\bar{\phi}$ | T | t | $\bar{h}$ | $p_r$ | $\bar{u}$ | $v_r$ | $\bar{\phi}$ |
|---|---|---|---|---|---|---|---|---|---|---|---|---|---|
| 100 | −173.15 | 2902.5 | .0453 | 2071.1 | 18.3578 | 165.716 | 600 | 326.85 | 17612.1 | 24.912 | 12623.5 | .2002 | 218.180 |
| 110 | −163.15 | 3193.6 | .0632 | 2279.0 | 14.4650 | 168.490 | 610 | 336.85 | 17916.8 | 26.468 | 12845.1 | .1916 | 218.684 |
| 120 | −153.15 | 3484.6 | .0857 | 2486.9 | 11.6366 | 171.022 | 620 | 346.85 | 18222.3 | 28.097 | 13067.4 | .1835 | 219.180 |
| 130 | −143.15 | 3775.7 | .1135 | 2694.8 | 9.5258 | 173.352 | 630 | 356.85 | 18528.5 | 29.803 | 13290.4 | .1758 | 219.670 |
| 140 | −133.15 | 4066.7 | .1471 | 2902.7 | 7.9144 | 175.509 | 640 | 366.85 | 18835.4 | 31.587 | 13514.2 | .1685 | 220.154 |
| 150 | −123.15 | 4357.8 | .1873 | 3110.6 | 6.6603 | 177.517 | 650 | 376.85 | 19143.0 | 33.452 | 13738.7 | .1616 | 220.631 |
| 160 | −113.15 | 4648.9 | .2347 | 3318.5 | 5.6676 | 179.396 | 660 | 386.85 | 19451.4 | 35.400 | 13963.9 | .1550 | 221.101 |
| 170 | −103.15 | 4939.9 | .2902 | 3526.5 | 4.8703 | 181.160 | 670 | 396.85 | 19760.5 | 37.436 | 14189.9 | .1488 | 221.566 |
| 180 | −93.15 | 5231.0 | .3545 | 3734.4 | 4.2216 | 182.824 | 680 | 406.85 | 20070.4 | 39.561 | 14416.6 | .1429 | 222.025 |
| 190 | −83.15 | 5522.1 | .4284 | 3942.3 | 3.6877 | 184.398 | 690 | 416.85 | 20381.0 | 41.778 | 14644.0 | .1373 | 222.479 |
| 200 | −73.15 | 5813.2 | .5126 | 4150.3 | 3.2437 | 185.891 | 700 | 426.85 | 20692.3 | 44.091 | 14872.2 | .1320 | 222.927 |
| 210 | −63.15 | 6104.2 | .6081 | 4358.2 | 2.8711 | 187.311 | 710 | 436.85 | 21004.4 | 46.502 | 15101.1 | .1269 | 223.369 |
| 220 | −53.15 | 6395.3 | .7157 | 4566.2 | 2.5557 | 188.665 | 720 | 446.85 | 21317.2 | 49.014 | 15330.8 | .1221 | 223.807 |
| 230 | −43.15 | 6686.4 | .8362 | 4774.1 | 2.2868 | 189.959 | 730 | 456.85 | 21630.7 | 51.631 | 15561.2 | .1176 | 224.239 |
| 240 | −33.15 | 6977.6 | .9706 | 4982.1 | 2.0559 | 191.198 | 740 | 466.85 | 21945.0 | 54.356 | 15792.3 | .1132 | 224.667 |
| 250 | −23.15 | 7268.7 | 1.1198 | 5190.1 | 1.8563 | 192.387 | 750 | 476.85 | 22260.0 | 57.192 | 16024.2 | .1090 | 225.090 |
| 260 | −13.15 | 7559.9 | 1.2846 | 5398.1 | 1.6828 | 193.529 | 760 | 486.85 | 22575.7 | 60.142 | 16256.8 | .1051 | 225.508 |
| 270 | −3.15 | 7851.1 | 1.4662 | 5606.2 | 1.5311 | 194.628 | 770 | 496.85 | 22892.2 | 63.210 | 16490.1 | .1013 | 225.922 |
| 280 | 6.85 | 8142.3 | 1.6654 | 5814.3 | 1.3979 | 195.687 | 780 | 506.85 | 23209.4 | 66.400 | 16724.2 | .0977 | 226.331 |
| 290 | 16.85 | 8433.7 | 1.8833 | 6022.5 | 1.2803 | 196.709 | 790 | 516.85 | 23527.3 | 69.714 | 16958.9 | .0942 | 226.736 |
| 300 | 26.85 | 8725.0 | 2.1208 | 6230.7 | 1.1761 | 197.697 | 800 | 526.85 | 23845.9 | 73.157 | 17194.4 | .0909 | 227.137 |
| 310 | 36.85 | 9016.5 | 2.3792 | 6439.0 | 1.0833 | 198.653 | 810 | 536.85 | 24165.2 | 76.732 | 17430.6 | .0878 | 227.533 |
| 320 | 46.85 | 9308.1 | 2.6594 | 6647.5 | 1.0005 | 199.579 | 820 | 546.85 | 24485.3 | 80.442 | 17667.4 | .0848 | 227.926 |
| 330 | 56.85 | 9599.8 | 2.9626 | 6856.0 | .9261 | 200.476 | 830 | 556.85 | 24806.0 | 84.293 | 17905.0 | .0819 | 228.315 |
| 340 | 66.85 | 9891.6 | 3.2899 | 7064.7 | .8593 | 201.347 | 840 | 566.85 | 25127.4 | 88.287 | 18143.3 | .0791 | 228.700 |
| 350 | 76.85 | 10183.7 | 3.6425 | 7273.6 | .7989 | 202.194 | 850 | 576.85 | 25449.5 | 92.429 | 18382.2 | .0765 | 229.081 |
| 360 | 86.85 | 10475.9 | 4.0215 | 7482.7 | .7443 | 203.017 | 860 | 586.85 | 25772.2 | 96.722 | 18621.8 | .0739 | 229.458 |
| 370 | 96.85 | 10768.3 | 4.4284 | 7692.0 | .6947 | 203.818 | 870 | 596.85 | 26095.7 | 101.171 | 18862.1 | .0715 | 229.832 |
| 380 | 106.85 | 11061.0 | 4.8642 | 7901.5 | .6495 | 204.599 | 880 | 606.85 | 26419.8 | 105.780 | 19103.1 | .0692 | 230.203 |
| 390 | 116.85 | 11353.9 | 5.3304 | 8111.3 | .6083 | 205.360 | 890 | 616.85 | 26744.5 | 110.553 | 19344.7 | .0669 | 230.570 |
| 400 | 126.85 | 11647.1 | 5.8282 | 8321.4 | .5706 | 206.102 | 900 | 626.85 | 27069.9 | 115.495 | 19586.9 | .0648 | 230.933 |
| 410 | 136.85 | 11940.7 | 6.3592 | 8531.8 | .5361 | 206.827 | 910 | 636.85 | 27395.9 | 120.609 | 19829.8 | .0627 | 231.293 |
| 420 | 146.85 | 12234.6 | 6.9246 | 8742.6 | .5043 | 207.535 | 920 | 646.85 | 27722.6 | 125.901 | 20073.4 | .0608 | 231.650 |
| 430 | 156.85 | 12528.9 | 7.5261 | 8953.7 | .4750 | 208.228 | 930 | 656.85 | 28049.9 | 131.375 | 20317.5 | .0589 | 232.004 |
| 440 | 166.85 | 12823.6 | 8.1650 | 9165.3 | .4481 | 208.905 | 940 | 666.85 | 28377.8 | 137.035 | 20562.3 | .0570 | 232.355 |
| 450 | 176.85 | 13118.7 | 8.8430 | 9377.3 | .4231 | 209.568 | 950 | 676.85 | 28706.4 | 142.886 | 20807.7 | .0553 | 232.703 |
| 460 | 186.85 | 13414.3 | 9.5617 | 9589.7 | .4000 | 210.218 | 960 | 686.85 | 29035.5 | 148.934 | 21053.7 | .0536 | 233.047 |
| 470 | 196.85 | 13710.4 | 10.3227 | 9802.6 | .3786 | 210.855 | 970 | 696.85 | 29365.2 | 155.182 | 21300.3 | .0520 | 233.389 |
| 480 | 206.85 | 14007.0 | 11.1277 | 10016.0 | .3586 | 211.479 | 980 | 706.85 | 29695.6 | 161.636 | 21547.4 | .0504 | 233.728 |
| 490 | 216.85 | 14304.0 | 11.9785 | 10230.0 | .3401 | 212.092 | 990 | 716.85 | 30026.5 | 168.300 | 21795.2 | .0489 | 234.064 |
| 500 | 226.85 | 14601.7 | 12.8769 | 10444.5 | .3228 | 212.693 | 1000 | 726.85 | 30357.9 | 175.181 | 22043.5 | .0475 | 234.397 |
| 510 | 236.85 | 14899.9 | 13.8247 | 10659.5 | .3067 | 213.284 | 1010 | 736.85 | 30690.0 | 182.282 | 22292.4 | .0461 | 234.727 |
| 520 | 246.85 | 15198.7 | 14.8239 | 10875.2 | .2917 | 213.864 | 1020 | 746.85 | 31022.6 | 189.610 | 22541.9 | .0447 | 235.055 |
| 530 | 256.85 | 15498.1 | 15.8764 | 11091.4 | .2776 | 214.434 | 1030 | 756.85 | 31355.7 | 197.169 | 22791.9 | .0434 | 235.380 |
| 540 | 266.85 | 15798.1 | 16.9842 | 11308.3 | .2644 | 214.995 | 1040 | 766.85 | 31689.4 | 204.965 | 23042.4 | .0422 | 235.702 |
| 550 | 276.85 | 16098.8 | 18.1494 | 11525.8 | .2520 | 215.547 | 1050 | 776.85 | 32023.7 | 213.003 | 23293.5 | .0410 | 236.022 |
| 560 | 286.85 | 16400.1 | 19.3740 | 11744.0 | .2403 | 216.090 | 1060 | 786.85 | 32358.4 | 221.289 | 23545.2 | .0398 | 236.340 |
| 570 | 296.85 | 16702.1 | 20.6604 | 11962.8 | .2294 | 216.624 | 1070 | 796.85 | 32693.7 | 229.829 | 23797.3 | .0387 | 236.654 |
| 580 | 306.85 | 17004.7 | 22.0106 | 12182.4 | .2191 | 217.150 | 1080 | 806.85 | 33029.5 | 238.628 | 24049.9 | .0376 | 236.967 |
| 590 | 316.85 | 17308.1 | 23.4271 | 12402.6 | .2094 | 217.669 | 1090 | 816.85 | 33365.8 | 247.692 | 24303.1 | .0366 | 237.277 |

# Table 21  Carbon monoxide at Low Pressures (for One Gram-Mole)

$\bar{m} = 28.0104$

| T | t | $\bar{h}$ | $p_r$ | $\bar{u}$ | $v_r$ | $\bar{\phi}$ | T | t | $\bar{h}$ | $p_r$ | $\bar{u}$ | $v_r$ | $\bar{\phi}$ |
|---|---|---|---|---|---|---|---|---|---|---|---|---|---|
| 1100 | 826.85 | 33702.6 | 257.03 | 24556.8 | .0356 | 237.584 | 1600 | 1326.85 | 51047.6 | 1224.2 | 37744.6 | .01087 | 250.562 |
| 1110 | 836.85 | 34039.9 | 266.64 | 24810.9 | .0346 | 237.890 | 1610 | 1336.85 | 51402.4 | 1257.2 | 38016.2 | .01065 | 250.783 |
| 1120 | 846.85 | 34377.7 | 276.53 | 25065.6 | .0337 | 238.193 | 1620 | 1346.85 | 51757.4 | 1290.9 | 38288.1 | .01043 | 251.003 |
| 1130 | 856.85 | 34716.0 | 286.72 | 25320.7 | .0328 | 238.493 | 1630 | 1356.85 | 52112.7 | 1325.2 | 38560.2 | .01023 | 251.221 |
| 1140 | 866.85 | 35054.7 | 297.20 | 25576.3 | .0319 | 238.792 | 1640 | 1366.85 | 52468.2 | 1360.4 | 38832.5 | .01002 | 251.439 |
| 1150 | 876.85 | 35393.9 | 307.98 | 25832.4 | .0310 | 239.088 | 1650 | 1376.85 | 52823.9 | 1396.2 | 39105.1 | .00983 | 251.655 |
| 1160 | 886.85 | 35733.6 | 319.06 | 26088.9 | .0302 | 239.382 | 1660 | 1386.85 | 53179.8 | 1432.8 | 39377.9 | .00963 | 251.870 |
| 1170 | 896.85 | 36073.7 | 330.47 | 26345.9 | .0294 | 239.674 | 1670 | 1396.85 | 53536.0 | 1470.1 | 39650.9 | .00944 | 252.084 |
| 1180 | 906.85 | 36414.3 | 342.19 | 26603.3 | .0287 | 239.964 | 1680 | 1406.85 | 53892.4 | 1508.2 | 39924.2 | .00926 | 252.297 |
| 1190 | 916.85 | 36755.3 | 354.24 | 26861.2 | .0279 | 240.252 | 1690 | 1416.85 | 54249.0 | 1547.1 | 40197.7 | .00908 | 252.508 |
| 1200 | 926.85 | 37096.7 | 366.63 | 27119.4 | .0272 | 240.537 | 1700 | 1426.85 | 54605.8 | 1586.8 | 40471.3 | .00891 | 252.719 |
| 1210 | 936.85 | 37438.6 | 379.35 | 27378.2 | .0265 | 240.821 | 1710 | 1436.85 | 54962.9 | 1627.3 | 40745.2 | .00874 | 252.928 |
| 1220 | 946.85 | 37780.9 | 392.43 | 27637.3 | .0258 | 241.103 | 1720 | 1446.85 | 55320.1 | 1668.6 | 41019.3 | .00857 | 253.137 |
| 1230 | 956.85 | 38123.6 | 405.86 | 27896.9 | .0252 | 241.382 | 1730 | 1456.85 | 55677.6 | 1710.7 | 41293.7 | .00841 | 253.344 |
| 1240 | 966.85 | 38466.7 | 419.65 | 28156.8 | .0246 | 241.660 | 1740 | 1466.85 | 56035.2 | 1753.6 | 41568.2 | .00825 | 253.550 |
| 1250 | 976.85 | 38810.2 | 433.81 | 28417.2 | .0240 | 241.936 | 1750 | 1476.85 | 56393.1 | 1797.4 | 41842.9 | .00810 | 253.755 |
| 1260 | 986.85 | 39154.1 | 448.34 | 28678.0 | .0234 | 242.210 | 1760 | 1486.85 | 56751.2 | 1842.1 | 42117.8 | .00794 | 253.959 |
| 1270 | 996.85 | 39498.4 | 463.26 | 28939.2 | .0228 | 242.482 | 1770 | 1496.85 | 57109.4 | 1887.6 | 42392.9 | .00780 | 254.162 |
| 1280 | 1006.85 | 39843.1 | 478.57 | 29200.7 | .0222 | 242.753 | 1780 | 1506.85 | 57467.9 | 1934.0 | 42668.2 | .00765 | 254.364 |
| 1290 | 1016.85 | 40188.2 | 494.28 | 29462.6 | .0217 | 243.021 | 1790 | 1516.85 | 57826.5 | 1981.3 | 42943.7 | .00751 | 254.565 |
| 1300 | 1026.85 | 40533.7 | 510.40 | 29724.9 | .0212 | 243.288 | 1800 | 1526.85 | 58185.4 | 2029.5 | 43219.4 | .00737 | 254.765 |
| 1310 | 1036.85 | 40879.5 | 526.93 | 29987.6 | .0207 | 243.553 | 1810 | 1536.85 | 58544.4 | 2078.6 | 43495.3 | .00724 | 254.964 |
| 1320 | 1046.85 | 41225.7 | 543.88 | 30250.7 | .0202 | 243.816 | 1820 | 1546.85 | 58903.6 | 2128.7 | 43771.4 | .00711 | 255.162 |
| 1330 | 1056.85 | 41572.2 | 561.26 | 30514.1 | .0197 | 244.078 | 1830 | 1556.85 | 59263.0 | 2179.7 | 44047.6 | .00698 | 255.359 |
| 1340 | 1066.85 | 41919.1 | 579.08 | 30777.8 | .0192 | 244.338 | 1840 | 1566.85 | 59622.6 | 2231.7 | 44324.1 | .00686 | 255.555 |
| 1350 | 1076.85 | 42266.3 | 597.34 | 31041.9 | .0188 | 244.596 | 1850 | 1576.85 | 59982.3 | 2284.7 | 44600.7 | .00673 | 255.750 |
| 1360 | 1086.85 | 42613.9 | 616.06 | 31306.3 | .0184 | 244.852 | 1860 | 1586.85 | 60342.2 | 2338.6 | 44877.5 | .00661 | 255.944 |
| 1370 | 1096.85 | 42961.9 | 635.24 | 31571.1 | .0179 | 245.107 | 1870 | 1596.85 | 60702.3 | 2393.6 | 45154.4 | .00650 | 256.137 |
| 1380 | 1106.85 | 43310.1 | 654.89 | 31836.2 | .0175 | 245.361 | 1880 | 1606.85 | 61062.6 | 2449.5 | 45431.5 | .00638 | 256.329 |
| 1390 | 1116.85 | 43658.7 | 675.01 | 32101.7 | .0171 | 245.612 | 1890 | 1616.85 | 61423.1 | 2506.5 | 45708.8 | .00627 | 256.520 |
| 1400 | 1126.85 | 44007.6 | 695.63 | 32367.5 | .0167 | 245.862 | 1900 | 1626.85 | 61783.7 | 2564.5 | 45986.3 | .00616 | 256.710 |
| 1410 | 1136.85 | 44356.9 | 716.74 | 32633.6 | .0164 | 246.111 | 1910 | 1636.85 | 62144.5 | 2623.6 | 46263.9 | .00605 | 256.900 |
| 1420 | 1146.85 | 44706.4 | 738.35 | 32900.0 | .0160 | 246.358 | 1920 | 1646.85 | 62505.4 | 2683.8 | 46541.7 | .00595 | 257.088 |
| 1430 | 1156.85 | 45056.3 | 760.48 | 33166.7 | .0156 | 246.604 | 1930 | 1656.85 | 62866.5 | 2745.0 | 46819.7 | .00585 | 257.276 |
| 1440 | 1166.85 | 45406.5 | 783.13 | 33433.7 | .0153 | 246.848 | 1940 | 1666.85 | 63227.8 | 2807.4 | 47097.8 | .00575 | 257.463 |
| 1450 | 1176.85 | 45756.9 | 806.31 | 33701.0 | .0150 | 247.090 | 1950 | 1676.85 | 63589.2 | 2870.8 | 47376.1 | .00565 | 257.648 |
| 1460 | 1186.85 | 46107.7 | 830.03 | 33968.7 | .0146 | 247.331 | 1960 | 1686.85 | 63950.8 | 2935.4 | 47654.5 | .00555 | 257.833 |
| 1470 | 1196.85 | 46458.8 | 854.31 | 34236.6 | .0143 | 247.571 | 1970 | 1696.85 | 64312.5 | 3001.1 | 47933.1 | .00546 | 258.017 |
| 1480 | 1206.85 | 46810.1 | 879.14 | 34504.8 | .0140 | 247.809 | 1980 | 1706.85 | 64674.4 | 3068.0 | 48211.9 | .00537 | 258.201 |
| 1490 | 1216.85 | 47161.8 | 904.53 | 34773.3 | .0137 | 248.046 | 1990 | 1716.85 | 65036.4 | 3136.0 | 48490.8 | .00528 | 258.383 |
| 1500 | 1226.85 | 47513.7 | 930.51 | 35042.1 | .0134 | 248.281 | 2000 | 1726.85 | 65398.6 | 3205.2 | 48769.8 | .00519 | 258.565 |
| 1510 | 1236.85 | 47865.9 | 957.07 | 35311.1 | .0131 | 248.515 | 2010 | 1736.85 | 65760.9 | 3275.7 | 49049.0 | .00510 | 258.745 |
| 1520 | 1246.85 | 48218.4 | 984.23 | 35580.5 | .0128 | 248.748 | 2020 | 1746.85 | 66123.4 | 3347.3 | 49328.3 | .00502 | 258.925 |
| 1530 | 1256.85 | 48571.1 | 1012.00 | 35850.1 | .0126 | 248.979 | 2030 | 1756.85 | 66486.0 | 3420.2 | 49607.8 | .00493 | 259.104 |
| 1540 | 1266.85 | 48924.1 | 1040.38 | 36119.9 | .0123 | 249.209 | 2040 | 1766.85 | 66848.8 | 3494.3 | 49887.4 | .00485 | 259.283 |
| 1550 | 1276.85 | 49277.4 | 1069.39 | 36390.1 | .0121 | 249.438 | 2050 | 1776.85 | 67211.7 | 3569.7 | 50167.2 | .00477 | 259.460 |
| 1560 | 1286.85 | 49630.9 | 1099.03 | 36660.5 | .0118 | 249.665 | 2060 | 1786.85 | 67574.8 | 3646.4 | 50447.1 | .00470 | 259.637 |
| 1570 | 1296.85 | 49984.7 | 1129.33 | 36931.1 | .0116 | 249.891 | 2070 | 1796.85 | 67938.0 | 3724.3 | 50727.2 | .00462 | 259.813 |
| 1580 | 1306.85 | 50338.8 | 1160.28 | 37202.0 | .0113 | 250.116 | 2080 | 1806.85 | 68301.3 | 3803.6 | 51007.3 | .00455 | 259.988 |
| 1590 | 1316.85 | 50693.1 | 1191.89 | 37473.2 | .0111 | 250.340 | 2090 | 1816.85 | 68664.7 | 3884.2 | 51287.6 | .00447 | 260.162 |

## Table 21  Carbon monoxide at Low Pressures (for One Gram-Mole)

$\bar{m} = 28.0104$

| T | t | $\bar{h}$ | $p_r$ | $\bar{u}$ | $v_r$ | $\bar{\phi}$ | T | t | $\bar{h}$ | $p_r$ | $\bar{u}$ | $v_r$ | $\bar{\phi}$ |
|---|---|---|---|---|---|---|---|---|---|---|---|---|---|
| 2100 | 1826.85 | 69028.3 | 3966 | 51568.1 | .00440 | 260.336 | 2600 | 2326.85 | 87351.7 | 10164 | 65734.2 | .00213 | 268.160 |
| 2110 | 1836.85 | 69392.1 | 4049 | 51848.7 | .00433 | 260.508 | 2610 | 2336.85 | 87720.6 | 10339 | 66020.0 | .00210 | 268.302 |
| 2120 | 1846.85 | 69755.9 | 4134 | 52129.4 | .00426 | 260.680 | 2620 | 2346.85 | 88089.6 | 10516 | 66305.8 | .00207 | 268.443 |
| 2130 | 1856.85 | 70119.9 | 4220 | 52410.2 | .00420 | 260.852 | 2630 | 2356.85 | 88458.6 | 10695 | 66591.7 | .00204 | 268.584 |
| 2140 | 1866.85 | 70484.0 | 4308 | 52691.2 | .00413 | 261.022 | 2640 | 2366.85 | 88827.8 | 10877 | 66877.7 | .00202 | 268.724 |
| 2150 | 1876.85 | 70848.3 | 4396 | 52972.3 | .00407 | 261.192 | 2650 | 2376.85 | 89197.0 | 11061 | 67163.8 | .00199 | 268.863 |
| 2160 | 1886.85 | 71212.6 | 4487 | 53253.5 | .00400 | 261.361 | 2660 | 2386.85 | 89566.3 | 11248 | 67450.0 | .00197 | 269.002 |
| 2170 | 1896.85 | 71577.1 | 4579 | 53534.9 | .00394 | 261.529 | 2670 | 2396.85 | 89935.6 | 11437 | 67736.2 | .00194 | 269.141 |
| 2180 | 1906.85 | 71941.7 | 4672 | 53816.3 | .00388 | 261.697 | 2680 | 2406.85 | 90305.1 | 11628 | 68022.5 | .00192 | 269.279 |
| 2190 | 1916.85 | 72306.5 | 4767 | 54097.9 | .00382 | 261.864 | 2690 | 2416.85 | 90674.6 | 11823 | 68308.9 | .00189 | 269.417 |
| 2200 | 1926.85 | 72671.3 | 4863 | 54379.6 | .00376 | 262.030 | 2700 | 2426.85 | 91044.2 | 12019 | 68595.3 | .00187 | 269.554 |
| 2210 | 1936.85 | 73036.3 | 4961 | 54661.5 | .00370 | 262.196 | 2710 | 2436.85 | 91413.9 | 12218 | 68881.9 | .00184 | 269.691 |
| 2220 | 1946.85 | 73401.4 | 5060 | 54943.4 | .00365 | 262.361 | 2720 | 2446.85 | 91783.7 | 12420 | 69168.5 | .00182 | 269.827 |
| 2230 | 1956.85 | 73766.6 | 5161 | 55225.5 | .00359 | 262.525 | 2730 | 2456.85 | 92153.5 | 12625 | 69455.2 | .00180 | 269.962 |
| 2240 | 1966.85 | 74131.9 | 5263 | 55507.7 | .00354 | 262.688 | 2740 | 2466.85 | 92523.4 | 12832 | 69741.9 | .00178 | 270.098 |
| 2250 | 1976.85 | 74497.4 | 5367 | 55789.9 | .00349 | 262.851 | 2750 | 2476.85 | 92893.4 | 13041 | 70028.7 | .00175 | 270.232 |
| 2260 | 1986.85 | 74862.9 | 5473 | 56072.4 | .00343 | 263.013 | 2760 | 2486.85 | 93263.4 | 13254 | 70315.6 | .00173 | 270.367 |
| 2270 | 1996.85 | 75228.6 | 5580 | 56354.9 | .00338 | 263.174 | 2770 | 2496.85 | 93633.5 | 13469 | 70602.6 | .00171 | 270.501 |
| 2280 | 2006.85 | 75594.3 | 5689 | 56637.5 | .00333 | 263.335 | 2780 | 2506.85 | 94003.7 | 13687 | 70889.7 | .00169 | 270.634 |
| 2290 | 2016.85 | 75960.2 | 5800 | 56920.2 | .00328 | 263.495 | 2790 | 2516.85 | 94374.0 | 13907 | 71176.8 | .00167 | 270.767 |
| 2300 | 2026.85 | 76326.2 | 5912 | 57203.1 | .00323 | 263.655 | 2800 | 2526.85 | 94744.3 | 14131 | 71464.0 | .00165 | 270.900 |
| 2310 | 2036.85 | 76692.3 | 6026 | 57486.0 | .00319 | 263.814 | 2810 | 2536.85 | 95114.7 | 14357 | 71751.2 | .00163 | 271.032 |
| 2320 | 2046.85 | 77058.5 | 6142 | 57769.1 | .00314 | 263.972 | 2820 | 2546.85 | 95485.2 | 14586 | 72038.5 | .00161 | 271.163 |
| 2330 | 2056.85 | 77424.8 | 6259 | 58052.3 | .00309 | 264.129 | 2830 | 2556.85 | 95855.7 | 14818 | 72325.9 | .00159 | 271.294 |
| 2340 | 2066.85 | 77791.2 | 6379 | 58335.5 | .00305 | 264.286 | 2840 | 2566.85 | 96226.3 | 15053 | 72613.4 | .00157 | 271.425 |
| 2350 | 2076.85 | 78157.8 | 6500 | 58618.9 | .00301 | 264.443 | 2850 | 2576.85 | 96597.0 | 15290 | 72900.9 | .00155 | 271.555 |
| 2360 | 2086.85 | 78524.4 | 6623 | 58902.4 | .00296 | 264.598 | 2860 | 2586.85 | 96967.7 | 15531 | 73188.5 | .00153 | 271.685 |
| 2370 | 2096.85 | 78891.1 | 6747 | 59186.0 | .00292 | 264.753 | 2870 | 2596.85 | 97338.5 | 15775 | 73476.2 | .00151 | 271.815 |
| 2380 | 2106.85 | 79257.9 | 6874 | 59469.7 | .00288 | 264.908 | 2880 | 2606.85 | 97709.4 | 16021 | 73763.9 | .00149 | 271.944 |
| 2390 | 2116.85 | 79624.9 | 7002 | 59753.4 | .00284 | 265.062 | 2890 | 2616.85 | 98080.3 | 16271 | 74051.7 | .00148 | 272.072 |
| 2400 | 2126.85 | 79991.9 | 7132 | 60037.3 | .00280 | 265.215 | 2900 | 2626.85 | 98451.3 | 16524 | 74339.6 | .00146 | 272.200 |
| 2410 | 2136.85 | 80359.0 | 7265 | 60321.3 | .00276 | 265.368 | 2910 | 2636.85 | 98822.4 | 16780 | 74627.5 | .00144 | 272.328 |
| 2420 | 2146.85 | 80726.2 | 7399 | 60605.4 | .00272 | 265.520 | 2920 | 2646.85 | 99193.5 | 17039 | 74915.5 | .00142 | 272.455 |
| 2430 | 2156.85 | 81093.5 | 7535 | 60889.5 | .00268 | 265.671 | 2930 | 2656.85 | 99564.7 | 17301 | 75203.5 | .00141 | 272.582 |
| 2440 | 2166.85 | 81460.9 | 7673 | 61173.8 | .00264 | 265.822 | 2940 | 2666.85 | 99936.0 | 17566 | 75491.6 | .00139 | 272.709 |
| 2450 | 2176.85 | 81828.4 | 7813 | 61458.1 | .00261 | 265.972 | 2950 | 2676.85 | 100307.3 | 17834 | 75779.8 | .00138 | 272.835 |
| 2460 | 2186.85 | 82196.0 | 7955 | 61742.6 | .00257 | 266.122 | 2960 | 2686.85 | 100678.7 | 18106 | 76068.0 | .00136 | 272.961 |
| 2470 | 2196.85 | 82563.7 | 8099 | 62027.1 | .00254 | 266.271 | 2970 | 2696.85 | 101050.1 | 18381 | 76356.3 | .00134 | 273.086 |
| 2480 | 2206.85 | 82931.5 | 8245 | 62311.8 | .00250 | 266.420 | 2980 | 2706.85 | 101421.6 | 18659 | 76644.7 | .00133 | 273.211 |
| 2490 | 2216.85 | 83299.4 | 8393 | 62596.5 | .00247 | 266.568 | 2990 | 2716.85 | 101793.2 | 18940 | 76933.1 | .00131 | 273.335 |
| 2500 | 2226.85 | 83667.3 | 8543 | 62881.3 | .00243 | 266.715 | 3000 | 2726.85 | 102164.8 | 19225 | 77221.6 | .00130 | 273.459 |
| 2510 | 2236.85 | 84035.4 | 8695 | 63166.2 | .00240 | 266.862 | 3010 | 2736.85 | 102536.5 | 19513 | 77510.1 | .00128 | 273.583 |
| 2520 | 2246.85 | 84403.5 | 8850 | 63451.2 | .00237 | 267.009 | 3020 | 2746.85 | 102908.2 | 19805 | 77798.7 | .00127 | 273.706 |
| 2530 | 2256.85 | 84771.7 | 9006 | 63736.3 | .00234 | 267.154 | 3030 | 2756.85 | 103280.0 | 20100 | 78087.4 | .00125 | 273.829 |
| 2540 | 2266.85 | 85140.0 | 9165 | 64021.4 | .00230 | 267.300 | 3040 | 2766.85 | 103651.9 | 20398 | 78376.1 | .00124 | 273.952 |
| 2550 | 2276.85 | 85508.4 | 9326 | 64306.7 | .00227 | 267.444 | 3050 | 2776.85 | 104023.8 | 20700 | 78664.9 | .00123 | 274.074 |
| 2560 | 2286.85 | 85876.9 | 9489 | 64592.0 | .00224 | 267.589 | 3060 | 2786.85 | 104395.8 | 21005 | 78953.7 | .00121 | 274.196 |
| 2570 | 2296.85 | 86245.5 | 9655 | 64877.5 | .00221 | 267.732 | 3070 | 2796.85 | 104767.9 | 21314 | 79242.6 | .00120 | 274.317 |
| 2580 | 2306.85 | 86614.1 | 9822 | 65163.0 | .00218 | 267.876 | 3080 | 2806.85 | 105140.0 | 21627 | 79531.6 | .00118 | 274.438 |
| 2590 | 2316.85 | 86982.9 | 9992 | 65448.6 | .00216 | 268.018 | 3090 | 2816.85 | 105512.1 | 21943 | 79820.6 | .00117 | 274.559 |

# Table 22    Carbon monoxide

$\bar{m} = 28.0104$

| T<br>K | t<br>C | $\bar{c}_p$<br>$\dfrac{J}{\text{g-mol K}}$ | $\bar{c}_v$<br>$\dfrac{J}{\text{g-mol K}}$ | $k = \dfrac{\bar{c}_p}{\bar{c}_v}$ | a<br>$\dfrac{m}{s}$ |
|---|---|---|---|---|---|
| 100 | −173.15 | 29.104 | 20.790 | 1.400 | 203.8 |
| 120 | −153.15 | 29.105 | 20.791 | 1.400 | 223.3 |
| 140 | −133.15 | 29.106 | 20.791 | 1.400 | 241.2 |
| 160 | −113.15 | 29.107 | 20.792 | 1.400 | 257.8 |
| 180 | −93.15 | 29.107 | 20.793 | 1.400 | 273.5 |
| 200 | −73.15 | 29.109 | 20.794 | 1.400 | 288.3 |
| 250 | −23.15 | 29.116 | 20.801 | 1.400 | 322.3 |
| 300 | 26.85 | 29.142 | 20.828 | 1.399 | 353.0 |
| 350 | 76.85 | 29.211 | 20.897 | 1.398 | 381.1 |
| 400 | 126.85 | 29.340 | 21.025 | 1.395 | 407.0 |
| 450 | 176.85 | 29.535 | 21.221 | 1.392 | 431.2 |
| 500 | 226.85 | 29.792 | 21.478 | 1.387 | 453.7 |
| 550 | 276.85 | 30.098 | 21.784 | 1.382 | 474.9 |
| 600 | 326.85 | 30.439 | 22.125 | 1.376 | 495.0 |
| 650 | 376.85 | 30.801 | 22.486 | 1.370 | 514.1 |
| 700 | 426.85 | 31.170 | 22.856 | 1.364 | 532.3 |
| 750 | 476.85 | 31.538 | 23.223 | 1.358 | 549.8 |
| 800 | 526.85 | 31.897 | 23.582 | 1.353 | 566.7 |
| 850 | 576.85 | 32.243 | 23.928 | 1.347 | 583.1 |
| 900 | 626.85 | 32.572 | 24.258 | 1.343 | 598.9 |
| 950 | 676.85 | 32.883 | 24.569 | 1.338 | 614.3 |
| 1000 | 726.85 | 33.176 | 24.862 | 1.334 | 629.4 |
| 1050 | 776.85 | 33.450 | 25.136 | 1.331 | 644.0 |
| 1100 | 826.85 | 33.706 | 25.391 | 1.327 | 658.4 |
| 1150 | 876.85 | 33.944 | 25.629 | 1.324 | 672.4 |
| 1200 | 926.85 | 34.166 | 25.851 | 1.322 | 686.1 |
| 1300 | 1026.85 | 34.563 | 26.249 | 1.317 | 712.8 |
| 1400 | 1126.85 | 34.908 | 26.593 | 1.313 | 738.6 |
| 1500 | 1226.85 | 35.206 | 26.892 | 1.309 | 763.5 |
| 1600 | 1326.85 | 35.466 | 27.152 | 1.306 | 787.6 |
| 1700 | 1426.85 | 35.693 | 27.379 | 1.304 | 811.1 |
| 1800 | 1526.85 | 35.893 | 27.579 | 1.301 | 833.9 |
| 1900 | 1626.85 | 36.070 | 27.755 | 1.300 | 856.1 |
| 2000 | 1726.85 | 36.226 | 27.912 | 1.298 | 877.8 |
| 2100 | 1826.85 | 36.366 | 28.052 | 1.296 | 898.9 |
| 2200 | 1926.85 | 36.491 | 28.177 | 1.295 | 919.6 |
| 2300 | 2026.85 | 36.605 | 28.290 | 1.294 | 939.9 |
| 2400 | 2126.85 | 36.707 | 28.393 | 1.293 | 959.7 |
| 2500 | 2226.85 | 36.800 | 28.486 | 1.292 | 979.1 |
| 2600 | 2326.85 | 36.886 | 28.571 | 1.291 | 998.2 |
| 2700 | 2426.85 | 36.964 | 28.650 | 1.290 | 1016.9 |
| 2800 | 2526.85 | 37.036 | 28.722 | 1.289 | 1035.2 |
| 2900 | 2626.85 | 37.103 | 28.789 | 1.289 | 1053.3 |
| 3000 | 2726.85 | 37.166 | 28.851 | 1.288 | 1071.0 |
| 3100 | 2826.85 | 37.224 | 28.909 | 1.288 | 1088.5 |
| 3200 | 2926.85 | 37.278 | 28.964 | 1.287 | 1105.7 |
| 3300 | 3026.85 | 37.329 | 29.015 | 1.287 | 1122.6 |
| 3400 | 3126.85 | 37.377 | 29.063 | 1.286 | 1139.3 |
| 3500 | 3226.85 | 37.423 | 29.109 | 1.286 | 1155.7 |
| 3600 | 3326.85 | 37.466 | 29.152 | 1.285 | 1171.9 |

# Table 23 Monatomic Gases, He, Ar, Hg, etc., at Low Pressures (for One Gram-Mole)

### Molecular weight, $\bar{m}$: He 4.0026, Ar 39.948, Hg 200.59

| T K | t C | $\bar{h}$ J/g-mole | $p_r$ | $\bar{u}$ J/g-mole | $v_r$ | $\bar{\phi}$ J/g-mole K |
|---|---|---|---|---|---|---|
| 100 | −173.15 | 2078.6 | .78767 | 1247.2 | 1.0556 | 132.028 |
| 120 | −153.15 | 2494.3 | 1.2425 | 1496.6 | .8030 | 135.818 |
| 140 | −133.15 | 2910.0 | 1.8267 | 1746.0 | .6372 | 139.022 |
| 160 | −113.15 | 3325.8 | 2.5506 | 1995.5 | .5216 | 141.797 |
| 180 | −93.15 | 3741.5 | 3.4239 | 2244.9 | .4371 | 144.246 |
| 200 | −73.15 | 4157.2 | 4.4557 | 2494.3 | .3732 | 146.436 |
| 250 | −23.15 | 5196.5 | 7.7839 | 3117.9 | .2670 | 151.074 |
| 300 | 26.85 | 6235.8 | 12.279 | 3741.5 | .2031 | 154.864 |
| 350 | 76.85 | 7275.1 | 18.052 | 4365.1 | .1612 | 158.068 |
| 400 | 126.85 | 8314.4 | 25.206 | 4988.6 | .1319 | 160.843 |
| 450 | 176.85 | 9353.7 | 33.836 | 5612.2 | .1106 | 163.292 |
| 500 | 226.85 | 10393.0 | 44.032 | 6235.8 | .0944 | 165.482 |
| 550 | 276.85 | 11432.3 | 55.879 | 6859.4 | .0818 | 167.463 |
| 600 | 326.85 | 12471.6 | 69.458 | 7483.0 | .07182 | 169.271 |
| 650 | 376.85 | 13510.9 | 84.845 | 8106.5 | .06370 | 170.935 |
| 700 | 426.85 | 14550.2 | 102.12 | 8730.1 | .05700 | 172.476 |
| 750 | 476.85 | 15589.5 | 121.34 | 9353.7 | .05139 | 173.910 |
| 800 | 526.85 | 16628.8 | 142.58 | 9977.3 | .04665 | 175.251 |
| 850 | 576.85 | 17668.1 | 165.92 | 10600.9 | .04259 | 176.511 |
| 900 | 626.85 | 18707.4 | 191.40 | 11224.4 | .03910 | 177.699 |
| 950 | 676.85 | 19746.7 | 219.11 | 11848.0 | .03605 | 178.823 |
| 1000 | 726.85 | 20786.0 | 249.08 | 12471.6 | .03338 | 179.889 |
| 1050 | 776.85 | 21825.3 | 281.40 | 13095.2 | .03102 | 180.904 |
| 1100 | 826.85 | 22864.6 | 316.10 | 13718.8 | .02893 | 181.871 |
| 1150 | 876.85 | 23903.9 | 353.26 | 14342.3 | .02707 | 182.795 |
| 1200 | 926.85 | 24943.2 | 392.91 | 14965.9 | .02539 | 183.679 |
| 1300 | 1026.85 | 27021.8 | 479.96 | 16213.1 | .02252 | 185.343 |
| 1400 | 1126.85 | 29100.4 | 577.65 | 17460.2 | .02015 | 186.883 |
| 1500 | 1226.85 | 31179.0 | 686.39 | 18707.4 | .01817 | 188.317 |
| 1600 | 1326.85 | 33257.6 | 806.58 | 19954.6 | .01649 | 189.659 |
| 1700 | 1426.85 | 35336.2 | 938.57 | 21201.7 | .01506 | 190.919 |
| 1800 | 1526.85 | 37414.8 | 1082.7 | 22448.9 | .01382 | 192.107 |
| 1900 | 1626.85 | 39493.4 | 1239.5 | 23696.0 | .01275 | 193.231 |
| 2000 | 1726.85 | 41572.0 | 1409.0 | 24943.2 | .01180 | 194.297 |
| 2100 | 1826.85 | 43650.6 | 1591.8 | 26190.4 | .01097 | 195.311 |
| 2200 | 1926.85 | 45729.2 | 1788.1 | 27437.5 | .01023 | 196.278 |
| 2300 | 2026.85 | 47807.8 | 1998.3 | 28684.7 | .00957 | 197.202 |
| 2400 | 2126.85 | 49886.4 | 2222.7 | 29931.9 | .00898 | 198.087 |
| 2500 | 2226.85 | 51965.0 | 2461.5 | 31179.0 | .00844 | 198.935 |
| 2600 | 2326.85 | 54043.6 | 2715.1 | 32426.2 | .00796 | 199.751 |
| 2700 | 2426.85 | 56122.2 | 2983.7 | 33673.3 | .00752 | 200.535 |
| 2800 | 2526.85 | 58200.8 | 3267.7 | 34920.5 | .00712 | 201.291 |
| 2900 | 2626.85 | 60279.4 | 3567.3 | 36167.7 | .00676 | 202.021 |
| 3000 | 2726.85 | 62358.0 | 3882.8 | 37414.8 | .00642 | 202.725 |
| 3100 | 2826.85 | 64436.6 | 4214.5 | 38662.0 | .00612 | 203.407 |
| 3200 | 2926.85 | 66515.2 | 4562.7 | 39909.1 | .00583 | 204.067 |
| 3300 | 3026.85 | 68593.8 | 4927.5 | 41156.3 | .00557 | 204.706 |
| 3400 | 3126.85 | 70672.4 | 5309.4 | 42403.5 | .00532 | 205.327 |
| 3500 | 3226.85 | 72751.0 | 5708.4 | 43650.6 | .00510 | 205.929 |
| 3600 | 3326.85 | 74829.6 | 6124.9 | 44897.8 | .00489 | 206.515 |

# SYMBOLS FOR POLYTROPIC FUNCTIONS AND COMPRESSIBLE-FLOW FUNCTIONS IN TABLES 24 TO 59

$A$    cross-sectional area

$a$    velocity of sound

$c_p$    specific heat at constant pressure

$c_v$    specific heat at constant volume

$D$    diameter of duct

$F$    wall-force function
$$= pA + \rho A V^2 = pA(1 + kM^2)$$

$f$    coefficient of friction $= \dfrac{2\tau_w}{\rho V^2}$

$G$    flow per unit area or mass velocity
$$= w/A$$

$h$    enthalpy per unit mass

$k$    ratio of specific heats $= c_p/c_v$

$L$    length of duct

$L_{\max}$    maximum length of duct for steady adiabatic constant-area flow with friction

$M$    Mach number $= V/a$

$\overline{m}$    molecular weight

$m$    length

$N$    force

$n$    exponent for polytropic process

$p$    static pressure

$\overline{R}$    universal gas constant $= 8.31441 \dfrac{\text{Nm}}{\text{Kg-mol}}$

$R$    gas constant $= \overline{R}/\overline{m}$

$r$    pressure ratio

$s$    entropy per unit mass

$T$    absolute thermodynamic temperature

$u$    internal energy per unit mass

$V$    velocity

$v$    specific volume

$w$    mass rate of flow

$\alpha$    Mach angle $= \text{arc sin}(1/M)$

$\alpha'$    angle between two-dimensional shock and incident flow, see sketch on p. 163

$\rho$    density

$\theta$    $= \alpha' - \omega'$

$\tau_w$    shear stress at wall

$\omega$    vector angle of characteristic curve in hodograph plane, see sketch on p. 158

$\omega'$    wedge angle for two-dimensional shock, see sketch on p. 163

$\omega'_m$    maximum wedge angle

$\omega'_a$    wedge angle for sonic flow downstream of two-dimensional shock

Superscript * refers to conditions in which $M = 1$; in particular, $M^*$ denotes the ratio of the velocity divided by the velocity of sound at the position at which $M = 1$.

Subscripts

0    refers to isentropic stagnation conditions

$x$    and $y$ refer to conditions upstream and downstream of a shock, respectively

# Formulas in Which the Quantities of Tables 25 to 29 Appear

General:

$$pv = \frac{\bar{R}T}{\bar{m}}. \quad r = \frac{p}{p_0}.$$

The Polytropic Process:

$$\frac{v_0}{v} = r^{\frac{1}{n}}. \qquad \frac{T}{T_0} = r^{\frac{n-1}{n}}. \qquad \int_{p_0}^{p} p\,dv = \frac{T_0}{\bar{m}}\frac{\bar{R}}{n-1}\left(1 - r^{\frac{n-1}{n}}\right) = \frac{T}{\bar{m}}\frac{\bar{R}}{n-1}\left(r^{-\left(\frac{n-1}{n}\right)} - 1\right).$$

$$\int_{p_0}^{p} v\,dp = -\frac{T_0}{\bar{m}}\frac{n\bar{R}}{n-1}\left(1 - r^{\frac{n-1}{n}}\right) = -\frac{T_n}{\bar{m}}\frac{\bar{R}}{n-1}\left(r^{-\left(\frac{n-1}{n}\right)} - 1\right).$$

Steady Flow through a Nozzle or a Diffuser:

$$\frac{V^2 - V_0^2}{2} = \frac{T_0}{\bar{m}}\frac{n\bar{R}}{n-1}\left(1 - r^{\frac{n-1}{n}}\right) = \frac{T}{\bar{m}}\frac{n\bar{R}}{n-1}\left(r^{-\left(\frac{n-1}{n}\right)} - 1\right).$$

$$V = \sqrt{\frac{T_0}{\bar{m}}}\sqrt{\frac{2n\bar{R}}{n-1}}\sqrt{\left(1 - r^{\frac{n-1}{n}}\right)} = \sqrt{\frac{T}{\bar{m}}}\sqrt{\frac{2n\bar{R}}{n-1}}\sqrt{r^{-\left(\frac{n-1}{n}\right)} - 1} \text{ for } V_0 = 0.$$

$$G = \frac{w}{A} = p_0\sqrt{\frac{\bar{m}}{T_0}}\sqrt{\frac{2}{\bar{R}}\frac{n}{n-1}}\,r^{\frac{1}{n}}\sqrt{1 - r^{\frac{n-1}{n}}} = p_0\sqrt{\frac{\bar{m}}{T}}\sqrt{\frac{2}{\bar{R}}\frac{n}{n-1}}\,r^{\frac{1}{n}}\sqrt{r^{\frac{n-1}{n}} - r^{2\left(\frac{n-1}{n}\right)}} \text{ for } V_0 = 0.$$

Reversible Adiabatic Process:

$$n = k = c_p/c_v. \quad u_0 - u = \int_{p_0}^{p} p\,dv = \frac{T_0}{\bar{m}}\frac{\bar{R}}{k-1}\left(1 - r^{\frac{k-1}{k}}\right) = \frac{T}{\bar{m}}\frac{\bar{R}}{k-1}\left(r^{-\left(\frac{k-1}{k}\right)} - 1\right).$$

$$h_0 - h = -\int_{p_0}^{p} v\,dp = \frac{T_0}{\bar{m}}\frac{k\bar{R}}{k-1}\left(1 - r^{\frac{k-1}{k}}\right) = \frac{T}{\bar{m}}\frac{k\bar{R}}{k-1}\left(r^{-\left(\frac{k-1}{k}\right)} - 1\right). \quad M = \sqrt{\frac{2}{k-1}\left(r^{-\left(\frac{k-1}{k}\right)} - 1\right)} \text{ for } M_0 = 0.$$

# Table 24  Functions of n in the Formulas Above

| $n$ | $\dfrac{1}{n}$ | $\dfrac{1}{n-1}$ | $\dfrac{n}{n-1}$ | $\dfrac{\bar{R}}{n-1}$ $\dfrac{\text{Nm}}{\text{K g-mol}}$ | $\dfrac{n\bar{R}}{n-1}$ $\dfrac{\text{Nm}}{\text{K g-mol}}$ | $\sqrt{\dfrac{2n\bar{R}}{n-1}}$ $\dfrac{\text{m}}{\text{s}}\left(\dfrac{\text{kg}}{\text{g-mol K}}\right)^{1/2}$ | $\sqrt{\dfrac{2}{\bar{R}}\dfrac{n}{n-1}}$ $\dfrac{(\text{kg g-mol K})^{1/2}}{\text{sN}}$ | $\dfrac{p^*}{p_0} =$ $\left(\dfrac{2}{n+1}\right)^{\frac{n}{n-1}}$ |
|---|---|---|---|---|---|---|---|---|
| 1.05 | 0.95238 | 20.000 | 21.000 | 166.29 | 174.60 | 18.687 | 2.2476 | 0.5954 |
| 1.10 | .90909 | 10.000 | 11.000 | 83.144 | 91.459 | 13.525 | 1.6267 | .5847 |
| 1.15 | .86957 | 6.6667 | 7.6667 | 55.429 | 63.744 | 11.291 | 1.3580 | .5744 |
| 1.20 | .83333 | 5.0000 | 6.0000 | 41.572 | 49.886 | 9.989 | 1.2014 | .5645 |
| 1.25 | .80000 | 4.0000 | 5.0000 | 33.258 | 41.572 | 9.118 | 1.0967 | .5549 |
| 1.30 | .76923 | 3.3333 | 4.3333 | 27.715 | 36.029 | 8.489 | 1.0210 | .5457 |
| 1.35 | .74074 | 2.8571 | 3.8571 | 23.755 | 32.070 | 8.009 | 0.9632 | .5369 |
| 1.40 | .71429 | 2.5000 | 3.5000 | 20.786 | 29.100 | 7.629 | 0.9176 | .5283 |
| 1.45 | .68966 | 2.2222 | 3.2222 | 18.476 | 26.791 | 7.320 | 0.8804 | .5200 |
| 1.50 | .66667 | 2.0000 | 3.0000 | 16.629 | 24.943 | 7.063 | 0.8495 | .5120 |
| 1.55 | .64516 | 1.8182 | 2.8182 | 15.117 | 23.432 | 6.846 | 0.8234 | .5043 |
| 1.60 | .62500 | 1.6667 | 2.6667 | 13.857 | 22.172 | 6.659 | 0.8009 | .4968 |
| 1.65 | .60606 | 1.5385 | 2.5385 | 12.791 | 21.106 | 6.497 | 0.7814 | .4895 |
| 1.70 | .58824 | 1.4286 | 2.4286 | 11.878 | 20.192 | 6.355 | 0.7643 | .4825 |
| 1.75 | .57143 | 1.3333 | 2.3333 | 11.086 | 19.400 | 6.229 | 0.7492 | .4757 |
| 1.80 | .55556 | 1.2500 | 2.2500 | 10.393 | 18.707 | 6.117 | 0.7357 | .4690 |

# Table 25 Functions of Pressure Ratio

n = 1.4

| $r$ | $\dfrac{1}{r}$ | $\dfrac{1}{r^{n}}$ | $r^{\frac{n-1}{n}}$ | $\sqrt{1-r^{\frac{n-1}{n}}}$ | $\dfrac{1}{r^{n}}\sqrt{1-r^{\frac{n-1}{n}}}$ |
|---|---|---|---|---|---|
| 1 | 1 | 1 | 1 | 0 | 0 |
| 0.999 | 1.001 | .99929 | .99971 | .01690 | .01689 |
| .998 | 1.002 | .99857 | .99943 | .02391 | .02388 |
| .997 | 1.003 | .99786 | .99914 | .02931 | .02924 |
| .996 | 1.004 | .99714 | .99886 | .03383 | .03373 |
| .995 | 1.005 | .99643 | .99857 | .03783 | .03770 |
| .994 | 1.006 | .99571 | .99828 | .04145 | .04127 |
| .993 | 1.007 | .99500 | .99800 | .04478 | .04455 |
| .992 | 1.008 | .99428 | .99771 | .04788 | .04760 |
| .991 | 1.009 | .99356 | .99742 | .05079 | .05046 |
| .990 | 1.010 | .99285 | .99713 | .05355 | .05316 |
| .989 | 1.011 | .99213 | .99684 | .05617 | .05573 |
| .988 | 1.012 | .99141 | .99656 | .05868 | .05818 |
| .987 | 1.013 | .99070 | .99627 | .06109 | .06052 |
| .986 | 1.014 | .98998 | .99598 | .06341 | .06277 |
| .985 | 1.015 | .98926 | .99569 | .06564 | .06494 |
| .984 | 1.016 | .98855 | .99540 | .06780 | .06703 |
| .983 | 1.017 | .98783 | .99511 | .06990 | .06906 |
| .982 | 1.018 | .98711 | .99482 | .07195 | .07102 |
| .981 | 1.019 | .98639 | .99453 | .07393 | .07293 |
| .980 | 1.020 | .98567 | .99424 | .07586 | .07478 |
| .979 | 1.021 | .98495 | .99395 | .07775 | .07658 |
| .978 | 1.022 | .98424 | .99366 | .07960 | .07834 |
| .977 | 1.024 | .98352 | .99337 | .08140 | .08006 |
| .976 | 1.025 | .98280 | .99308 | .08317 | .08174 |
| .975 | 1.026 | .98208 | .99279 | .08490 | .08338 |
| .974 | 1.027 | .98136 | .99250 | .08659 | .08498 |
| .973 | 1.028 | .98064 | .99221 | .08826 | .08655 |
| .972 | 1.029 | .97992 | .99192 | .08990 | .08809 |
| .971 | 1.030 | .97920 | .99163 | .09150 | .08960 |
| .970 | 1.031 | .97848 | .99134 | .09308 | .09108 |
| .969 | 1.032 | .97776 | .99104 | .09464 | .09253 |
| .968 | 1.033 | .97704 | .99075 | .09617 | .09396 |
| .967 | 1.034 | .97632 | .99046 | .09768 | .09537 |
| .966 | 1.035 | .97559 | .99017 | .09917 | .09675 |
| .965 | 1.036 | .97487 | .98987 | .10064 | .09811 |
| .964 | 1.037 | .97415 | .98958 | .10208 | .09944 |
| .963 | 1.038 | .97343 | .98929 | .10351 | .10076 |
| .962 | 1.040 | .97271 | .98899 | .10492 | .10205 |
| .961 | 1.041 | .97199 | .98870 | .10631 | .10333 |
| .960 | 1.042 | .97126 | .98840 | .10768 | .10459 |
| .958 | 1.044 | .96982 | .98782 | .11038 | .10705 |
| .956 | 1.046 | .96837 | .98723 | .11302 | .10945 |
| .954 | 1.048 | .96692 | .98664 | .11561 | .11178 |
| .952 | 1.050 | .96547 | .98604 | .11814 | .11406 |
| .950 | 1.053 | .96402 | .98545 | .12062 | .11628 |
| .948 | 1.055 | .96257 | .98486 | .12305 | .11844 |
| .946 | 1.057 | .96112 | .98426 | .12544 | .12056 |
| .944 | 1.059 | .95967 | .98367 | .12779 | .12264 |
| .942 | 1.062 | .95822 | .98307 | .13010 | .12467 |
| .940 | 1.064 | .95677 | .98248 | .13238 | .12665 |
| .938 | 1.066 | .95531 | .98188 | .13462 | .12860 |
| .936 | 1.068 | .95386 | .98128 | .13683 | .13051 |
| .934 | 1.071 | .95240 | .98068 | .13900 | .13238 |
| .932 | 1.073 | .95094 | .98008 | .14114 | .13421 |
| .930 | 1.075 | .94948 | .97948 | .14325 | .13601 |
| .928 | 1.078 | .94803 | .97888 | .14534 | .13778 |
| .926 | 1.080 | .94657 | .97827 | .14740 | .13952 |
| .924 | 1.082 | .94510 | .97767 | .14943 | .14123 |
| .922 | 1.085 | .94364 | .97706 | .15144 | .14291 |
| .920 | 1.087 | .94218 | .97646 | .15343 | .14456 |
| .918 | 1.089 | .94072 | .97585 | .15540 | .14618 |
| .916 | 1.092 | .93925 | .97524 | .15734 | .14778 |
| .914 | 1.094 | .93779 | .97463 | .15927 | .14936 |
| .912 | 1.096 | .93632 | .97402 | .16117 | .15091 |
| .910 | 1.099 | .93485 | .97341 | .16305 | .15243 |
| .908 | 1.101 | .93339 | .97280 | .16492 | .15393 |
| .906 | 1.104 | .93192 | .97219 | .16676 | .15541 |
| .904 | 1.106 | .93045 | .97158 | .16859 | .15687 |
| .902 | 1.109 | .92898 | .97096 | .17041 | .15830 |
| .900 | 1.111 | .92750 | .97035 | .17220 | .15972 |
| .895 | 1.117 | .92382 | .96880 | .17663 | .16317 |
| .890 | 1.124 | .92013 | .96725 | .18096 | .16651 |
| .885 | 1.130 | .91644 | .96570 | .18521 | .16973 |
| .880 | 1.136 | .91274 | .96414 | .18938 | .17285 |
| .875 | 1.143 | .90903 | .96257 | .19347 | .17588 |
| .870 | 1.149 | .90531 | .96099 | .19750 | .17881 |
| .865 | 1.156 | .90159 | .95941 | .20146 | .18165 |
| .860 | 1.163 | .89787 | .95782 | .20537 | .18440 |
| .855 | 1.170 | .89414 | .95623 | .20922 | .18707 |
| .850 | 1.176 | .89040 | .95463 | .21301 | .18966 |
| .845 | 1.183 | .88665 | .95302 | .21674 | .19218 |
| .840 | 1.190 | .88290 | .95141 | .22043 | .19463 |
| .835 | 1.198 | .87914 | .94979 | .22408 | .19701 |
| .830 | 1.205 | .87538 | .94816 | .22768 | .19932 |
| .825 | 1.212 | .87161 | .94652 | .23125 | .20157 |
| .820 | 1.220 | .86784 | .94488 | .23478 | .20375 |
| .815 | 1.227 | .86406 | .94323 | .23826 | .20588 |
| .810 | 1.235 | .86027 | .94157 | .24171 | .20795 |
| .805 | 1.242 | .85647 | .93990 | .24513 | .20996 |
| .800 | 1.250 | .85267 | .93823 | .24852 | .21191 |
| .795 | 1.258 | .84886 | .93656 | .25188 | .21381 |
| .790 | 1.266 | .84504 | .93487 | .25520 | .21566 |
| .785 | 1.274 | .84121 | .93317 | .25850 | .21746 |
| .780 | 1.282 | .83738 | .93147 | .26178 | .21921 |
| .775 | 1.290 | .83354 | .92976 | .26502 | .22091 |
| .770 | 1.299 | .82970 | .92804 | .26824 | .22256 |
| .765 | 1.307 | .82585 | .92631 | .27144 | .22417 |
| .760 | 1.316 | .82199 | .92458 | .27462 | .22573 |
| .755 | 1.324 | .81812 | .92284 | .27777 | .22725 |
| .750 | 1.333 | .81425 | .92109 | .28090 | .22873 |
| .745 | 1.342 | .81037 | .91933 | .28401 | .23016 |
| .740 | 1.351 | .80648 | .91757 | .28711 | .23155 |
| .735 | 1.361 | .80258 | .91580 | .29019 | .23290 |
| .730 | 1.370 | .79868 | .91401 | .29325 | .23421 |
| .725 | 1.379 | .79477 | .91221 | .29629 | .23548 |
| .720 | 1.389 | .79085 | .91041 | .29931 | .23671 |
| .715 | 1.399 | .78692 | .90860 | .30233 | .23791 |
| .710 | 1.408 | .78299 | .90678 | .30532 | .23906 |
| .705 | 1.418 | .77905 | .90495 | .30830 | .24018 |

## Table 25   $n = 1.4$

| $r$ | $\frac{1}{r}$ | $\frac{1}{r^n}$ | $r^{\frac{n-1}{n}}$ | $\sqrt{1-r^{\frac{n-1}{n}}}$ | $\frac{1}{r^n}\sqrt{1-r^{\frac{n-1}{n}}}$ | $r$ | $\frac{1}{r}$ | $\frac{1}{r^n}$ | $r^{\frac{n-1}{n}}$ | $\sqrt{1-r^{\frac{n-1}{n}}}$ | $\frac{1}{r^n}\sqrt{1-r^{\frac{n-1}{n}}}$ |
|---|---|---|---|---|---|---|---|---|---|---|---|
| .700 | 1.429 | .77510 | .90311 | .31127 | .24126 | .425 | 2.353 | .54271 | .78311 | .46571 | .25274 |
| .695 | 1.439 | .77114 | .90126 | .31422 | .24231 | .420 | 2.381 | .53814 | .78047 | .46854 | .25214 |
| .690 | 1.449 | .76717 | .89940 | .31716 | .24332 | .415 | 2.410 | .53355 | .77781 | .47138 | .25150 |
| .685 | 1.460 | .76319 | .89753 | .32009 | .24429 | .410 | 2.439 | .52895 | .77512 | .47422 | .25084 |
| .680 | 1.471 | .75921 | .89566 | .32301 | .24523 | .405 | 2.469 | .52434 | .77240 | .47707 | .25015 |
| .675 | 1.481 | .75522 | .89377 | .32592 | .24614 | .400 | 2.500 | .51971 | .76967 | .47993 | .24942 |
| .670 | 1.493 | .75122 | .89188 | .32882 | .24701 | .395 | 2.532 | .51506 | .76691 | .48280 | .24867 |
| .665 | 1.504 | .74721 | .88998 | .33170 | .24785 | .390 | 2.564 | .51039 | .76412 | .48567 | .24789 |
| .660 | 1.515 | .74320 | .88806 | .33458 | .24866 | .385 | 2.597 | .50571 | .76131 | .48856 | .24707 |
| .655 | 1.527 | .73917 | .88613 | .33745 | .24943 | .380 | 2.632 | .50101 | .75847 | .49146 | .24622 |
| .650 | 1.538 | .73514 | .88419 | .34031 | .25017 | .375 | 2.667 | 49629 | .75560 | .49436 | .24535 |
| .645 | 1.550 | .73109 | .88224 | .34316 | .25088 | .370 | 2.703 | .49155 | .75271 | .49728 | .24444 |
| .640 | 1.562 | .72704 | .88028 | .34600 | .25155 | .365 | 2.740 | .48680 | .74979 | .50021 | .24350 |
| .635 | 1.575 | .72298 | .87831 | .34883 | .25220 | .360 | 2.778 | .48203 | .74684 | .50315 | .24253 |
| .630 | 1.587 | .71891 | .87633 | .35166 | .25281 | .355 | 2.817 | .47724 | .74386 | .50610 | .24153 |
| .625 | 1.600 | .71483 | .87434 | .35448 | .25340 | .350 | 2.857 | .47243 | .74086 | .50906 | .24049 |
| .620 | 1.613 | .71074 | .87233 | .35730 | .25395 | .345 | 2.896 | .46760 | .73782 | .51204 | .23943 |
| .615 | 1.626 | .70664 | .87032 | .36011 | .25447 | .340 | 2.941 | .46275 | .73475 | .51503 | .23833 |
| .610 | 1.639 | .70253 | .86829 | .36291 | .25496 | .335 | 2.985 | .45788 | .73164 | .51803 | .23719 |
| .605 | 1.653 | .69841 | .86625 | .36571 | .25542 | .330 | 3.030 | .45299 | .72850 | .52105 | .23603 |
| .600 | 1.667 | .69428 | .86420 | .36851 | .25585 | .325 | 3.077 | .44807 | .72533 | .52409 | .23483 |
| .595 | 1.681 | .69015 | .86214 | .37130 | .25625 | .320 | 3.125 | .44313 | .72213 | .52714 | .23359 |
| .590 | 1.695 | .68600 | .86006 | .37409 | .25662 | .315 | 3.175 | .43818 | .71889 | .53020 | .23232 |
| .585 | 1.709 | .68184 | .85797 | .37687 | .25696 | .310 | 3.226 | .43320 | .71561 | .53328 | .23102 |
| .580 | 1.724 | .67767 | .85587 | .37965 | .25727 | .305 | 3.279 | .42820 | .71229 | .53638 | .22968 |
| .575 | 1.739 | .67350 | .85376 | .38242 | .25756 | .300 | 3.333 | .42317 | .70893 | .53950 | .22830 |
| .570 | 1.754 | .66931 | .85163 | .38519 | .25781 | .295 | 3.390 | .41812 | .70554 | .54264 | .22689 |
| .565 | 1.770 | .66511 | .84949 | .38796 | .25803 | .290 | 3.448 | .41305 | .70210 | .54580 | .22544 |
| .560 | 1.786 | .66090 | .84733 | .39073 | .25823 | .285 | 3.509 | .40795 | .69862 | .54898 | .22396 |
| .555 | 1.802 | .65668 | .84516 | .39350 | .25840 | .280 | 3.571 | .40282 | .69510 | .55218 | .22243 |
| .550 | 1.818 | .65245 | .84298 | .39626 | .25854 | .275 | 3.636 | .39767 | .69153 | .55540 | .22087 |
| .545 | 1.835 | .64820 | .84078 | .39902 | .25865 | .270 | 3.704 | .39249 | .68791 | .55865 | .21927 |
| .540 | 1.852 | .64395 | .83857 | .40178 | .25873 | .265 | 3.774 | .38728 | .68425 | .56192 | .21762 |
| .535 | 1.869 | .63969 | .83635 | .40454 | .25878 | .260 | 3.846 | .38205 | .68053 | .56521 | .21594 |
| .530 | 1.887 | .63541 | .83411 | .40730 | .25880 | .255 | 3.922 | .37679 | .67677 | .56853 | .21422 |
| .525 | 1.905 | .63112 | .83185 | .41006 | .25880 | .250 | 4.000 | .37150 | .67295 | .57188 | .21245 |
| .520 | 1.923 | .62682 | .82958 | .41282 | .25877 | .245 | 4.082 | .36618 | .66908 | .57526 | .21065 |
| .515 | 1.942 | .62251 | .82729 | .41558 | .25870 | .240 | 4.167 | .36082 | .66515 | .57867 | .20880 |
| .510 | 1.961 | .61819 | .82499 | .41834 | .25861 | .235 | 4.255 | .35544 | .66116 | .58211 | .20690 |
| .505 | 1.980 | .61386 | .82267 | .42110 | .25850 | .230 | 4.348 | .35002 | .65711 | .58558 | .20496 |
| .500 | 2.000 | .60951 | .82034 | .42387 | .25835 | .225 | 4.444 | .34457 | .65299 | .58908 | .20298 |
| .495 | 2.020 | .60515 | .81798 | .42663 | .25818 | .220 | 4.545 | .33908 | .64881 | .59261 | .20094 |
| .490 | 2.041 | .60078 | .81561 | .42940 | .25797 | .215 | 4.651 | .33356 | .64457 | .59619 | .19886 |
| .485 | 2.062 | .59639 | .81323 | .43217 | .25774 | .210 | 4.762 | .32800 | .64025 | .59980 | .19673 |
| .480 | 2.083 | .59199 | .81082 | .43494 | .25748 | .205 | 4.878 | .32240 | .63585 | .60345 | .19455 |
| .475 | 2.105 | .58758 | .80840 | .43772 | .25720 | .200 | 5.000 | .31676 | .63139 | .60714 | .19232 |
| .470 | 2.128 | .58316 | .80596 | .44050 | .25688 | .198 | 5.051 | .31450 | .62957 | .60863 | .19141 |
| .465 | 2.151 | .57872 | .80350 | .44328 | .25654 | .196 | 5.102 | .31223 | .62775 | .61012 | .19050 |
| .460 | 2.174 | .57427 | .80102 | .44607 | .25616 | .194 | 5.155 | .30995 | .62591 | .61162 | .18957 |
| .455 | 2.198 | .56980 | .79853 | .44886 | .25576 | .192 | 5.208 | .30766 | .62406 | .61313 | .18864 |
| .450 | 2.222 | .56532 | .79601 | .45165 | .25533 | .190 | 5.263 | .30537 | .62220 | .61465 | .18770 |
| .445 | 2.247 | .56083 | .79347 | .45445 | .25487 | .188 | 5.319 | .30307 | .62032 | .61618 | .18675 |
| .440 | 2.273 | .55632 | .79091 | .45726 | .25438 | .186 | 5.376 | .30076 | .61843 | .61771 | .18579 |
| .435 | 2.299 | .55180 | .78834 | .46007 | .25386 | .184 | 5.435 | .29845 | .61652 | .61926 | .18482 |
| .430 | 2.326 | .54726 | .78574 | .46289 | .25332 | .182 | 5.495 | .29613 | .61460 | .62081 | .18384 |

# Table 25    n = 1.4

| r | $\frac{1}{r}$ | $r^{\frac{1}{n}}$ | $r^{\frac{n-1}{n}}$ | $\sqrt{1-r^{\frac{n-1}{n}}}$ | $r^{\frac{1}{n}}\sqrt{1-r^{\frac{n-1}{n}}}$ | r | $\frac{1}{r}$ | $r^{\frac{1}{n}}$ | $r^{\frac{n-1}{n}}$ | $\sqrt{1-r^{\frac{n-1}{n}}}$ | $r^{\frac{1}{n}}\sqrt{1-r^{\frac{n-1}{n}}}$ |
|---|---|---|---|---|---|---|---|---|---|---|---|
| .180 | 5.556 | .29380 | .61266 | .62237 | .18285 | .070 | 14.3 | .14965 | .46777 | .72954 | .10918 |
| .178 | 5.618 | .29146 | .61071 | .62394 | .18185 | .069 | 14.5 | .14812 | .46585 | .73086 | .10825 |
| .176 | 5.682 | .28912 | .60874 | .62551 | .18085 | .068 | 14.7 | .14658 | .46391 | .73218 | .10732 |
| .174 | 5.747 | .28677 | .60676 | .62709 | .17983 | .067 | 14.9 | .14504 | .46195 | .73352 | .10639 |
| .172 | 5.814 | .28441 | .60476 | .62869 | .17881 | .066 | 15.2 | .14349 | .45997 | .73487 | .10545 |
| .170 | 5.882 | .28205 | .60274 | .63029 | .17777 | .065 | 15.4 | .14193 | .45797 | .73623 | .10450 |
| .168 | 5.952 | .27967 | .60070 | .63190 | .17672 | .064 | 15.6 | .14037 | .45594 | .73760 | .10354 |
| .166 | 6.024 | .27729 | .59865 | .63352 | .17567 | .063 | 15.9 | .13880 | .45389 | .73899 | .10257 |
| .164 | 6.098 | .27490 | .59658 | .63516 | .17460 | .062 | 16.1 | .13722 | .45182 | .74039 | .10160 |
| .162 | 6.173 | .27250 | .59449 | .63680 | .17353 | .061 | 16.4 | .13564 | .44973 | .74180 | .10062 |
| .160 | 6.250 | .27009 | .59239 | .63845 | .17244 | .060 | 16.7 | .13405 | .44761 | .74323 | .09963 |
| .158 | 6.329 | .26768 | .59027 | .64010 | .17134 | .059 | 16.9 | .13245 | .44547 | .74467 | .09863 |
| .156 | 6.410 | .26525 | .58812 | .64177 | .17023 | .058 | 17.2 | .13084 | .44330 | .74613 | .09762 |
| .154 | 6.494 | .26282 | .58595 | .64346 | .16911 | .057 | 17.5 | .12922 | .44110 | .74760 | .09660 |
| .152 | 6.579 | .26038 | .58376 | .64516 | .16798 | .056 | 17.9 | .12760 | .43887 | .74908 | .09558 |
| .150 | 6.667 | .25793 | .58156 | .64687 | .16684 | .055 | 18.2 | .12597 | .43662 | .75059 | .09455 |
| .148 | 6.757 | .25546 | .57934 | .64858 | .16569 | .054 | 18.5 | .12433 | .43434 | .75211 | .09351 |
| .146 | 6.849 | .25299 | .57709 | .65031 | .16453 | .053 | 18.9 | .12268 | .43202 | .75364 | .09246 |
| .144 | 6.944 | .25051 | .57482 | .65205 | .16335 | .052 | 19.2 | .12102 | .42968 | .75520 | .09139 |
| .142 | 7.042 | .24802 | .57253 | .65381 | .16216 | .051 | 19.6 | .11935 | .42730 | .75677 | .09032 |
| .140 | 7.143 | .24552 | .57021 | .65558 | .16096 | .050 | 20.0 | .11768 | .42489 | .75836 | .08924 |
| .138 | 7.246 | .24301 | .56787 | .65736 | .15975 | .049 | 20.4 | .11599 | .42245 | .75997 | .08815 |
| .136 | 7.353 | .24049 | .56551 | .65916 | .15852 | .048 | 20.8 | .11430 | .41997 | .76160 | .08705 |
| .134 | 7.463 | .23796 | .56312 | .66097 | .15728 | .047 | 21.3 | .11259 | .41745 | .76325 | .08593 |
| .132 | 7.576 | .23542 | .56071 | .66279 | .15603 | .046 | 21.7 | .11087 | .41489 | .76493 | .08481 |
| .130 | 7.692 | .23286 | .55827 | .66463 | .15477 | .045 | 22.2 | .10915 | .41229 | .76662 | .08367 |
| .128 | 7.812 | .23030 | .55580 | .66648 | .15349 | .044 | 22.7 | .10741 | .40965 | .76834 | .08253 |
| .126 | 7.937 | .22773 | .55330 | .66835 | .15220 | .043 | 23.3 | .10566 | .40697 | .77008 | .08137 |
| .124 | 8.065 | .22514 | .55078 | .67024 | .15090 | .042 | 23.8 | .10390 | .40424 | .77185 | .08019 |
| .122 | 8.197 | .22254 | .54823 | .67214 | .14958 | .041 | 24.4 | .10212 | .40147 | .77364 | .07901 |
| .120 | 8.333 | .21992 | .54564 | .67406 | .14824 | .0400 | 25.00 | .10034 | .39865 | .77547 | .07781 |
| .118 | 8.475 | .21730 | .54303 | .67600 | .14689 | .0395 | 25.32 | .09944 | .39722 | .77639 | .07721 |
| .116 | 8.621 | .21466 | .54038 | .67795 | .14553 | .0390 | 25.64 | .09854 | .39577 | .77732 | .07660 |
| .114 | 8.772 | .21201 | .53770 | .67992 | .14415 | .0385 | 25.97 | .09764 | .39432 | .77826 | .07599 |
| .112 | 8.929 | .20935 | .53499 | .68191 | .14276 | .0380 | 26.32 | .09673 | .39285 | .77920 | .07537 |
| .110 | 9.091 | .20667 | .53224 | .68393 | .14135 | .0375 | 26.67 | .09582 | .39136 | .78015 | .07475 |
| .108 | 9.259 | .20398 | .52946 | .68596 | .13992 | .0370 | 27.03 | .09490 | .38986 | .78111 | .07413 |
| .106 | 9.434 | .20128 | .52664 | .68801 | .13848 | .0365 | 27.40 | .09399 | .38835 | .78208 | .07350 |
| .104 | 9.615 | .19856 | .52379 | .69008 | .13702 | .0360 | 27.78 | .09307 | .38682 | .78305 | .07288 |
| .102 | 9.804 | .19582 | .52089 | .69218 | .13554 | .0355 | 28.17 | .09214 | .38528 | .78404 | .07224 |
| .100 | 10.0 | .19307 | .51795 | .69430 | .13405 | .0350 | 28.57 | .09121 | .38372 | .78503 | .07160 |
| .098 | 10.2 | .19030 | .51497 | .69644 | .13254 | .0345 | 28.99 | .09028 | .38215 | .78603 | .07096 |
| .096 | 10.4 | .18752 | .51194 | .69861 | .13100 | .0340 | 29.41 | .08934 | .38056 | .78704 | .07032 |
| .094 | 10.6 | .18472 | .50887 | .70080 | .12945 | .0335 | 29.85 | .08840 | .37895 | .78807 | .06967 |
| .092 | 10.9 | .18191 | .50576 | .70302 | .12788 | .0330 | 30.30 | .08746 | .37733 | .78910 | .06901 |
| .090 | 11.1 | .17907 | .50259 | .70527 | .12630 | .0325 | 30.77 | .08651 | .37568 | .79014 | .06835 |
| .088 | 11.4 | .17622 | .49938 | .70755 | .12469 | .0320 | 31.25 | .08556 | .37402 | .79119 | .06769 |
| .086 | 11.6 | .17335 | .49611 | .70986 | .12305 | .0315 | 31.75 | .08460 | .37235 | .79225 | .06702 |
| .084 | 11.9 | .17046 | .49278 | .71219 | .12140 | .0310 | 32.26 | .08364 | .37065 | .79332 | .06635 |
| .082 | 12.2 | .16755 | .48940 | .71456 | .11973 | .0305 | 32.79 | .08267 | .36893 | .79440 | .06567 |
| .080 | 12.5 | .16462 | .48596 | .71697 | .11803 | .0300 | 33.33 | .08170 | .36719 | .79549 | .06499 |
| .078 | 12.8 | .16167 | .48246 | .71941 | .11631 | .0295 | 33.89 | .08073 | .36543 | .79660 | .06431 |
| .076 | 13.2 | .15870 | .47889 | .72188 | .11456 | .0290 | 34.48 | .07975 | .36365 | .79772 | .06362 |
| .074 | 13.5 | .15571 | .47525 | .72439 | .11279 | .0285 | 35.09 | .07876 | .36185 | .79884 | .06292 |
| .072 | 13.9 | .15269 | .47155 | .72695 | .11100 | .0280 | 35.71 | .07777 | .36002 | .79998 | .06222 |

# Table 25   n = 1.4

| r | $\frac{1}{r}$ | $r^{\frac{1}{n}}$ | $r^{\frac{n-1}{n}}$ | $\sqrt{1-r^{\frac{n-1}{n}}}$ | $r^{\frac{1}{n}}\sqrt{1-r^{\frac{n-1}{n}}}$ | r | $\frac{1}{r}$ | $r^{\frac{1}{n}}$ | $r^{\frac{n-1}{n}}$ | $\sqrt{1-r^{\frac{n-1}{n}}}$ | $r^{\frac{1}{n}}\sqrt{1-r^{\frac{n-1}{n}}}$ |
|---|---|---|---|---|---|---|---|---|---|---|---|
| .0275 | 36.36 | .07678 | .35817 | .80114 | .06151 | .0060 | 167 | .02588 | .23184 | .87645 | .02268 |
| .0270 | 37.04 | .07578 | .35630 | .80231 | .06080 | .0058 | 172 | .02526 | .22960 | .87772 | .02217 |
| .0265 | 37.74 | .07477 | .35440 | .80349 | .06008 | .0056 | 179 | .02464 | .22731 | .87902 | .02166 |
| .0260 | 38.46 | .07376 | .35248 | .80469 | .05936 | .0054 | 185 | .02400 | .22496 | .88036 | .02113 |
| .0255 | 39.22 | .07275 | .35053 | .80590 | .05863 | .0052 | 192 | .02337 | .22255 | .88173 | .02060 |
| .0250 | 40.00 | .07173 | .34855 | .80712 | .05789 | .0050 | 200 | .02272 | .22007 | .88314 | .02006 |
| .0245 | 40.82 | .07070 | .34655 | .80836 | .05715 | .0048 | 208 | .02207 | .21752 | .88458 | .01952 |
| .0240 | 41.67 | .06966 | .34451 | .80962 | .05640 | .0046 | 217 | .02141 | .21489 | .88606 | .01897 |
| .0235 | 42.55 | .06862 | .34244 | .81090 | .05565 | .0044 | 227 | .02074 | .21218 | .88759 | .01841 |
| .0230 | 43.48 | .06758 | .34035 | .81219 | .05489 | .0042 | 238 | .02006 | .20938 | .88917 | .01784 |
| .0225 | 44.44 | .06653 | .33822 | .81350 | .05412 | .0040 | 250 | .01937 | .20648 | .89080 | .01726 |
| .0220 | 45.45 | .06547 | .33605 | .81483 | .05334 | .0039 | 256 | .01903 | .20499 | .89163 | .01696 |
| .0215 | 46.51 | .06440 | .33385 | .81618 | .05256 | .0038 | 263 | .01868 | .20347 | .89248 | .01667 |
| .0210 | 47.62 | .06333 | .33162 | .81755 | .05177 | .0037 | 270 | .01832 | .20193 | .89335 | .01637 |
| .0205 | 48.78 | .06225 | .32934 | .81894 | .05098 | .0036 | 278 | .01797 | .20036 | .89423 | .01607 |
| .0200 | 50.00 | .06116 | .32702 | .82035 | .05017 | .0035 | 286 | .01761 | .19875 | .89513 | .01576 |
| .0195 | 51.28 | .06006 | .32467 | .82179 | .04936 | .0034 | 294 | .01725 | .19711 | .89604 | .01546 |
| .0190 | 52.63 | .05896 | .32227 | .82325 | .04854 | .0033 | 303 | .01688 | .19544 | .89698 | .01515 |
| .0185 | 54.05 | .05784 | .31982 | .82473 | .04771 | .0032 | 312 | .01652 | .19373 | .89793 | .01483 |
| .0180 | 55.56 | .05672 | .31733 | .82624 | .04687 | .0031 | 323 | .01615 | .19198 | .89890 | .01452 |
| .0175 | 57.14 | .05559 | .31479 | .82778 | .04602 | .0030 | 333 | .01577 | .19019 | .89989 | .01420 |
| .0170 | 58.82 | .05445 | .31219 | .82934 | .04516 | .0029 | 345 | .01540 | .18835 | .90091 | .01387 |
| .0165 | 60.61 | .05331 | .30953 | .83094 | .04429 | .0028 | 357 | .01502 | .18647 | .90196 | .01354 |
| .0160 | 62.50 | .05215 | .30682 | .83257 | .04342 | .0027 | 370 | .01463 | .18455 | .90303 | .01321 |
| .0155 | 64.52 | .05098 | .30405 | .83423 | .04253 | .0026 | 385 | .01424 | .18257 | .90412 | .01288 |
| .0150 | 66.67 | .04980 | .30122 | .83593 | .04163 | .0025 | 400 | .01385 | .18053 | .90524 | .01254 |
| .0145 | 68.97 | .04861 | .29832 | .83767 | .04072 | .0024 | 417 | .01345 | .17844 | .90640 | .01219 |
| .0140 | 71.43 | .04740 | .29534 | .83944 | .03979 | .0023 | 435 | .01305 | .17628 | .90759 | .01184 |
| .0135 | 74.07 | .04619 | .29229 | .84126 | .03886 | .0022 | 455 | .01264 | .17406 | .90881 | .01149 |
| .0130 | 76.92 | .04496 | .28915 | .84312 | .03791 | .0021 | 476 | .01223 | .17176 | .91008 | .01113 |
| .0125 | 80.00 | .04372 | .28593 | .84503 | .03694 | .0020 | 500 | .01181 | .16938 | .91138 | .01076 |
| .0120 | 83.33 | .04246 | .28261 | .84699 | .03596 | .0019 | 526 | .01138 | .16692 | .91273 | .01039 |
| .0115 | 86.96 | .04119 | .27920 | .84900 | .03497 | .0018 | 556 | .01095 | .16436 | .91413 | .01001 |
| .0110 | 90.91 | .03990 | .27568 | .85107 | .03396 | .0017 | 588 | .01051 | .16170 | .91559 | .00963 |
| .0105 | 95.24 | .03860 | .27204 | .85321 | .03293 | .0016 | 625 | .01007 | .15892 | .91710 | .00923 |
| .0100 | 100 | .03728 | .26827 | .85541 | .03189 | .0015 | 667 | .00961 | .15602 | .91868 | .00883 |
| .0098 | 102 | .03674 | .26673 | .85631 | .03146 | .0014 | 714 | .00915 | .15298 | .92034 | .00842 |
| .0096 | 104 | .03620 | .26516 | .85723 | .03104 | .0013 | 769 | .00868 | .14977 | .92208 | .00800 |
| .0094 | 106 | .03566 | .26357 | .85816 | .03061 | .0012 | 833 | .00820 | .14638 | .92392 | .00757 |
| .0092 | 109 | .03512 | .26195 | .85910 | .03017 | .0011 | 909 | .00770 | .14279 | .92586 | .00713 |
| .0090 | 111 | .03457 | .26031 | .86005 | .02974 | .0010 | 1000 | .00720 | .13895 | .92793 | .00668 |
| .0088 | 114 | .03402 | .25865 | .86102 | .02929 | .0009 | 1111 | .00668 | .13483 | .93015 | .00621 |
| .0086 | 116 | .03347 | .25696 | .86200 | .02885 | .0008 | 1250 | .00614 | .13037 | .93254 | .00572 |
| .0084 | 119 | .03291 | .25523 | .86300 | .02840 | .0007 | 1429 | .00558 | .12549 | .93515 | .00522 |
| .0082 | 122 | .03235 | .25348 | .86401 | .02795 | .0006 | 1667 | .00500 | .12008 | .93804 | .00469 |
| .0080 | 125 | .03178 | .25170 | .86504 | .02749 | .0005 | 2000 | .00439 | .11399 | .94128 | .00413 |
| .0078 | 128 | .03121 | .24989 | .86609 | .02703 | .0004 | 2500 | .00374 | .10694 | .94502 | .00354 |
| .0076 | 132 | .03064 | .24804 | .86716 | .02657 | .0003 | 3333 | .00304 | .09851 | .94947 | .00289 |
| .0074 | 135 | .03006 | .24616 | .86824 | .02610 | .0002 | 5000 | .00228 | .08773 | .95513 | .00218 |
| .0072 | 139 | .02948 | .24424 | .86935 | .02563 | .0001 | 10,000 | .00139 | .07197 | .96334 | .00134 |
| .0070 | 143 | .02889 | .24228 | .87047 | .02515 | 0 | ∞ | 0 | 0 | 1 | 0 |
| .0068 | 147 | .02830 | .24028 | .87162 | .02467 | | | | | | |
| .0066 | 152 | .02770 | .23824 | .87279 | .02418 | | | | | | |
| .0064 | 156 | .02710 | .23616 | .87398 | .02369 | | | | | | |
| .0062 | 161 | .02649 | .23402 | .87520 | .02319 | | | | | | |

# Table 26  $r^{\frac{1}{n}}$

**Values of n**

| r | $\frac{1}{r}$ | 1.05 | 1.10 | 1.15 | 1.20 | 1.25 | 1.30 | 1.35 | 1.40 | 1.45 | 1.50 | 1.55 | 1.60 | 1.65 | 1.70 | 1.75 | 1.80 | r |
|---|---|---|---|---|---|---|---|---|---|---|---|---|---|---|---|---|---|---|
| 1 | 1 | 1 | 1 | 1 | 1 | 1 | 1 | 1 | 1 | 1 | 1 | 1 | 1 | 1 | 1 | 1 | 1 | 1 |
| 0.98 | 1.02 | 0.9809 | .9818 | .9826 | .9833 | .9840 | .9846 | .9851 | .9857 | .9862 | .9866 | .9871 | .9875 | .9878 | .9882 | .9885 | .9888 | 0.98 |
| .96 | 1.04 | 9619 | 9636 | 9651 | 9666 | 9679 | 9691 | 9702 | 9713 | 9722 | 9732 | 9740 | 9748 | 9756 | 9763 | 9769 | 9776 | .96 |
| .94 | 1.06 | 9428 | 9453 | 9476 | 9497 | 9517 | 9535 | 9552 | 9568 | 9582 | 9596 | 9609 | 9621 | 9632 | 9643 | 9653 | 9662 | .94 |
| .92 | 1.09 | 9237 | 9270 | 9301 | 9329 | 9355 | 9379 | 9401 | 9422 | 9441 | 9459 | 9476 | 9492 | 9507 | 9521 | 9535 | 9547 | .92 |
| .90 | 1.11 | 9045 | 9087 | 9125 | 9159 | 9192 | 9222 | 9249 | 9275 | 9299 | 9322 | 9343 | 9363 | 9381 | 9399 | 9416 | 9431 | .90 |
| .88 | 1.14 | 8854 | 8903 | 8948 | 8990 | 9028 | 9063 | 9097 | 9127 | 9156 | 9183 | 9208 | 9232 | 9255 | 9276 | 9296 | 9314 | .88 |
| .86 | 1.16 | 8662 | 8719 | 8771 | 8819 | 8863 | 8905 | 8943 | 8979 | 9012 | 9043 | 9073 | 9100 | 9126 | 9151 | 9174 | 9196 | .86 |
| .84 | 1.19 | 8470 | 8534 | 8593 | 8648 | 8698 | 8745 | 8788 | 8829 | 8867 | 8903 | 8936 | 8968 | 8997 | 9025 | 9052 | 9077 | .84 |
| .82 | 1.22 | 8278 | 8349 | 8415 | 8476 | 8532 | 8584 | 8633 | 8678 | 8721 | 8761 | 8798 | 8834 | 8867 | 8898 | 8928 | 8956 | .82 |
| .80 | 1.25 | 8085 | 8164 | 8236 | 8303 | 8365 | 8423 | 8476 | 8527 | 8574 | 8618 | 8659 | 8698 | 8735 | 8770 | 8803 | 8834 | .80 |
| .78 | 1.28 | 7893 | 7978 | 8057 | 8130 | 8197 | 8260 | 8319 | 8374 | 8425 | 8474 | 8519 | 8562 | 8602 | 8640 | 8676 | 8711 | .78 |
| .76 | 1.32 | 7700 | 7792 | 7877 | 7956 | 8029 | 8097 | 8160 | 8220 | 8276 | 8328 | 8377 | 8424 | 8468 | 8509 | 8549 | 8586 | .76 |
| .74 | 1.35 | 7507 | 7605 | 7696 | 7781 | 7859 | 7932 | 8001 | 8065 | 8125 | 8181 | 8234 | 8285 | 8332 | 8377 | 8419 | 8460 | .74 |
| .72 | 1.39 | 7314 | 7418 | 7515 | 7605 | 7689 | 7767 | 7840 | 7909 | 7973 | 8033 | 8090 | 8144 | 8195 | 8243 | 8288 | 8332 | .72 |
| .70 | 1.43 | 7120 | 7231 | 7333 | 7429 | 7518 | 7601 | 7678 | 7751 | 7819 | 7884 | 7944 | 8002 | 8056 | 8107 | 8156 | 8202 | .70 |
| .68 | 1.47 | 6926 | 7043 | 7151 | 7251 | 7345 | 7433 | 7515 | 7592 | 7665 | 7733 | 7797 | 7858 | 7916 | 7970 | 8022 | 8071 | .68 |
| .66 | 1.52 | 6732 | 6854 | 6968 | 7073 | 7172 | 7264 | 7351 | 7432 | 7508 | 7580 | 7649 | 7713 | 7774 | 7832 | 7886 | 7939 | .66 |
| .64 | 1.56 | 6537 | 6665 | 6784 | 6894 | 6998 | 7094 | 7185 | 7270 | 7351 | 7427 | 7498 | 7566 | 7630 | 7691 | 7749 | 7804 | .64 |
| .62 | 1.61 | 6343 | 6475 | 6599 | 6714 | 6822 | 6923 | 7018 | 7107 | 7192 | 7271 | 7346 | 7417 | 7485 | 7549 | 7610 | 7668 | .62 |
| .60 | 1.67 | 6148 | 6285 | 6413 | 6533 | 6645 | 6751 | 6850 | 6943 | 7031 | 7114 | 7192 | 7267 | 7337 | 7405 | 7468 | 7529 | .60 |
| .58 | 1.72 | 5952 | 6094 | 6227 | 6351 | 6468 | 6577 | 6680 | 6777 | 6868 | 6955 | 7037 | 7114 | 7188 | 7258 | 7325 | 7389 | .58 |
| .56 | 1.79 | 5757 | 5903 | 6040 | 6168 | 6289 | 6402 | 6508 | 6609 | 6704 | 6794 | 6879 | 6960 | 7037 | 7110 | 7180 | 7246 | .56 |
| .54 | 1.85 | 5561 | 5711 | 5852 | 5984 | 6108 | 6225 | 6335 | 6440 | 6538 | 6631 | 6720 | 6804 | 6884 | 6960 | 7032 | 7101 | .54 |
| .52 | 1.92 | 5364 | 5519 | 5663 | 5799 | 5927 | 6047 | 6161 | 6268 | 6370 | 6466 | 6558 | 6645 | 6728 | 6807 | 6882 | 6954 | .52 |
| .50 | 2.00 | 5168 | 5325 | 5473 | 5612 | 5743 | 5867 | 5984 | 6095 | 6200 | 6300 | 6394 | 6484 | 6570 | 6652 | 6730 | 6804 | .50 |
| .48 | 2.08 | 4971 | 5131 | 5282 | 5425 | 5559 | 5686 | 5806 | 5920 | 6028 | 6130 | 6228 | 6321 | 6409 | 6494 | 6574 | 6651 | .48 |
| .46 | 2.17 | 4773 | 4936 | 5090 | 5236 | 5373 | 5503 | 5626 | 5743 | 5854 | 5959 | 6059 | 6155 | 6246 | 6333 | 6416 | 6496 | .46 |
| .44 | 2.27 | 4575 | 4741 | 4897 | 5045 | 5185 | 5318 | 5444 | 5563 | 5677 | 5785 | 5888 | 5986 | 6080 | 6170 | 6255 | 6338 | .44 |
| .42 | 2.38 | 4377 | 4545 | 4703 | 4853 | 4996 | 5131 | 5259 | 5381 | 5498 | 5608 | 5714 | 5815 | 5911 | 6003 | 6091 | 6176 | .42 |
| .40 | 2.50 | 4178 | 4347 | 4508 | 4660 | 4804 | 4942 | 5073 | 5197 | 5316 | 5429 | 5537 | 5640 | 5739 | 5833 | 5924 | 6011 | .40 |
| .38 | 2.63 | 3979 | 4149 | 4311 | 4465 | 4611 | 4751 | 4883 | 5010 | 5131 | 5246 | 5357 | 5462 | 5563 | 5660 | 5753 | 5842 | .38 |
| .36 | 2.78 | 3779 | 3950 | 4113 | 4268 | 4416 | 4557 | 4692 | 4820 | 4943 | 5061 | 5173 | 5281 | 5384 | 5483 | 5578 | 5669 | .36 |
| .34 | 2.94 | 3579 | 3750 | 3914 | 4070 | 4219 | 4361 | 4497 | 4627 | 4752 | 4871 | 4986 | 5095 | 5201 | 5302 | 5399 | 5492 | .34 |
| .32 | 3.12 | 3378 | 3549 | 3713 | 3869 | 4019 | 4162 | 4300 | 4431 | 4557 | 4678 | 4794 | 4906 | 5013 | 5116 | 5215 | 5310 | .32 |
| .30 | 3.33 | 3177 | 3347 | 3510 | 3667 | 3817 | 3961 | 4099 | 4232 | 4359 | 4481 | 4599 | 4712 | 4821 | 4925 | 5026 | 5123 | .30 |
| .28 | 3.57 | 2975 | 3144 | 3306 | 3462 | 3612 | 3756 | 3895 | 4028 | 4157 | 4280 | 4399 | 4513 | 4623 | 4729 | 4832 | 4930 | .28 |
| .26 | 3.85 | 2772 | 2939 | 3099 | 3254 | 3404 | 3548 | 3687 | 3821 | 3949 | 4074 | 4193 | 4309 | 4420 | 4528 | 4631 | 4731 | .26 |
| .24 | 4.17 | 2569 | 2732 | 2891 | 3044 | 3193 | 3336 | 3475 | 3608 | 3737 | 3862 | 3982 | 4099 | 4211 | 4319 | 4424 | 4526 | .24 |
| .22 | 4.55 | 2364 | 2525 | 2680 | 2832 | 2978 | 3120 | 3258 | 3391 | 3520 | 3644 | 3765 | 3882 | 3995 | 4104 | 4210 | 4312 | .22 |
| .20 | 5.00 | 2159 | 2315 | 2467 | 2615 | 2759 | 2900 | 3036 | 3168 | 3296 | 3420 | 3540 | 3657 | 3770 | 3880 | 3986 | 4090 | .20 |
| .18 | 5.56 | 1953 | 2104 | 2251 | 2395 | 2536 | 2674 | 2808 | 2938 | 3065 | 3188 | 3308 | 3424 | 3537 | 3647 | 3754 | 3857 | .18 |
| .16 | 6.25 | 1746 | 1890 | 2032 | 2172 | 2308 | 2442 | 2573 | 2701 | 2826 | 2947 | 3066 | 3181 | 3293 | 3403 | 3509 | 3613 | .16 |
| .14 | 7.14 | 1537 | 1674 | 1809 | 1943 | 2074 | 2204 | 2331 | 2455 | 2577 | 2696 | 2813 | 2926 | 3037 | 3146 | 3251 | 3354 | .14 |
| .12 | 8.33 | 1327 | 1455 | 1582 | 1709 | 1834 | 1957 | 2079 | 2199 | 2317 | 2433 | 2546 | 2658 | 2766 | 2873 | 2977 | 3079 | .12 |
| .10 | 10.0 | 1116 | 1233 | 1350 | 1468 | 1585 | 1701 | 1817 | 1931 | 2043 | 2154 | 2264 | 2371 | 2477 | 2581 | 2683 | 2783 | .10 |
| .08 | 12.5 | 0902 | 1006 | 1112 | 1219 | 1326 | 1433 | 1540 | 1646 | 1752 | 1857 | 1960 | 2063 | 2164 | 2263 | 2362 | 2458 | .08 |
| .06 | 16.7 | 0686 | 0775 | 0866 | 0959 | 1053 | 1148 | 1244 | 1340 | 1437 | 1533 | 1628 | 1723 | 1818 | 1911 | 2004 | 2095 | .06 |
| .04 | 25.0 | 0466 | 0536 | 0609 | 0684 | 0761 | 0841 | 0921 | 1003 | 1086 | 1170 | 1253 | 1337 | 1422 | 1506 | 1590 | 1673 | .04 |
| .02 | 50.0 | 0241 | 0285 | 0333 | 0384 | 0437 | 0493 | 0551 | 0612 | 0673 | 0737 | 0801 | 0867 | 0934 | 1001 | 1069 | 1138 | .02 |
| 0 | ∞ | 0 | 0 | 0 | 0 | 0 | 0 | 0 | 0 | 0 | 0 | 0 | 0 | 0 | 0 | 0 | 0 | 0 |

# Table 27  $r^{\frac{n-1}{n}}$

**Values of n**

| r | $\frac{1}{r}$ | 1.05 | 1.10 | 1.15 | 1.20 | 1.25 | 1.30 | 1.35 | 1.40 | 1.45 | 1.50 | 1.55 | 1.60 | 1.65 | 1.70 | 1.75 | 1.80 | r |
|---|---|---|---|---|---|---|---|---|---|---|---|---|---|---|---|---|---|---|
| 1 | 1 | 1 | 1 | 1 | 1 | 1 | 1 | 1 | 1 | 1 | 1 | 1 | 1 | 1 | 1 | 1 | 1 | 1 |
| 0.98 | 1.02 | 0.9990 | .9982 | .9974 | .9966 | .9960 | .9953 | .9948 | .9942 | .9938 | .9933 | .9929 | .9925 | .9921 | .9917 | .9914 | .9911 | 0.98 |
| .96 | 1.04 | 9981 | 9963 | 9947 | 9932 | 9919 | 9906 | 9895 | 9884 | 9874 | 9865 | 9856 | 9848 | 9840 | 9833 | 9827 | 9820 | .96 |
| .94 | 1.06 | 9971 | 9944 | 9920 | 9897 | 9877 | 9858 | 9841 | 9825 | 9810 | 9796 | 9783 | 9771 | 9759 | 9748 | 9738 | 9729 | .94 |
| .92 | 1.09 | 9960 | 9924 | 9892 | 9862 | 9835 | 9809 | 9786 | 9765 | 9745 | 9726 | 9708 | 9692 | 9677 | 9662 | 9649 | 9636 | .92 |
| .90 | 1.11 | 9950 | 9905 | 9864 | 9826 | 9791 | 9760 | 9731 | 9703 | 9678 | 9655 | 9633 | 9613 | 9593 | 9575 | 9558 | 9543 | .90 |
| .88 | 1.14 | 9939 | 9884 | 9835 | 9789 | 9748 | 9709 | 9674 | 9641 | 9611 | 9583 | 9557 | 9532 | 9509 | 9487 | 9467 | 9448 | .88 |
| .86 | 1.16 | 9928 | 9864 | 9805 | 9752 | 9703 | 9658 | 9617 | 9578 | 9543 | 9510 | 9479 | 9450 | 9423 | 9398 | 9374 | 9352 | .86 |
| .84 | 1.19 | 9917 | 9843 | 9775 | 9714 | 9657 | 9606 | 9558 | 9514 | 9473 | 9435 | 9400 | 9367 | 9336 | 9307 | 9280 | 9254 | .84 |
| .82 | 1.22 | 9906 | 9821 | 9744 | 9675 | 9611 | 9552 | 9499 | 9449 | 9403 | 9360 | 9320 | 9283 | 9248 | 9215 | 9185 | 9156 | .82 |
| .80 | 1.25 | 9894 | 9799 | 9713 | 9635 | 9564 | 9498 | 9438 | 9382 | 9331 | 9283 | 9239 | 9197 | 9158 | 9122 | 9088 | 9056 | .80 |
| .78 | 1.28 | 9882 | 9777 | 9681 | 9594 | 9515 | 9443 | 9376 | 9315 | 9258 | 9205 | 9156 | 9110 | 9068 | 9028 | 8990 | 8955 | .78 |
| .76 | 1.32 | 9870 | 9754 | 9648 | 9553 | 9466 | 9386 | 9313 | 9246 | 9184 | 9126 | 9072 | 9022 | 8975 | 8931 | 8890 | 8852 | .76 |
| .74 | 1.35 | 9858 | 9730 | 9615 | 9511 | 9416 | 9329 | 9249 | 9176 | 9108 | 9045 | 8987 | 8932 | 8881 | 8834 | 8789 | 8747 | .74 |
| .72 | 1.39 | 9845 | 9706 | 9581 | 9467 | 9364 | 9270 | 9184 | 9104 | 9031 | 8963 | 8900 | 8841 | 8786 | 8735 | 8687 | 8642 | .72 |
| .70 | 1.43 | 9832 | 9681 | 9545 | 9423 | 9312 | 9210 | 9117 | 9031 | 8952 | 8879 | 8811 | 8748 | 8689 | 8634 | 8582 | 8534 | .70 |
| .68 | 1.47 | 9818 | 9655 | 9509 | 9377 | 9258 | 9148 | 9048 | 8957 | 8872 | 8794 | 8721 | 8653 | 8591 | 8532 | 8477 | 8425 | .68 |
| .66 | 1.52 | 9804 | 9629 | 9472 | 9331 | 9203 | 9086 | 8979 | 8881 | 8790 | 8707 | 8629 | 8557 | 8490 | 8427 | 8369 | 8314 | .66 |
| .64 | 1.56 | 9790 | 9602 | 9435 | 9283 | 9146 | 9021 | 8907 | 8803 | 8707 | 8618 | 8535 | 8459 | 8388 | 8321 | 8259 | 8201 | .64 |
| .62 | 1.61 | 9775 | 9575 | 9396 | 9234 | 9088 | 8956 | 8834 | 8723 | 8621 | 8527 | 8440 | 8359 | 8284 | 8213 | 8148 | 8086 | .62 |
| .60 | 1.67 | 9760 | 9546 | 9355 | 9184 | 9029 | 8888 | 8760 | 8642 | 8534 | 8434 | 8342 | 8257 | 8177 | 8103 | 8034 | 7969 | .60 |
| .58 | 1.72 | 9744 | 9517 | 9314 | 9132 | 8968 | 8819 | 8683 | 8559 | 8445 | 8340 | 8242 | 8152 | 8069 | 7991 | 7918 | 7850 | .58 |
| .56 | 1.79 | 9728 | 9487 | 9272 | 9079 | 8905 | 8748 | 8604 | 8473 | 8353 | 8243 | 8140 | 8046 | 7958 | 7876 | 7800 | 7728 | .56 |
| .54 | 1.85 | 9711 | 9455 | 9228 | 9024 | 8841 | 8675 | 8524 | 8386 | 8259 | 8143 | 8036 | 7937 | 7845 | 7759 | 7679 | 7604 | .54 |
| .52 | 1.92 | 9693 | 9423 | 9182 | 8967 | 8774 | 8599 | 8441 | 8296 | 8163 | 8041 | 7929 | 7825 | 7729 | 7639 | 7556 | 7478 | .52 |
| .50 | 2.00 | 9675 | 9389 | 9136 | 8909 | 8706 | 8522 | 8355 | 8203 | 8064 | 7937 | 7820 | 7711 | 7610 | 7517 | 7430 | 7349 | .50 |
| .48 | 2.08 | 9657 | 9355 | 9087 | 8849 | 8635 | 8442 | 8267 | 8108 | 7963 | 7830 | 7707 | 7594 | 7489 | 7392 | 7301 | 7217 | .48 |
| .46 | 2.17 | 9637 | 9318 | 9037 | 8786 | 8562 | 8359 | 8176 | 8010 | 7858 | 7719 | 7592 | 7474 | 7365 | 7263 | 7169 | 7081 | .46 |
| .44 | 2.27 | 9617 | 9281 | 8984 | 8721 | 8486 | 8274 | 8083 | 7909 | 7751 | 7606 | 7473 | 7350 | 7237 | 7132 | 7034 | 6943 | .44 |
| .42 | 2.38 | 9595 | 9242 | 8930 | 8654 | 8407 | 8186 | 7986 | 7805 | 7640 | 7489 | 7350 | 7223 | 7105 | 6996 | 6895 | 6801 | .42 |
| .40 | 2.50 | 9573 | 9201 | 8873 | 8584 | 8326 | 8094 | 7886 | 7697 | 7525 | 7368 | 7224 | 7092 | 6970 | 6857 | 6752 | 6655 | .40 |
| .38 | 2.63 | 9550 | 9158 | 8814 | 8511 | 8241 | 7999 | 7781 | 7585 | 7406 | 7243 | 7094 | 6957 | 6831 | 6714 | 6606 | 6505 | .38 |
| .36 | 2.78 | 9525 | 9113 | 8752 | 8434 | 8152 | 7900 | 7673 | 7468 | 7283 | 7114 | 6959 | 6817 | 6687 | 6566 | 6454 | 6350 | .36 |
| .34 | 2.94 | 9499 | 9066 | 8687 | 8354 | 8059 | 7796 | 7560 | 7347 | 7155 | 6980 | 6819 | 6673 | 6538 | 6413 | 6298 | 6191 | .34 |
| .32 | 3.12 | 9472 | 9016 | 8619 | 8270 | 7962 | 7688 | 7442 | 7221 | 7021 | 6840 | 6674 | 6523 | 6384 | 6255 | 6137 | 6027 | .32 |
| .30 | 3.33 | 9443 | 8963 | 8547 | 8182 | 7860 | 7574 | 7319 | 7089 | 6882 | 6694 | 6523 | 6367 | 6223 | 6091 | 5969 | 5856 | .30 |
| .28 | 3.57 | 9412 | 8907 | 8470 | 8088 | 7752 | 7455 | 7189 | 6951 | 6736 | 6542 | 6366 | 6204 | 6056 | 5921 | 5795 | 5679 | .28 |
| .26 | 3.85 | 9379 | 8847 | 8389 | 7989 | 7638 | 7328 | 7052 | 6805 | 6583 | 6383 | 6200 | 6034 | 5882 | 5743 | 5614 | 5494 | .26 |
| .24 | 4.17 | 9343 | 8783 | 8302 | 7883 | 7518 | 7194 | 6907 | 6651 | 6422 | 6214 | 6027 | 5856 | 5700 | 5556 | 5425 | 5303 | .24 |
| .22 | 4.55 | 9304 | 8714 | 8208 | 7770 | 7387 | 7051 | 6753 | 6488 | 6251 | 6037 | 5843 | 5668 | 5507 | 5361 | 5226 | 5102 | .22 |
| .20 | 5.00 | 9262 | 8639 | 8106 | 7647 | 7248 | 6898 | 6589 | 6314 | 6068 | 5848 | 5649 | 5469 | 5305 | 5155 | 5017 | 4890 | .20 |
| .18 | 5.56 | 9216 | 8557 | 7996 | 7514 | 7097 | 6732 | 6411 | 6127 | 5873 | 5646 | 5442 | 5257 | 5089 | 4936 | 4795 | 4667 | .18 |
| .16 | 6.25 | 9164 | 8465 | 7874 | 7368 | 6931 | 6551 | 6218 | 5924 | 5662 | 5429 | 5219 | 5030 | 4858 | 4702 | 4559 | 4429 | .16 |
| .14 | 7.14 | 9106 | 8363 | 7738 | 7206 | 6749 | 6353 | 6007 | 5702 | 5433 | 5192 | 4978 | 4784 | 4609 | 4450 | 4306 | 4174 | .14 |
| .12 | 8.33 | 9040 | 8247 | 7584 | 7023 | 6544 | 6131 | 5771 | 5456 | 5179 | 4932 | 4713 | 4515 | 4338 | 4177 | 4031 | 3897 | .12 |
| .10 | 10.0 | 8962 | 8111 | 7406 | 6813 | 6310 | 5878 | 5505 | 5179 | 4894 | 4642 | 4417 | 4217 | 4037 | 3875 | 3728 | 3594 | .10 |
| .08 | 12.5 | 8867 | 7948 | 7193 | 6564 | 6034 | 5583 | 5195 | 4860 | 4566 | 4309 | 4081 | 3878 | 3697 | 3535 | 3388 | 3255 | .08 |
| .06 | 16.7 | 8746 | 7743 | 6928 | 6257 | 5697 | 5224 | 4822 | 4476 | 4176 | 3915 | 3685 | 3482 | 3301 | 3140 | 2995 | 2864 | .06 |
| .04 | 25.0 | 8579 | 7463 | 6571 | 5848 | 5253 | 4758 | 4341 | 3986 | 3683 | 3420 | 3192 | 2991 | 2814 | 2657 | 2517 | 2392 | .04 |
| .02 | 50.0 | 8300 | 7007 | 6003 | 5210 | 4573 | 4054 | 3627 | 3270 | 2970 | 2714 | 2495 | 2306 | 2141 | 1997 | 1870 | 1758 | .02 |
| 0 | ∞ | 0 | 0 | 0 | 0 | 0 | 0 | 0 | 0 | 0 | 0 | 0 | 0 | 0 | 0 | 0 | 0 | 0 |

# Table 28 $\sqrt{1-r^{\frac{n-1}{n}}}$

## Values of n

| r | $\frac{1}{r}$ | 1.05 | 1.10 | 1.15 | 1.20 | 1.25 | 1.30 | 1.35 | 1.40 | 1.45 | 1.50 | 1.55 | 1.60 | 1.65 | 1.70 | 1.75 | 1.80 | r |
|---|---|---|---|---|---|---|---|---|---|---|---|---|---|---|---|---|---|---|
| 1 | 1 | 0 | 0 | 0 | 0 | 0 | 0 | 0 | 0 | 0 | 0 | 0 | 0 | 0 | 0 | 0 | 0 | 1 |
| 0.98 | 1.02 | 0.0310 | .0428 | .0513 | .0580 | .0635 | .0682 | .0723 | .0759 | .0791 | .0819 | .0845 | .0869 | .0890 | .0910 | .0928 | .0945 | 0.98 |
| .96 | 1.04 | 0441 | 0609 | 0729 | 0823 | 0902 | 0968 | 1026 | 1077 | 1122 | 1163 | 1199 | 1233 | 1263 | 1291 | 1317 | 1341 | .96 |
| .94 | 1.06 | 0542 | 0749 | 0897 | 1013 | 1109 | 1191 | 1261 | 1324 | 1379 | 1429 | 1474 | 1514 | 1552 | 1586 | 1618 | 1647 | .94 |
| .92 | 1.09 | 0630 | 0869 | 1040 | 1175 | 1286 | 1381 | 1462 | 1534 | 1598 | 1656 | 1707 | 1755 | 1798 | 1837 | 1874 | 1907 | .92 |
| .90 | 1.11 | 0707 | 0976 | 1168 | 1319 | 1444 | 1550 | 1642 | 1722 | 1794 | 1858 | 1916 | 1968 | 2016 | 2060 | 2101 | 2139 | .90 |
| .88 | 1.14 | 0779 | 1075 | 1286 | 1452 | 1589 | 1705 | 1806 | 1894 | 1972 | 2042 | 2106 | 2163 | 2216 | 2264 | 2309 | 2350 | .88 |
| .86 | 1.16 | 0846 | 1167 | 1396 | 1576 | 1724 | 1849 | 1958 | 2054 | 2138 | 2214 | 2283 | 2345 | 2402 | 2454 | 2502 | 2546 | .86 |
| .84 | 1.19 | 0909 | 1254 | 1500 | 1692 | 1851 | 1986 | 2102 | 2204 | 2295 | 2376 | 2449 | 2516 | 2576 | 2632 | 2683 | 2731 | .84 |
| .82 | 1.22 | 0970 | 1337 | 1599 | 1804 | 1973 | 2116 | 2239 | 2348 | 2444 | 2530 | 2608 | 2678 | 2742 | 2801 | 2855 | 2906 | .82 |
| .80 | 1.25 | 1028 | 1417 | 1694 | 1911 | 2089 | 2240 | 2371 | 2485 | 2587 | 2677 | 2759 | 2833 | 2901 | 2963 | 3020 | 3073 | .80 |
| .78 | 1.28 | 1085 | 1494 | 1786 | 2014 | 2202 | 2361 | 2498 | 2618 | 2724 | 2819 | 2905 | 2983 | 3054 | 3118 | 3178 | 3233 | .78 |
| .76 | 1.32 | 1139 | 1570 | 1875 | 2114 | 2311 | 2477 | 2621 | 2746 | 2857 | 2957 | 3046 | 3127 | 3201 | 3269 | 3331 | 3389 | .76 |
| .74 | 1.35 | 1193 | 1643 | 1962 | 2212 | 2418 | 2591 | 2740 | 2871 | 2987 | 3090 | 3183 | 3268 | 3344 | 3415 | 3479 | 3539 | .74 |
| .72 | 1.39 | 1246 | 1715 | 2048 | 2308 | 2522 | 2702 | 2857 | 2993 | 3113 | 3221 | 3317 | 3404 | 3484 | 3557 | 3624 | 3686 | .72 |
| .70 | 1.43 | 1298 | 1786 | 2132 | 2402 | 2624 | 2811 | 2972 | 3113 | 3237 | 3348 | 3448 | 3538 | 3621 | 3696 | 3765 | 3829 | .70 |
| .68 | 1.47 | 1349 | 1856 | 2215 | 2495 | 2725 | 2918 | 3085 | 3230 | 3359 | 3473 | 3576 | 3669 | 3754 | 3832 | 3903 | 3969 | .68 |
| .66 | 1.52 | 1400 | 1925 | 2297 | 2587 | 2824 | 3024 | 3196 | 3346 | 3478 | 3596 | 3703 | 3798 | 3886 | 3966 | 4039 | 4106 | .66 |
| .64 | 1.56 | 1450 | 1994 | 2378 | 2677 | 2922 | 3128 | 3305 | 3460 | 3596 | 3718 | 3827 | 3926 | 4015 | 4097 | 4172 | 4242 | .64 |
| .62 | 1.61 | 1500 | 2062 | 2459 | 2767 | 3020 | 3232 | 3414 | 3573 | 3713 | 3838 | 3950 | 4051 | 4143 | 4227 | 4304 | 4375 | .62 |
| .60 | 1.67 | 1550 | 2130 | 2539 | 2857 | 3116 | 3335 | 3522 | 3685 | 3829 | 3957 | 4072 | 4175 | 4269 | 4355 | 4434 | 4507 | .60 |
| .58 | 1.72 | 1600 | 2198 | 2619 | 2946 | 3213 | 3437 | 3629 | 3796 | 3944 | 4075 | 4192 | 4298 | 4395 | 4482 | 4563 | 4637 | .58 |
| .56 | 1.79 | 1650 | 2266 | 2699 | 3035 | 3309 | 3539 | 3736 | 3907 | 4058 | 4192 | 4312 | 4421 | 4519 | 4609 | 4691 | 4766 | .56 |
| .54 | 1.85 | 1700 | 2334 | 2779 | 3124 | 3405 | 3641 | 3842 | 4018 | 4172 | 4309 | 4432 | 4542 | 4642 | 4734 | 4818 | 4895 | .54 |
| .52 | 1.92 | 1751 | 2402 | 2859 | 3213 | 3501 | 3743 | 3949 | 4128 | 4286 | 4426 | 4551 | 4663 | 4766 | 4859 | 4944 | 5022 | .52 |
| .50 | 2.00 | 1802 | 2471 | 2940 | 3303 | 3598 | 3845 | 4056 | 4239 | 4399 | 4542 | 4670 | 4784 | 4888 | 4983 | 5070 | 5149 | .50 |
| .48 | 2.08 | 1853 | 2541 | 3022 | 3393 | 3695 | 3947 | 4163 | 4349 | 4513 | 4659 | 4788 | 4905 | 5011 | 5107 | 5195 | 5276 | .48 |
| .46 | 2.17 | 1905 | 2611 | 3104 | 3484 | 3793 | 4050 | 4270 | 4461 | 4628 | 4776 | 4908 | 5026 | 5134 | 5231 | 5321 | 5402 | .46 |
| .44 | 2.27 | 1958 | 2682 | 3187 | 3576 | 3891 | 4154 | 4379 | 4573 | 4743 | 4893 | 5027 | 5148 | 5257 | 5356 | 5446 | 5529 | .44 |
| .42 | 2.38 | 2012 | 2754 | 3271 | 3669 | 3991 | 4259 | 4488 | 4685 | 4858 | 5011 | 5147 | 5270 | 5380 | 5481 | 5572 | 5656 | .42 |
| .40 | 2.50 | 2066 | 2827 | 3356 | 3763 | 4092 | 4366 | 4598 | 4799 | 4975 | 5130 | 5268 | 5393 | 5505 | 5606 | 5699 | 5784 | .40 |
| .38 | 2.63 | 2122 | 2902 | 3443 | 3859 | 4195 | 4473 | 4710 | 4915 | 5093 | 5251 | 5391 | 5516 | 5630 | 5733 | 5826 | 5912 | .38 |
| .36 | 2.78 | 2179 | 2978 | 3532 | 3957 | 4299 | 4583 | 4824 | 5031 | 5213 | 5372 | 5514 | 5642 | 5756 | 5860 | 5955 | 6041 | .36 |
| .34 | 2.94 | 2238 | 3056 | 3623 | 4057 | 4405 | 4695 | 4939 | 5150 | 5334 | 5496 | 5640 | 5768 | 5884 | 5989 | 6084 | 6172 | .34 |
| .32 | 3.12 | 2298 | 3137 | 3716 | 4159 | 4514 | 4809 | 5057 | 5271 | 5458 | 5621 | 5767 | 5897 | 6014 | 6120 | 6216 | 6304 | .32 |
| .30 | 3.33 | 2361 | 3220 | 3812 | 4264 | 4626 | 4925 | 5178 | 5395 | 5584 | 5749 | 5896 | 6028 | 6146 | 6252 | 6349 | 6437 | .30 |
| .28 | 3.57 | 2425 | 3306 | 3911 | 4372 | 4741 | 5045 | 5302 | 5522 | 5713 | 5880 | 6029 | 6161 | 6280 | 6387 | 6484 | 6573 | .28 |
| .26 | 3.85 | 2493 | 3395 | 4014 | 4484 | 4860 | 5169 | 5429 | 5652 | 5845 | 6015 | 6164 | 6298 | 6417 | 6525 | 6623 | 6712 | .26 |
| .24 | 4.17 | 2563 | 3488 | 4121 | 4601 | 4982 | 5297 | 5561 | 5787 | 5982 | 6153 | 6303 | 6438 | 6558 | 6666 | 6764 | 6853 | .24 |
| .22 | 4.55 | 2637 | 3586 | 4233 | 4723 | 5111 | 5430 | 5698 | 5926 | 6123 | 6295 | 6447 | 6582 | 6703 | 6811 | 6909 | 6999 | .22 |
| .20 | 5.00 | 2716 | 3689 | 4352 | 4851 | 5246 | 5570 | 5841 | 6071 | 6270 | 6444 | 6596 | 6731 | 6852 | 6961 | 7059 | 7148 | .20 |
| .18 | 5.56 | 2800 | 3799 | 4477 | 4986 | 5388 | 5717 | 5991 | 6224 | 6424 | 6598 | 6751 | 6887 | 7008 | 7116 | 7214 | 7303 | .18 |
| .16 | 6.25 | 2891 | 3917 | 4611 | 5130 | 5539 | 5872 | 6150 | 6384 | 6586 | 6761 | 6914 | 7050 | 7171 | 7279 | 7376 | 7464 | .16 |
| .14 | 7.14 | 2990 | 4046 | 4756 | 5286 | 5702 | 6039 | 6319 | 6556 | 6758 | 6934 | 7087 | 7222 | 7342 | 7450 | 7546 | 7633 | .14 |
| .12 | 8.33 | 3099 | 4187 | 4915 | 5456 | 5879 | 6220 | 6503 | 6741 | 6943 | 7119 | 7271 | 7406 | 7525 | 7631 | 7726 | 7812 | .12 |
| .10 | 10.0 | 3223 | 4346 | 5093 | 5645 | 6075 | 6420 | 6705 | 6943 | 7146 | 7320 | 7472 | 7605 | 7722 | 7826 | 7920 | 8004 | .10 |
| .08 | 12.5 | 3366 | 4529 | 5298 | 5862 | 6297 | 6646 | 6932 | 7170 | 7371 | 7544 | 7693 | 7824 | 7939 | 8041 | 8132 | 8213 | .08 |
| .06 | 16.7 | 3541 | 4751 | 5542 | 6118 | 6560 | 6911 | 7196 | 7432 | 7631 | 7801 | 7947 | 8074 | 8185 | 8283 | 8370 | 8448 | .06 |
| .04 | 25.0 | 3770 | 5037 | 5855 | 6444 | 6890 | 7240 | 7523 | 7755 | 7948 | 8112 | 8252 | 8372 | 8477 | 8569 | 8650 | 8723 | .04 |
| .02 | 50.0 | 4123 | 5471 | 6322 | 6921 | 7367 | 7711 | 7983 | 8204 | 8385 | 8536 | 8663 | 8771 | 8865 | 8946 | 9017 | 9079 | .02 |
| 0 | ∞ | 1 | 1 | 1 | 1 | 1 | 1 | 1 | 1 | 1 | 1 | 1 | 1 | 1 | 1 | 1 | 1 | 0 |

# Table 29   $r^{\frac{1}{n}}\sqrt{1-r^{\frac{n-1}{n}}}$

## Values of n

| r | $\frac{1}{r}$ | 1.05 | 1.10 | 1.15 | 1.20 | 1.25 | 1.30 | 1.35 | 1.40 | 1.45 | 1.50 | 1.55 | 1.60 | 1.65 | 1.70 | 1.75 | 1.80 | r |
|---|---|---|---|---|---|---|---|---|---|---|---|---|---|---|---|---|---|---|
| 1 | 1 | 0 | 0 | 0 | 0 | 0 | 0 | 0 | 0 | 0 | 0 | 0 | 0 | 0 | 0 | 0 | 0 | 1 |
| 0.98 | 1.02 | 0.0304 | .0421 | .0504 | .0570 | .0625 | .0672 | .0713 | .0749 | .0780 | .0808 | .0834 | .0858 | .0880 | .0900 | .0918 | .0935 | 0.98 |
| .96 | 1.04 | 0424 | 0586 | 0703 | 0796 | 0873 | 0938 | 0995 | 1046 | 1091 | 1131 | 1168 | 1201 | 1232 | 1260 | 1287 | 1311 | .96 |
| .94 | 1.06 | 0511 | 0708 | 0850 | 0962 | 1055 | 1135 | 1205 | 1267 | 1322 | 1371 | 1416 | 1457 | 1495 | 1529 | 1561 | 1591 | .94 |
| .92 | 1.09 | 0581 | 0806 | 0967 | 1096 | 1203 | 1295 | 1375 | 1446 | 1509 | 1566 | 1618 | 1665 | 1709 | 1749 | 1786 | 1821 | .92 |
| .90 | 1.11 | 0640 | 0887 | 1066 | 1208 | 1327 | 1429 | 1518 | 1597 | 1668 | 1732 | 1790 | 1843 | 1892 | 1937 | 1978 | 2017 | .90 |
| .88 | 1.14 | 0690 | 0957 | 1151 | 1305 | 1434 | 1545 | 1642 | 1729 | 1806 | 1876 | 1939 | 1997 | 2051 | 2100 | 2146 | 2189 | .88 |
| .86 | 1.16 | 0733 | 1017 | 1224 | 1389 | 1528 | 1647 | 1751 | 1844 | 1927 | 2002 | 2071 | 2134 | 2192 | 2246 | 2295 | 2342 | .86 |
| .84 | 1.19 | 0770 | 1070 | 1289 | 1463 | 1610 | 1737 | 1848 | 1946 | 2035 | 2116 | 2189 | 2256 | 2318 | 2375 | 2429 | 2479 | .84 |
| .82 | 1.22 | 0803 | 1116 | 1345 | 1529 | 1683 | 1816 | 1934 | 2038 | 2131 | 2216 | 2294 | 2366 | 2432 | 2493 | 2549 | 2602 | .82 |
| .80 | 1.25 | 0831 | 1157 | 1395 | 1586 | 1748 | 1887 | 2010 | 2119 | 2218 | 2307 | 2389 | 2464 | 2534 | 2598 | 2658 | 2714 | .80 |
| .78 | 1.28 | 0856 | 1192 | 1439 | 1637 | 1805 | 1950 | 2078 | 2192 | 2295 | 2389 | 2475 | 2554 | 2627 | 2694 | 2758 | 2817 | .78 |
| .76 | 1.32 | 0877 | 1223 | 1477 | 1682 | 1855 | 2006 | 2139 | 2257 | 2365 | 2463 | 2552 | 2634 | 2711 | 2782 | 2848 | 2909 | .76 |
| .74 | 1.35 | 0896 | 1250 | 1510 | 1721 | 1900 | 2055 | 2192 | 2316 | 2427 | 2528 | 2621 | 2707 | 2787 | 2860 | 2929 | 2994 | .74 |
| .72 | 1.39 | 0911 | 1272 | 1539 | 1755 | 1939 | 2099 | 2240 | 2367 | 2482 | 2587 | 2684 | 2773 | 2855 | 2932 | 3004 | 3071 | .72 |
| .70 | 1.43 | 0924 | 1292 | 1563 | 1785 | 1973 | 2136 | 2282 | 2413 | 2532 | 2640 | 2739 | 2831 | 2917 | 2997 | 3071 | 3141 | .70 |
| .68 | 1.47 | 0934 | 1307 | 1584 | 1809 | 2001 | 2169 | 2318 | 2452 | 2574 | 2686 | 2789 | 2884 | 2972 | 3054 | 3131 | 3203 | .68 |
| .66 | 1.52 | 0942 | 1320 | 1600 | 1830 | 2025 | 2197 | 2349 | 2487 | 2612 | 2726 | 2832 | 2930 | 3021 | 3106 | 3185 | 3260 | .66 |
| .64 | 1.56 | 0948 | 1329 | 1613 | 1846 | 2045 | 2219 | 2375 | 2516 | 2644 | 2761 | 2870 | 2971 | 3064 | 3151 | 3233 | 3310 | .64 |
| .62 | 1.61 | 0952 | 1335 | 1622 | 1858 | 2060 | 2237 | 2396 | 2539 | 2670 | 2791 | 2902 | 3005 | 3101 | 3191 | 3275 | 3355 | .62 |
| .60 | 1.67 | 0953 | 1339 | 1628 | 1866 | 2071 | 2251 | 2412 | 2558 | 2692 | 2815 | 2928 | 3034 | 3133 | 3225 | 3312 | 3393 | .60 |
| .58 | 1.72 | 0952 | 1340 | 1631 | 1871 | 2078 | 2260 | 2424 | 2573 | 2709 | 2834 | 2950 | 3058 | 3159 | 3253 | 3342 | 3426 | .58 |
| .56 | 1.79 | 0950 | 1338 | 1630 | 1872 | 2081 | 2266 | 2431 | 2582 | 2721 | 2848 | 2967 | 3077 | 3180 | 3277 | 3368 | 3454 | .56 |
| .54 | 1.85 | 0946 | 1333 | 1626 | 1869 | 2080 | 2266 | 2434 | 2587 | 2728 | 2857 | 2978 | 3090 | 3196 | 3295 | 3388 | 3476 | .54 |
| .52 | 1.92 | 0939 | 1326 | 1619 | 1863 | 2075 | 2263 | 2433 | 2588 | 2730 | 2862 | 2985 | 3099 | 3206 | 3307 | 3402 | 3492 | .52 |
| .50 | 2.00 | 0931 | 1316 | 1609 | 1854 | 2066 | 2256 | 2427 | 2584 | 2728 | 2861 | 2986 | 3103 | 3212 | 3315 | 3412 | 3503 | .50 |
| .48 | 2.08 | 0921 | 1304 | 1596 | 1841 | 2054 | 2244 | 2417 | 2575 | 2721 | 2856 | 2982 | 3101 | 3212 | 3316 | 3415 | 3509 | .48 |
| .46 | 2.17 | 0909 | 1289 | 1580 | 1824 | 2038 | 2229 | 2402 | 2562 | 2709 | 2846 | 2974 | 3094 | 3207 | 3313 | 3414 | 3509 | .46 |
| .44 | 2.27 | 0896 | 1271 | 1561 | 1804 | 2018 | 2209 | 2384 | 2544 | 2692 | 2831 | 2960 | 3082 | 3196 | 3304 | 3407 | 3504 | .44 |
| .42 | 2.38 | 0881 | 1252 | 1538 | 1781 | 1994 | 2185 | 2360 | 2521 | 2671 | 2810 | 2941 | 3064 | 3180 | 3290 | 3394 | 3493 | .42 |
| .40 | 2.50 | 0863 | 1229 | 1513 | 1754 | 1966 | 2157 | 2333 | 2494 | 2645 | 2785 | 2917 | 3041 | 3159 | 3270 | 3376 | 3476 | .40 |
| .38 | 2.63 | 0844 | 1204 | 1484 | 1723 | 1934 | 2125 | 2300 | 2462 | 2613 | 2755 | 2888 | 3013 | 3132 | 3245 | 3352 | 3454 | .38 |
| .36 | 2.78 | 0824 | 1176 | 1453 | 1689 | 1898 | 2089 | 2263 | 2425 | 2577 | 2719 | 2853 | 2979 | 3099 | 3213 | 3321 | 3425 | .36 |
| .34 | 2.94 | 0801 | 1146 | 1418 | 1651 | 1859 | 2047 | 2221 | 2383 | 2535 | 2677 | 2812 | 2939 | 3060 | 3175 | 3285 | 3389 | .34 |
| .32 | 3.12 | 0776 | 1113 | 1380 | 1609 | 1814 | 2002 | 2175 | 2336 | 2487 | 2630 | 2765 | 2893 | 3015 | 3131 | 3241 | 3347 | .32 |
| .30 | 3.33 | 0750 | 1078 | 1338 | 1563 | 1766 | 1951 | 2123 | 2283 | 2434 | 2577 | 2712 | 2840 | 2963 | 3079 | 3191 | 3298 | .30 |
| .28 | 3.57 | 0721 | 1039 | 1293 | 1514 | 1712 | 1895 | 2065 | 2224 | 2375 | 2517 | 2652 | 2781 | 2904 | 3021 | 3133 | 3241 | .28 |
| .26 | 3.85 | 0691 | 0998 | 1244 | 1459 | 1654 | 1834 | 2002 | 2159 | 2308 | 2450 | 2585 | 2713 | 2836 | 2954 | 3067 | 3176 | .26 |
| .24 | 4.17 | 0658 | 0953 | 1191 | 1401 | 1591 | 1767 | 1932 | 2088 | 2236 | 2376 | 2510 | 2638 | 2761 | 2879 | 2993 | 3102 | .24 |
| .22 | 4.55 | 0624 | 0905 | 1135 | 1337 | 1522 | 1694 | 1856 | 2009 | 2155 | 2294 | 2427 | 2555 | 2677 | 2795 | 2909 | 3018 | .22 |
| .20 | 5.00 | 0587 | 0854 | 1074 | 1269 | 1448 | 1615 | 1773 | 1923 | 2066 | 2204 | 2335 | 2462 | 2584 | 2701 | 2814 | 2923 | .20 |
| .18 | 5.56 | 0547 | 0799 | 1008 | 1194 | 1367 | 1529 | 1682 | 1829 | 1969 | 2104 | 2233 | 2358 | 2479 | 2595 | 2708 | 2817 | .18 |
| .16 | 6.25 | 0505 | 0740 | 0937 | 1114 | 1279 | 1434 | 1582 | 1724 | 1861 | 1993 | 2120 | 2243 | 2362 | 2477 | 2588 | 2697 | .16 |
| .14 | 7.14 | 0460 | 0677 | 0861 | 1027 | 1183 | 1331 | 1473 | 1610 | 1742 | 1869 | 1993 | 2113 | 2230 | 2343 | 2454 | 2561 | .14 |
| .12 | 8.33 | 0411 | 0609 | 0778 | 0932 | 1078 | 1218 | 1352 | 1482 | 1609 | 1732 | 1852 | 1968 | 2082 | 2192 | 2300 | 2405 | .12 |
| .10 | 10.0 | 0360 | 0536 | 0688 | 0829 | 0963 | 1092 | 1218 | 1340 | 1460 | 1577 | 1691 | 1803 | 1913 | 2020 | 2125 | 2227 | .10 |
| .08 | 12.5 | 0304 | 0456 | 0589 | 0714 | 0835 | 0952 | 1067 | 1180 | 1291 | 1401 | 1508 | 1614 | 1718 | 1820 | 1920 | 2019 | .08 |
| .06 | 16.7 | 0243 | 0368 | 0480 | 0587 | 0691 | 0794 | 0895 | 0996 | 1096 | 1196 | 1294 | 1391 | 1488 | 1583 | 1677 | 1770 | .06 |
| .04 | 25.0 | 0176 | 0270 | 0356 | 0441 | 0525 | 0609 | 0693 | 0778 | 0863 | 0949 | 1034 | 1120 | 1205 | 1290 | 1375 | 1459 | .04 |
| .02 | 50.0 | 0099 | 0156 | 0211 | 0266 | 0322 | 0380 | 0440 | 0502 | 0565 | 0629 | 0694 | 0761 | 0828 | 0896 | 0964 | 1033 | .02 |
| 0 | ∞ | 0 | 0 | 0 | 0 | 0 | 0 | 0 | 0 | 0 | 0 | 0 | 0 | 0 | 0 | 0 | 0 | 0 |

# Table 30 One-Dimensional Isentropic Compressible-Flow Functions

For a perfect gas with constant specific heat and molecular weight
$k = 1.4$

| M | M* | $\frac{A}{A^*}$ | $\frac{p}{p_0}$ | $\frac{\rho}{\rho_0}$ | $\frac{T}{T_0}$ | $\frac{F}{F^*}$ | $\left(\frac{A}{A^*}\right)\left(\frac{p}{p_0}\right)$ |
|---|---|---|---|---|---|---|---|
| 0 | 0 | ∞ | 1.00000 | 1.00000 | 1.00000 | ∞ | ∞ |
| 0.01 | 0.01096 | 57.874 | .99993 | .99995 | .99998 | 45.650 | 57.870 |
| .02 | .02191 | 28.942 | .99972 | .99980 | .99992 | 22.834 | 28.934 |
| .03 | .03286 | 19.300 | .99937 | .99955 | .99982 | 15.232 | 19.288 |
| .04 | .04381 | 14.482 | .99888 | .99920 | .99968 | 11.435 | 14.465 |
| .05 | .05476 | 11.592 | .99825 | .99875 | .99950 | 9.1584 | 11.571 |
| .06 | .06570 | 9.6659 | .99748 | .99820 | .99928 | 7.6428 | 9.6415 |
| .07 | .07664 | 8.2915 | .99658 | .99755 | .99902 | 6.5620 | 8.2631 |
| .08 | .08758 | 7.2616 | .99553 | .99680 | .99872 | 5.7529 | 7.2291 |
| .09 | .09851 | 6.4613 | .99435 | .99596 | .99838 | 5.1249 | 6.4248 |
| .10 | .10943 | 5.8218 | .99303 | .99502 | .99800 | 4.6236 | 5.7812 |
| .11 | .12035 | 5.2992 | .99157 | .99398 | .99758 | 4.2146 | 5.2546 |
| .12 | .13126 | 4.8643 | .98998 | .99284 | .99714 | 3.8747 | 4.8157 |
| .13 | .14216 | 4.4968 | .98826 | .99160 | .99664 | 3.5880 | 4.4440 |
| .14 | .15306 | 4.1824 | .98640 | .99027 | .99610 | 3.3432 | 4.1255 |
| .15 | .16395 | 3.9103 | .98441 | .98884 | .99552 | 3.1317 | 3.8493 |
| .16 | .17483 | 3.6727 | .98228 | .98731 | .99490 | 2.9474 | 3.6076 |
| .17 | .18569 | 3.4635 | .98003 | .98569 | .99425 | 2.7855 | 3.3943 |
| .18 | .19654 | 3.2779 | .97765 | .98398 | .99356 | 2.6422 | 3.2046 |
| .19 | .20738 | 3.1122 | .97514 | .98217 | .99283 | 2.5146 | 3.0348 |
| .20 | .21822 | 2.9635 | .97250 | .98027 | .99206 | 2.4004 | 2.8820 |
| .21 | .22904 | 2.8293 | .96973 | .97828 | .99125 | 2.2976 | 2.7437 |
| .22 | .23984 | 2.7076 | .96685 | .97621 | .99041 | 2.2046 | 2.6178 |
| .23 | .25063 | 2.5968 | .96383 | .97403 | .98953 | 2.1203 | 2.5029 |
| .24 | .26141 | 2.4956 | .96070 | .97177 | .98861 | 2.0434 | 2.3975 |
| .25 | .27216 | 2.4027 | .95745 | .96942 | .98765 | 1.9732 | 2.3005 |
| .26 | .28291 | 2.3173 | .95408 | .96699 | .98666 | 1.9088 | 2.2109 |
| .27 | .29364 | 2.2385 | .95060 | .96446 | .98563 | 1.8496 | 2.1279 |
| .28 | .30435 | 2.1656 | .94700 | .96185 | .98456 | 1.7950 | 2.0508 |
| .29 | .31504 | 2.0979 | .94329 | .95916 | .98346 | 1.7446 | 1.9789 |
| .30 | .32572 | 2.0351 | .93947 | .95638 | .98232 | 1.6979 | 1.9119 |
| .31 | .33638 | 1.9765 | .93554 | .95352 | .98114 | 1.6546 | 1.8491 |
| .32 | .34701 | 1.9218 | .93150 | .95058 | .97993 | 1.6144 | 1.7902 |
| .33 | .35762 | 1.8707 | .92736 | .94756 | .97868 | 1.5769 | 1.7348 |
| .34 | .36821 | 1.8229 | .92312 | .94446 | .97740 | 1.5420 | 1.6828 |
| .35 | .37879 | 1.7780 | .91877 | .94128 | .97608 | 1.5094 | 1.6336 |
| .36 | .38935 | 1.7358 | .91433 | .93803 | .97473 | 1.4789 | 1.5871 |
| .37 | .39988 | 1.6961 | .90979 | .93470 | .97335 | 1.4503 | 1.5431 |
| .38 | .41039 | 1.6587 | .90516 | .93129 | .97193 | 1.4236 | 1.5014 |
| .39 | .42087 | 1.6234 | .90044 | .92782 | .97048 | 1.3985 | 1.4618 |
| 0.40 | .43133 | 1.5901 | .89562 | .92428 | .96899 | 1.3749 | 1.4241 |
| .41 | .44177 | 1.5587 | .89071 | .92066 | .96747 | 1.3527 | 1.3883 |
| .42 | .45218 | 1.5289 | .88572 | .91697 | .96592 | 1.3318 | 1.3542 |
| .43 | .46256 | 1.5007 | .88065 | .91322 | .96434 | 1.3122 | 1.3216 |
| .44 | .47292 | 1.4740 | .87550 | .90940 | .96272 | 1.2937 | 1.2905 |
| .45 | .48326 | 1.4487 | .87027 | .90552 | .96108 | 1.2763 | 1.2607 |
| .46 | .49357 | 1.4246 | .86496 | .90157 | .95940 | 1.2598 | 1.2322 |
| .47 | .50385 | 1.4018 | .85958 | .89756 | .95769 | 1.2443 | 1.2050 |
| .48 | .51410 | 1.3801 | .85413 | .89349 | .95595 | 1.2296 | 1.1788 |
| .49 | .52432 | 1.3594 | .84861 | .88936 | .95418 | 1.2158 | 1.1537 |
| .50 | .53452 | 1.3398 | .84302 | .88517 | .95238 | 1.2027 | 1.12951 |
| .51 | .54469 | 1.3212 | .83737 | .88092 | .95055 | 1.1903 | 1.10631 |
| .52 | .55482 | 1.3034 | .83166 | .87662 | .94869 | 1.1786 | 1.08397 |
| .53 | .56493 | 1.2864 | .82589 | .87227 | .94681 | 1.1675 | 1.06245 |
| .54 | .57501 | 1.2703 | .82005 | .86788 | .94489 | 1.1571 | 1.04173 |
| .55 | .58506 | 1.2550 | .81416 | .86342 | .94295 | 1.1472 | 1.02174 |
| .56 | .59508 | 1.2403 | .80822 | .85892 | .94098 | 1.1378 | 1.00244 |
| .57 | .60506 | 1.2263 | .80224 | .85437 | .93898 | 1.1289 | .98381 |
| .58 | .61500 | 1.2130 | .79621 | .84977 | .93696 | 1.1205 | .96581 |
| .59 | .62491 | 1.2003 | .79012 | .84513 | .93491 | 1.1126 | .94839 |
| .60 | .63480 | 1.1882 | .78400 | .84045 | .93284 | 1.10504 | .93155 |
| .61 | .64466 | 1.1766 | .77784 | .83573 | .93074 | 1.09793 | .91525 |
| .62 | .65448 | 1.1656 | .77164 | .83096 | .92861 | 1.09120 | .89946 |
| .63 | .66427 | 1.1551 | .76540 | .82616 | .92646 | 1.08485 | .88416 |
| .64 | .67402 | 1.1451 | .75913 | .82132 | .92428 | 1.07883 | .86932 |
| .65 | .68374 | 1.1356 | .75283 | .81644 | .92208 | 1.07314 | .85493 |
| .66 | .69342 | 1.1265 | .74650 | .81153 | .91986 | 1.06777 | .84096 |
| .67 | .70307 | 1.1178 | .74014 | .80659 | .91762 | 1.06271 | .82740 |
| .68 | .71268 | 1.1096 | .73376 | .80162 | .91535 | 1.05792 | .81421 |
| .69 | .72225 | 1.1018 | .72735 | .79662 | .91306 | 1.05340 | .80141 |
| .70 | .73179 | 1.09437 | .72092 | .79158 | .91075 | 1.04915 | .78896 |
| .71 | .74129 | 1.08729 | .71448 | .78652 | .90842 | 1.04514 | .77685 |
| .72 | .75076 | 1.08057 | .70802 | .78143 | .90606 | 1.04137 | .76507 |
| .73 | .76019 | 1.07419 | .70155 | .77632 | .90368 | 1.03783 | .75360 |
| .74 | .76958 | 1.06814 | .69507 | .77119 | .90129 | 1.03450 | .74243 |
| .75 | .77893 | 1.06242 | .68857 | .76603 | .89888 | 1.03137 | .73155 |
| .76 | .78825 | 1.05700 | .68207 | .76086 | .89644 | 1.02844 | .72095 |
| .77 | .79753 | 1.05188 | .67556 | .75567 | .89399 | 1.02570 | .71062 |
| .78 | .80677 | 1.04705 | .66905 | .75046 | .89152 | 1.02314 | .70054 |
| .79 | .81597 | 1.04250 | .66254 | .74524 | .88903 | 1.02075 | .69070 |

See page 113 for definitions of symbols.

## Table 30   One-Dimensional Isentropic Compressible-Flow Functions

**For a perfect gas with constant specific heat and molecular weight**
**k = 1.4**

| M | M* | $\frac{A}{A^*}$ | $\frac{p}{p_0}$ | $\frac{\rho}{\rho_0}$ | $\frac{T}{T_0}$ | $\frac{F}{F^*}$ | $\left(\frac{A}{A^*}\right)\left(\frac{p}{p_0}\right)$ | M | M* | $\frac{A}{A^*}$ | $\frac{p}{p_0}$ | $\frac{\rho}{\rho_0}$ | $\frac{T}{T_0}$ | $\frac{F}{F^*}$ | $\left(\frac{A}{A^*}\right)\left(\frac{p}{p_0}\right)$ |
|---|---|---|---|---|---|---|---|---|---|---|---|---|---|---|---|
| 0.80 | .82514 | 1.03823 | .65602 | .74000 | .88652 | 1.01853 | .68110 | 1.20 | 1.1583 | 1.03044 | .41238 | .53114 | .77640 | 1.01082 | .42493 |
| .81 | .83426 | 1.03422 | .64951 | .73474 | .88400 | 1.01646 | .67173 | 1.21 | 1.1658 | 1.03344 | .40702 | .52620 | .77350 | 1.01178 | .42063 |
| .82 | .84334 | 1.03046 | .64300 | .72947 | .88146 | 1.01455 | .66259 | 1.22 | 1.1732 | 1.03657 | .40171 | .52129 | .77061 | 1.01278 | .41640 |
| .83 | .85239 | 1.02696 | .63650 | .72419 | .87890 | 1.01278 | .65366 | 1.23 | 1.1806 | 1.03983 | .39645 | .51640 | .76771 | 1.01381 | .41224 |
| .84 | .86140 | 1.02370 | .63000 | .71890 | .87633 | 1.01115 | .64493 | 1.24 | 1.1879 | 1.04323 | .39123 | .51154 | .76481 | 1.01486 | .40814 |
| .85 | .87037 | 1.02067 | .62351 | .71361 | .87374 | 1.00966 | .63640 | 1.25 | 1.1952 | 1.04676 | .38606 | .50670 | .76190 | 1.01594 | .40411 |
| .86 | .87929 | 1.01787 | .61703 | .70831 | .87114 | 1.00829 | .62806 | 1.26 | 1.2025 | 1.05041 | .38094 | .50189 | .75900 | 1.01705 | .40014 |
| .87 | .88817 | 1.01530 | .61057 | .70300 | .86852 | 1.00704 | .61991 | 1.27 | 1.2097 | 1.05419 | .37586 | .49710 | .75610 | 1.01818 | .39622 |
| .88 | .89702 | 1.01294 | 60412 | .69769 | .86589 | 1.00591 | .61193 | 1.28 | 1.2169 | 1.05810 | .37083 | .49234 | .75319 | 1.01933 | .39237 |
| .89 | .90583 | 1.01080 | .59768 | 69237 | .86324 | 1.00490 | .60413 | 1.29 | 1.2240 | 1.06214 | .36585 | .48761 | .75029 | 1.02050 | .38858 |
| .90 | .91460 | 1.00886 | .59126 | .68704 | .86058 | 1.00399 | .59650 | 1.30 | 1.2311 | 1.06631 | .36092 | .48291 | .74738 | 1.02170 | .38484 |
| .91 | .92333 | 1.00713 | .58486 | .68171 | .85791 | 1.00318 | .58903 | 1.31 | 1.2382 | 1.07060 | .35603 | .47823 | .74448 | 1.02292 | .38116 |
| .92 | .93201 | 1.00560 | .57848 | .67639 | .85523 | 1.00248 | .58171 | 1.32 | 1.2452 | 1.07502 | .35119 | .47358 | .74158 | 1.02415 | .37754 |
| .93 | .94065 | 1.00426 | .57212 | .67107 | .85253 | 1.00188 | .57455 | 1.33 | 1.2522 | 1.07957 | .34640 | .46895 | .73867 | 1.02540 | .37397 |
| .94 | .94925 | 1.00311 | .56578 | .66575 | .84982 | 1.00136 | .56754 | 1.34 | 1.2591 | 1.08424 | .34166 | .46436 | .73577 | 1.02666 | .37044 |
| .95 | .95781 | 1.00214 | .55946 | .66044 | .84710 | 1.00093 | .56066 | 1.35 | 1.2660 | 1.08904 | .33697 | .45980 | .73287 | 1.02794 | 36697 |
| .96 | .96633 | 1.00136 | .55317 | .65513 | .84437 | 1.00059 | .55392 | 1.36 | 1.2729 | 1.09397 | .33233 | .45527 | .72997 | 1.02924 | .36355 |
| .97 | .97481 | 1.00076 | .54691 | .64982 | .84162 | 1.00033 | .54732 | 1.37 | 1.2797 | 1.09902 | .32774 | .45076 | .72707 | 1.03056 | .36018 |
| .98 | .98325 | 1.00033 | .54067 | .64452 | .83887 | 1.00014 | .54085 | 1.38 | 1.2865 | 1.10420 | .32319 | .44628 | .72418 | 1.03189 | .35686 |
| .99 | .99165 | 1.00008 | .53446 | .63923 | .83611 | 1.00003 | .53450 | 1.39 | 1.2932 | 1.10950 | .31869 | .44183 | .72128 | 1.03323 | .35359 |
| 1.00 | 1.00000 | 1.00000 | .52828 | .63394 | .83333 | 1.00000 | .52828 | 1.40 | 1.2999 | 1.1149 | .31424 | .43742 | .71839 | 1.03458 | .35036 |
| 1.01 | 1.00831 | 1.00008 | .52213 | .62866 | .83055 | 1.00003 | .52218 | 1.41 | 1.3065 | 1.1205 | .30984 | .43304 | .71550 | 1.03595 | .34717 |
| 1.02 | 1.01658 | 1.00033 | .51602 | .62339 | .82776 | 1.00013 | .51619 | 1.42 | 1.3131 | 1.1262 | .30549 | .42869 | .71261 | 1.03733 | .34403 |
| 1.03 | 1.02481 | 1.00074 | .50994 | .61813 | .82496 | 1.00030 | .51031 | 1.43 | 1.3197 | 1.1320 | .30119 | .42436 | .70973 | 1.03872 | .34093 |
| 1.04 | 1.03300 | 1.00130 | .50389 | .61288 | .82215 | 1.00053 | .50454 | 1.44 | 1.3262 | 1.1379 | .29693 | .42007 | .70685 | 1.04012 | .33787 |
| 1.05 | 1.04114 | 1.00202 | .49787 | .60765 | .81933 | 1.00082 | .49888 | 1.45 | 1.3327 | 1.1440 | .29272 | .41581 | .70397 | 1.04153 | .33486 |
| 1.06 | 1.04924 | 1.00290 | .49189 | .60243 | .81651 | 1.00116 | .49332 | 1.46 | 1.3392 | 1.1502 | .28856 | .41158 | .70110 | 1.04295 | .33189 |
| 1.07 | 1.05730 | 1.00394 | .48595 | .59722 | .81368 | 1.00155 | .48787 | 1.47 | 1.3456 | 1.1565 | .28445 | .40738 | .69823 | 1.04438 | .32896 |
| 1.08 | 1.06532 | 1.00512 | .48005 | .59203 | .81084 | 1.00200 | .48251 | 1.48 | 1.3520 | 1.1629 | .28039 | .40322 | .69537 | 1.04581 | .32607 |
| 1.09 | 1.07330 | 1.00645 | .47418 | .58685 | .80800 | 1.00250 | .47724 | 1.49 | 1.3583 | 1.1695 | .27637 | .39909 | .69251 | 1.04725 | .32321 |
| 1.10 | 1.08124 | 1.00793 | .46835 | .58169 | .80515 | 1.00305 | .47206 | 1.50 | 1.3646 | 1.1762 | .27240 | .39498 | .68965 | 1.04870 | .32039 |
| 1.11 | 1.08914 | 1.00955 | .46256 | .57655 | .80230 | 1.00365 | .46698 | 1.51 | 1.3708 | 1.1830 | .26848 | .39091 | .68680 | 1.05016 | .31761 |
| 1.12 | 1.09699 | 1.01131 | .45682 | .57143 | .79944 | 1.00429 | .46199 | 1.52 | 1.3770 | 1.1899 | .26461 | .38637 | .68396 | 1.05162 | .31487 |
| 1.13 | 1.10480 | 1.01322 | .45112 | .56632 | .79657 | 1.00497 | .45708 | 1.53 | 1.3832 | 1.1970 | .26078 | .38287 | .68112 | 1.05309 | .31216 |
| 1.14 | 1.11256 | 1.01527 | .44545 | .56123 | .79370 | 1.00569 | .45225 | 1.54 | 1.3894 | 1.2042 | .25700 | .37890 | .67828 | 1.05456 | .30948 |
| 1.15 | 1.1203 | 1.01746 | .43983 | .55616 | .79083 | 1.00646 | .44751 | 1.55 | 1.3955 | 1.2115 | .25326 | .37496 | .67545 | 1.05604 | .30685 |
| 1.16 | 1.1280 | 1.01978 | .43425 | .55112 | .78795 | 1.00726 | .44284 | 1.56 | 1.4016 | 1.2190 | .24957 | .37105 | .67262 | 1.05752 | .30424 |
| 1.17 | 1.1356 | 1.02224 | .42872 | .54609 | .78507 | 1.00810 | .43825 | 1.57 | 1.4076 | 1.2266 | .24593 | .36717 | .66980 | 1.05900 | .30167 |
| 1.18 | 1.1432 | 1.02484 | .42323 | .54108 | .78218 | 1.00897 | .43374 | 1.58 | 1.4135 | 1.2343 | .24233 | .36332 | .66699 | 1.06049 | .29913 |
| 1.19 | 1.1508 | 1.02757 | .41778 | .53610 | .77929 | 1.00988 | .42930 | 1.59 | 1.4195 | 1.2422 | .23878 | .35951 | .66418 | 1.06198 | .29662 |

# Table 30 One-Dimensional Isentropic Compressible-Flow Functions

## For a perfect gas with constant specific heat and molecular weight
## k = 1.4

| M | M* | $\frac{A}{A^*}$ | $\frac{p}{p_0}$ | $\frac{\rho}{\rho_0}$ | $\frac{T}{T_0}$ | $\frac{F}{F^*}$ | $\left(\frac{A}{A^*}\right)\left(\frac{p}{p_0}\right)$ | M | M* | $\frac{A}{A^*}$ | $\frac{p}{p_0}$ | $\frac{\rho}{\rho_0}$ | $\frac{T}{T_0}$ | $\frac{F}{F^*}$ | $\left(\frac{A}{A^*}\right)\left(\frac{p}{p_0}\right)$ |
|---|---|---|---|---|---|---|---|---|---|---|---|---|---|---|---|
| 1.60 | 1.4254 | 1.2502 | .23527 | .35573 | .66138 | 1.06348 | .29414 | 2.00 | 1.6330 | 1.6875 | .12780 | .23005 | .55556 | 1.1227 | .21567 |
| 1.61 | 1.4313 | 1.2583 | .23181 | .35198 | .65858 | 1.06498 | .29169 | 2.01 | 1.6375 | 1.7017 | .12583 | .22751 | .55310 | 1.1241 | .21412 |
| 1.62 | 1.4371 | 1.2666 | .22839 | .34826 | .65579 | 1.06648 | .28928 | 2.02 | 1.6420 | 1.7160 | .12389 | .22499 | .55064 | 1.1255 | .21259 |
| 1.63 | 1.4429 | 1.2750 | .22501 | .34458 | .65301 | 1.06798 | .28690 | 2.03 | 1.6465 | 1.7305 | .12198 | .22250 | .54819 | 1.1269 | .21107 |
| 1.64 | 1.4487 | 1.2835 | .22168 | .34093 | .65023 | 1.06948 | .28454 | 2.04 | 1.6509 | 1.7452 | .12009 | .22004 | .54576 | 1.1283 | .20957 |
| 1.65 | 1.4544 | 1.2922 | .21839 | .33731 | .64746 | 1.07098 | .28221 | 2.05 | 1.6553 | 1.7600 | .11823 | .21760 | .54333 | 1.1297 | .20808 |
| 1.66 | 1.4601 | 1.3010 | .21515 | .33372 | .64470 | 1.07249 | .27991 | 2.06 | 1.6597 | 1.7750 | .11640 | .21519 | .54091 | 1.1311 | .20661 |
| 1.67 | 1.4657 | 1.3099 | .21195 | .33016 | .64194 | 1.07399 | .27764 | 2.07 | 1.6640 | 1.7902 | .11460 | .21281 | .53850 | 1.1325 | .20515 |
| 1.68 | 1.4713 | 1.3190 | .20879 | .32664 | .63919 | 1.07550 | .27540 | 2.08 | 1.6683 | 1.8056 | .11282 | .21045 | .53611 | 1.1339 | .20371 |
| 1.69 | 1.4769 | 1.3282 | .20567 | .32315 | .63645 | 1.07701 | .27318 | 2.09 | 1.6726 | 1.8212 | .11107 | .20811 | .53373 | 1.1352 | .20228 |
| 1.70 | 1.4825 | 1.3376 | .20259 | .31969 | .63372 | 1.07851 | .27099 | 2.10 | 1.6769 | 1.8369 | .10935 | .20580 | .53135 | 1.1366 | .20087 |
| 1.71 | 1.4880 | 1.3471 | .19955 | .31626 | .63099 | 1.08002 | .26882 | 2.11 | 1.6811 | 1.8529 | .10766 | .20352 | .52898 | 1.1380 | .19947 |
| 1.72 | 1.4935 | 1.3567 | .19656 | .31286 | .62827 | 1.08152 | .26668 | 2.12 | 1.6853 | 1.8690 | .10599 | .20126 | .52663 | 1.1393 | .19809 |
| 1.73 | 1.4989 | 1.3665 | .19361 | .30950 | .62556 | 1.08302 | .26457 | 2.13 | 1.6895 | 1.8853 | .10434 | .19902 | .52428 | 1.1407 | .19672 |
| 1.74 | 1.5043 | 1.3764 | .19070 | .30617 | .62286 | 1.08453 | .26248 | 2.14 | 1.6936 | 1.9018 | .10272 | .19681 | .52194 | 1.1420 | .19537 |
| 1.75 | 1.5097 | 1.3865 | .18782 | .30287 | .62016 | 1.08603 | .26042 | 2.15 | 1.6977 | 1.9185 | .10113 | .19463 | .51962 | 1.1434 | .19403 |
| 1.76 | 1.5150 | 1.3967 | .18499 | .29959 | .61747 | 1.08753 | .25838 | 2.16 | 1.7018 | 1.9354 | .09956 | .19247 | .51730 | 1.1447 | .19270 |
| 1.77 | 1.5203 | 1.4071 | .18220 | .29635 | .61479 | 1.08903 | .25636 | 2.17 | 1.7059 | 1.9525 | .09802 | .19033 | .51499 | 1.1460 | .19138 |
| 1.78 | 1.5256 | 1.4176 | .17944 | .29314 | .61211 | 1.09053 | .25436 | 2.18 | 1.7099 | 1.9698 | .09650 | .18821 | .51269 | 1.1474 | .19008 |
| 1.79 | 1.5308 | 1.4282 | .17672 | .28997 | .60945 | 1.09202 | .25239 | 2.19 | 1.7139 | 1.9873 | .09500 | .18612 | .51041 | 1.1487 | .18879 |
| 1.80 | 1.5360 | 1.4390 | .17404 | .28682 | .60680 | 1.09352 | .25044 | 2.20 | 1.7179 | 2.0050 | .09352 | .18405 | .50813 | 1.1500 | .18751 |
| 1.81 | 1.5412 | 1.4499 | .17140 | .28370 | .60415 | 1.09500 | .24851 | 2.21 | 1.7219 | 2.0229 | .09207 | .18200 | .50586 | 1.1513 | .18624 |
| 1.82 | 1.5463 | 1.4610 | .16879 | .28061 | .60151 | 1.09649 | .24660 | 2.22 | 1.7258 | 2.0409 | .09064 | .17998 | .50361 | 1.1526 | .18499 |
| 1.83 | 1.5514 | 1.4723 | .16622 | .27756 | .59888 | 1.09798 | .24472 | 2.23 | 1.7297 | 2.0592 | .08923 | .17798 | .50136 | 1.1539 | .18375 |
| 1.84 | 1.5564 | 1.4837 | .16369 | .27453 | .59626 | 1.09946 | .24286 | 2.24 | 1.7336 | 2.0777 | .08784 | .17600 | .49912 | 1.1552 | .18252 |
| 1.85 | 1.5614 | 1.4952 | .16120 | .27153 | .59365 | 1.1009 | .24102 | 2.25 | 1.7374 | 2.0964 | .08648 | .17404 | .49689 | 1.1565 | .18130 |
| 1.86 | 1.5664 | 1.5069 | .15874 | .26857 | .59105 | 1.1024 | .23919 | 2.26 | 1.7412 | 2.1154 | .08514 | .17211 | .49468 | 1.1578 | .18009 |
| 1.87 | 1.5714 | 1.5188 | .15631 | .26563 | .58845 | 1.1039 | .23739 | 2.27 | 1.7450 | 2.1345 | .08382 | .17020 | .49247 | 1.1590 | .17890 |
| 1.88 | 1.5763 | 1.5308 | .15392 | .26272 | .58586 | 1.1054 | .23561 | 2.28 | 1.7488 | 2.1538 | .08252 | .16830 | .49027 | 1.1603 | .17772 |
| 1.89 | 1.5812 | 1.5429 | .15156 | .25984 | .58329 | 1.1068 | .23385 | 2.29 | 1.7526 | 2.1734 | .08123 | .16643 | .48809 | 1.1616 | .17655 |
| 1.90 | 1.5861 | 1.5552 | .14924 | .25699 | .58072 | 1.1083 | .23211 | 2.30 | 1.7563 | 2.1931 | .07997 | .16458 | .48591 | 1.1629 | .17539 |
| 1.91 | 1.5909 | 1.5677 | .14695 | .25417 | .57816 | 1.1097 | .23039 | 2.31 | 1.7600 | 2.2131 | .07873 | .16275 | .48374 | 1.1641 | .17424 |
| 1.92 | 1.5957 | 1.5804 | .14469 | .25138 | .57561 | 1.1112 | .22868 | 2.32 | 1.7637 | 2.2333 | .07751 | .16095 | .48158 | 1.1653 | .17310 |
| 1.93 | 1.6005 | 1.5932 | .14247 | .24862 | .57307 | 1.1126 | .22699 | 2.33 | 1.7673 | 2.2537 | .07631 | .15916 | .47944 | 1.1666 | .17197 |
| 1.94 | 1.6052 | 1.6062 | .14028 | .24588 | .57054 | 1.1141 | .22532 | 2.34 | 1.7709 | 2.2744 | .07513 | .15739 | .47730 | 1.1678 | .17085 |
| 1.95 | 1.6099 | 1.6193 | .13813 | .24317 | .56802 | 1.1155 | .22367 | 2.35 | 1.7745 | 2.2953 | .07396 | .15564 | .47517 | 1.1690 | .16975 |
| 1.96 | 1.6146 | 1.6326 | .13600 | .24049 | .56551 | 1.1170 | .22204 | 2.36 | 1.7781 | 2.3164 | .07281 | .15391 | .47305 | 1.1703 | .16866 |
| 1.97 | 1.6193 | 1.6461 | .13390 | .23784 | .56301 | 1.1184 | .22042 | 2.37 | 1.7817 | 2.3377 | .07168 | .15220 | .47095 | 1.1715 | .16757 |
| 1.98 | 1.6239 | 1.6597 | .13184 | .23522 | .56051 | 1.1198 | .21882 | 2.38 | 1.7852 | 2.3593 | .07057 | .15052 | .46885 | 1.1727 | .16649 |
| 1.99 | 1.6285 | 1.6735 | .12981 | .23262 | .55803 | 1.1213 | .21724 | 2.39 | 1.7887 | 2.3811 | .06948 | .14885 | .46676 | 1.1739 | .16543 |

# Table 30  One-Dimensional Isentropic Compressible-Flow Functions

**For a perfect gas with constant specific heat and molecular weight**
**k = 1.4**

| M | M* | $\frac{A}{A^*}$ | $\frac{p}{p_0}$ | $\frac{\rho}{\rho_0}$ | $\frac{T}{T_0}$ | $\frac{F}{F^*}$ | $\left(\frac{A}{A^*}\right)\left(\frac{p}{p_0}\right)$ | M | M* | $\frac{A}{A^*}$ | $\frac{p}{p_0}$ | $\frac{\rho}{\rho_0}$ | $\frac{T}{T_0}$ | $\frac{F}{F^*}$ | $\left(\frac{A}{A^*}\right)\left(\frac{p}{p_0}\right)$ |
|---|---|---|---|---|---|---|---|---|---|---|---|---|---|---|---|
| 2.40 | 1.7922 | 2.4031 | .06840 | .14720 | .46468 | 1.1751 | .16437 | 2.80 | 1.9140 | 3.5001 | .03685 | .09462 | .38941 | 1.2182 | .12897 |
| 2.41 | 1.7957 | 2.4254 | .06734 | .14557 | .46262 | 1.1763 | .16332 | 2.81 | 1.9167 | 3.5336 | .03629 | .09360 | .38771 | 1.2192 | .12823 |
| 2.42 | 1.7991 | 2.4479 | .06630 | .14395 | .46056 | 1.1775 | .16229 | 2.82 | 1.9193 | 3.5674 | .03574 | .09259 | .38603 | 1.2202 | .12750 |
| 2.43 | 1.8025 | 2.4706 | .06527 | .14235 | .45851 | 1.1786 | .16126 | 2.83 | 1.9220 | 3.6015 | .03520 | .09158 | .38435 | 1.2211 | .12678 |
| 2.44 | 1.8059 | 2.4936 | .06426 | .14078 | .45647 | 1.1798 | .16024 | 2.84 | 1.9246 | 3.6359 | .03467 | .09059 | .38268 | 1.2221 | .12605 |
| 2.45 | 1.8093 | 2.5168 | .06327 | .13922 | .45444 | 1.1810 | .15923 | 2.85 | 1.9271 | 3.6707 | .03415 | .08962 | .38102 | 1.2230 | .12534 |
| 2.46 | 1.8126 | 2.5403 | .06229 | .13768 | .45242 | 1.1821 | .15823 | 2.86 | 1.9297 | 3.7058 | .03363 | .08865 | .37937 | 1.2240 | .12463 |
| 2.47 | 1.8159 | 2.5640 | .06133 | .13616 | .45041 | 1.1833 | .15724 | 2.87 | 1.9322 | 3.7413 | .03312 | .08769 | .37773 | 1.2249 | .12393 |
| 2.48 | 1.8192 | 2.5880 | .06038 | .13465 | .44841 | 1.1844 | .15626 | 2.88 | 1.9348 | 3.7771 | .03262 | .08674 | .37610 | 1.2258 | .12323 |
| 2.49 | 1.8225 | 2.6122 | .05945 | .13316 | .44642 | 1.1856 | .15528 | 2.89 | 1.9373 | 3.8133 | .03213 | .08581 | .37448 | 1.2268 | .12254 |
| 2.50 | 1.8258 | 2.6367 | .05853 | .13169 | .44444 | 1.1867 | .15432 | 2.90 | 1.9398 | 3.8498 | .03165 | .08489 | .37286 | 1.2277 | .12185 |
| 2.51 | 1.8290 | 2.6615 | .05763 | .13023 | .44247 | 1.1879 | .15337 | 2.91 | 1.9423 | 3.8866 | .03118 | .08398 | .37125 | 1.2286 | .12117 |
| 2.52 | 1.8322 | 2.6865 | .05674 | .12879 | .44051 | 1.1890 | .15242 | 2.92 | 1.9448 | 3.9238 | .03071 | .08308 | .36965 | 1.2295 | .12049 |
| 2.53 | 1.8354 | 2.7117 | .05586 | .12737 | .43856 | 1.1901 | .15148 | 2.93 | 1.9472 | 3.9614 | .03025 | .08218 | .36806 | 1.2304 | .11982 |
| 2.54 | 1.8386 | 2.7372 | .05500 | .12597 | .43662 | 1.1912 | .15055 | 2.94 | 1.9497 | 3.9993 | .02980 | .08130 | .36648 | 1.2313 | .11916 |
| 2.55 | 1.8417 | 2.7630 | .05415 | .12458 | .43469 | 1.1923 | .14963 | 2.95 | 1.9521 | 4.0376 | .02935 | .08043 | .36490 | 1.2322 | .11850 |
| 2.56 | 1.8448 | 2.7891 | .05332 | .12321 | .43277 | 1.1934 | .14871 | 2.96 | 1.9545 | 4.0763 | .02891 | .07957 | .36333 | 1.2331 | .11785 |
| 2.57 | 1.8479 | 2.8154 | .05250 | .12185 | .43085 | 1.1945 | .14780 | 2.97 | 1.9569 | 4.1153 | .02848 | .07872 | .36177 | 1.2340 | .11720 |
| 2.58 | 1.8510 | 2.8420 | .05169 | .12051 | .42894 | 1.1956 | .14691 | 2.98 | 1.9593 | 4.1547 | .02805 | .07788 | .36022 | 1.2348 | .11656 |
| 2.59 | 1.8541 | 2.8689 | .05090 | .11918 | .42705 | 1.1967 | .14601 | 2.99 | 1.9616 | 4.1944 | .02764 | .07705 | .35868 | 1.2357 | .11591 |
| 2.60 | 1.8572 | 2.8960 | .05012 | .11787 | .42517 | 1.1978 | .14513 | 3.00 | 1.9640 | 4.2346 | .02722 | .07623 | .35714 | 1.2366 | .11528 |
| 2.61 | 1.8602 | 2.9234 | .04935 | .11658 | .42330 | 1.1989 | .14426 | 3.10 | 1.9866 | 4.6573 | .02345 | .06852 | .34223 | 1.2450 | .10921 |
| 2.62 | 1.8632 | 2.9511 | .04859 | .11530 | .42143 | 1.2000 | .14339 | 3.20 | 2.0079 | 5.1210 | .02023 | .06165 | .32808 | 1.2530 | .10359 |
| 2.63 | 1.8662 | 2.9791 | .04784 | .11403 | .41957 | 1.2011 | .14253 | 3.30 | 2.0279 | 5.6287 | .01748 | .05554 | .31466 | 1.2605 | .09837 |
| 2.64 | 1.8692 | 3.0074 | .04711 | .11278 | .41772 | 1.2021 | .14168 | 3.40 | 2.0466 | 6.1837 | .01512 | .05009 | .30193 | 1.2676 | .09353 |
| 2.65 | 1.8721 | 3.0359 | .04639 | .11154 | .41589 | 1.2031 | .14083 | 3.50 | 2.0642 | 6.7896 | .01311 | .04523 | .28986 | 1.2743 | .08902 |
| 2.66 | 1.8750 | 3.0647 | .04568 | .11032 | .41406 | 1.2042 | .13999 | 3.60 | 2.0808 | 7.4501 | .01138 | .04089 | .27840 | 1.2807 | .08482 |
| 2.67 | 1.8779 | 3.0938 | .04498 | .10911 | .41224 | 1.2052 | .13916 | 3.70 | 2.0964 | 8.1691 | .00990 | .03702 | .26752 | 1.2867 | .08090 |
| 2.68 | 1.8808 | 3.1233 | .04429 | .10792 | .41043 | 1.2062 | .13834 | 3.80 | 2.1111 | 8.9506 | .00863 | .03355 | .25720 | 1.2924 | .07723 |
| 2.69 | 1.8837 | 3.1530 | .04361 | .10674 | .40863 | 1.2073 | .13752 | 3.90 | 2.1250 | 9.7990 | .00753 | .03044 | .24740 | 1.2978 | .07380 |
| 2.70 | 1.8865 | 3.1830 | .04295 | .10557 | .40684 | 1.2083 | .13671 | 4.00 | 2.1381 | 10.719 | .00658 | .02766 | .23810 | 1.3029 | .07059 |
| 2.71 | 1.8894 | 3.2133 | .04230 | .10442 | .40505 | 1.2093 | .13591 | 4.10 | 2.1505 | 11.715 | .00577 | .02516 | .22925 | 1.3077 | .06758 |
| 2.72 | 1.8922 | 3.2440 | .04166 | .10328 | .40327 | 1.2103 | .13511 | 4.20 | 2.1622 | 12.792 | .00506 | .02292 | .22085 | 1.3123 | .06475 |
| 2.73 | 1.8950 | 3.2749 | .04102 | .10215 | .40151 | 1.2113 | .13432 | 4.30 | 2.1732 | 13.955 | .00445 | .02090 | .21286 | 1.3167 | .06209 |
| 2.74 | 1.8978 | 3.3061 | .04039 | .10104 | .39976 | 1.2123 | .13354 | 4.40 | 2.1837 | 15.210 | .00392 | .01909 | .20525 | 1.3208 | .05959 |
| 2.75 | 1.9005 | 3.3376 | .03977 | .09994 | .39801 | 1.2133 | .13276 | 4.50 | 2.1936 | 16.562 | .00346 | .01745 | .19802 | 1.3247 | .05723 |
| 2.76 | 1.9032 | 3.3695 | .03917 | .09885 | .39627 | 1.2143 | .13199 | 4.60 | 2.2030 | 18.018 | .00305 | .01597 | .19113 | 1.3284 | .05500 |
| 2.77 | 1.9060 | 3.4017 | .03858 | .09777 | .39454 | 1.2153 | .13123 | 4.70 | 2.2119 | 19.583 | .00270 | .01463 | .18457 | 1.3320 | .05289 |
| 2.78 | 1.9087 | 3.4342 | .03800 | .09671 | .39282 | 1.2163 | .13047 | 4.80 | 2.2204 | 21.264 | .00240 | .01343 | .17832 | 1.3354 | .05091 |
| 2.79 | 1.9114 | 3.4670 | .03742 | .09566 | .39111 | 1.2173 | .12972 | 4.90 | 2.2284 | 23.067 | .00213 | .01233 | .17235 | 1.3386 | .04904 |
| | | | | | | | | 5.00 | 2.2361 | 25.000 | $189(10)^{-5}$ | .01134 | .16667 | 1.3416 | .04725 |
| | | | | | | | | 6.00 | 2.2953 | 53.180 | $633(10)^{-6}$ | .00519 | .12195 | 1.3655 | .03368 |
| | | | | | | | | 7.00 | 2.3333 | 104.143 | $242(10)^{-6}$ | .00261 | .09259 | 1.3810 | .02516 |
| | | | | | | | | 8.00 | 2.3591 | 190.109 | $102(10)^{-6}$ | .00141 | .07246 | 1.3915 | .01947 |
| | | | | | | | | 9.00 | 2.3772 | 327.189 | $474(10)^{-7}$ | .000815 | .05814 | 1.3989 | .01550 |
| | | | | | | | | 10.00 | 2.3904 | 535.938 | $236(10)^{-7}$ | .000495 | .04762 | 1.4044 | .01263 |
| | | | | | | | | ∞ | 2.4495 | ∞ | 0 | 0 | 0 | 1.4289 | 0 |

# Table 31   One-Dimensional Isentropic Compressible-Flow Functions

For a perfect gas with constant specific heat and molecular weight
k = 1.0

| M | M* | $\frac{A}{A^*}$ | $\frac{p}{p_0}=\frac{\rho}{\rho_0}$ | $\frac{T}{T_0}$ | $\frac{F}{F^*}$ | $\left(\frac{A}{A^*}\right)\left(\frac{p}{p_0}\right)$ | M | M* | $\frac{A}{A^*}$ | $\frac{p}{p_0}=\frac{\rho}{\rho_0}$ | $\frac{T}{T_0}$ | $\frac{F}{F^*}$ | $\left(\frac{A}{A^*}\right)\left(\frac{p}{p_0}\right)$ |
|---|---|---|---|---|---|---|---|---|---|---|---|---|---|
| 0 | 0 | ∞ | 1.0000 | 1.000 | ∞ | ∞ | 1.75 | 1.75 | 1.603 | .2163 | 1.000 | 1.161 | .3466 |
| 0.05 | .05 | 12.146 | .9989 | | 10.025 | 12.136 | 1.80 | 1.80 | 1.703 | .1979 | | 1.178 | .3370 |
| .10 | .10 | 6.096 | .9951 | | 5.050 | 6.065 | 1.85 | 1.85 | 1.815 | .1806 | | 1.195 | .3279 |
| .15 | .15 | 4.089 | .9888 | | 3.408 | 4.044 | 1.90 | 1.90 | 1.941 | .1645 | | 1.213 | .3192 |
| .20 | .20 | 3.094 | .9802 | | 2.600 | 3.033 | 1.95 | 1.95 | 2.082 | .1494 | | 1.231 | .3110 |
| .25 | .25 | 2.503 | .9693 | | 2.125 | 2.426 | 2.00 | 2.00 | 2.241 | .1353 | | 1.250 | .3033 |
| .30 | .30 | 2.115 | .9561 | | 1.817 | 2.022 | 2.05 | 2.05 | 2.419 | .1223 | | 1.269 | .2959 |
| .35 | .35 | 1.842 | .9406 | | 1.604 | 1.733 | 2.10 | 2.10 | 2.620 | .1102 | | 1.288 | .2888 |
| .40 | .40 | 1.643 | .9231 | | 1.450 | 1.516 | 2.15 | 2.15 | 2.846 | .09914 | | 1.307 | .2821 |
| .45 | .45 | 1.491 | .9037 | | 1.336 | 1.348 | 2.20 | 2.20 | 3.100 | .08892 | | 1.327 | .2757 |
| .50 | .50 | 1.375 | .8825 | | 1.250 | 1.2131 | 2.25 | 2.25 | 3.388 | .07956 | | 1.347 | .2696 |
| .55 | .55 | 1.283 | .8597 | | 1.184 | 1.1028 | 2.30 | 2.30 | 3.714 | .07100 | | 1.367 | .2637 |
| .60 | .60 | 1.210 | .8353 | | 1.133 | 1.0109 | 2.35 | 2.35 | 4.083 | .06321 | | 1.387 | .2581 |
| .65 | .65 | 1.153 | .8096 | | 1.0942 | .9331 | 2.40 | 2.40 | 4.502 | .05614 | | 1.408 | .2527 |
| .70 | .70 | 1.107 | .7827 | | 1.0643 | .8665 | 2.45 | 2.45 | 4.979 | .04973 | | 1.429 | .2475 |
| .75 | .75 | 1.0714 | .7549 | | 1.0417 | .8087 | 2.50 | 2.50 | 5.522 | .04394 | | 1.450 | .2426 |
| .80 | .80 | 1.0441 | .7262 | | 1.0250 | .7582 | 2.55 | 2.55 | 6.142 | .03873 | | 1.471 | .2379 |
| .85 | .85 | 1.0240 | .6968 | | 1.0132 | .7136 | 2.60 | 2.60 | 6.852 | .03405 | | 1.492 | .2333 |
| .90 | .90 | 1.0104 | .6670 | | 1.0056 | .6739 | 2.65 | 2.65 | 7.665 | .02986 | | 1.513 | .2289 |
| .95 | .95 | 1.0025 | .6369 | | 1.0013 | .6385 | 2.70 | 2.70 | 8.600 | .02612 | | 1.535 | .2247 |
| 1.00 | 1.00 | 1.0000 | .6065 | | 1.0000 | .6065 | 2.75 | 2.75 | 9.676 | .02279 | | 1.557 | .2206 |
| 1.05 | 1.05 | 1.0025 | .5762 | | 1.0012 | .5777 | 2.80 | 2.80 | 10.92 | .01984 | | 1.579 | .2166 |
| 1.10 | 1.10 | 1.0097 | .5461 | | 1.0045 | .5514 | 2.85 | 2.85 | 12.35 | .01723 | | 1.600 | .2128 |
| 1.15 | 1.15 | 1.0217 | .5163 | | 1.0098 | .5274 | 2.90 | 2.90 | 14.02 | .01492 | | 1.622 | .2092 |
| 1.20 | 1.20 | 1.0384 | .4868 | | 1.0167 | .5054 | 2.95 | 2.95 | 15.95 | .01289 | | 1.644 | .2056 |
| 1.25 | 1.25 | 1.0598 | .4578 | | 1.0250 | .4852 | 3.00 | 3.00 | 18.20 | $1111(10)^{-5}$ | | 1.667 | .2022 |
| 1.30 | 1.30 | 1.0861 | .4295 | | 1.0346 | .4666 | 3.50 | 3.50 | 79.22 | $219(10)^{-5}$ | | 1.893 | .1733 |
| 1.35 | 1.35 | 1.117 | .4020 | | 1.0453 | .4493 | 4.00 | 4.00 | 452.0 | $335(10)^{-6}$ | | 2.125 | .1516 |
| 1.40 | 1.40 | 1.154 | .3753 | | 1.0571 | .4332 | 4.50 | 4.50 | 3364 | $401(10)^{-7}$ | | 2.361 | .1348 |
| 1.45 | 1.45 | 1.197 | .3495 | | 1.0698 | .4183 | 5.00 | 5.00 | 32550 | $373(10)^{-8}$ | | 2.600 | .1213 |
| 1.50 | 1.50 | 1.245 | .3247 | | 1.0833 | .4044 | 6.00 | 6.00 | $664(10)^4$ | $152(10)^{-10}$ | | 3.083 | .1010 |
| 1.55 | 1.55 | 1.300 | .3008 | | 1.0976 | .3913 | 7.00 | 7.00 | $378(10)^7$ | $229(10)^{-13}$ | | 3.571 | .08665 |
| 1.60 | 1.60 | 1.363 | .2780 | | 1.113 | .3791 | 8.00 | 8.00 | $599(10)^{10}$ | $127(10)^{-16}$ | | 4.062 | .07578 |
| 1.65 | 1.65 | 1.434 | .2563 | | 1.128 | .3676 | 9.00 | 9.00 | $262(10)^{14}$ | $258(10)^{-20}$ | | 4.556 | .06741 |
| 1.70 | 1.70 | 1.514 | .2357 | | 1.144 | .3568 | 10.00 | 10.00 | $314(10)^{18}$ | $193(10)^{-24}$ | | 5.050 | .06065 |
| | | | | | | | ∞ | ∞ | ∞ | 0 | 1.000 | ∞ | 0 |

# Table 32  One-Dimensional Isentropic Compressible-Flow Functions

For a perfect gas with constant specific heat and molecular weight
k = 1.1

| M | $M^*$ | $\frac{A}{A^*}$ | $\frac{p}{p_0}$ | $\frac{\rho}{\rho_0}$ | $\frac{T}{T_0}$ | $\frac{F}{F^*}$ | $\left(\frac{A}{A^*}\right)\left(\frac{p}{p_0}\right)$ | M | $M^*$ | $\frac{A}{A^*}$ | $\frac{p}{p_0}$ | $\frac{\rho}{\rho_0}$ | $\frac{T}{T_0}$ | $\frac{F}{F^*}$ | $\left(\frac{A}{A^*}\right)\left(\frac{p}{p_0}\right)$ |
|---|---|---|---|---|---|---|---|---|---|---|---|---|---|---|---|
| 0 | 0 | ∞ | 1.0000 | 1.0000 | 1.0000 | ∞ | ∞ | 1.75 | 1.670 | 1.528 | .2086 | .2406 | .8672 | 1.134 | .3188 |
| 0.05 | 0.05123 | 11.999 | .9986 | .9988 | .9999 | 9.785 | 11.982 | 1.80 | 1.711 | 1.610 | .1917 | .2228 | .8606 | 1.147 | .3088 |
| .10 | .1024 | 6.023 | .9945 | .9950 | .9995 | 4.931 | 5.990 | 1.85 | 1.752 | 1.701 | .1759 | .2060 | .8539 | 1.161 | .2993 |
| .15 | .1536 | 4.042 | .9877 | .9888 | .9989 | 3.332 | 3.992 | 1.90 | 1.792 | 1.801 | .1612 | .1902 | .8471 | 1.175 | .2902 |
| .20 | .2047 | 3.059 | .9783 | .9802 | .9980 | 2.545 | 2.993 | 1.95 | 1.832 | 1.911 | .1474 | .1754 | .8402 | 1.189 | .2816 |
| .25 | .2558 | 2.476 | .9663 | .9693 | .9969 | 2.083 | 2.393 | 2.00 | 1.871 | 2.032 | .1346 | .1615 | .8333 | 1.203 | .2735 |
| .30 | .3067 | 2.094 | .9518 | .9561 | .9955 | 1.784 | 1.993 | 2.05 | 1.910 | 2.165 | .1227 | .1485 | .8264 | 1.217 | .2657 |
| .35 | .3575 | 1.825 | .9350 | .9408 | .9939 | 1.577 | 1.707 | 2.10 | 1.948 | 2.312 | .1117 | .1363 | .8194 | 1.231 | .2582 |
| .40 | .4082 | 1.628 | .9161 | .9234 | .9921 | 1.429 | 1.492 | 2.15 | 1.986 | 2.473 | .1015 | .1250 | .8123 | 1.245 | .2511 |
| .45 | .4588 | 1.480 | .8951 | .9042 | .9900 | 1.319 | 1.325 | 2.20 | 2.023 | 2.651 | .09218 | .1145 | .8052 | 1.259 | .2444 |
| .50 | .5092 | 1.365 | .8723 | .8832 | .9877 | 1.237 | 1.1908 | 2.25 | 2.060 | 2.846 | .08357 | .10473 | .7980 | 1.273 | .2379 |
| .55 | .5594 | 1.275 | .8478 | .8606 | .9851 | 1.174 | 1.0812 | 2.30 | 2.096 | 3.061 | .07566 | .09568 | .7908 | 1.286 | .2317 |
| .60 | .6094 | 1.204 | .8218 | .8366 | .9823 | 1.125 | .9897 | 2.35 | 2.132 | 3.299 | .06842 | .08731 | .7836 | 1.300 | .2257 |
| .65 | .6592 | 1.148 | .7945 | .8113 | .9793 | 1.0882 | .9121 | 2.40 | 2.167 | 3.560 | .06179 | .07959 | .7764 | 1.314 | .2200 |
| .70 | .7087 | 1.104 | .7662 | .7850 | .9761 | 1.0599 | .8456 | 2.45 | 2.202 | 3.848 | .05574 | .07247 | .7692 | 1.328 | .2145 |
| .75 | .7579 | 1.0689 | .7370 | .7578 | .9727 | 1.0387 | .7878 | 2.50 | 2.236 | 4.165 | .05022 | .06592 | .7619 | 1.342 | .2092 |
| .80 | .8069 | 1.0425 | .7071 | .7298 | .9690 | 1.0231 | .7372 | 2.55 | 2.270 | 4.515 | .04520 | .05990 | .7546 | 1.356 | .2041 |
| .85 | .8557 | 1.0231 | .6768 | .7013 | .9651 | 1.0122 | .6924 | 2.60 | 2.303 | 4.902 | .04064 | .05438 | .7473 | 1.369 | .1992 |
| .90 | .9041 | 1.0100 | .6462 | .6724 | .9610 | 1.0051 | .6526 | 2.65 | 2.336 | 5.328 | .03650 | .04932 | .7401 | 1.382 | .1945 |
| .95 | .9522 | 1.0024 | .6154 | .6431 | .9568 | 1.0012 | .6169 | 2.70 | 2.368 | 5.799 | .03276 | .04470 | .7329 | 1.395 | .1900 |
| 1.00 | 1.0000 | 1.0000 | .5847 | .6139 | .9524 | 1.0000 | .5847 | 2.75 | 2.400 | 6.320 | .02936 | .04047 | .7256 | 1.408 | .1856 |
| 1.05 | 1.0474 | 1.0023 | .5542 | .5847 | .9478 | 1.0011 | .5555 | 2.80 | 2.432 | 6.895 | .02630 | .03661 | .7184 | 1.422 | .1814 |
| 1.10 | 1.0945 | 1.0092 | .5240 | .5557 | .9430 | 1.0041 | .5289 | 2.85 | 2.463 | 7.532 | .02354 | .03310 | .7112 | 1.435 | .1773 |
| 1.15 | 1.141 | 1.0204 | .4944 | .5271 | .9380 | 1.0087 | .5046 | 2.90 | 2.493 | 8.237 | .02104 | .02989 | .7040 | 1.448 | .1733 |
| 1.20 | 1.188 | 1.0360 | .4654 | .4989 | .9328 | 1.0148 | .4822 | 2.95 | 2.523 | 9.016 | .01880 | .02698 | .6968 | 1.460 | .1695 |
| 1.25 | 1.234 | 1.0559 | .4371 | .4713 | .9275 | 1.0221 | .4616 | 3.0 | 2.553 | 9.880 | $1679(10)^{-5}$ | .02434 | .6897 | 1.472 | .1658 |
| 1.30 | 1.279 | 1.0801 | .4097 | .4443 | .9221 | 1.0305 | .4425 | 3.5 | 2.824 | 25.83 | $522(10)^{-5}$ | .008414 | .6202 | 1.589 | .1348 |
| 1.35 | 1.324 | 1.1088 | .3832 | .4180 | .9165 | 1.0397 | .4249 | 4.0 | 3.055 | 71.75 | $156(10)^{-5}$ | .002801 | .5556 | 1.691 | .1116 |
| 1.40 | 1.369 | 1.1421 | .3576 | .3926 | .9107 | 1.0497 | .4084 | 4.5 | 3.250 | 205.8 | $456(10)^{-6}$ | .000918 | .4969 | 1.779 | .09385 |
| 1.45 | 1.413 | 1.1802 | .3330 | .3680 | .9048 | 1.0604 | .3930 | 5.0 | 3.416 | 597.7 | $134(10)^{-6}$ | .000301 | .4444 | 1.854 | .07988 |
| 1.50 | 1.457 | 1.223 | .3095 | .3443 | .8989 | 1.0717 | .3787 | 6.0 | 3.674 | 4949 | $121(10)^{-7}$ | $338(10)^{-7}$ | .3571 | 1.973 | .05967 |
| 1.55 | 1.501 | 1.272 | .2871 | .3216 | .8928 | 1.0836 | .3652 | 7.0 | 3.862 | 37976 | $121(10)^{-8}$ | $419(10)^{-8}$ | .2899 | 2.060 | .04608 |
| 1.60 | 1.544 | 1.326 | .2658 | .2999 | .8865 | 1.0958 | .3526 | 8.0 | 4.000 | $262(10)^{3}$ | $139(10)^{-9}$ | $585(10)^{-9}$ | .2381 | 2.125 | .03654 |
| 1.65 | 1.586 | 1.387 | .2456 | .2791 | .8801 | 1.108 | .3407 | 9.0 | 4.104 | $161(10)^{4}$ | $184(10)^{-10}$ | $927(10)^{-10}$ | .1980 | 2.174 | .02962 |
| 1.70 | 1.628 | 1.454 | .2266 | .2593 | .8737 | 1.121 | .3294 | 10.0 | 4.183 | $887(10)^{4}$ | $275(10)^{-11}$ | $165(10)^{-10}$ | .1667 | 2.211 | .02446 |
| | | | | | | | | ∞ | 4.583 | ∞ | 0 | 0 | 0 | 2.400 | 0 |

# Table 33   One-Dimensional Isentropic Compressible-Flow Functions

For a perfect gas with constant specific heat and molecular weight
k = 1.2

| M | M* | $\frac{A}{A^*}$ | $\frac{p}{p_0}$ | $\frac{\rho}{\rho_0}$ | $\frac{T}{T_0}$ | $\frac{F}{F^*}$ | $\left(\frac{A}{A^*}\right)\left(\frac{p}{p_0}\right)$ | M | M* | $\frac{A}{A^*}$ | $\frac{p}{p_0}$ | $\frac{\rho}{\rho_0}$ | $\frac{T}{T_0}$ | $\frac{F}{F^*}$ | $\left(\frac{A}{A^*}\right)\left(\frac{p}{p_0}\right)$ |
|---|---|---|---|---|---|---|---|---|---|---|---|---|---|---|---|
| 0 | 0 | ∞ | 1.0000 | 1.0000 | 1.0000 | ∞ | ∞ | 1.75 | 1.606 | 1.470 | .2013 | .2629 | .7656 | 1.114 | .2960 |
| 0.05 | 0.05243 | 11.857 | .9985 | .9988 | .9998 | 9.562 | 11.839 | 1.80 | 1.641 | 1.539 | .1856 | .2458 | .7553 | 1.125 | .2858 |
| .10 | .1048 | 5.953 | .9940 | .9950 | .9991 | 4.822 | 5.917 | 1.85 | 1.675 | 1.615 | .1710 | .2295 | .7450 | 1.136 | .2762 |
| .15 | .1571 | 3.996 | .9866 | .9888 | .9978 | 3.260 | 3.942 | 1.90 | 1.708 | 1.698 | .1573 | .2141 | .7347 | 1.147 | .2671 |
| .20 | .2093 | 3.026 | .9763 | .9802 | .9960 | 2.493 | 2.954 | 1.95 | 1.741 | 1.787 | .1446 | .1996 | .7245 | 1.158 | .2584 |
| .25 | .2614 | 2.451 | .9633 | .9693 | .9938 | 2.044 | 2.361 | 2.00 | 1.773 | 1.884 | .1328 | .1859 | .7143 | 1.168 | .2502 |
| .30 | .3133 | 2.073 | .9477 | .9562 | .9911 | 1.753 | 1.965 | 2.05 | 1.804 | 1.989 | .1218 | .1730 | .7041 | 1.179 | .2423 |
| .35 | .3649 | 1.809 | .9296 | .9409 | .9879 | 1.553 | 1.681 | 2.10 | 1.835 | 2.103 | .1117 | .1609 | .6940 | 1.190 | .2348 |
| .40 | .4162 | 1.615 | .9092 | .9236 | .9843 | 1.409 | 1.468 | 2.15 | 1.865 | 2.226 | .1023 | .1496 | .6839 | 1.201 | .2277 |
| .45 | .4672 | 1.469 | .8867 | .9046 | .9802 | 1.304 | 1.302 | 2.20 | 1.894 | 2.359 | .09362 | .1390 | .6739 | 1.212 | .2209 |
| .50 | .5179 | 1.356 | .8623 | .8839 | .9756 | 1.224 | 1.170 | 2.25 | 1.923 | 2.504 | .08563 | .1290 | .6639 | 1.222 | .2144 |
| .55 | .5683 | 1.268 | .8363 | .8616 | .9706 | 1.164 | 1.0606 | 2.30 | 1.951 | 2.660 | .07826 | .1197 | .6540 | 1.232 | .2082 |
| .60 | .6183 | 1.199 | .8088 | .8379 | .9653 | 1.118 | .9694 | 2.35 | 1.978 | 2.829 | .07148 | .1110 | .6442 | 1.242 | .2022 |
| .65 | .6678 | 1.144 | .7801 | .8131 | .9595 | 1.0826 | .8922 | 2.40 | 2.005 | 3.011 | .06526 | .1029 | .6345 | 1.252 | .1965 |
| .70 | .7168 | 1.100 | .7505 | .7873 | .9533 | 1.0559 | .8258 | 2.45 | 2.031 | 3.208 | .05955 | .09529 | .6249 | 1.262 | .1910 |
| 75 | .7654 | 1.0666 | .7201 | .7606 | .9467 | 1.0360 | .7681 | 2.50 | 2.057 | 3.421 | .05431 | .08825 | .6154 | 1.272 | .1858 |
| .80 | .8134 | 1.0410 | .6892 | .7333 | .9398 | 1.0214 | .7174 | 2.55 | 2.082 | 3.650 | .04951 | .08170 | .6060 | 1.282 | .1808 |
| .85 | .8609 | 1.0222 | .6580 | .7055 | .9326 | 1.0112 | .6726 | 2.60 | 2.106 | 3.898 | .04512 | .07562 | .5967 | 1.291 | .1759 |
| .90 | .9078 | 1.0096 | .6267 | .6774 | .9251 | 1.0047 | .6327 | 2.65 | 2.130 | 4.166 | .04110 | .06997 | .5875 | 1.300 | .1712 |
| .95 | .9542 | 1.0023 | .5954 | .6492 | .9172 | 1.0011 | .5968 | 2.70 | 2.154 | 4.455 | .03743 | .06472 | .5784 | 1.309 | .1667 |
| 1.00 | 1.0000 | 1.0000 | .5644 | .6209 | .9091 | 1.0000 | .5644 | 2.75 | 2.177 | 4.767 | .03408 | .05985 | .5694 | 1.318 | .1624 |
| 1.05 | 1.0451 | 1.0022 | .5339 | .5928 | .9007 | 1.0010 | .5351 | 2.80 | 2.199 | 5.103 | .03102 | .05534 | .5605 | 1.327 | .1583 |
| 1.10 | 1.0896 | 1.0087 | .5039 | .5649 | .8921 | 1.0037 | .5083 | 2.85 | 2.220 | 5.466 | .02823 | .05116 | .5518 | 1.335 | .1543 |
| 1.15 | 1.134 | 1.0194 | .4746 | .5374 | .8832 | 1.0079 | .4838 | 2.90 | 2.241 | 5.858 | .02569 | .04729 | .5432 | 1.343 | .1505 |
| 1.20 | 1.177 | 1.0340 | .4461 | .5104 | .8741 | 1.0133 | .4613 | 2.95 | 2.262 | 6.280 | .02337 | .04370 | .5347 | 1.352 | .1467 |
| 1.25 | 1.219 | 1.0525 | .4185 | .4839 | .8648 | 1.0197 | .4405 | 3.00 | 2.283 | 6.735 | .02126 | .04039 | .5263 | 1.360 | .1432 |
| 1.30 | 1.261 | 1.0749 | .3918 | .4581 | .8554 | 1.0270 | .4212 | 3.5 | 2.461 | 13.76 | .008242 | .01834 | .4494 | 1.434 | .1134 |
| 1.35 | 1.302 | 1.101 | .3662 | .4330 | .8458 | 1.0350 | .4033 | 4.0 | 2.602 | 28.35 | .003237 | .008417 | .3846 | 1.493 | .09179 |
| 1.40 | 1.342 | 1.131 | .3417 | .4087 | .8361 | 1.0437 | .3867 | 4.5 | 2.714 | 57.96 | .001305 | .003948 | .3306 | 1.541 | .07564 |
| 1.45 | 1.382 | 1.166 | .3182 | .3852 | .8263 | 1.0529 | .3712 | 5.0 | 2.803 | 116.3 | .000544 | .001904 | .2857 | 1.580 | .06329 |
| 1.50 | 1.421 | 1.205 | .2959 | .3625 | .8163 | 1.0625 | .3566 | 6.0 | 2.934 | 435.9 | $106(10)^{-6}$ | $486(10)^{-6}$ | .2174 | 1.637 | .04601 |
| 1.55 | 1.459 | 1.248 | .2747 | .3407 | .8063 | 1.0724 | .3430 | 7.0 | 3.023 | 1469 | $237(10)^{-7}$ | $140(10)^{-6}$ | .1695 | 1.677 | .03482 |
| 1.60 | 1.497 | 1.296 | .2547 | .3199 | .7962 | 1.0826 | .3302 | 8.0 | 3.084 | 4467 | $609(10)^{-8}$ | $451(10)^{-7}$ | .1351 | 1.704 | .02720 |
| 1.65 | 1.534 | 1.349 | .2358 | .3000 | .7860 | 1.0930 | .3181 | 9.0 | 3.129 | 12383 | $176(10)^{-8}$ | $160(10)^{-7}$ | .1099 | 1.724 | .02181 |
| 1.70 | 1.570 | 1.407 | .2180 | .2810 | .7758 | 1.1036 | .3067 | 10.0 | 3.162 | 31623 | $564(10)^{-9}$ | $621(10)^{-8}$ | .09091 | 1.739 | .01785 |
| | | | | | | | | ∞ | 3.317 | ∞ | 0 | 0 | 0 | 1.809 | 0 |

# Table 34  One-Dimensional Isentropic Compressible-Flow Functions

For a perfect gas with constant specific heat and molecular weight
k = 1.3

| M | M* | $\frac{A}{A*}$ | $\frac{p}{p_0}$ | $\frac{\rho}{\rho_0}$ | $\frac{T}{T_0}$ | $\frac{F}{F*}$ | $\left(\frac{A}{A*}\right)\left(\frac{p}{p_0}\right)$ | M | M* | $\frac{A}{A*}$ | $\frac{p}{p_0}$ | $\frac{\rho}{\rho_0}$ | $\frac{T}{T_0}$ | $\frac{F}{F*}$ | $\left(\frac{A}{A*}\right)\left(\frac{p}{p_0}\right)$ |
|---|---|---|---|---|---|---|---|---|---|---|---|---|---|---|---|
| 0 | 0 | ∞ | 1.0000 | 1.0000 | 1.0000 | ∞ | ∞ | 1.75 | 1.554 | 1.424 | .1944 | .2836 | .6852 | 1.0986 | .2768 |
| 0.05 | 0.0536 | 11.721 | .9984 | .9988 | .9996 | 9.354 | 11.702 | 1.80 | 1.584 | 1.484 | .1797 | .2670 | .6729 | 1.108 | .2667 |
| .10 | .1072 | 5.885 | .9936 | .9951 | .9985 | 4.720 | 5.848 | 1.85 | 1.613 | 1.549 | .1660 | .2513 | .6607 | 1.116 | .2571 |
| .15 | .1606 | 3.952 | .9855 | .9889 | .9966 | 3.194 | 3.895 | 1.90 | 1.641 | 1.618 | .1533 | .2364 | .6487 | 1.125 | .2481 |
| .20 | .2138 | 2.994 | .9744 | .9803 | .9940 | 2.445 | 2.917 | 1.95 | 1.669 | 1.693 | .1415 | .2222 | .6368 | 1.134 | .2395 |
| .25 | .2668 | 2.426 | .9603 | .9694 | .9907 | 2.007 | 2.330 | 2.00 | 1.696 | 1.773 | .1305 | .2087 | .6250 | 1.143 | .2313 |
| .30 | .3195 | 2.054 | .9435 | .9563 | .9867 | 1.724 | 1.938 | 2.05 | 1.722 | 1.859 | .1203 | .1960 | .6134 | 1.152 | .2236 |
| .35 | .3719 | 1.793 | .9241 | .9411 | .9820 | 1.530 | 1.657 | 2.10 | 1.747 | 1.951 | .1108 | .1841 | .6019 | 1.160 | .2162 |
| .40 | .4239 | 1.602 | .9023 | .9240 | .9766 | 1.391 | 1.446 | 2.15 | 1.772 | 2.050 | .1020 | .1728 | .5905 | 1.168 | .2092 |
| .45 | .4754 | 1.459 | .8784 | .9051 | .9705 | 1.289 | 1.281 | 2.20 | 1.796 | 2.156 | .0939 | .1621 | .5793 | 1.176 | .2025 |
| .50 | .5264 | 1.348 | .8526 | .8845 | .9638 | 1.213 | 1.1491 | 2.25 | 1.819 | 2.268 | .08645 | .1521 | .5684 | 1.184 | .1961 |
| .55 | .5769 | 1.261 | .8251 | .8625 | .9566 | 1.155 | 1.0407 | 2.30 | 1.842 | 2.388 | .07955 | .1427 | .5576 | 1.192 | .1900 |
| .60 | .6267 | 1.193 | .7962 | .8392 | .9488 | 1.111 | .9501 | 2.35 | 1.864 | 2.517 | .07318 | .1338 | .5470 | 1.200 | .1842 |
| .65 | .6759 | 1.139 | .7662 | .8148 | .9404 | 1.0777 | .8731 | 2.40 | 1.885 | 2.654 | .06731 | .1254 | .5365 | 1.208 | .1786 |
| .70 | .7245 | 1.0972 | .7354 | .7895 | .9315 | 1.0524 | .8069 | 2.45 | 1.906 | 2.799 | .06190 | .1176 | .5262 | 1.216 | .1733 |
| .75 | .7724 | 1.0644 | .7040 | .7634 | .9222 | 1.0336 | .7493 | 2.50 | 1.926 | 2.954 | .05692 | .1103 | .5161 | 1.223 | .1682 |
| .80 | .8195 | 1.0395 | .6723 | .7367 | .9124 | 1.0199 | .6988 | 2.55 | 1.946 | 3.119 | .05234 | .1034 | .5062 | 1.230 | .1633 |
| .85 | .8658 | 1.0214 | .6403 | .7096 | .9022 | 1.0104 | .6540 | 2.60 | 1.965 | 3.295 | .04813 | .09693 | .4965 | 1.237 | .1586 |
| .90 | .9113 | 1.0092 | .6084 | .6823 | .8917 | 1.0043 | .6140 | 2.65 | 1.983 | 3.482 | .04426 | .09087 | .4870 | 1.244 | .1541 |
| .95 | .9561 | 1.0022 | .5768 | .6549 | .8808 | 1.0010 | .5781 | 2.70 | 2.001 | 3.681 | .04070 | .08520 | .4777 | 1.250 | .1498 |
| 1.00 | 1.0000 | 1.0000 | .5457 | .6276 | .8696 | 1.0000 | .5457 | 2.75 | 2.019 | 3.892 | .03743 | .07988 | .4686 | 1.257 | .1457 |
| 1.05 | 1.0430 | 1.0021 | .5152 | .6004 | .8581 | 1.0009 | .5163 | 2.80 | 2.036 | 4.116 | .03442 | .07490 | .4596 | 1.264 | .1417 |
| 1.10 | 1.0852 | 1.0083 | .4854 | .5735 | .8464 | 1.0034 | .4895 | 2.85 | 2.052 | 4.354 | .03166 | .07024 | .4508 | 1.270 | .1379 |
| 1.15 | 1.127 | 1.0183 | .4565 | .5470 | .8345 | 1.0072 | .4649 | 2.90 | 2.068 | 4.607 | .02913 | .06587 | .4422 | 1.276 | .1342 |
| 1.20 | 1.167 | 1.0321 | .4285 | .5210 | .8224 | 1.0120 | .4423 | 2.95 | 2.084 | 4.875 | .02680 | .06178 | .4338 | 1.282 | .1307 |
| 1.25 | 1.206 | 1.0495 | .4015 | .4956 | .8102 | 1.0177 | .4214 | 3.0 | 2.099 | 5.160 | .02466 | .05796 | .4255 | 1.288 | .12725 |
| 1.30 | 1.245 | 1.0704 | .3756 | .4709 | .7978 | 1.0241 | .4021 | 3.5 | 2.228 | 9.110 | .01090 | .03092 | .3524 | 1.338 | .09926 |
| 1.35 | 1.283 | 1.0948 | .3509 | .4468 | .7853 | 1.0312 | .3842 | 4.0 | 2.326 | 15.94 | .00498 | .01692 | .2941 | 1.378 | .07935 |
| 1.40 | 1.320 | 1.123 | .3273 | .4235 | .7728 | 1.0388 | .3675 | 4.5 | 2.402 | 27.39 | .00236 | .00954 | .2477 | 1.409 | .06472 |
| 1.45 | 1.356 | 1.154 | .3049 | .4010 | .7603 | 1.0467 | .3519 | 5.0 | 2.460 | 45.96 | .00117 | .00555 | .2105 | 1.433 | .05370 |
| 1.50 | 1.391 | 1.189 | .2836 | .3793 | .7477 | 1.0549 | .3374 | 6.0 | 2.543 | 120.1 | $321(10)^{-6}$ | .00206 | .15625 | 1.468 | .03856 |
| 1.55 | 1.425 | 1.228 | .2635 | .3585 | .7351 | 1.0634 | .3237 | 7.0 | 2.598 | 285.3 | $101(10)^{-6}$ | $847(10)^{-6}$ | .11976 | 1.491 | .02893 |
| 1.60 | 1.458 | 1.271 | .2446 | .3385 | .7225 | 1.0720 | .3109 | 8.0 | 2.635 | 623.1 | $361(10)^{-7}$ | $382(10)^{-6}$ | .09434 | 1.507 | .02247 |
| 1.65 | 1.491 | 1.318 | .2268 | .3194 | .7100 | 1.0808 | .2989 | 9.0 | 2.662 | 1266 | $142(10)^{-7}$ | $186(10)^{-6}$ | .07605 | 1.519 | .01793 |
| 1.70 | 1.523 | 1.369 | .2101 | .3011 | .6976 | 1.0897 | .2875 | 10.0 | 2.681 | 2416 | $606(10)^{-8}$ | $969(10)^{-7}$ | .06250 | 1.527 | .01463 |
| | | | | | | | | ∞ | 2.769 | ∞ | 0 | 0 | 0 | 1.565 | 0 |

(130)

# Table 35   One-Dimensional Isentropic Compressible-Flow Functions

**For a perfect gas with constant specific heat and molecular weight**
**k = 1.67**

| M | M* | $\frac{A}{A^*}$ | $\frac{p}{p_0}$ | $\frac{\rho}{\rho_0}$ | $\frac{T}{T_0}$ | $\frac{F}{F^*}$ | $\left(\frac{A}{A^*}\right)\left(\frac{p}{p_0}\right)$ |
|---|---|---|---|---|---|---|---|
| 0 | 0 | ∞ | 1.0000 | 1.0000 | 1.0000 | 0 | ∞ |
| 0.05 | 0.05775 | 11.265 | .9979 | .9988 | .9992 | 8.687 | 11.242 |
| .10 | .1154 | 5.661 | .9917 | .9950 | .9967 | 4.392 | 5.614 |
| .15 | .1727 | 3.805 | .9815 | .9888 | .9925 | 2.982 | 3.735 |
| .20 | .2296 | 2.887 | .9674 | .9803 | .9868 | 2.293 | 2.793 |
| .25 | .2859 | 2.344 | .9497 | .9695 | .9795 | 1.892 | 2.226 |
| .30 | .3415 | 1.989 | .9286 | .9566 | .9708 | 1.635 | 1.847 |
| .35 | .3963 | 1.741 | .9046 | .9417 | .9606 | 1.460 | 1.575 |
| .40 | .4502 | 1.560 | .8780 | .9250 | .9491 | 1.336 | 1.370 |
| .45 | .5031 | 1.424 | .8491 | .9067 | .9364 | 1.245 | 1.209 |
| .50 | .5549 | 1.320 | .8184 | .8869 | .9227 | 1.178 | 1.0803 |
| .55 | .6055 | 1.239 | .7862 | .8658 | .9080 | 1.128 | .9742 |
| .60 | .6548 | 1.176 | .7529 | .8437 | .8924 | 1.0909 | .8853 |
| .65 | .7029 | 1.126 | .7190 | .8207 | .8760 | 1.0628 | .8097 |
| .70 | .7496 | 1.0874 | .6847 | .7970 | .8590 | 1.0422 | .7445 |
| .75 | .7949 | 1.0576 | .6503 | .7728 | .8414 | 1.0265 | .6877 |
| .80 | .8388 | 1.0351 | .6162 | .7483 | .8234 | 1.0155 | .6378 |
| .85 | .8812 | 1.0189 | .5826 | .7236 | .8051 | 1.0080 | .5936 |
| .90 | .9222 | 1.0080 | .5497 | .6988 | .7866 | 1.0033 | .5541 |
| .95 | .9618 | 1.0019 | .5177 | .6742 | .7679 | 1.0008 | .5187 |
| 1.00 | 1.0000 | 1.0000 | .4867 | .6497 | .7491 | 1.0000 | .4867 |
| 1.05 | 1.0368 | 1.0018 | .4568 | .6255 | .7303 | 1.0007 | .4576 |
| 1.10 | 1.0721 | 1.0071 | .4282 | .6017 | .7116 | 1.0024 | .4312 |
| 1.15 | 1.106 | 1.0154 | .4009 | .5784 | .6930 | 1.0051 | .4070 |
| 1.20 | 1.139 | 1.0266 | .3749 | .5557 | .6746 | 1.0085 | .3849 |
| 1.25 | 1.170 | 1.0406 | .3502 | .5335 | .6564 | 1.0124 | .3646 |
| 1.30 | 1.200 | 1.0573 | .3269 | .5119 | .6385 | 1.0167 | .3457 |
| 1.35 | 1.229 | 1.0765 | .3049 | .4910 | .6209 | 1.0213 | .3282 |
| 1.40 | 1.257 | 1.0981 | .2842 | .4707 | .6036 | 1.0262 | .3121 |
| 1.45 | 1.283 | 1.122 | .2647 | .4511 | .5867 | 1.0313 | .2972 |
| 1.50 | 1.309 | 1.148 | .2465 | .4323 | .5702 | 1.0364 | .2830 |
| 1.55 | 1.333 | 1.176 | .2295 | .4142 | .5541 | 1.0416 | .2700 |
| 1.60 | 1.356 | 1.207 | .2136 | .3968 | .5383 | 1.0468 | .2579 |
| 1.65 | 1.379 | 1.240 | .1988 | .3801 | .5230 | 1.0520 | .2465 |
| 1.70 | 1.400 | 1.275 | .1850 | .3640 | .5081 | 1.0572 | .2358 |
| 1.75 | 1.420 | 1.312 | .1721 | .3486 | .4936 | 1.0623 | .2257 |
| 1.80 | 1.440 | 1.351 | .1601 | .3339 | .4795 | 1.0673 | .2163 |
| 1.85 | 1.459 | 1.392 | .1490 | .3198 | .4658 | 1.0722 | .2075 |
| 1.90 | 1.477 | 1.436 | .1386 | .3063 | .4526 | 1.0770 | .1991 |
| 1.95 | 1.494 | 1.482 | .1290 | .2934 | .4398 | 1.0817 | .1912 |
| 2.00 | 1.511 | 1.530 | .1201 | .2811 | .4274 | 1.0863 | .1838 |
| 2.05 | 1.527 | 1.580 | .1119 | .2694 | .4153 | 1.0908 | .1768 |
| 2.10 | 1.542 | 1.632 | .1042 | .2582 | .4036 | 1.0952 | .1701 |
| 2.15 | 1.556 | 1.687 | .09712 | .2475 | .3923 | 1.0994 | .1638 |
| 2.20 | 1.570 | 1.744 | .09053 | .2373 | .3814 | 1.1035 | .1579 |
| 2.25 | 1.583 | 1.803 | .08442 | .2276 | .3709 | 1.107 | .1522 |
| 2.30 | 1.596 | 1.865 | .07875 | .2183 | .3607 | 1.111 | .1468 |
| 2.35 | 1.608 | 1.929 | .07349 | .2094 | .3508 | 1.115 | .1417 |
| 2.40 | 1.620 | 1.995 | .06862 | .2010 | .3413 | 1.119 | .1369 |
| 2.45 | 1.631 | 2 064 | .06410 | .1930 | .3321 | 1.123 | .1323 |
| 2.50 | 1.642 | 2.135 | .05990 | .1853 | .3232 | 1.126 | .1279 |
| 2.55 | 1.653 | 2.209 | .05601 | .1780 | .3146 | 1.129 | .1237 |
| 2.60 | 1.663 | 2.285 | .05239 | .1710 | .3063 | 1.132 | .1197 |
| 2.65 | 1.673 | 2.364 | .04903 | .1644 | .2983 | 1.135 | .1159 |
| 2.70 | 1.682 | 2.445 | .04591 | .1581 | .2905 | 1.138 | .1123 |
| 2.75 | 1.691 | 2.529 | .04301 | .1520 | .2830 | 1.141 | .1088 |
| 2.80 | 1.699 | 2.616 | .04032 | .1462 | .2757 | 1.144 | .1055 |
| 2.85 | 1.707 | 2.705 | .03781 | .1407 | .2687 | 1.146 | .1023 |
| 2.90 | 1.715 | 2.797 | .03547 | .1354 | .2620 | 1.149 | .09924 |
| 2.95 | 1.723 | 2.892 | .03330 | .1304 | .2554 | 1.152 | .09633 |
| 3.0 | 1.730 | 2.990 | .03128 | .12560 | .2491 | 1.154 | .09354 |
| 3.5 | 1.790 | 4.134 | .01720 | .08779 | .1959 | 1.174 | .07111 |
| 4.0 | 1.833 | 5.608 | .009939 | .06321 | .1572 | 1.189 | .05574 |
| 4.5 | 1.864 | 7.456 | .006007 | .04676 | .1285 | 1.200 | .04479 |
| 5.0 | 1.887 | 9.721 | .003779 | .03542 | .1067 | 1.208 | .03673 |
| 6.0 | 1.918 | 15.68 | $165(10)^{-5}$ | .02160 | .07657 | 1.220 | .02593 |
| 7.0 | 1.938 | 23.85 | $807(10)^{-6}$ | .01406 | .05742 | 1.227 | .01925 |
| 8.0 | 1.951 | 34.58 | $429(10)^{-6}$ | .00963 | .04456 | 1.232 | .01484 |
| 9.0 | 1.960 | 48.24 | $244(10)^{-6}$ | .00687 | .03554 | 1.235 | .01178 |
| 10.0 | 1.967 | 65.18 | $147(10)^{-6}$ | .00507 | .02898 | 1.238 | .00958 |
| ∞ | 1.996 | ∞ | 0 | 0 | 0 | 1.249 | 0 |

# Table 36 Rayleigh Line—One-Dimensional Compressible-Flow Functions for Stagnation-Temperature Change in the Absence of Friction and Area Change

**For a perfect gas with constant specific heat and molecular weight**
**k = 1.4**

| M | $\dfrac{T_0}{T_0^*}$ | $\dfrac{T}{T^*}$ | $\dfrac{p}{p^*}$ | $\dfrac{p_0}{p_0^*}$ | $\dfrac{V}{V^*}$ | M | $\dfrac{T_0}{T_0^*}$ | $\dfrac{T}{T^*}$ | $\dfrac{p}{p^*}$ | $\dfrac{p_0}{p_0^*}$ | $\dfrac{V}{V^*}$ |
|---|---|---|---|---|---|---|---|---|---|---|---|
| 0 | 0 | 0 | 2.4000 | 1.2679 | 0 | 0.40 | .52903 | .61515 | 1.9608 | 1.1566 | .31372 |
| 0.01 | .000480 | .000576 | 2.3997 | 1.2678 | .000240 | .41 | .54651 | .63448 | 1.9428 | 1.1523 | .32658 |
| .02 | .00192 | .00230 | 2.3987 | 1.2675 | .000959 | .42 | .56376 | .65345 | 1.9247 | 1.1480 | .33951 |
| .03 | .00431 | .00516 | 2.3970 | 1.2671 | .00216 | .43 | .58075 | .67205 | 1.9065 | 1.1437 | .35251 |
| .04 | .00765 | .00917 | 2.3946 | 1.2665 | .00383 | .44 | .59748 | .69025 | 1.8882 | 1.1394 | .36556 |
| .05 | .01192 | .01430 | 2.3916 | 1.2657 | .00598 | .45 | .61393 | .70803 | 1.8699 | 1.1351 | .37865 |
| .06 | .01712 | .02053 | 2.3880 | 1.2647 | .00860 | .46 | .63007 | .72538 | 1.8515 | 1.1308 | .39178 |
| .07 | .02322 | .02784 | 2.3837 | 1.2636 | .01168 | .47 | .64589 | .74228 | 1.8331 | 1.1266 | .40493 |
| .08 | .03021 | .03621 | 2.3787 | 1.2623 | .01522 | .48 | .66139 | .75871 | 1.8147 | 1.1224 | .41810 |
| .09 | .03807 | .04562 | 2.3731 | 1.2608 | .01922 | .49 | .67655 | .77466 | 1.7962 | 1.1182 | .43127 |
| .10 | .04678 | .05602 | 2.3669 | 1.2591 | .02367 | .50 | .69136 | .79012 | 1.7778 | 1.1140 | .44445 |
| .11 | .05630 | .06739 | 2.3600 | 1.2573 | .02856 | .51 | .70581 | .80509 | 1.7594 | 1.1099 | .45761 |
| .12 | .06661 | .07970 | 2.3526 | 1.2554 | .03388 | .52 | .71990 | .81955 | 1.7410 | 1.1059 | .47075 |
| .13 | .07768 | .09290 | 2.3445 | 1.2533 | .03962 | .53 | .73361 | .83351 | 1.7226 | 1.1019 | .48387 |
| .14 | .08947 | .10695 | 2.3359 | 1.2510 | .04578 | .54 | .74695 | .84695 | 1.7043 | 1.0979 | .49696 |
| .15 | .10196 | .12181 | 2.3267 | 1.2486 | .05235 | .55 | .75991 | .85987 | 1.6860 | 1.09397 | .51001 |
| .16 | .11511 | .13743 | 2.3170 | 1.2461 | .05931 | .56 | .77248 | .87227 | 1.6678 | 1.09010 | .52302 |
| .17 | .12888 | .15377 | 2.3067 | 1.2434 | .06666 | .57 | .78467 | .88415 | 1.6496 | 1.08630 | .53597 |
| .18 | .14324 | .17078 | 2.2959 | 1.2406 | .07438 | .58 | .79647 | .89552 | 1.6316 | 1.08255 | .54887 |
| .19 | .15814 | .18841 | 2.2845 | 1.2377 | .08247 | .59 | .80789 | .90637 | 1.6136 | 1.07887 | .56170 |
| .20 | .17355 | .20661 | 2.2727 | 1.2346 | .09091 | .60 | .81892 | .91670 | 1.5957 | 1.07525 | .57447 |
| .21 | .18943 | .22533 | 2.2604 | 1.2314 | .09969 | .61 | .82956 | .92653 | 1.5780 | 1.07170 | .58716 |
| .22 | .20574 | .24452 | 2.2477 | 1.2281 | .10879 | .62 | .83982 | .93585 | 1.5603 | 1.06821 | .59978 |
| .23 | .22244 | .26413 | 2.2345 | 1.2248 | .11820 | .63 | .84970 | .94466 | 1.5427 | 1.06480 | .61232 |
| .24 | .23948 | .28411 | 2.2209 | 1.2213 | .12792 | .64 | .85920 | .95298 | 1.5253 | 1.06146 | .62477 |
| .25 | .25684 | .30440 | 2.2069 | 1.2177 | .13793 | .65 | .86833 | .96081 | 1.5080 | 1.05820 | .63713 |
| .26 | .27446 | .32496 | 2.1925 | 1.2140 | .14821 | .66 | .87709 | .96816 | 1.4908 | 1.05502 | .64941 |
| .27 | .29231 | .34573 | 2.1777 | 1.2102 | .15876 | .67 | .88548 | .97503 | 1.4738 | 1.05192 | .66159 |
| .28 | .31035 | .36667 | 2.1626 | 1.2064 | .16955 | .68 | .89350 | .98144 | 1.4569 | 1.04890 | .67367 |
| .29 | .32855 | .38773 | 2.1472 | 1.2025 | .18058 | .69 | .90117 | .98739 | 1.4401 | 1.04596 | .68564 |
| .30 | .34686 | .40887 | 2.1314 | 1.1985 | .19183 | .70 | .90850 | .99289 | 1.4235 | 1.04310 | .69751 |
| .31 | .36525 | .43004 | 2.1154 | 1.1945 | .20329 | .71 | .91548 | .99796 | 1.4070 | 1.04033 | .70927 |
| .32 | .38369 | .45119 | 2.0991 | 1.1904 | .21494 | .72 | .92212 | 1.00260 | 1.3907 | 1.03764 | .72093 |
| .33 | .40214 | .47228 | 2.0825 | 1.1863 | .22678 | .73 | .92843 | 1.00682 | 1.3745 | 1.03504 | .73248 |
| .34 | .42057 | .49327 | 2.0657 | 1.1821 | .23879 | .74 | .93442 | 1.01062 | 1.3585 | 1.03253 | .74392 |
| .35 | .43894 | .51413 | 2.0487 | 1.1779 | .25096 | .75 | .94009 | 1.01403 | 1.3427 | 1.03010 | .75525 |
| .36 | .45723 | .53482 | 2.0314 | 1.1737 | .26327 | .76 | .94546 | 1.01706 | 1.3270 | 1.02776 | .76646 |
| .37 | .47541 | .55530 | 2.0140 | 1.1695 | .27572 | .77 | .95052 | 1.01971 | 1.3115 | 1.02552 | .77755 |
| .38 | .49346 | .57553 | 1.9964 | 1.1652 | .28828 | .78 | .95528 | 1.02198 | 1.2961 | 1.02337 | .78852 |
| .39 | .51134 | .59549 | 1.9787 | 1.1609 | .30095 | .79 | .95975 | 1.02390 | 1.2809 | 1.02131 | .79938 |

See page 113 for definitions of symbols.

# Table 36  Rayleigh Line—One-Dimensional Compressible-Flow Functions for Stagnation-Temperature Change in the Absence of Friction and Area Change

**For a perfect gas with constant specific heat and molecular weight**
**k = 1.4**

| M | $\dfrac{T_0}{T_0^*}$ | $\dfrac{T}{T^*}$ | $\dfrac{p}{p^*}$ | $\dfrac{p_0}{p_0^*}$ | $\dfrac{V}{V^*}$ | M | $\dfrac{T_0}{T_0^*}$ | $\dfrac{T}{T^*}$ | $\dfrac{p}{p^*}$ | $\dfrac{p_0}{p_0^*}$ | $\dfrac{V}{V^*}$ |
|---|---|---|---|---|---|---|---|---|---|---|---|
| 0.80 | .96394 | 1.02548 | 1.2658 | 1.01934 | .81012 | 1.20 | .97872 | .91185 | .79576 | 1.01941 | 1.1459 |
| .81 | .96786 | 1.02672 | 1.2509 | 1.01746 | .82075 | 1.21 | .97685 | .90671 | .78695 | 1.02140 | 1.1522 |
| .82 | .97152 | 1.02763 | 1.2362 | 1.01569 | .83126 | 1.22 | .97492 | .90153 | .77827 | 1.02348 | 1.1584 |
| .83 | .97492 | 1.02823 | 1.2217 | 1.01399 | .84164 | 1.23 | .97294 | .89632 | .76971 | 1.02566 | 1.1645 |
| .84 | .97807 | 1.02853 | 1.2073 | 1.01240 | .85190 | 1.24 | .97092 | .89108 | .76127 | 1.02794 | 1.1705 |
| .85 | .98097 | 1.02854 | 1.1931 | 1.01091 | .86204 | 1.25 | .96886 | .88581 | .75294 | 1.03032 | 1.1764 |
| .86 | .98363 | 1.02826 | 1.1791 | 1.00951 | .87206 | 1.26 | .96675 | .88052 | .74473 | 1.03280 | 1.1823 |
| .87 | .98607 | 1.02771 | 1.1652 | 1.00819 | .88196 | 1.27 | .96461 | .87521 | .73663 | 1.03536 | 1.1881 |
| .88 | .98828 | 1.02690 | 1.1515 | 1.00698 | .89175 | 1.28 | .96243 | .86988 | .72865 | 1.03803 | 1.1938 |
| .89 | .99028 | 1.02583 | 1.1380 | 1.00587 | .90142 | 1.29 | .96022 | .86453 | .72078 | 1.04080 | 1.1994 |
| .90 | .99207 | 1.02451 | 1.1246 | 1.04485 | .91097 | 1.30 | .95798 | .85917 | .71301 | 1.04365 | 1.2050 |
| .91 | .99366 | 1.02297 | 1.1114 | 1.00393 | .92039 | 1.31 | .95571 | .85380 | .70535 | 1.04661 | 1.2105 |
| .92 | .99506 | 1.02120 | 1.09842 | 1.00310 | .92970 | 1.32 | .95341 | .84843 | .69780 | 1.04967 | 1.2159 |
| .93 | .99627 | 1.01921 | 1.08555 | 1.00237 | .93889 | 1.33 | .95108 | .84305 | .69035 | 1.05283 | 1.2212 |
| .94 | .99729 | 1.01702 | 1.07285 | 1.00174 | .94796 | 1.34 | .94873 | .83766 | .68301 | 1.05608 | 1.2264 |
| .95 | .99814 | 1.01463 | 1.06030 | 1.00121 | .95692 | 1.35 | .94636 | .83227 | .67577 | 1.05943 | 1.2316 |
| .96 | .99883 | 1.01205 | 1.04792 | 1.00077 | .96576 | 1.36 | .94397 | .82698 | .66863 | 1.06288 | 1.2367 |
| .97 | .99935 | 1.00929 | 1.03570 | 1.00043 | .97449 | 1.37 | .94157 | .82151 | .66159 | 1.06642 | 1.2417 |
| .98 | .99972 | 1.00636 | 1.02364 | 1.00019 | .98311 | 1.38 | .93915 | .81613 | .65464 | 1.07006 | 1.2467 |
| .99 | .99993 | 1.00326 | 1.01174 | 1.00004 | .99161 | 1.39 | .93671 | .81076 | .64778 | 1.07380 | 1.2516 |
| 1.00 | 1.00000 | 1.00000 | 1.00000 | 1.00000 | 1.00000 | 1.40 | .93425 | .80540 | .64102 | 1.07765 | 1.2564 |
| 1.01 | .99993 | .99659 | .98841 | 1.00004 | 1.00828 | 1.41 | .93178 | .80004 | .63436 | 1.08159 | 1.2612 |
| 1.02 | .99973 | .99304 | .97697 | 1.00019 | 1.01644 | 1.42 | .92931 | .79469 | .62779 | 1.08563 | 1.2659 |
| 1.03 | .99940 | .98936 | .96569 | 1.00043 | 1.02450 | 1.43 | .92683 | .78936 | .62131 | 1.08977 | 1.2705 |
| 1.04 | .99895 | .98553 | .95456 | 1.00077 | 1.03246 | 1.44 | .92434 | .78405 | .61491 | 1.09400 | 1.2751 |
| 1.05 | .99838 | .98161 | .94358 | 1.00121 | 1.04030 | 1.45 | .92184 | .77875 | .60860 | 1.0983 | 1.2796 |
| 1.06 | .99769 | .97755 | .93275 | 1.00175 | 1.04804 | 1.46 | .91933 | .77346 | .60237 | 1.1028 | 1.2840 |
| 1.07 | .99690 | .97339 | .92206 | 1.00238 | 1.05567 | 1.47 | .91682 | .76819 | .59623 | 1.1073 | 1.2884 |
| 1.08 | .99600 | .96913 | .91152 | 1.00311 | 1.06320 | 1.48 | .91431 | .76294 | .59018 | 1.1120 | 1.2927 |
| 1.09 | .99501 | .96477 | .90112 | 1.00394 | 1.07062 | 1.49 | .91179 | .75771 | .58421 | 1.1167 | 1.2970 |
| 1.10 | .99392 | .96031 | .89086 | 1.00486 | 1.07795 | 1.50 | .90928 | .75250 | .57831 | 1.1215 | 1.3012 |
| 1.11 | .99274 | .95577 | .88075 | 1.00588 | 1.08518 | 1.51 | .90676 | .74731 | .57250 | 1.1264 | 1.3054 |
| 1.12 | .99148 | .95115 | .87078 | 1.00699 | 1.09230 | 1.52 | .90424 | .74215 | .56677 | 1.1315 | 1.3095 |
| 1.13 | .99013 | .94646 | .86094 | 1.00820 | 1.09933 | 1.53 | .90172 | .73701 | .56111 | 1.1367 | 1.3135 |
| 1.14 | .98871 | .94169 | .85123 | 1.00951 | 1.10626 | 1.54 | .89920 | .73189 | .55553 | 1.1420 | 1.3175 |
| 1.15 | .98721 | .93685 | .84166 | 1.01092 | 1.1131 | 1.55 | .89669 | .72680 | .55002 | 1.1473 | 1.3214 |
| 1.16 | .98564 | .93195 | .83222 | 1.01243 | 1.1198 | 1.56 | .89418 | .72173 | .54458 | 1.1527 | 1.3253 |
| 1.17 | .98400 | .92700 | .82292 | 1.01403 | 1.1264 | 1.57 | .89167 | .71669 | .53922 | 1.1582 | 1.3291 |
| 1.18 | .98230 | .92200 | .81374 | 1.01572 | 1.1330 | 1.58 | .88917 | .71168 | .53393 | 1.1639 | 1.3329 |
| 1.19 | .98054 | .91695 | .80468 | 1.01752 | 1.1395 | 1.59 | .88668 | .70669 | .52871 | 1.1697 | 1.3366 |

**Table 36  Rayleigh Line—One-Dimensional Compressible-Flow Functions for Stagnation-Temperature Change in the Absence of Friction and Area Change**

**For a perfect gas with constant specific heat and molecular weight**
**k = 1.4**

| M | $\dfrac{T_0}{T_0{}^*}$ | $\dfrac{T}{T^*}$ | $\dfrac{p}{p^*}$ | $\dfrac{p_0}{p_0{}^*}$ | $\dfrac{V}{V^*}$ | M | $\dfrac{T_0}{T_0{}^*}$ | $\dfrac{T}{T^*}$ | $\dfrac{p}{p^*}$ | $\dfrac{p_0}{p_0{}^*}$ | $\dfrac{V}{V^*}$ |
|---|---|---|---|---|---|---|---|---|---|---|---|
| 1.60 | .88419 | .70173 | .52356 | 1.1756 | 1.3403 | 2.00 | .79339 | .52893 | .36364 | 1.5031 | 1.4545 |
| 1.61 | .88170 | .69680 | .51848 | 1.1816 | 1.3439 | 2.01 | .79139 | .52526 | .36057 | 1.5138 | 1.4567 |
| 1.62 | .87922 | .69190 | .51346 | 1.1877 | 1.3475 | 2.02 | .78941 | .52161 | .35754 | 1.5246 | 1.4589 |
| 1.63 | .87675 | .68703 | .50851 | 1.1939 | 1.3511 | 2.03 | .78744 | .51800 | .35454 | 1.5356 | 1.4610 |
| 1.64 | .87429 | .68219 | .50363 | 1.2002 | 1.3546 | 2.04 | .78549 | .51442 | .35158 | 1.5467 | 1.4631 |
| 1.65 | .87184 | .67738 | .49881 | 1.2066 | 1.3580 | 2.05 | .78355 | .51087 | .34866 | 1.5579 | 1.4652 |
| 1.66 | .86940 | .67259 | .49405 | 1.2131 | 1.3614 | 2.06 | .78162 | .50735 | .34577 | 1.5693 | 1.4673 |
| 1.67 | .86696 | .66784 | .48935 | 1.2197 | 1.3648 | 2.07 | .77971 | .50386 | .34291 | 1.5808 | 1.4694 |
| 1.68 | .86453 | .66312 | .48471 | 1.2264 | 1.3681 | 2.08 | .77781 | .50040 | .34009 | 1.5924 | 1.4714 |
| 1.69 | .86211 | .65843 | .48014 | 1.2332 | 1.3713 | 2.09 | .77593 | .49697 | .33730 | 1.6042 | 1.4734 |
| 1.70 | .85970 | .65377 | .47563 | 1.2402 | 1.3745 | 2.10 | .77406 | .49356 | .33454 | 1.6161 | 1.4753 |
| 1.71 | .85731 | .64914 | .47117 | 1.2473 | 1.3777 | 2.11 | .77221 | .49018 | .33181 | 1.6282 | 1.4773 |
| 1.72 | .85493 | .64455 | .46677 | 1.2545 | 1.3809 | 2.12 | .77037 | .48683 | .32912 | 1.6404 | 1.4792 |
| 1.73 | .85256 | .63999 | .46242 | 1.2618 | 1.3840 | 2.13 | .76854 | .48351 | .32646 | 1.6528 | 1.4811 |
| 1.74 | .85020 | .63546 | .45813 | 1.2692 | 1.3871 | 2.14 | .76673 | .48022 | .32383 | 1.6653 | 1.4830 |
| 1.75 | .84785 | .63096 | .45390 | 1.2767 | 1.3901 | 2.15 | .76493 | .47696 | .32122 | 1.6780 | 1.4849 |
| 1.76 | .84551 | .62649 | .44972 | 1.2843 | 1.3931 | 2.16 | .76314 | .47373 | .31864 | 1.6908 | 1.4867 |
| 1.77 | .84318 | .62205 | .44559 | 1.2920 | 1.3960 | 2.17 | .76137 | .47052 | .31610 | 1.7037 | 1.4885 |
| 1.78 | .84087 | .61765 | .44152 | 1.2998 | 1.3989 | 2.18 | .75961 | .46734 | .31359 | 1.7168 | 1.4903 |
| 1.79 | .83857 | .61328 | .43750 | 1.3078 | 1.4018 | 2.19 | .75787 | .46419 | .31110 | 1.7300 | 1.4921 |
| 1.80 | .83628 | .60894 | .43353 | 1.3159 | 1.4046 | 2.20 | .75614 | .46106 | .30864 | 1.7434 | 1.4939 |
| 1.81 | .83400 | .60463 | .42960 | 1.3241 | 1.4074 | 2.21 | .75442 | .45796 | .30621 | 1.7570 | 1.4956 |
| 1.82 | .83174 | .60036 | .42573 | 1.3324 | 1.4102 | 2.22 | .75271 | .45489 | .30381 | 1.7707 | 1.4973 |
| 1.83 | .82949 | .59612 | .42191 | 1.3408 | 1.4129 | 2.23 | .75102 | .45184 | .30143 | 1.7846 | 1.4990 |
| 1.84 | .82726 | .59191 | .41813 | 1.3494 | 1.4156 | 2.24 | .74934 | .44882 | .29908 | 1.7986 | 1.5007 |
| 1.85 | .82504 | .58773 | .41440 | 1.3581 | 1.4183 | 2.25 | .74767 | .44582 | .29675 | 1.8128 | 1.5024 |
| 1.86 | .82283 | .58359 | .41072 | 1.3669 | 1.4209 | 2.26 | .74602 | .44285 | .29445 | 1.8271 | 1.5040 |
| 1.87 | .82064 | .57948 | .40708 | 1.3758 | 1.4235 | 2.27 | .74438 | .43990 | .29218 | 1.8416 | 1.5056 |
| 1.88 | .81846 | .57540 | .40349 | 1.3848 | 1.4261 | 2.28 | .74275 | .43698 | .28993 | 1.8562 | 1.5072 |
| 1.89 | .81629 | .57135 | .39994 | 1.3940 | 1.4286 | 2.29 | .74114 | .43409 | .28771 | 1.8710 | 1.5088 |
| 1.90 | .81414 | .56734 | .39643 | 1.4033 | 1.4311 | 2.30 | .73954 | .43122 | .28551 | 1.8860 | 1.5104 |
| 1.91 | .81200 | .56336 | .39297 | 1.4127 | 1.4336 | 2.31 | .73795 | .42837 | .28333 | 1.9012 | 1.5119 |
| 1.92 | .80987 | .55941 | .38955 | 1.4222 | 1.4360 | 2.32 | .73638 | .42555 | .28118 | 1.9165 | 1.5134 |
| 1.93 | .80776 | .55549 | .38617 | 1.4319 | 1.4384 | 2.33 | .73482 | .42276 | .27905 | 1.9320 | 1.5150 |
| 1.94 | .80567 | .55160 | .38283 | 1.4417 | 1.4408 | 2.34 | .73327 | .41999 | .27695 | 1.9476 | 1.5165 |
| 1.95 | .80359 | .54774 | .37954 | 1.4516 | 1.4432 | 2.35 | .73173 | .41724 | .27487 | 1.9634 | 1.5180 |
| 1.96 | .80152 | .54391 | .37628 | 1.4616 | 1.4455 | 2.36 | .73020 | .41451 | .27281 | 1.9794 | 1.5195 |
| 1.97 | .79946 | .54012 | .37306 | 1.4718 | 1.4478 | 2.37 | .72868 | .41181 | .27077 | 1.9955 | 1.5209 |
| 1.98 | .79742 | .53636 | .36988 | 1.4821 | 1.4501 | 2.38 | .72718 | .40913 | .26875 | 2.0118 | 1.5223 |
| 1.99 | .79540 | .53263 | .36674 | 1.4925 | 1.4523 | 2.39 | .72569 | .40647 | .26675 | 2.0283 | 1.5237 |

## Table 36 Rayleigh Line—One-Dimensional Compressible-Flow Functions for Stagnation-Temperature Change in the Absence of Friction and Area Change

For a perfect gas with constant specific heat and molecular weight
k = 1.4

| M | $\frac{T_0}{T_0*}$ | $\frac{T}{T*}$ | $\frac{p}{p*}$ | $\frac{p_0}{p_0*}$ | $\frac{V}{V*}$ | M | $\frac{T_0}{T_0*}$ | $\frac{T}{T*}$ | $\frac{p}{p*}$ | $\frac{p_0}{p_0*}$ | $\frac{V}{V*}$ |
|---|---|---|---|---|---|---|---|---|---|---|---|
| 2.40 | .72421 | .40383 | .26478 | 2.0450 | 1.5252 | 2.80 | .67380 | .31486 | .20040 | 2.8731 | 1.5711 |
| 2.41 | .72274 | .40122 | .26283 | 2.0619 | 1.5266 | 2.81 | .67273 | .31299 | .19909 | 2.8982 | 1.5721 |
| 2.42 | .72129 | .39863 | .26090 | 2.0789 | 1.5279 | 2.82 | .67167 | .31114 | .19780 | 2.9236 | 1.5730 |
| 2.43 | .71985 | .39606 | .25899 | 2.0961 | 1.5293 | 2.83 | .67062 | .30931 | .19652 | 2.9493 | 1.5739 |
| 2.44 | .71842 | .39352 | .25710 | 2.1135 | 1.5306 | 2.84 | .66958 | .30749 | .19525 | 2.9752 | 1.5748 |
| 2.45 | .71700 | .39100 | .25523 | 2.1311 | 1.5320 | 2.85 | .66855 | .30568 | .19399 | 3.0013 | 1.5757 |
| 2.46 | .71559 | .38850 | .25337 | 2.1489 | 1.5333 | 2.86 | .66752 | .30389 | .19274 | 3.0277 | 1.5766 |
| 2.47 | .71419 | .38602 | .25153 | 2.1669 | 1.5346 | 2.87 | .66650 | .30211 | .19151 | 3.0544 | 1.5775 |
| 2.48 | .71280 | .38356 | .24972 | 2.1850 | 1.5359 | 2.88 | .66549 | .30035 | .19029 | 3.0813 | 1.5784 |
| 2.49 | .71142 | .38112 | .24793 | 2.2033 | 1.5372 | 2.89 | .66449 | .29860 | .18908 | 3.1084 | 1.5792 |
| 2.50 | .71005 | .37870 | .24616 | 2.2218 | 1.5385 | 2.90 | .66350 | .29687 | .18788 | 3.1358 | 1.5801 |
| 2.51 | .70870 | .37630 | .24440 | 2.2405 | 1.5398 | 2.91 | .66252 | .29515 | .18669 | 3.1635 | 1.5809 |
| 2.52 | .70736 | .37392 | .24266 | 2.2594 | 1.5410 | 2.92 | .66154 | .29344 | .18551 | 3.1914 | 1.5818 |
| 2.53 | .70603 | .37157 | .24094 | 2.2785 | 1.5422 | 2.93 | .66057 | .29175 | .18435 | 3.2196 | 1.5826 |
| 2.54 | .70471 | .36923 | .23923 | 2.2978 | 1.5434 | 2.94 | .65961 | .29007 | .18320 | 3.2481 | 1.5834 |
| 2.55 | .70340 | .36691 | .23754 | 2.3173 | 1.5446 | 2.95 | .65865 | .28841 | .18205 | 3.2768 | 1.5843 |
| 2.56 | .70210 | .36461 | .23587 | 2.3370 | 1.5458 | 2.96 | .65770 | .28676 | .18091 | 3.3058 | 1.5851 |
| 2.57 | .70081 | .36233 | .23422 | 2.3569 | 1.5470 | 2.97 | .65676 | .28512 | .17978 | 3.3351 | 1.5859 |
| 2.58 | .69953 | .36007 | .23258 | 2.3770 | 1.5482 | 2.98 | .65583 | .28349 | .17867 | 3.3646 | 1.5867 |
| 2.59 | .69825 | .35783 | .23096 | 2.3972 | 1.5494 | 2.99 | .65490 | .28188 | .17757 | 3.3944 | 1.5875 |
| 2.60 | .69699 | .35561 | .22936 | 2.4177 | 1.5505 | 3.00 | .65398 | .28028 | .17647 | 3.4244 | 1.5882 |
| 2.61 | .69574 | .35341 | .22777 | 2.4384 | 1.5516 | 3.50 | .61580 | .21419 | .13223 | 5.3280 | 1.6198 |
| 2.62 | .69450 | .35123 | .22620 | 2.4593 | 1.5527 | 4.00 | .58909 | .16831 | .10256 | 8.2268 | 1.6410 |
| 2.63 | .69327 | .34906 | .22464 | 2.4804 | 1.5538 | 4.50 | .56983 | .13540 | .08177 | 12.502 | 1.6559 |
| 2.64 | .69205 | .34691 | .22310 | 2.5017 | 1.5549 | 5.00 | .55555 | .11111 | .06667 | 18.634 | 1.6667 |
| 2.65 | .69084 | .34478 | .22158 | 2.5233 | 1.5560 | 6.00 | .53633 | .07849 | .04669 | 38.946 | 1.6809 |
| 2.66 | .68964 | .34267 | .22007 | 2.5451 | 1.5571 | 7.00 | .52437 | .05826 | .03448 | 75.414 | 1.6897 |
| 2.67 | .68845 | .34057 | .21857 | 2.5671 | 1.5582 | 8.00 | .51646 | .04491 | .02649 | 136.62 | 1.6954 |
| 2.68 | .68727 | .33849 | .21709 | 2.5892 | 1.5593 | 9.00 | .51098 | .03565 | .02098 | 233.88 | 1.6993 |
| 2.69 | .68610 | .33643 | .21562 | 2.6116 | 1.5603 | 10.00 | .50702 | .02897 | .01702 | 381.61 | 1.7021 |
| 2.70 | .68494 | .33439 | .21417 | 2.6342 | 1.5613 | ∞ | .48980 | 0 | 0 | ∞ | 1.7143 |
| 2.71 | .68378 | .33236 | .21273 | 2.6571 | 1.5623 | | | | | | |
| 2.72 | .68263 | .33035 | .21131 | 2.6802 | 1.5633 | | | | | | |
| 2.73 | .68150 | .32836 | .20990 | 2.7035 | 1.5644 | | | | | | |
| 2.74 | .68038 | .32638 | .20850 | 2.7270 | 1.5654 | | | | | | |
| 2.75 | .67926 | .32442 | .20712 | 2.7508 | 1.5663 | | | | | | |
| 2.76 | .67815 | .32248 | .20575 | 2.7748 | 1.5673 | | | | | | |
| 2.77 | .67704 | .32055 | .20439 | 2.7990 | 1.5683 | | | | | | |
| 2.78 | .67595 | .31864 | .20305 | 2.8235 | 1.5692 | | | | | | |
| 2.79 | .67487 | .31674 | .20172 | 2.8482 | 1.5702 | | | | | | |

# Table 37  Rayleigh Line—One-Dimensional Compressible-Flow Functions for Stagnation-Temperature Change in the Absence of Friction and Area Change

**For a perfect gas with constant specific heat and molecular weight**
**k = 1.0**

| M | $\dfrac{T_0}{T_0^*}=\dfrac{T}{T^*}$ | $\dfrac{p}{p^*}$ | $\dfrac{p_0}{p_0^*}$ | $\dfrac{V}{V^*}$ | M | $\dfrac{T_0}{T_0^*}=\dfrac{T}{T^*}$ | $\dfrac{p}{p^*}$ | $\dfrac{p_0}{p_0^*}$ | $\dfrac{V}{V^*}$ |
|---|---|---|---|---|---|---|---|---|---|
| 0 | 0 | 2.000 | 1.213 | 0 | 1.75 | .7422 | .4923 | 1.381 | 1.508 |
| 0.05 | .00995 | 1.995 | 1.212 | .00499 | 1.80 | .7209 | .4717 | 1.446 | 1.528 |
| .10 | .03921 | 1.980 | 1.207 | .01980 | 1.85 | .7000 | .4522 | 1.519 | 1.547 |
| .15 | .08608 | 1.956 | 1.200 | .04401 | 1.90 | .6795 | .4338 | 1.600 | 1.566 |
| .20 | .14793 | 1.923 | 1.190 | .07692 | 1.95 | .6595 | .4164 | 1.691 | 1.584 |
| .25 | .2215 | 1.882 | 1.178 | .1176 | 2.00 | .6400 | .4000 | 1.793 | 1.601 |
| .30 | .3030 | 1.835 | 1.164 | .1651 | 2.05 | .6211 | .3844 | 1.907 | 1.616 |
| .35 | .3889 | 1.782 | 1.149 | .2183 | 2.10 | .6027 | .3697 | 2.034 | 1.630 |
| .40 | .4756 | 1.724 | 1.133 | .2758 | 2.15 | .5849 | .3557 | 2.176 | 1.644 |
| .45 | .5602 | 1.663 | 1.116 | .3368 | 2.20 | .5677 | .3425 | 2.336 | 1.657 |
| .50 | .6400 | 1.600 | 1.0997 | .4000 | 2.25 | .5510 | .3299 | 2.515 | 1.670 |
| .55 | .7132 | 1.536 | 1.0834 | .4645 | 2.30 | .5348 | .3179 | 2.716 | 1.682 |
| .60 | .7785 | 1.471 | 1.0679 | .5294 | 2.35 | .5192 | .3066 | 2.942 | 1.693 |
| .65 | .8352 | 1.406 | 1.0534 | .5940 | 2.40 | .5042 | .2959 | 3.197 | 1.704 |
| .70 | .8828 | 1.342 | 1.0402 | .6577 | 2.45 | .4897 | .2857 | 3.484 | 1.714 |
| .75 | .9216 | 1.280 | 1.0285 | .7200 | 2.50 | .4757 | .2759 | 3.808 | 1.724 |
| .80 | .9518 | 1.220 | 1.0186 | .7805 | 2.55 | .4621 | .2666 | 4.175 | 1.733 |
| .85 | .9740 | 1.161 | 1.0107 | .8389 | 2.60 | .4490 | .2577 | 4.591 | 1.742 |
| .90 | .9890 | 1.105 | 1.0048 | .8950 | 2.65 | .4364 | .2493 | 5.064 | 1.751 |
| .95 | .9974 | 1.0512 | 1.0012 | .9488 | 2.70 | .4243 | .2413 | 5.602 | 1.759 |
| 1.00 | 1.0000 | 1.0000 | 1.0000 | 1.0000 | 2.75 | .4126 | .2336 | 6.215 | 1.766 |
| 1.05 | .9976 | .9512 | 1.0013 | 1.0488 | 2.80 | .4013 | .2262 | 6.916 | 1.774 |
| 1.10 | .9910 | .9049 | 1.0052 | 1.0951 | 2.85 | .3904 | .2192 | 7.719 | 1.781 |
| 1.15 | .9807 | .8611 | 1.0118 | 1.1389 | 2.90 | .3799 | .2125 | 8.640 | 1.787 |
| 1.20 | .9675 | .8197 | 1.0214 | 1.1802 | 2.95 | .3698 | .2061 | 9.699 | 1.794 |
| 1.25 | .9518 | .7805 | 1.0340 | 1.220 | 3.00 | .3600 | .2000 | 10.92 | 1.800 |
| 1.30 | .9342 | .7435 | 1.0498 | 1.257 | 3.50 | .2791 | .1509 | 41.85 | 1.849 |
| 1.35 | .9151 | .7086 | 1.0690 | 1.291 | 4.00 | .2215 | .1176 | 212.71 | 1.882 |
| 1.40 | .8948 | .6757 | 1.0919 | 1.324 | 4.50 | .1794 | .09412 | 1425 | 1.906 |
| 1.45 | .8737 | .6447 | 1.1187 | 1.355 | 5.00 | .1479 | .07692 | 12519 | 1.923 |
| 1.50 | .8521 | .6154 | 1.150 | 1.384 | 6.00 | .10519 | .05405 | $215(10)^4$ | 1.946 |
| 1.55 | .8301 | .5878 | 1.186 | 1.412 | 7.00 | .07840 | .04000 | $106(10)^7$ | 1.960 |
| 1.60 | .8080 | .5618 | 1.226 | 1.438 | 8.00 | .06059 | .03077 | $147(10)^{10}$ | 1.969 |
| 1.65 | .7859 | .5373 | 1.271 | 1.463 | 9.00 | .04818 | .02439 | $574(10)^{13}$ | 1.976 |
| 1.70 | .7639 | .5141 | 1.323 | 1.486 | 10.00 | .03921 | .01980 | $623(10)^{17}$ | 1.980 |
|  |  |  |  |  | $\infty$ | 0 | 0 | $\infty$ | 2.000 |

# Table 38  Rayleigh Line—One-Dimensional Compressible-Flow Functions for Stagnation-Temperature Change in the Absence of Friction and Area Change

For a perfect gas with constant specific heat and molecular weight
k = 1.1

| M | $\dfrac{T_0}{T_0{}^*}$ | $\dfrac{T}{T^*}$ | $\dfrac{p}{p^*}$ | $\dfrac{p_0}{p_0{}^*}$ | $\dfrac{V}{V^*}$ | M | $\dfrac{T_0}{T_0{}^*}$ | $\dfrac{T}{T^*}$ | $\dfrac{p}{p^*}$ | $\dfrac{p_0}{p_0{}^*}$ | $\dfrac{V}{V^*}$ |
|---|---|---|---|---|---|---|---|---|---|---|---|
| 0 | 0 | 0 | 2.100 | 1.228 | 0 | 1.75 | .7771 | .7076 | .4807 | 1.347 | 1.472 |
| 0.05 | .01044 | .01097 | 2.094 | 1.226 | .00524 | 1.80 | .7591 | .6859 | .4601 | 1.403 | 1.491 |
| .10 | .04111 | .04315 | 2.077 | 1.221 | .02077 | 1.85 | .7415 | .6648 | .4407 | 1.465 | 1.508 |
| .15 | .09009 | .09449 | 2.049 | 1.213 | .04611 | 1.90 | .7243 | .6443 | .4224 | 1.532 | 1.525 |
| .20 | .15444 | .16184 | 2.011 | 1.203 | .08046 | 1.95 | .7076 | .6243 | .4052 | 1.607 | 1.541 |
| .25 | .2305 | .2413 | 1.965 | 1.190 | .1228 | 2.00 | .6914 | .6049 | .3889 | 1.689 | 1.556 |
| .30 | .3144 | .3286 | 1.911 | 1.174 | .1720 | 2.05 | .6756 | .5862 | .3735 | 1.780 | 1.570 |
| .35 | .4020 | .4195 | 1.851 | 1.157 | .2267 | 2.10 | .6603 | .5681 | .3589 | 1.879 | 1.583 |
| .40 | .4898 | .5102 | 1.786 | 1.140 | .2857 | 2.15 | .6456 | .5506 | .3451 | 1.987 | 1.595 |
| .45 | .5746 | .5973 | 1.717 | 1.122 | .3478 | 2.20 | .6313 | .5337 | .3321 | 2.106 | 1.607 |
| .50 | .6540 | .6782 | 1.647 | 1.1040 | .4118 | 2.25 | .6175 | .5174 | .3197 | 2.237 | 1.618 |
| .55 | .7261 | .7510 | 1.576 | 1.0867 | .4766 | 2.30 | .6042 | .5017 | .3079 | 2.380 | 1.629 |
| .60 | .7898 | .8147 | 1.504 | 1.0702 | .5416 | 2.35 | .5914 | .4866 | .2968 | 2.537 | 1.639 |
| .65 | .8446 | .8684 | 1.434 | 1.0550 | .6057 | 2.40 | .5790 | .4720 | .2863 | 2.709 | 1.649 |
| .70 | .8902 | .9123 | 1.365 | 1.0412 | .6686 | 2.45 | .5671 | .4580 | .2763 | 2.897 | 1.658 |
| .75 | .9270 | .9467 | 1.297 | 1.0291 | .7297 | 2.50 | .5556 | .4444 | .2667 | 3.104 | 1.667 |
| .80 | .9554 | .9720 | 1.232 | 1.0189 | .7887 | 2.55 | .5445 | .4314 | .2576 | 3.332 | 1.675 |
| .85 | .9761 | .9892 | 1.170 | 1.0109 | .8453 | 2.60 | .5338 | .4189 | .2489 | 3.581 | 1.683 |
| .90 | .9899 | .9989 | 1.111 | 1.0050 | .8995 | 2.65 | .5235 | .4068 | .2406 | 3.855 | 1.690 |
| .95 | .9976 | 1.0023 | 1.0538 | 1.0013 | .9511 | 2.70 | .5136 | .3952 | .2328 | 4.156 | 1.697 |
| 1.00 | 1.0000 | 1.0000 | 1.0000 | 1.0000 | 1.0000 | 2.75 | .5041 | .3840 | .2253 | 4.487 | 1.704 |
| 1.05 | .9979 | .9930 | .9490 | 1.0013 | 1.0463 | 2.80 | .4949 | .3733 | .2182 | 4.851 | 1.711 |
| 1.10 | .9919 | .9821 | .9009 | 1.0051 | 1.0901 | 2.85 | .4860 | .3629 | .2114 | 5.251 | 1.717 |
| 1.15 | .9827 | .9679 | .8555 | 1.0116 | 1.1314 | 2.90 | .4775 | .3529 | .2049 | 5.692 | 1.723 |
| 1.20 | .9710 | .9511 | .8127 | 1.0209 | 1.1703 | 2.95 | .4693 | .3433 | .1986 | 6.176 | 1.729 |
| 1.25 | .9572 | .9322 | .7724 | 1.0331 | 1.207 | 3.00 | .4613 | .3341 | .1927 | 6.710 | 1.734 |
| 1.30 | .9418 | .9118 | .7345 | 1.0483 | 1.241 | 3.50 | .3960 | .2578 | .1451 | 16.26 | 1.777 |
| 1.35 | .9251 | .8902 | .6989 | 1.0665 | 1.273 | 4.00 | .3496 | .2040 | .1129 | 42.42 | 1.806 |
| 1.40 | .9074 | .8678 | .6654 | 1.0880 | 1.304 | 4.50 | .3160 | .1648 | .0902 | 115.70 | 1.827 |
| 1.45 | .8892 | .8449 | .6339 | 1.1130 | 1.333 | 5.00 | .2909 | .1357 | .0737 | 322.33 | 1.842 |
| 1.50 | .8706 | .8217 | .6043 | 1.141 | 1.360 | 6.00 | .2568 | .09631 | .05172 | 2508 | 1.862 |
| 1.55 | .8518 | .7984 | .5765 | 1.173 | 1.385 | 7.00 | .2356 | .07169 | .03825 | 18430 | 1.874 |
| 1.60 | .8329 | .7753 | .5503 | 1.210 | 1.409 | 8.00 | .2215 | .05536 | .02941 | $123(10)^3$ | 1.882 |
| 1.65 | .8141 | .7524 | .5257 | 1.251 | 1.431 | 9.00 | .2116 | .04400 | .02331 | $743(10)^3$ | 1.888 |
| 1.70 | .7955 | .7298 | .5025 | 1.297 | 1.452 | 10.00 | .2045 | .03579 | .01892 | $401(10)^4$ | 1.892 |
| | | | | | | $\infty$ | .1736 | 0 | 0 | $\infty$ | 1.909 |

# Table 39 Rayleigh Line—One-Dimensional Compressible-Flow Functions for Stagnation-Temperature Change in the Absence of Friction and Area Change

**For a perfect gas with constant specific heat and molecular weight**
**k = 1.2**

| M | $\dfrac{T_0}{T_0{}^*}$ | $\dfrac{T}{T^*}$ | $\dfrac{p}{p^*}$ | $\dfrac{p_0}{p_0{}^*}$ | $\dfrac{V}{V^*}$ | M | $\dfrac{T_0}{T_0{}^*}$ | $\dfrac{T}{T^*}$ | $\dfrac{p}{p^*}$ | $\dfrac{p_0}{p_0{}^*}$ | $\dfrac{V}{V^*}$ |
|---|---|---|---|---|---|---|---|---|---|---|---|
| 0 | 0 | 0 | 2.200 | 1.242 | 0 | 1.75 | .8054 | .6782 | .4706 | 1.320 | 1.441 |
| 0.05 | .01094 | .01203 | 2.193 | 1.239 | .00548 | 1.80 | .7900 | .6563 | .4501 | 1.369 | 1.458 |
| .10 | .04301 | .04726 | 2.173 | 1.234 | .02174 | 1.85 | .7750 | .6351 | .4308 | 1.422 | 1.474 |
| .15 | .09408 | .10325 | 2.141 | 1.226 | .04820 | 1.90 | .7604 | .6146 | .4126 | 1.480 | 1.490 |
| .20 | .16089 | .17627 | 2.099 | 1.214 | .08397 | 1.95 | .7462 | .5947 | .3955 | 1.543 | 1.504 |
| .25 | .2395 | .2618 | 2.047 | 1.199 | .1279 | 2.00 | .7325 | .5755 | .3793 | 1.612 | 1.517 |
| .30 | .3255 | .3548 | 1.986 | 1.183 | .1787 | 2.05 | .7192 | .5570 | .3641 | 1.687 | 1.530 |
| .35 | .4147 | .4507 | 1.918 | 1.165 | .2350 | 2.10 | .7063 | .5391 | .3497 | 1.767 | 1.542 |
| .40 | .5034 | .5450 | 1.846 | 1.146 | .2953 | 2.15 | .6939 | .5219 | .3360 | 1.854 | 1.553 |
| .45 | .5884 | .6343 | 1.770 | 1.127 | .3584 | 2.20 | .6819 | .5054 | .3231 | 1.948 | 1.564 |
| .50 | .6672 | .7160 | 1.692 | 1.1078 | .4231 | 2.25 | .6703 | .4895 | .3109 | 2.050 | 1.574 |
| .55 | .7381 | .7881 | 1.614 | 1.0895 | .4884 | 2.30 | .6591 | .4742 | .2994 | 2.159 | 1.584 |
| .60 | .8003 | .8497 | 1.536 | 1.0722 | .5531 | 2.35 | .6484 | .4595 | .2884 | 2.277 | 1.593 |
| .65 | .8531 | .9004 | 1.460 | 1.0563 | .6168 | 2.40 | .6381 | .4453 | .2780 | 2.405 | 1.602 |
| .70 | .8969 | .9405 | 1.385 | 1.0420 | .6788 | 2.45 | .6281 | .4317 | .2682 | 2.542 | 1.610 |
| .75 | .9318 | .9704 | 1.313 | 1.0296 | .7388 | 2.50 | .6185 | .4187 | .2588 | 2.690 | 1.618 |
| .80 | .9585 | .9910 | 1.244 | 1.0191 | .7964 | 2.55 | .6093 | .4062 | .2499 | 2.849 | 1.625 |
| .85 | .9779 | 1.0032 | 1.178 | 1.0109 | .8514 | 2.60 | .6004 | .3941 | .2414 | 3.021 | 1.632 |
| .90 | .9907 | 1.0081 | 1.115 | 1.0049 | .9037 | 2.65 | .5918 | .3825 | .2334 | 3.205 | 1.639 |
| .95 | .9978 | 1.0067 | 1.0562 | 1.0012 | .9532 | 2.70 | .5836 | .3713 | .2257 | 3.403 | 1.645 |
| 1.00 | 1.0000 | 1.0000 | 1.0000 | 1.0000 | 1.0000 | 2.75 | .5757 | .3606 | .2184 | 3.617 | 1.651 |
| 1.05 | .9981 | .9888 | .9471 | 1.0013 | 1.0441 | 2.80 | .5681 | .3503 | .2114 | 3.847 | 1.657 |
| 1.10 | .9927 | .9741 | .8972 | 1.0050 | 1.0856 | 2.85 | .5608 | .3404 | .2047 | 4.094 | 1.663 |
| 1.15 | .9845 | .9564 | .8504 | 1.0114 | 1.1247 | 2.90 | .5537 | .3309 | .1983 | 4.359 | 1.668 |
| 1.20 | .9740 | .9365 | .8065 | 1.0204 | 1.1613 | 2.95 | .5469 | .3217 | .1923 | 4.644 | 1.673 |
| 1.25 | .9617 | .9149 | .7653 | 1.0322 | 1.196 | 3.00 | .5404 | .3128 | .1864 | 4.951 | 1.678 |
| 1.30 | .9481 | .8921 | .7266 | 1.0467 | 1.228 | 3.50 | .4865 | .2405 | .1401 | 9.597 | 1.717 |
| 1.35 | .9334 | .8685 | .6903 | 1.0640 | 1.258 | 4.00 | .4486 | .1898 | .1089 | 18.99 | 1.743 |
| 1.40 | .9180 | .8443 | .6563 | 1.0843 | 1.286 | 4.50 | .4211 | .1531 | .08696 | 37.61 | 1.761 |
| 1.45 | .9021 | .8199 | .6245 | 1.1077 | 1.313 | 5.00 | .4006 | .1259 | .07097 | 73.64 | 1.774 |
| 1.50 | .8859 | .7955 | .5946 | 1.134 | 1.338 | 6.00 | .3730 | .08919 | .04977 | 266.2 | 1.792 |
| 1.55 | .8695 | .7712 | .5666 | 1.164 | 1.361 | 7.00 | .3557 | .06632 | .03679 | 875.9 | 1.803 |
| 1.60 | .8532 | .7473 | .5403 | 1.197 | 1.383 | 8.00 | .3443 | .05118 | .02828 | 2621 | 1.810 |
| 1.65 | .8370 | .7237 | .5156 | 1.234 | 1.404 | 9.00 | .3363 | .04065 | .02240 | 7181 | 1.815 |
| 1.70 | .8211 | .7007 | .4924 | 1.275 | 1.423 | 10.00 | .3306 | .03306 | .01818 | 18182 | 1.818 |
|  |  |  |  |  |  | $\infty$ | .3056 | 0 | 0 | $\infty$ | 1.833 |

(138)

# Table 40 Rayleigh Line—One-Dimensional Compressible-Flow Functions for Stagnation-Temperature Change in the Absence of Friction and Area Change

**For a perfect gas with constant specific heat and molecular weight**
**k = 1.3**

| M | $\dfrac{T_0}{T_0{}^*}$ | $\dfrac{T}{T^*}$ | $\dfrac{p}{p^*}$ | $\dfrac{p_0}{p_0{}^*}$ | $\dfrac{V}{V^*}$ | M | $\dfrac{T_0}{T_0{}^*}$ | $\dfrac{T}{T^*}$ | $\dfrac{p}{p^*}$ | $\dfrac{p_0}{p_0{}^*}$ | $\dfrac{V}{V^*}$ |
|---|---|---|---|---|---|---|---|---|---|---|---|
| 0 | 0 | 0 | 2.300 | 1.255 | 0 | 1.75 | .8285 | .6529 | .4617 | 1.296 | 1.414 |
| 0.05 | .01143 | .01314 | 2.293 | 1.253 | .00573 | 1.80 | .8153 | .6309 | .4413 | 1.340 | 1.430 |
| .10 | .04489 | .05155 | 2.270 | 1.247 | .02270 | 1.85 | .8024 | .6097 | .4221 | 1.387 | 1.445 |
| .15 | .09803 | .11236 | 2.234 | 1.237 | .05028 | 1.90 | .7898 | .5892 | .4040 | 1.438 | 1.459 |
| .20 | .16726 | .19120 | 2.186 | 1.224 | .08745 | 1.95 | .7776 | .5695 | .3870 | 1.493 | 1.472 |
| .25 | .2482 | .2828 | 2.127 | 1.209 | .1329 | 2.00 | .7659 | .5505 | .3710 | 1.552 | 1.484 |
| .30 | .3363 | .3816 | 2.059 | 1.191 | .1853 | 2.05 | .7545 | .5322 | .3559 | 1.615 | 1.495 |
| .35 | .4270 | .4822 | 1.984 | 1.172 | .2430 | 2.10 | .7435 | .5146 | .3416 | 1.683 | 1.506 |
| .40 | .5165 | .5800 | 1.904 | 1.152 | .3046 | 2.15 | .7329 | .4977 | .3281 | 1.755 | 1.517 |
| .45 | .6015 | .6713 | 1.821 | 1.131 | .3687 | 2.20 | .7227 | .4815 | .3154 | 1.832 | 1.527 |
| .50 | .6796 | .7533 | 1.736 | 1.1112 | .4340 | 2.25 | .7129 | .4659 | .3034 | 1.915 | 1.536 |
| .55 | .7494 | .8244 | 1.651 | 1.0919 | .4994 | 2.30 | .7034 | .4510 | .2920 | 2.003 | 1.545 |
| .60 | .8099 | .8837 | 1.567 | 1.0739 | .5640 | 2.35 | .6943 | .4367 | .2812 | 2.097 | 1.553 |
| .65 | .8611 | .9312 | 1.485 | 1.0574 | .6272 | 2.40 | .6855 | .4229 | .2710 | 2.197 | 1.561 |
| .70 | .9029 | .9673 | 1.405 | 1.0426 | .6885 | 2.45 | .6771 | .4097 | .2613 | 2.303 | 1.568 |
| .75 | .9361 | .9928 | 1.328 | 1.0299 | .7473 | 2.50 | .6690 | .3971 | .2521 | 2.416 | 1.575 |
| .80 | .9614 | 1.0088 | 1.255 | 1.0193 | .8035 | 2.55 | .6612 | .3850 | .2433 | 2.536 | 1.582 |
| .85 | .9795 | 1.0163 | 1.186 | 1.0109 | .8569 | 2.60 | .6537 | .3733 | .2350 | 2.664 | 1.588 |
| .90 | .9914 | 1.0166 | 1.120 | 1.0049 | .9075 | 2.65 | .6465 | .3621 | .2271 | 2.800 | 1.594 |
| .95 | .9980 | 1.0108 | 1.0583 | 1.0012 | .9552 | 2.70 | .6396 | .3513 | .2195 | 2.944 | 1.600 |
| 1.00 | 1.0000 | 1.0000 | 1.0000 | 1.0000 | 1.0000 | 2.75 | .6329 | .3410 | .2123 | 3.096 | 1.606 |
| 1.05 | .9982 | .9851 | .9452 | 1.0012 | 1.0421 | 2.80 | .6265 | .3311 | .2055 | 3.258 | 1.611 |
| 1.10 | .9933 | .9669 | .8939 | 1.0049 | 1.0816 | 2.85 | .6203 | .3216 | .1990 | 3.429 | 1.616 |
| 1.15 | .9859 | .9461 | .8458 | 1.0111 | 1.1186 | 2.90 | .6144 | .3124 | .1928 | 3.611 | 1.621 |
| 1.20 | .9765 | .9235 | .8008 | 1.0199 | 1.1532 | 2.95 | .6087 | .3036 | .1868 | 3.804 | 1.626 |
| 1.25 | .9656 | .8996 | .7588 | 1.0312 | 1.186 | 3.00 | .6032 | .2952 | .1811 | 4.007 | 1.630 |
| 1.30 | .9534 | .8747 | .7194 | 1.0451 | 1.216 | 3.50 | .5582 | .2262 | .1359 | 6.806 | 1.665 |
| 1.35 | .9404 | .8493 | .6826 | .1.0617 | 1.244 | 4.00 | .5265 | .1781 | .1055 | 11.57 | 1.688 |
| 1.40 | .9268 | .8237 | .6483 | 1.0809 | 1.270 | 4.50 | .5037 | .1435 | .08417 | 19.44 | 1.704 |
| 1.45 | .9128 | .7980 | .6161 | 1.1028 | 1.295 | 5.00 | .4867 | .1178 | .06866 | 32.06 | 1.716 |
| 1.50 | .8986 | .7726 | .5860 | 1.128 | 1.318 | 6.00 | .4639 | .08335 | .04812 | 81.79 | 1.732 |
| 1.55 | .8843 | .7475 | .5578 | 1.155 | 1.340 | 7.00 | .4496 | .06192 | .03555 | 191.3 | 1.742 |
| 1.60 | .8701 | .7230 | .5314 | 1.185 | 1.360 | 8.00 | .4402 | .04775 | .02732 | 413.4 | 1.748 |
| 1.65 | .8560 | .6990 | .5067 | 1.219 | 1.379 | 9.00 | .4336 | .03792 | .02164 | 833.4 | 1.753 |
| 1.70 | .8421 | .6756 | .4835 | 1.256 | 1.397 | 10.00 | .4289 | .03082 | .01756 | 1582 | 1.756 |
| | | | | | | ∞ | .4083 | 0 | 0 | ∞ | 1.769 |

# Table 41 Rayleigh Line—One-Dimensional Compressible-Flow Functions for Stagnation-Temperature Change in the Absence of Friction and Area Change

**For a perfect gas with constant specific heat and molecular weight**
**k = 1.67**

| M | $\dfrac{T_0}{T_0{}^*}$ | $\dfrac{T}{T^*}$ | $\dfrac{p}{p^*}$ | $\dfrac{p_0}{p_0{}^*}$ | $\dfrac{V}{V^*}$ | M | $\dfrac{T_0}{T_0{}^*}$ | $\dfrac{T}{T^*}$ | $\dfrac{p}{p^*}$ | $\dfrac{p_0}{p_0{}^*}$ | $\dfrac{V}{V^*}$ |
|---|---|---|---|---|---|---|---|---|---|---|---|
| 0 | 0 | 0 | 2.670 | 1.299 | 0 | 1.75 | .8862 | .5840 | .4367 | 1.235 | 1.337 |
| 0.05 | .01325 | .01767 | 2.659 | 1.297 | .00665 | 1.80 | .8779 | .5620 | .4165 | 1.266 | 1.349 |
| .10 | .05183 | .06896 | 2.626 | 1.289 | .02626 | 1.85 | .8699 | .5410 | .3976 | 1.299 | 1.360 |
| .15 | .11243 | .1490 | 2.573 | 1.276 | .05790 | 1.90 | .8621 | .5209 | .3799 | 1.334 | 1.371 |
| .20 | .19020 | .2506 | 2.503 | 1.259 | .10011 | 1.95 | .8546 | .5018 | .3633 | 1.370 | 1.381 |
| .25 | .2794 | .3653 | 2.418 | 1.239 | .1511 | 2.00 | .8474 | .4835 | .3477 | 1.408 | 1.391 |
| .30 | .3742 | .4849 | 2.321 | 1.216 | .2089 | 2.05 | .8405 | .4660 | .3330 | 1.448 | 1.400 |
| .35 | .4693 | .6018 | 2.216 | 1.192 | .2715 | 2.10 | .8338 | .4493 | .3192 | 1.490 | 1.408 |
| .40 | .5606 | .7103 | 2.107 | 1.168 | .3371 | 2.15 | .8274 | .4334 | .3062 | 1.534 | 1.415 |
| .45 | .6448 | .8062 | 1.995 | 1.144 | .4040 | 2.20 | .8213 | .4183 | .2940 | 1.580 | 1.423 |
| .50 | .7201 | .8870 | 1.884 | 1.1202 | .4709 | 2.25 | .8154 | .4038 | .2824 | 1.628 | 1.430 |
| .55 | .7853 | .9519 | 1.774 | 1.0981 | .5366 | 2.30 | .8097 | .3899 | .2715 | 1.678 | 1.436 |
| .60 | .8402 | 1.0010 | 1.667 | 1.0778 | .6003 | 2.35 | .8043 | .3767 | .2612 | 1.729 | 1.442 |
| .65 | .8853 | 1.0354 | 1.565 | 1.0597 | .6614 | 2.40 | .7991 | .3641 | .2514 | 1.783 | 1.448 |
| .70 | .9213 | 1.0565 | 1.468 | 1.0438 | .7195 | 2.45 | .7941 | .3521 | .2422 | 1.839 | 1.454 |
| .75 | .9491 | 1.0662 | 1.377 | 1.0303 | .7744 | 2.50 | .7893 | .3406 | .2334 | 1.897 | 1.459 |
| .80 | .9697 | 1.0660 | 1.291 | 1.0193 | .8260 | 2.55 | .7847 | .3296 | .2251 | 1.956 | 1.464 |
| .85 | .9842 | 1.0578 | 1.210 | 1.0108 | .8742 | 2.60 | .7803 | .3191 | .2173 | 2.018 | 1.469 |
| .90 | .9935 | 1.0432 | 1.135 | 1.0048 | .9192 | 2.65 | .7761 | .3090 | .2098 | 2.082 | 1.473 |
| .95 | .9985 | 1.0235 | 1.0649 | 1.0012 | .9611 | 2.70 | .7721 | .2994 | .2027 | 2.148 | 1.477 |
| 1.00 | 1.0000 | 1.0000 | 1.0000 | 1.0000 | 1.0000 | 2.75 | .7682 | .2902 | .1959 | 2.216 | 1.481 |
| 1.05 | .9987 | .9736 | .9398 | 1.0012 | 1.0361 | 2.80 | .7644 | .2814 | .1895 | 2.287 | 1.485 |
| 1.10 | .9952 | .9454 | .8839 | 1.0046 | 1.0695 | 2.85 | .7608 | .2730 | .1834 | 2.360 | 1.489 |
| 1.15 | .9899 | .9158 | .8321 | 1.0103 | 1.1005 | 2.90 | .7574 | .2649 | .1775 | 2.435 | 1.493 |
| 1.20 | .9833 | .8855 | .7842 | 1.0181 | 1.1292 | 2.95 | .7541 | .2571 | .1719 | 2.512 | 1.496 |
| 1.25 | .9757 | .8550 | .7397 | 1.0280 | 1.156 | 3.00 | .7509 | .2497 | .1666 | 2.587 | 1.499 |
| 1.30 | .9674 | .8246 | .6985 | 1.0400 | 1.181 | 3.50 | .7251 | .1897 | .1244 | 3.521 | 1.524 |
| 1.35 | .9586 | .7946 | .6603 | 1.0540 | 1.204 | 4.00 | .7072 | .1484 | .09632 | 4.716 | 1.541 |
| 1.40 | .9495 | .7652 | .6249 | 1.0700 | 1.225 | 4.50 | .6943 | .1191 | .07669 | 6.213 | 1.553 |
| 1.45 | .9403 | .7365 | .5919 | 1.0880 | 1.245 | 5.00 | .6848 | .0975 | .06246 | 8.044 | 1.561 |
| 1.50 | .9310 | .7087 | .5612 | 1.108 | 1.263 | 6.00 | .6721 | .06870 | .04368 | 12.86 | 1.573 |
| 1.55 | .9217 | .6818 | .5327 | 1.130 | 1.280 | 7.00 | .6642 | .05092 | .03224 | 19.44 | 1.580 |
| 1.60 | .9125 | .6559 | .5062 | 1.154 | 1.296 | 8.00 | .6590 | .03920 | .02475 | 28.07 | 1.584 |
| 1.65 | .9035 | .6309 | .4814 | 1.179 | 1.311 | 9.00 | .6553 | .03110 | .01959 | 39.05 | 1.587 |
| 1.70 | .8947 | .6069 | .4583 | 1.206 | 1.324 | 10.00 | .6527 | .02526 | .01589 | 52.66 | 1.589 |
| | | | | | | $\infty$ | .6414 | 0 | 0 | $\infty$ | 1.599 |

# Table 42  Fanno Line—One-Dimensional Compressible-Flow Functions for Adiabatic Flow at Constant Area with Friction

**For a perfect gas with constant specific heat and molecular weight**
**k = 1.4**

| M | $\frac{T}{T^*}$ | $\frac{p}{p^*}$ | $\frac{p_0}{p_0{}^*}$ | $\frac{V}{V^*}$ | $\frac{F}{F^*}$ | $4\frac{fL_{max}}{D}$ | M | $\frac{T}{T^*}$ | $\frac{p}{p^*}$ | $\frac{p_0}{p_0{}^*}$ | $\frac{V}{V^*}$ | $\frac{F}{F^*}$ | $4\frac{fL_{max}}{D}$ |
|---|---|---|---|---|---|---|---|---|---|---|---|---|---|
| 0 | 1.2000 | ∞ | ∞ | 0 | ∞ | ∞ | 0.40 | 1.1628 | 2.6958 | 1.5901 | .43133 | 1.3749 | 2.3085 |
| 0.01 | 1.2000 | 109.544 | 57.874 | .01095 | 45.650 | 7134.40 | .41 | 1.1610 | 2.6280 | 1.5587 | .44177 | 1.3527 | 2.1344 |
| .02 | 1.1999 | 54.770 | 28.942 | .02191 | 22.834 | 1778.45 | .42 | 1.1591 | 2.5634 | 1.5289 | .45218 | 1.3318 | 1.9744 |
| .03 | 1.1998 | 36.511 | 19.300 | .03286 | 15.232 | 787.08 | .43 | 1.1572 | 2.5017 | 1.5007 | .46257 | 1.3122 | 1.8272 |
| .04 | 1.1996 | 27.382 | 14.482 | .04381 | 11.435 | 440.35 | .44 | 1.1553 | 2.4428 | 1.4739 | .47293 | 1.2937 | 1.6915 |
| .05 | 1.1994 | 21.903 | 11.5914 | .05476 | 9.1584 | 280.02 | .45 | 1.1533 | 2.3865 | 1.4486 | .48326 | 1.2763 | 1.5664 |
| .06 | 1.1991 | 18.251 | 9.6659 | .06570 | 7.6428 | 193.03 | .46 | 1.1513 | 2.3326 | 1.4246 | .49357 | 1.2598 | 1.4509 |
| .07 | 1.1988 | 15.642 | 8.2915 | .07664 | 6.5620 | 140.66 | .47 | 1.1492 | 2.2809 | 1.4018 | .50385 | 1.2443 | 1.3442 |
| .08 | 1.1985 | 13.684 | 7.2616 | .08758 | 5.7529 | 106.72 | .48 | 1.1471 | 2.2314 | 1.3801 | .51410 | 1.2296 | 1.2453 |
| .09 | 1.1981 | 12.162 | 6.4614 | .09851 | 5.1249 | 83.496 | .49 | 1.1450 | 2.1838 | 1.3595 | .52433 | 1.2158 | 1.1539 |
| .10 | 1.1976 | 10.9435 | 5.8218 | .10943 | 4.6236 | 66.922 | .50 | 1.1429 | 2.1381 | 1.3399 | .53453 | 1.2027 | 1.06908 |
| .11 | 1.1971 | 9.9465 | 5.2992 | .12035 | 4.2146 | 54.688 | .51 | 1.1407 | 2.0942 | 1.3212 | .54469 | 1.1903 | .99042 |
| .12 | 1.1966 | 9.1156 | 4.8643 | .13126 | 3.8747 | 45.408 | .52 | 1.1384 | 2.0519 | 1.3034 | .55482 | 1.1786 | .91741 |
| .13 | 1.1960 | 8.4123 | 4.4968 | .14216 | 3.5880 | 38.207 | .53 | 1.1362 | 2.0112 | 1.2864 | .56493 | 1.1675 | .84963 |
| .14 | 1.1953 | 7.8093 | 4.1824 | .15306 | 3.3432 | 32.511 | .54 | 1.1339 | 1.9719 | 1.2702 | .57501 | 1.1571 | .78662 |
| .15 | 1.1946 | 7.2866 | 3.9103 | .16395 | 3.1317 | 27.932 | .55 | 1.1315 | 1.9341 | 1.2549 | .58506 | 1.1472 | .72805 |
| .16 | 1.1939 | 6.8291 | 3.6727 | .17482 | 2.9474 | 24.198 | .56 | 1.1292 | 1.8976 | 1.2403 | .59507 | 1.1378 | .67357 |
| .17 | 1.1931 | 6.4252 | 3.4635 | .18568 | 2.7855 | 21.115 | .57 | 1.1268 | 1.8623 | 1.2263 | .60505 | 1.1289 | .62286 |
| .18 | 1.1923 | 6.0662 | 3.2779 | .19654 | 2.6422 | 18.543 | .58 | 1.1244 | 1.8282 | 1.2130 | .61500 | 1.1205 | .57568 |
| .19 | 1.1914 | 5.7448 | 3.1123 | .20739 | 2.5146 | 16.375 | .59 | 1.1219 | 1.7952 | 1.2003 | .62492 | 1.1126 | .53174 |
| .20 | 1.1905 | 5.4555 | 2.9635 | .21822 | 2.4004 | 14.533 | .60 | 1.1194 | 1.7634 | 1.1882 | .63481 | 1.10504 | .49081 |
| .21 | 1.1895 | 5.1936 | 2.8293 | .22904 | 2.2976 | 12.956 | .61 | 1.1169 | 1.7325 | 1.1766 | .64467 | 1.09793 | .45270 |
| .22 | 1.1885 | 4.9554 | 2.7076 | .23984 | 2.2046 | 11.596 | .62 | 1.1144 | 1.7026 | 1.1656 | .65449 | 1.09120 | .41720 |
| .23 | 1.1874 | 4.7378 | 2.5968 | .25063 | 2.1203 | 10.416 | .63 | 1.1118 | 1.6737 | 1.1551 | .66427 | 1.08485 | .38411 |
| .24 | 1.1863 | 4.5383 | 2.4956 | .26141 | 2.0434 | 9.3865 | .64 | 1.1091 | 1.6456 | 1.1451 | .67402 | 1.07883 | .35330 |
| .25 | 1.1852 | 4.3546 | 2.4027 | .27217 | 1.9732 | 8.4834 | .65 | 1.10650 | 1.6183 | 1.1356 | .68374 | 1.07314 | .32460 |
| .26 | 1.1840 | 4.1850 | 2.3173 | .28291 | 1.9088 | 7.6876 | .66 | 1.10383 | 1.5919 | 1.1265 | .69342 | 1.06777 | .29785 |
| .27 | 1.1828 | 4.0280 | 2.2385 | .29364 | 1.8496 | 6.9832 | .67 | 1.10114 | 1.5662 | 1.1179 | .70306 | 1.06271 | .27295 |
| .28 | 1.1815 | 3.8820 | 2.1656 | .30435 | 1.7950 | 6.3572 | .68 | 1.09842 | 1.5413 | 1.1097 | .71267 | 1.05792 | .24978 |
| .29 | 1.1802 | 3.7460 | 2.0979 | .31504 | 1.7446 | 5.7989 | .69 | 1.09567 | 1.5170 | 1.1018 | .72225 | 1.05340 | .22821 |
| .30 | 1.1788 | 3.6190 | 2.0351 | .32572 | 1.6979 | 5.2992 | .70 | 1.09290 | 1.4934 | 1.09436 | .73179 | 1.04915 | .20814 |
| .31 | 1.1774 | 3.5002 | 1.9765 | .33637 | 1.6546 | 4.8507 | .71 | 1.09010 | 1.4705 | 1.08729 | .74129 | 1.04514 | .18949 |
| .32 | 1.1759 | 3.3888 | 1.9219 | .34700 | 1.6144 | 4.4468 | .72 | 1.08727 | 1.4482 | 1.08057 | .75076 | 1.04137 | .17215 |
| .33 | 1.1744 | 3.2840 | 1.8708 | .35762 | 1.5769 | 4.0821 | .73 | 1.08442 | 1.4265 | 1.07419 | .76019 | 1.03783 | .15606 |
| .34 | 1.1729 | 3.1853 | 1.8229 | .36822 | 1.5420 | 3.7520 | .74 | 1.08155 | 1.4054 | 1.06815 | .76958 | 1.03450 | .14113 |
| .35 | 1.1713 | 3.0922 | 1.7780 | .37880 | 1.5094 | 3.4525 | .75 | 1.07865 | 1.3848 | 1.06242 | .77893 | 1.03137 | .12728 |
| .36 | 1.1697 | 3.0042 | 1.7358 | .38935 | 1.4789 | 3.1801 | .76 | 1.07573 | 1.3647 | 1.05700 | .78825 | 1.02844 | .11446 |
| .37 | 1.1680 | 2.9209 | 1.6961 | .39988 | 1.4503 | 2.9320 | .77 | 1.07279 | 1.3451 | 1.05188 | .79753 | 1.02570 | .10262 |
| .38 | 1.1663 | 2.8420 | 1.6587 | .41039 | 1.4236 | 2.7055 | .78 | 1.06982 | 1.3260 | 1.04705 | .80677 | 1.02314 | .09167 |
| .39 | 1.1646 | 2.7671 | 1.6234 | .42087 | 1.3985 | 2.4983 | .79 | 1.06684 | 1.3074 | 1.04250 | .81598 | 1.02075 | .08159 |

See page 113 for definitions of symbols.

# Table 42 Fanno Line—One-Dimensional Compressible-Flow Functions for Adiabatic Flow at Constant Area with Friction

**For a perfect gas with constant specific heat and molecular weight**
$$k = 1.4$$

| M | $\frac{T}{T^*}$ | $\frac{p}{p^*}$ | $\frac{p_0}{p_0^*}$ | $\frac{V}{V^*}$ | $\frac{F}{F^*}$ | $4\frac{fL_{max}}{D}$ | M | $\frac{T}{T^*}$ | $\frac{p}{p^*}$ | $\frac{p_0}{p_0^*}$ | $\frac{V}{V^*}$ | $\frac{F}{F^*}$ | $4\frac{fL_{max}}{D}$ |
|---|---|---|---|---|---|---|---|---|---|---|---|---|---|
| 0.80 | 1.06383 | 1.2892 | 1.03823 | .82514 | 1.01853 | .07229 | 1.20 | .93168 | .80436 | 1.03044 | 1.1583 | 1.01082 | .03364 |
| .81 | 1.06080 | 1.2715 | 1.03422 | .83426 | 1.01646 | .06375 | 1.21 | .92820 | .79623 | 1.03344 | 1.1658 | 1.01178 | .03650 |
| .82 | 1.05775 | 1.2542 | 1.03047 | .84334 | 1.01455 | .05593 | 1.22 | .92473 | .78822 | 1.03657 | 1.1732 | 1.01278 | .03942 |
| .83 | 1.05468 | 1.2373 | 1.02696 | .85239 | 1.01278 | .04878 | 1.23 | .92125 | .78034 | 1.03983 | 1.1806 | 1.01381 | .04241 |
| .84 | 1.05160 | 1.2208 | 1.02370 | .86140 | 1.01115 | .04226 | 1.24 | .91777 | .77258 | 1.04323 | 1.1879 | 1.01486 | .04547 |
| .85 | 1.04849 | 1.2047 | 1.02067 | .87037 | 1.00966 | .03632 | 1.25 | .91429 | .76495 | 1.04676 | 1.1952 | 1.01594 | .04858 |
| .86 | 1.04537 | 1.1889 | 1.01787 | .87929 | 1.00829 | .03097 | 1.26 | .91080 | .75743 | 1.05041 | 1.2025 | 1.01705 | .05174 |
| .87 | 1.04223 | 1.1735 | 1.01529 | .88818 | 1.00704 | .02613 | 1.27 | .90732 | .75003 | 1.05419 | 1.2097 | 1.01818 | .05494 |
| .88 | 1.03907 | 1.1584 | 1.01294 | .89703 | 1.00591 | .02180 | 1.28 | .90383 | .74274 | 1.05809 | 1.2169 | 1.01933 | .05820 |
| .89 | 1.03589 | 1.1436 | 1.01080 | .90583 | 1.00490 | .01793 | 1.29 | .90035 | .73556 | 1.06213 | 1.2240 | 1.02050 | .06150 |
| .90 | 1.03270 | 1.12913 | 1.00887 | .91459 | 1.00399 | .014513 | 1.30 | .89686 | .72848 | 1.06630 | 1.2311 | 1.02169 | .06483 |
| .91 | 1.02950 | 1.11500 | 1.00714 | .92332 | 1.00318 | .011519 | 1.31 | .89338 | .72152 | 1.07060 | 1.2382 | 1.02291 | .06820 |
| .92 | 1.02627 | 1.10114 | 1.00560 | .93201 | 1.00248 | .008916 | 1.32 | .88989 | .71465 | 1.07502 | 1.2452 | 1.02415 | .07161 |
| .93 | 1.02304 | 1.08758 | 1.00426 | .94065 | 1.00188 | .006694 | 1.33 | .88641 | .70789 | 1.07957 | 1.2522 | 1.02540 | .07504 |
| .94 | 1.01978 | 1.07430 | 1.00311 | .94925 | 1.00136 | .004815 | 1.34 | .88292 | .70123 | 1.08424 | 1.2591 | 1.02666 | .07850 |
| .95 | 1.01652 | 1.06129 | 1.00215 | .95782 | 1.00093 | .003280 | 1.35 | .87944 | .69466 | 1.08904 | 1.2660 | 1.02794 | .08199 |
| .96 | 1.01324 | 1.04854 | 1.00137 | .96634 | 1.00059 | .002056 | 1.36 | .87596 | .68818 | 1.09397 | 1.2729 | 1.02924 | .08550 |
| .97 | 1.00995 | 1.03605 | 1.00076 | .97481 | 1.00033 | .001135 | 1.37 | .87249 | .68180 | 1.09902 | 1.2797 | 1.03056 | .08904 |
| .98 | 1.00664 | 1.02379 | 1.00033 | .98324 | 1.00014 | .000493 | 1.38 | .86901 | .67551 | 1.10419 | 1.2864 | 1.03189 | .09259 |
| .99 | 1.00333 | 1.01178 | 1.00008 | .99164 | 1.00003 | .000120 | 1.39 | .86554 | .66931 | 1.10948 | 1.2932 | 1.03323 | .09616 |
| 1.00 | 1.00000 | 1.00000 | 1.00000 | 1.00000 | 1.00000 | 0 | 1.40 | .86207 | .66320 | 1.1149 | 1.2999 | 1.03458 | .09974 |
| 1.01 | .99666 | .98844 | 1.00008 | 1.00831 | 1.00003 | .000114 | 1.41 | .85860 | .65717 | 1.1205 | 1.3065 | 1.03595 | .10333 |
| 1.02 | .99331 | .97711 | 1.00033 | 1.01658 | 1.00013 | .000458 | 1.42 | .85514 | .65122 | 1.1262 | 1.3131 | 1.03733 | .10694 |
| 1.03 | .98995 | .96598 | 1.00073 | 1.02481 | 1.00030 | .001013 | 1.43 | .85168 | .64536 | 1.1320 | 1.3197 | 1.03872 | .11056 |
| 1.04 | .98658 | .95506 | 1.00130 | 1.03300 | 1.00053 | .001771 | 1.44 | .84822 | .63958 | 1.1379 | 1.3262 | 1.04012 | .11419 |
| 1.05 | .98320 | .94435 | 1.00203 | 1.04115 | 1.00082 | .002712 | 1.45 | .84477 | 63387 | 1.1440 | 1.3327 | 1.04153 | .11782 |
| 1.06 | .97982 | .93383 | 1.00291 | 1.04925 | 1.00116 | .003837 | 1.46 | .84133 | .62824 | 1.1502 | 1.3392 | 1.04295 | .12146 |
| 1.07 | .97642 | .92350 | 1.00394 | 1.05731 | 1.00155 | .005129 | 1.47 | .83788 | .62269 | 1.1565 | 1.3456 | 1.04438 | .12510 |
| 1.08 | .97302 | .91335 | 1.00512 | 1.06533 | 1.00200 | .006582 | 1.48 | .83445 | .61722 | 1.1629 | 1.3520 | 1.04581 | .12875 |
| 1.09 | .96960 | .90338 | 1.00645 | 1.07331 | 1.00250 | .008185 | 1.49 | .83101 | .61181 | 1.1695 | 1.3583 | 1.04725 | .13240 |
| 1.10 | .96618 | .89359 | 1.00793 | 1.08124 | 1.00305 | .009933 | 1.50 | .82759 | .60648 | 1.1762 | 1.3646 | 1.04870 | .13605 |
| 1.11 | .96276 | .88397 | 1.00955 | 1.08913 | 1.00365 | .011813 | 1.51 | .82416 | .60122 | 1.1830 | 1.3708 | 1.05016 | .13970 |
| 1.12 | .95933 | .87451 | 1.01131 | 1.09698 | 1.00429 | .013824 | 1.52 | .82075 | .59602 | 1.1899 | 1.3770 | 1.05162 | .14335 |
| 1.13 | .95589 | .86522 | 1.01322 | 1.10479 | 1.00497 | .015949 | 1.53 | .81734 | .59089 | 1.1970 | 1.3832 | 1.05309 | .14699 |
| 1.14 | .95244 | .85608 | 1.01527 | 1.11256 | 1.00569 | .018187 | 1.54 | .81394 | .58583 | 1.2043 | 1.3894 | 1.05456 | .15063 |
| 1.15 | .94899 | .84710 | 1.01746 | 1.1203 | 1.00646 | .02053 | 1.55 | .81054 | .58084 | 1.2116 | 1.3955 | 1.05604 | .15427 |
| 1.16 | .94554 | .83827 | 1.01978 | 1.1280 | 1.00726 | .02298 | 1.56 | .80715 | .57591 | 1.2190 | 1.4015 | 1.05752 | .15790 |
| 1.17 | .94208 | .82958 | 1.02224 | 1.1356 | 1.00810 | .02552 | 1.57 | .80376 | .57104 | 1.2266 | 1.4075 | 1.05900 | .16152 |
| 1.18 | .93862 | .82104 | 1.02484 | 1.1432 | 1.00897 | .02814 | 1.58 | .80038 | .56623 | 1.2343 | 1.4135 | 1.06049 | .16514 |
| 1.19 | .93515 | .81263 | 1.02757 | 1.1508 | 1.00988 | .03085 | 1.59 | .79701 | .56148 | 1.2422 | 1.4195 | 1.06198 | .16876 |

# Table 42 Fanno Line—One-Dimensional Compressible-Flow Functions for Adiabatic Flow at Constant Area with Friction

**For a perfect gas with constant specific heat and molecular weight**
**k = 1.4**

| M | $\dfrac{T}{T^*}$ | $\dfrac{p}{p^*}$ | $\dfrac{p_0}{p_0^*}$ | $\dfrac{V}{V^*}$ | $\dfrac{F}{F^*}$ | $4\dfrac{fL_{max}}{D}$ | M | $\dfrac{T}{T^*}$ | $\dfrac{p}{p^*}$ | $\dfrac{p_0}{p_0^*}$ | $\dfrac{V}{V^*}$ | $\dfrac{F}{F^*}$ | $4\dfrac{fL_{max}}{D}$ |
|---|---|---|---|---|---|---|---|---|---|---|---|---|---|
| 1.60 | .79365 | .55679 | 1.2502 | 1.4254 | 1.06348 | .17236 | 2.00 | .66667 | .40825 | 1.6875 | 1.6330 | 1.1227 | .30499 |
| 1.61 | .79030 | .55216 | 1.2583 | 1.4313 | 1.06498 | .17595 | 2.01 | .66371 | .40532 | 1.7017 | 1.6375 | 1.1241 | .30796 |
| 1.62 | .78695 | .54759 | 1.2666 | 1.4371 | 1.06648 | .17953 | 2.02 | .66076 | .40241 | 1.7160 | 1.6420 | 1.1255 | .31091 |
| 1.63 | .78361 | .54308 | 1.2750 | 1.4429 | 1.06798 | .18311 | 2.03 | .65783 | .39954 | 1.7305 | 1.6465 | 1.1269 | .31384 |
| 1.64 | .78028 | .53862 | 1.2835 | 1.4487 | 1.06948 | .18667 | 2.04 | .65491 | .39670 | 1.7452 | 1.6509 | 1.1283 | .31675 |
| 1.65 | .77695 | .53421 | 1.2922 | 1.4544 | 1.07098 | .19022 | 2.05 | .65200 | .39389 | 1.7600 | 1.6553 | 1.1297 | .31965 |
| 1.66 | .77363 | .52986 | 1.3010 | 1.4601 | 1.07249 | .19376 | 2.06 | .64910 | .39110 | 1.7750 | 1.6597 | 1.1311 | .32253 |
| 1.67 | .77033 | .52556 | 1.3099 | 1.4657 | 1.07399 | .19729 | 2.07 | .64621 | .38834 | 1.7902 | 1.6640 | 1.1325 | .32538 |
| 1.68 | .76703 | .52131 | 1.3190 | 1.4713 | 1.07550 | .20081 | 2.08 | .64333 | .38562 | 1.8056 | 1.6683 | 1.1339 | .32822 |
| 1.69 | .76374 | .51711 | 1.3282 | 1.4769 | 1.07701 | .20431 | 2.09 | .64047 | .38292 | 1.8212 | 1.6726 | 1.1352 | .33104 |
| 1.70 | .76046 | .51297 | 1.3376 | 1.4825 | 1.07851 | .20780 | 2.10 | .63762 | .38024 | 1.8369 | 1.6769 | 1.1366 | .33385 |
| 1.71 | .75718 | .50887 | 1.3471 | 1.4880 | 1.08002 | .21128 | 2.11 | .63478 | .37760 | 1.8528 | 1.6811 | 1.1380 | .33664 |
| 1.72 | .75392 | .50482 | 1.3567 | 1.4935 | 1.08152 | .21474 | 2.12 | .63195 | .37498 | 1.8690 | 1.6853 | 1.1393 | .33940 |
| 1.73 | .75067 | .50082 | 1.3665 | 1.4989 | 1.08302 | .21819 | 2.13 | .62914 | .37239 | 1.8853 | 1.6895 | 1.1407 | .34215 |
| 1.74 | .74742 | .49686 | 1.3764 | 1.5043 | 1.08453 | .22162 | 2.14 | .62633 | .36982 | 1.9018 | 1.6936 | 1.1420 | .34488 |
| 1.75 | .74419 | .49295 | 1.3865 | 1.5097 | 1.08603 | .22504 | 2.15 | .62354 | .36728 | 1.9185 | 1.6977 | 1.1434 | .34760 |
| 1.76 | .74096 | .48909 | 1.3967 | 1.5150 | 1.08753 | .22844 | 2.16 | .62076 | .36476 | 1.9354 | 1.7018 | 1.1447 | .35030 |
| 1.77 | .73774 | .48527 | 1.4070 | 1.5203 | 1.08903 | .23183 | 2.17 | .61799 | .36227 | 1.9525 | 1.7059 | 1.1460 | .35298 |
| 1.78 | .73453 | .48149 | 1.4175 | 1.5256 | 1.09053 | .23520 | 2.18 | .61523 | .35980 | 1.9698 | 1.7099 | 1.1474 | .35564 |
| 1.79 | .73134 | .47776 | 1.4282 | 1.5308 | 1.09202 | .23855 | 2.19 | .61249 | .35736 | 1.9873 | 1.7139 | 1.1487 | .35828 |
| 1.80 | .72816 | .47407 | 1.4390 | 1.5360 | 1.09352 | .24189 | 2.20 | .60976 | .35494 | 2.0050 | 1.7179 | 1.1500 | .36091 |
| 1.81 | .72498 | .47042 | 1.4499 | 1.5412 | 1.09500 | .24521 | 2.21 | .60704 | .35254 | 2.0228 | 1.7219 | 1.1513 | .36352 |
| 1.82 | .72181 | .46681 | 1.4610 | 1.5463 | 1.09649 | .24851 | 2.22 | .60433 | .35017 | 2.0409 | 1.7258 | 1.1526 | .36611 |
| 1.83 | .71865 | .46324 | 1.4723 | 1.5514 | 1.09798 | .25180 | 2.23 | .60163 | .34782 | 2.0592 | 1.7297 | 1.1539 | .36868 |
| 1.84 | .71551 | .45972 | 1.4837 | 1.5564 | 1.09946 | .25507 | 2.24 | .59895 | .34550 | 2.0777 | 1.7336 | 1.1552 | .37124 |
| 1.85 | .71238 | .45623 | 1.4952 | 1.5614 | 1.1009 | .25832 | 2.25 | .59627 | .34319 | 2.0964 | 1.7374 | 1.1565 | .37378 |
| 1.86 | .70925 | .45278 | 1.5069 | 1.5664 | 1.1024 | .26156 | 2.26 | .59361 | .34091 | 2.1154 | 1.7412 | 1.1578 | .37630 |
| 1.87 | .70614 | .44937 | 1.5188 | 1.5714 | 1.1039 | .26478 | 2.27 | .59096 | .33865 | 2.1345 | 1.7450 | 1.1590 | .37881 |
| 1.88 | .70304 | .44600 | 1.5308 | 1.5763 | 1.1054 | .26798 | 2.28 | .58833 | .33641 | 2.1538 | 1.7488 | 1.1603 | .38130 |
| 1.89 | .69995 | .44266 | 1.5429 | 1.5812 | 1.1068 | .27116 | 2.29 | .58570 | .33420 | 2.1733 | 1.7526 | 1.1616 | .38377 |
| 1.90 | .69686 | .43936 | 1.5552 | 1.5861 | 1.1083 | .27433 | 2.30 | .58309 | .33200 | 2.1931 | 1.7563 | 1.1629 | .38623 |
| 1.91 | .69379 | .43610 | 1.5677 | 1.5909 | 1.1097 | .27744 | 2.31 | .58049 | .32983 | 2.2131 | 1.7600 | 1.1641 | .38867 |
| 1.92 | .69074 | .43287 | 1.5804 | 1.5957 | 1.1112 | .28061 | 2.32 | .57790 | .32767 | 2.2333 | 1.7637 | 1.1653 | .39109 |
| 1.93 | .68769 | .42967 | 1.5932 | 1.6005 | 1.1126 | .28372 | 2.33 | .57532 | .32554 | 2.2537 | 1.7673 | 1.1666 | .39350 |
| 1.94 | .68465 | .42651 | 1.6062 | 1.6052 | 1.1141 | .28681 | 2.34 | .57276 | .32342 | 2.2744 | 1.7709 | 1.1678 | .39589 |
| 1.95 | .68162 | .42339 | 1.6193 | 1.6099 | 1.1155 | .28989 | 2.35 | .57021 | .32133 | 2.2953 | 1.7745 | 1.1690 | .39826 |
| 1.96 | .67861 | .42030 | 1.6326 | 1.6146 | 1.1170 | .29295 | 2.36 | .56767 | .31925 | 2.3164 | 1.7781 | 1.1703 | .40062 |
| 1.97 | .67561 | .41724 | 1.6461 | 1.6193 | 1.1184 | .29599 | 2.37 | .56514 | .31720 | 2.3377 | 1.7817 | 1.1715 | .40296 |
| 1.98 | .67262 | .41421 | 1.6597 | 1.6239 | 1.1198 | .29901 | 2.38 | .56262 | .31516 | 2.3593 | 1.7852 | 1.1727 | .40528 |
| 1.99 | .66964 | .41121 | 1.6735 | 1.6284 | 1.1213 | .30201 | 2.39 | .56011 | .31314 | 2.3811 | 1.7887 | 1.1739 | .40760 |

# Table 42 Fanno Line—One-Dimensional Compressible-Flow Functions for Adiabatic Flow at Constant Area with Friction

For a perfect gas with constant specific heat and molecular weight
k = 1.4

| M | $\frac{T}{T^*}$ | $\frac{p}{p^*}$ | $\frac{p_0}{p_0{}^*}$ | $\frac{V}{V^*}$ | $\frac{F}{F^*}$ | $4\frac{fL_{max}}{D}$ | M | $\frac{T}{T^*}$ | $\frac{p}{p^*}$ | $\frac{p_0}{p_0{}^*}$ | $\frac{V}{V^*}$ | $\frac{F}{F^*}$ | $4\frac{fL_{max}}{D}$ |
|---|---|---|---|---|---|---|---|---|---|---|---|---|---|
| 2.40 | .55762 | .31114 | 2.4031 | 1.7922 | 1.1751 | .40989 | 2.80 | .46729 | .24414 | 3.5001 | 1.9140 | 1.2182 | .48976 |
| 2.41 | .55514 | .30916 | 2.4254 | 1.7956 | 1.1763 | .41216 | 2.81 | .46526 | .24274 | 3.5336 | 1.9167 | 1.2192 | .49148 |
| 2.42 | .55267 | .30720 | 2.4479 | 1.7991 | 1.1775 | .41442 | 2.82 | .46324 | .24135 | 3.5674 | 1.9193 | 1.2202 | .49321 |
| 2.43 | .55021 | .30525 | 2.4706 | 1.8025 | 1.1786 | .41667 | 2.83 | .46122 | .23997 | 3.6015 | 1.9220 | 1.2211 | .49491 |
| 2.44 | .54776 | .30332 | 2.4936 | 1.8059 | 1.1798 | .41891 | 2.84 | .45922 | .23861 | 3.6359 | 1.9246 | 1.2221 | .49660 |
| 2.45 | .54533 | .30141 | 2.5168 | 1.8092 | 1.1810 | .42113 | 2.85 | .45723 | .23726 | 3.6707 | 1.9271 | 1.2230 | .49828 |
| 2.46 | .54291 | .29952 | 2.5403 | 1.8126 | 1.1821 | .42333 | 2.86 | .45525 | .23592 | 3.7058 | 1.9297 | 1.2240 | .49995 |
| 2.47 | .54050 | .29765 | 2.5640 | 1.8159 | 1.1833 | .42551 | 2.87 | .45328 | .23458 | 3.7413 | 1.9322 | 1.2249 | .50161 |
| 2.48 | .53810 | .29579 | 2.5880 | 1.8192 | 1.1844 | .42768 | 2.88 | .45132 | .23326 | 3.7771 | 1.9348 | 1.2258 | .50326 |
| 2.49 | .53571 | .29395 | 2.6122 | 1.8225 | 1.1856 | .42983 | 2.89 | .44937 | .23196 | 3.8133 | 1.9373 | 1.2268 | .50489 |
| 2.50 | .53333 | .29212 | 2.6367 | 1.8257 | 1.1867 | .43197 | 2.90 | .44743 | .23066 | 3.8498 | 1.9398 | 1.2277 | .50651 |
| 2.51 | .53097 | .29031 | 2.6615 | 1.8290 | 1.1879 | .43410 | 2.91 | .44550 | .22937 | 3.8866 | 1.9423 | 1.2286 | .50812 |
| 2.52 | .52862 | .28852 | 2.6865 | 1.8322 | 1.1890 | .43621 | 2.92 | .44358 | .22809 | 3.9238 | 1.9448 | 1.2295 | .50973 |
| 2.53 | .52627 | .28674 | 2.7117 | 1.8354 | 1.1910 | .43831 | 2.93 | .44167 | .22682 | 3.9614 | 1.9472 | 1.2304 | .51133 |
| 2.54 | .52394 | .28498 | 2.7372 | 1.8386 | 1.1912 | .44040 | 2.94 | .43977 | .22556 | 3.9993 | 1.9497 | 1.2313 | .51291 |
| 2.55 | .52163 | .28323 | 2.7630 | 1.8417 | 1.1923 | .44247 | 2.95 | .43788 | .22431 | 4.0376 | 1.9521 | 1.2322 | .51447 |
| 2.56 | .51932 | .28150 | 2.7891 | 1.8448 | 1.1934 | .44452 | 2.96 | .43600 | .22307 | 4.0763 | 1.9545 | 1.2331 | .51603 |
| 2.57 | .51702 | .27978 | 2.8154 | 1.8479 | 1.1945 | .44655 | 2.97 | .43413 | .22185 | 4.1153 | 1.9569 | 1.2340 | .51758 |
| 2.58 | .51474 | .27808 | 2.8420 | 1.8510 | 1.1956 | .44857 | 2.98 | .43226 | .22063 | 4.1547 | 1.9592 | 1.2348 | .51912 |
| 2.59 | .51247 | .27640 | 2.8689 | 1.8541 | 1.1967 | .45059 | 2.99 | .43041 | .21942 | 4.1944 | 1.9616 | 1.2357 | .52064 |
| 2.60 | .51020 | .27473 | 2.8960 | 1.8571 | 1.1978 | .45259 | 3.0 | .42857 | .21822 | 4.2346 | 1.9640 | 1.2366 | .52216 |
| 2.61 | .50795 | .27307 | 2.9234 | 1.8602 | 1.1989 | .45457 | 3.5 | .34783 | .16850 | 6.7896 | 2.0642 | 1.2743 | .58643 |
| 2.62 | .50571 | .27143 | 2.9511 | 1.8632 | 1.2000 | .45654 | 4.0 | .28571 | .13363 | 10.719 | 2.1381 | 1.3029 | .63306 |
| 2.63 | .50349 | .26980 | 2.9791 | 1.8662 | 1.2011 | .45850 | 4.5 | .23762 | .10833 | 16.562 | 2.1936 | 1.3247 | .66764 |
| 2.64 | .50127 | .26818 | 3.0074 | 1.8691 | 1.2021 | .46044 | 5.0 | .20000 | .08944 | 25.000 | 2.2361 | 1.3416 | .69380 |
| 2.65 | .49906 | .26658 | 3.0359 | 1.8721 | 1.2031 | .46237 | 6.0 | .14634 | .06376 | 53.180 | 2.2953 | 1.3655 | .72987 |
| 2.66 | .49687 | .26499 | 3.0647 | 1.8750 | 1.2042 | .46429 | 7.0 | .11111 | .04762 | 104.14 | 2.3333 | 1.3810 | .75280 |
| 2.67 | .49469 | .26342 | 3.0938 | 1.8779 | 1.2052 | .46619 | 8.0 | .08696 | .03686 | 190.11 | 2.3591 | 1.3915 | .76819 |
| 2.68 | .49251 | .26186 | 3.1234 | 1.8808 | 1.2062 | .46807 | 9.0 | .06977 | .02935 | 327.19 | 2.3772 | 1.3989 | .77898 |
| 2.69 | .49035 | .26032 | 3.1530 | 1.8837 | 1.2073 | .46996 | 10.0 | .05714 | .02390 | 535.94 | 2.3905 | 1.4044 | .78683 |
| 2.70 | .48820 | .25878 | 3.1830 | 1.8865 | 1.2083 | .47182 | ∞ | 0 | 0 | ∞ | 2.4495 | 1.4289 | .82153 |
| 2.71 | .48606 | .25726 | 3.2133 | 1.8894 | 1.2093 | .47367 | | | | | | | |
| 2.72 | .48393 | .25575 | 3.2440 | 1.8922 | 1.2103 | .47551 | | | | | | | |
| 2.73 | .48182 | .25426 | 3.2749 | 1.8950 | 1.2113 | .47734 | | | | | | | |
| 2.74 | .47971 | .25278 | 3.3061 | 1.8978 | 1.2123 | .47915 | | | | | | | |
| 2.75 | .47761 | .25131 | 3.3376 | 1.9005 | 1.2133 | .48095 | | | | | | | |
| 2.76 | .47553 | .24985 | 3.3695 | 1.9032 | 1.2143 | .48274 | | | | | | | |
| 2.77 | .47346 | .24840 | 3.4017 | 1.9060 | 1.2153 | .48452 | | | | | | | |
| 2.78 | .47139 | .24697 | 3.4342 | 1.9087 | 1.2163 | .48628 | | | | | | | |
| 2.79 | .46933 | .24555 | 3.4670 | 1.9114 | 1.2173 | .48803 | | | | | | | |

# Table 43 Fanno Line—One-Dimensional Compressible-Flow Functions for Adiabatic Flow at Constant Area with Friction

For a perfect gas with constant specific heat and molecular weight
k = 1.0

| M | $\frac{T}{T^*}$ | $\frac{p}{p^*}$ | $\frac{p_0}{p_0^*}$ | $\frac{V}{V^*}$ | $\frac{F}{F^*}$ | $4\frac{fL_{max}}{D}$ |
|---|---|---|---|---|---|---|
| 0 | 1.000 | ∞ | ∞ | 0 | ∞ | ∞ |
| 0.05 | | 20.000 | 12.146 | .0500 | 10.025 | 393.01 |
| .10 | | 10.000 | 6.096 | .1000 | 5.050 | 94.39 |
| .15 | | 6.667 | 4.089 | .1500 | 3.408 | 39.65 |
| .20 | | 5.000 | 3.094 | .2000 | 2.600 | 20.78 |
| .25 | | 4.000 | 2.503 | .2500 | 2.125 | 12.227 |
| .30 | | 3.333 | 2.115 | .3000 | 1.817 | 7.703 |
| .35 | | 2.857 | 1.842 | .3500 | 1.604 | 5.064 |
| .40 | | 2.500 | 1.643 | .4000 | 1.450 | 3.417 |
| .45 | | 2.222 | 1.492 | .4500 | 1.336 | 2.341 |
| .50 | | 2.000 | 1.375 | .5000 | 1.250 | 1.614 |
| .55 | | 1.818 | 1.283 | .5500 | 1.184 | 1.110 |
| .60 | | 1.667 | 1.210 | .6000 | 1.133 | .7561 |
| .65 | | 1.539 | 1.153 | .6500 | 1.0942 | .5053 |
| .70 | | 1.429 | 1.107 | .7000 | 1.0643 | .3275 |
| .75 | | 1.333 | 1.0714 | .7500 | 1.0417 | .2024 |
| .80 | | 1.250 | 1.0441 | .8000 | 1.0250 | .1162 |
| .85 | | 1.176 | 1.0240 | .8500 | 1.0132 | .05904 |
| .90 | | 1.111 | 1.0104 | .9000 | 1.0056 | .02385 |
| .95 | | 1.0526 | 1.0026 | .9500 | 1.0013 | .00545 |
| 1.00 | | 1.0000 | 1.0000 | 1.0000 | 1.0000 | 0 |
| 1.05 | | .9524 | 1.0025 | 1.0500 | 1.0012 | .00461 |
| 1.10 | | .9091 | 1.0097 | 1.100 | 1.0045 | .01707 |
| 1.15 | | .8695 | 1.0217 | 1.150 | 1.0098 | .03567 |
| 1.20 | | .8333 | 1.0384 | 1.200 | 1.0167 | .05909 |
| 1.25 | | .8000 | 1.0598 | 1.250 | 1.0250 | .08629 |
| 1.30 | | .7692 | 1.0862 | 1.300 | 1.0346 | .1164 |
| 1.35 | | .7407 | 1.118 | 1.350 | 1.0453 | .1489 |
| 1.40 | | .7143 | 1.154 | 1.400 | 1.0571 | .1831 |
| 1.45 | | .6897 | 1.196 | 1.450 | 1.0698 | .2188 |
| 1.50 | | .6667 | 1.245 | 1.500 | 1.0833 | .2554 |
| 1.55 | | .6452 | 1.300 | 1.550 | 1.0976 | .2927 |
| 1.60 | | .6250 | 1.363 | 1.600 | 1.112 | .3306 |
| 1.65 | | .6061 | 1.434 | 1.650 | 1.128 | .3689 |
| 1.70 | | .5882 | 1.514 | 1.700 | 1.144 | .4073 |

| M | $\frac{T}{T^*}$ | $\frac{p}{p^*}$ | $\frac{p_0}{p_0^*}$ | $\frac{V}{V^*}$ | $\frac{F}{F^*}$ | $4\frac{fL_{max}}{D}$ |
|---|---|---|---|---|---|---|
| 1.75 | 1.000 | .5714 | 1.603 | 1.750 | 1.161 | .4458 |
| 1.80 | | .5556 | 1.703 | 1.800 | 1.178 | .4842 |
| 1.85 | | .5406 | 1.815 | 1.850 | 1.195 | .5225 |
| 1.90 | | .5263 | 1.941 | 1.900 | 1.213 | .5607 |
| 1.95 | | .5128 | 2.082 | 1.950 | 1.231 | .5986 |
| 2.00 | | .5000 | 2.241 | 2.000 | 1.250 | .6363 |
| 2.05 | | .4878 | 2.419 | 2.050 | 1.269 | .6736 |
| 2.10 | | .4762 | 2.620 | 2.100 | 1.288 | .7106 |
| 2.15 | | .4651 | 2.846 | 2.150 | 1.308 | .7472 |
| 2.20 | | .4545 | 3.100 | 2.200 | 1.327 | .7835 |
| 2.25 | | .4444 | 3.388 | 2.250 | 1.347 | .8194 |
| 2.30 | | .4348 | 3.714 | 2.300 | 1.367 | .8549 |
| 2.35 | | .4256 | 4.083 | 2.350 | 1.388 | .8900 |
| 2.40 | | .4167 | 4.502 | 2.400 | 1.408 | .9246 |
| 2.45 | | .4082 | 4.979 | 2.450 | 1.429 | .9588 |
| 2.50 | | .4000 | 5.522 | 2.500 | 1.450 | .9926 |
| 2.55 | | .3922 | 6.142 | 2.550 | 1.471 | 1.0260 |
| 2.60 | | .3847 | 6.852 | 2.600 | 1.492 | 1.0590 |
| 2.65 | | .3774 | 7.665 | 2.650 | 1.514 | 1.0916 |
| 2.70 | | .3704 | 8.600 | 2.700 | 1.535 | 1.1237 |
| 2.75 | | .3636 | 9.676 | 2.750 | 1.557 | 1.155 |
| 2.80 | | .3571 | 10.92 | 2.800 | 1.579 | 1.187 |
| 2.85 | | .3509 | 12.35 | 2.850 | 1.600 | 1.218 |
| 2.90 | | .3449 | 14.02 | 2.900 | 1.622 | 1.248 |
| 2.95 | | .3390 | 15.95 | 2.950 | 1.644 | 1.279 |
| 3.00 | | .3333 | 18.20 | 3.000 | 1.667 | 1.308 |
| 3.50 | | .2857 | 79.22 | 3.500 | 1.893 | 1.587 |
| 4.00 | | .2500 | 452.01 | 4.000 | 2.125 | 1.835 |
| 4.50 | | .2222 | 3364 | 4.500 | 2.361 | 2.058 |
| 5.00 | | .2000 | 32550 | 5.000 | 2.600 | 2.259 |
| 6.00 | | .1667 | $664(10)^4$ | 6.000 | 3.083 | 2.611 |
| 7.00 | | .1429 | $378(10)^7$ | 7.000 | 3.571 | 2.912 |
| 8.00 | | .1250 | $599(10)^{10}$ | 8.000 | 4.062 | 3.174 |
| 9.00 | | .1111 | $262(10)^{14}$ | 9.000 | 4.556 | 3.407 |
| 10.00 | | .1000 | $314(10)^{18}$ | 10.000 | 5.050 | 3.615 |
| ∞ | 1.000 | 0 | ∞ | ∞ | ∞ | ∞ |

# Table 44  Fanno Line—One-Dimensional Compressible-Flow Functions for Adiabatic Flow at Constant Area with Friction

**For a perfect gas with constant specific heat and molecular weight**
**k = 1.1**

| M | $\frac{T}{T^*}$ | $\frac{p}{p^*}$ | $\frac{p_0}{p_0{}^*}$ | $\frac{V}{V^*}$ | $\frac{F}{F^*}$ | $4\frac{fL_{max}}{D}$ |
|---|---|---|---|---|---|---|
| 0 | 1.0500 | ∞ | ∞ | 0 | ∞ | ∞ |
| 0.05 | 1.0499 | 20.493 | 11.999 | .05123 | 9.785 | 357.05 |
| .10 | 1.0495 | 10.244 | 6.023 | .1024 | 4.932 | 85.65 |
| .15 | 1.0488 | 6.828 | 4.042 | .1536 | 3.332 | 35.92 |
| .20 | 1.0479 | 5.118 | 3.059 | .2047 | 2.545 | 18.79 |
| .25 | 1.0467 | 4.092 | 2.476 | .2558 | 2.083 | 11.03 |
| .30 | 1.0453 | 3.408 | 2.094 | .3067 | 1.784 | 6.936 |
| .35 | 1.0436 | 2.919 | 1.825 | .3575 | 1.577 | 4.549 |
| .40 | 1.0417 | 2.552 | 1.628 | .4082 | 1.429 | 3.062 |
| .45 | 1.0395 | 2.266 | 1.480 | .4588 | 1.319 | 2.093 |
| .50 | 1.0370 | 2.037 | 1.365 | .5092 | 1.237 | 1.439 |
| .55 | 1.0343 | 1.849 | 1.275 | .5594 | 1.174 | .9871 |
| .60 | 1.0314 | 1.693 | 1.204 | .6094 | 1.125 | .6705 |
| .65 | 1.0283 | 1.560 | 1.148 | .6591 | 1.0882 | .4468 |
| .70 | 1.0249 | 1.446 | 1.104 | .7086 | 1.0599 | .2887 |
| .75 | 1.0213 | 1.347 | 1.0689 | .7579 | 1.0386 | .1780 |
| .80 | 1.0174 | 1.261 | 1.0425 | .8069 | 1.0231 | .1019 |
| .85 | 1.0133 | 1.184 | 1.0231 | .8557 | 1.0122 | .05160 |
| .90 | 1.0091 | 1.116 | 1.0100 | .9041 | 1.0051 | .02078 |
| .95 | 1.0047 | 1.0551 | 1.0024 | .9522 | 1.0012 | .00472 |
| 1.00 | 1.0000 | 1.0000 | 1.0000 | 1.0000 | 1.0000 | 0 |
| 1.05 | .9951 | .9501 | 1.0023 | 1.0474 | 1.0011 | .00398 |
| 1.10 | .9901 | .9046 | 1.0092 | 1.0945 | 1.0041 | .01468 |
| 1.15 | .9849 | .8630 | 1.0204 | 1.1412 | 1.0087 | .03058 |
| 1.20 | .9795 | .8247 | 1.0360 | 1.1876 | 1.0148 | .05050 |
| 1.25 | .9739 | .7895 | 1.0559 | 1.234 | 1.0221 | .07350 |
| 1.30 | .9682 | .7569 | 1.0801 | 1.279 | 1.0304 | .09885 |
| 1.35 | .9623 | .7266 | 1.109 | 1.324 | 1.0397 | .1260 |
| 1.40 | .9563 | .6985 | 1.142 | 1.369 | 1.0498 | .1544 |
| 1.45 | .9501 | .6722 | 1.180 | 1.413 | 1.0605 | .1838 |
| 1.50 | .9438 | .6476 | 1.223 | 1.457 | 1.0717 | .2138 |
| 1.55 | .9374 | .6246 | 1.272 | 1.501 | 1.0835 | .2443 |
| 1.60 | .9309 | .6030 | 1.326 | 1.544 | 1.0958 | .2749 |
| 1.65 | .9242 | .5826 | 1.387 | 1.586 | 1.108 | .3056 |
| 1.70 | .9174 | .5634 | 1.454 | 1.628 | 1.121 | .3362 |
| 1.75 | .9105 | .5453 | 1.528 | 1.670 | 1.134 | .3667 |
| 1.80 | .9036 | .5281 | 1.610 | 1.711 | 1.148 | .3969 |
| 1.85 | .8966 | .5118 | 1.701 | 1.752 | 1.161 | .4268 |
| 1.90 | .8895 | .4964 | 1.801 | 1.792 | 1.175 | .4563 |
| 1.95 | .8823 | .4817 | 1.911 | 1.832 | 1.189 | .4854 |
| 2.00 | .8750 | .4677 | 2.032 | 1.871 | 1.203 | .5140 |
| 2.05 | .8677 | .4544 | 2.165 | 1.910 | 1.217 | .5422 |
| 2.10 | .8603 | .4417 | 2.312 | 1.948 | 1.231 | .5698 |
| 2.15 | .8529 | .4295 | 2.473 | 1.986 | 1.245 | .5970 |
| 2.20 | .8454 | .4179 | 2.651 | 2.023 | 1.259 | .6237 |
| 2.25 | .8379 | .4068 | 2.846 | 2.060 | 1.273 | .6498 |
| 2.30 | .8304 | .3962 | 3.061 | 2.096 | 1.286 | .6754 |
| 2.35 | .8228 | .3860 | 3.299 | 2.132 | 1.300 | .7005 |
| 2.40 | .8152 | .3762 | 3.560 | 2.167 | 1.314 | .7251 |
| 2.45 | .8076 | .3668 | 3.848 | 2.202 | 1.328 | .7491 |
| 2.50 | .8000 | .3578 | 4.165 | 2.236 | 1.342 | .7726 |
| 2.55 | .7924 | .3491 | 4.515 | 2.270 | 1.355 | .7957 |
| 2.60 | .7848 | .3407 | 4.902 | 2.303 | 1.369 | .8182 |
| 2.65 | .7771 | .3327 | 5.328 | 2.336 | 1.382 | .8402 |
| 2.70 | .7695 | .3249 | 5.799 | 2.368 | 1.395 | .8617 |
| 2.75 | .7619 | .3174 | 6.320 | 2.400 | 1.409 | .8828 |
| 2.80 | .7543 | .3102 | 6.895 | 2.432 | 1.422 | .9034 |
| 2.85 | .7467 | .3032 | 7.532 | 2.463 | 1.434 | .9235 |
| 2.90 | .7392 | .2965 | 8.237 | 2.493 | 1.447 | .9432 |
| 2.95 | .7316 | .2900 | 9.016 | 2.523 | 1.460 | .9624 |
| 3.00 | .7241 | .2837 | 9.880 | 2.553 | 1.472 | .9812 |
| 3.50 | .6512 | .2305 | 25.83 | 2.824 | 1.589 | 1.147 |
| 4.00 | .5833 | .1909 | 71.74 | 3.055 | 1.691 | 1.280 |
| 4.50 | .5217 | .1605 | 205.7 | 3.250 | 1.779 | 1.386 |
| 5.00 | .4667 | .1366 | 597.7 | 3.416 | 1.854 | 1.472 |
| 6.00 | .3750 | .1021 | 4949 | 3.674 | 1.973 | 1.601 |
| 7.00 | .3043 | .07881 | 37976 | 3.862 | 2.060 | 1.689 |
| 8.00 | .2500 | .06250 | $262(10)^3$ | 4.000 | 2.125 | 1.752 |
| 9.00 | .2079 | .05067 | $161(10)^4$ | 4.104 | 2.174 | 1.798 |
| 10.00 | .1750 | .04183 | $887(10)^4$ | 4.183 | 2.211 | 1.832 |
| ∞ | 0 | 0 | ∞ | 4.583 | 2.400 | 1.997 |

# Table 45 Fanno Line—One-Dimensional Compressible-Flow Functions for Adiabatic Flow at Constant Area with Friction

**For a perfect gas with constant specific heat and molecular weight**
**k = 1.2**

| M | $\dfrac{T}{T^*}$ | $\dfrac{p}{p^*}$ | $\dfrac{p_0}{p_0{}^*}$ | $\dfrac{V}{V^*}$ | $\dfrac{F}{F^*}$ | $4\dfrac{fL_{max}}{D}$ | M | $\dfrac{T}{T^*}$ | $\dfrac{p}{p^*}$ | $\dfrac{p_0}{p_0{}^*}$ | $\dfrac{V}{V^*}$ | $\dfrac{F}{F^*}$ | $4\dfrac{fL_{max}}{D}$ |
|---|---|---|---|---|---|---|---|---|---|---|---|---|---|
| 0 | 1.1000 | ∞ | ∞ | 0 | ∞ | ∞ | 1.75 | .8421 | .5244 | 1.471 | 1.606 | 1.114 | .3072 |
| 0.05 | 1.0997 | 20.974 | 11.857 | .05243 | 9.562 | 327.09 | 1.80 | .8308 | .5064 | 1.540 | 1.641 | 1.125 | .3316 |
| .10 | 1.0989 | 10.483 | 5.953 | .1048 | 4.822 | 78.36 | 1.85 | .8195 | .4894 | 1.615 | 1.675 | 1.136 | .3556 |
| .15 | 1.0975 | 6.984 | 3.996 | .1571 | 3.260 | 32.81 | 1.90 | .8082 | .4732 | 1.697 | 1.708 | 1.147 | .3791 |
| .20 | 1.0956 | 5.234 | 3.026 | .2093 | 2.493 | 17.13 | 1.95 | .7970 | .4578 | 1.787 | 1.741 | 1.158 | .4021 |
| .25 | 1.0932 | 4.182 | 2.451 | .2614 | 2.044 | 10.04 | 2.00 | .7857 | .4432 | 1.884 | 1.773 | 1.168 | .4247 |
| .30 | 1.0902 | 3.480 | 2.073 | .3133 | 1.753 | 6.298 | 2.05 | .7745 | .4293 | 1.989 | 1.804 | 1.179 | .4468 |
| .35 | 1.0867 | 2.978 | 1.809 | .3649 | 1.553 | 4.121 | 2.10 | .7634 | .4160 | 2.103 | 1.835 | 1.190 | .4684 |
| .40 | 1.0827 | 2.601 | 1.615 | .4162 | 1.409 | 2.768 | 2.15 | .7523 | .4034 | 2.226 | 1.865 | 1.201 | .4894 |
| .45 | 1.0782 | 2.307 | 1.469 | .4672 | 1.304 | 1.887 | 2.20 | .7413 | .3913 | 2.359 | 1.894 | 1.211 | .5099 |
| .50 | 1.0732 | 2.072 | 1.356 | .5179 | 1.224 | 1.294 | 2.25 | .7303 | .3798 | 2.504 | 1.923 | 1.221 | .5299 |
| .55 | 1.0677 | 1.879 | 1.268 | .5683 | 1.164 | .8855 | 2.30 | .7194 | .3688 | 2.660 | 1.951 | 1.232 | .5493 |
| .60 | 1.0618 | 1.717 | 1.199 | .6183 | 1.118 | .5999 | 2.35 | .7086 | .3582 | 2.829 | 1.978 | 1.242 | .5683 |
| .65 | 1.0554 | 1.581 | 1.144 | .6678 | 1.0826 | .3987 | 2.40 | .6980 | .3481 | 3.011 | 2.005 | 1.252 | .5868 |
| .70 | 1.0486 | 1.463 | 1.100 | .7168 | 1.0561 | .2570 | 2.45 | .6874 | .3384 | 3.208 | 2.031 | 1.262 | .6047 |
| .75 | 1.0414 | 1.361 | 1.0666 | .7654 | 1.0360 | .1579 | 2.50 | .6769 | .3291 | 3.420 | 2.057 | 1.272 | .6222 |
| .80 | 1.0338 | 1.271 | 1.0410 | .8134 | 1.0214 | .09016 | 2.55 | .6665 | .3202 | 3.650 | 2.082 | 1.281 | .6392 |
| .85 | 1.0259 | 1.192 | 1.0222 | .8609 | 1.0112 | .04554 | 2.60 | .6563 | .3116 | 3.898 | 2.106 | 1.291 | .6557 |
| .90 | 1.0176 | 1.121 | 1.0096 | .9078 | 1.0047 | .01829 | 2.65 | .6462 | .3033 | 4.166 | 2.130 | 1.300 | .6718 |
| .95 | 1.0089 | 1.0573 | 1.0023 | .9542 | 1.0011 | .00414 | 2.70 | .6362 | .2954 | 4.455 | 2.154 | 1.309 | .6874 |
| 1.00 | 1.0000 | 1.0000 | 1.0000 | 1.0000 | 1.0000 | 0 | 2.75 | .6263 | .2878 | 4.767 | 2.176 | 1.318 | .7026 |
| 1.05 | .9908 | .9480 | 1.0022 | 1.0451 | 1.0010 | .00347 | 2.80 | .6166 | .2804 | 5.103 | 2.199 | 1.327 | .7173 |
| 1.10 | .9813 | .9005 | 1.0087 | 1.0896 | 1.0037 | .01277 | 2.85 | .6070 | .2733 | 5.466 | 2.220 | 1.335 | .7316 |
| 1.15 | .9715 | .8571 | 1.0194 | 1.134 | 1.0079 | .02657 | 2.90 | .5975 | .2665 | 5.858 | 2.242 | 1.344 | .7456 |
| 1.20 | .9615 | .8172 | 1.0340 | 1.177 | 1.0134 | .04368 | 2.95 | .5882 | .2600 | 6.280 | 2.263 | 1.352 | .7592 |
| 1.25 | .9514 | .7803 | 1.0525 | 1.219 | 1.0197 | .06338 | 3.00 | .5789 | .2536 | 6.735 | 2.283 | 1.360 | .7724 |
| 1.30 | .9410 | .7462 | 1.0749 | 1.261 | 1.0270 | .08500 | 3.50 | .4944 | .2009 | 13.76 | 2.461 | 1.434 | .8857 |
| 1.35 | .9304 | .7145 | 1.101 | 1.302 | 1.0351 | .1080 | 4.00 | .4231 | .1626 | 28.35 | 2.602 | 1.493 | .9718 |
| 1.40 | .9197 | .6850 | 1.132 | 1.342 | 1.0437 | .1320 | 4.50 | .3636 | .1340 | 57.96 | 2.714 | 1.541 | 1.0380 |
| 1.45 | .9089 | .6575 | 1.166 | 1.382 | 1.0529 | .1567 | 5.00 | .3143 | .1121 | 116.34 | 2.803 | 1.580 | 1.0896 |
| 1.50 | .8980 | .6317 | 1.205 | 1.421 | 1.0625 | .1817 | 6.00 | .2391 | .08150 | 435.9 | 2.934 | 1.637 | 1.163 |
| 1.55 | .8869 | .6076 | 1.248 | 1.459 | 1.0724 | .2069 | 7.00 | .1864 | .06168 | 1469 | 3.023 | 1.677 | 1.212 |
| 1.60 | .8758 | .5849 | 1.296 | 1.497 | 1.0826 | .2323 | 8.00 | .1486 | .04819 | 4467 | 3.084 | 1.704 | 1.245 |
| 1.65 | .8646 | .5635 | 1.349 | 1.534 | 1.0930 | .2575 | 9.00 | .1209 | .03863 | 12383 | 3.129 | 1.724 | 1.268 |
| 1.70 | .8534 | .5434 | 1.407 | 1.570 | 1.1036 | .2825 | 10.00 | .1000 | .03162 | 31623 | 3.162 | 1.739 | 1.286 |
| | | | | | | | ∞ | 0 | 0 | ∞ | 3.317 | 1.809 | 1.365 |

# Table 46 Fanno Line—One-Dimensional Compressible-Flow Functions for Adiabatic Flow at Constant Area with Friction

For a perfect gas with constant specific heat and molecular weight
k = 1.3

| M | $\frac{T}{T^*}$ | $\frac{p}{p^*}$ | $\frac{p_0}{p_0{}^*}$ | $\frac{V}{V^*}$ | $\frac{F}{F^*}$ | $4\frac{fL_{max}}{D}$ | M | $\frac{T}{T^*}$ | $\frac{p}{p^*}$ | $\frac{p_0}{p_0{}^*}$ | $\frac{V}{V^*}$ | $\frac{F}{F^*}$ | $4\frac{fL_{max}}{D}$ |
|---|---|---|---|---|---|---|---|---|---|---|---|---|---|
| 0 | 1.150 | ∞ | ∞ | 0 | ∞ | ∞ | 1.75 | .7880 | .5073 | 1.424 | 1.554 | 1.0986 | .2613 |
| 0.05 | 1.149 | 21.444 | 11.721 | .05361 | 9.354 | 301.74 | 1.80 | .7739 | .4887 | 1.484 | 1.584 | 1.108 | .2814 |
| .10 | 1.148 | 10.716 | 5.885 | .1072 | 4.720 | 72.20 | 1.85 | .7599 | .4712 | 1.549 | 1.613 | 1.116 | .3010 |
| .15 | 1.146 | 7.137 | 3.952 | .1606 | 3.194 | 30.18 | 1.90 | .7460 | .4546 | 1.618 | 1.641 | 1.125 | .3202 |
| .20 | 1.143 | 5.346 | 2.994 | .2138 | 2.445 | 15.73 | 1.95 | .7323 | .4388 | 1.693 | 1.669 | 1.134 | .3390 |
| .25 | 1.139 | 4.270 | 2.426 | .2668 | 2.007 | 9.201 | 2.00 | .7188 | .4239 | 1.773 | 1.696 | 1.143 | .3573 |
| .30 | 1.134 | 3.551 | 2.054 | .3195 | 1.724 | 5.759 | 2.05 | .7054 | .4097 | 1.859 | 1.722 | 1.151 | .3751 |
| .35 | 1.129 | 3.036 | 1.793 | .3719 | 1.530 | 3.760 | 2.10 | .6922 | .3962 | 1.951 | 1.747 | 1.160 | .3924 |
| .40 | 1.123 | 2.649 | 1.602 | .4239 | 1.391 | 2.520 | 2.15 | .6791 | .3833 | 2.050 | 1.772 | 1.168 | .4092 |
| .45 | 1.116 | 2.348 | 1.459 | .4754 | 1.289 | 1.714 | 2.20 | .6662 | .3710 | 2.156 | 1.796 | 1.176 | .4255 |
| .50 | 1.1084 | 2.106 | 1.348 | .5264 | 1.213 | 1.172 | 2.25 | .6536 | .3593 | 2.268 | 1.819 | 1.184 | .4413 |
| .55 | 1.1001 | 1.907 | 1.261 | .5769 | 1.155 | .8004 | 2.30 | .6412 | .3482 | 2.388 | 1.842 | 1.192 | .4566 |
| .60 | 1.0911 | 1.741 | 1.193 | .6267 | 1.111 | .5409 | 2.35 | .6290 | .3375 | 2.517 | 1.864 | 1.200 | .4715 |
| .65 | 1.0815 | 1.600 | 1.140 | .6759 | 1.0777 | .3586 | 2.40 | .6170 | .3273 | 2.654 | 1.885 | 1.208 | .4860 |
| .70 | 1.0713 | 1.479 | 1.0972 | .7245 | 1.0524 | .2305 | 2.45 | .6051 | .3175 | 2.800 | 1.906 | 1.215 | .5000 |
| .75 | 1.0605 | 1.373 | 1.0644 | .7724 | 1.0336 | .14131 | 2.50 | .5935 | .3082 | 2.954 | 1.926 | 1.223 | .5136 |
| .80 | 1.0493 | 1.280 | 1.0395 | .8195 | 1.0199 | .08044 | 2.55 | .5822 | .2992 | 3.119 | 1.946 | 1.230 | .5267 |
| .85 | 1.0376 | 1.198 | 1.0214 | .8658 | 1.0104 | .04053 | 2.60 | .5711 | .2906 | 3.295 | 1.965 | 1.237 | .5394 |
| .90 | 1.0254 | 1.125 | 1.0092 | .9113 | 1.0043 | .01623 | 2.65 | .5601 | .2824 | 3.482 | 1.983 | 1.244 | .5517 |
| .95 | 1.0129 | 1.0594 | 1.0022 | .9561 | 1.0010 | .00367 | 2.70 | .5493 | .2745 | 3.681 | 2.001 | 1.250 | .5636 |
| 1.00 | 1.0000 | 1.0000 | 1.0000 | 1.0000 | 1.0000 | 0 | 2.75 | .5388 | .2669 | 3.892 | 2.019 | 1.257 | .5752 |
| 1.05 | .9868 | .9461 | 1.0021 | 1.0430 | 1.0009 | .00305 | 2.80 | .5285 | .2596 | 4.116 | 2.036 | 1.263 | .5864 |
| 1.10 | .9733 | .8969 | 1.0083 | 1.0852 | 1.0033 | .01122 | 2.85 | .5184 | .2526 | 4.354 | 2.052 | 1.270 | .5972 |
| 1.15 | .9596 | .8518 | 1.0183 | 1.1266 | 1.0071 | .02324 | 2.90 | .5085 | .2459 | 4.607 | 2.068 | 1.276 | .6077 |
| 1.20 | .9457 | .8104 | 1.0321 | 1.1670 | 1.0120 | .03820 | 2.95 | .4988 | .2394 | 4.875 | 2.084 | 1.282 | .6179 |
| 1.25 | .9316 | .7722 | 1.0495 | 1.206 | 1.0177 | .05524 | 3.00 | .4894 | .2332 | 5.160 | 2.099 | 1.288 | .6277 |
| 1.30 | .9174 | .7368 | 1.0704 | 1.245 | 1.0241 | .07388 | 3.50 | .4053 | .1819 | 9.110 | 2.228 | 1.338 | .7110 |
| 1.35 | .9031 | .7039 | 1.0948 | 1.283 | 1.0312 | .09365 | 4.00 | .3382 | .1454 | 15.94 | 2.326 | 1.378 | .7726 |
| 1.40 | .8887 | .6734 | 1.1227 | 1.320 | 1.0388 | .11417 | 4.50 | .2848 | .1186 | 27.39 | 2.402 | 1.409 | .8189 |
| 1.45 | .8743 | .6448 | 1.1543 | 1.356 | 1.0467 | .13513 | 5.00 | .2421 | .09841 | 45.95 | 2.460 | 1.433 | .8543 |
| 1.50 | .8598 | .6182 | 1.189 | 1.391 | 1.0549 | .1564 | 6.00 | .1797 | .07065 | 120.1 | 2.543 | 1.468 | .9037 |
| 1.55 | .8454 | .5932 | 1.228 | 1.425 | 1.0634 | .1777 | 7.00 | .1377 | .05302 | 285.3 | 2.598 | 1.491 | .9355 |
| 1.60 | .8309 | .5697 | 1.271 | 1.458 | 1.0721 | .1989 | 8.00 | .1085 | .04117 | 623.1 | 2.635 | 1.507 | .9570 |
| 1.65 | .8165 | .5477 | 1.318 | 1.491 | 1.0808 | .2200 | 9.00 | .08745 | .03286 | 1266 | 2.662 | 1.519 | .9722 |
| 1.70 | .8022 | .5269 | 1.369 | 1.523 | 1.0897 | .2408 | 10.00 | .07188 | .02681 | 2416 | 2.681 | 1.527 | .9832 |
| | | | | | | | ∞ | 0 | 0 | ∞ | 2.769 | 1.565 | 1.0326 |

# Table 47 Fanno Line—One-Dimensional Compressible-Flow Functions for Adiabatic Flow at Constant Area with Friction

**For a perfect gas with constant specific heat and molecular weight**
**k = 1.67**

| M | $\frac{T}{T^*}$ | $\frac{p}{p^*}$ | $\frac{p_0}{p_0{}^*}$ | $\frac{V}{V^*}$ | $\frac{F}{F^*}$ | $4\frac{fL_{max}}{D}$ | M | $\frac{T}{T^*}$ | $\frac{p}{p^*}$ | $\frac{p_0}{p_0{}^*}$ | $\frac{V}{V^*}$ | $\frac{F}{F^*}$ | $4\frac{fL_{max}}{D}$ |
|---|---|---|---|---|---|---|---|---|---|---|---|---|---|
| 0 | 1.335 | ∞ | ∞ | 0 | ∞ | ∞ | 1.75 | .6590 | .4639 | 1.312 | 1.421 | 1.0623 | .1580 |
| 0.05 | 1.334 | 23.099 | 11.265 | .05775 | 8.687 | 234.36 | 1.80 | .6402 | .4445 | 1.351 | 1.440 | 1.0673 | .1692 |
| .10 | 1.331 | 11.535 | 5.661 | .1154 | 4.392 | 55.83 | 1.85 | .6219 | .4263 | 1.392 | 1.459 | 1.0722 | .1800 |
| .15 | 1.325 | 7.674 | 3.805 | .1727 | 2.982 | 23.21 | 1.90 | .6042 | .4091 | 1.436 | 1.477 | 1.0770 | .1905 |
| .20 | 1.317 | 5.739 | 2.887 | .2296 | 2.293 | 12.11 | 1.95 | .5871 | .3929 | 1.482 | 1.494 | 1.0817 | .2007 |
| .25 | 1.308 | 4.574 | 2.344 | .2859 | 1.892 | 6.980 | 2.00 | .5705 | .3776 | 1.530 | 1.510 | 1.0863 | .2105 |
| .30 | 1.296 | 3.795 | 1.989 | .3415 | 1.635 | 4.337 | 2.05 | .5544 | .3632 | 1.580 | 1.526 | 1.0908 | .2199 |
| .35 | 1.282 | 3.235 | 1.741 | .3963 | 1.460 | 2.810 | 2.10 | .5388 | .3496 | 1.632 | 1.541 | 1.0952 | .2290 |
| .40 | 1.267 | 2.814 | 1.560 | .4502 | 1.336 | 1.868 | 2.15 | .5238 | .3367 | 1.687 | 1.556 | 1.0994 | .2377 |
| .45 | 1.250 | 2.485 | 1.424 | .5031 | 1.245 | 1.260 | 2.20 | .5093 | .3244 | 1.744 | 1.570 | 1.1035 | .2461 |
| .50 | 1.232 | 2.220 | 1.320 | .5549 | 1.178 | .8549 | 2.25 | .4952 | .3128 | 1.803 | 1.583 | 1.107 | .2542 |
| .55 | 1.212 | 2.002 | 1.239 | .6056 | 1.128 | .5787 | 2.30 | .4816 | .3017 | 1.865 | 1.596 | 1.111 | .2620 |
| .60 | 1.191 | 1.819 | 1.176 | .6548 | 1.0909 | .3877 | 2.35 | .4684 | .2912 | 1.929 | 1.608 | 1.115 | .2694 |
| .65 | 1.169 | 1.664 | 1.126 | .7029 | 1.0628 | .2548 | 2.40 | .4557 | .2813 | 1.995 | 1.620 | 1.119 | .2766 |
| .70 | 1.146 | 1.530 | 1.0874 | .7496 | 1.0418 | .1625 | 2.45 | .4434 | .2718 | 2.064 | 1.631 | 1.122 | .2835 |
| .75 | 1.1233 | 1.413 | 1.0576 | .7949 | 1.0265 | .09870 | 2.50 | .4315 | .2628 | 2.135 | 1.642 | 1.126 | .2901 |
| .80 | 1.0993 | 1.311 | 1.0351 | .8388 | 1.0155 | .05576 | 2.55 | .4200 | .2542 | 2.209 | 1.653 | 1.129 | .2965 |
| .85 | 1.0748 | 1.220 | 1.0189 | .8812 | 1.0080 | .02780 | 2.60 | .4089 | .2460 | 2.285 | 1.663 | 1.132 | .3026 |
| .90 | 1.0501 | 1.139 | 1.0081 | .9222 | 1.0033 | .01106 | 2.65 | .3982 | .2381 | 2.364 | 1.672 | 1.135 | .3085 |
| .95 | 1.0251 | 1.0657 | 1.0019 | .9618 | 1.0008 | .00248 | 2.70 | .3878 | .2306 | 2.445 | 1.682 | 1.138 | .3141 |
| 1.00 | 1.0000 | 1.0000 | 1.0000 | 1.0000 | 1.0000 | 0 | 2.75 | .3778 | .2235 | 2.529 | 1.691 | 1.141 | .3196 |
| 1.05 | .9749 | .9404 | 1.0018 | 1.0368 | 1.0006 | .00203 | 2.80 | .3681 | .2167 | 2.616 | 1.699 | 1.144 | .3248 |
| 1.10 | .9499 | .8860 | 1.0070 | 1.0721 | 1.0024 | .00740 | 2.85 | .3587 | .2102 | 2.705 | 1.707 | 1.146 | .3299 |
| 1.15 | .9251 | .8364 | 1.0154 | 1.1061 | 1.0051 | .01522 | 2.90 | .3497 | .2039 | 2.797 | 1.715 | 1.149 | .3348 |
| 1.20 | .9006 | .7908 | 1.0266 | 1.1388 | 1.0084 | .02481 | 2.95 | .3410 | .1979 | 2.892 | 1.723 | 1.152 | .3395 |
| 1.25 | .8763 | .7489 | 1.0406 | 1.170 | 1.0124 | .03564 | 3.00 | .3325 | .1922 | 2.990 | 1.730 | 1.154 | .3440 |
| 1.30 | .8524 | .7102 | 1.0573 | 1.200 | 1.0167 | .04733 | 3.50 | .2616 | .1461 | 4.134 | 1.790 | 1.174 | .3810 |
| 1.35 | .8289 | .6744 | 1.0765 | 1.229 | 1.0213 | .05957 | 4.00 | .2099 | .1145 | 5.608 | 1.833 | 1.189 | .4071 |
| 1.40 | .8059 | .6412 | 1.0981 | 1.257 | 1.0262 | .07212 | 4.50 | .1715 | .09203 | 7.456 | 1.864 | 1.200 | .4261 |
| 1.45 | .7833 | .6104 | 1.1220 | 1.284 | 1.0313 | .08481 | 5.00 | .1424 | .07547 | 9.721 | 1.887 | 1.208 | .4402 |
| 1.50 | .7612 | .5817 | 1.148 | 1.309 | 1.0364 | .09749 | 6.00 | .10222 | .05329 | 15.68 | 1.918 | 1.220 | .4594 |
| 1.55 | .7397 | .5549 | 1.176 | 1.333 | 1.0416 | .1101 | 7.00 | .07666 | .03955 | 23.85 | 1.938 | 1.227 | .4714 |
| 1.60 | .7187 | .5298 | 1.207 | 1.356 | 1.0468 | .1225 | 8.00 | .05949 | .03049 | 34.58 | 1.951 | 1.232 | .4793 |
| 1.65 | .6982 | .5064 | 1.240 | 1.378 | 1.0520 | .1346 | 9.00 | .04745 | .02420 | 48.24 | 1.960 | 1.235 | .4849 |
| 1.70 | .6783 | .4845 | 1.275 | 1.400 | 1.0572 | .1465 | 10.00 | .03870 | .01967 | 65.18 | 1.967 | 1.238 | .4889 |
| | | | | | | | ∞ | 0 | 0 | ∞ | 1.996 | 1.249 | .5064 |

# Table 48   One-Dimensional Normal-Shock Functions

For a perfect gas with constant specific heat and molecular weight
k = 1.4

| $M_x$ | $M_y$ | $\frac{p_y}{p_x}$ | $\frac{\rho_y}{\rho_x}$ | $\frac{T_y}{T_x}$ | $\frac{p_{0y}}{p_{0x}}$ | $\frac{p_{0y}}{p_x}$ | $M_x$ | $M_y$ | $\frac{p_y}{p_x}$ | $\frac{\rho_y}{\rho_x}$ | $\frac{T_y}{T_x}$ | $\frac{p_{0y}}{p_{0x}}$ | $\frac{p_{0y}}{p_x}$ |
|---|---|---|---|---|---|---|---|---|---|---|---|---|---|
| 1.00 | 1.00000 | 1.00000 | 1.00000 | 1.00000 | 1.00000 | 1.8929 | 1.40 | .73971 | 2.1200 | 1.6896 | 1.2547 | .95819 | 3.0493 |
| 1.01 | .99013 | 1.02345 | 1.01669 | 1.00665 | .99999 | 1.9152 | 1.41 | .73554 | 2.1528 | 1.7070 | 1.2612 | .95566 | 3.0844 |
| 1.02 | .98052 | 1.04713 | 1.03344 | 1.01325 | .99998 | 1.9379 | 1.42 | .73144 | 2.1858 | 1.7243 | 1.2676 | .95306 | 3.1198 |
| 1.03 | .97115 | 1.07105 | 1.05024 | 1.01981 | .99997 | 1.9610 | 1.43 | .72741 | 2.2190 | 1.7416 | 1.2742 | .95039 | 3.1555 |
| 1.04 | .96202 | 1.09520 | 1.06709 | 1.02634 | .99994 | 1.9845 | 1.44 | .72345 | 2.2525 | 1.7589 | 1.2807 | .94765 | 3.1915 |
| 1.05 | .95312 | 1.1196 | 1.08398 | 1.03284 | .99987 | 2.0083 | 1.45 | .71956 | 2.2862 | 1.7761 | 1.2872 | .94483 | 3.2278 |
| 1.06 | .94444 | 1.1442 | 1.10092 | 1.03931 | .99976 | 2.0325 | 1.46 | .71574 | 2.3202 | 1.7934 | 1.2938 | .94196 | 3.2643 |
| 1.07 | .93598 | 1.1690 | 1.11790 | 1.04575 | .99962 | 2.0570 | 1.47 | .71198 | 2.3544 | 1.8106 | 1.3004 | .93901 | 3.3011 |
| 1.08 | .92772 | 1.1941 | 1.13492 | 1.05217 | .99944 | 2.0819 | 1.48 | .70829 | 2.3888 | 1.8278 | 1.3070 | .93600 | 3.3382 |
| 1.09 | .91965 | 1.2194 | 1.15199 | 1.05856 | .99921 | 2.1072 | 1.49 | .70466 | 2.4234 | 1.8449 | 1.3136 | .93292 | 3.3756 |
| 1.10 | .91177 | 1.2450 | 1.1691 | 1.06494 | .99892 | 2.1328 | 1.50 | .70109 | 2.4583 | 1.8621 | 1.3202 | .92978 | 3.4133 |
| 1.11 | .90408 | 1.2708 | 1.1862 | 1.07130 | .99858 | 2.1588 | 1.51 | .69758 | 2.4934 | 1.8792 | 1.3269 | .92658 | 3.4512 |
| 1.12 | .89656 | 1.2968 | 1.2034 | 1.07764 | .99820 | 2.1851 | 1.52 | .69413 | 2.5288 | 1.8962 | 1.3336 | .92331 | 3.4894 |
| 1.13 | .88922 | 1.3230 | 1.2206 | 1.08396 | .99776 | 2.2118 | 1.53 | .69073 | 2.5644 | 1.9133 | 1.3403 | .91999 | 3.5279 |
| 1.14 | .88204 | 1.3495 | 1.2378 | 1.09027 | .99726 | 2.2388 | 1.54 | .68739 | 2.6003 | 1.9303 | 1.3470 | .91662 | 3.5667 |
| 1.15 | .87502 | 1.3762 | 1.2550 | 1.09657 | .99669 | 2.2661 | 1.55 | .68410 | 2.6363 | 1.9473 | 1.3538 | .91319 | 3.6058 |
| 1.16 | .86816 | 1.4032 | 1.2723 | 1.10287 | .99605 | 2.2937 | 1.56 | .68086 | 2.6725 | 1.9643 | 1.3606 | .90970 | 3.6451 |
| 1.17 | .86145 | 1.4304 | 1.2896 | 1.10916 | .99534 | 2.3217 | 1.57 | .67768 | 2.7090 | 1.9812 | 1.3674 | .90615 | 3.6847 |
| 1.18 | .85488 | 1.4578 | 1.3069 | 1.11544 | .99455 | 2.3499 | 1.58 | .67455 | 2.7458 | 1.9981 | 1.3742 | .90255 | 3.7245 |
| 1.19 | .84846 | 1.4854 | 1.3243 | 1.12172 | .99371 | 2.3786 | 1.59 | .67147 | 2.7828 | 2.0149 | 1.3811 | .89889 | 3.7645 |
| 1.20 | .84217 | 1.5133 | 1.3416 | 1.1280 | .99280 | 2.4075 | 1.60 | .66844 | 2.8201 | 2.0317 | 1.3880 | .89520 | 3.8049 |
| 1.21 | .83601 | 1.5414 | 1.3590 | 1.1343 | .99180 | 2.4367 | 1.61 | .66545 | 2.8575 | 2.0485 | 1.3949 | .89144 | 3.8456 |
| 1.22 | .82998 | 1.5698 | 1.3764 | 1.1405 | .99073 | 2.4662 | 1.62 | .66251 | 2.8951 | 2.0652 | 1.4018 | .88764 | 3.8866 |
| 1.23 | .82408 | 1.5984 | 1.3938 | 1.1468 | .98957 | 2.4961 | 1.63 | .65962 | 2.9330 | 2.0820 | 1.4088 | .88380 | 3.9278 |
| 1.24 | .81830 | 1.6272 | 1.4112 | 1.1531 | .98835 | 2.5263 | 1.64 | .65677 | 2.9712 | 2.0986 | 1.4158 | .87992 | 3.9693 |
| 1.25 | .81264 | 1.6562 | 1.4286 | 1.1594 | .98706 | 2.5568 | 1.65 | .65396 | 3.0096 | 2.1152 | 1.4228 | .87598 | 4.0111 |
| 1.26 | .80709 | 1.6855 | 1.4460 | 1.1657 | .98568 | 2.5876 | 1.66 | .65119 | 3.0482 | 2.1318 | 1.4298 | .87201 | 4.0531 |
| 1.27 | .80165 | 1.7150 | 1.4634 | 1.1720 | .98422 | 2.6187 | 1.67 | .64847 | 3.0870 | 2.1484 | 1.4369 | .86800 | 4.0954 |
| 1.28 | .79631 | 1.7448 | 1.4808 | 1.1782 | .98268 | 2.6500 | 1.68 | .64579 | 3.1261 | 2.1649 | 1.4440 | .86396 | 4.1379 |
| 1.29 | .79108 | 1.7748 | 1.4983 | 1.1846 | .98106 | 2.6816 | 1.69 | .64315 | 3.1654 | 2.1813 | 1.4512 | .85987 | 4.1807 |
| 1.30 | .78596 | 1.8050 | 1.5157 | 1.1909 | .97935 | 2.7135 | 1.70 | .64055 | 3.2050 | 2.1977 | 1.4583 | .85573 | 4.2238 |
| 1.31 | .78093 | 1.8354 | 1.5331 | 1.1972 | .97758 | 2.7457 | 1.71 | .63798 | 3.2448 | 2.2141 | 1.4655 | .85155 | 4.2672 |
| 1.32 | .77600 | 1.8661 | 1.5505 | 1.2035 | .97574 | 2.7783 | 1.72 | .63545 | 3.2848 | 2.2304 | 1.4727 | .84735 | 4.3108 |
| 1.33 | .77116 | 1.8970 | 1.5680 | 1.2099 | .97382 | 2.8112 | 1.73 | .63296 | 3.3250 | 2.2467 | 1.4800 | .84312 | 4.3547 |
| 1.34 | .76641 | 1.9282 | 1.5854 | 1.2162 | .97181 | 2.8444 | 1.74 | .63051 | 3.3655 | 2.2629 | 1.4873 | .83886 | 4.3989 |
| 1.35 | .76175 | 1.9596 | 1.6028 | 1.2226 | .96972 | 2.8778 | 1.75 | .62809 | 3.4062 | 2.2791 | 1.4946 | .83456 | 4.4433 |
| 1.36 | .75718 | 1.9912 | 1.6202 | 1.2290 | .96756 | 2.9115 | 1.76 | .62570 | 3.4472 | 2.2952 | 1.5019 | .83024 | 4.4880 |
| 1.37 | .75269 | 2.0230 | 1.6376 | 1.2354 | .96534 | 2.9455 | 1.77 | .62335 | 3.4884 | 2.3113 | 1.5093 | .82589 | 4.5330 |
| 1.38 | .74828 | 2.0551 | 1.6550 | 1.2418 | .96304 | 2.9798 | 1.78 | .62104 | 3.5298 | 2.3273 | 1.5167 | .82152 | 4.5783 |
| 1.39 | .74396 | 2.0874 | 1.6723 | 1.2482 | .96065 | 3.0144 | 1.79 | .61875 | 3.5714 | 2.3433 | 1.5241 | .81711 | 4.6238 |

See page 113 for definitions of symbols.

# Table 48 One-Dimensional Normal-Shock Functions

**For a perfect gas with constant specific heat and molecular weight**
**k = 1.4**

| $M_x$ | $M_y$ | $\dfrac{p_y}{p_x}$ | $\dfrac{\rho_y}{\rho_x}$ | $\dfrac{T_y}{T_x}$ | $\dfrac{p_{0y}}{p_{0x}}$ | $\dfrac{p_{0y}}{p_x}$ | $M_x$ | $M_y$ | $\dfrac{p_y}{p_x}$ | $\dfrac{\rho_y}{\rho_x}$ | $\dfrac{T_y}{T_x}$ | $\dfrac{p_{0y}}{p_{0x}}$ | $\dfrac{p_{0y}}{p_x}$ |
|---|---|---|---|---|---|---|---|---|---|---|---|---|---|
| 1.80 | .61650 | 3.6133 | 2.3592 | 1.5316 | .81268 | 4.6695 | 2.20 | .54706 | 5.4800 | 2.9512 | 1.8569 | .62812 | 6.7163 |
| 1.81 | .61428 | 3.6554 | 2.3751 | 1.5391 | .80823 | 4.7155 | 2.21 | .54572 | 5.5314 | 2.9648 | 1.8657 | .62358 | 6.7730 |
| 1.82 | .61209 | 3.6978 | 2.3909 | 1.5466 | .80376 | 4.7618 | 2.22 | .54440 | 5.5831 | 2.9783 | 1.8746 | .61905 | 6.8299 |
| 1.83 | .60993 | 3.7404 | 2.4067 | 1.5542 | .79926 | 4.8083 | 2.23 | .54310 | 5.6350 | 2.9918 | 1.8835 | .61453 | 6.8869 |
| 1.84 | .60780 | 3.7832 | 2.4224 | 1.5617 | .79474 | 4.8551 | 2.24 | .54182 | 5.6872 | 3.0052 | 1.8924 | .61002 | 6.9442 |
| 1.85 | .60570 | 3.8262 | 2.4381 | 1.5694 | .79021 | 4.9022 | 2.25 | .54055 | 5.7396 | 3.0186 | 1.9014 | .60554 | 7.0018 |
| 1.86 | .60363 | 3.8695 | 2.4537 | 1.5770 | .78567 | 4.9498 | 2.26 | .53929 | 5.7922 | 3.0319 | 1.9104 | .60106 | 7.0597 |
| 1.87 | .60159 | 3.9130 | 2.4693 | 1.5847 | .78112 | 4.9974 | 2.27 | .53805 | 5.8451 | 3.0452 | 1.9194 | .59659 | 7.1178 |
| 1.88 | .59957 | 3.9568 | 2.4848 | 1.5924 | .77656 | 5.0453 | 2.28 | .53683 | 5.8982 | 3.0584 | 1.9285 | .59214 | 7.1762 |
| 1.89 | .59758 | 4.0008 | 2.5003 | 1.6001 | .77197 | 5.0934 | 2.29 | .53561 | 5.9515 | 3.0715 | 1.9376 | .58772 | 7.2348 |
| 1.90 | .59562 | 4.0450 | 2.5157 | 1.6079 | .76735 | 5.1417 | 2.30 | .53441 | 6.0050 | 3.0846 | 1.9468 | .58331 | 7.2937 |
| 1.91 | .59368 | 4.0894 | 2.5310 | 1.6157 | .76273 | 5.1904 | 2.31 | .53322 | 6.0588 | 3.0976 | 1.9560 | .57891 | 7.3529 |
| 1.92 | .59177 | 4.1341 | 2.5463 | 1.6236 | .75812 | 5.2394 | 2.32 | .53205 | 6.1128 | 3.1105 | 1.9652 | .57452 | 7.4123 |
| 1.93 | .58988 | 4.1790 | 2.5615 | 1.6314 | .75347 | 5.2886 | 2.33 | .53089 | 6.1670 | 3.1234 | 1.9745 | .57015 | 7.4720 |
| 1.94 | .58802 | 4.2242 | 2.5767 | 1.6394 | .74883 | 5.3381 | 2.34 | .52974 | 6.2215 | 3.1362 | 1.9838 | .56580 | 7.5319 |
| 1.95 | .58618 | 4.2696 | 2.5919 | 1.6473 | .74418 | 5.3878 | 2.35 | .52861 | 6.2762 | 3.1490 | 1.9931 | .56148 | 7.5920 |
| 1.96 | .58437 | 4.3152 | 2.6070 | 1.6553 | .73954 | 5.4378 | 2.36 | .52749 | 6.3312 | 3.1617 | 2.0025 | .55717 | 7.6524 |
| 1.97 | .58258 | 4.3610 | 2.6220 | 1.6633 | .73487 | 5.4880 | 2.37 | .52638 | 6.3864 | 3.1743 | 2.0119 | .55288 | 7.7131 |
| 1.98 | .58081 | 4.4071 | 2.6369 | 1.6713 | .73021 | 5.5385 | 2.38 | .52528 | 6.4418 | 3.1869 | 2.0213 | .54862 | 7.7741 |
| 1.99 | .57907 | 4.4534 | 2.6518 | 1.6794 | .72554 | 5.5894 | 2.39 | .52419 | 6.4974 | 3.1994 | 2.0308 | .54438 | 7.8354 |
| 2.00 | .57735 | 4.5000 | 2.6666 | 1.6875 | .72088 | 5.6405 | 2.40 | .52312 | 6.5533 | 3.2119 | 2.0403 | .54015 | 7.8969 |
| 2.01 | .57565 | 4.5468 | 2.6814 | 1.6956 | .71619 | 5.6918 | 2.41 | .52206 | 6.6094 | 3.2243 | 2.0499 | .53594 | 7.9587 |
| 2.02 | .57397 | 4.5938 | 2.6962 | 1.7038 | .71152 | 5.7434 | 2.42 | .52100 | 6.6658 | 3.2366 | 2.0595 | .53175 | 8.0207 |
| 2.03 | .57231 | 4.6411 | 2.7109 | 1.7120 | .70686 | 5.7952 | 2.43 | .51996 | 6.7224 | 3.2489 | 2.0691 | .52758 | 8.0830 |
| 2.04 | .57068 | 4.6886 | 2.7255 | 1.7203 | .70218 | 5.8473 | 2.44 | .51894 | 6.7792 | 3.2611 | 2.0788 | .52344 | 8.1455 |
| 2.05 | .56907 | 4.7363 | 2.7400 | 1.7286 | .69752 | 5.8997 | 2.45 | .51792 | 6.8362 | 3.2733 | 2.0885 | .51932 | 8.2083 |
| 2.06 | .56747 | 4.7842 | 2.7545 | 1.7369 | .69284 | 5.9523 | 2.46 | .51691 | 6.8935 | 3.2854 | 2.0982 | .51521 | 8.2714 |
| 2.07 | .56589 | 4.8324 | 2.7690 | 1.7452 | .68817 | 6.0052 | 2.47 | .51592 | 6.9510 | 3.2975 | 2.1080 | .51112 | 8.3347 |
| 2.08 | .56433 | 4.8808 | 2.7834 | 1.7536 | .68351 | 6.0584 | 2.48 | .51493 | 7.0088 | 3.3095 | 2.1178 | .50706 | 8.3983 |
| 2.09 | .56280 | 4.9295 | 2.7977 | 1.7620 | .67886 | 6.1118 | 2.49 | .51395 | 7.0668 | 3.3214 | 2.1276 | .50303 | 8.4622 |
| 2.10 | .56128 | 4.9784 | 2.8119 | 1.7704 | .67422 | 6.1655 | 2.50 | .51299 | 7.1250 | 3.3333 | 2.1375 | .49902 | 8.5262 |
| 2.11 | .55978 | 5.0275 | 2.8216 | 1.7789 | .66957 | 6.2194 | 2.51 | .51204 | 7.1834 | 3.3451 | 2.1474 | .49502 | 8.5904 |
| 2.12 | .55830 | 5.0768 | 2.8402 | 1.7874 | .66492 | 6.2736 | 2.52 | .51109 | 7.2421 | 3.3569 | 2.1574 | .49104 | 8.6549 |
| 2.13 | .55683 | 5.1264 | 2.8543 | 1.7960 | .66029 | 6.3280 | 2.53 | .51015 | 7.3010 | 3.3686 | 2.1674 | .48709 | 8.7198 |
| 2.14 | .55538 | 5.1762 | 2.8683 | 1.8046 | .65567 | 6.3827 | 2.54 | .50923 | 7.3602 | 3.3802 | 2.1774 | .48317 | 8.7850 |
| 2.15 | .55395 | 5.2262 | 2.8823 | 1.8132 | .65105 | 6.4377 | 2.55 | .50831 | 7.4196 | 3.3918 | 2.1875 | .47927 | 8.8505 |
| 2.16 | .55254 | 5.2765 | 2.8962 | 1.8219 | .64644 | 6.4929 | 2.56 | .50740 | 7.4792 | 3.4034 | 2.1976 | .47540 | 8.9162 |
| 2.17 | .55114 | 5.3270 | 2.9100 | 1.8306 | .64185 | 6.5484 | 2.57 | .50651 | 7.5391 | 3.4149 | 2.2077 | .47155 | 8.9821 |
| 2.18 | .54976 | 5.3778 | 2.9238 | 1.8393 | .63728 | 6.6042 | 2.58 | .50562 | 7.5992 | 3.4263 | 2.2179 | .46772 | 9.0482 |
| 2.19 | .54841 | 5.4288 | 2.9376 | 1.8481 | .63270 | 6.6602 | 2.59 | .50474 | 7.6595 | 3.4376 | 2.2281 | .46391 | 9.1146 |

## Table 48  One-Dimensional Normal-Shock Functions

For a perfect gas with constant specific heat and molecular weight
k = 1.4

| $M_x$ | $M_y$ | $\dfrac{p_y}{p_x}$ | $\dfrac{\rho_y}{\rho_x}$ | $\dfrac{T_y}{T_x}$ | $\dfrac{p_{0y}}{p_{0x}}$ | $\dfrac{p_{0y}}{p_x}$ |
|---|---|---|---|---|---|---|
| 2.60 | .50387 | 7.7200 | 3.4489 | 2.2383 | .46012 | 9.1813 |
| 2.61 | .50301 | 7.7808 | 3.4602 | 2.2486 | .45636 | 9.2481 |
| 2.62 | .50216 | 7.8418 | 3.4714 | 2.2589 | .45262 | 9.3154 |
| 2.63 | .50132 | 7.9030 | 3.4825 | 2.2693 | .44891 | 9.3829 |
| 2.64 | .50048 | 7.9645 | 3.4936 | 2.2797 | .44522 | 9.4507 |
| 2.65 | .49965 | 8.0262 | 3.5047 | 2.2901 | .44155 | 9.5187 |
| 2.66 | .49883 | 8.0882 | 3.5157 | 2.3006 | .43791 | 9.5869 |
| 2.67 | .49802 | 8.1504 | 3.5266 | 2.3111 | .43429 | 9.6553 |
| 2.68 | .49722 | 8.2128 | 3.5374 | 2.3217 | .43070 | 9.7241 |
| 2.69 | .49642 | 8.2754 | 3.5482 | 2.3323 | .42713 | 9.7932 |
| 2.70 | .49563 | 8.3383 | 3.5590 | 2.3429 | .42359 | 9.8625 |
| 2.71 | .49485 | 8.4014 | 3.5697 | 2.3536 | .42007 | 9.9320 |
| 2.72 | .49408 | 8.4648 | 3.5803 | 2.3643 | .41657 | 10.002 |
| 2.73 | .49332 | 8.5284 | 3.5909 | 2.3750 | .41310 | 10.072 |
| 2.74 | .49256 | 8.5922 | 3.6014 | 2.3858 | .40965 | 10.142 |
| 2.75 | .49181 | 8.6562 | 3.6119 | 2.3966 | .40622 | 10.212 |
| 2.76 | .49107 | 8.7205 | 3.6224 | 2.4074 | .40282 | 10.283 |
| 2.77 | .49033 | 8.7850 | 3.6328 | 2.4183 | .39945 | 10.354 |
| 2.78 | .48960 | 8.8497 | 3.6431 | 2.4292 | .39610 | 10.426 |
| 2.79 | .48888 | 8.9147 | 3.6533 | 2.4402 | .39276 | 10.498 |
| 2.80 | .48817 | 8.9800 | 3.6635 | 2.4512 | .38946 | 10.569 |
| 2.81 | .48746 | 9.0454 | 3.6737 | 2.4622 | .38618 | 10.641 |
| 2.82 | .48676 | 9.1111 | 3.6838 | 2.4733 | .38293 | 10.714 |
| 2.83 | .48607 | 9.1770 | 3.6939 | 2.4844 | .37970 | 10.787 |
| 2.84 | .48538 | 9.2432 | 3.7039 | 2.4955 | .37649 | 10.860 |
| 2.85 | .48470 | 9.3096 | 3.7139 | 2.5067 | .37330 | 10.933 |
| 2.86 | .48402 | 9.3762 | 3.7238 | 2.5179 | .37013 | 11.006 |
| 2.87 | .48334 | 9.4431 | 3.7336 | 2.5292 | .36700 | 11.080 |
| 2.88 | .48268 | 9.5102 | 3.7434 | 2.5405 | .36389 | 11.154 |
| 2.89 | .48203 | 9.5775 | 3.7532 | 2.5518 | .36080 | 11.228 |
| 2.90 | .48138 | 9.6450 | 3.7629 | 2.5632 | .35773 | 11.302 |
| 2.91 | .48074 | 9.7127 | 3.7725 | 2.5746 | .35469 | 11.377 |
| 2.92 | .48010 | 9.7808 | 3.7821 | 2.5860 | .35167 | 11.452 |
| 2.93 | .47946 | 9.8491 | 3.7917 | 2.5975 | .34867 | 11.527 |
| 2.94 | .47883 | 9.9176 | 3.8012 | 2.6090 | .34570 | 11.603 |
| 2.95 | .47821 | 9.986 | 3.8106 | 2.6206 | .34275 | 11.679 |
| 2.96 | .47760 | 10.055 | 3.8200 | 2.6322 | .33982 | 11.755 |
| 2.97 | .47699 | 10.124 | 3.8294 | 2.6438 | .33692 | 11.831 |
| 2.98 | .47638 | 10.194 | 3.8387 | 2.6555 | .33404 | 11.907 |
| 2.99 | .47578 | 10.263 | 3.8479 | 2.6672 | .33118 | 11.984 |
| 3.00 | .47519 | 10.333 | 3.8571 | 2.6790 | .32834 | 12.061 |
| 3.50 | .45115 | 14.125 | 4.2608 | 3.3150 | .21295 | 16.242 |
| 4.00 | .43496 | 18.500 | 4.5714 | 4.0469 | .13876 | 21.068 |
| 4.50 | .42355 | 23.458 | 4.8119 | 4.8751 | .09170 | 26.539 |
| 5.00 | .41523 | 29.000 | 5.0000 | 5.8000 | .06172 | 32.654 |
| 6.00 | .40416 | 41.833 | 5.2683 | 7.941 | .02965 | 46.815 |
| 7.00 | .39736 | 57.000 | 5.4444 | 10.469 | .01535 | 63.552 |
| 8.00 | .39289 | 74.500 | 5.5652 | 13.387 | .00849 | 82.865 |
| 9.00 | .38980 | 94.333 | 5.6512 | 16.693 | .00496 | 104.753 |
| 10.00 | .38757 | 116.50 | 5.7143 | 20.388 | .00304 | 129.217 |
| ∞ | .37796 | ∞ | 6.000 | ∞ | 0 | ∞ |

(152)

# Table 49  One-Dimensional Normal-Shock Functions

For a perfect gas with constant specific heat and molecular weight
k = 1.0

| $M_x$ | $M_y$ | $\dfrac{p_y}{p_x} = \dfrac{\rho_y}{\rho_x}$ | $\dfrac{T_y}{T_x}$ | $\dfrac{p_{0y}}{p_{0x}}$ | $\dfrac{p_{0y}}{p_x}$ |
|---|---|---|---|---|---|
| 1.00 | 1.0000 | 1.000 | 1.000 | 1.0000 | 1.649 |
| 1.05 | .9524 | 1.103 | | .9998 | 1.735 |
| 1.10 | .9091 | 1.210 | | .9988 | 1.829 |
| 1.15 | .8696 | 1.322 | | .9964 | 1.930 |
| 1.20 | .8333 | 1.440 | | .9919 | 2.038 |
| 1.25 | .8000 | 1.563 | | .9851 | 2.152 |
| 1.30 | .7692 | 1.690 | | .9759 | 2.272 |
| 1.35 | .7407 | 1.822 | | .9640 | 2.398 |
| 1.40 | .7143 | 1.960 | | .9494 | 2.530 |
| 1.45 | .6897 | 2.103 | | .9321 | 2.667 |
| 1.50 | .6667 | 2.250 | | .9122 | 2.810 |
| 1.55 | .6452 | 2.402 | | .8899 | 2.958 |
| 1.60 | .6250 | 2.560 | | .8653 | 3.112 |
| 1.65 | .6061 | 2.723 | | .8386 | 3.271 |
| 1.70 | .5882 | 2.890 | | .8100 | 3.436 |
| 1.75 | .5714 | 3.062 | | .7798 | 3.606 |
| 1.80 | .5556 | 3.240 | | .7482 | 3.781 |
| 1.85 | .5406 | 3.423 | | .7155 | 3.961 |
| 1.90 | .5263 | 3.610 | | .6819 | 4.146 |
| 1.95 | .5128 | 3.802 | | .6478 | 4.337 |
| 2.00 | .5000 | 4.000 | | .6134 | 4.532 |
| 2.05 | .4878 | 4.203 | | .5789 | 4.733 |
| 2.10 | .4762 | 4.410 | | .5446 | 4.940 |
| 2.15 | .4651 | 4.622 | | .5106 | 5.151 |
| 2.20 | .4545 | 4.840 | | .4772 | 5.367 |
| 2.25 | .4444 | 5.063 | | .4446 | 5.588 |
| 2.30 | .4347 | 5.290 | | .4129 | 5.814 |
| 2.35 | .4255 | 5.522 | | .3822 | 6.045 |
| 2.40 | .4167 | 5.760 | | .3527 | 6.282 |
| 2.45 | .4082 | 6.003 | | .3244 | 6.524 |
| 2.50 | .4000 | 6.250 | | .2975 | 6.771 |
| 2.55 | .3921 | 6.502 | | .2720 | 7.022 |
| 2.60 | .3846 | 6.760 | | .2479 | 7.279 |
| 2.65 | .3774 | 7.023 | | .2252 | 7.541 |
| 2.70 | .3704 | 7.290 | | .2040 | 7.807 |
| 2.75 | .3636 | 7.562 | | .1842 | 8.079 |
| 2.80 | .3571 | 7.840 | | .1658 | 8.356 |
| 2.85 | .3508 | 8.123 | | .1488 | 8.638 |
| 2.90 | .3448 | 8.410 | | .1332 | 8.925 |
| 2.95 | .3390 | 8.702 | | .1188 | 9.217 |
| 3.00 | .3333 | 9.000 | | $1055(10)^{-4}$ | 9.514 |
| 3.50 | .2857 | 12.25 | | $2791(10)^{-5}$ | 12.76 |
| 4.00 | .2500 | 16.00 | | $554(10)^{-5}$ | 16.51 |
| 4.50 | .2222 | 20.25 | | $832(10)^{-6}$ | 20.76 |
| 5.00 | .2000 | 25.00 | | $951(10)^{-7}$ | 25.51 |
| 6.00 | .1667 | 36.00 | | $556(10)^{-9}$ | 36.50 |
| 7.00 | .1429 | 49.00 | | $113(10)^{-11}$ | 49.50 |
| 8.00 | .1250 | 64.00 | | $817(10)^{-15}$ | 64.50 |
| 9.00 | .1111 | 81.00 | | $210(10)^{-18}$ | 81.50 |
| 10.00 | .1000 | 100.00 | | $194(10)^{-22}$ | 100.50 |
| $\infty$ | 0 | $\infty$ | 1.000 | 0 | $\infty$ |

(153)

# Table 50　One-Dimensional Normal-Shock Functions

For a perfect gas with constant specific heat and molecular weight
k = 1.1

| $M_x$ | $M_y$ | $\dfrac{p_y}{p_x}$ | $\dfrac{\rho_y}{\rho_x}$ | $\dfrac{T_y}{T_x}$ | $\dfrac{p_{0y}}{p_{0x}}$ | $\dfrac{p_{0y}}{p_x}$ |
|---|---|---|---|---|---|---|
| 1.00 | 1.0000 | 1.0000 | 1.000 | 1.0000 | 1.0000 | 1.710 |
| 1.05 | .9526 | 1.107 | 1.097 | 1.0093 | .9998 | 1.804 |
| 1.10 | .9099 | 1.220 | 1.198 | 1.0183 | .9988 | 1.906 |
| 1.15 | .8712 | 1.338 | 1.303 | 1.0271 | .9965 | 2.015 |
| 1.20 | .8360 | 1.461 | 1.410 | 1.0358 | .9921 | 2.132 |
| 1.25 | .8038 | 1.589 | 1.521 | 1.0444 | .9856 | 2.255 |
| 1.30 | .7743 | 1.723 | 1.636 | 1.0529 | .9769 | 2.384 |
| 1.35 | .7471 | 1.862 | 1.754 | 1.0615 | .9657 | 2.520 |
| 1.40 | .7221 | 2.006 | 1.875 | 1.0701 | .9519 | 2.662 |
| 1.45 | .6989 | 2.155 | 1.998 | 1.0788 | .9358 | 2.810 |
| 1.50 | .6773 | 2.309 | 2.124 | 1.0876 | .9174 | 2.964 |
| 1.55 | .6573 | 2.469 | 2.252 | 1.0965 | .8969 | 3.124 |
| 1.60 | .6386 | 2.634 | 2.383 | 1.1055 | .8744 | 3.289 |
| 1.65 | .6211 | 2.804 | 2.516 | 1.1146 | .8501 | 3.460 |
| 1.70 | .6048 | 2.980 | 2.651 | 1.1239 | .8242 | 3.637 |
| 1.75 | .5895 | 3.161 | 2.789 | 1.133 | .7970 | 3.820 |
| 1.80 | .5751 | 3.347 | 2.928 | 1.143 | .7686 | 4.008 |
| 1.85 | .5615 | 3.538 | 3.069 | 1.153 | .7393 | 4.202 |
| 1.90 | .5487 | 3.734 | 3.211 | 1.163 | .7093 | 4.401 |
| 1.95 | .5366 | 3.936 | 3.355 | 1.173 | .6789 | 4.606 |
| 2.00 | .5252 | 4.143 | 3.500 | 1.184 | .6483 | 4.817 |
| 2.05 | .5144 | 4.355 | 3.646 | 1.194 | .6175 | 5.033 |
| 2.10 | .5042 | 4.572 | 3.793 | 1.205 | .5869 | 5.254 |
| 2.15 | .4945 | 4.795 | 3.942 | 1.216 | .5566 | 5.481 |
| 2.20 | .4853 | 5.023 | 4.092 | 1.228 | .5267 | 5.713 |
| 2.25 | .4765 | 5.256 | 4.242 | 1.239 | .4974 | 5.951 |
| 2.30 | .4682 | 5.494 | 4.393 | 1.251 | .4687 | 6.194 |
| 2.35 | .4603 | 5.738 | 4.544 | 1.263 | .4408 | 6.443 |
| 2.40 | .4527 | 5.987 | 4.696 | 1.275 | .4138 | 6.697 |
| 2.45 | .4454 | 6.241 | 4.848 | 1.287 | .3878 | 6.957 |
| 2.50 | .4385 | 6.500 | 5.000 | 1.300 | .3627 | 7.222 |
| 2.55 | .4319 | 6.764 | 5.152 | 1.313 | .3387 | 7.492 |
| 2.60 | .4256 | 7.034 | 5.304 | 1.326 | .3157 | 7.768 |
| 2.65 | .4196 | 7.309 | 5.457 | 1.339 | .2938 | 8.049 |
| 2.70 | .4138 | 7.589 | 5.610 | 1.353 | .2730 | 8.335 |
| 2.75 | .4082 | 7.875 | 5.762 | 1.367 | .2533 | 8.627 |
| 2.80 | .4029 | 8.166 | 5.914 | 1.381 | .2347 | 8.925 |
| 2.85 | .3978 | 8.462 | 6.065 | 1.395 | .2172 | 9.228 |
| 2.90 | .3929 | 8.763 | 6.216 | 1.410 | .2007 | 9.536 |
| 2.95 | .3882 | 9.069 | 6.367 | 1.424 | .1852 | 9.850 |
| 3.00 | .3837 | 9.381 | 6.517 | 1.439 | .17070 | 10.17 |
| 3.50 | .3466 | 12.786 | 7.977 | 1.603 | .07126 | 13.66 |
| 4.00 | .3203 | 16.714 | 9.333 | 1.791 | .02750 | 17.68 |
| 4.50 | .3009 | 21.167 | 10.565 | 2.003 | .01014 | 22.24 |
| 5.00 | .2863 | 26.143 | 11.667 | 2.241 | .00366 | 27.35 |
| 6.00 | .2661 | 37.67 | 13.50 | 2.790 | $472(10)^{-6}$ | 39.16 |
| 7.00 | .2531 | 51.29 | 14.91 | 3.439 | $645(10)^{-7}$ | 53.12 |
| 8.00 | .2443 | 67.00 | 16.00 | 4.188 | $965(10)^{-8}$ | 69.23 |
| 9.00 | .2381 | 84.81 | 16.84 | 5.036 | $161(10)^{-8}$ | 87.49 |
| 10.00 | .2336 | 104.71 | 17.50 | 5.984 | $297(10)^{-9}$ | 107.90 |
| $\infty$ | .2132 | $\infty$ | 21.00 | $\infty$ | 0 | $\infty$ |

# Table 51   One-Dimensional Normal-Shock Functions

For a perfect gas with constant specific heat and molecular weight
k = 1.2

| $M_x$ | $M_y$ | $\dfrac{p_y}{p_x}$ | $\dfrac{\rho_y}{\rho_x}$ | $\dfrac{T_y}{T_x}$ | $\dfrac{p_{0y}}{p_{0x}}$ | $\dfrac{p_{0y}}{p_x}$ |
|---|---|---|---|---|---|---|
| 1.00 | 1.0000 | 1.000 | 1.000 | 1.0000 | 1.0000 | 1.772 |
| 1.05 | .9528 | 1.112 | 1.092 | 1.0178 | .9998 | 1.873 |
| 1.10 | .9106 | 1.229 | 1.187 | 1.0351 | .9989 | 1.982 |
| 1.15 | .8726 | 1.352 | 1.285 | 1.0521 | .9965 | 2.099 |
| 1.20 | .8383 | 1.480 | 1.385 | 1.0689 | .9923 | 2.224 |
| 1.25 | .8071 | 1.614 | 1.486 | 1.0855 | .9861 | 2.356 |
| 1.30 | .7787 | 1.753 | 1.590 | 1.1022 | .9777 | 2.495 |
| 1.35 | .7527 | 1.897 | 1.696 | 1.1189 | .9671 | 2.641 |
| 1.40 | .7288 | 2.047 | 1.803 | 1.1357 | .9542 | 2.793 |
| 1.45 | .7067 | 2.203 | 1.911 | 1.1527 | .9391 | 2.951 |
| 1.50 | .6864 | 2.364 | 2.020 | 1.170 | .9220 | 3.115 |
| 1.55 | .6676 | 2.530 | 2.131 | 1.187 | .9030 | 3.286 |
| 1.60 | .6501 | 2.702 | 2.242 | 1.205 | .8822 | 3.463 |
| 1.65 | .6338 | 2.879 | 2.354 | 1.223 | .8599 | 3.646 |
| 1.70 | .6186 | 3.062 | 2.466 | 1.241 | .8362 | 3.836 |
| 1.75 | .6044 | 3.250 | 2.579 | 1.260 | .8114 | 4.031 |
| 1.80 | .5912 | 3.444 | 2.692 | 1.279 | .7856 | 4.232 |
| 1.85 | .5788 | 3.643 | 2.805 | 1.299 | .7591 | 4.439 |
| 1.90 | .5671 | 3.847 | 2.918 | 1.319 | .7320 | 4.652 |
| 1.95 | .5561 | 4.057 | 3.031 | 1.339 | .7045 | 4.871 |
| 2.00 | .5458 | 4.273 | 3.143 | 1.360 | .6768 | 5.096 |
| 2.05 | .5360 | 4.494 | 3.255 | 1.381 | .6490 | 5.326 |
| 2.10 | .5268 | 4.720 | 3.366 | 1.402 | .6213 | 5.562 |
| 2.15 | .5181 | 4.952 | 3.477 | 1.424 | .5938 | 5.805 |
| 2.20 | .5099 | 5.189 | 3.587 | 1.446 | .5667 | 6.053 |
| 2.25 | .5021 | 5.432 | 3.697 | 1.469 | .5400 | 6.307 |
| 2.30 | .4947 | 5.680 | 3.806 | 1.492 | .5139 | 6.567 |
| 2.35 | .4877 | 5.934 | 3.914 | 1.516 | .4884 | 6.832 |
| 2.40 | .4810 | 6.193 | 4.020 | 1.540 | .4636 | 7.104 |
| 2.45 | .4746 | 6.457 | 4.126 | 1.565 | .4397 | 7.383 |
| 2.50 | .4686 | 6.727 | 4.231 | 1.590 | .4162 | 7.664 |
| 2.55 | .4629 | 7.003 | 4.335 | 1.616 | .3937 | 7.952 |
| 2.60 | .4574 | 7.284 | 4.437 | 1.642 | .3721 | 8.247 |
| 2.65 | .4521 | 7.570 | 4.538 | 1.668 | .3513 | 8.547 |
| 2.70 | .4471 | 7.862 | 4.638 | 1.695 | .3314 | 8.853 |
| 2.75 | .4424 | 8.159 | 4.737 | 1.723 | .3124 | 9.165 |
| 2.80 | .4378 | 8.462 | 4.834 | 1.751 | .2942 | 9.483 |
| 2.85 | .4334 | 8.770 | 4.930 | 1.779 | .2768 | 9.806 |
| 2.90 | .4292 | 9.084 | 5.025 | 1.808 | .2603 | 10.135 |
| 2.95 | .4252 | 9.403 | 5.118 | 1.837 | .2446 | 10.470 |
| 3.00 | .4214 | 9.727 | 5.211 | 1.867 | .22980 | 10.81 |
| 3.50 | .3904 | 13.273 | 6.056 | 2.192 | .11978 | 14.53 |
| 4.00 | .3690 | 17.364 | 6.769 | 2.565 | .06096 | 18.83 |
| 4.50 | .3536 | 22.000 | 7.364 | 2.988 | .03093 | 23.70 |
| 5.00 | .3421 | 27.182 | 7.857 | 3.459 | .01586 | 29.15 |
| 6.00 | .3267 | 39.18 | 8.609 | 4.551 | $441(10)^{-5}$ | 41.76 |
| 7.00 | .3170 | 53.36 | 9.136 | 5.841 | $134(10)^{-5}$ | 56.66 |
| 8.00 | .3106 | 69.73 | 9.513 | 7.329 | $450(10)^{-6}$ | 73.86 |
| 9.00 | .3061 | 88.27 | 9.791 | 9.016 | $164(10)^{-6}$ | 93.35 |
| 10.00 | .3029 | 109.00 | 10.000 | 10.900 | $650(10)^{-7}$ | 115.14 |
| $\infty$ | .2887 | $\infty$ | 11.00 | $\infty$ | 0 | $\infty$ |

# Table 52  One-Dimensional Normal-Shock Functions

For a perfect gas with constant specific heat and molecular weight
k = 1.3

| $M_x$ | $M_y$ | $\dfrac{p_y}{p_x}$ | $\dfrac{\rho_y}{\rho_x}$ | $\dfrac{T_y}{T_x}$ | $\dfrac{p_{0y}}{p_{0x}}$ | $\dfrac{p_{0y}}{p_x}$ |
|---|---|---|---|---|---|---|
| 1.00 | 1.0000 | 1.000 | 1.000 | 1.0000 | 1.0000 | 1.832 |
| 1.05 | .9530 | 1.116 | 1.088 | 1.0257 | .9998 | 1.941 |
| 1.10 | .9112 | 1.237 | 1.178 | 1.0507 | .9989 | 2.058 |
| 1.15 | .8739 | 1.364 | 1.269 | 1.0752 | .9966 | 2.183 |
| 1.20 | .8403 | 1.497 | 1.362 | 1.0995 | .9925 | 2.316 |
| 1.25 | .8100 | 1.636 | 1.456 | 1.124 | .9866 | 2.457 |
| 1.30 | .7825 | 1.780 | 1.551 | 1.148 | .9786 | 2.605 |
| 1.35 | .7575 | 1.930 | 1.646 | 1.172 | .9684 | 2.760 |
| 1.40 | .7346 | 2.085 | 1.742 | 1.197 | .9562 | 2.922 |
| 1.45 | .7136 | 2.246 | 1.838 | 1.222 | .9421 | 3.090 |
| 1.50 | .6942 | 2.413 | 1.935 | 1.247 | .9261 | 3.265 |
| 1.55 | .6764 | 2.585 | 2.031 | 1.273 | .9084 | 3.447 |
| 1.60 | .6599 | 2.763 | 2.127 | 1.299 | .8891 | 3.635 |
| 1.65 | .6446 | 2.947 | 2.223 | 1.326 | .8684 | 3.830 |
| 1.70 | .6304 | 3.137 | 2.318 | 1.353 | .8466 | 4.031 |
| 1.75 | .6172 | 3.332 | 2.413 | 1.380 | .8238 | 4.238 |
| 1.80 | .6048 | 3.532 | 2.507 | 1.408 | .8001 | 4.452 |
| 1.85 | .5933 | 3.738 | 2.601 | 1.437 | .7758 | 4.672 |
| 1.90 | .5825 | 3.950 | 2.694 | 1.467 | .7510 | 4.898 |
| 1.95 | .5724 | 4.168 | 2.785 | 1.497 | .7259 | 5.131 |
| 2.00 | .5629 | 4.391 | 2.875 | 1.527 | .7006 | 5.370 |
| 2.05 | .5539 | 4.620 | 2.964 | 1.558 | .6752 | 5.615 |
| 2.10 | .5455 | 4.855 | 3.052 | 1.590 | .6499 | 5.866 |
| 2.15 | .5376 | 5.095 | 3.139 | 1.623 | .6248 | 6.123 |
| 2.20 | .5301 | 5.341 | 3.225 | 1.656 | .6000 | 6.387 |
| 2.25 | .5230 | 5.592 | 3.309 | 1.690 | .5755 | 6.657 |
| 2.30 | .5163 | 5.849 | 3.392 | 1.725 | .5515 | 6.933 |
| 2.35 | .5100 | 6.112 | 3.474 | 1.760 | .5280 | 7.215 |
| 2.40 | .5040 | 6.381 | 3.554 | 1.796 | .5050 | 7.503 |
| 2.45 | .4983 | 6.655 | 3.633 | 1.832 | .4827 | 7.798 |
| 2.50 | .4929 | 6.935 | 3.710 | 1.869 | .4610 | 8.098 |
| 2.55 | .4878 | 7.220 | 3.786 | 1.907 | .4400 | 8.405 |
| 2.60 | .4829 | 7.511 | 3.860 | 1.946 | .4196 | 8.718 |
| 2.65 | .4782 | 7.808 | 3.933 | 1.985 | .3999 | 9.037 |
| 2.70 | .4738 | 8.110 | 4.005 | 2.025 | .3810 | 9.362 |
| 2.75 | .4696 | 8.418 | 4.075 | 2.066 | .3628 | 9.693 |
| 2.80 | .4655 | 8.732 | 4.144 | 2.108 | .3452 | 10.030 |
| 2.85 | .4616 | 9.052 | 4.211 | 2.150 | .3284 | 10.373 |
| 2.90 | .4579 | 9.377 | 4.277 | 2.193 | .3123 | 10.723 |
| 2.95 | .4544 | 9.708 | 4.341 | 2.236 | .2969 | 11.079 |
| 3.00 | .4511 | 10.04 | 4.404 | 2.280 | .28216 | 11.44 |
| 3.50 | .4241 | 13.72 | 4.964 | 2.763 | .16774 | 15.39 |
| 4.00 | .4058 | 17.96 | 5.412 | 3.318 | .09933 | 19.96 |
| 4.50 | .3927 | 22.76 | 5.768 | 3.946 | .05939 | 25.13 |
| 5.00 | .3832 | 28.13 | 6.053 | 4.648 | .03613 | 30.92 |
| 6.00 | .3704 | 40.57 | 6.469 | 6.271 | $1422(10)^{-5}$ | 44.31 |
| 7.00 | .3625 | 55.26 | 6.749 | 8.189 | $610(10)^{-5}$ | 60.14 |
| 8.00 | .3573 | 72.22 | 6.943 | 10.401 | $283(10)^{-5}$ | 78.40 |
| 9.00 | .3536 | 91.43 | 7.084 | 12.908 | $140(10)^{-5}$ | 99.10 |
| 10.00 | .3510 | 112.91 | 7.188 | 15.710 | $740(10)^{-6}$ | 122.24 |
| $\infty$ | .3397 | $\infty$ | 7.667 | $\infty$ | 0 | $\infty$ |

(156)

# Table 53   One-Dimensional Normal-Shock Functions

For a perfect gas with constant specific heat and molecular weight
k = 1.67

| $M_x$ | $M_y$ | $\dfrac{p_y}{p_x}$ | $\dfrac{\rho_y}{\rho_x}$ | $\dfrac{T_y}{T_x}$ | $\dfrac{p_{0y}}{p_{0x}}$ | $\dfrac{p_{0y}}{p_x}$ |
|---|---|---|---|---|---|---|
| 1.00 | 1.0000 | 1.000 | 1.000 | 1.0000 | 1.0000 | 2.055 |
| 1.05 | .9535 | 1.128 | 1.075 | 1.0496 | .9998 | 2.189 |
| 1.10 | .9131 | 1.262 | 1.149 | 1.0985 | .9990 | 2.333 |
| 1.15 | .8776 | 1.403 | 1.223 | 1.1471 | .9969 | 2.486 |
| 1.20 | .8463 | 1.550 | 1.297 | 1.195 | .9934 | 2.649 |
| 1.25 | .8184 | 1.703 | 1.370 | 1.244 | .9883 | 2.821 |
| 1.30 | .7935 | 1.863 | 1.441 | 1.293 | .9813 | 3.002 |
| 1.35 | .7711 | 2.029 | 1.511 | 1.343 | .9728 | 3.191 |
| 1.40 | .7509 | 2.201 | 1.580 | 1.394 | .9627 | 3.388 |
| 1.45 | .7325 | 2.379 | 1.647 | 1.445 | .9511 | 3.592 |
| 1.50 | .7158 | 2.563 | 1.713 | 1.497 | .9381 | 3.804 |
| 1.55 | .7006 | 2.754 | 1.777 | 1.550 | .9239 | 4.025 |
| 1.60 | .6866 | 2.951 | 1.840 | 1.604 | .9086 | 4.254 |
| 1.65 | .6738 | 3.154 | 1.901 | 1.659 | .8924 | 4.490 |
| 1.70 | .6620 | 3.364 | 1.960 | 1.716 | .8754 | 4.733 |
| 1.75 | .6511 | 3.580 | 2.018 | 1.774 | .8577 | 4.984 |
| 1.80 | .6410 | 3.802 | 2.074 | 1.833 | .8395 | 5.243 |
| 1.85 | .6316 | 4.030 | 2.128 | 1.893 | .8209 | 5.510 |
| 1.90 | .6229 | 4.265 | 2.181 | 1.955 | .8019 | 5.784 |
| 1.95 | .6148 | 4.506 | 2.232 | 2.018 | .7827 | 6.065 |
| 2.00 | .6073 | 4.753 | 2.282 | 2.083 | .7634 | 6.354 |
| 2.05 | .6002 | 5.006 | 2.330 | 2.149 | .7441 | 6.650 |
| 2.10 | .5936 | 5.266 | 2.376 | 2.216 | .7248 | 6.954 |
| 2.15 | .5875 | 5.532 | 2.421 | 2.284 | .7056 | 7.266 |
| 2.20 | .5817 | 5.804 | 2.465 | 2.354 | .6866 | 7.585 |
| 2.25 | .5762 | 6.082 | 2.507 | 2.426 | .6678 | 7.911 |
| 2.30 | .5711 | 6.366 | 2.548 | 2.499 | .6493 | 8.244 |
| 2.35 | .5663 | 6.657 | 2.587 | 2.574 | .6310 | 8.585 |
| 2.40 | .5617 | 6.954 | 2.625 | 2.650 | .6130 | 8.934 |
| 2.45 | .5574 | 7.257 | 2.662 | 2.727 | .5954 | 9.290 |
| 2.50 | .5534 | 7.567 | 2.697 | 2.806 | .5783 | 9.65 |
| 2.55 | .5495 | 7.883 | 2.731 | 2.886 | .5615 | 10.02 |
| 2.60 | .5459 | 8.205 | 2.764 | 2.968 | .5450 | 10.40 |
| 2.65 | .5425 | 8.533 | 2.796 | 3.052 | .5289 | 10.79 |
| 2.70 | .5392 | 8.868 | 2.827 | 3.137 | .5133 | 11.18 |
| 2.75 | .5361 | 9.209 | 2.857 | 3.223 | .4981 | 11.58 |
| 2.80 | .5331 | 9.556 | 2.886 | 3.311 | .4833 | 11.99 |
| 2.85 | .5303 | 9.909 | 2.914 | 3.401 | .4689 | 12.41 |
| 2.90 | .5276 | 10.269 | 2.941 | 3.492 | .4550 | 12.83 |
| 2.95 | .5251 | 10.635 | 2.967 | 3.584 | .4415 | 13.26 |
| 3.00 | .5227 | 11.01 | 2.992 | 3.678 | .4283 | 13.69 |
| 3.50 | .5036 | 15.07 | 3.204 | 4.704 | .3177 | 18.47 |
| 4.00 | .4910 | 19.76 | 3.358 | 5.885 | .2384 | 23.99 |
| 4.50 | .4822 | 25.08 | 3.473 | 7.221 | .1816 | 30.24 |
| 5.00 | .4758 | 31.02 | 3.560 | 8.714 | .1406 | 37.23 |
| 6.00 | .4674 | 44.78 | 3.680 | 12.17 | .08831 | 53.40 |
| 7.00 | .4623 | 61.04 | 3.756 | 16.25 | .05854 | 72.53 |
| 8.00 | .4589 | 79.81 | 3.807 | 20.96 | .04059 | 94.59 |
| 9.00 | .4566 | 101.08 | 3.843 | 26.30 | .02920 | 119.6 |
| 10.00 | .4550 | 124.84 | 3.870 | 32.26 | .02167 | 147.6 |
| $\infty$ | .4479 | $\infty$ | 3.985 | $\infty$ | 0 | $\infty$ |

# Table 54  Two-Dimensional Isentropic Compressible-Flow Functions for Method of Characteristics

**For a perfect gas with constant specific heat and molecular weight**
**k = 1.4**

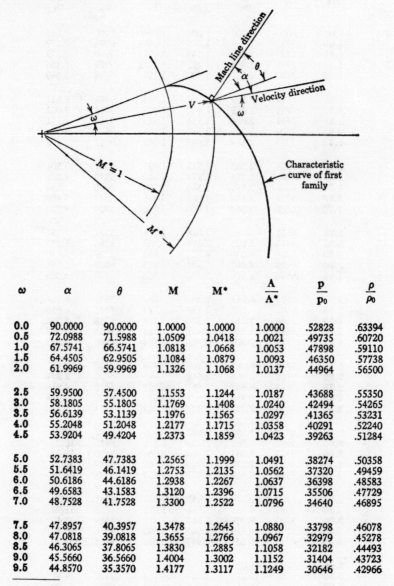

| $\omega$ | $\alpha$ | $\theta$ | M | M* | $\dfrac{A}{A^*}$ | $\dfrac{p}{p_0}$ | $\dfrac{\rho}{\rho_0}$ |
|---|---|---|---|---|---|---|---|
| 0.0 | 90.0000 | 90.0000 | 1.0000 | 1.0000 | 1.0000 | .52828 | .63394 |
| 0.5 | 72.0988 | 71.5988 | 1.0509 | 1.0418 | 1.0021 | .49735 | .60720 |
| 1.0 | 67.5741 | 66.5741 | 1.0818 | 1.0668 | 1.0053 | .47898 | .59110 |
| 1.5 | 64.4505 | 62.9505 | 1.1084 | 1.0879 | 1.0093 | .46350 | .57738 |
| 2.0 | 61.9969 | 59.9969 | 1.1326 | 1.1068 | 1.0137 | .44964 | .56500 |
| 2.5 | 59.9500 | 57.4500 | 1.1553 | 1.1244 | 1.0187 | .43688 | .55350 |
| 3.0 | 58.1805 | 55.1805 | 1.1769 | 1.1408 | 1.0240 | .42494 | .54265 |
| 3.5 | 56.6139 | 53.1139 | 1.1976 | 1.1565 | 1.0297 | .41365 | .53231 |
| 4.0 | 55.2048 | 51.2048 | 1.2177 | 1.1715 | 1.0358 | .40291 | .52240 |
| 4.5 | 53.9204 | 49.4204 | 1.2373 | 1.1859 | 1.0423 | .39263 | .51284 |
| 5.0 | 52.7383 | 47.7383 | 1.2565 | 1.1999 | 1.0491 | .38274 | .50358 |
| 5.5 | 51.6419 | 46.1419 | 1.2753 | 1.2135 | 1.0562 | .37320 | .49459 |
| 6.0 | 50.6186 | 44.6186 | 1.2938 | 1.2267 | 1.0637 | .36398 | .48583 |
| 6.5 | 49.6583 | 43.1583 | 1.3120 | 1.2396 | 1.0715 | .35506 | .47729 |
| 7.0 | 48.7528 | 41.7528 | 1.3300 | 1.2522 | 1.0796 | .34640 | .46895 |
| 7.5 | 47.8957 | 40.3957 | 1.3478 | 1.2645 | 1.0880 | .33798 | .46078 |
| 8.0 | 47.0818 | 39.0818 | 1.3655 | 1.2766 | 1.0967 | .32979 | .45278 |
| 8.5 | 46.3065 | 37.8065 | 1.3830 | 1.2885 | 1.1058 | .32182 | .44493 |
| 9.0 | 45.5660 | 36.5660 | 1.4004 | 1.3002 | 1.1152 | .31404 | .43723 |
| 9.5 | 44.8570 | 35.3570 | 1.4177 | 1.3117 | 1.1249 | .30646 | .42966 |

See page 113 for definitions of symbols.

# Table 54   Method of Characteristics

For a perfect gas with constant specific heat and molecular weight
k = 1.4

| ω | α | θ | M | M* | $\frac{A}{A^*}$ | $\frac{p}{p_0}$ | $\frac{\rho}{\rho_0}$ |
|---|---|---|---|---|---|---|---|
| 10.0 | 44.1770 | 34.1770 | 1.4349 | 1.3230 | 1.1349 | .29906 | .42222 |
| 10.5 | 43.5233 | 33.0233 | 1.4521 | 1.3341 | 1.1453 | .29184 | .41491 |
| 11.0 | 42.8940 | 31.8940 | 1.4692 | 1.3451 | 1.1560 | .28478 | .40772 |
| 11.5 | 42.2869 | 30.7869 | 1.4862 | 1.3559 | 1.1670 | .27788 | .40064 |
| 12.0 | 41.7007 | 29.7007 | 1.5032 | 1.3666 | 1.1783 | .27114 | .39367 |
| 12.5 | 41.1338 | 28.6338 | 1.5202 | 1.3772 | 1.1900 | .26454 | .38680 |
| 13.0 | 40.5849 | 27.5849 | 1.5371 | 1.3876 | 1.2021 | .25809 | .38004 |
| 13.5 | 40.0529 | 26.5529 | 1.5540 | 1.3979 | 1.2145 | .25178 | .37338 |
| 14.0 | 39.5366 | 25.5366 | 1.5709 | 1.4081 | 1.2273 | .24560 | .36681 |
| 14.5 | 39.0350 | 24.5350 | 1.5878 | 1.4182 | 1.2405 | .23955 | .36034 |
| 15.0 | 38.5474 | 23.5474 | 1.6047 | 1.4282 | 1.2541 | .23363 | .35396 |
| 15.5 | 38.0730 | 22.5730 | 1.6216 | 1.4380 | 1.2680 | .22783 | .34766 |
| 16.0 | 37.6108 | 21.6108 | 1.6385 | 1.4478 | 1.2823 | .22216 | .34145 |
| 16.5 | 37.1605 | 20.6605 | 1.6555 | 1.4575 | 1.2970 | .21661 | .33533 |
| 17.0 | 36.7212 | 19.7212 | 1.6725 | 1.4671 | 1.3121 | .21117 | .32929 |
| 17.5 | 36.2925 | 18.7925 | 1.6895 | 1.4766 | 1.3277 | .20584 | .32334 |
| 18.0 | 35.8739 | 17.8739 | 1.7065 | 1.4860 | 1.3437 | .20062 | .31747 |
| 18.5 | 35.4648 | 16.9648 | 1.7235 | 1.4953 | 1.3602 | .19551 | .31168 |
| 19.0 | 35.0648 | 16.0648 | 1.7406 | 1.5046 | 1.3771 | .19051 | .30596 |
| 19.5 | 34.6735 | 15.1735 | 1.7578 | 1.5138 | 1.3945 | .18562 | .30032 |
| 20.0 | 34.2904 | 14.2904 | 1.7750 | 1.5229 | 1.4123 | .18082 | .29475 |
| 20.5 | 33.9153 | 13.4153 | 1.7922 | 1.5319 | 1.4306 | .17612 | .28926 |
| 21.0 | 33.5479 | 12.5479 | 1.8095 | 1.5409 | 1.4494 | .17152 | .28385 |
| 21.5 | 33.1877 | 11.6877 | 1.8269 | 1.5498 | 1.4687 | .16702 | .27851 |
| 22.0 | 32.8344 | 10.8344 | 1.8443 | 1.5586 | 1.4886 | .16261 | .27324 |
| 22.5 | 32.4879 | 9.9879 | 1.8618 | 1.5673 | 1.5090 | .15830 | .26804 |
| 23.0 | 32.1478 | 9.1478 | 1.8793 | 1.5760 | 1.5300 | .15408 | .26291 |
| 23.5 | 31.8138 | 8.3138 | 1.8969 | 1.5846 | 1.5515 | .14995 | .25786 |
| 24.0 | 31.4859 | 7.4859 | 1.9146 | 1.5932 | 1.5736 | .14590 | .25287 |
| 24.5 | 31.1637 | 6.6637 | 1.9324 | 1.6017 | 1.5963 | .14194 | .24795 |
| 25.0 | 30.8469 | 5.8469 | 1.9503 | 1.6101 | 1.6197 | .13806 | .24310 |
| 25.5 | 30.5355 | 5.0355 | 1.9682 | 1.6184 | 1.6437 | .13427 | .23831 |
| 26.0 | 30.2293 | 4.2293 | 1.9862 | 1.6267 | 1.6683 | .13057 | .23359 |
| 26.5 | 29.9281 | 3.4281 | 2.0044 | 1.6350 | 1.6936 | .12694 | .22894 |
| 27.0 | 29.6316 | 2.6316 | 2.0226 | 1.6432 | 1.7196 | .12339 | .22435 |
| 27.5 | 29.3397 | 1.8397 | 2.0409 | 1.6513 | 1.7464 | .11992 | .21982 |
| 28.0 | 29.0524 | 1.0524 | 2.0593 | 1.6594 | 1.7739 | .11653 | .21536 |
| 28.5 | 28.7694 | 0.2694 | 2.0778 | 1.6674 | 1.8022 | .11321 | .21097 |
| 29.0 | 28.4906 | −0.5094 | 2.0964 | 1.6753 | 1.8312 | .10997 | .20664 |
| 29.5 | 28.2158 | −1.2842 | 2.1151 | 1.6832 | 1.8611 | .10680 | .20237 |

(159)

# Table 54   Method of Characteristics

For a perfect gas with constant specific heat and molecular weight
k = 1.4

| $\omega$ | $\alpha$ | $\theta$ | M | M* | $\dfrac{A}{A^*}$ | $\dfrac{p}{p_0}$ | $\dfrac{\rho}{\rho_0}$ |
|---|---|---|---|---|---|---|---|
| 30.0 | 27.9451 | −2.0549 | 2.1339 | 1.6911 | 1.8918 | .10370 | .19816 |
| 30.5 | 27.6782 | −2.8218 | 2.1528 | 1.6989 | 1.9233 | .10068 | .19401 |
| 31.0 | 27.4149 | −3.5851 | 2.1718 | 1.7066 | 1.9557 | .09773 | .18992 |
| 31.5 | 27.1552 | −4.3448 | 2.1910 | 1.7143 | 1.9891 | .09484 | .18590 |
| 32.0 | 26.8991 | −5.1009 | 2.2103 | 1.7220 | 2.0235 | .09202 | .18194 |
| 32.5 | 26.6464 | −5.8536 | 2.2297 | 1.7296 | 2.0588 | .08927 | .17804 |
| 33.0 | 26.3970 | −6.6030 | 2.2492 | 1.7371 | 2.0951 | .08658 | .17419 |
| 33.5 | 26.1507 | −7.3493 | 2.2689 | 1.7446 | 2.1324 | .08395 | .17040 |
| 34.0 | 25.9076 | −8.0924 | 2.2887 | 1.7521 | 2.1709 | .08139 | .16667 |
| 34.5 | 25.6675 | −8.8325 | 2.3086 | 1.7595 | 2.2105 | .07889 | .16300 |
| 35.0 | 25.4304 | −9.5696 | 2.3287 | 1.7669 | 2.2512 | .07646 | .15938 |
| 35.5 | 25.1962 | −10.3038 | 2.3489 | 1.7742 | 2.2931 | .07408 | .15582 |
| 36.0 | 24.9648 | −11.0352 | 2.3693 | 1.7814 | 2.3363 | .07176 | .15232 |
| 36.5 | 24.7361 | −11.7639 | 2.3898 | 1.7886 | 2.3807 | .06950 | .14887 |
| 37.0 | 24.5101 | −12.4899 | 2.4105 | 1.7958 | 2.4264 | .06729 | .14548 |
| 37.5 | 24.2866 | −13.2134 | 2.4313 | 1.8029 | 2.4736 | .06514 | .14215 |
| 38.0 | 24.0657 | −13.9343 | 2.4523 | 1.8100 | 2.5222 | .06304 | .13886 |
| 38.5 | 23.8473 | −14.6527 | 2.4734 | 1.8170 | 2.5722 | .06100 | .13563 |
| 39.0 | 23.6313 | −15.3687 | 2.4947 | 1.8240 | 2.6237 | .05901 | .13246 |
| 39.5 | 23.4176 | −16.0824 | 2.5162 | 1.8310 | 2.6768 | .05707 | .12934 |
| 40.0 | 23.2061 | −16.7939 | 2.5378 | 1.8379 | 2.7316 | .05519 | .12627 |
| 40.5 | 22.9969 | −17.5031 | 2.5596 | 1.8447 | 2.7881 | .05335 | .12325 |
| 41.0 | 22.7900 | −18.2100 | 2.5816 | 1.8515 | 2.8463 | .05156 | .12029 |
| 41.5 | 22.5852 | −18.9148 | 2.6038 | 1.8583 | 2.9063 | .04982 | .11738 |
| 42.0 | 22.3824 | −19.6176 | 2.6261 | 1.8650 | 2.9682 | .04813 | .11452 |
| 42.5 | 22.1816 | −20.3184 | 2.6487 | 1.8717 | 3.0321 | .04648 | .11170 |
| 43.0 | 21.9828 | −21.0172 | 2.6714 | 1.8783 | 3.0981 | .04488 | .10894 |
| 43.5 | 21.7860 | −21.7140 | 2.6944 | 1.8849 | 3.1662 | .04332 | .10622 |
| 44.0 | 21.5911 | −22.4089 | 2.7176 | 1.8915 | 3.2364 | .04181 | .10356 |
| 44.5 | 21.3980 | −23.1020 | 2.7409 | 1.8980 | 3.3089 | .04034 | .10094 |
| 45.0 | 21.2068 | −23.7932 | 2.7644 | 1.9045 | 3.3838 | .03891 | .09837 |
| 45.5 | 21.0174 | −24.4826 | 2.7882 | 1.9109 | 3.4611 | .03752 | .09585 |
| 46.0 | 20.8297 | −25.1703 | 2.8122 | 1.9173 | 3.5410 | .03617 | .09338 |
| 46.5 | 20.6437 | −25.8563 | 2.8364 | 1.9236 | 3.6236 | .03486 | .09095 |
| 47.0 | 20.4594 | −26.5406 | 2.8609 | 1.9299 | 3.7089 | .03359 | .08856 |
| 47.5 | 20.2767 | −27.2233 | 2.8856 | 1.9362 | 3.7971 | .03235 | .08622 |
| 48.0 | 20.0956 | −27.9044 | 2.9105 | 1.9424 | 3.8883 | .03115 | .08393 |
| 48.5 | 19.9160 | −28.5840 | 2.9356 | 1.9486 | 3.9827 | .02999 | .08168 |
| 49.0 | 19.7380 | −29.2620 | 2.9610 | 1.9547 | 4.0803 | .02886 | .07948 |
| 49.5 | 19.5615 | −29.9385 | 2.9867 | 1.9608 | 4.1812 | .02777 | .07732 |

# Table 54    Method of Characteristics

**For a perfect gas with constant specific heat and molecular weight**
**k = 1.4**

| $\omega$ | $\alpha$ | $\theta$ | $M$ | $M^*$ | $\dfrac{A}{A^*}$ | $\dfrac{p}{p_0}$ | $\dfrac{\rho}{\rho_0}$ |
|---|---|---|---|---|---|---|---|
| 50.0 | 19.3865 | −30.6135 | 3.0126 | 1.9669 | 4.2857 | .02671 | .07520 |
| 50.5 | 19.2129 | −31.2871 | 3.0388 | 1.9729 | 4.3938 | .02568 | .07313 |
| 51.0 | 19.0408 | −31.9592 | 3.0652 | 1.9789 | 4.5058 | .02469 | .07110 |
| 51.5 | 18.8700 | −32.6300 | 3.0919 | 1.9848 | 4.6218 | .02373 | .06911 |
| 52.0 | 18.7005 | −33.2995 | 3.1189 | 1.9907 | 4.7419 | .02280 | .06716 |
| 52.5 | 18.5324 | −33.9676 | 3.1462 | 1.9966 | 4.8663 | .02190 | .06525 |
| 53.0 | 18.3657 | −34.6343 | 3.1738 | 2.0024 | 4.9953 | .02103 | .06338 |
| 53.5 | 18.2002 | −35.2998 | 3.2016 | 2.0082 | 5.1290 | .02018 | .06155 |
| 54.0 | 18.0360 | −35.9640 | 3.2298 | 2.0139 | 5.2676 | .01936 | .05976 |
| 54.5 | 17.8730 | −36.6270 | 3.2583 | 2.0196 | 5.4114 | .01857 | .05801 |
| 55.0 | 17.7112 | −37.2888 | 3.2871 | 2.0253 | 5.5606 | .01781 | .05629 |
| 55.5 | 17.5506 | −37.9494 | 3.3162 | 2.0309 | 5.7154 | .01707 | .05461 |
| 56.0 | 17.3911 | −38.6089 | 3.3457 | 2.0365 | 5.8761 | .01636 | .05297 |
| 56.5 | 17.2328 | −39.2672 | 3.3755 | 2.0421 | 6.0429 | .01567 | .05137 |
| 57.0 | 17.0757 | −39.9243 | 3.4056 | 2.0476 | 6.2162 | .01500 | .04981 |
| 57.5 | 16.9196 | −40.5804 | 3.4361 | 2.0531 | 6.3963 | .01436 | .04828 |
| 58.0 | 16.7646 | −41.2354 | 3.4669 | 2.0585 | 6.5834 | .01374 | .04678 |
| 58.5 | 16.6107 | −41.8893 | 3.4981 | 2.0639 | 6.7778 | .01314 | .04532 |
| 59.0 | 16.4579 | −42.5421 | 3.5297 | 2.0692 | 6.9799 | .01257 | .04389 |
| 59.5 | 16.3061 | −43.1939 | 3.5616 | 2.0745 | 7.1902 | .01202 | .04250 |
| 60.0 | 16.1552 | −43.8448 | 3.5940 | 2.0798 | 7.4090 | .01148 | .04114 |
| 60.5 | 16.0053 | −44.4947 | 3.6268 | 2.0850 | 7.6368 | .01096 | .03981 |
| 61.0 | 15.8564 | −45.1436 | 3.6600 | 2.0902 | 7.8739 | .01047 | .03851 |
| 61.5 | 15.7085 | −45.7915 | 3.6936 | 2.0954 | 8.1208 | .00999 | .03725 |
| 62.0 | 15.5615 | −46.4385 | 3.7276 | 2.1005 | 8.3780 | .00953 | .03602 |
| 62.5 | 15.4154 | −47.0846 | 3.7620 | 2.1056 | 8.6460 | .00909 | .03482 |
| 63.0 | 15.2703 | −47.7297 | 3.7969 | 2.1107 | 8.9254 | .00867 | .03365 |
| 63.5 | 15.1260 | −48.3740 | 3.8323 | 2.1157 | 9.2168 | .00826 | .03251 |
| 64.0 | 14.9826 | −49.0174 | 3.8681 | 2.1207 | 9.5208 | .00786 | .03140 |
| 64.5 | 14.8400 | −49.6600 | 3.9044 | 2.1256 | 9.8380 | .00749 | .03032 |
| 65.0 | 14.6983 | −50.3017 | 3.9412 | 2.1305 | 10.169 | .00712 | .02926 |
| 65.5 | 14.5574 | −50.9426 | 3.9785 | 2.1353 | 10.515 | .00678 | .02823 |
| 66.0 | 14.4174 | −51.5826 | 4.0163 | 2.1401 | 10.876 | .00644 | .02723 |
| 66.5 | 14.2781 | −52.2219 | 4.0547 | 2.1449 | 11.253 | .00612 | .02626 |
| 67.0 | 14.1396 | −52.8604 | 4.0936 | 2.1497 | 11.648 | .00582 | .02532 |
| 67.5 | 14.0019 | −53.4981 | 4.1330 | 2.1544 | 12.061 | .00552 | .02440 |
| 68.0 | 13.8650 | −54.1350 | 4.1730 | 2.1591 | 12.493 | .00524 | .02350 |
| 68.5 | 13.7288 | −54.7712 | 4.2136 | 2.1637 | 12.945 | .00497 | .02263 |
| 69.0 | 13.5934 | −55.4066 | 4.2548 | 2.1683 | 13.418 | .00472 | .02179 |
| 69.5 | 13.4587 | −56.0413 | 4.2966 | 2.1728 | 13.913 | .00447 | .02097 |

# Table 54    Method of Characteristics

**For a perfect gas with constant specific heat and molecular weight**
**k = 1.4**

| $\omega$ | $\alpha$ | $\theta$ | M | M* | $\dfrac{A}{A^*}$ | $\dfrac{p}{p_0}$ | $\dfrac{\rho}{\rho_0}$ |
|---|---|---|---|---|---|---|---|
| 70.0 | 13.3247 | −56.6753 | 4.3390 | 2.1773 | 14.433 | .00423 | .02017 |
| 70.5 | 13.1913 | −57.3087 | 4.3821 | 2.1818 | 14.978 | .00401 | .01940 |
| 71.0 | 13.0587 | −57.9413 | 4.4258 | 2.1863 | 15.549 | .00379 | .01865 |
| 71.5 | 12.9268 | −58.5732 | 4.4702 | 2.1907 | 16.148 | .00359 | .01792 |
| 72.0 | 12.7955 | −59.2045 | 4.5152 | 2.1951 | 16.777 | .00339 | .01721 |
| 72.5 | 12.6649 | −59.8351 | 4.5610 | 2.1994 | 17.438 | .00320 | .01652 |
| 73.0 | 12.5349 | −60.4651 | 4.6076 | 2.2037 | 18.132 | .00302 | .01586 |
| 73.5 | 12.4055 | −61.0945 | 4.6549 | 2.2080 | 18.862 | .00285 | .01522 |
| 74.0 | 12.2768 | −61.7232 | 4.7029 | 2.2122 | 19.630 | .00269 | .01460 |
| 74.5 | 12.1487 | −62.3513 | 4.7517 | 2.2164 | 20.438 | .00254 | .01400 |
| 75.0 | 12.0212 | −62.9788 | 4.8014 | 2.2205 | 21.288 | .00239 | .01342 |
| 75.5 | 11.8943 | −63.6057 | 4.8519 | 2.2246 | 22.183 | .00225 | .01285 |
| 76.0 | 11.7680 | −64.2320 | 4.9032 | 2.2287 | 23.126 | .00212 | .01230 |
| 76.5 | 11.6422 | −64.8578 | 4.9554 | 2.2327 | 24.121 | .00199 | .01177 |
| 77.0 | 11.5170 | −65.4830 | 5.0085 | 2.2367 | 25.171 | .00187 | .01126 |
| 77.5 | 11.3924 | −66.1076 | 5.0626 | 2.2407 | 26.279 | .00176 | .01077 |
| 78.0 | 11.2683 | −66.7317 | 5.1176 | 2.2446 | 27.449 | .00165 | .01029 |
| 78.5 | 11.1447 | −67.3553 | 5.1736 | 2.2485 | 28.685 | .00155 | .00983 |
| 79.0 | 11.0217 | −67.9783 | 5.2306 | 2.2523 | 29.992 | .00145 | .00939 |
| 79.5 | 10.8992 | −68.6008 | 5.2887 | 2.2561 | 31.374 | .00136 | .00896 |
| 80.0 | 10.7772 | −69.2228 | 5.3479 | 2.2599 | 32.837 | .00127 | .00854 |
| 80.5 | 10.6558 | −69.8442 | 5.4081 | 2.2636 | 34.387 | .00119 | .00814 |
| 81.0 | 10.5348 | −70.4652 | 5.4694 | 2.2673 | 36.029 | .00111 | .00776 |
| 81.5 | 10.4143 | −71.0857 | 5.5320 | 2.2710 | 37.770 | .00104 | .00739 |
| 82.0 | 10.2942 | −71.7058 | 5.5959 | 2.2746 | 39.618 | .00097 | .00703 |
| 82.5 | 10.1746 | −72.3254 | 5.6610 | 2.2782 | 41.580 | .00090 | .00669 |
| 83.0 | 10.0555 | −72.9445 | 5.7274 | 2.2818 | 43.665 | .00084 | .00636 |
| 83.5 | 9.9369 | −73.5631 | 5.7950 | 2.2853 | 45.882 | .00078 | .00604 |
| 84.0 | 9.8187 | −74.1813 | 5.8640 | 2.2888 | 48.240 | .00073 | .00574 |
| 84.5 | 9.7010 | −74.7990 | 5.9345 | 2.2922 | 50.750 | .00068 | .00545 |
| 85.0 | 9.5837 | −75.4163 | 6.0064 | 2.2956 | 53.424 | .00063 | .00517 |
| 85.5 | 9.4668 | −76.0332 | 6.0799 | 2.2990 | 56.276 | .00058 | .00490 |
| 86.0 | 9.3503 | −76.6497 | 6.1550 | 2.3023 | 59.320 | .00054 | .00464 |
| 86.5 | 9.2342 | −77.2658 | 6.2317 | 2.3056 | 62.570 | .00050 | .00439 |
| 87.0 | 9.1185 | −77.8815 | 6.3101 | 2.3088 | 66.043 | .00046 | .00416 |
| 87.5 | 9.0032 | −78.4968 | 6.3902 | 2.3120 | 69.758 | .00043 | .00393 |
| 88.0 | 8.8884 | −79.1116 | 6.4720 | 2.3152 | 73.734 | .00040 | .00371 |
| 88.5 | 8.7740 | −79.7260 | 6.5558 | 2.3183 | 77.994 | .00037 | .00351 |
| 89.0 | 8.6599 | −80.3401 | 6.6415 | 2.3214 | 82.561 | .00034 | .00331 |
| 89.5 | 8.5462 | −80.9538 | 6.7292 | 2.3245 | 87.463 | .00031 | .00312 |
| 90.0 | 8.4328 | −81.5672 | 6.8190 | 2.3275 | 92.730 | .00029 | .00294 |

# Table 55  Wedge Angle for Sonic Flow Downstream of and Maximum Wedge Angle for Two-Dimensional Shock

**For a perfect gas with constant specific heat and molecular weight**
**k = 1.4**

| $M_x$ | $\omega_a'$ | $\omega_m'$ | $M_x$ | $\omega_a'$ | $\omega_m'$ | $M_x$ | $\omega_a'$ | $\omega_m'$ | $M_x$ | $\omega_a'$ | $\omega_m'$ |
|---|---|---|---|---|---|---|---|---|---|---|---|
| 1.01 | 0.05° | 0.05° | 1.36 | 7.94° | 8.32° | 1.71 | 16.86° | 17.24° | 2.06 | 23.73° | 23.98° |
| 1.02 | 0.13 | 0.15 | 1.37 | 8.21 | 8.60 | 1.72 | 17.09 | 17.46 | 2.07 | 23.90 | 24.14 |
| 1.03 | 0.24 | 0.26 | 1.38 | 8.48 | 8.88 | 1.73 | 17.31 | 17.68 | 2.08 | 24.06 | 24.30 |
| 1.04 | 0.37 | 0.40 | 1.39 | 8.76 | 9.15 | 1.74 | 17.54 | 17.90 | 2.09 | 24.22 | 24.46 |
| 1.05 | 0.52 | 0.56 | 1.40 | 9.03 | 9.43 | 1.75 | 17.76 | 18.12 | 2.10 | 24.38 | 24.61 |
| 1.06 | 0.67 | 0.73 | 1.41 | 9.30 | 9.70 | 1.76 | 17.98 | 18.34 | 2.11 | 24.54 | 24.77 |
| 1.07 | 0.84 | 0.91 | 1.42 | 9.57 | 9.97 | 1.77 | 18.20 | 18.55 | 2.12 | 24.70 | 24.92 |
| 1.08 | 1.02 | 1.10 | 1.43 | 9.84 | 10.25 | 1.78 | 18.41 | 18.76 | 2.13 | 24.85 | 25.08 |
| 1.09 | 1.21 | 1.30 | 1.44 | 10.10 | 10.52 | 1.79 | 18.63 | 18.97 | 2.14 | 25.01 | 25.23 |
| 1.10 | 1.41 | 1.51 | 1.45 | 10.37 | 10.79 | 1.80 | 18.84 | 19.18 | 2.15 | 25.16 | 25.38 |
| 1.11 | 1.61 | 1.73 | 1.46 | 10.64 | 11.05 | 1.81 | 19.05 | 19.39 | 2.16 | 25.31 | 25.52 |
| 1.12 | 1.82 | 1.96 | 1.47 | 10.90 | 11.32 | 1.82 | 19.26 | 19.59 | 2.17 | 25.46 | 25.67 |
| 1.13 | 2.04 | 2.19 | 1.48 | 11.17 | 11.59 | 1.83 | 19.46 | 19.80 | 2.18 | 25.61 | 25.82 |
| 1.14 | 2.26 | 2.43 | 1.49 | 11.43 | 11.85 | 1.84 | 19.67 | 20.00 | 2.19 | 25.76 | 25.96 |
| 1.15 | 2.49 | 2.67 | 1.50 | 11.69 | 12.11 | 1.85 | 19.87 | 20.20 | 2.20 | 25.90 | 26.10 |
| 1.16 | 2.73 | 2.92 | 1.51 | 11.96 | 12.37 | 1.86 | 20.07 | 20.40 | 2.21 | 26.05 | 26.24 |
| 1.17 | 2.97 | 3.17 | 1.52 | 12.21 | 12.63 | 1.87 | 20.27 | 20.59 | 2.22 | 26.19 | 26.38 |
| 1.18 | 3.21 | 3.42 | 1.53 | 12.47 | 12.89 | 1.88 | 20.47 | 20.78 | 2.23 | 26.33 | 26.52 |
| 1.19 | 3.45 | 3.68 | 1.54 | 12.73 | 13.15 | 1.89 | 20.67 | 20.98 | 2.24 | 26.47 | 26.66 |
| 1.20 | 3.70 | 3.94 | 1.55 | 12.99 | 13.40 | 1.90 | 20.86 | 21.17 | 2.25 | 26.61 | 26.80 |
| 1.21 | 3.95 | 4.21 | 1.56 | 13.24 | 13.66 | 1.91 | 21.05 | 21.36 | 2.3 | 27.28 | 27.45 |
| 1.22 | 4.21 | 4.48 | 1.57 | 13.49 | 13.91 | 1.92 | 21.24 | 21.54 | 2.4 | 28.53 | 28.68 |
| 1.23 | 4.46 | 4.74 | 1.58 | 13.74 | 14.16 | 1.93 | 21.43 | 21.73 | 2.5 | 29.67 | 29.80 |
| 1.24 | 4.72 | 5.01 | 1.59 | 13.99 | 14.41 | 1.94 | 21.62 | 21.91 | 2.6 | 30.70 | 30.81 |
| 1.25 | 4.99 | 5.29 | 1.60 | 14.24 | 14.65 | 1.95 | 21.81 | 22.09 | | | |
| 1.26 | 5.25 | 5.56 | 1.61 | 14.49 | 14.90 | 1.96 | 21.99 | 22.27 | 2.7 | 31.64 | 31.74 |
| 1.27 | 5.52 | 5.83 | 1.62 | 14.73 | 15.14 | 1.97 | 22.17 | 22.45 | 2.8 | 32.50 | 32.59 |
| 1.28 | 5.78 | 6.11 | 1.63 | 14.98 | 15.38 | 1.98 | 22.35 | 22.63 | 2.9 | 33.29 | 33.36 |
| 1.29 | 6.05 | 6.39 | 1.64 | 15.22 | 15.62 | 1.99 | 22.53 | 22.80 | 3.0 | 34.01 | 34.07 |
| 1.30 | 6.32 | 6.66 | 1.65 | 15.46 | 15.86 | 2.00 | 22.71 | 22.97 | | | |
| 1.31 | 6.59 | 6.94 | 1.66 | 15.70 | 16.09 | 2.01 | 22.88 | 23.14 | 4 | 38.75 | 38.77 |
| 1.32 | 6.86 | 7.22 | 1.67 | 15.93 | 16.32 | 2.02 | 23.05 | 23.31 | 5 | 41.11 | 41.12 |
| 1.33 | 7.13 | 7.49 | 1.68 | 16.17 | 16.55 | 2.03 | 23.23 | 23.48 | 6 | 42.44 | 42.44 |
| 1.34 | 7.40 | 7.77 | 1.69 | 16.40 | 16.78 | 2.04 | 23.40 | 23.65 | 8 | 43.79 | 43.79 |
| 1.35 | 7.67 | 8.05 | 1.70 | 16.63 | 17.01 | 2.05 | 23.56 | 23.81 | | | |
| | | | | | | | | | 10 | 44.43 | 44.43 |
| | | | | | | | | | 15 | 45.07 | 45.07 |
| | | | | | | | | | 20 | 45.29 | 45.29 |
| | | | | | | | | | ∞ | 45.58 | 45.58 |

See page 113 for definitions of symbols.

# Table 56 Upstream Mach Number, $M_x$, for Two-Dimensional Shock

**For a perfect gas with constant specific heat and molecular weight**
**k = 1.4**

| $\alpha' \rightarrow$ $\omega' \downarrow$ | 15° | 16° | 17° | 18° | 19° | 20° | 21° | 22° | 23° | 24° | 25° | 26° | 27° | 28° | 29° | 30° | 31° |
|---|---|---|---|---|---|---|---|---|---|---|---|---|---|---|---|---|---|
| 1° | 4.036 | 3.779 | 3.555 | 3.358 | 3.182 | 3.023 | 2.882 | 2.754 | 2.637 | 2.531 | 2.433 | 2.344 | 2.262 | 2.186 | 2.116 | 2.050 | 1.989 |
| 2° | 4.232 | 3.950 | 3.706 | 3.491 | 3.301 | 3.132 | 2.981 | 2.844 | 2.721 | 2.608 | 2.505 | 2.410 | 2.325 | 2.244 | 2.171 | 2.102 | 2.039 |
| 3° | 4.457 | 4.145 | 3.876 | 3.643 | 3.436 | 3.252 | 3.089 | 2.943 | 2.811 | 2.691 | 2.582 | 2.482 | 2.391 | 2.307 | 2.230 | 2.158 | 2.091 |
| 4° | 4.717 | 4.369 | 4.070 | 3.811 | 3.586 | 3.386 | 3.211 | 3.051 | 2.910 | 2.782 | 2.666 | 2.560 | 2.463 | 2.374 | 2.293 | 2.217 | 2.147 |
| 5° | 5.032 | 4.632 | 4.294 | 4.005 | 3.755 | 3.536 | 3.344 | 3.173 | 3.019 | 2.881 | 2.757 | 2.644 | 2.541 | 2.447 | 2.360 | 2.280 | 2.207 |
| 6° | 5.414 | 4.946 | 4.559 | 4.232 | 3.949 | 3.708 | 3.495 | 3.307 | 3.139 | 2.990 | 2.856 | 2.735 | 2.625 | 2.525 | 2.433 | 2.349 | 2.271 |
| 7° | 5.897 | 5.332 | 4.876 | 4.498 | 4.177 | 3.904 | 3.666 | 3.459 | 3.275 | 3.112 | 2.966 | 2.835 | 2.717 | 2.609 | 2.511 | 2.421 | 2.339 |
| 8° | 6.531 | 5.823 | 5.267 | 4.818 | 4.445 | 4.131 | 3.864 | 3.630 | 3.427 | 3.249 | 3.089 | 2.946 | 2.818 | 2.702 | 2.597 | 2.500 | 2.413 |
| 9° | 7.425 | 6.475 | 5.767 | 5.215 | 4.771 | 4.403 | 4.094 | 3.830 | 3.601 | 3.403 | 3.226 | 3.069 | 2.930 | 2.804 | 2.690 | 2.587 | 2.492 |
| 10° | 8.823 | 7.407 | 6.442 | 5.729 | 5.177 | 4.734 | 4.371 | 4.065 | 3.805 | 3.579 | 3.381 | 3.209 | 3.055 | 2.916 | 2.793 | 2.681 | 2.579 |
| 11° | 11.486 | 8.902 | 7.419 | 6.427 | 5.706 | 5.151 | 4.709 | 4.346 | 4.042 | 3.783 | 3.560 | 3.366 | 3.195 | 3.043 | 2.907 | 2.785 | 2.675 |
| 12° | 20.743 | 11.921 | 9.034 | 7.459 | 6.433 | 5.698 | 5.137 | 4.691 | 4.329 | 4.027 | 3.769 | 3.547 | 3.356 | 3.185 | 3.035 | 2.901 | 2.780 |
| 13° | | 26.089 | 12.533 | 9.223 | 7.532 | 6.460 | 5.703 | 5.134 | 4.684 | 4.321 | 4.017 | 3.761 | 3.540 | 3.348 | 3.180 | 3.030 | 2.898 |
| 14° | | | 45.340 | 13.410 | 9.485 | 7.638 | 6.507 | 5.725 | 5.142 | 4.686 | 4.319 | 4.015 | 3.757 | 3.537 | 3.346 | 3.178 | 3.030 |
| 15° | | | | | 14.693 | 9.830 | 7.782 | 6.579 | 5.761 | 5.161 | 4.698 | 4.324 | 4.017 | 3.759 | 3.538 | 3.348 | 3.180 |
| 16° | | | | | | 16.729 | 10.295 | 7.970 | 6.672 | 5.815 | 5.193 | 4.717 | 4.338 | 4.027 | 3.767 | 3.546 | 3.353 |
| 17° | | | | | | | 20.472 | 10.911 | 8.213 | 6.794 | 5.885 | 5.237 | 4.748 | 4.359 | 4.044 | 3.781 | 3.557 |
| 18° | | | | | | | | 30.452 | 11.765 | 8.519 | 6.949 | 5.974 | 5.295 | 4.787 | 4.390 | 4.067 | 3.800 |
| 19° | | | | | | | | | | 12.994 | 8.915 | 7.139 | 6.085 | 5.368 | 4.839 | 4.428 | 4.098 |
| 20° | | | | | | | | | | | 14.923 | 9.422 | 7.377 | 6.222 | 5.457 | 4.901 | 4.475 |
| 21° | | | | | | | | | | | | 18.474 | 10.104 | 7.669 | 6.388 | 5.564 | 4.978 |
| 22° | | | | | | | | | | | | | 28.833 | 11.044 | 9.039 | 6.590 | 5.694 |
| 23° | | | | | | | | | | | | | | | 12.431 | 8.512 | 6.835 |
| 24° | | | | | | | | | | | | | | | | 14.720 | 9.133 |
| 25° | | | | | | | | | | | | | | | | | 19.514 |
| 26° | | | | | | | | | | | | | | | | | |
| 27° | | | | | | | | | | | | | | | | | |
| 28° | | | | | | | | | | | | | | | | | |
| 29° | | | | | | | | | | | | | | | | | |
| 30° | | | | | | | | | | | | | | | | | |

| $\alpha' \rightarrow$ $\omega' \downarrow$ | 48° | 49° | 50° | 51° | 52° | 53° | 54° | 55° | 56° | 57° | 58° | 59° | 60° | 61° | 62° | 63° | 64° |
|---|---|---|---|---|---|---|---|---|---|---|---|---|---|---|---|---|---|
| 1° | 1.375 | 1.354 | 1.334 | 1.314 | 1.296 | 1.280 | 1.263 | 1.249 | 1.234 | 1.220 | 1.207 | 1.195 | 1.183 | 1.172 | 1.162 | 1.152 | 1.142 |
| 2° | 1.404 | 1.383 | 1.362 | 1.343 | 1.325 | 1.308 | 1.291 | 1.276 | 1.262 | 1.248 | 1.235 | 1.223 | 1.211 | 1.201 | 1.191 | 1.181 | 1.172 |
| 3° | 1.434 | 1.412 | 1.391 | 1.372 | 1.354 | 1.337 | 1.320 | 1.305 | 1.290 | 1.277 | 1.264 | 1.252 | 1.240 | 1.230 | 1.220 | 1.211 | 1.203 |
| 4° | 1.465 | 1.443 | 1.422 | 1.403 | 1.383 | 1.366 | 1.349 | 1.334 | 1.319 | 1.305 | 1.293 | 1.281 | 1.269 | 1.260 | 1.250 | 1.241 | 1.234 |
| 5° | 1.497 | 1.474 | 1.453 | 1.433 | 1.414 | 1.396 | 1.379 | 1.363 | 1.349 | 1.335 | 1.322 | 1.310 | 1.300 | 1.290 | 1.280 | 1.272 | 1.265 |
| 6° | 1.530 | 1.508 | 1.485 | 1.465 | 1.445 | 1.427 | 1.410 | 1.394 | 1.379 | 1.366 | 1.353 | 1.341 | 1.330 | 1.320 | 1.312 | 1.304 | 1.296 |
| 7° | 1.565 | 1.541 | 1.519 | 1.498 | 1.478 | 1.459 | 1.441 | 1.426 | 1.411 | 1.397 | 1.384 | 1.372 | 1.361 | 1.352 | 1.343 | 1.335 | 1.328 |
| 8° | 1.601 | 1.576 | 1.553 | 1.531 | 1.511 | 1.492 | 1.474 | 1.458 | 1.443 | 1.429 | 1.416 | 1.404 | 1.393 | 1.383 | 1.375 | 1.368 | 1.362 |
| 9° | 1.638 | 1.612 | 1.589 | 1.566 | 1.546 | 1.526 | 1.508 | 1.491 | 1.476 | 1.462 | 1.449 | 1.436 | 1.426 | 1.416 | 1.408 | 1.401 | 1.395 |
| 10° | 1.677 | 1.651 | 1.625 | 1.603 | 1.581 | 1.561 | 1.543 | 1.526 | 1.510 | 1.496 | 1.482 | 1.471 | 1.460 | 1.450 | 1.442 | 1.435 | 1.430 |
| 11° | 1.717 | 1.689 | 1.664 | 1.641 | 1.618 | 1.598 | 1.579 | 1.561 | 1.545 | 1.530 | 1.517 | 1.505 | 1.495 | 1.485 | 1.477 | 1.470 | 1.465 |
| 12° | 1.760 | 1.731 | 1.704 | 1.680 | 1.656 | 1.636 | 1.616 | 1.598 | 1.581 | 1.567 | 1.553 | 1.541 | 1.530 | 1.521 | 1.513 | 1.506 | 1.502 |
| 13° | 1.804 | 1.774 | 1.747 | 1.721 | 1.697 | 1.675 | 1.655 | 1.636 | 1.620 | 1.604 | 1.590 | 1.578 | 1.567 | 1.557 | 1.550 | 1.544 | 1.539 |
| 14° | 1.851 | 1.819 | 1.790 | 1.764 | 1.739 | 1.716 | 1.695 | 1.676 | 1.659 | 1.643 | 1.629 | 1.616 | 1.605 | 1.596 | 1.588 | 1.582 | 1.577 |
| 15° | 1.900 | 1.867 | 1.837 | 1.809 | 1.783 | 1.760 | 1.738 | 1.718 | 1.699 | 1.683 | 1.669 | 1.656 | 1.645 | 1.635 | 1.628 | 1.621 | 1.617 |
| 16° | 1.952 | 1.917 | 1.886 | 1.857 | 1.830 | 1.805 | 1.782 | 1.762 | 1.743 | 1.725 | 1.710 | 1.698 | 1.686 | 1.676 | 1.668 | 1.662 | 1.658 |
| 17° | 2.007 | 1.971 | 1.937 | 1.907 | 1.878 | 1.853 | 1.829 | 1.807 | 1.787 | 1.770 | 1.754 | 1.740 | 1.729 | 1.719 | 1.711 | 1.705 | 1.701 |
| 18° | 2.067 | 2.029 | 1.992 | 1.960 | 1.930 | 1.902 | 1.877 | 1.855 | 1.835 | 1.816 | 1.800 | 1.786 | 1.774 | 1.764 | 1.756 | 1.750 | 1.746 |
| 19° | 2.131 | 2.089 | 2.051 | 2.016 | 1.985 | 1.955 | 1.930 | 1.906 | 1.884 | 1.865 | 1.848 | 1.833 | 1.820 | 1.811 | 1.802 | 1.796 | 1.792 |
| 20° | 2.199 | 2.154 | 2.114 | 2.077 | 2.043 | 2.012 | 1.985 | 1.959 | 1.937 | 1.916 | 1.898 | 1.883 | 1.870 | 1.859 | 1.851 | 1.845 | 1.841 |
| 21° | 2.273 | 2.225 | 2.181 | 2.142 | 2.105 | 2.073 | 2.042 | 2.016 | 1.992 | 1.970 | 1.952 | 1.935 | 1.922 | 1.911 | 1.902 | 1.896 | 1.892 |
| 22° | 2.353 | 2.302 | 2.254 | 2.211 | 2.172 | 2.137 | 2.105 | 2.076 | 2.051 | 2.028 | 2.008 | 1.991 | 1.977 | 1.965 | 1.955 | 1.949 | 1.946 |
| 23° | 2.441 | 2.385 | 2.333 | 2.287 | 2.244 | 2.207 | 2.172 | 2.142 | 2.114 | 2.090 | 2.069 | 2.050 | 2.035 | 2.022 | 2.013 | 2.006 | 2.003 |
| 24° | 2.538 | 2.476 | 2.419 | 2.369 | 2.323 | 2.281 | 2.244 | 2.211 | 2.182 | 2.155 | 2.133 | 2.113 | 2.096 | 2.083 | 2.073 | 2.066 | 2.062 |
| 25° | 2.646 | 2.577 | 2.515 | 2.459 | 2.408 | 2.363 | 2.322 | 2.287 | 2.254 | 2.226 | 2.201 | 2.180 | 2.163 | 2.148 | 2.137 | 2.130 | 2.126 |
| 26° | 2.767 | 2.689 | 2.620 | 2.557 | 2.502 | 2.452 | 2.408 | 2.368 | 2.333 | 2.302 | 2.275 | 2.253 | 2.233 | 2.218 | 2.206 | 2.198 | 2.193 |
| 27° | 2.903 | 2.815 | 2.738 | 2.668 | 2.606 | 2.551 | 2.501 | 2.458 | 2.419 | 2.385 | 2.355 | 2.330 | 2.310 | 2.293 | 2.280 | 2.271 | 2.266 |
| 28° | 3.061 | 2.960 | 2.871 | 2.791 | 2.722 | 2.660 | 2.605 | 2.556 | 2.513 | 2.476 | 2.443 | 2.416 | 2.393 | 2.374 | 2.359 | 2.350 | 2.344 |
| 29° | 3.244 | 3.126 | 3.023 | 2.932 | 2.852 | 2.782 | 2.720 | 2.665 | 2.617 | 2.575 | 2.539 | 2.509 | 2.483 | 2.462 | 2.446 | 2.435 | 2.429 |
| 30° | 3.462 | 3.321 | 3.199 | 3.094 | 3.001 | 2.920 | 2.849 | 2.788 | 2.733 | 2.686 | 2.646 | 2.610 | 2.582 | 2.558 | 2.541 | 2.528 | 2.521 |

See page 113 for definitions of symbols.

# Table 56  Upstream Mach Number, $M_x$, for Two-Dimensional Shock

**For a perfect gas with constant specific heat and molecular weight**
**k = 1.4**

| $\alpha' \rightarrow$ $\omega' \downarrow$ | 32° | 33° | 34° | 35° | 36° | 37° | 38° | 39° | 40° | 41° | 42° | 43° | 44° | 45° | 46° | 47° |
|---|---|---|---|---|---|---|---|---|---|---|---|---|---|---|---|---|
| 1° | 1.932 | 1.879 | 1.830 | 1.783 | 1.740 | 1.698 | 1.660 | 1.623 | 1.589 | 1.557 | 1.527 | 1.498 | 1.471 | 1.445 | 1.420 | 1.397 |
| 2° | 1.979 | 1.925 | 1.872 | 1.824 | 1.779 | 1.737 | 1.697 | 1.660 | 1.624 | 1.591 | 1.559 | 1.530 | 1.502 | 1.475 | 1.450 | 1.426 |
| 3° | 2.029 | 1.972 | 1.919 | 1.869 | 1.821 | 1.777 | 1.736 | 1.698 | 1.661 | 1.627 | 1.594 | 1.563 | 1.534 | 1.507 | 1.481 | 1.457 |
| 4° | 2.082 | 2.023 | 1.966 | 1.914 | 1.865 | 1.819 | 1.776 | 1.736 | 1.698 | 1.663 | 1.629 | 1.598 | 1.568 | 1.540 | 1.514 | 1.489 |
| 5° | 2.138 | 2.075 | 2.017 | 1.962 | 1.911 | 1.864 | 1.819 | 1.777 | 1.738 | 1.701 | 1.667 | 1.634 | 1.604 | 1.575 | 1.547 | 1.522 |
| 6° | 2.198 | 2.133 | 2.071 | 2.013 | 1.959 | 1.910 | 1.863 | 1.820 | 1.779 | 1.740 | 1.705 | 1.671 | 1.640 | 1.610 | 1.582 | 1.556 |
| 7° | 2.263 | 2.192 | 2.128 | 2.067 | 2.012 | 1.959 | 1.910 | 1.865 | 1.822 | 1.783 | 1.745 | 1.710 | 1.678 | 1.647 | 1.618 | 1.591 |
| 8° | 2.331 | 2.258 | 2.199 | 2.125 | 2.065 | 2.011 | 1.960 | 1.913 | 1.868 | 1.827 | 1.788 | 1.751 | 1.717 | 1.685 | 1.655 | 1.627 |
| 9° | 2.406 | 2.327 | 2.253 | 2.187 | 2.124 | 2.066 | 2.012 | 1.963 | 1.916 | 1.872 | 1.832 | 1.794 | 1.758 | 1.725 | 1.694 | 1.665 |
| 10° | 2.487 | 2.402 | 2.324 | 2.253 | 2.187 | 2.125 | 2.068 | 2.016 | 1.966 | 1.921 | 1.879 | 1.839 | 1.802 | 1.767 | 1.735 | 1.705 |
| 11° | 2.575 | 2.483 | 2.400 | 2.324 | 2.253 | 2.188 | 2.128 | 2.072 | 2.021 | 1.973 | 1.928 | 1.886 | 1.848 | 1.812 | 1.778 | 1.746 |
| 12° | 2.672 | 2.573 | 2.482 | 2.401 | 2.325 | 2.256 | 2.192 | 2.133 | 2.078 | 2.027 | 1.981 | 1.937 | 1.896 | 1.858 | 1.823 | 1.790 |
| 13° | 2.779 | 2.671 | 2.573 | 2.484 | 2.404 | 2.330 | 2.260 | 2.197 | 2.140 | 2.086 | 2.036 | 1.990 | 1.947 | 1.907 | 1.871 | 1.836 |
| 14° | 2.898 | 2.780 | 2.673 | 2.576 | 2.488 | 2.408 | 2.335 | 2.268 | 2.205 | 2.148 | 2.095 | 2.046 | 2.002 | 1.959 | 1.920 | 1.884 |
| 15° | 3.033 | 2.902 | 2.784 | 2.678 | 2.583 | 2.495 | 2.416 | 2.344 | 2.276 | 2.215 | 2.159 | 2.107 | 2.059 | 2.014 | 1.973 | 1.935 |
| 16° | 3.187 | 3.039 | 2.909 | 2.792 | 2.686 | 2.591 | 2.505 | 2.426 | 2.354 | 2.288 | 2.228 | 2.172 | 2.121 | 2.073 | 2.030 | 1.990 |
| 17° | 3.365 | 3.197 | 3.049 | 2.919 | 2.802 | 2.697 | 2.602 | 2.516 | 2.439 | 2.367 | 2.302 | 2.242 | 2.167 | 2.137 | 2.090 | 2.047 |
| 18° | 3.574 | 3.380 | 3.211 | 3.063 | 2.932 | 2.815 | 2.711 | 2.617 | 2.531 | 2.454 | 2.382 | 2.318 | 2.259 | 2.205 | 2.155 | 2.109 |
| 19° | 3.826 | 3.596 | 3.399 | 3.230 | 3.080 | 2.949 | 2.832 | 2.728 | 2.634 | 2.549 | 2.471 | 2.401 | 2.337 | 2.278 | 2.224 | 2.176 |
| 20° | 4.136 | 3.857 | 3.623 | 3.424 | 3.253 | 3.103 | 2.971 | 2.853 | 2.748 | 2.654 | 2.569 | 2.492 | 2.422 | 2.359 | 2.300 | 2.248 |
| 21° | 4.534 | 4.182 | 3.896 | 3.656 | 3.454 | 3.280 | 3.128 | 2.995 | 2.877 | 2.771 | 2.678 | 2.592 | 2.516 | 2.446 | 2.383 | 2.326 |
| 22° | 5.069 | 4.602 | 4.236 | 3.942 | 3.697 | 3.490 | 3.313 | 3.159 | 3.023 | 2.905 | 2.799 | 2.704 | 2.619 | 2.543 | 2.473 | 2.411 |
| 23° | 5.846 | 5.176 | 4.683 | 4.301 | 3.995 | 3.743 | 3.532 | 3.350 | 3.195 | 3.057 | 2.937 | 2.830 | 2.735 | 2.650 | 2.573 | 2.504 |
| 24° | 7.137 | 6.031 | 5.303 | 4.778 | 4.377 | 4.058 | 3.798 | 3.579 | 3.394 | 3.234 | 3.096 | 2.974 | 2.866 | 2.770 | 2.684 | 2.607 |
| 25° | 9.973 | 7.516 | 6.254 | 5.454 | 4.890 | 4.465 | 4.131 | 3.860 | 3.635 | 3.445 | 3.281 | 3.140 | 3.016 | 2.906 | 2.810 | 2.723 |
| 26° | 43.967 | 11.198 | 7.993 | 6.523 | 5.632 | 5.021 | 4.568 | 4.216 | 3.931 | 3.698 | 3.502 | 3.334 | 3.189 | 3.063 | 2.952 | 2.854 |
| 27° | | | 13.162 | 8.624 | 6.854 | 5.847 | 5.173 | 4.687 | 4.313 | 4.015 | 3.770 | 3.567 | 3.394 | 3.246 | 3.116 | 3.003 |
| 28° | | | | 17.083 | 9.495 | 7.272 | 6.104 | 5.365 | 4.825 | 4.424 | 4.108 | 3.853 | 3.641 | 3.462 | 3.309 | 3.176 |
| 29° | | | | | 32.120 | 10.771 | 7.809 | 6.420 | 5.571 | 4.988 | 4.554 | 4.218 | 3.948 | 3.725 | 3.539 | 3.380 |
| 30° | | | | | | | 12.895 | 8.531 | 6.812 | 5.831 | 5.178 | 4.705 | 4.344 | 4.056 | 3.822 | 3.628 |

| $\alpha' \rightarrow$ $\omega' \downarrow$ | 65° | 66° | 67° | 68° | 69° | 70° | 71° | 72° | 73° | 74° | 75° | 76° | 77° | 78° | 79° | 80° |
|---|---|---|---|---|---|---|---|---|---|---|---|---|---|---|---|---|
| 1° | 1.134 | 1.125 | 1.118 | 1.111 | 1.105 | 1.099 | 1.094 | 1.089 | 1.085 | 1.081 | 1.078 | 1.077 | 1.075 | 1.075 | 1.075 | 1.077 |
| 2° | 1.164 | 1.156 | 1.150 | 1.144 | 1.139 | 1.134 | 1.130 | 1.126 | 1.124 | 1.122 | 1.122 | 1.122 | 1.124 | | | |
| 3° | 1.195 | 1.188 | 1.182 | 1.177 | 1.173 | 1.169 | 1.166 | 1.164 | 1.163 | 1.164 | 1.165 | 1.168 | | | | |
| 4° | 1.227 | 1.220 | 1.215 | 1.210 | 1.207 | 1.204 | 1.203 | 1.202 | 1.203 | 1.205 | 1.208 | 1.214 | | | | |
| 5° | 1.258 | 1.252 | 1.248 | 1.244 | 1.242 | 1.240 | 1.239 | 1.240 | 1.243 | 1.247 | 1.252 | 1.260 | | | | |
| 6° | 1.291 | 1.285 | 1.281 | 1.278 | 1.276 | 1.276 | 1.277 | 1.279 | 1.283 | 1.289 | 1.297 | 1.307 | | | | |
| 7° | 1.323 | 1.319 | 1.315 | 1.313 | 1.312 | 1.313 | 1.314 | 1.318 | 1.324 | 1.331 | 1.341 | 1.354 | | | | |
| 8° | 1.357 | 1.352 | 1.350 | 1.348 | 1.348 | 1.350 | 1.353 | 1.358 | 1.365 | 1.375 | 1.386 | 1.402 | | | | |
| 9° | 1.390 | 1.387 | 1.385 | 1.384 | 1.385 | 1.388 | 1.392 | 1.398 | 1.407 | 1.418 | 1.433 | 1.450 | | | | |
| 10° | 1.426 | 1.423 | 1.421 | 1.421 | 1.423 | 1.426 | 1.432 | 1.440 | 1.450 | 1.463 | 1.479 | 1.500 | | | | |
| 11° | 1.461 | 1.459 | 1.458 | 1.459 | 1.462 | 1.466 | 1.473 | 1.482 | 1.494 | 1.509 | 1.528 | 1.551 | | | | |
| 12° | 1.498 | 1.496 | 1.496 | 1.497 | 1.501 | 1.507 | 1.514 | 1.525 | 1.539 | 1.556 | 1.577 | 1.603 | | | | |
| 13° | 1.535 | 1.535 | 1.535 | 1.537 | 1.541 | 1.548 | 1.557 | 1.569 | 1.584 | 1.604 | 1.627 | 1.657 | | | | |
| 14° | 1.575 | 1.574 | 1.575 | 1.578 | 1.583 | 1.591 | 1.601 | 1.615 | 1.632 | 1.653 | 1.679 | 1.712 | | | | |
| 15° | 1.615 | 1.614 | 1.616 | 1.620 | 1.626 | 1.635 | 1.647 | 1.662 | 1.681 | 1.704 | 1.733 | 1.768 | | | | |
| 16° | 1.656 | 1.657 | 1.659 | 1.663 | 1.670 | 1.680 | 1.694 | 1.710 | 1.731 | 1.757 | 1.788 | 1.827 | | | | |
| 17° | 1.700 | 1.700 | 1.703 | 1.708 | 1.716 | 1.727 | 1.742 | 1.760 | 1.783 | 1.811 | 1.846 | 1.888 | | | | |
| 18° | 1.745 | 1.745 | 1.749 | 1.755 | 1.764 | 1.776 | 1.792 | 1.812 | 1.837 | 1.868 | 1.906 | 1.952 | | | | |
| 19° | 1.791 | 1.792 | 1.796 | 1.804 | 1.813 | 1.827 | 1.844 | 1.866 | 1.893 | 1.927 | 1.968 | 2.019 | | | | |
| 20° | 1.840 | 1.842 | 1.846 | 1.854 | 1.865 | 1.879 | 1.899 | 1.923 | 1.953 | 1.989 | 2.033 | 2.088 | | | | |
| 21° | 1.891 | 1.893 | 1.898 | 1.907 | 1.919 | 1.935 | 1.956 | 1.982 | 2.014 | 2.054 | 2.102 | 2.161 | | | | |
| 22° | 1.945 | 1.948 | 1.953 | 1.962 | 1.975 | 1.993 | 2.016 | 2.043 | 2.079 | 2.122 | 2.174 | 2.239 | | | | |
| 23° | 2.002 | 2.005 | 2.011 | 2.021 | 2.035 | 2.054 | 2.079 | 2.109 | 2.147 | 2.193 | 2.251 | 2.321 | | | | |
| 24° | 2.062 | 2.064 | 2.071 | 2.083 | 2.098 | 2.119 | 2.145 | 2.178 | 2.219 | 2.269 | 2.332 | 2.410 | | | | |
| 25° | 2.125 | 2.129 | 2.136 | 2.148 | 2.165 | 2.187 | 2.216 | 2.251 | 2.296 | 2.351 | 2.419 | 2.504 | | | | |
| 26° | 2.193 | 2.197 | 2.205 | 2.218 | 2.236 | 2.260 | 2.290 | 2.329 | 2.378 | 2.437 | 2.512 | 2.604 | | | | |
| 27° | 2.266 | 2.270 | 2.278 | 2.292 | 2.311 | 2.337 | 2.370 | 2.413 | 2.466 | 2.531 | 2.612 | 2.715 | | | | |
| 28° | 2.344 | 2.348 | 2.357 | 2.372 | 2.393 | 2.420 | 2.457 | 2.503 | 2.560 | 2.631 | 2.720 | 2.833 | | | | |
| 29° | 2.428 | 2.432 | 2.441 | 2.457 | 2.481 | 2.511 | 2.549 | 2.599 | 2.662 | 2.740 | 2.839 | 2.965 | | | | |
| 30° | 2.520 | 2.523 | 2.533 | 2.551 | 2.575 | 2.608 | 2.651 | 2.705 | 2.774 | 2.860 | 2.970 | 3.110 | | | | |

# Table 57 Downstream Mach Number, $M_y$, for Two-Dimensional Shock

For a perfect gas with constant specific heat and molecular weight
k = 1.4

| α' → ω' ↓ | 24° | 25° | 26° | 27° | 28° | .29° | 30°.. | 31° | 32° | 33° | 34° | 35° | 36° | 37° | 38° |
|---|---|---|---|---|---|---|---|---|---|---|---|---|---|---|---|
| 1° | 2.487 | 2.392 | 2.303 | 2.222 | 2.147 | 2.077 | 2.014 | 1.953 | 1.897 | 1.844 | 1.795 | 1.749 | 1.705 | 1.665 | 1.626 |
| 2° | | 2.420 | 2.329 | 2.244 | 2.167 | 2.095 | 2.028 | 1.966 | 1.908 | 1.854 | 1.804 | 1.756 | 1.711 | 1.669 | 1.630 |
| 3° | | | 2.357 | 2.270 | 2.189 | 2.114 | 2.045 | 1.981 | 1.921 | 1.865 | 1.812 | 1.763 | 1.718 | 1.675 | 1.633 |
| 4° | | | | 2.298 | 2.214 | 2.136 | 2.065 | 1.998 | 1.936 | 1.878 | 1.825 | 1.774 | 1.726 | 1.682 | 1.640 |
| 5° | | | | | 2.242 | 2.161 | 2.087 | 2.018 | 1.954 | 1.894 | 1.837 | 1.785 | 1.736 | 1.690 | 1.646 |
| 6° | | | | | | 2.189 | 2.111 | 2.039 | 1.973 | 1.910 | 1.852 | 1.798 | 1.748 | 1.700 | 1.655 |
| 7° | | | | | | | 2.139 | 2.063 | 1.994 | 1.930 | 1.869 | 1.813 | 1.760 | 1.711 | 1.665 |
| 8° | | | | | | | | 2.090 | 2.018 | 1.951 | 1.888 | 1.830 | 1.776 | 1.724 | 1.676 |
| 9° | | | | | | | | | 2.045 | 1.974 | 1.910 | 1.848 | 1.792 | 1.739 | 1.689 |
| 10° | | | | | | | | | | 2.001 | 1.933 | 1.869 | 1.810 | 1.755 | 1.704 |
| 11° | | | | | | | | | | | 1.959 | 1.893 | 1.831 | 1.774 | 1.720 |
| 12° | | | | | | | | | | | | 1.918 | 1.854 | 1.794 | 1.738 |
| 13° | | | | | | | | | | | | | 1.879 | 1.817 | 1.759 |
| 14° | | | | | | | | | | | | | | 1.842 | 1.781 |
| 15° | | | | | | | | | | | | | | | 1.806 |
| 16° | | | | | | | | | | | | | | | |
| 17° | | | | | | | | | | | | | | | |
| 18° | | | | | | | | | | | | | | | |
| 19° | | | | | | | | | | | | | | | |
| 20° | | | | | | | | | | | | | | | |
| 21° | | | | | | | | | | | | | | | |
| 22° | | | | | | | | | | | | | | | |
| 23° | | | | | | | | | | | | | | | |
| 24° | | | | | | | | | | | | | | | |
| 25° | | | | | | | | | | | | | | | |
| 26° | | | | | | | | | | | | | | | |
| 27° | | | | | | | | | | | | | | | |
| 28° | | | | | | | | | | | | | | | |
| 29° | | | | | | | | | | | | | | | |
| 30° | | | | | | | | | | | | | | | |

| α' → ω' ↓ | 54° | 55° | 56° | 57° | 58° | 59° | 60° | 61° | 62° | 63° | 64° | 65° | 66° | 67° | 68° |
|---|---|---|---|---|---|---|---|---|---|---|---|---|---|---|---|
| 1° | 1.225 | 1.209 | 1.194 | 1.179 | 1.165 | 1.152 | 1.139 | 1.127 | 1.115 | 1.104 | 1.093 | 1.083 | 1.074 | 1.064 | 1.055 |
| 2° | 1.215 | 1.199 | 1.183 | 1.167 | 1.153 | 1.138 | 1.125 | 1.112 | 1.099 | 1.087 | 1.076 | 1.064 | 1.054 | 1.043 | 1.033 |
| 3° | 1.206 | 1.189 | 1.173 | 1.156 | 1.141 | 1.126 | 1.112 | 1.098 | 1.085 | 1.072 | 1.060 | 1.048 | 1.036 | 1.025 | 1.014 |
| 4° | 1.199 | 1.181 | 1.164 | 1.147 | 1.130 | 1.115 | 1.100 | 1.086 | 1.072 | 1.058 | 1.045 | 1.032 | 1.020 | 1.008 | .9958 |
| 5° | 1.192 | 1.173 | 1.155 | 1.138 | 1.121 | 1.105 | 1.089 | 1.074 | 1.060 | 1.046 | 1.032 | 1.018 | 1.005 | .9920 | .9794 |
| 6° | 1.186 | 1.167 | 1.148 | 1.130 | 1.113 | 1.096 | 1.080 | 1.064 | 1.049 | 1.034 | 1.020 | 1.005 | .9915 | .9778 | .9644 |
| 7° | 1.181 | 1.161 | 1.142 | 1.123 | 1.105 | 1.088 | 1.071 | 1.055 | 1.039 | 1.024 | 1.009 | .9938 | .9790 | .9649 | .9507 |
| 8° | 1.177 | 1.156 | 1.136 | 1.117 | 1.099 | 1.081 | 1.064 | 1.047 | 1.030 | 1.014 | .9984 | .9829 | .9681 | .9528 | .9382 |
| 9° | 1.174 | 1.153 | 1.132 | 1.112 | 1.093 | 1.075 | 1.056 | 1.039 | 1.022 | 1.005 | .9891 | .9735 | .9575 | .9422 | .9268 |
| 10° | 1.171 | 1.150 | 1.128 | 1.108 | 1.088 | 1.069 | 1.051 | 1.032 | 1.015 | .9978 | .9809 | .9644 | .9483 | .9322 | .9162 |
| 11° | 1.170 | 1.147 | 1.125 | 1.104 | 1.084 | 1.064 | 1.045 | 1.027 | 1.009 | .9911 | .9736 | .9566 | .9397 | .9233 | .9070 |
| 12° | 1.169 | 1.146 | 1.123 | 1.102 | 1.081 | 1.060 | 1.041 | 1.022 | 1.003 | .9849 | .9669 | .9495 | .9321 | .9150 | .8984 |
| 13° | 1.169 | 1.145 | 1.122 | 1.100 | 1.078 | 1.057 | 1.037 | 1.018 | .9982 | .9794 | .9612 | .9433 | .9252 | .9077 | .8904 |
| 14° | 1.170 | 1.145 | 1.121 | 1.099 | 1.076 | 1.055 | 1.034 | 1.014 | .9943 | .9749 | .9561 | .9374 | .9192 | .9013 | .8834 |
| 15° | 1.171 | 1.146 | 1.122 | 1.098 | 1.075 | 1.053 | 1.032 | 1.011 | .9907 | .9710 | .9517 | .9327 | .9138 | .8953 | .8771 |
| 16° | 1.174 | 1.148 | 1.123 | 1.099 | 1.075 | 1.052 | 1.030 | 1.009 | .9881 | .9678 | .9479 | .9284 | .9091 | .8902 | .8714 |
| 17° | 1.177 | 1.150 | 1.124 | 1.100 | 1.075 | 1.052 | 1.030 | 1.007 | .9860 | .9653 | .9448 | .9248 | .9053 | .8857 | .8666 |
| 18° | 1.182 | 1.154 | 1.127 | 1.101 | 1.077 | 1.052 | 1.029 | 1.007 | .9846 | .9632 | .9424 | .9219 | .9018 | .8819 | .8622 |
| 19° | 1.186 | 1.158 | 1.130 | 1.104 | 1.078 | 1.054 | 1.030 | 1.006 | .9840 | .9622 | .9407 | .9196 | .8991 | .8787 | .8584 |
| 20° | 1.192 | 1.163 | 1.134 | 1.107 | 1.081 | 1.056 | 1.031 | 1.007 | .9840 | .9614 | .9395 | .9179 | .8968 | .8760 | .8554 |
| 21° | 1.200 | 1.169 | 1.140 | 1.112 | 1.084 | 1.058 | 1.033 | 1.008 | .9846 | .9615 | .9389 | .9169 | .8953 | .8740 | .8530 |
| 22° | 1.207 | 1.176 | 1.145 | 1.116 | 1.088 | 1.061 | 1.036 | 1.010 | .9861 | .9622 | .9391 | .9164 | .8943 | .8725 | .8511 |
| 23° | 1.216 | 1.183 | 1.152 | 1.122 | 1.093 | 1.066 | 1.039 | 1.013 | .9878 | .9636 | .9398 | .9166 | .8939 | .8716 | .9497 |
| 24° | 1.226 | 1.192 | 1.160 | 1.129 | 1.099 | 1.071 | 1.043 | 1.016 | .9906 | .9655 | .9413 | .9174 | .8943 | .8714 | .8489 |
| 25° | 1.237 | 1.202 | 1.169 | 1.137 | 1.106 | 1.076 | 1.048 | 1.020 | .9939 | .9682 | .9432 | .9188 | .8949 | .8717 | .8487 |
| 26° | 1.249 | 1.213 | 1.178 | 1.145 | 1.113 | 1.083 | 1.053 | 1.025 | .9979 | .9716 | .9459 | .9209 | .8963 | .8725 | .8490 |
| 27° | 1.263 | 1.225 | 1.189 | 1.155 | 1.122 | 1.090 | 1.060 | 1.031 | 1.003 | .9756 | .9491 | .9234 | .8983 | .8738 | .8499 |
| 28° | 1.278 | 1.238 | 1.201 | 1.165 | 1.131 | 1.099 | 1.067 | 1.037 | 1.008 | .9802 | .9531 | .9267 | .9010 | .8759 | .8513 |
| 29° | 1.294 | 1.253 | 1.214 | 1.177 | 1.142 | 1.108 | 1.075 | 1.044 | 1.015 | .9856 | .9578 | .9306 | .9042 | .8786 | .8534 |
| 30° | 1.312 | 1.269 | 1.228 | 1.190 | 1.153 | 1.118 | 1.085 | 1.053 | 1.022 | .9919 | .9631 | .9352 | .9082 | .8818 | .8559 |

# Table 57  Downstream Mach Number, $M_y$, for Two-Dimensional Shock

**For a perfect gas with constant specific heat and molecular weight**
**k = 1.4**

| $\alpha' \rightarrow$ $\omega' \downarrow$ | 39° | 40° | 41° | 42° | 43° | 44° | 45° | 46° | 47° | 48° | 49° | 50° | 51° | 52° | 53° |
|---|---|---|---|---|---|---|---|---|---|---|---|---|---|---|---|
| 1° | 1.590 | 1.556 | 1.523 | 1.492 | 1.463 | 1.436 | 1.409 | 1.385 | 1.361 | 1.339 | 1.317 | 1.297 | 1.278 | 1.260 | 1.242 |
| 2° | 1.592 | 1.557 | 1.524 | 1.492 | 1.461 | 1.433 | 1.407 | 1.381 | 1.357 | 1.333 | 1.311 | 1.290 | 1.271 | 1.251 | 1.232 |
| 3° | 1.595 | 1.558 | 1.524 | 1.492 | 1.461 | 1.432 | 1.404 | 1.378 | 1.353 | 1.329 | 1.306 | 1.284 | 1.263 | 1.243 | 1.224 |
| 4° | 1.600 | 1.562 | 1.527 | 1.494 | 1.461 | 1.431 | 1.403 | 1.376 | 1.350 | 1.325 | 1.302 | 1.279 | 1.257 | 1.237 | 1.217 |
| 5° | 1.605 | 1.567 | 1.531 | 1.496 | 1.463 | 1.432 | 1.403 | 1.375 | 1.348 | 1.323 | 1.298 | 1.275 | 1.253 | 1.231 | 1.211 |
| 6° | 1.613 | 1.573 | 1.535 | 1.500 | 1.466 | 1.434 | 1.403 | 1.375 | 1.347 | 1.321 | 1.296 | 1.272 | 1.249 | 1.227 | 1.206 |
| 7° | 1.621 | 1.580 | 1.541 | 1.504 | 1.470 | 1.436 | 1.405 | 1.375 | 1.347 | 1.320 | 1.294 | 1.270 | 1.246 | 1.224 | 1.202 |
| 8° | 1.631 | 1.588 | 1.548 | 1.510 | 1.474 | 1.441 | 1.408 | 1.378 | 1.348 | 1.320 | 1.294 | 1.268 | 1.244 | 1.221 | 1.198 |
| 9° | 1.642 | 1.599 | 1.557 | 1.517 | 1.480 | 1.445 | 1.412 | 1.380 | 1.350 | 1.322 | 1.294 | 1.268 | 1.243 | 1.219 | 1.196 |
| 10° | 1.655 | 1.610 | 1.566 | 1.526 | 1.488 | 1.451 | 1.417 | 1.384 | 1.353 | 1.323 | 1.295 | 1.269 | 1.242 | 1.218 | 1.194 |
| 11° | 1.670 | 1.622 | 1.578 | 1.536 | 1.496 | 1.459 | 1.423 | 1.389 | 1.357 | 1.327 | 1.298 | 1.270 | 1.243 | 1.218 | 1.193 |
| 12° | 1.686 | 1.637 | 1.591 | 1.547 | 1.506 | 1.467 | 1.430 | 1.396 | 1.362 | 1.331 | 1.301 | 1.272 | 1.245 | 1.219 | 1.193 |
| 13° | 1.704 | 1.653 | 1.605 | 1.560 | 1.517 | 1.477 | 1.439 | 1.402 | 1.368 | 1.336 | 1.305 | 1.275 | 1.247 | 1.220 | 1.194 |
| 14° | 1.724 | 1.671 | 1.621 | 1.574 | 1.530 | 1.488 | 1.448 | 1.411 | 1.376 | 1.342 | 1.310 | 1.279 | 1.250 | 1.222 | 1.195 |
| 15° | 1.746 | 1.691 | 1.638 | 1.589 | 1.543 | 1.500 | 1.459 | 1.421 | 1.384 | 1.349 | 1.316 | 1.284 | 1.254 | 1.225 | 1.198 |
| 16° | 1.770 | 1.712 | 1.658 | 1.607 | 1.559 | 1.514 | 1.472 | 1.431 | 1.393 | 1.358 | 1.323 | 1.290 | 1.259 | 1.230 | 1.201 |
| 17° |  | 1.736 | 1.680 | 1.626 | 1.576 | 1.529 | 1.485 | 1.444 | 1.404 | 1.367 | 1.331 | 1.298 | 1.265 | 1.235 | 1.205 |
| 18° |  |  | 1.703 | 1.648 | 1.595 | 1.546 | 1.500 | 1.457 | 1.416 | 1.377 | 1.340 | 1.306 | 1.272 | 1.241 | 1.211 |
| 19° |  |  | 1.729 | 1.671 | 1.617 | 1.565 | 1.517 | 1.472 | 1.429 | 1.389 | 1.351 | 1.315 | 1.281 | 1.248 | 1.217 |
| 20° |  |  | 1.759 | 1.697 | 1.640 | 1.586 | 1.536 | 1.489 | 1.444 | 1.402 | 1.363 | 1.325 | 1.290 | 1.256 | 1.224 |
| 21° |  |  | 1.791 | 1.726 | 1.665 | 1.609 | 1.556 | 1.507 | 1.461 | 1.417 | 1.376 | 1.337 | 1.300 | 1.265 | 1.231 |
| 22° |  |  | 1.826 | 1.757 | 1.694 | 1.634 | 1.579 | 1.527 | 1.479 | 1.433 | 1.390 | 1.350 | 1.311 | 1.275 | 1.240 |
| 23° |  |  | 1.865 | 1.792 | 1.725 | 1.662 | 1.604 | 1.550 | 1.499 | 1.451 | 1.406 | 1.364 | 1.324 | 1.286 | 1.250 |
| 24° |  |  | 1.909 | 1.831 | 1.759 | 1.693 | 1.631 | 1.574 | 1.521 | 1.471 | 1.424 | 1.380 | 1.338 | 1.299 | 1.262 |
| 25° |  |  | 1.957 | 1.873 | 1.797 | 1.726 | 1.662 | 1.601 | 1.545 | 1.493 | 1.444 | 1.397 | 1.354 | 1.313 | 1.274 |
| 26° |  |  | 2.011 | 1.921 | 1.839 | 1.764 | 1.695 | 1.631 | 1.572 | 1.516 | 1.465 | 1.417 | 1.371 | 1.328 | 1.288 |
| 27° |  |  | 2.071 | 1.974 | 1.885 | 1.805 | 1.731 | 1.664 | 1.601 | 1.543 | 1.488 | 1.438 | 1.390 | 1.345 | 1.303 |
| 28° |  |  | 2.139 | 2.033 | 1.938 | 1.851 | 1.772 | 1.700 | 1.633 | 1.572 | 1.514 | 1.461 | 1.411 | 1.364 | 1.320 |
| 29° |  |  | 2.217 | 2.100 | 1.996 | 1.902 | 1.818 | 1.740 | 1.669 | 1.603 | 1.542 | 1.486 | 1.434 | 1.384 | 1.338 |
| 30° |  |  | 2.306 | 2.177 | 2.062 | 1.960 | 1.868 | 1.785 | 1.708 | 1.638 | 1.574 | 1.514 | 1.459 | 1.407 | 1.358 |

| $\alpha' \rightarrow$ $\omega' \downarrow$ | 69° | 70° | 71° | 72° | 73° | 74° | 75° | 76° | 77° | 78° | 79° | 80° |
|---|---|---|---|---|---|---|---|---|---|---|---|---|
| 1° | 1.046 | 1.037 | 1.030 | 1.021 | 1.014 | 1.007 | .9992 | .9916 | .9844 | .9771 | .9697 |  |
| 2° | 1.023 | 1.014 | 1.004 | .9949 | .9857 | .9764 | .9671 | .9577 |  |  |  |  |
| 3° | 1.003 | .9920 | .9815 | .9709 | .9603 | .9493 | .9384 | .9274 |  |  |  |  |
| 4° | .9837 | .9722 | .9604 | .9490 | .9369 | .9253 | .9130 | .9004 |  |  |  |  |
| 5° | .9665 | .9542 | .9416 | .9291 | .9160 | .9031 | .8900 | .8761 |  |  |  |  |
| 6° | .9511 | .9376 | .9244 | .9109 | .8974 | .8833 | .8689 | .8542 |  |  |  |  |
| 7° | .9367 | .9225 | .9087 | .8944 | .8800 | .8654 | .8501 | .8344 |  |  |  |  |
| 8° | .9235 | .9088 | .8940 | .8793 | .8642 | .8487 | .8330 | .8167 |  |  |  |  |
| 9° | .9115 | .8962 | .8810 | .8655 | .8498 | .8338 | .8172 | .8004 |  |  |  |  |
| 10° | .9006 | .8848 | .8688 | .8529 | .8364 | .8200 | .8030 | .7853 |  |  |  |  |
| 11° | .8905 | .8741 | .8577 | .8412 | .8245 | .8074 | .7899 | .7718 |  |  |  |  |
| 12° | .8813 | .8645 | .8477 | .8307 | .8135 | .7959 | .7777 | .7593 |  |  |  |  |
| 13° | .8732 | .8559 | .8384 | .8211 | .8034 | .7853 | .7669 | .7478 |  |  |  |  |
| 14° | .8656 | .8479 | .8301 | .8122 | .7942 | .7758 | .7567 | .7373 |  |  |  |  |
| 15° | .8588 | .8407 | .8225 | .8041 | .7856 | .7669 | .7475 | .7279 |  |  |  |  |
| 16° | .8528 | .8343 | .8155 | .7969 | .7780 | .7588 | .7393 | .7190 |  |  |  |  |
| 17° | .8475 | .8285 | .8094 | .7903 | .7711 | .7514 | .7315 | .7110 |  |  |  |  |
| 18° | .8426 | .8232 | .8040 | .7844 | .7648 | .7449 | .7245 | .7038 |  |  |  |  |
| 19° | .8386 | .8187 | .7990 | .7791 | .7591 | .7388 | .7182 | .6971 |  |  |  |  |
| 20° | .8350 | .8149 | .7946 | .7743 | .7540 | .7334 | .7125 | .6911 |  |  |  |  |
| 21° | .8321 | .8116 | .7909 | .7703 | .7496 | .7286 | .7074 | .6857 |  |  |  |  |
| 22° | .8299 | .8087 | .7876 | .7668 | .7455 | .7243 | .7028 | .6808 |  |  |  |  |
| 23° | .8279 | .8065 | .7851 | .7637 | .7421 | .7206 | .6987 | .6764 |  |  |  |  |
| 24° | .8268 | .8048 | .7830 | .7612 | .7394 | .7174 | .6951 | .6725 |  |  |  |  |
| 25° | .8260 | .8037 | .7813 | .7592 | .7370 | .7146 | .6920 | .6692 |  |  |  |  |
| 26° | .8258 | .8030 | .7803 | .7577 | .7351 | .7124 | .6895 | .6663 |  |  |  |  |
| 27° | .8262 | .8029 | .7797 | .7566 | .7336 | .7106 | .6874 | .6638 |  |  |  |  |
| 28° | .8271 | .8033 | .7797 | .7562 | .7327 | .7093 | .6857 | .6619 |  |  |  |  |
| 29° | .8285 | .8041 | .7801 | .7562 | .7323 | .7085 | .6845 | .6603 |  |  |  |  |
| 30° | .8305 | .8057 | .7810 | .7566 | .7323 | .7081 | .6837 | .6592 |  |  |  |  |

# Table 58  Ratio of Downstream to Upstream Pressure for Two-Dimensional Shock

**For a perfect gas with constant specific heat and molecular weight**
**k = 1.4**

| α' → ω' ↓ | 1° | 2° | 3° | 4° | 5° | 6° | 7° | 8° | 9° | 10° | 11° | 12° | 13° | 14° | 15° | 16° | 17° | 18° | 19° |
|---|---|---|---|---|---|---|---|---|---|---|---|---|---|---|---|---|---|---|---|
| 1° | | 2.753 | 1.780 | 1.501 | 1.370 | 1.293 | 1.243 | 1.209 | 1.182 | 1.162 | 1.146 | 1.134 | 1.122 | 1.114 | 1.107 | 1.099 | 1.093 | 1.089 | 1.085 |
| 2° | | | 5.674 | 2.754 | 2.081 | 1.783 | 1.614 | 1.506 | 1.431 | 1.376 | 1.333 | 1.299 | 1.273 | 1.251 | 1.233 | 1.216 | 1.203 | 1.191 | 1.181 |
| 3° | | | | 11.553 | 4.014 | 2.763 | 2.247 | 1.965 | 1.790 | 1.669 | 1.581 | 1.514 | 1.461 | 1.419 | 1.386 | 1.356 | 1.331 | 1.311 | 1.293 |
| 4° | | | | | 29.425 | 5.706 | 3.571 | 2.772 | 2.353 | 2.098 | 1.924 | 1.799 | 1.706 | 1.632 | 1.575 | 1.525 | 1.485 | 1.452 | 1.423 |
| 5° | | | | | | | 8.092 | 4.548 | 3.371 | 2.783 | 2.432 | 2.199 | 2.033 | 1.908 | 1.812 | 1.735 | 1.672 | 1.620 | 1.577 |
| 6° | | | | | | | | 11.722 | 5.752 | 4.063 | 3.264 | 2.798 | 2.495 | 2.282 | 2.124 | 2.002 | 1.906 | 1.828 | 1.762 |
| 7° | | | | | | | | | 17.907 | 7.283 | 4.248 | 3.804 | 3.203 | 2.819 | 2.551 | 2.353 | 2.204 | 2.087 | 1.991 |
| 8° | | | | | | | | | | 30.803 | 9.275 | 5.824 | 4.412 | 3.647 | 3.167 | 2.838 | 2.600 | 2.420 | 2.276 |
| 9° | | | | | | | | | | | 74.645 | 12.004 | 6.972 | 5.111 | 4.047 | 3.550 | 3.150 | 2.864 | 2.649 |
| 10° | | | | | | | | | | | | | 15.944 | 8.381 | 5.918 | 4.697 | 3.971 | 3.490 | 3.147 |
| 11° | | | | | | | | | | | | | | 22.158 | 10.144 | 6.859 | 5.322 | 4.437 | 3.859 |
| 12° | | | | | | | | | | | | | | | 33.459 | 12.431 | 7.972 | 6.033 | 4.950 |
| 13° | | | | | | | | | | | | | | | | 60.163 | 15.499 | 9.309 | 6.850 |
| 14° | | | | | | | | | | | | | | | | | 204.843 | 19.868 | 10.957 |
| 15° | | | | | | | | | | | | | | | | | | | 26.528 |
| 16° | | | | | | | | | | | | | | | | | | | |
| 17° | | | | | | | | | | | | | | | | | | | |
| 18° | | | | | | | | | | | | | | | | | | | |
| 19° | | | | | | | | | | | | | | | | | | | |
| 20° | | | | | | | | | | | | | | | | | | | |
| 21° | | | | | | | | | | | | | | | | | | | |
| 22° | | | | | | | | | | | | | | | | | | | |
| 23° | | | | | | | | | | | | | | | | | | | |
| 24° | | | | | | | | | | | | | | | | | | | |
| 25° | | | | | | | | | | | | | | | | | | | |
| 26° | | | | | | | | | | | | | | | | | | | |
| 27° | | | | | | | | | | | | | | | | | | | |
| 28° | | | | | | | | | | | | | | | | | | | |
| 29° | | | | | | | | | | | | | | | | | | | |
| 30° | | | | | | | | | | | | | | | | | | | |

| α' → ω' ↓ | 39° | 40° | 41° | 42° | 43° | 44° | 45° | 46° | 47° | 48° | 49° | 50° | 51° | 52° | 53° | 54° | 55° | 56° | 57° |
|---|---|---|---|---|---|---|---|---|---|---|---|---|---|---|---|---|---|---|---|
| 1° | 1.051 | 1.051 | 1.051 | 1.051 | 1.051 | 1.051 | 1.051 | 1.051 | 1.051 | 1.051 | 1.051 | 1.051 | 1.051 | 1.051 | 1.052 | 1.052 | 1.054 | 1.054 | 1.055 |
| 2° | 1.107 | 1.105 | 1.104 | 1.104 | 1.104 | 1.104 | 1.102 | 1.102 | 1.102 | 1.104 | 1.104 | 1.104 | 1.104 | 1.105 | 1.107 | 1.107 | 1.108 | 1.109 | 1.111 |
| 3° | 1.165 | 1.163 | 1.162 | 1.160 | 1.159 | 1.159 | 1.159 | 1.157 | 1.157 | 1.159 | 1.159 | 1.159 | 1.160 | 1.162 | 1.163 | 1.165 | 1.166 | 1.168 | 1.171 |
| 4° | 1.225 | 1.224 | 1.222 | 1.219 | 1.219 | 1.218 | 1.216 | 1.216 | 1.216 | 1.216 | 1.216 | 1.218 | 1.219 | 1.219 | 1.222 | 1.224 | 1.225 | 1.228 | 1.231 |
| 5° | 1.293 | 1.289 | 1.286 | 1.284 | 1.283 | 1.281 | 1.280 | 1.278 | 1.278 | 1.278 | 1.278 | 1.280 | 1.281 | 1.283 | 1.284 | 1.286 | 1.289 | 1.293 | 1.296 |
| 6° | 1.364 | 1.359 | 1.354 | 1.351 | 1.348 | 1.347 | 1.345 | 1.344 | 1.344 | 1.342 | 1.344 | 1.344 | 1.345 | 1.347 | 1.348 | 1.351 | 1.354 | 1.359 | 1.364 |
| 7° | 1.440 | 1.434 | 1.429 | 1.425 | 1.420 | 1.418 | 1.415 | 1.414 | 1.412 | 1.412 | 1.412 | 1.412 | 1.414 | 1.415 | 1.418 | 1.420 | 1.425 | 1.429 | 1.434 |
| 8° | 1.524 | 1.516 | 1.509 | 1.503 | 1.498 | 1.493 | 1.490 | 1.487 | 1.485 | 1.485 | 1.483 | 1.485 | 1.485 | 1.487 | 1.490 | 1.493 | 1.498 | 1.503 | 1.509 |
| 9° | 1.614 | 1.602 | 1.594 | 1.587 | 1.581 | 1.574 | 1.569 | 1.566 | 1.564 | 1.561 | 1.561 | 1.561 | 1.561 | 1.564 | 1.566 | 1.569 | 1.574 | 1.581 | 1.587 |
| 10° | 1.711 | 1.697 | 1.687 | 1.677 | 1.669 | 1.662 | 1.655 | 1.650 | 1.647 | 1.645 | 1.644 | 1.642 | 1.644 | 1.645 | 1.647 | 1.650 | 1.655 | 1.662 | 1.669 |
| 11° | 1.818 | 1.801 | 1.788 | 1.776 | 1.764 | 1.755 | 1.749 | 1.742 | 1.737 | 1.733 | 1.730 | 1.730 | 1.730 | 1.730 | 1.733 | 1.737 | 1.742 | 1.749 | 1.755 |
| 12° | 1.935 | 1.915 | 1.897 | 1.883 | 1.869 | 1.856 | 1.848 | 1.839 | 1.833 | 1.828 | 1.825 | 1.821 | 1.821 | 1.821 | 1.825 | 1.828 | 1.833 | 1.839 | 1.848 |
| 13° | 2.065 | 2.040 | 2.018 | 1.998 | 1.982 | 1.967 | 1.955 | 1.945 | 1.936 | 1.929 | 1.924 | 1.922 | 1.920 | 1.920 | 1.924 | 1.924 | 1.929 | 1.936 | 1.945 |
| 14° | 2.210 | 2.178 | 2.151 | 2.126 | 2.106 | 2.089 | 2.072 | 2.059 | 2.048 | 2.040 | 2.033 | 2.027 | 2.026 | 2.024 | 2.025 | 2.027 | 2.033 | 2.040 | 2.048 |
| 15° | 2.371 | 2.331 | 2.298 | 2.269 | 2.243 | 2.220 | 2.201 | 2.183 | 2.170 | 2.158 | 2.149 | 2.143 | 2.140 | 2.138 | 2.138 | 2.140 | 2.143 | 2.149 | 2.158 |
| 16° | 2.524 | 2.505 | 2.461 | 2.427 | 2.393 | 2.365 | 2.341 | 2.321 | 2.304 | 2.288 | 2.276 | 2.268 | 2.262 | 2.259 | 2.257 | 2.259 | 2.262 | 2.268 | 2.276 |
| 17° | 2.759 | 2.700 | 2.646 | 2.602 | 2.562 | 2.526 | 2.497 | 2.470 | 2.448 | 2.430 | 2.415 | 2.403 | 2.395 | 2.389 | 2.387 | 2.387 | 2.389 | 2.395 | 2.403 |
| 18° | 2.997 | 2.922 | 2.856 | 2.798 | 2.750 | 2.706 | 2.670 | 2.638 | 2.610 | 2.587 | 2.568 | 2.551 | 2.541 | 2.532 | 2.526 | 2.524 | 2.526 | 2.532 | 2.541 |
| 19° | 3.272 | 3.177 | 3.096 | 3.023 | 2.961 | 2.908 | 2.861 | 2.821 | 2.787 | 2.759 | 2.733 | 2.713 | 2.698 | 2.687 | 2.678 | 2.676 | 2.676 | 2.678 | 2.687 |
| 20° | 3.594 | 3.472 | 3.371 | 3.281 | 3.203 | 3.136 | 3.079 | 3.028 | 2.986 | 2.949 | 2.917 | 2.892 | 2.872 | 2.856 | 2.845 | 2.841 | 2.838 | 2.841 | 2.845 |
| 21° | 3.977 | 3.823 | 3.690 | 3.579 | 3.480 | 3.396 | 3.323 | 3.262 | 3.208 | 3.162 | 3.124 | 3.091 | 3.065 | 3.044 | 3.030 | 3.018 | 3.014 | 3.014 | 3.018 |
| 22° | 4.445 | 4.240 | 4.072 | 3.926 | 3.801 | 3.695 | 3.605 | 3.526 | 3.459 | 3.401 | 3.353 | 3.311 | 3.279 | 3.252 | 3.232 | 3.218 | 3.208 | 3.207 | 3.208 |
| 23° | 5.020 | 4.753 | 4.526 | 4.340 | 4.180 | 4.046 | 3.929 | 3.831 | 3.747 | 3.674 | 3.613 | 3.560 | 3.519 | 3.483 | 3.457 | 3.437 | 3.424 | 3.416 | 3.416 |
| 24° | 5.753 | 5.386 | 5.086 | 4.839 | 4.632 | 4.457 | 4.309 | 4.183 | 4.075 | 3.983 | 3.907 | 3.840 | 3.787 | 3.741 | 3.706 | 3.679 | 3.661 | 3.650 | 3.645 |
| 25° | 6.718 | 6.204 | 5.792 | 5.457 | 5.182 | 4.955 | 4.759 | 4.600 | 4.460 | 4.343 | 4.246 | 4.163 | 4.092 | 4.034 | 3.989 | 3.952 | 3.926 | 3.907 | 3.898 |
| 26° | 8.045 | 7.283 | 6.700 | 6.238 | 5.864 | 5.559 | 5.305 | 5.094 | 4.917 | 4.766 | 4.638 | 4.533 | 4.442 | 4.367 | 4.306 | 4.261 | 4.222 | 4.198 | 4.183 |
| 27° | 9.982 | 8.799 | 7.928 | 7.258 | 6.737 | 6.320 | 5.978 | 5.694 | 5.460 | 5.265 | 5.101 | 4.965 | 4.849 | 4.753 | 4.674 | 4.609 | 4.561 | 4.526 | 4.501 |
| 28° | 13.084 | 11.057 | 9.660 | 8.650 | 7.889 | 7.298 | 6.825 | 6.442 | 6.128 | 5.868 | 5.655 | 5.475 | 5.323 | 5.200 | 5.097 | 5.013 | 4.948 | 4.897 | 4.863 |
| 29° | 18.876 | 14.796 | 12.325 | 10.667 | 9.487 | 8.608 | 7.928 | 7.395 | 6.963 | 6.613 | 6.328 | 6.090 | 5.889 | 5.725 | 5.593 | 5.483 | 5.393 | 5.327 | 5.276 |
| 30° | 33.459 | 22.203 | 16.906 | 13.841 | 11.848 | 10.455 | 9.432 | 8.650 | 8.045 | 7.555 | 7.163 | 6.839 | 6.576 | 6.359 | 6.178 | 6.032 | 5.917 | 5.824 | 5.753 |

## Table 58  Ratio of Downstream to Upstream Pressure for Two-Dimensional Shock

**For a perfect gas with constant specific heat and molecular weight**
**k = 1.4**

| $\omega'$ \ $\alpha'\rightarrow$ | 20° | 21° | 22° | 23° | 24° | 25° | 26° | 27° | 28° | 29° | 30° | 31° | 32° | 33° | 34° | 35° | 36° | 37° | 38° |
|---|---|---|---|---|---|---|---|---|---|---|---|---|---|---|---|---|---|---|---|
| 1° | 1.081 | 1.078 | 1.075 | 1.072 | 1.069 | 1.066 | 1.065 | 1.064 | 1.062 | 1.061 | 1.059 | 1.058 | 1.056 | 1.055 | 1.055 | 1.054 | 1.054 | 1.052 | 1.052 |
| 2° | 1.172 | 1.165 | 1.157 | 1.152 | 1.146 | 1.141 | 1.136 | 1.133 | 1.128 | 1.125 | 1.122 | 1.120 | 1.117 | 1.115 | 1.112 | 1.111 | 1.109 | 1.108 | 1.107 |
| 3° | 1.277 | 1.263 | 1.249 | 1.240 | 1.231 | 1.222 | 1.215 | 1.208 | 1.202 | 1.197 | 1.192 | 1.187 | 1.182 | 1.180 | 1.177 | 1.174 | 1.171 | 1.168 | 1.166 |
| 4° | 1.398 | 1.378 | 1.358 | 1.342 | 1.327 | 1.315 | 1.302 | 1.292 | 1.283 | 1.275 | 1.267 | 1.260 | 1.254 | 1.249 | 1.243 | 1.239 | 1.236 | 1.231 | 1.228 |
| 5° | 1.540 | 1.509 | 1.482 | 1.456 | 1.436 | 1.417 | 1.401 | 1.386 | 1.373 | 1.361 | 1.350 | 1.341 | 1.331 | 1.324 | 1.318 | 1.312 | 1.305 | 1.301 | 1.296 |
| 6° | 1.709 | 1.662 | 1.624 | 1.589 | 1.559 | 1.533 | 1.511 | 1.490 | 1.472 | 1.456 | 1.442 | 1.429 | 1.417 | 1.407 | 1.398 | 1.389 | 1.381 | 1.375 | 1.368 |
| 7° | 1.913 | 1.848 | 1.792 | 1.743 | 1.702 | 1.667 | 1.635 | 1.609 | 1.584 | 1.563 | 1.543 | 1.527 | 1.511 | 1.496 | 1.485 | 1.474 | 1.464 | 1.455 | 1.447 |
| 8° | 2.162 | 2.070 | 1.991 | 1.926 | 1.870 | 1.821 | 1.780 | 1.743 | 1.711 | 1.682 | 1.658 | 1.635 | 1.614 | 1.597 | 1.581 | 1.566 | 1.553 | 1.541 | 1.532 |
| 9° | 2.478 | 2.345 | 2.235 | 2.143 | 2.068 | 2.002 | 1.946 | 1.897 | 1.855 | 1.818 | 1.785 | 1.755 | 1.730 | 1.708 | 1.686 | 1.669 | 1.652 | 1.637 | 1.624 |
| 10° | 2.891 | 2.696 | 2.540 | 2.410 | 2.306 | 2.216 | 2.141 | 2.078 | 2.020 | 1.973 | 1.929 | 1.892 | 1.860 | 1.830 | 1.804 | 1.781 | 1.761 | 1.742 | 1.725 |
| 11° | 3.455 | 3.155 | 2.915 | 2.743 | 2.595 | 2.475 | 2.373 | 2.288 | 2.214 | 2.151 | 2.096 | 2.048 | 2.005 | 1.967 | 1.935 | 1.906 | 1.879 | 1.856 | 1.835 |
| 12° | 4.264 | 3.786 | 3.438 | 3.171 | 2.963 | 2.794 | 2.655 | 2.541 | 2.442 | 2.359 | 2.288 | 2.226 | 2.172 | 2.124 | 2.081 | 2.046 | 2.013 | 1.984 | 1.958 |
| 13° | 5.527 | 4.708 | 4.149 | 3.741 | 3.437 | 3.196 | 3.004 | 2.847 | 2.715 | 2.606 | 2.512 | 2.432 | 2.363 | 2.302 | 2.249 | 2.202 | 2.162 | 2.126 | 2.092 |
| 14° | 7.797 | 6.179 | 5.200 | 4.542 | 4.072 | 3.720 | 3.447 | 3.228 | 3.051 | 2.904 | 2.779 | 2.674 | 2.585 | 2.508 | 2.440 | 2.381 | 2.329 | 2.284 | 2.245 |
| 15° | 13.022 | 8.905 | 6.921 | 5.745 | 4.977 | 4.432 | 4.026 | 3.714 | 3.467 | 3.267 | 3.103 | 2.963 | 2.847 | 2.748 | 2.661 | 2.587 | 2.522 | 2.464 | 2.415 |
| 16° | 38.025 | 15.712 | 10.234 | 7.763 | 6.359 | 5.453 | 4.823 | 4.358 | 4.003 | 3.725 | 3.501 | 3.313 | 3.160 | 3.030 | 2.919 | 2.825 | 2.741 | 2.670 | 2.608 |
| 17° | | 62.627 | 19.326 | 11.848 | 8.743 | 7.050 | 5.982 | 5.254 | 4.720 | 4.318 | 4.003 | 3.749 | 3.542 | 3.371 | 3.225 | 3.103 | 2.997 | 2.906 | 2.827 |
| 18° | | | 151.658 | 24.488 | 13.841 | 9.895 | 7.834 | 6.576 | 5.725 | 5.118 | 4.658 | 4.303 | 4.017 | 3.787 | 3.594 | 3.434 | 3.299 | 3.182 | 3.084 |
| 19° | | | | | 32.419 | 16.393 | 11.258 | 8.737 | 7.243 | 6.255 | 5.551 | 5.031 | 4.629 | 4.309 | 4.049 | 3.837 | 3.658 | 3.508 | 3.381 |
| 20° | | | | | | 46.239 | 19.735 | 12.919 | 9.787 | 8.000 | 6.839 | 6.032 | 5.438 | 4.982 | 4.622 | 4.334 | 4.098 | 3.901 | 3.736 |
| 21° | | | | | | | 76.353 | 24.382 | 14.958 | 11.023 | 8.862 | 7.503 | 6.567 | 5.885 | 5.371 | 4.965 | 4.642 | 4.380 | 4.160 |
| 22° | | | | | | | | 192.864 | 31.194 | 17.553 | 12.499 | 9.866 | 8.251 | 7.163 | 6.381 | 5.796 | 5.342 | 4.979 | 4.687 |
| 23° | | | | | | | | | | 42.209 | 20.965 | 14.290 | 11.032 | 9.106 | 7.834 | 6.934 | 6.268 | 5.753 | 5.349 |
| 24° | | | | | | | | | | | 63.034 | 25.646 | 16.523 | 12.421 | 10.092 | 8.595 | 7.555 | 6.792 | 6.212 |
| 25° | | | | | | | | | | | | 117.675 | 32.419 | 19.381 | 14.103 | 11.250 | 9.473 | 8.257 | 7.379 |
| 26° | | | | | | | | | | | | | 630.364 | 43.226 | 23.142 | 16.166 | 12.617 | 10.486 | 9.060 |
| 27° | | | | | | | | | | | | | | | 63.034 | 28.381 | 18.770 | 14.278 | 11.668 |
| 28° | | | | | | | | | | | | | | | | 111.845 | 36.169 | 22.180 | 16.307 |
| 29° | | | | | | | | | | | | | | | | | 415.687 | 48.859 | 26.802 |
| 30° | | | | | | | | | | | | | | | | | | | 73.365 |

| $\omega'$ \ $\alpha'\rightarrow$ | 58° | 59° | 60° | 61° | 62° | 63° | 64° | 65° | 66° | 67° | 68° | 69° | 70° | 71° | 72° | 73° | 74° | 75° | 76° |
|---|---|---|---|---|---|---|---|---|---|---|---|---|---|---|---|---|---|---|---|
| 1° | 1.055 | 1.056 | 1.058 | 1.059 | 1.061 | 1.062 | 1.064 | 1.065 | 1.066 | 1.069 | 1.072 | 1.075 | 1.078 | 1.081 | 1.085 | 1.089 | 1.093 | 1.099 | 1.107 |
| 2° | 1.112 | 1.115 | 1.117 | 1.120 | 1.122 | 1.125 | 1.128 | 1.133 | 1.136 | 1.141 | 1.146 | 1.152 | 1.157 | 1.165 | 1.172 | 1.181 | 1.191 | 1.203 | 1.216 |
| 3° | 1.174 | 1.177 | 1.180 | 1.182 | 1.187 | 1.192 | 1.197 | 1.202 | 1.208 | 1.215 | 1.222 | 1.231 | 1.240 | 1.249 | 1.263 | 1.277 | 1.293 | 1.311 | 1.331 |
| 4° | 1.236 | 1.239 | 1.243 | 1.249 | 1.254 | 1.260 | 1.267 | 1.275 | 1.283 | 1.292 | 1.302 | 1.315 | 1.327 | 1.342 | 1.358 | 1.378 | 1.398 | 1.423 | 1.452 |
| 5° | 1.301 | 1.305 | 1.312 | 1.318 | 1.324 | 1.331 | 1.341 | 1.350 | 1.361 | 1.373 | 1.386 | 1.401 | 1.417 | 1.436 | 1.456 | 1.482 | 1.509 | 1.540 | 1.577 |
| 6° | 1.368 | 1.375 | 1.381 | 1.389 | 1.398 | 1.407 | 1.417 | 1.429 | 1.442 | 1.456 | 1.472 | 1.490 | 1.511 | 1.533 | 1.559 | 1.589 | 1.624 | 1.662 | 1.709 |
| 7° | 1.440 | 1.447 | 1.455 | 1.464 | 1.474 | 1.485 | 1.496 | 1.511 | 1.527 | 1.543 | 1.563 | 1.584 | 1.609 | 1.635 | 1.667 | 1.702 | 1.743 | 1.792 | 1.848 |
| 8° | 1.516 | 1.524 | 1.532 | 1.541 | 1.553 | 1.566 | 1.581 | 1.597 | 1.614 | 1.635 | 1.658 | 1.682 | 1.711 | 1.743 | 1.780 | 1.821 | 1.870 | 1.926 | 1.991 |
| 9° | 1.594 | 1.602 | 1.614 | 1.624 | 1.637 | 1.652 | 1.669 | 1.686 | 1.708 | 1.730 | 1.755 | 1.785 | 1.818 | 1.855 | 1.897 | 1.946 | 2.002 | 2.068 | 2.143 |
| 10° | 1.677 | 1.687 | 1.697 | 1.711 | 1.725 | 1.742 | 1.761 | 1.781 | 1.804 | 1.830 | 1.860 | 1.892 | 1.929 | 1.973 | 2.020 | 2.078 | 2.141 | 2.216 | 2.306 |
| 11° | 1.764 | 1.776 | 1.788 | 1.801 | 1.818 | 1.835 | 1.856 | 1.879 | 1.906 | 1.935 | 1.967 | 2.005 | 2.048 | 2.096 | 2.151 | 2.214 | 2.288 | 2.373 | 2.475 |
| 12° | 1.856 | 1.869 | 1.883 | 1.897 | 1.915 | 1.935 | 1.958 | 1.984 | 2.013 | 2.046 | 2.081 | 2.124 | 2.172 | 2.226 | 2.288 | 2.359 | 2.442 | 2.541 | 2.655 |
| 13° | 1.955 | 1.967 | 1.982 | 1.998 | 2.018 | 2.040 | 2.065 | 2.092 | 2.126 | 2.162 | 2.202 | 2.249 | 2.302 | 2.363 | 2.432 | 2.512 | 2.606 | 2.715 | 2.847 |
| 14° | 2.059 | 2.072 | 2.089 | 2.106 | 2.126 | 2.151 | 2.178 | 2.210 | 2.245 | 2.284 | 2.329 | 2.381 | 2.440 | 2.508 | 2.585 | 2.674 | 2.779 | 2.904 | 3.051 |
| 15° | 2.170 | 2.183 | 2.201 | 2.220 | 2.243 | 2.269 | 2.298 | 2.331 | 2.371 | 2.415 | 2.464 | 2.522 | 2.587 | 2.661 | 2.748 | 2.847 | 2.963 | 3.103 | 3.267 |
| 16° | 2.288 | 2.304 | 2.321 | 2.341 | 2.365 | 2.393 | 2.427 | 2.461 | 2.505 | 2.524 | 2.608 | 2.670 | 2.741 | 2.825 | 2.919 | 3.030 | 3.160 | 3.313 | 3.501 |
| 17° | 2.415 | 2.430 | 2.448 | 2.470 | 2.497 | 2.526 | 2.562 | 2.602 | 2.646 | 2.700 | 2.759 | 2.827 | 2.906 | 2.997 | 3.103 | 3.225 | 3.371 | 3.542 | 3.749 |
| 18° | 2.551 | 2.568 | 2.587 | 2.610 | 2.638 | 2.670 | 2.706 | 2.750 | 2.798 | 2.856 | 2.922 | 2.997 | 3.084 | 3.182 | 3.299 | 3.434 | 3.594 | 3.787 | 4.017 |
| 19° | 2.698 | 2.713 | 2.737 | 2.759 | 2.787 | 2.821 | 2.861 | 2.908 | 2.963 | 3.023 | 3.096 | 3.177 | 3.272 | 3.381 | 3.508 | 3.658 | 3.837 | 4.049 | 4.309 |
| 20° | 2.856 | 2.872 | 2.892 | 2.917 | 2.949 | 2.986 | 3.028 | 3.079 | 3.136 | 3.203 | 3.281 | 3.371 | 3.472 | 3.594 | 3.736 | 3.901 | 4.098 | 4.334 | 4.622 |
| 21° | 3.030 | 3.044 | 3.065 | 3.091 | 3.124 | 3.162 | 3.208 | 3.262 | 3.323 | 3.396 | 3.480 | 3.579 | 3.690 | 3.824 | 3.977 | 4.160 | 4.380 | 4.642 | 4.965 |
| 22° | 3.218 | 3.232 | 3.252 | 3.252 | 3.279 | 3.311 | 3.353 | 3.401 | 3.459 | 3.526 | 3.605 | 3.695 | 3.801 | 3.926 | 4.240 | 4.445 | 4.687 | 4.979 | 5.342 |
| 23° | 3.424 | 3.437 | 3.457 | 3.483 | 3.519 | 3.560 | 3.613 | 3.674 | 3.747 | 3.831 | 3.929 | 4.046 | 4.180 | 4.340 | 4.526 | 4.753 | 5.020 | 5.349 | 5.753 |
| 24° | 3.650 | 3.661 | 3.679 | 3.706 | 3.741 | 3.787 | 3.840 | 3.907 | 3.983 | 4.075 | 4.183 | 4.309 | 4.457 | 4.632 | 4.839 | 5.086 | 5.386 | 5.753 | 6.212 |
| 25° | 3.898 | 3.907 | 3.926 | 3.952 | 3.989 | 4.034 | 4.092 | 4.163 | 4.246 | 4.343 | 4.460 | 4.600 | 4.759 | 4.955 | 5.182 | 5.457 | 5.792 | 6.204 | 6.718 |
| 26° | 4.177 | 4.183 | 4.198 | 4.222 | 4.261 | 4.306 | 4.367 | 4.442 | 4.533 | 4.638 | 4.766 | 4.917 | 5.094 | 5.305 | 5.559 | 5.864 | 6.238 | 6.700 | 7.283 |
| 27° | 4.488 | 4.488 | 4.501 | 4.526 | 4.561 | 4.609 | 4.674 | 4.753 | 4.849 | 4.965 | 5.101 | 5.265 | 5.460 | 5.694 | 5.978 | 6.320 | 6.737 | 7.258 | 7.928 |
| 28° | 4.843 | 4.836 | 4.843 | 4.863 | 4.897 | 4.948 | 5.013 | 5.097 | 5.200 | 5.323 | 5.475 | 5.655 | 5.868 | 6.128 | 6.442 | 6.825 | 7.298 | 7.889 | 8.650 |
| 29° | 5.243 | 5.229 | 5.229 | 5.243 | 5.276 | 5.327 | 5.393 | 5.483 | 5.593 | 5.725 | 5.889 | 6.090 | 6.328 | 6.613 | 6.963 | 7.395 | 7.928 | 8.608 | 9.487 |
| 30° | 5.706 | 5.674 | 5.667 | 5.674 | 5.706 | 5.753 | 5.824 | 5.917 | 6.032 | 6.178 | 6.359 | 6.576 | 6.839 | 7.163 | 7.555 | 8.045 | 8.650 | 9.432 | 10.455 |

# Table 59  Ratio of Downstream to Upstream Density for Two-Dimensional Shock

For a perfect gas with constant specific heat and molecular weight
k = 1.4

| ω' ↓ \ α' → | 1° | 2° | 3° | 4° | 5° | 6° | 7° | 8° | 9° | 10° | 11° | 12° | 13° | 14° | 15° | 16° | 17° | 18° | 19° |
|---|---|---|---|---|---|---|---|---|---|---|---|---|---|---|---|---|---|---|---|
| 1° | | 2.001 | 1.501 | 1.334 | 1.251 | 1.201 | 1.168 | 1.145 | 1.127 | 1.113 | 1.102 | 1.094 | 1.086 | 1.080 | 1.075 | 1.070 | 1.066 | 1.063 | 1.060 |
| 2° | | | 3.002 | 2.002 | 1.669 | 1.503 | 1.403 | 1.337 | 1.290 | 1.255 | 1.227 | 1.205 | 1.188 | 1.173 | 1.161 | 1.150 | 1.141 | 1.133 | 1.126 |
| 3° | | | | 4.006 | 2.505 | 2.006 | 1.756 | 1.606 | 1.507 | 1.436 | 1.383 | 1.342 | 1.309 | 1.283 | 1.261 | 1.242 | 1.226 | 1.213 | 1.201 |
| 4° | | | | | 5.012 | 3.010 | 2.343 | 2.010 | 1.810 | 1.678 | 1.583 | 1.512 | 1.458 | 1.414 | 1.378 | 1.349 | 1.324 | 1.303 | 1.285 |
| 5° | | | | | | 6.021 | 3.516 | 2.682 | 2.265 | 2.015 | 1.849 | 1.731 | 1.643 | 1.574 | 1.520 | 1.475 | 1.438 | 1.407 | 1.381 |
| 6° | | | | | | | 7.034 | 4.025 | 3.022 | 2.522 | 2.222 | 2.022 | 1.880 | 1.774 | 1.692 | 1.626 | 1.573 | 1.529 | 1.491 |
| 7° | | | | | | | | 8.052 | 4.536 | 3.365 | 2.780 | 2.430 | 2.197 | 2.031 | 1.907 | 1.810 | 1.734 | 1.672 | 1.620 |
| 8° | | | | | | | | | 9.074 | 5.049 | 3.709 | 3.040 | 2.639 | 2.372 | 2.182 | 2.040 | 1.930 | 1.843 | 1.771 |
| 9° | | | | | | | | | | 10.102 | 5.566 | 4.056 | 3.302 | 2.850 | 2.549 | 2.335 | 2.175 | 2.051 | 1.953 |
| 10° | | | | | | | | | | | 11.136 | 6.087 | 4.405 | 3.566 | 3.063 | 2.728 | 2.490 | 2.312 | 2.174 |
| 11° | | | | | | | | | | | | 12.177 | 6.611 | 4.757 | 3.832 | 3.278 | 2.909 | 2.646 | 2.450 |
| 12° | | | | | | | | | | | | | 13.226 | 7.140 | 5.113 | 4.101 | 3.495 | 3.091 | 2.804 |
| 13° | | | | | | | | | | | | | | | 7.673 | 5.471 | 4.372 | 3.714 | 3.276 |
| 14° | | | | | | | | | | | | | | | | 8.211 | 5.834 | 4.647 | 3.936 |
| 15° | | | | | | | | | | | | | | | | | 8.755 | 6.200 | 4.924 |
| 16° | | | | | | | | | | | | | | | | | | 9.304 | 6.570 |
| 17° | | | | | | | | | | | | | | | | | | | 9.860 |
| 18° | | | | | | | | | | | | | | | | | | | |
| 19° | | | | | | | | | | | | | | | | | | | |
| 20° | | | | | | | | | | | | | | | | | | | |
| 21° | | | | | | | | | | | | | | | | | | | |
| 22° | | | | | | | | | | | | | | | | | | | |
| 23° | | | | | | | | | | | | | | | | | | | |
| 24° | | | | | | | | | | | | | | | | | | | |
| 25° | | | | | | | | | | | | | | | | | | | |
| 26° | | | | | | | | | | | | | | | | | | | |
| 27° | | | | | | | | | | | | | | | | | | | |
| 28° | | | | | | | | | | | | | | | | | | | |
| 29° | | | | | | | | | | | | | | | | | | | |
| 30° | | | | | | | | | | | | | | | | | | | |

| ω' ↓ \ α' → | 39° | 40° | 41° | 42° | 43° | 44° | 45° | 46° | 47° | 48° | 49° | 50° | 51° | 52° | 53° | 54° | 55° | 56° | 57° |
|---|---|---|---|---|---|---|---|---|---|---|---|---|---|---|---|---|---|---|---|
| 1° | 1.036 | 1.036 | 1.036 | 1.036 | 1.036 | 1.036 | 1.036 | 1.036 | 1.036 | 1.036 | 1.036 | 1.036 | 1.036 | 1.036 | 1.037 | 1.037 | 1.038 | 1.038 | 1.039 |
| 2° | 1.075 | 1.074 | 1.073 | 1.073 | 1.073 | 1.073 | 1.072 | 1.072 | 1.072 | 1.073 | 1.073 | 1.073 | 1.073 | 1.074 | 1.075 | 1.075 | 1.076 | 1.077 | 1.078 |
| 3° | 1.115 | 1.114 | 1.113 | 1.112 | 1.111 | 1.111 | 1.111 | 1.110 | 1.110 | 1.111 | 1.111 | 1.111 | 1.112 | 1.113 | 1.114 | 1.115 | 1.116 | 1.117 | 1.119 |
| 4° | 1.156 | 1.155 | 1.154 | 1.152 | 1.152 | 1.151 | 1.150 | 1.150 | 1.150 | 1.150 | 1.150 | 1.151 | 1.152 | 1.152 | 1.154 | 1.155 | 1.156 | 1.158 | 1.160 |
| 5° | 1.201 | 1.198 | 1.196 | 1.195 | 1.194 | 1.193 | 1.192 | 1.191 | 1.191 | 1.191 | 1.191 | 1.192 | 1.193 | 1.194 | 1.195 | 1.196 | 1.198 | 1.201 | 1.203 |
| 6° | 1.247 | 1.244 | 1.241 | 1.239 | 1.237 | 1.236 | 1.235 | 1.234 | 1.234 | 1.233 | 1.234 | 1.234 | 1.235 | 1.236 | 1.237 | 1.239 | 1.241 | 1.244 | 1.247 |
| 7° | 1.296 | 1.292 | 1.289 | 1.286 | 1.283 | 1.282 | 1.280 | 1.279 | 1.278 | 1.278 | 1.278 | 1.278 | 1.279 | 1.280 | 1.282 | 1.283 | 1.286 | 1.289 | 1.292 |
| 8° | 1.348 | 1.343 | 1.339 | 1.335 | 1.332 | 1.329 | 1.327 | 1.325 | 1.324 | 1.324 | 1.323 | 1.324 | 1.324 | 1.325 | 1.327 | 1.329 | 1.332 | 1.335 | 1.339 |
| 9° | 1.403 | 1.396 | 1.391 | 1.387 | 1.383 | 1.379 | 1.376 | 1.374 | 1.373 | 1.371 | 1.371 | 1.371 | 1.371 | 1.373 | 1.374 | 1.376 | 1.379 | 1.383 | 1.387 |
| 10° | 1.461 | 1.453 | 1.447 | 1.441 | 1.436 | 1.432 | 1.428 | 1.425 | 1.423 | 1.422 | 1.421 | 1.420 | 1.421 | 1.422 | 1.423 | 1.425 | 1.428 | 1.432 | 1.436 |
| 11° | 1.523 | 1.514 | 1.506 | 1.499 | 1.492 | 1.487 | 1.483 | 1.479 | 1.476 | 1.474 | 1.472 | 1.472 | 1.472 | 1.472 | 1.474 | 1.476 | 1.479 | 1.483 | 1.487 |
| 12° | 1.589 | 1.578 | 1.568 | 1.560 | 1.552 | 1.545 | 1.540 | 1.535 | 1.532 | 1.529 | 1.527 | 1.525 | 1.525 | 1.525 | 1.527 | 1.529 | 1.532 | 1.535 | 1.540 |
| 13° | 1.660 | 1.647 | 1.635 | 1.624 | 1.615 | 1.607 | 1.600 | 1.595 | 1.590 | 1.586 | 1.583 | 1.582 | 1.581 | 1.581 | 1.582 | 1.583 | 1.586 | 1.590 | 1.595 |
| 14° | 1.737 | 1.720 | 1.706 | 1.693 | 1.682 | 1.673 | 1.664 | 1.657 | 1.651 | 1.647 | 1.643 | 1.640 | 1.639 | 1.638 | 1.639 | 1.640 | 1.643 | 1.647 | 1.651 |
| 15° | 1.819 | 1.799 | 1.782 | 1.767 | 1.754 | 1.742 | 1.732 | 1.723 | 1.716 | 1.710 | 1.705 | 1.702 | 1.700 | 1.699 | 1.699 | 1.700 | 1.702 | 1.705 | 1.710 |
| 16° | 1.908 | 1.885 | 1.864 | 1.846 | 1.830 | 1.816 | 1.804 | 1.794 | 1.785 | 1.777 | 1.771 | 1.767 | 1.764 | 1.762 | 1.761 | 1.762 | 1.764 | 1.767 | 1.771 |
| 17° | 2.004 | 1.977 | 1.952 | 1.931 | 1.912 | 1.895 | 1.881 | 1.868 | 1.857 | 1.848 | 1.841 | 1.835 | 1.831 | 1.828 | 1.827 | 1.827 | 1.828 | 1.831 | 1.835 |
| 18° | 2.110 | 2.077 | 2.048 | 2.022 | 2.000 | 1.980 | 1.963 | 1.948 | 1.935 | 1.924 | 1.915 | 1.907 | 1.902 | 1.898 | 1.895 | 1.894 | 1.895 | 1.898 | 1.902 |
| 19° | 2.225 | 2.186 | 2.152 | 2.121 | 2.094 | 2.071 | 2.050 | 2.032 | 2.017 | 2.004 | 1.992 | 1.983 | 1.976 | 1.971 | 1.967 | 1.966 | 1.966 | 1.967 | 1.971 |
| 20° | 2.352 | 2.305 | 2.265 | 2.229 | 2.197 | 2.169 | 2.145 | 2.123 | 2.105 | 2.089 | 2.075 | 2.064 | 2.055 | 2.048 | 2.043 | 2.041 | 2.040 | 2.041 | 2.043 |
| 21° | 2.492 | 2.437 | 2.388 | 2.346 | 2.308 | 2.275 | 2.246 | 2.221 | 2.199 | 2.180 | 2.164 | 2.150 | 2.139 | 2.130 | 2.124 | 2.119 | 2.117 | 2.117 | 2.119 |
| 22° | 2.649 | 2.582 | 2.525 | 2.474 | 2.429 | 2.390 | 2.356 | 2.326 | 2.300 | 2.277 | 2.258 | 2.241 | 2.228 | 2.217 | 2.209 | 2.203 | 2.199 | 2.198 | 2.199 |
| 23° | 2.824 | 2.745 | 2.675 | 2.615 | 2.562 | 2.516 | 2.475 | 2.440 | 2.409 | 2.382 | 2.359 | 2.339 | 2.323 | 2.309 | 2.299 | 2.291 | 2.286 | 2.283 | 2.283 |
| 24° | 3.022 | 2.926 | 2.843 | 2.771 | 2.708 | 2.653 | 2.605 | 2.563 | 2.526 | 2.494 | 2.467 | 2.443 | 2.424 | 2.407 | 2.394 | 2.384 | 2.377 | 2.373 | 2.371 |
| 25° | 3.248 | 3.132 | 3.032 | 2.945 | 2.870 | 2.805 | 2.747 | 2.698 | 2.654 | 2.616 | 2.584 | 2.556 | 2.532 | 2.512 | 2.496 | 2.483 | 2.474 | 2.467 | 2.464 |
| 26° | 3.508 | 3.365 | 3.244 | 3.140 | 3.050 | 2.972 | 2.904 | 2.845 | 2.794 | 2.749 | 2.710 | 2.677 | 2.648 | 2.624 | 2.604 | 2.589 | 2.576 | 2.568 | 2.563 |
| 27° | 3.810 | 3.635 | 3.487 | 3.360 | 3.252 | 3.159 | 3.078 | 3.007 | 2.946 | 2.893 | 2.847 | 2.808 | 2.774 | 2.745 | 2.721 | 2.701 | 2.686 | 2.675 | 2.667 |
| 28° | 4.166 | 3.948 | 3.765 | 3.611 | 3.480 | 3.368 | 3.271 | 3.187 | 3.114 | 3.051 | 2.997 | 2.950 | 2.909 | 2.875 | 2.846 | 2.822 | 2.803 | 2.788 | 2.778 |
| 29° | 4.593 | 4.317 | 4.090 | 3.900 | 3.740 | 3.604 | 3.487 | 3.387 | 3.300 | 3.225 | 3.161 | 3.105 | 3.056 | 3.015 | 2.981 | 2.952 | 2.928 | 2.910 | 2.896 |
| 30° | 5.113 | 4.759 | 4.472 | 4.236 | 4.039 | 3.873 | 3.732 | 3.611 | 3.508 | 3.418 | 3.341 | 3.274 | 3.217 | 3.168 | 3.126 | 3.091 | 3.063 | 3.040 | 3.022 |

## Table 59  Ratio of Downstream to Upstream Density for Two-Dimensional Shock

**For a perfect gas with constant specific heat and molecular weight**
**k = 1.4**

| α′ → / ω′ ↓ | 20° | 21° | 22° | 23° | 24° | 25° | 26° | 27° | 28° | 29° | 30° | 31° | 32° | 33° | 34° | 35° | 36° | 37° | 38° |
|---|---|---|---|---|---|---|---|---|---|---|---|---|---|---|---|---|---|---|---|
| 1° | 1.057 | 1.055 | 1.053 | 1.051 | 1.049 | 1.047 | 1.046 | 1.045 | 1.044 | 1.043 | 1.042 | 1.041 | 1.040 | 1.039 | 1.039 | 1.038 | 1.038 | 1.037 | 1.037 |
| 2° | 1.120 | 1.115 | 1.110 | 1.106 | 1.102 | 1.099 | 1.095 | 1.093 | 1.090 | 1.088 | 1.086 | 1.084 | 1.082 | 1.081 | 1.079 | 1.078 | 1.077 | 1.076 | 1.075 |
| 3° | 1.190 | 1.181 | 1.173 | 1.166 | 1.160 | 1.154 | 1.149 | 1.144 | 1.140 | 1.137 | 1.133 | 1.130 | 1.127 | 1.125 | 1.123 | 1.121 | 1.119 | 1.117 | 1.116 |
| 4° | 1.269 | 1.256 | 1.243 | 1.233 | 1.223 | 1.215 | 1.207 | 1.200 | 1.194 | 1.189 | 1.184 | 1.179 | 1.175 | 1.172 | 1.168 | 1.165 | 1.163 | 1.160 | 1.158 |
| 5° | 1.358 | 1.339 | 1.322 | 1.306 | 1.293 | 1.281 | 1.271 | 1.261 | 1.253 | 1.245 | 1.238 | 1.232 | 1.226 | 1.221 | 1.217 | 1.213 | 1.209 | 1.206 | 1.203 |
| 6° | 1.460 | 1.433 | 1.409 | 1.388 | 1.370 | 1.354 | 1.340 | 1.327 | 1.316 | 1.306 | 1.297 | 1.289 | 1.281 | 1.275 | 1.269 | 1.263 | 1.258 | 1.254 | 1.250 |
| 7° | 1.577 | 1.540 | 1.508 | 1.480 | 1.456 | 1.435 | 1.416 | 1.400 | 1.385 | 1.372 | 1.360 | 1.350 | 1.340 | 1.331 | 1.324 | 1.317 | 1.311 | 1.305 | 1.300 |
| 8° | 1.712 | 1.663 | 1.620 | 1.584 | 1.553 | 1.525 | 1.501 | 1.480 | 1.461 | 1.444 | 1.429 | 1.416 | 1.403 | 1.393 | 1.383 | 1.374 | 1.366 | 1.359 | 1.353 |
| 9° | 1.872 | 1.806 | 1.750 | 1.702 | 1.662 | 1.626 | 1.595 | 1.568 | 1.544 | 1.523 | 1.504 | 1.487 | 1.472 | 1.459 | 1.446 | 1.436 | 1.426 | 1.417 | 1.409 |
| 10° | 2.064 | 1.975 | 1.901 | 1.839 | 1.786 | 1.740 | 1.701 | 1.667 | 1.636 | 1.610 | 1.586 | 1.565 | 1.547 | 1.530 | 1.515 | 1.502 | 1.490 | 1.479 | 1.469 |
| 11° | 2.298 | 2.177 | 2.079 | 1.997 | 1.928 | 1.870 | 1.820 | 1.777 | 1.739 | 1.706 | 1.677 | 1.651 | 1.628 | 1.607 | 1.589 | 1.573 | 1.558 | 1.545 | 1.533 |
| 12° | 2.590 | 2.424 | 2.291 | 2.184 | 2.095 | 2.020 | 1.956 | 1.902 | 1.854 | 1.813 | 1.777 | 1.745 | 1.717 | 1.692 | 1.669 | 1.650 | 1.632 | 1.616 | 1.602 |
| 13° | 2.964 | 2.731 | 2.551 | 2.407 | 2.291 | 2.194 | 2.113 | 2.044 | 1.984 | 1.933 | 1.888 | 1.849 | 1.815 | 1.784 | 1.757 | 1.733 | 1.712 | 1.693 | 1.675 |
| 14° | 3.463 | 3.126 | 2.875 | 2.680 | 2.525 | 2.399 | 2.295 | 2.207 | 2.133 | 2.069 | 2.013 | 1.965 | 1.923 | 1.886 | 1.853 | 1.824 | 1.798 | 1.775 | 1.755 |
| 15° | 4.160 | 3.652 | 3.291 | 3.020 | 2.811 | 2.645 | 2.509 | 2.397 | 2.303 | 2.223 | 2.155 | 2.095 | 2.044 | 1.999 | 1.959 | 1.924 | 1.893 | 1.865 | 1.841 |
| 16° | 5.205 | 4.388 | 3.844 | 3.457 | 3.168 | 2.944 | 2.766 | 2.621 | 2.501 | 2.401 | 2.316 | 2.242 | 2.179 | 2.124 | 2.076 | 2.034 | 1.996 | 1.963 | 1.934 |
| 17° | 6.945 | 5.490 | 4.618 | 4.039 | 3.626 | 3.318 | 3.079 | 2.890 | 2.735 | 2.608 | 2.501 | 2.410 | 2.332 | 2.265 | 2.206 | 2.155 | 2.110 | 2.070 | 2.035 |
| 18° | 10.423 | 7.325 | 5.778 | 4.852 | 4.236 | 3.798 | 3.470 | 3.217 | 3.015 | 2.852 | 2.716 | 2.603 | 2.506 | 2.424 | 2.352 | 2.290 | 2.236 | 2.188 | 2.147 |
| 19° | | | 7.709 | 6.070 | 5.089 | 4.437 | 3.972 | 3.625 | 3.357 | 3.144 | 2.970 | 2.827 | 2.707 | 2.605 | 2.517 | 2.442 | 2.376 | 2.319 | 2.269 |
| 20° | | | | 8.099 | 6.367 | 5.330 | 4.640 | 4.150 | 3.783 | 3.500 | 3.274 | 3.091 | 2.940 | 2.813 | 2.705 | 2.613 | 2.534 | 2.465 | 2.405 |
| 21° | | | | | 8.495 | 6.669 | 5.575 | 4.848 | 4.330 | 3.944 | 3.645 | 3.408 | 3.215 | 3.055 | 2.922 | 2.808 | 2.711 | 2.628 | 2.555 |
| 22° | | | | | | 8.898 | 6.975 | 5.824 | 5.059 | 4.514 | 4.108 | 3.794 | 3.544 | 3.341 | 3.173 | 3.033 | 2.914 | 2.812 | 2.725 |
| 23° | | | | | | | 9.306 | 7.287 | 6.077 | 5.274 | 4.702 | 4.275 | 3.945 | 3.683 | 3.470 | 3.294 | 3.147 | 3.022 | 2.916 |
| 24° | | | | | | | | | 7.604 | 6.336 | 5.493 | 4.894 | 4.446 | 4.100 | 3.825 | 3.602 | 3.418 | 3.264 | 3.134 |
| 25° | | | | | | | | | | 7.927 | 6.599 | 5.717 | 5.089 | 4.621 | 4.259 | 3.971 | 3.738 | 3.545 | 3.384 |
| 26° | | | | | | | | | | | 8.256 | 6.868 | 5.945 | 5.289 | 4.799 | 4.421 | 4.120 | 3.877 | 3.676 |
| 27° | | | | | | | | | | | | 8.593 | 7.142 | 6.179 | 5.493 | 4.982 | 4.587 | 4.274 | 4.019 |
| 28° | | | | | | | | | | | | | | 7.423 | 6.418 | 5.703 | 5.170 | 4.758 | 4.431 |
| 29° | | | | | | | | | | | | | | | | 6.662 | 5.917 | 5.362 | 4.933 |
| 30° | | | | | | | | | | | | | | | | | 6.913 | 6.137 | 5.559 |

| α′ → / ω′ ↓ | 58° | 59° | 60° | 61° | 62° | 63° | 64° | 65° | 66° | 67° | 68° | 69° | 70° | 71° | 72° | 73° | 74° | 75° | 76° |
|---|---|---|---|---|---|---|---|---|---|---|---|---|---|---|---|---|---|---|---|
| 1° | 1.039 | 1.040 | 1.041 | 1.042 | 1.043 | 1.044 | 1.045 | 1.046 | 1.047 | 1.049 | 1.051 | 1.053 | 1.055 | 1.057 | 1.060 | 1.063 | 1.066 | 1.070 | 1.075 |
| 2° | 1.079 | 1.081 | 1.082 | 1.084 | 1.086 | 1.088 | 1.090 | 1.093 | 1.095 | 1.099 | 1.102 | 1.106 | 1.110 | 1.115 | 1.120 | 1.126 | 1.133 | 1.141 | 1.150 |
| 3° | 1.121 | 1.123 | 1.125 | 1.127 | 1.130 | 1.133 | 1.137 | 1.140 | 1.144 | 1.149 | 1.154 | 1.160 | 1.166 | 1.173 | 1.181 | 1.190 | 1.201 | 1.213 | 1.226 |
| 4° | 1.163 | 1.165 | 1.168 | 1.172 | 1.175 | 1.179 | 1.184 | 1.189 | 1.194 | 1.200 | 1.207 | 1.215 | 1.223 | 1.233 | 1.243 | 1.256 | 1.269 | 1.285 | 1.303 |
| 5° | 1.206 | 1.209 | 1.213 | 1.217 | 1.221 | 1.226 | 1.232 | 1.238 | 1.245 | 1.253 | 1.261 | 1.271 | 1.281 | 1.293 | 1.306 | 1.322 | 1.339 | 1.358 | 1.381 |
| 6° | 1.250 | 1.254 | 1.258 | 1.263 | 1.269 | 1.275 | 1.281 | 1.289 | 1.297 | 1.306 | 1.316 | 1.327 | 1.340 | 1.354 | 1.370 | 1.388 | 1.409 | 1.433 | 1.460 |
| 7° | 1.296 | 1.300 | 1.305 | 1.311 | 1.317 | 1.324 | 1.331 | 1.340 | 1.350 | 1.360 | 1.372 | 1.385 | 1.400 | 1.416 | 1.435 | 1.456 | 1.480 | 1.508 | 1.540 |
| 8° | 1.343 | 1.348 | 1.353 | 1.359 | 1.366 | 1.374 | 1.383 | 1.393 | 1.403 | 1.416 | 1.429 | 1.444 | 1.461 | 1.480 | 1.501 | 1.525 | 1.553 | 1.584 | 1.620 |
| 9° | 1.391 | 1.396 | 1.403 | 1.409 | 1.417 | 1.426 | 1.436 | 1.446 | 1.459 | 1.472 | 1.487 | 1.504 | 1.523 | 1.544 | 1.568 | 1.595 | 1.626 | 1.662 | 1.702 |
| 10° | 1.441 | 1.447 | 1.453 | 1.461 | 1.469 | 1.479 | 1.490 | 1.502 | 1.515 | 1.530 | 1.547 | 1.565 | 1.586 | 1.610 | 1.636 | 1.667 | 1.701 | 1.740 | 1.786 |
| 11° | 1.492 | 1.499 | 1.506 | 1.514 | 1.523 | 1.533 | 1.545 | 1.558 | 1.573 | 1.589 | 1.607 | 1.628 | 1.651 | 1.677 | 1.706 | 1.739 | 1.777 | 1.820 | 1.870 |
| 12° | 1.545 | 1.552 | 1.560 | 1.568 | 1.578 | 1.589 | 1.602 | 1.616 | 1.632 | 1.650 | 1.669 | 1.692 | 1.717 | 1.745 | 1.777 | 1.813 | 1.854 | 1.902 | 1.956 |
| 13° | 1.600 | 1.607 | 1.615 | 1.624 | 1.635 | 1.647 | 1.660 | 1.675 | 1.693 | 1.712 | 1.733 | 1.757 | 1.784 | 1.815 | 1.849 | 1.888 | 1.933 | 1.984 | 2.044 |
| 14° | 1.657 | 1.664 | 1.673 | 1.682 | 1.693 | 1.706 | 1.720 | 1.737 | 1.755 | 1.775 | 1.798 | 1.824 | 1.853 | 1.886 | 1.923 | 1.965 | 2.013 | 2.069 | 2.133 |
| 15° | 1.716 | 1.723 | 1.732 | 1.742 | 1.754 | 1.767 | 1.782 | 1.799 | 1.819 | 1.841 | 1.865 | 1.893 | 1.924 | 1.959 | 1.999 | 2.044 | 2.095 | 2.155 | 2.223 |
| 16° | 1.777 | 1.785 | 1.794 | 1.804 | 1.816 | 1.830 | 1.846 | 1.864 | 1.885 | 1.908 | 1.934 | 1.963 | 1.996 | 2.034 | 2.076 | 2.124 | 2.179 | 2.242 | 2.316 |
| 17° | 1.841 | 1.848 | 1.857 | 1.868 | 1.881 | 1.895 | 1.912 | 1.931 | 1.952 | 1.977 | 2.004 | 2.035 | 2.070 | 2.110 | 2.155 | 2.205 | 2.265 | 2.332 | 2.410 |
| 18° | 1.907 | 1.915 | 1.924 | 1.935 | 1.948 | 1.963 | 1.980 | 2.000 | 2.022 | 2.048 | 2.077 | 2.110 | 2.147 | 2.188 | 2.236 | 2.290 | 2.352 | 2.424 | 2.506 |
| 19° | 1.976 | 1.983 | 1.992 | 2.004 | 2.017 | 2.032 | 2.050 | 2.071 | 2.094 | 2.121 | 2.152 | 2.186 | 2.225 | 2.269 | 2.319 | 2.376 | 2.442 | 2.517 | 2.605 |
| 20° | 2.048 | 2.055 | 2.064 | 2.075 | 2.089 | 2.105 | 2.123 | 2.145 | 2.169 | 2.197 | 2.229 | 2.265 | 2.305 | 2.352 | 2.405 | 2.465 | 2.534 | 2.613 | 2.705 |
| 21° | 2.124 | 2.130 | 2.139 | 2.150 | 2.164 | 2.180 | 2.199 | 2.221 | 2.246 | 2.275 | 2.308 | 2.346 | 2.388 | 2.437 | 2.492 | 2.555 | 2.628 | 2.711 | 2.803 |
| 22° | 2.203 | 2.209 | 2.217 | 2.228 | 2.241 | 2.258 | 2.277 | 2.300 | 2.326 | 2.356 | 2.390 | 2.429 | 2.474 | 2.525 | 2.582 | 2.649 | 2.725 | 2.812 | 2.914 |
| 23° | 2.286 | 2.291 | 2.299 | 2.309 | 2.323 | 2.339 | 2.359 | 2.382 | 2.409 | 2.440 | 2.475 | 2.516 | 2.562 | 2.615 | 2.675 | 2.745 | 2.824 | 2.916 | 3.022 |
| 24° | 2.373 | 2.377 | 2.384 | 2.394 | 2.407 | 2.424 | 2.443 | 2.467 | 2.494 | 2.526 | 2.563 | 2.605 | 2.653 | 2.708 | 2.771 | 2.843 | 2.926 | 3.022 | 3.134 |
| 25° | 2.464 | 2.467 | 2.474 | 2.483 | 2.496 | 2.512 | 2.532 | 2.556 | 2.584 | 2.616 | 2.654 | 2.698 | 2.747 | 2.805 | 2.870 | 2.945 | 3.032 | 3.132 | 3.248 |
| 26° | 2.561 | 2.563 | 2.568 | 2.576 | 2.589 | 2.604 | 2.624 | 2.648 | 2.677 | 2.710 | 2.749 | 2.794 | 2.845 | 2.904 | 2.972 | 3.050 | 3.140 | 3.244 | 3.365 |
| 27° | 2.663 | 2.663 | 2.667 | 2.675 | 2.686 | 2.701 | 2.721 | 2.745 | 2.774 | 2.808 | 2.847 | 2.893 | 2.946 | 3.007 | 3.078 | 3.159 | 3.252 | 3.360 | 3.487 |
| 28° | 2.772 | 2.770 | 2.772 | 2.778 | 2.788 | 2.803 | 2.822 | 2.846 | 2.875 | 2.909 | 2.950 | 2.997 | 3.051 | 3.114 | 3.187 | 3.271 | 3.368 | 3.480 | 3.611 |
| 29° | 2.887 | 2.883 | 2.883 | 2.887 | 2.896 | 2.910 | 2.928 | 2.952 | 2.981 | 3.015 | 3.056 | 3.105 | 3.161 | 3.225 | 3.300 | 3.387 | 3.487 | 3.604 | 3.740 |
| 30° | 3.010 | 3.002 | 3.000 | 3.002 | 3.010 | 3.022 | 3.040 | 3.063 | 3.091 | 3.126 | 3.168 | 3.217 | 3.274 | 3.341 | 3.418 | 3.508 | 3.611 | 3.732 | 3.873 |

# Table 60    Standard Atmosphere

| Altitude m | Temperature K | Temperature °C | Pressure kPa | Pressure torr | Density kg m$^{-3}$ | Density ratio, $\rho / \rho_0$ | Speed of Sound m s$^{-1}$ | Absolute Viscosity Ns m$^{-2}$ | Kinematic Viscosity m$^2$ s$^{-1}$ |
|---|---|---|---|---|---|---|---|---|---|
| 0 | 288.15 | 15.0 | 101.325 | 760.0 | 1.2255 | 1.0000 | 341.0 | $1.796 \times 10^{-5}$ | $1.466 \times 10^{-5}$ |
| 1000 | 281.65 | 8.5 | 89.87 | 674.1 | 1.1120 | 0.9074 | 337.1 | 1.765 | 1.587 |
| 2000 | 275.15 | +2.0 | 79.49 | 596.2 | 1.0068 | 0.8215 | 333.4 | 1.733 | 1.721 |
| 3000 | 268.65 | −4.5 | 70.10 | 525.8 | 0.9094 | 0.7421 | 329.3 | 1.700 | 1.869 |
| 4000 | 262.15 | −11.0 | 61.63 | 462.3 | 0.8193 | 0.6685 | 325.4 | 1.668 | 2.036 |
| 5000 | 255.65 | −17.5 | 54.01 | 405.1 | 0.7363 | 0.6008 | 321.1 | 1.635 | 2.221 |
| 6000 | 249.15 | −24.0 | 47.17 | 353.8 | 0.6598 | 0.5384 | 317.1 | 1.602 | 2.428 |
| 7000 | 242.65 | −30.5 | 41.05 | 307.9 | 0.5896 | 0.4811 | 312.7 | 1.568 | 2.659 |
| 8000 | 236.15 | −37.0 | 35.58 | 266.9 | 0.5252 | 0.4286 | 308.4 | 1.534 | 2.921 |
| 9000 | 229.65 | −43.5 | 30.72 | 230.4 | 0.4664 | 0.3806 | 304.3 | 1.499 | 3.214 |
| 10000 | 223.15 | −50.0 | 26.42 | 198.2 | 0.4127 | 0.3368 | 297.5 | 1.464 | 3.547 |
| 11000 | 218.15 | −55.0 | 22.62 | 169.7 | 0.3614 | 0.2949 | 296.7 | 1.437 | 3.976 |
| 12000 | 218.15 | −55.0 | 19.33 | 145.0 | 0.3090 | 0.2521 | 296.7 | 1.437 | 4.650 |
| 13000 | 218.15 | −55.0 | 16.53 | 124.0 | 0.2642 | 0.2156 | 296.7 | 1.437 | 5.439 |
| 14000 | 218.15 | −55.0 | 14.13 | 106.0 | 0.2259 | 0.1843 | 296.7 | 1.437 | 6.361 |
| 15000 | 218.15 | −55.0 | 12.08 | 90.6 | 0.1931 | 0.1576 | 296.7 | 1.437 | 7.442 |
| 16000 | 218.15 | −55.0 | 10.47 | 78.5 | 0.1670 | 0.1363 | 296.7 | 1.437 | 8.605 |
| 17000 | 218.15 | −55.0 | 8.97 | 67.3 | 0.1431 | 0.1168 | 296.8 | 1.437 | $1.004 \times 10^{-4}$ |
| 18000 | 218.15 | −55.0 | 7.68 | 57.6 | 0.1225 | 0.1000 | 296.8 | 1.437 | 1.173 |
| 19000 | 218.15 | −55.0 | 6.58 | 49.4 | 0.1049 | 0.0856 | 296.8 | 1.437 | 1.370 |
| 20000 | 218.15 | −55.0 | 5.64 | 42.3 | 0.0899 | 0.0734 | 296.8 | 1.437 | 1.598 |
| 21000 | 218.15 | −55.0 | 4.83 | 36.2 | 0.0770 | 0.0628 | 296.8 | 1.437 | 1.866 |
| 22000 | 218.15 | −55.0 | 4.14 | 31.1 | 0.0660 | 0.0539 | 296.8 | 1.437 | 2.177 |
| 23000 | 218.15 | −55.0 | 3.55 | 26.6 | 0.0565 | 0.0461 | 296.8 | 1.437 | 2.543 |
| 24000 | 218.15 | −55.0 | 3.04 | 22.8 | 0.0484 | 0.0395 | 296.8 | 1.437 | 2.969 |
| 25000 | 218.15 | −55.0 | 2.60 | 19.5 | 0.0414 | 0.0338 | 296.9 | 1.437 | 3.471 |
| 26000 | 218.15 | −55.0 | 2.23 | 16.7 | 0.0355 | 0.0290 | 296.9 | 1.437 | 4.048 |
| 27000 | 218.15 | −55.0 | 1.91 | 14.3 | 0.0304 | 0.0248 | 297.0 | 1.437 | 4.727 |
| 28000 | 218.15 | −55.0 | 1.64 | 12.3 | 0.0260 | 0.0212 | 297.0 | 1.437 | 5.527 |
| 29000 | 218.15 | −55.0 | 1.40 | 10.5 | 0.0223 | 0.0182 | 297.1 | 1.437 | 6.444 |
| 30000 | 218.15 | −55.0 | 1.20 | 9.00 | 0.0191 | 0.0156 | 297.1 | 1.437 | 7.524 |
| 32000 | 218.15 | −55.0 | 0.890 | 6.68 | 0.0141 | 0.0115 | 297.2 | 1.437 | $1.019 \times 10^{-3}$ |
| 34000 | 233.15 | −40.0 | 0.605 | 4.54 | 0.0096 | 0.0078 | 304.4 | 1.518 | 1.581 |
| 36000 | 248.35 | −24.8 | 0.417 | 3.13 | 0.0066 | 0.0054 | 323.2 | 1.597 | 2.420 |
| 38000 | 263.65 | −9.5 | 0.288 | 2.16 | 0.0046 | 0.0038 | 329.4 | 1.675 | 3.641 |
| 40000 | 278.85 | +5.7 | 0.200 | 1.50 | 0.0032 | 0.0026 | 336.7 | 1.751 | 5.472 |

# Table 61  Physical Constants[a]

| Quantity | Symbol | Value | Uncertainty ppm |
|----------|--------|-------|-----------------|
| Speed of light in vacuum | $c$ | 299792458 m/s | 0.004 |
| Avogadro number | $N$ | $6.022045 \times 10^{23}$/g-mol | 5.1 |
| Faraday constant | $F$ | 96484.56 C/g-mol | 2.8 |
| Planck constant | $h$ | $6.626176 \times 10^{-34}$ J/Hz | 5.4 |
| Molar gas constant | $R$ | 8.31441 J/K g-mol | 31. |
| Elementary charge | $e$ | $1.6021892 \times 10^{-19}$ C | 2.9 |
| Boltzmann constant | $k$ | $1.380662 \times 10^{-23}$ J/K | 32. |
| First radiation constant | $c_1 = 2\pi hc^2$ | $3.741832 \times 10^{-16}$ W m$^2$ | 5.4 |
| Second radiation constant | $c_2 = hc/k$ | 0.01438786 m/K | 31. |
| Molar volume, ideal gas at $T = 273.15$ K, $p = 1$ atm | $V_0$ | 0.02241383 m$^3$/g-mol | 31. |
| Gravitational constant | $G$ | $6.6720 \times 10^{-11}$ N m$^2$/kg$^2$ | 615. |

[a]Data from Reference 41.

# Table 62   Conversion Factors

To convert the numerical value of a property expressed in one of the units in the left-hand column of a table to the numerical value of the same property given in one of the units in the top row of the same table, multiply the former value by the factor in the block common to both units. Numbers followed by an asterisk (*) are definitions of the relation between the two units. SI is the abbreviation of the International System of Units.

## Table 62a   Length

| Units | in. | ft | cm | m(SI) |
|---|---|---|---|---|
| 1 in. = | 1 | 0.0833333 | 2.54* | 0.0254 |
| 1 ft = | 12* | 1 | 30.48 | 0.3048 |
| 1 cm = | 0.3937008 | 0.0328084 | 1 | 0.01* |
| 1 m(SI) = | 39.37008 | 3.280840 | 100 | 1 |

## Table 62b   Volume

| Units | in.$^3$ | ft$^3$ | gal | liter | cm$^3$ | m$^3$(SI) |
|---|---|---|---|---|---|---|
| 1 in.$^3$ = | 1 | $5.787037 \times 10^{-4}$ | $4.329004 \times 10^{-3}$ | 0.01638706 | 16.38706* | $1.638706 \times 10^{-5}$ |
| 1 ft$^3$ = | 1728* | 1 | 7.480519 | 28.31684 | 28316.84 | 0.02831684 |
| 1 gal = | 231* | 0.1336806 | 1 | 3.785411 | 3785.411 | $3.785411 \times 10^{-3}$ |
| 1 liter = | 61.02374 | 0.03531467 | 0.2641721 | 1 | 1000* | $10^{-3}$ |
| 1 cm$^3$ = | 0.06102374 | $3.531467 \times 10^{-5}$ | $2.641721 \times 10^{-4}$ | $10^{-3}$ | 1 | $10^{-6}$ |
| 1 m$^3$(SI) = | 61023.74 | 35.31467 | 264.1721 | $10^3$ | $10^6$ | 1 |

## Table 62c   Mass

| Units | lb | g | kg(SI) | ton | metric ton |
|---|---|---|---|---|---|
| 1 lb = | 1 | 453.59237* | 0.45359237 | 0.0005 | $4.5359237 \times 10^{-4}$ |
| 1 g = | $2.204623 \times 10^{-3}$ | 1 | $10^{-3}$ | $1.1023115 \times 10^{-6}$ | $10^{-6}$ |
| 1 kg(SI) = | 2.204623 | 1000 | 1 | $1.1023115 \times 10^{-3}$ | $10^{-3}$ |
| 1 ton = | 2000* | 907184.74 | 907.18474 | 1 | 0.9071846 |
| 1 metric ton = | 2204.623 | $10^6$ | 1000* | 1.1023115 | 1 |

## Table 62d   Density

| Units | lb/ft$^3$ | lb/gal | g/cm$^3$ | kg/m$^3$(SI) |
|---|---|---|---|---|
| 1 lb/ft$^3$ = | 1 | 0.13368056 | 0.01601847 | 16.01847 |
| 1 lb/gal = | 7.480519 | 1 | 0.11982646 | 119.82646 |
| 1 g/cm$^3$ = | 62.42793 | 8.345402 | 1 | $10^3$ |
| 1 kg/m$^3$(SI) = | 0.06242793 | $8.345402 \times 10^{-3}$ | $10^{-3}$ | 1 |

## Table 62e  Force

| Unit | lbf | pdl[a] | dyn | g[b] | N(SI) |
|---|---|---|---|---|---|
| 1 lbf = | 1 | 32.1740 | $4.44822 \times 10^5$ | 453.59237 | 4.448222 |
| 1 pdl = | 0.0310809 | 1 | $1.38255 \times 10^4$ | 14.0981 | 0.138255 |
| 1 dyn = | $2.24809 \times 10^{-6}$ | $7.23301 \times 10^{-5}$ | 1 | $1.01972 \times 10^{-3}$ | $10^{-5}$ |
| 1 g = | $2.20462 \times 10^{-3}$ | 0.070931 | 980.665 | 1 | $9.80665 \times 10^{-3}$ |
| 1 N(SI) = | 0.224809 | 7.23298 | $10^5$ | 101.972 | 1 |

[a]Poundal.
[b]Gram force.


## Table 62f  Pressure

| Unit | lb/in.$^2$ | ft H$_2$O 60°F | atm | kg/cm$^2$ | torr | bar | N/m$^2$(SI) |
|---|---|---|---|---|---|---|---|
| 1 lb/in.$^2$ = | 1 | 2.30897 | 0.06804596 | 0.07030669 | 51.71495 | 0.06894757 | $6.894757 \times 10^3$ |
| 1 ft H$_2$O, 60°F = | 0.43309 | 1 | 0.0294703 | 0.0304495 | 22.3974 | 0.0298608 | $2.98608 \times 10^3$ |
| 1 atm = | 14.69595 | 33.9325 | 1 | 1.033227 | 760* | 1.013250 | $1.013250 \times 10^5$ |
| 1 kg/cm$^2$ = | 14.22335 | 32.8413 | 0.9678411 | 1 | 735.5596 | 0.980665 | $9.80665 \times 10^4$ |
| 1 torr = | 0.0193368 | 0.0446480 | $1.315789 \times 10^{-3}$ | $1.359509 \times 10^{-3}$ | 1 | $1.333224 \times 10^{-3}$ | 133.3224 |
| 1 bar = | 14.503775 | 33.4887 | 0.9869233 | 1.019716 | 750.0617 | 1 | $10^5$ |
| 1 N/m$^2$(SI) = | $1.450378 \times 10^{-4}$ | $3.34887 \times 10^{-4}$ | $9.869233 \times 10^{-6}$ | $1.019716 \times 10^{-5}$ | $7.500617 \times 10^{-3}$ | $10^{-5}$ | 1 |


## Table 62g  Energy

| Units | Btu$_{IT}$ | cal$_{IT}$ | ft-lb | ft$^3$-lb/in.$^2$ | hp hr | kg m | liter atm | cal | J(SI) |
|---|---|---|---|---|---|---|---|---|---|
| 1 Btu$_{IT}$ = | 1 | 251.9958* | 778.1693 | 5.403953 | $3.930148 \times 10^{-4}$ | 107.5858 | 10.41259 | 252.1644 | 1055.056 |
| 1 cal$_{IT}$ = | $3.968320 \times 10^{-3}$ | 1 | 3.088025 | 0.02144462 | $1.559608 \times 10^{-6}$ | 0.4269349 | 0.04132048 | 1.000669 | 4.1868* |
| 1 ft-lb = | $1.285067 \times 10^{-3}$ | 0.3238316 | 1 | $6.944444 \times 10^{-3}$ | $5.050505 \times 10^{-7}$ | 0.1382550 | 0.01338088 | 0.3240482 | 1.355818 |
| ft$^3$-lb/in.$^2$ = | 0.1850497 | 46.63175 | 144* | 1 | $7.272727 \times 10^{-5}$ | 19.90873 | 1.926847 | 46.66294 | 195.2378 |
| 1 hp hr = | 2544.433 | $6.411866 \times 10^5$ | 1980000* | 13750 | 1 | $2.73745 \times 10^5$ | $2.649414 \times 10^4$ | $6.41615 \times 10^5$ | $2.684520 \times 10^6$ |
| 1 kg m = | $9.294906 \times 10^{-3}$ | 2.342277 | 7.233009 | 0.0502292 | $3.653035 \times 10^{-6}$ | 1 | 0.0967840 | 2.343844 | 9.80665 |
| 1 liter atm = | 0.09603759 | 24.20107 | 74.73351 | 0.5189827 | $3.774420 \times 10^{-5}$ | 10.33229 | 1 | 24.21726 | 101.3250 |
| 1 cal = | $3.965667 \times 10^{-3}$ | 0.9993314 | 3.085960 | $2.143028 \times 10^{-2}$ | $1.558566 \times 10^{-6}$ | 0.4266495 | 0.04129286 | 1 | 4.184* |
| 1 J(SI) = | $9.47817 \times 10^{-4}$ | 0.2388459 | 0.7375621 | $5.121959 \times 10^{-3}$ | $3.725062 \times 10^{-7}$ | 0.1019716 | $9.869233 \times 10^{-3}$ | 0.2390057 | 1 |


## Table 62h  Power

| Units | ft-lb/sec | Btu$_{IT}$/hr | hp | W(SI) |
|---|---|---|---|---|
| 1 ft-lb/sec = | 1 | 4.626243 | $1.818182 \times 10^{-3}$ | 1.355818 |
| 1 Btu$_{IT}$/hr = | 0.216158 | 1 | $3.930148 \times 10^{-4}$ | 0.2930711 |
| 1 hp = | 550 * | 2544.43 | 1 | 745.6999 |
| 1 W(SI) = | 0.737562 | 3.412142 | $1.341022 \times 10^{-3}$ | 1 |


## Table 62i  Specific Energy

| Units | Btu$_{IT}$/lb | cal$_{IT}$/g | cal/g | J/g | J/kg(SI) |
|---|---|---|---|---|---|
| 1 Btu$_{IT}$/lb = | 1 | 0.55555556 | 0.5559272 | 2.326000 | 2326 |
| 1 cal$_{IT}$/g = | 1.8* | 1 | 1.000669 | 4.1868* | 4186.8 |
| 1 cal/g = | 1.798797 | 0.9993314 | 1 | 4.1840* | 4184 |
| 1 J/g = | 0.4299226 | 0.2388459 | 0.2390057 | 1 | 1000 |
| 1 J/kg(SI) = | $4.299226 \times 10^{-4}$ | $2.388459 \times 10^{-4}$ | $2.390057 \times 10^{-4}$ | $10^{-3}$ | 1 |

## Table 62j  Specific Energy per Degree

| Units | $Btu_{IT}/R\,lb$ | $cal_{IT}/K\,g$ | $cal/K\,g$ | $J/K\,g$ | $J/K\,kg(SI)$ |
|---|---|---|---|---|---|
| 1 $Btu_{IT}/R\,lb =$ | 1 | 1 | 1.000669 | 4.1868 | 4186.8 |
| 1 $cal_{IT}/K\,g =$ | 1 | 1 | 1.000669 | 4.1868* | 4186.8 |
| 1 $cal/K\,g =$ | 0.9993314 | 0.9993314 | 1 | 4.1840* | 4184 |
| 1 $J/K\,g =$ | 0.2388459 | 0.2388459 | 0.2390057 | 1 | 1000 |
| 1 $J/K\,kg(SI) =$ | $2.388459 \times 10^{-4}$ | $2.388459 \times 10^{-4}$ | $2.390057 \times 10^{-4}$ | $10^{-3}$ | 1 |

## Table 62k  Absolute Viscosity

| Unit | centipoise | $lbm/ft\,sec$ | $lbf\,sec/ft^2$ | $N\,sec/m^2(SI)$ |
|---|---|---|---|---|
| 1 centipoise = | 1 | $6.71971 \times 10^{-4}$ | $2.08855 \times 10^{-5}$ | $10^{-3}$* |
| 1 $lbm/ft\,sec =$ | 1488.16 | 1 | 0.031081 | 1.48816 |
| 1 $lbf\,sec/ft^2 =$ | 47880.3 | 32.1741 | 1 | 47.8803 |
| 1 $N\,sec/m^2(SI) =$ | $10^3$ | 0.671971 | 0.0208855 | 1 |

## Table 62l  Kinematic Viscosity

| Units | centistoke | $ft^2/sec$ | $cm^2/hr$ | $m^2/sec(SI)$ |
|---|---|---|---|---|
| 1 centistoke = | 1 | $1.076391 \times 10^{-5}$ | 36.0000 | $10^{-6}$* |
| 1 $ft^2/sec =$ | 92903.04 | 1 | 3344509. | 0.09290304* |
| 1 $cm^2/hr =$ | 0.02777778 | $2.989975 \times 10^{-7}$ | 1 | $2.777778 \times 10^{-8}$ |
| 1 $m^2/sec(SI) =$ | $10^6$ | 10.76391 | $3.6 \times 10^7$ | 1 |

## Table 62m  Thermal Conductivity

| Units | $Btu_{IT}/ft\,hr\,R$ | $cal_{IT}/cm\,sec\,K$ | $kcal_{IT}/m\,hr\,K$ | $W/m\,K(SI)$ |
|---|---|---|---|---|
| 1 $Btu_{IT}/ft\,hr\,R =$ | 1 | 0.0041336 | 1.4881 | 1.7307 |
| 1 $cal_{IT}/cm\,sec\,K =$ | 241.92 | 1 | 360 | 418.68 |
| 1 $kcal_{IT}/m\,hr\,K =$ | 0.67199 | 0.0027778 | 1 | 1.1630 |
| 1 $W/m\,K(SI) =$ | 0.57781 | 0.0023885 | 0.85985 | 1 |

## Table 62n  Molar Gas Constant

| $\bar{R} =$ | | |
|---|---|---|
| 1.98586 | $cal_{IT}/K\,g$-mole |
| 1.98586 | $Btu_{IT}/R\,lb$-mole |
| 10.73150 | $ft^3/lb$-in.$^2$ R lb-mole |
| 0.730235 | $ft^3$ atm/R lb-mole |
| 62363.2 | $cm^3$ torr/K g-mole |
| 82.0568 | $cm^3$ atm/K g-mole |
| 0.0820568 | liter atm/K g-mole |
| 1.98719 | cal/K g-mole |
| 8.31441 | J/K g-mole(SI) |

## Table 62o    Miscellaneous

**Temperature**

Triple point of water = 273.16 K*
$t(°C) = T(K) - 273.15$
$T(R) = 1.8T(K)$
$t(°F) = T(R) - 459.67$*
$t(°F) = 1.8t(°C) + 32$

**Force and Mass Conversion**

$g_c = 32.1740$ lbm ft/lbf sec$^2$

$\quad = 1.0$ kg m/N sec$^2$(SI)

**Gravity**

$g = 980.665$ cm/sec$^2$

$\quad = 9.80665$ m/sec$^2$(SI)

$(PV)^{P=0}_{T=0°C}$

5276.36 ft$^3$ lb/in.$^2$ lb-mole
22.41383 liter atm/g-mole
542.8015 cal/g-mole
2271.082 J/g-mole(SI)

# Sources of Data and Calculation Methods

## 1 GENERAL DESCRIPTION

For decades engineers and scientists have sought to determine with precision the thermodynamic properties of the air around us and the simple gases from which it is composed. A great many experimental data across a broad range of variables have been accumulated, but accurate measurements have proved elusive, perhaps unobtainable, at certain critical extremes of pressure and temperature.

Where experimental techniques and the methods for deriving additional data produced unreliable results, researchers turned to theoretical analysis in an effort to learn even more about the properties of gases. It was well known, for instance, that "real" gases begin to acquire the characteristics attributed to an "ideal" gas when pressures are reduced to critically low levels. The theoretical concept of an ideal gas, therefore, became a vital link in the understanding of gas properties and in the practical application of this knowledge.

### 1.1 Ideal Gas Properties

As pressure is reduced to zero, the properties of real gases approach those of an ideal gas. The latter is assumed to exhibit two characteristic properties—its molecules occupy no space and they produce no molecular interactions. Consequently, the equation of state for an ideal gas is

$$pv = RT \qquad (1)$$

where $p$ denotes the pressure, $v$ the specific volume, $R$ the gas constant, and $T$ the absolute thermodynamic temperature.[1]*

From (1) the relations between internal energy $u$ and enthalpy $h$, for unit mass, and specific heat capacity at constant volume $c_v$ and specific heat capacity at constant pressure $c_p$ are

$$u = h - RT \qquad (2)$$

*Superscript numbers refer to items in the Bibliography, which begins on page 210.

and

$$c_v = c_p - R. \qquad (3)$$

It may be shown that $h$, $u$, $c_p$, and $c_v$ are functions of temperature only.[2] Hence, complete presentation of these quantities will consist of a table with a single variable, the temperature.

The entropy, on the other hand, proves to be a function of both temperature and pressure, so that an equally simple presentation is not possible. In an isentropic (constant-entropy) process the ratio of the pressures corresponding to a given pair of temperatures is the same for all isentropics. Therefore the pressure ratio of any two pressures $p_a$ and $p_b$, corresponding to the given temperatures $T_a$ and $T_b$ of a gas in an isentropic process, can be obtained from the relative pressures $p_{ra}$ and $p_{rb}$ as tabulated for $T_a$ and $T_b$ in the property table under the heading of the given gas. It is also true that the ratio of the volumes corresponding to a given pair of temperatures is the same for all isentropics. These two statements may be proved as shown in the following.

For a homogeneous system in the absence of gravity, electricity, capillarity, magnetism, and chemical change, we may write for changes between equilibrium states

$$T\,ds = dh - v\,dp$$

where $s$ denotes the entropy, $h$ the enthalpy, $v$ the specific volume—all for unit mass—and $p$ the pressure. Since $h$ is a function of $T$ only, we have

$$T\,ds = c_p\,dT - v\,dp \qquad (4)$$

where $c_p$ is the specific heat capacity at constant pressure, $(\partial h/\partial T)_P = dh/dt$. For an infinitesimal step in an isentropic process we have

$$0 = c_p\,dT - v\,dp$$

and, upon substituting from (1) and transposing,

$$\frac{dp}{p} = \frac{c_P}{R} \cdot \frac{dT}{T}.$$

Because $c_p$, like the enthalpy, is a function of temperature only we get, upon integrating between temperatures $T_0$ and $T$,

$$\ln\frac{p}{p_0} = \frac{1}{R}\int_{T_0}^{T}\frac{c_P}{T}\,dT. \qquad (5)$$

When a base temperature $T_0$ is selected the ratio $p/p_0$ becomes a single function of the temperature $T$ regardless of the value of the entropy.

The corresponding volume ratio is a single function of the temperature $T$ also and is given by

$$\ln\frac{v}{v_0} = -\frac{1}{R}\int_{T_0}^{T}\frac{c_v}{T}\,dT \qquad (6)$$

or, alternatively,

$$\frac{v}{v_0} = \frac{p_0}{p}\cdot\frac{T}{T_0}. \qquad (7)$$

Solving (4) for $ds$ and substituting from (1), we get

$$ds = \frac{c_P}{T}\,dT - R\frac{dp}{p}.$$

The entropy $s$ at any state reckoned from $T_0 = 0$ and unit pressure is then

$$s = \int_{T_0}^{T}\frac{c_P}{T}\,dT - R\ln p = \phi - R\ln p \qquad (8)$$

in which

$$\phi = \int_{T_0}^{T}\frac{c_P}{T}\,dT. \qquad (9)$$

The change in entropy between states 1 and 2 is then

$$s_2 - s_1 = \phi_2 - \phi_1 - R\ln\frac{p_2}{p_1}. \qquad (10)$$

From (5), it is evident that the ratio of the pressures $p_a$ and $p_b$, corresponding to the temperatures $T_a$ and $T_b$, respectively, along a given isentropic, is equal to the ratio of the relative pressures $p_{ra}$ and $p_{rb}$ as tabulated for $T_a$ and $T_b$, respectively. Thus

$$\left(\frac{p_a}{p_b}\right)_s = \text{constant} = \frac{p_{ra}}{p_{rb}}.$$

Similarly, from (6),

$$\left(\frac{v_a}{v_b}\right)_s = \text{constant} = \frac{v_{ra}}{v_{rb}}.$$

In terms of an ideal gas, thermodynamic properties represented by $h$, $p_r$, $u$, $v_r$, $\phi$, $c_p$, $c_v$, and $k$ $(= c_p/c_v)$ are presented as functions of temperatures, K and °C. Included in this book are tables for air, fuel combustion products with 400%, 200%, and 100% theoretical air, $N_2$, $O_2$, $H_2O$, $CO_2$, $H_2$, CO, and Ar.

Velocity of sound, $\alpha$, as a function of temperatures was included in all these tables except those for the fuel combustion products. In Table 2, the transport properties for air at low pressures, for example

mass velocity $(G)$, viscosity $(\mu)$, thermal conductivity $(\lambda)$, and Prandtl number (Pr) at selected temperatures were also listed.

The sources of data and methods for evaluating the tabulated properties are described in the remaining sections.

## 1.2 Calculation of Ideal Gas Thermodynamic Properties

The thermodynamic properties like $c_p$, $s$, $h$, and so on may be calculated for simple gases in the ideal gaseous state over a wide range of temperatures with high reliability by applying the method of statistical mechanics to basic empirical data.[3] The basic data are the energy levels of various molecular motions, which include both external motions (translation of the molecules) and internal motions (rotational, vibrational, etc.). Based on such energy levels ($\varepsilon_i$) and their respective statistical weights ($g_i$), the molecular partition function

$$q = \sum_i g_i e^{-\varepsilon_i/kt}$$

is obtained for each of the $3n$ degrees of freedom for a gaseous molecule having $n$ atoms. From these partition functions, the thermodynamic properties of the given substance in the ideal gaseous state at 101.325 kPa(1 atm) pressure can be evaluated by use of the standard method of statistical mechanics.[3]

The internal energy of a gaseous molecule is composed of rotational kinetic energy $\varepsilon_{rot}$, which is produced by collisions of the gaseous molecules, and interatomic energies or vibrational energy $\varepsilon_{vib}$ of the molecule. The atoms within the molecule possess kinetic energy due to vibration, as well as potential energy arising from forces acting between the atoms. In complex molecules certain groups of atoms can rotate against the remaining part of the molecule to produce "internal rotation." This kind of motion inside the molecule is treated as a special kind of vibration in statistical calculations.

For monatomic substances, such as argon, neon, and so on, there are no rotational and vibrational degrees of freedom. The total energy of such substances contains only translational and electronic energies. The total energy of diatomic and polyatomic molecules, on the other hand, includes energy contributions from $3n$ degrees of freedom as shown in Table A. For polyatomic substances the contributions from electronic energy are usually neglected

## Table A  Distribution of Energy in Gaseous Molecule[a]

| Type | Translation | Rotation | Vibration | Internal Rotation | Total |
|---|---|---|---|---|---|
| Monatomic | 3 | 0 | 0 | 0 | 3 |
| Diatomic | 3 | 2 | 1 | 0 | 6 |
| Triatomic | | | | | |
| linear | 3 | 2 | 4 | 0 | 9 |
| nonlinear | 3 | 3 | 3 | 0 | 9 |
| Polyatomic | | | | | |
| linear | 3 | 2 | $3n-a-5$ | a | $3n$ |
| nonlinear | 3 | 3 | $3n-b-6$ | b | $3n$ |

[a]In degrees of freedom.

since the electronic state of the molecule is undisturbed under ordinary conditions (i.e., it remains at the electronic ground state). Similarly the nuclear energy inside the atoms of the gaseous molecules is not considered in statistical calculations. Some simple molecules ($CClO$, $TiCl_3$, etc.) have low-lying electronic levels, the contributions to the ideal gas thermodynamic properties from these electronic energy levels are included.

In Table B are summarized the partition functions that can be employed for evaluation of thermodynamic properties for chemical substances in the ideal gaseous state at given temperatures. The derivation of these partition functions may be found in standard textbooks on statistical mechanics.[4] In this table: $V$ denotes volume of the given gas; $m$, mass of the molecule; $k$, Boltzmann's constant; $T$, absolute thermodynamic temperature; $h$, Planck's constant; $I$, moment of inertia of the linear molecule; $\sigma$, symmetry number; $I_x I_y I_z$, product of the principal moments of inertia of a nonlinear molecule; $\omega_i$, wavenumber of $i$th fundamental vibration; $g_0$, statistical weight of electronic ground state; $g_i$, statistical weight of electronic energy level $\varepsilon_i$; $I_r$, reduced moment of inertia of the rotating top; and $\sigma_{IR}$, symmetry number of the top.

The thermodynamic properties of a gaseous substance can be calculated by substituting each of the partition functions listed in Table B into the equations shown in Table C, where $R$ denotes the gas constant; $q'$, $T(\partial q/\partial T)_v$; $E_0^\circ$, energy of the substance at absolute zero temperature; and $N$, Avogadro's number. The calculated contributions from different molecular motions for each of the thermodynamic properties presented in Table C are added to yield the final property values at the given temperature $T$.

## Table C  Equations for Calculating Thermodynamic Properties

$$E^\circ - E_0^\circ = RT^2\left(\frac{\partial \ln q}{\partial T}\right)_v = RT\left(\frac{q'}{q}\right)$$

$$H^\circ - E_0^\circ = E^\circ - E_0^\circ + RT = RT^2\left(\frac{\partial \ln q}{\partial T}\right)_P$$

$$C_P^\circ = \left(\frac{\partial E}{\partial T}\right)_v + R = \frac{R}{T^2}\left[\frac{\partial^2 \ln q}{\partial(1/T)^2}\right]_P$$

$$C_v^\circ = C_P^\circ - R = \frac{R}{T^2}\left[\frac{\partial^2 \ln q}{\partial(1/T)^2}\right]_v$$

$$S^\circ = \frac{H^\circ - E_0^\circ}{T} + R\ln\frac{q}{N}$$

## Table B  Partition Functions of Molecular Motion

| Motion | Partition Function | Degrees of Freedom Included |
|---|---|---|
| Translation | $q_{\text{trans}} = V(2\pi mkT)^{3/2}h^{-3}$ | 3 |
| Rotation | $q_{\text{rot}} = \dfrac{8\pi^2 IkT}{\sigma h^2}$  (linear molecule) | 2 |
| | $q_{\text{rot}} = \dfrac{(8\pi^2 kT)^{3/2}(\pi I_x I_y I_z)^{1/2}}{\sigma h^3}$  (nonlinear molecule) | 3 |
| Vibration | $q_{\text{vib}} = (1 - e^{-h\omega_i/kT})^{-1}$ | 1 |
| Electronic | $q_{\text{el}} = g_0 + \sum_i g_i e^{-\varepsilon_i/kT}$ | — |
| Free internal rotation | $q_f = (8\pi^3 I_r kT)^{1/2}(h\sigma_{IR})^{-1}$ | 1 |

(181)

## 1.3 Equations for Calculating Ideal Gas Thermodynamic Properties

In Table D a summary of equations for calculation of thermodynamic properties for monatomic substances in the ideal gaseous state at 101.325 kPa pressure is presented. The input data needed for evaluation of translational contributions to thermodynamic properties are molecular weight ($M$) of the given substance and the absolute temperature in degrees Kelvin (K) at which the thermodynamic properties are required. For evaluation of thermodynamic properties due to electronic contributions, the values of electronic energy levels $\varepsilon_i$ and the corresponding statistical weight $g_i$ for each of the $i$ energy levels are employed. These equations were used for calculating the ideal gas thermodynamic properties for Ar.

In calculating the thermodynamic properties for diatomic gases, for example $N_2$, $O_2$, and CO, in the ideal gaseous state at one atmosphere pressure, the equations listed in Table E were employed. These equations were derived based on a nonrigid-rotor and anharmonic-oscillator molecular model.[3] In Table E $u$ denotes $(\omega_e - 2\omega_e X_e)hc/kT$, where $\omega_e$ denotes vibrational constant; $\omega_e X_e$, anharmonicity constant; $h$, Planck's constant; $c$, speed of light; $k$, Boltzmann's constant; $T$, absolute temperature in K; $I$, moment of inertia of the given molecule; $X$, $\omega_e X_e/\omega_e$; and $\delta$, $\alpha_e B_e$, where $\alpha_e$ = vibrational-rotational coupling constant, $B_e$ = rotational constant; and $\gamma = B_e/\omega_e$.

As mentioned in Section 1.2, the total thermodynamic property $G$ of a diatomic substance is obtained as

$$G = G_{\text{trans}} + G_{\text{rot}} + G_{\text{vib}} + G_{\text{el}} + G_{\text{anh}}$$

where $G_{\text{trans}}$ denotes contributions from 3 translational degrees of freedom; $G_{\text{rot}}$, contributions from 2 rotational degrees of freedom; $G_{\text{vib}}$, vibrational contribution; $G_{\text{el}}$, electronic contribution; and $G_{\text{anh}}$, contributions from anharmonicity corrections.

The input data, such as moment of inertia $I$, may be derived from molecular structure of the given substance, which can be determined by microwave spectroscopy or X-ray diffraction. The vibrational constants are usually obtained from infrared and Raman spectroscopy.[5] The thermodynamic properties for $H_2$, $H_2O$, and $CO_2$ were adopted from literature.[6] These have been the best values available. Based on the calculated values of $h$, $u$, $p_r$, $v_r$, and $\phi$, and $c_p$, $c_v$, and $k(= c_p/c_v)$ for $N_2$, $O_2$, and Ar, the corresponding properties for air at low pressures (for one kilogram) were evaluated at close temperature ranges. The thermodynamic properties at the intermediate temperatures can be obtained by linear interpolation of the tabulated property values without significant error.

## 2 SOURCES AND METHODS FOR INDIVIDUAL TABLES

### 2.1 Tables 1 and 2—Air at Low Pressures (for one kilogram)

In principle, a mathematical model of the equation of state for air could be developed based upon the experimental $P$–$V$–$T$ measurements. By combining this model with the standard thermodynamic rela-

**Table D  Equations for Calculating Ideal Gas Thermodynamic Properties for Monatomic Molecules at a Pressure of 1 atm[a]**

| Contribution | Property | Equation |
|---|---|---|
| Translation | $C_p^\circ$ | 4.967975 |
| | $H^\circ - H_0^\circ$ | $4.967975T$ |
| | $S^\circ$ | $6.863511 \log M + 11.439185 \log T - 2.314820$ |
| Electronic | $C_p^\circ$ | $\dfrac{4.113692}{T^2}\left[\dfrac{\sum \varepsilon_i^2 g_i e^{-1.438786\varepsilon_i/T}}{\sum g_i e^{-1.438786\varepsilon_i/T}} - \left(\dfrac{\sum \varepsilon_i g_i e^{-1.438786\varepsilon_i/T}}{\sum g_i e^{-1.438786\varepsilon_i/T}}\right)^2\right]$ |
| | $H^\circ - H_0^\circ$ | $2.859141\left(\dfrac{\sum \varepsilon_i g_i e^{-1.438786\varepsilon_i/T}}{\sum g_i e^{-1.438786\varepsilon_i/T}}\right)$ |
| | $S^\circ$ | $\dfrac{2.859141}{T}\left(\dfrac{\sum \varepsilon_i g_i e^{-1.438786\varepsilon_i/T}}{\sum g_i e^{-1.438786\varepsilon_i/T}}\right) + 4.575674 \log \sum g_i e^{-1.438786\varepsilon_i/T}$ |

[a]Units: cal/mole for $H^\circ - H_0^\circ$ and cal/K mole for the remaining quantities, where 1 cal = 4.184 J.

(182)

**Table E  Equations for Calculating Ideal Gas Thermodynamic Properties for Diatomic Molecules at a Pressure of 1 atm[a]**

| Contribution | Property | Equation |
|---|---|---|
| **Translation** | $C_p^\circ$ | 4.967975 |
| | $H^\circ - H_0^\circ$ | $4.967975T$ |
| | $S^\circ$ | $6.863511 \log M + 11.439185 \log T - 2.314820$ |
| **Rotation** | $C_p^\circ$ | 1.98719 |
| | $H^\circ - H_0^\circ$ | $1.98719T$ |
| | $S^\circ$ | $4.575674 \log[(IT \times 10^{39})/\sigma] - 4.349171$ |
| **Vibration** | $C_p^\circ$ | $1.98719 u^2 e^{-u}/(1 - e^{-u})^2$ |
| | $H^\circ - H_0^\circ$ | $1.98719 T u e^{-u}/(1 - e^{-u})$ |
| | $S^\circ$ | $1.98719 u e^{-u}/(1 - e^{-u}) - 4.575674 \log(1 - e^{-u})$ |
| **Electronic** | $C_p^\circ$ | $\dfrac{4.113692}{T^2}\left[\dfrac{\sum \varepsilon_i^2 g_i e^{-1.438786\varepsilon_i/T}}{\sum g_i e^{-1.438786\varepsilon_i/T}} - \left(\dfrac{\sum \varepsilon_i g_i e^{-1.438786\varepsilon_i/T}}{\sum g_i e^{-1.438786\varepsilon_i/T}}\right)^2\right]$ |
| | $H^\circ - H_0^\circ$ | $2.859141\left(\dfrac{\sum \varepsilon_i g_i e^{-1.438786\varepsilon_i/T}}{\sum g_i e^{-1.438786\varepsilon_i/T}}\right)$ |
| | $S^\circ$ | $\dfrac{2.859141}{T}\left(\dfrac{\sum \varepsilon_i g_i e^{-1.438786\varepsilon_i/T}}{\sum g_i e^{-1.438786\varepsilon_i/T}}\right) + 4.575674 \log \sum g_i e^{-1.438786\varepsilon_i/T}$ |
| **Anharmonicity Corrections** | $C_p^\circ$ | $1.98719\left[\dfrac{16\gamma}{u} - \dfrac{\delta u^2 e^u}{(e^u-1)^2} + \dfrac{u^2 e^u(2\delta e^u - 4Xu - 8X)}{(e^u-1)^3} + \dfrac{12Xu^3 e^{2u}}{(e^u-1)^4}\right]$ |
| | $H^\circ - H_0^\circ$ | $1.98719 T\left[\dfrac{8\gamma}{u} + \dfrac{u(\delta e^u - 2X)}{(e^u-1)^2} + \dfrac{4Xu^2 e^u}{(e^u-1)^3}\right]$ |
| | $S^\circ$ | $1.98719\left[\dfrac{16\gamma}{u} + \dfrac{\delta}{(e^u-1)} + \dfrac{\delta u e^u}{(e^u-1)^2} + \dfrac{4Xu^2 e^u}{(e^u-1)^3}\right]$ |

[a]Units: cal/mole for $H^\circ - H_0^\circ$ and cal/K mole for the remaining quantities, where 1 cal = 4.184 J.

tionships, it appears likely that certain properties, such as heat capacity, enthalpy, entropy, and internal energy, might be derived as a result. In fact, the accuracy of the experimentally measured property values is less than those obtained by theoretical calculations, because an adequate amount of high-precision, experimental data is lacking.

More than 100 sources of information on the specific heat of air at selected pressures are found in the literature. These results were obtained by theoretical calculations, correlations, calorimetry, adiabatic expansion, Joule-Thomson experiments, the velocity of sound measurements, and heat transfer measurements.[8] Several sets of extensive values were derived by use of the method of statistical mechanics.[6,9,10] These results are more reliable than those evaluated from correlations, earlier statistical calculations, and extrapolated values to zero pressures for $P$-$V$-$T$ measurements.

This work employed the modern physical constants,[11] atomic weights,[12] and the recent molecular and spectroscopic constants for nitrogen,[13] oxygen,[13] and argon[14] for reevaluation of the thermodynamic properties of air by using statistical mechanical method.[3–5] In these calculations a molecular model with a nonrigid rotor and anharmonic oscillator was adopted for both nitrogen and oxygen molecules. As mentioned previously, the calculated results, as given in Table 1, are adequate for industrial applications, where the pressure of air is low and the temperature is moderate.

For computation of the thermodynamic properties of air, the composition of air was assumed to be as follows:

| Gas | Molecular Weight | Percentage by Volume |
|---|---|---|
| Nitrogen | 28.0134 | 78.03 |
| Oxygen | 31.9988 | 20.99 |
| Argon | 39.948 | 0.98 |

The molecular weight and the gas constant for air are calculated as 28.9669 and 287.031 $\text{Nm K}^{-1}\text{kg}^{-1}$ or 287.031 $\text{J K}^{-1}\text{kg}^{-1}$, respectively.

The enthalpy, entropy, and relative pressure of air were derived from the corresponding values of the three constituents in air in accordance with the Gibbs-Dalton law as follows:

$$h = \sum_i x_i h_i \qquad (11)$$

$$\phi = \sum_i x_i \phi_i \qquad (12)$$

$$\ln p_r = \sum_i x_i \ln p_{ri} \qquad (13)$$

where $x_i$ denotes the mole fraction of gas $i$, and the summation takes place over all the constituent gases. The values of $h$, $p_r$, $u$, $v_r$ and $\phi$, evaluated over the temperature range from 100 to 3595 K at low pressures for one kilogram of air are presented in Table 1.

The values of the internal energy were found from the equation

$$u = h - RT$$

$$= h - 0.287031T. \qquad (14)$$

The relative pressure $p_r$ which is the $p/p_0$ of (5), is given by

$$\ln p_r = \frac{\phi}{R}$$

if $T_0$ is the same for $\phi$ and $p_r$. The relative pressure was determined in this way. Since the resulting numbers were inconveniently large they were divided by $10^n$ before tabulating. The magnitude of $n$, which was always an integer, was selected for each table so as to give a convenient range of numbers.

The relative volume $v_r$, which is the $v/v_0$ of (6), could be evaluated from

$$-\frac{1}{R} \int_{T_0}^{T} \frac{c_v}{T} dT \qquad (a)$$

or from

$$\frac{T}{p_r T_0} \qquad (b)$$

indifferently, provided that $T_0$ is not too near zero. The fact, however, that both $c_p$ and $c_v$ become zero at zero temperature (and that $c_p - c_v \neq p_r$ near zero

temperature) precludes the relation

$$pv = RT$$

near zero temperature. With a $T_0$ of zero, expression (a) remains finite, whereas (b) becomes infinite.

The primary requirement that

$$\frac{v_2}{v_1} = \frac{v_{r2}}{v_{r1}},$$

where states 1 and 2 are at the same entropy, can be fulfilled by the definition

$$v_r = \frac{RT}{p_r}, \qquad (15)$$

which has the further advantage that $v_r$ is equal to the specific volume when $p_r$ is equal to the pressure. This definition was therefore adopted, with $R$ having the value 0.287031, making $v_r$ numerically equal to the specific volume in $m^3$/kg when $p_r$ is numerically equal to the pressure in kPa.

Each value of $\phi$ listed in Table 1 is the value computed from (12) minus unity.

In Table 2, the specific heat at constant volume and the velocity of sound for air were computed from the relations

$$c_v = c_p - R \qquad (16)$$

and

$$a = \sqrt{kRT}. \qquad (17)$$

The maximum flow per unit area is the maximum value for isentropic expansion from temperature $T$, unit pressure, and zero velocity. Its value for any other pressure is that for unit pressure multiplied by the pressure in pounds per square inch absolute. The values given in Table 2 were obtained from Table 1 by trial for the maximum value of the quantity

$$\frac{G_x}{p_i} = \frac{p_{rx} \sqrt{2(h_i - h_x)}}{p_{r_i} R T_x} \qquad (18)$$

where subscript $i$ refers to the initial state of the expansion and temperature $T$ in Table 2, and subscript $x$ refers to any state at the same entropy as state $i$ but at lower pressures.

(184)

**Table F   Comparison of Calculated Ideal Gas Properties With Real Gas Properties[a]**

| | T | | | P, psia | | |
|---|---|---|---|---|---|---|
| | R | 0.147 | 14.70 | 147.0 | 588.0 | 1470.0 |
| $\Delta(h - h_0)$ | | | | | | |
| | 540 | 0.01 | −0.06 | −0.74 | −3.02 | −7.33 |
| | 720 | 0.01 | −0.02 | −0.27 | −1.08 | −2.51 |
| | 1080 | 0.02 | 0.01 | 0.00 | −0.06 | −0.08 |
| | 1800 | 0.04 | 0.04 | 0.08 | 0.26 | 0.60 |
| $\Delta C_p$ | | | | | | |
| | 540 | 0.04 | 0.17 | 1.56 | 5.98 | 13.52 |
| | 720 | 0.00 | 0.08 | 0.78 | 3.02 | 6.82 |
| | 1080 | 0.04 | 0.08 | 0.36 | 1.22 | 2.75 |
| | 1800 | 0.14 | 0.14 | 0.24 | 0.49 | 0.96 |
| $\Delta\phi$ | | | | | | |
| | 540 | 2.16 | 2.38 | 2.44 | 2.20 | 1.60 |
| | 720 | 2.09 | 2.29 | 2.38 | 2.30 | 2.06 |
| | 1080 | 1.99 | 2.02 | 2.27 | 2.28 | 2.23 |
| | 1800 | 1.88 | 2.02 | 2.10 | 2.15 | 2.16 |

[a]Percent = (NBS value − calculated value) × 100/NBS value.

More than fifty sets of experimental measurements on viscosity of gaseous air have been reported.[6,17,18] Most of them were made close to room temperature. Touloukian et al.[18] critically evaluated the reported experimental data and obtained a consistent set of air viscosities in the temperature range from 80 to 2000 K (−193.15 to 1726.85°C). Their recommended values were adopted to recalculate the $\eta$ values for air, as shown in Table 2. The uncertainty of the calculated results is estimated as ±2% in the whole temperature range.

Numerous thermal conductivity ($\lambda$) measurements for air were made, only few covered extensive ranges of temperatures.[6,7] The listed values of $\lambda$ in Table 2 were based on the thermal conductivity data selected for air in the temperature range from 50 to 1500 K (−223.15 to 1226.85°C) by Touloukian et al.[19] They obtained these data from large scale graphs of different sets of measurements and checked by differencing. Below 400K the recommended $\lambda$ values should be accurate to within about 1%; the uncertainty then increased to about 5% at 1500K. The Prandtl numbers, Pr, for air were computed in accordance with the following relationship: $Pr = c_p\eta/\lambda$ where the effect of pressure on $\eta$ and $\lambda$ is less than the uncertainty in the given values for pressures up to 1400 kPa.

Real gas thermodynamic properties of air have been derived by several authors.[6,15,16] Hilsenrath et al.[6] divided the range of the calculated thermodynamic properties for air into two regions. In the region below 1500 K the composition of air was considered fixed and the corrections for gas imperfection were significant. Above 1500 K, the corrections for gas imperfection were small and the predominant influence on the thermodynamic properties was the effect of the dissociation of the constituents of air at high temperatures. In this region, the properties of air were based on the contributions from each of the molecular and atomic species present in the equilibrium composition at each temperature and pressure. The properties of ideal gas for each of the molecular and atomic species in air were obtained by use of the method of statistical mechanic calculations.

A comparison of calculated values of $h$, $c_p$, and $\phi$ with those derived for air in the real gaseous state by Hilsenrath et al.,[6] at temperatures 540, 720, 1080, 1800$R$ and pressures 0.147, 14.70, 147.0, 588.0, 1470 psia, is shown in Table F. In general, the average deviation in enthalpy is less than five percent at these conditions. The agreement is better at lower pressures and higher temperatures. The differences in $\phi$ between the real gas and ideal gas for air are 2 to 3%, in the temperature range 300 to 1000 K and at pressures up to $10^4$ kPa.

## 2.2   Tables 3 to 8—Products of Combustion with 400, 200, and 100% of Theoretical Air

Tables 3, 5, and 7 are for one gram-mole of products of combustion of a hydrocarbon fuel of composition $(CH_2)_n$ with 400, 200, and 100% of theoretical air,

### Table G Molecular Weight and Composition of Combustion Gases

| Gases in the Combustion Products | Molecular Weight | Percentage by Volume | | |
|---|---|---|---|---|
| | | 400% Theoretical Air | 200% Theoretical Air | 100% Theoretical Air |
| Nitrogen ($N_2$) | 28.0134 | 76.6886 | 75.3925 | 72.9275 |
| Oxygen ($O_2$) | 31.9988 | 15.4719 | 10.1403 | 0.0000 |
| Carbon dioxide ($CO_2$) | 44.0098 | 3.4382 | 6.7602 | 13.0783 |
| Water ($H_2O$) | 18.0152 | 3.4382 | 6.7602 | 13.0783 |
| Argon (Ar) | 39.948 | 0.9631 | 0.9468 | 0.9159 |

respectively. The composition of products of combustion employed for evaluation of their thermodynamic properties are presented in Table G. The molecular weights of these mixtures are 28.9512, 28.9360, 28.9072; and their gas constants are 287.187, 287.338, and 287.624 newton-meter $K^{-1}kg^{-1}$ or $JK^{-1}kg^{-1}$, respectively, where 1 newton-meter = 1 J.

The molar enthalpy, entropy $\phi$, and relative pressure were computed by combining the corresponding values of the constituents as shown in Table G in accordance with equations (11), (12), and (13). The molar internal energy was derived from the equation

$$\bar{u} = \bar{h} - 8.31441T, \qquad (19)$$

where the bar over the symbol indicates that the quantity is per gram-mole.

The relative volume was calculated by the equation

$$v_r = 0.0083144\frac{T}{P_r}, \qquad (20)$$

making $v_r$ numerically equal to the molar volume in $m^3$/g-mole when $P_r$ is numerically equal to the pressure in kPa.

Although the thermodynamic properties given in Tables 3, 5, and 7 were calculated for hydrocarbon fuel of composition $(CH_2)_n$, it has been shown[20] that they represent with high precision the properties of fuel combustion products over a wide range of composition with the same percent of theoretical air.

Evidence to this effect appear in Tables 4, 6, and 8, which show specific heats and ratios of specific heats for three different compositions of hydrocarbon fuel, namely, $(CH)_n$, $(CH_2)_n$, and $(CH_3)_n$. In Tables 4, 6, and 8, the molar specific heats at constant pressure were calculated by means of the Gibbs-Dalton law from the composition of combustion products and from the specific heats of the individual component gases.

The molar specific heat at constant volume was then computed from the equation

$$\bar{c}_v = \bar{c}_p - \bar{R}$$

The ratio of the specific heats was obtained from

$$k = \frac{\bar{c}_p}{\bar{c}_v}.$$

### 2.3 Use of Tables 1, 3, 5, and 7 for Other Mixtures

Table 1 for air may be converted to a table for 1 mole merely by multiplication of the enthalpy, internal energy, and ($\phi + 1$) by the molecular weight of air. The resultant table for air is also a molar products table for infinite percent of theoretical air or zero percent of theoretical fuel. It has been shown[20] that Table 1 converted to the molar form and the molar Tables 3, 5, and 7 have utility for mixtures other than those for which they were calculated. They represent combustion products of hydrocarbon fuels ranging from $(CH)_n$ to $(CH_3)_n$ for the percentages of theoretical air as indicated. They also represent certain mixtures of air and octane vapor and of air and water vapor in accordance with Table H.

### Table H Application of Evaluated Combustion Product Properties for Other Gas Mixtures

| Table Number | Products | | Reactants | Air and Water Vapor |
|---|---|---|---|---|
| | % Theoretical Air | % Theoretical Fuel | % Theoretical Fuel | Mass % Water |
| 1, 2 | ∞ | 0 | 0 | 0 |
| 3, 4 | 400 | 25 | 14 | 6.7 |
| 5, 6 | 200 | 50 | 28 | 13 |
| 7, 8 | 100 | 100 | 54 | 26 |

Problems involving mixtures intermediate between those indicated in Table H may be solved by linear interpolation. These extensions of the tables yield a precision in general better than one part in five hundred.

## 2.4  Tables 9 and 10

In Table 9 the molecular weight and gas constant are listed for the combustion products of hydrocarbon fuels containing $(CH_x)_n$ and air. The values of $x$, atomic ratio of H/C, were chosen in the range 0.715 to 4.052 which corresponds to 0.06 to 0.34 kg of H/kg of C in the fuel. The percent of theoretical fuel for combustion increases from 0 (i.e., pure air) to 100 in 10 steps with increment of 10% for each step.

Table 10 presents the enthalpy of combustion of three classes of pure hydrocarbons related to liquid and gaseous fuels. These values were taken from the Thermodynamic Research Center Hydrocarbon Project Tables.[21] The enthalpy of combustion of hydrocarbons other than those listed in Table 10 can be calculated from the chemical equation of combustion reaction with the enthalpy of formation values for all the reactants and products known at the given temperature. In complete combustion at 298.15 K, the products formed are $CO_2$ (gas) and $H_2O$ (liquid), of which the enthalpy of formation is reported[22] to be $-393.51 \pm 0.13$ kJ g-mol and $-285.830 \pm 0.042$ kJ g-mol$^{-1}$, respectively. Also available are the enthalpies of formation at 298.15 K for hundreds of organic substances related to petroleum and coal products.[21]

The enthalpy of formation of higher members of an homologous series can be estimated by the addition of appropriate numbers of methylene ($CH_2$) increments to known values for a lower homolog. The $n$-alkyl chain of the lower homolog must be long enough so that the increment for the addition of another $CH_2$ group has approached the limiting value for the long chains.[23]

## 2.5  Tables 11 to 23—$N_2$, $O_2$, $H_2O$, $CO_2$, $H_2$, CO, and Ar

Tables 11 to 23 are for 1 g-mol of gas. The ideal gas thermodynamic properties for $N_2$, $O_2$, $H_2O$, $CO_2$, $H_2$, and CO are listed for each compound in two consecutive tables. The values of $\bar{h}$, $p_r$, $\bar{u}$, $v_r$, and $\bar{\phi}$ are given in the first table. Those of $\bar{c}_p$, $\bar{c}_v$, $\bar{c}_p/\bar{c}_v$, and $a$

are presented in the following one. For a monatomic species such as argon, only the first table is included since $\bar{c}_p$ and $\bar{c}_v$ are constants and independent of temperature. Using the statistical mechanical method, the molecular and spectroscopic constants reported by Rosen[13] were employed in computing the thermodynamic properties for diatomic molecules such as $N_2$, $O_2$, and CO.

The thermodynamic properties for argon in the ideal gaseous state were derived from statistical mechanical calculation. The required spectroscopic data for argon were obtained from C. E. Moore.[14] In calculating the required thermodynamic properties of atomic argon at each given temperature 265 electronic energy levels (up to 127970 cm$^{-1}$) were used. The ideal gas thermodynamic properties of $CO_2$ (gas) were taken from those reported by J. Hilsenrath et al.[6] The properties for hydrogen (75% ortho and 25% para) were adopted from those reported by Woolley et al.[24] and those for $H_2O$ (gas) were adopted from a recent work recommended by L. Haar,[25] where the thermodynamic properties were computed by statistical mechanical method, employing direct summation of electronic energy levels. The above calculated results are considered to be the most reliable ones.

The formulas employed to derive the other related thermodynamic quantities as given in Tables 11 to 23 are similar to those described in Section 2.1 for air. For ease of presentation, the tabulated values of the relative pressure have been decreased from the calculated values by a constant factor of $10^{10}$ for $N_2$, $O_2$, $H_2O$, and CO; $10^{12}$ for $CO_2$; $10^{15}$ for $H_2$; and $10^7$ for Ar. The low critical temperatures of $N_2$, $O_2$, $H_2$, CO, and Ar insure that the equation, $p\bar{v} = \bar{R}T$, is a good approximation to the true equation of state in each instance over a wide range of pressures. Therefore, Tables 11, 13, 19, 21, and 23, have validity over a range comparable to that of Table 1.

The critical temperature of $CO_2$ is close to room temperature, far below the critical point encountered in power plants. Consequently the range of validity for Table 17 should compare favorably with that of Table 1, although it is doubtless slightly less extensive. The critical temperature of water being the highest of all the gases considered here, the range of validity found in Table 15 is at the low end of the scale. Nevertheless, the range is considerable, owing largely to the high critical pressure of water. The pressures of combustion products in power plants

# Table I  Comparison of the Ideal Gas Thermodynamic Data at 101.325 kPa and Selected Temperatures[a]

| | $\bar{h}$, cal / g-mol | | | $\bar{c}_p$, cal / K g-mol | | | $\bar{\phi}$, cal / K g-mol | | |
|---|---|---|---|---|---|---|---|---|---|
| | 298.15 K | 1000 K | 3000 K | 298.15 K | 1000 K | 3000 K | 298.15 K | 1000 K | 3000 K |
| **Nitrogen ($N_2$)** | | | | | | | | | |
| CODATA,[22] 1975 | 2072 | | | | | | 45.770 | | |
| JANAF,[26] 1977 | 2072 | 7202 | 24231 | 6.961 | 7.815 | 8.850 | 45.770 | 54.508 | 63.762 |
| TRCDP,[27] 1971 | 2072 | 7202 | 24233 | 6.961 | 7.815 | 8.852 | 45.761 | 54.500 | 63.755 |
| Gurvich et al.,[28] 1962 | 2072 | 7202 | 24236 | | | | 45.771 | 54.510 | 63.766 |
| NBS 564,[6] 1955 | | 7202 | 24234 | | 7.815 | 8.852 | | 54.501 | 63.756 |
| This work | 2072 | 7202 | 24225 | 6.961 | 7.814 | 8.842 | 45.761 | 54.500 | 63.751 |
| **Oxygen ($O_2$)** | | | | | | | | | |
| CODATA,[22] 1975 | 2075 | | | | | | 49.005 | | |
| JANAF,[26] 1977 | 2075 | 7501 | 25501 | 7.021 | 8.334 | 9.528 | 49.005 | 58.190 | 67.963 |
| TRCDP,[27] 1968 | 2075 | 7501 | 25519 | 7.020 | 8.336 | 9.551 | 49.003 | 58.191 | 67.971 |
| Gurvich et al.,[28] 1962 | 2075 | 7502 | 25525 | | | | 49.006 | 58.193 | 67.976 |
| NBS 564,[6] 1955 | | 7501 | 25521 | | 8.336 | 9.551 | | 58.200 | 67.981 |
| This work | 2075 | 7501 | 25424 | 7.021 | 8.334 | 9.496 | 48.994 | 58.180 | 67.924 |
| **Argon (Ar)** | | | | | | | | | |
| CODATA,[22] 1975 | 1481 | | | | | | 36.982 | | |
| JANAF,[26] 1977 | 1481 | 4968 | 14904 | 4.968 | 4.968 | 4.968 | 36.983 | 42.995 | 48.453 |
| Gurvich et al.[28] 1962 | 1481 | 4968 | 14905 | | | | 36.983 | 42.995 | 48.454 |
| NBS 564,[6] 1955 | | 4968 | 14904 | | 4.968 | 4.968 | | 42.997 | 48.454 |
| This work | 1481 | 4968 | 14904 | 4.968 | 4.968 | 4.968 | 36.983 | 42.995 | 48.452 |
| **Water ($H_2O$)** | | | | | | | | | |
| CODATA,[22] 1975 | 2368 | | | | | | 45.106 | | |
| JANAF,[26] 1961 | 2367 | 8576 | 32568 | 8.025 | 9.851 | 13.304 | 45.106 | 55.592 | 68.421 |
| TRCDP,[27] 1969 | 2367 | 8576 | 32567 | 8.025 | 9.850 | 13.303 | 45.103 | 55.589 | 68.418 |
| Gurvich et al.,[28] 1962 | 2368 | 8591 | 32910 | | | | 45.108 | 55.614 | 68.594 |
| NBS 564,[6] 1955 | | 8576 | 32567 | | 9.850 | 13.303 | | 55.590 | 68.419 |
| This work | 2367 | 8624 | 32773 | 8.028 | 9.864 | 13.331 | 45.104 | 55.598 | 68.446 |
| **Carbon Dioxide ($CO_2$)** | | | | | | | | | |
| CODATA,[22] 1975 | 2238 | | | | | | 51.070 | | |
| JANAF,[26] 1965 | 2238 | 10222 | 38773 | 8.874 | 12.980 | 14.873 | 51.072 | 64.344 | 79.848 |
| TRCDP,[27] 1966 | 2238 | 10221 | 38762 | 8.874 | 12.980 | 14.855 | 51.070 | 64.337 | 79.837 |
| Gurvich et al.,[28] 1962 | 2239 | 10220 | 38781 | | | | 51.071 | 64.333 | 79.838 |
| NBS 564,[6] 1955 | | 10221 | 38770 | | 12.980 | 14.872 | | 64.337 | 79.841 |
| This work | 2238 | 10221 | 38771 | 8.874 | 12.980 | 14.872 | 51.070 | 64.337 | 79.841 |
| **Hydrogen ($H_2$)** | | | | | | | | | |
| CODATA,[22] 1975 | 2024 | | | | | | 31.207 | | |
| JANAF,[26] 1977 | 2024 | 6967 | 23233 | 6.892 | 7.219 | 8.864 | 31.207 | 39.700 | 48.466 |
| TRCDP,[27] 1968 | 2024 | 6966 | 23231 | 6.889 | 7.219 | 8.859 | 31.206 | 39.701 | 48.465 |
| Gurvich et al.,[28] 1962 | 2019 | 6967 | 23232 | | | | 31.195 | 39.702 | 48.467 |
| NBS 564,[6] 1955 | 2037 | 6967 | 23232 | 6.894 | 7.219 | 8.859 | 31.251 | 39.700 | 48.466 |
| This work | 2024 | 6966 | 23231 | 6.890 | 7.220 | 8.860 | 31.251 | 39.700 | 48.466 |
| **Carbon Monoxide (CO)** | | | | | | | | | |
| CODATA,[22] 1975 | 2073 | | | | | | 47.217 | | |
| JANAF,[26] 1965 | 2072 | 7255 | 24429 | 6.965 | 7.931 | 8.895 | 47.214 | 56.028 | 65.370 |
| TRCDP,[27] 1966 | 2073 | 7256 | 24423 | 6.965 | 7.931 | 8.886 | 47.219 | 56.033 | 65.372 |
| Gurvich et al.,[28] 1962 | 2073 | 7256 | 24429 | | | | 47.217 | 56.032 | 65.373 |
| NBS 564,[6] 1955 | 2085 | 7256 | 24430 | 6.965 | 7.931 | 8.895 | 47.259 | 56.031 | 65.371 |
| This work | 2072 | 7256 | 24418 | 6.965 | 7.929 | 8.883 | 47.208 | 56.022 | 65.358 |

[a] 1 cal = 4.184 J.

(188)

seldom exceed 1400 kPa. For 200% of theoretical air, the corresponding partial pressures are: nitrogen 15, oxygen 20, carbon dioxide 13.5, and argon 2 psi. It follows from this example that the present tables are reliable for pressures several times as large as those now employed in power plants.

Presented in Table I is a comparison of our calculated results with those reported in the literature for molar enthalpy, heat capacity at constant pressure, and entropy at selected temperatures, i.e., 298.15, 1000, and 3000 K, and 101.325 kPa (1 atm) pressure. The agreement between them is excellent. Some slight discrepancies are caused by the use of different values of physical, molecular, and spectroscopic constants for calculations. Nevertheless, in most cases, these discrepancies are well within their respective assigned uncertainties. The thermodynamic properties of nitrogen, oxygen, water, carbon dioxide, hydrogen, carbon monoxide, and argon in the real gaseous state have been critically evaluated by J. Hilsenrath et al.[6] The reported values are internally consistent, very reliable, and highly recommended for use. Comprehensive reports on argon,[29] carbon dioxide,[30] and air components[36] are available. V. J. Johnson has published a review[31] of recent data compiled on hydrogen, argon, nitrogen, and oxygen. A survey of existing data on thermodynamic properties of nitrogen and carbon dioxide was reported by G. M. Wilson et al.[32] F. Din surveyed the existing data, calculated the thermodynamic functions, and constructed the thermodynamic diagrams for $CO_2$,[33] $CO$,[33] $Ar$,[34] and $N_2$.[35]

## 2.6 Tables 30 to 53—One-Dimensional Compressible-Flow Functions

Developments in high-speed propulsion have led to a better understanding of the flow of compressible fluids. Generalized[37]* treatments of one-dimensional flow are available in both analytical and numerical formulations. Tables 30 to 53 represent a portion of the numerical formulations prepared by A. H. Shapiro and G. M. Edelman. Some of this material has been published in the *Journal of Applied Mechanics*.[38] These tables contain the functions useful in many engineering problems in the one-dimensional flow of a perfect gas with constant specific heat and molecular weight.

*References to earlier papers are given in reference 37.

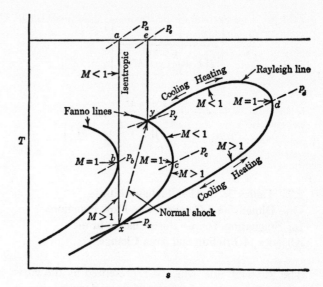

**Figure 1**

The isentropic, which is represented by the line *abx* in Figure 1, is the locus of states for a frictionless adiabatic process. Segments *ab* and *bx* represent subsonic and supersonic flows, respectively. At point *b* the Mach number is unity, and this condition is denoted by a superscript asterisk.

The following relations hold along the isentropic:

$$T_0 = \text{constant} = T_0{}^* = T_a$$

$$T^* = \text{constant} = T_b$$

$$p_0 = \text{constant} = p_0{}^* = p_a$$

$$p^* = \text{constant} = p_b$$

$$M^* = \frac{V}{V^*} = M\sqrt{\frac{k+1}{2\left(1 + \frac{k-1}{2}M^2\right)}}$$

$$\frac{A}{A^*} = \frac{1}{M}\left[\frac{2\left(1 + \frac{k-1}{2}M^2\right)}{k+1}\right]^{\frac{k+1}{2(k-1)}}$$

$$\frac{T}{T^*} = \frac{k+1}{2\left(1 + \frac{k-1}{2}M^2\right)}$$

$$\frac{\rho}{\rho^*} = \left[\frac{k+1}{2\left(1 + \frac{k-1}{2}M^2\right)}\right]^{\frac{1}{k-1}}$$

$$\frac{p}{p^*} = \left[ \frac{k+1}{2\left(1 + \frac{k-1}{2}M^2\right)} \right]^{\frac{k}{k-1}}$$

$$\frac{F}{F^*} = \frac{1 + kM^2}{m\sqrt{2(k+1)\left(1 + \frac{k-1}{2}M^2\right)}} \; .$$

## 2.7 Tables 36 to 41—Rayleigh Line—One-Dimensional Compressible-Flow Functions for Stagnation-Temperature Change in the Absence of Friction and Area Change

The Rayleigh line, which is represented by the curve $ydx$ of Figure 1, is the locus of states for a process at constant flow per unit area—that is, constant $A$—and constant impulse $F$. The process being reversible, increasing entropy corresponds to heat flow to the stream, decreasing entropy to heat flow from the stream. It follows that the stagnation temperature increases with increasing entropy and decreases with decreasing entropy. At point $d$ both the entropy and the stagnation temperature are at a maximum and the Mach number is unity. This condition is denoted by a superscript asterisk.

The following relations hold along the Rayleigh line:

$$A = A^* = \text{constant}$$

$$F = F^* = \text{constant}$$

$$\frac{T_0}{T_0^*} = \frac{2(k+1)M^2\left(1 + \frac{k-1}{2}M^2\right)}{(1 + kM^2)^2}$$

$$\frac{T}{T^*} = \frac{(k+1)^2 M^2}{(1 + kM^2)^2}$$

$$\frac{p_0}{p_0^*} = \frac{k+1}{1 + kM^2} \left[ \frac{2\left(1 + \frac{k-1}{2}M^2\right)}{k+1} \right]^{\frac{k}{k-1}}$$

$$\frac{p}{p^*} = \frac{k+1}{1 + kM^2}$$

$$\frac{\rho}{\rho^*} = \frac{V^*}{V} = \frac{1}{M^*} = \frac{1 + kM^2}{(k+1)M^2} \; .$$

## 2.8 Tables 42 to 47—Fanno Line—One-Dimensional Compressible-Flow Functions for Adiabatic Flow at Constant Area with Friction

The Fanno line, which is represented by the curve $ycx$ in Figure 1, is the locus of states for an adiabatic process at constant flow per unit area, that is, constant $A$. A second Fanno line, passing through the point $b$ in Figure 1, is for a higher flow per unit area. At points $b$ and $c$ the Mach number is unity, and this condition is denoted by a superscript asterisk. At higher points on the line the velocities are subsonic; at lower points supersonic.

It should be noted that the superscript asterisk always refers to a state where the Mach number is unity, but that this state corresponding to a given state $x$ (Figure 1) is state $b$ for the isentropic, state $c$ for the Fanno line, and state $d$ for the Rayleigh line.

The following relations hold along the Fanno line:

$$A = A^*$$

$$T_0 = T_0^* = T_a$$

$$\frac{T}{T^*} = \frac{k+1}{2\left(1 + \frac{k-1}{2}M^2\right)}$$

$$\frac{p_0}{p_0^*} = \frac{1}{M} \left[ \frac{2\left(1 + \frac{k-1}{2}M^2\right)}{k+1} \right]^{\frac{k+1}{2(k-1)}}$$

$$\frac{p}{p^*} = \frac{1}{M} \sqrt{\frac{k+1}{2\left(1 + \frac{k-1}{2}M^2\right)}}$$

$$\frac{\rho}{\rho^*} = \frac{V^*}{V} = \frac{1}{M^*} = \frac{1}{M} \sqrt{\frac{2\left(1 + \frac{k-1}{2}M^2\right)}{k+1}}$$

$$\frac{F}{F^*} = \frac{1 + kM^2}{M\sqrt{2(k+1)\left(1 + \frac{k-1}{2}M^2\right)}}$$

$$\frac{4fL_{\max}}{D} = \frac{1 - M^2}{kM^2} + \frac{k+1}{2k} \log_e \frac{(k+1)M^2}{2\left(1 + \frac{k-1}{2}M^2\right)} \; .$$

(190)

## 2.9 Tables 48 to 53—One-Dimensional Normal Shock Functions

States $x$ and $y$ in Figure 1, which have the same flow per unit area, the same stagnation temperature and the same impulse, represent the initial and final states of a normal shock.

The following relations hold for the normal shock:

$$T_{0x} = T_{0y}$$

$$M_y^2 = \frac{M_x^2 + \dfrac{2}{k-1}}{\dfrac{2k}{k-1}M_x^2 - 1}$$

$$\frac{T_y}{T_x} = \frac{\left(1 + \dfrac{k-1}{2}M_x^2\right)\left(\dfrac{2k}{k-1}M_x^2 - 1\right)}{\dfrac{(k+1)^2}{2(k-1)}M_x^2}$$

$$\frac{p_y}{p_x} = \frac{2k}{k+1}M_x^2 - \frac{k-1}{k+1}$$

$$\frac{\rho_y}{\rho_x} = \left(\frac{p_y}{p_x}\right)\left(\frac{T_x}{T_y}\right)$$

$$\frac{p_{0y}}{p_{0x}} = \left[\frac{\dfrac{k+1}{2}M_x^2}{1 + \dfrac{k-1}{2}M_x^2}\right]^{\frac{k}{k-1}}\left[\frac{2k}{k+1}M_x^2 - \frac{k-1}{k+1}\right]^{\frac{1}{1-k}}$$

$$\frac{p_{0y}}{p_x} = \left[\frac{k+1}{2}M_x^2\right]^{\frac{k}{k-1}}\left[\frac{2k}{k+1}M_x^2 - \frac{k-1}{k+1}\right]^{\frac{1}{1-k}}$$

## 2.10 Tables 54 to 59—Two-Dimensional Compressible-Flow Functions

Table 54, taken from the data of Shapiro and Edelman,[38] presents functions useful in the application of the method of characteristics to two-dimensional, isentropic, supersonic-flow problems. Under these conditions the following relations hold:

$$\frac{p}{p_0} = \left(1 + \frac{k-1}{2}M^2\right)^{\frac{k}{1-k}}$$

$$\frac{\rho}{\rho_0} = \left(1 + \frac{k-1}{2}M^2\right)^{\frac{1}{1-k}}$$

$$M^* = \frac{V}{V^*} = M\sqrt{\frac{k+1}{2\left(1 + \dfrac{k-1}{2}M^2\right)}}$$

$$\frac{A}{A^*} = \frac{1}{M}\left[\frac{2\left(1 + \dfrac{k-1}{2}M^2\right)}{k+1}\right]^{\frac{k+1}{2(k-1)}}$$

$$\alpha = \arcsin\frac{1}{M}.$$

Tables 55 to 59, taken from the data of Edmonson, Murnaghan, and Snow,[39] present functions useful in applications of the theory of two-dimensional shocks, or oblique shock waves. For these the following relations hold:

$$\frac{\rho_y}{\rho_x} = X = \frac{\tan\alpha}{\tan(\alpha - \omega)}$$

$$M_x = \left(\frac{5X}{6-X}\right)^{1/2}\csc\alpha$$

$$M_y = \left(\frac{5}{6X-1}\right)^{1/2}\csc(\alpha - \omega)$$

$$\frac{p_y}{p_x} = \frac{6X-1}{6-X}.$$

## 2.11 Tables 60 to 62—Standard Atmosphere, Physical Constants, and Conversion Factors

The contents in Table 60 represent a combination of two tables found in the handbook, *Smithsonian Physical Tables*.[40] The values of physical constants in Table 61 are those of the 1973 International Physical Constants recommended by CODATA-ICSU.[41] As the most reliable values presently available, these constants were employed throughout this work in computing the ideal gas thermodynamic properties of chemical compounds by means of statistical mechanics.

The conversion factor tables demonstrate the numerical relationships among English, cgs, and SI units for such quantities as length, volume, mass, density, force, pressure, energy, power, specific energy, specific energy per degree, absolute viscosity, kinematic viscosity, thermal conductivity, and molar gas constants. Of course, the trend of the future is towards international conformity in the use of SI units.[42]

# EXAMPLES

## Illustrating the Use of the Tables

### Example 1 Compression of Air in Steady Flow

Air at a pressure of 101.325 kPa (1 atm abs) and a temperature of 300 K is compressed in steady flow to a pressure of 607.95 kPa (6 atm abs). Find the work of compression and the temperature after compression for (a) 100% efficiency of compression and (b) 60% efficiency of compression. The efficiency of compression is here defined as the ratio of the isentropic work of compression to the actual work of compression.

*Solution.* (a) From Table 1 we get for $T_1 = 300$ K

$$p_{r1} = 1.3801, \qquad h_1 = 300.43 \text{ kJ/kg},$$

where subscript 1 refers to the state at the compressor inlet. To determine the properties at the compressor outlet for isentropic compression we compute the relative pressure there

$$p_{r2} = \tfrac{6}{1} \times 1.3801 = 8.2806.$$

Entering Table 1 with this value of $p_r$, we find, for $h_{2s}$ and $T_{2s}$, the enthalpy and temperature at the compressor outlet for isentropic compression

$$h_{2s} = 501.62 \text{ kJ/kg}, \qquad T_{2s} = 498.37 \text{ K}.$$

The work of compression for 100% efficiency is then

$$h_{2s} - h_1 = 201.19 \text{ kJ/kg}.$$

(b) Since the efficiency of compression $\eta$ is defined by the equation $\eta = (h_{2s} - h_1)/\text{work per kg}$, we have for 60% efficiency

$$\text{work per kilogram} = \frac{201.19}{0.60} = 335.32 \text{ kJ/kg}.$$

For the enthalpy at state 2, the state at the compressor outlet, we have

$$h_2 = h_1 + \text{work per kilogram}$$

$$= 300.43 + 335.32 = 635.75 \text{ kJ/kg}.$$

Entering Table 1 with this value of the enthalpy, we get the temperature at the compressor outlet

$$T_2 = 627.04 \text{ K}.$$

(The value of $p_{r2}$ is irrelevant because the process is not isentropic.)

If in this problem the definition of the efficiency is altered to be the ratio of the reversible isothermal work of compression to the actual work of compression, Table 1 is not necessary to the solution, for it is readily shown that the work of reversible isothermal compression in steady flow is given by

$$RT \ln \frac{p_2}{p_1},$$

provided only that

$$pv = RT.$$

### Example 2 Change in Entropy

Find the increase in entropy of each kilogram of air in Example 1(b).

*Solution.*

$$s_2 - s_1 = \phi_2 - \phi_1 - R \ln \frac{p_2}{p_1}.$$

From Table 1,

$T_1 = 300$ K, $p_1 = 101.325$ kPa, $\phi_1 = 5.7016$ kJ/K kg,

$T_2 = 627.04$ K, $p_2 = 607.95$ kPa, $\phi_2 = 6.4552$ kJ/K kg.

$$-R \ln \frac{p_2}{p_1} = -0.51429.$$

Hence

$$s_2 - s_1 = 6.4552 - 5.7016 - 0.51429$$

$$= 0.2393 \text{ kJ/K kg}.$$

An alternative method consists of determining the increase in entropy between $T_{2s}$ and $T_2$ at $p_2$. Then

$$s_2 - s_1 = \phi_2 - \phi_{2s}$$

$$= 6.4552 - 6.2159$$

$$= 0.2393 \text{ kJ/K kg}.$$

### Example 3    The Pressure in a Turbine Stage

A steady stream of air approaches a ring of convergent nozzles at a pressure of 350 kPa, a temperature of 800 K, and at a rate of flow per unit area of 400 kg/s m$^2$ of nozzle exit area. Assuming a discharge coefficient of unity, find the pressure in the exit plane of the nozzle.

*Solution.* The alternative methods (*a*) and (*b*) given below are based on Tables 1 and 29 respectively.

(*a*) From the equation

$$G = \frac{w}{A} = \frac{V}{v} = \frac{p\sqrt{2(h_0 - h)}}{RT},$$

where $h$ denotes exit enthalpy and $h_0$, entrance enthalpy (see page 113 for definition of other symbols) and from Table 1, the exit pressure is determined by a trial-and-error procedure to be 284 kPa.

(*b*) The value of $k$ at 800 K from Table 2 is 1.354. The rate of flow per unit area is given by the equation (page 114)

$$G = \frac{p_0\sqrt{m}}{\sqrt{T_0}} \sqrt{\frac{2}{R} \frac{n}{n-1}} \; r^{\frac{1}{n}} \sqrt{1 - r^{\frac{n-1}{n}}}$$

where $n = k$. Substituting given values and obtaining the value for $\sqrt{(2/\overline{R})[n/(n-1)]}$ from Table 24, we get

$$400 = \frac{350{,}000\sqrt{28.9669}}{\sqrt{800}} (0.030332) r^{\frac{1}{n}} \sqrt{1 - r^{\frac{n-1}{n}}}$$

or

$$0.1980 = r^{\frac{1}{n}} \sqrt{1 - r^{\frac{n-1}{n}}}$$

From Table 29, $r = 0.810$. It follows that the exit pressure is 284 kPa.

### Example 4    Reversible Adiabatic Steady Flow

A steady stream of air expands reversibly and adiabatically in a nozzle passage from a pressure of 1000 kPa, a temperature of 300 K, and negligible velocity. Find the specific volume, velocity, Mach number, and mass velocity at the cross sections of the stream where the values of the pressure are 200, 900, and 995 kPa respectively.

*Solution.* A solution may be found from Table 1 upon noting that

$$\frac{v}{v_0} = \frac{v_r}{v_{r0}},$$

$$V = \sqrt{2(h_0 - h)}$$

(where the subscript 0 refers to the section where the velocity is zero),

$$M = \frac{V}{a}$$

(where $a$ denotes the velocity of sound as given in Table 2), and

$$G = \frac{V}{v}.$$

Since the ratio of the specific heats $k$ is constant at a value of 1.4 throughout this expansion, according to the data of Table 2, a second solution may be found from Table 25 with the aid of the relations given on page 114.

For small changes in pressure, a third method of solution yields precise results. Noting that

$$\left( \frac{\partial h}{\partial p} \right)_s = v,$$

we get

$$h_0 - h \cong \frac{v_0 + v}{2}(p_0 - p),$$

in which $v_0$ and $v$ may be obtained from Table 1.

The results of these three methods are summarized in Table J. The precision of Table 1, as indicated by

(193)

## Table J

| $\frac{p}{p_0}$ | Method | T K | $\Delta h_s$ kJ / kg | v $m^3$ / kg | V m / s | M | G kg / $m^2$s |
|---|---|---|---|---|---|---|---|
| **0.20** | Table 1 | 189.26 | 111.06 | 0.27162 | 471.3 | 1.709 | 1734 |
| | Table 25 | 189.42 | 111.09 | 0.27184 | 471.4 | 1.709 | 1734 |
| | $\tilde{v}\,\Delta p$ | 189.26 | 143.09 | 0.27162 | 535.0 | 1.939 | 1969 |
| **0.90** | Table 1 | 291.10 | 8.94 | 0.09284 | 133.7 | 0.3911 | 1440 |
| | Table 25 | 291.10 | 8.937 | 0.09284 | 133.7 | 0.3909 | 1440 |
| | $\tilde{v}\,\Delta p$ | 291.10 | 8.947 | 0.09284 | 133.8 | 0.3912 | 1441 |
| **0.995** | Table 1 | 299.57 | 0.43 | 0.08642 | 29.37 | 0.0847 | 339.9 |
| | Table 25 | 299.57 | 0.4313 | 0.08642 | 29.37 | 0.0847 | 339.9 |
| | $\tilde{v}\,\Delta \dot{p}$ | 299.57 | 0.4313 | 0.08642 | 29.37 | 0.0847 | 339.9 |

the value of $\Delta h_s$ when $p/p_0$ is 0.995, becomes inadequate for small changes in pressure. The precision of the method of Table 25 is good in all three instances. It deteriorates as the pressure change is reduced still further. For large pressure changes it is precise provided that a mean value of $k$ is used and the variation in $k$ is not large. The method employing $\tilde{v}\,\Delta p$ is good for small changes in pressure and is distinguished from the other methods in that it improves in precision as the pressure change approaches zero.

If the value of $k$ were different from 1.4, Tables 26 to 29 would be used instead of Table 25 in the second method.

### Example 5   Polytropic Process

Air at 1000 kPa and 1500 K expands according to the relation

$$pv^{1.3} = \text{constant}$$

to a pressure of 40 kPa. Find (a) the temperature at the end of the expansion, (b) the change in entropy, (c) the work done on a piston by a kilogram of air expanding slowly, (d) the heat flow to a kilogram of air expanding slowly, (e) the work delivered to a turbine shaft per kilogram of air entering the turbine for reversible expansion in steady flow, and the corresponding flow of heat.

*Solution.* (a)

$$\frac{T_2}{T_1} = \left(\frac{p_2}{p_1}\right)^{\frac{n-1}{n}} = 0.4758, \text{ according to Table 27.}$$

$$T_2 = 0.4758 \times 1500 = 713.7 \text{ K.}$$

(b) Referring to Table 1, we get

$$s_2 - s_1 = \phi_2 - \phi_1 + R \ln \frac{p_1}{p_2}$$

$$= 6.5934 - 7.4444 + 0.9239$$

$$= 0.0729 \text{ kJ/K kg.}$$

(c)

$$W = \int_1^2 p\,dv = \frac{RT_1}{n-1}\left[1 - \left(\frac{p_2}{p_1}\right)^{\frac{n-1}{n}}\right]$$

$$= \frac{0.28703 \times 1500}{0.3}[1 - 0.4758] = 752.3 \text{ kJ/kg.}$$

(d) $Q = u_2 - u_1 + W$, which, upon substitution of values from Table 1 and from above, becomes

$$Q = 523.40 - 1205.14 + 752.3 = 70.6 \text{ kJ/kg.}$$

(e) For an infinitesimal step between states of zero velocity in a steady-flow process, the shaft work $dW_x$ is given by

$$dW_x = -dh + dQ,$$

which for reversibility becomes

$$dW_x = -dh + T\,ds = -v\,dp,$$

and, therefore,

$$W_x = -\int_1^2 v\,dp = \frac{nR}{n-1}(T_1 - T_2) = 978.0 \text{ kJ/kg.}$$

$$Q = h_2 - h_1 + W_x,$$

(194)

which, upon substitution of values from Table 1 and from above, becomes

$$Q = 728.26 - 1635.68 + 978.0$$

$$= 70.6 \text{ kJ/kg},$$

a value identical with that found in ($d$) above, both being equal to the integral of $T\,ds$.

### Example 6   Compression of Air and Water Vapor

A mixture of air and water vapor at 300 K and 100 kPa has a specific humidity of 0.030 kilogram of water per kilogram of dry air. The mixture is compressed isentropically in steady flow to 400 kPa. Calculate the work of compression per kilogram of mixture.

*Solution.* The molar products table for 400% of theoretical air corresponds to a mixture of air and 6.70% by mass of water vapor (see Table H). Linear interpolation is necessary to obtain the solution.

*From Table 3: Air with 6.70%*
*water vapor*
$$T_1 = 300 \text{ K}$$
$$p_{r1} = 1.3688$$
$$\bar{h}_1 = 8769.3 \text{ J/g-mole}$$
$$p_{r2} = 4 \times p_{r1} = 5.4752$$
$$T_2 = 441.91 \text{ K}$$
$$\bar{h}_2 = 12,995.0 \text{ J/g-mole}$$
$$(\bar{h}_2 - \bar{h}_1)_s = 4225.7 \text{ J/g-mole}$$

*From Table 1: Air with 0%*
*water vapor*
$$T_1 = 300 \text{ K}$$
$$p_{r1} = 1.3801$$
$$\bar{h}_1 = 8702.5 \text{ J/g-mole}$$
$$p_{r2} = 5.5204$$
$$T_2 = 444.83$$
$$\bar{h}_2 = 12,942.3 \text{ J/g-mole}$$
$$(\bar{h}_2 - \bar{h}_1) = 4239.8 \text{ J/g-mole}$$

Since the mixture is $(0.03/1.03) \times 100$ or 2.91% by mass of water vapor, linear interpolation with respect to the percentage of water vapor yields the following expression for the work of compression per kilogram of mixture:

$$W = \frac{\dfrac{2.91}{6.70}(4225.7 - 4239.8) + 4239.8}{28.466}$$

$$= 148.73 \text{ kJ/kg},$$

where 28.466 is the molecular weight of the mixture.

### Example 7   Compression of Air and Octane Vapor

A mixture of air and octane vapor, corresponding to 25% of theoretical fuel, is compressed isentropically in steady flow from 300 K and 100 kPa to 600 kPa. Calculate the work of compression per kilogram of reactants.

*Solution.* The molar products tables for 200 and 400% of theoretical air correspond to a mixture of air and octane vapor for 28 and 14% of theoretical fuel, respectively (see Table H). Linear interpolation is necessary to obtain the solution.

*From Table 5: Reactants for 28%*
*theoretical fuel*
$$T_1 = 300 \text{ K}, \ p_{r1} = 1.3580$$
$$\bar{h}_1 = 8833.8 \text{ J/g-mole}$$
$$p_{r2} = 6 \times p_{r1} = 8.1480$$
$$T_2 = 490.01 \text{ K}$$
$$\bar{h}_2 = 14,611.3 \text{ J/g-mole}$$
$$(\bar{h}_2 - \bar{h}_1)_s = 5777.5 \text{ J/g-mole}$$

*From Table 3: Reactants for 14%*
*theoretical fuel*
$$T_1 = 300 \text{ K}, \ p_{r1} = 1.3688$$
$$\bar{h}_1 = 8769.3 \text{ J/g-mole}$$
$$p_{r2} = 6 \times p_{r1} = 8.2128$$
$$T_2 = 494.02 \text{ K}$$
$$\bar{h}_2 = 14,571.0 \text{ J/g-mole}$$
$$(\bar{h}_2 - \bar{h}_1)_s = 5801.7 \text{ J/g-mole}$$

Interpolating linearly with respect to the percentage of theoretical fuel, we get for the work of compression per kilogram of reactants

$$W = \frac{\dfrac{0.25 - 0.14}{0.28 - 0.14}(5777.5 - 5801.7) + 5801.7}{29.326}$$

$$= 197.19 \text{ kJ/kg}$$

where 29.326 is the molecular weight of the mixture.

## Example 8  Expansion of Products of Combustion of Benzene

The products of combustion of benzene with 200% of theoretical air expand in steady flow in a turbine from an initial temperature of 800 K and an initial pressure of 1013.25 kPa (10 atm) to an exit pressure of 101.325 kPa (1 atm). The efficiency of the turbine is 80% based on the isentropic work of expansion. Calculate the work to the turbine shaft per kilogram of products.

*Solution.* The composition of benzene is $C_6H_6$. It has been shown[20] that a molar products table based on a fuel composition of $(CH_2)_n$ will yield precise results for the products of combustion of a hydrocarbon with the same composition as benzene, provided that the percentage of theoretical air is the same for the benzene as for the $(CH_2)_n$. Using Table 5 for 200% of theoretical air, we obtain for the isentropic expansion

$$T_1 = 800 \text{ K}, \qquad \bar{h}_1 = 24,585.4 \text{ J/g-mole},$$

$$p_{r1} = 53.992, \qquad p_{r2} = 53.992 \times \tfrac{1}{10} = 5.3992$$

$$T_{2s} = 438.49 \text{ K}, \qquad \bar{h}_{2s} = 13,024.3 \text{ J/g-mole},$$

where subscript 1 refers to the inlet of the turbine and subscript $2s$ refers to the state at the exit pressure for isentropic expansion.

The work per kilogram of products is given by

$$W = 0.8 \frac{h_1 - h_{2s}}{\bar{m}} = 0.8 \frac{24,585.4 - 13,024.3}{29.445}$$

$$= 314.11 \text{ kJ/kg},$$

where $\bar{m}$ is the molecular weight of the products of combustion for benzene with 200% of theoretical air.

## Example 9  Heat Transfer to Products of Combustion of Benzene

The products of combustion of benzene with 300% of theoretical air flow steadily through a heat exchanger and drop in temperature from 600 to 400 K at a constant pressure of 101.325 kPa (1 atm). Calculate the heat transfer per kilogram of products.

*Solution.* The molar products table based on a fuel composition of $(CH_2)_n$ may be used for this mixture (see Table H). Linear interpolation between Tables 3 and 5 is necessary to obtain the answer for 300% of theoretical air.

*From Table 5: Products for 50%
theoretical fuel*

$$T_1 = 600 \text{ K}$$
$$\bar{h}_1 = 18,063.2 \text{ J/g-mole}$$
$$T_2 = 400 \text{ K}$$
$$\bar{h}_2 = 11,849.3 \text{ J/g-mole}$$
$$\bar{h}_1 - \bar{h}_2 = 6213.9 \text{ J/g-mole}$$

*From Table 3: Products for 25%
theoretical fuel*

$$T_1 = 600 \text{ K}$$
$$\bar{h}_1 = 17,830.9 \text{ J/g-mole}$$
$$T_2 = 400 \text{ K}$$
$$\bar{h}_2 = 11,738.1 \text{ J/g-mole}$$
$$\bar{h}_1 - \bar{h}_2 = 6092.8 \text{ J/g-mole}$$

Interpolating linearly with respect to percentage of theoretical fuel for 33.33% of theoretical fuel we get for the heat transfer per kilogram of mixture

$$Q = \frac{\dfrac{33.33 - 25.0}{50.0 - 25.0}(6213.9 - 6092.8) + 6092.8}{29.289}$$

$$= 209.40 \text{ kJ/kg},$$

where 29.289 is the molecular weight of the mixture.

## Example 10  Adiabatic Combustion at Constant Pressure

Liquid octane originally at 15 C is burned at constant pressure in steady flow in 200% of theoretical air originally at 200 C. Find the flame temperature for complete adiabatic combustion.

*Solution.* Application of the First Law indicates equality of the enthalpy of reactants, $H_R$, and that of products, $H_P$; thus

$$H_R = H_P,$$

where $H_R$ and $H_P$ are rederived from the same base state. Expressing this in terms of the enthalpy of combustion of liquid octane to gaseous products at

25 °C (which is given in Table 10), we have

$$[H_R - H_R'] = H_P - H_P' + (H_P' - H_R')$$

$$= [H_P - H_P'] + H_{RP}'. \qquad [a]$$

In general,

$$Q = H_P - H_R = (H_P - H_P') - (H_R - H_R')$$

$$+ (H_P' - H_R')$$

$$= (H_P - H_P') - (H_R - H_R') + H_{RP}'$$

where the primed symbols refer to a temperature of 25 °C (298.15 K) and $H_{RP}$ denotes enthalpy of combustion. The quantities in brackets, being differences in enthalpy between two states of the same chemical aggregation, are independent of the base state selected and therefore may be taken from any tables of reactants and products.

The enthalpy of the reactants may be considered to be the sum of the enthalpies of liquid octane and of air. The former is given satisfactorily by the equation

$$h_f = 2.1T - 668$$

where $h_f$ is the enthalpy in kJ per kilogram of liquid octane and $T$ is the absolute temperature in degrees Kelvin. (The state of zero enthalpy, which is irrelevant in this analysis, is octane vapor at 0 K.) The enthalpy of air may be taken from Table 1 and that of products from Table 5. The enthalpy of combustion at 25 °C is taken from Table 10.

For 200% of theoretical air the chemical equation is

$$C_8H_{18} + 25O_2 + 94.10N_2$$

$$= 8CO_2 + 9H_2O + 12.5O_2 + 94.10N_2,$$

where $N_2$ represents nitrogen and argon. Per g-mole of liquid octane we have

$$n_A = 119.10$$

$$m_A = 119.10 \times 28.9669 = 3450.0 \text{ g}$$

$$n_P = 123.60,$$

where $n_A$ and $n_P$ denote numbers of moles of air and products, respectively, and $m_A$ denotes the number of grams of air.

Expressing equation [a] in terms of specific enthalpies and solving for the enthalpy per mole of products, we have

$$\bar{h}_P = \frac{1}{n_P}\left[\bar{m}_f\left(h_f - h_f' - h_{RP}'\right) + m_A(h_A - h_A')\right] + \bar{h}_P',$$

where $\bar{m}_f$ denotes the molecular weight of octane. Substituting numbers, we get

$$\bar{h}_P = \frac{1}{123.60}\{114.22[2.1(288.2 - 298.2) + 44,420]$$

$$+ 3450.0(475.73 - 298.57)\} + 8778$$

$$= 54,750 \text{ J/g-mole.}$$

From Table 5 we get the flame temperature

$$T_P = 1639.6 \text{ K.}$$

It is evident from equation [a] that a simplification results if the enthalpy of combustion at 0 K, $H_{RP}^0$, is used, for we get

$$H_P = H_R - H_{RP}^0,$$

or, in terms of the specific enthalpies and the masses of air and fuel,

$$(m_A + m_F)h_P = (m_A + m_F)h_R - m_F h_{RP}^0$$

and

$$h_P = h_R - \frac{m_F}{m_A + m_F}h_{RP}^0. \qquad [b]$$

Equation [b] is often employed in combustion calculations.

## Example 11 Adiabatic Combustion at Constant Volume

A mixture of octane vapor and 200% of theoretical air at a pressure of 101.325 kPa (1 atm) and a temperature of 500 K is burned adiabatically and completely at constant volume. Find the temperature and pressure after combustion.

*Solution.* Application of the First Law indicates equality of the internal energy of reactants, $U_R$, and that of products, $U_P$; thus

$$U_R = U_P.$$

Expressing this in terms of the internal energy of combustion $U_{RP}$ at 25 °C, we have

$$[U_R - U_R'] = U_P - U_P' + (U_P' - U_R')$$

$$= [U_P - U_P'] + U_{RP}', \qquad [a]$$

where the primed symbols refer to a temperature of 25 °C (298.15 K). The quantities in brackets, being differences in internal energy between two states of the same chemical aggregation, are independent of the base state selected. That for reactants may be taken from a table for reactants, and that for products from a table for products.

The values of the properties of a mixture of octane vapor and 200% of theoretical air (or 50% of theoretical fuel) are obtained by linear interpolation, on the basis of percentage of theoretical fuel, with the aid of Tables 5 and 7, and with the data given in Table H on page 186.

Values of the properties of the products may be obtained directly from Table 5.

The value of $U_{RP}'$ is obtained from $H_{RP}'$ through the relation

$$U_{RP} = H_{RP} + [(pV)_R - (pV)_P],$$

or

$$U_{RP} = H_{RP} + \bar{R}T(n_R - n_P),$$

where $T$ denotes the absolute temperature and $n_R$ and $n_P$ the number of moles of reactants and products, respectively. The value of $H_{RP}$ at 25 °C is given in Table 10.

For 200% of theoretical air the chemical equation is

$$C_8H_{18} + 25O_2 + 25\frac{0.7901}{0.2099}N_2$$

$$= 8CO_2 + 9H_2O + 12.5O_2 + 94.10N_2,$$

where $N_2$ represents nitrogen and argon.

Per mole of gaseous octane we have

$$n_R = 120.10$$

and

$$n_P = 123.60,$$

so that

$$U_R = 120.10\bar{u}_R,$$

$$U_P = 123.60\bar{u}_P,$$

and, at 25 °C or 298.15 K,

$$U_{RP} = H_{RP} + 8.31441 \times 298.15[120.10 - 123.60]$$

$$= -44,782 \times 114.22 - 8676$$

$$= -5,115,000 - 8676$$

$$= -5,123,676 \text{ J/g-mole octane,}$$

and per kilogram of octane

$$u_{RP} = -\frac{5,123,676}{114.22} = -44,858 \text{ kJ/kg.}$$

The value of $\bar{u}_R$ at 500 K is obtained as follows: From Table 5, for 28% of theoretical fuel,

$$\bar{u}_R = 10,763.9 \text{ J/g-mole.}$$

From Table 7, for 54% of theoretical fuel,

$$\bar{u}_R = 11,083.5 \text{ J/g-mole.}$$

Hence, for the mixture of air and octane vapor having 50% of theoretical fuel, we have by interpolation

$$\bar{u}_R = \frac{50 - 28}{54 - 28}(11,083.5 - 10,763.9) + 10,763.9$$

$$= 11,034.3 \text{ J/g-mole.}$$

Similarly,

$$\bar{u}_R' = 6402.2 \text{ J/g-mole.}$$

Introducing these numbers and a value for $\bar{u}_P'$ from Table 5 into equation [a] and solving for $\bar{u}_P$, we get

$$\bar{u}_P = \frac{1}{123.60}\left[120.10(\bar{u}_R - \bar{u}_R') - U_{RP}'\right] + \bar{u}_P'$$

$$= \frac{1}{123.60}\left[120.10(11{,}034.3 - 6402.2)\right.$$

$$\left. + 5{,}123{,}676\right] + 6299.5$$

$$= 52{,}254 \text{ J/g-mole}.$$

The last entry in Table 5 lists an internal energy of 51,874.1 J/g-mole corresponding to a temperature of 1999 K. Thus we must use Table 6 in conjunction with Table 5. The change in internal energy from state 1 to state 2 may be expressed as

$$u_2 - u_1 = \int_{T_1}^{T_2} c_v \, dT.$$

Letting $u_1 = 51{,}874.1$ J/g-mole, $T_1 = 1999$ K, and $u_2 = 52{,}254$ J/g-mole and using Table 6 we get $52{,}254 - 51{,}874.1 = 30.355 \, (T_P - 1999)$ or

$$T_P = 2012 \text{ K}.$$

The ratio of the pressure of the products to that of the reactants is given by

$$\frac{p_P}{p_R} = \frac{n_P T_P}{n_R T_R}$$

since

$$V_P = V_R.$$

Thus

$$p_P = 101.325 \times \frac{123.60}{120.10} \times \frac{2012}{500}$$

$$= 419.6 \text{ kPa (4.14 atm)}.$$

### Example 12  Enthalpy of Combustion of *n*-Octane

The enthalpy of combustion at 25 °C of gaseous *n*-octane is given in Table 10 as −44,782 kJ per kilogram of octane for gaseous products of combustion. Correct this value to 0 K.

*Solution.* The chemical equation may be given in the form

$$C_8H_{18}(g) + 12.5O_2(g) + D \rightarrow 8CO_2(g)$$

$$+ 9H_2O(g) + D,$$

where $D$ denotes diluent gases which are unaffected by the reaction.

For a given temperature $T$ the enthalpy of combustion per mole of octane is given by

$$\bar{h}_{RP} = -\bar{h}_A - 12.5\bar{h}_B + 8\bar{h}_M + 9\bar{h}_N,$$

where $\bar{h}_A$, $\bar{h}_B$, $\bar{h}_M$ and $\bar{h}_N$ denote the enthalpy per mole of $C_8H_{18}$, $O_2$, $CO_2$, and $H_2O$, respectively, all at the temperature $T$ and all reckoned from the same base (for example, zero enthalpy for each element at 0 K).

The relation between the enthalpies of combustion per kilogram of octane at two temperatures $T'$ and $T''$ is then given by

$$h_{RP}' = h_{RP}'' + \frac{1}{\bar{m}_A}\left[-(\bar{h} - \bar{h}'')_A - 12.5(\bar{h} - \bar{h}'')_B\right.$$

$$\left. + 8(\bar{h} - \bar{h}'')_M + 9(\bar{h} - \bar{h}'')_N\right],$$

where the superscript refers to the corresponding temperature, and $\bar{m}_A$ denotes the molecular weight of octane. In this equation the enthalpies for each substance appear only in the difference between two values for that substance, and the base selected is, therefore, of no consequence. Values for $O_2$, $CO_2$, and $H_2O$ may be taken from Tables 13, 17, and 15, respectively. Values for octane must be found elsewhere and are here taken from Table 3u-E (part I) of "Tables of Selected Values of Chemical Thermodynamic Properties" (December 31, 1944), published by the National Bureau of Standards. The numerical solution for 0 K in terms of values at that temperature and at 25 °C is as follows.

$$h_{RP} \text{ at } 0 \text{ K} = -44{,}782 + \frac{1}{114.22}\left[-(0 - 114.22\right.$$

$$\times 320.01) - 12.5(0 - 8683.3)$$

$$+ 8(0 - 9365.3) + 9(0 - 9904.0)\right]$$

$$= -44{,}782 - 166$$

$$= -44{,}948 \text{ kJ/kg of octane}.$$

## Example 13   Discharge of Hydrogen from a Closed Container

Hydrogen in a tank having a volume of 10 m³ is initially at a pressure of 1000 kPa and at a temperature of 400 K. The hydrogen discharges to a pressure of 101.325 kPa (1 atm) through a small nozzle until one half the original mass remains in the tank. No heat is exchanged between the hydrogen and the walls of the tank. Find the final pressure and temperature in the tank.

*Solution.*  The process experienced by the hydrogen which finally remains in the tank is an isentropic expansion. Using subscripts 1 and 2 to denote, respectively, the initial and final states of hydrogen inside the tank, we have from Table 19

$$T_1 = 400 \text{ K}, \qquad v_{r1} = 0.180, \qquad p_{r1} = 18.4763$$

Moreover,

$$v_2 = 2v_1$$

and, therefore,

$$v_{r2} = 2v_{r1} = 0.360.$$

From Table 19,

$$T_2 = 303.1 \text{ K}, \qquad p_{r2} = 7.0093$$

$$p_2 = p_1 \times \frac{p_{r2}}{p_{r1}} = 379.4 \text{ kPa}.$$

## Example 14   Gas Turbine

Air is compressed isentropically in steady flow from 101.325 kPa (1 atm) and 300 K to 506.625 kPa (5 atm). Liquid octane at 300 K is introduced into the stream of compressed air at such a rate that the resultant mixture contains 300% of theoretical air. The octane burns completely at constant pressure, and the products of combustion expand isentropically to 101.325 kPa (1 atm). Find the efficiency.

*Solution.*  Compression: Let subscripts 1 and 2 refer to states at inlet and outlet of the compressor, respectively. Then from Table 1 we have

$$T_1 = 300 \text{ K}, \qquad h_1 = 300.43 \text{ kJ/kg}, \qquad p_{r1} = 1.3801$$

$$p_{r2} = \tfrac{5}{1} \times 1.3801 = 6.9005, \qquad T_2 = 473.6 \text{ K},$$

$$h_2 = 476.20 \text{ kJ/kg}.$$

Mixing: Application of the First Law to the adiabatic process of mixing liquid octane and air shows that the enthalpy of the resultant air-fuel mixture is equal to the sum of the enthalpies of the air and fuel before mixing. To evaluate the temperature of the mixture of air and octane vapor, a common base state is selected from which the enthalpies of the mixture and of the components may be reckoned.

In Table 1 the enthalpy of air is zero at 0 K. Using an analogous convention, we may fix the enthalpy of octane vapor at zero at 0 K. Data relative to this base have been published by the Bureau of Standards for hydrocarbons. The enthalpy of liquid octane as obtained from these data is given with satisfactory precision over a range of temperatures centering on 278 K by the equation

$$h_f = 2.1T - 668 \text{ kJ/kg},$$

where $h_f$ denotes the enthalpy of a kilogram of liquid octane at the temperature $T$ K. The effect of pressure

## Table K   States in the Gas Turbine, Figure 2

| State | Table Number | T K | h kJ / kg | $\bar{h}$ J / g-mole | $p_r$ | p kPa |
|---|---|---|---|---|---|---|
| 1 | 1 | 300 | 300.43* | 8702.5* | 1.3801 | 101.3 |
| 2 | 1 | 473.6 | 476.20* | 13,794.0* | 6.9005 | 506.6 |
| 3 | 5,7 | 458.5 | 465.09* | 13,694.6* | | 506.6 |
| 4 | 3,5 | 1297.4 | 1435.97 | 41,475.1 | 381.7 | 506.6 |
| 5 | 3,5 | 881.8 | 936.36 | 27,045.0 | 76.3 | 101.3 |

*Base state is a fuel-air mixture. Where values of these properties are not so marked the base state is a mixture of air and products of complete combustion.

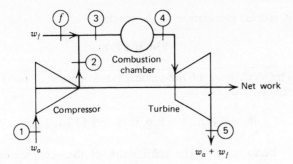

**Figure 1**

on enthalpy may be safely ignored in problems involving combustion.

The theoretical fuel-air ratio for octane ($C_8H_{18}$) is 0.06621, based on the composition of air used in constructing Table 1. Then for 300% of theoretical air, each kilogram of fuel-air mixture contains 0.0216 kg of fuel and 0.9784 kg of air. The enthalpy of each kilogram of liquid fuel at 300 K, based on vapor at 0 K, is

$$h_f = 2.1 \times 300 - 668 = -38 \text{ kJ/kg}.$$

The enthalpy of the fuel-air mixture based on a state having the same chemical aggregation is then

$$h_3 = 0.0216(-38) + 0.9784 \times 476.20$$

$$= 465.09 \text{ kJ/kg},$$

and

$$\bar{h}_3 = 465.09 \times 29.445 = 13{,}694.6 \text{ J/g-mole},$$

where subscript 3 refers to the state immediately after mixing and 29.445 is the molecular weight of the mixture. The temperature of this mixture is obtained by linear interpolation, on the basis of percentage of theoretical fuel, with the aid of Tables 5 and 7. From Table H on page 186, a mixture of air and octane vapor for 28% of theoretical fuel corresponds closely to a products table for 50% of theoretical fuel. From the value of $\bar{h}_3$ we get:

From Table 5, for 28% of theoretical fuel, $T_3 = 460.3$ K.

From Table 7, for 54% of theoretical fuel, $T_3 = 451.7$.

Hence, for 33.33% of theoretical fuel, we have by interpolation

$$T_3 = 460.3 + \frac{33.33 - 28}{54 - 28}(451.7 - 460.3) = 458.5 \text{ K}.$$

Combustion: Application of the First Law to the process of adiabatic combustion between states 3 and 4 (see Figure 1) results in the equation

$$h_3 = h_4,$$

where $h_3$ and $h_4$ represent the enthalpy per kilogram for reactants and products of combustion, respectively. The base states for $h_3$ and $h_4$ must, of course, be the same as explained on page 196.

The values of enthalpy used above in the calculation of the mixing process are based on an enthalpy of zero for gaseous reactants at 0 K. The values of enthalpy given in Tables 5 and 7, on the other hand, are based on an enthalpy of zero for gaseous products of combustion at 0 K. If the enthalpy of the reactants is to be reckoned from this latter or products base, the values on the reactants base must be augmented by the increase in enthalpy when products at 0 K are changed to reactants at 0 K. This quantity is the negative of the *enthalpy of combustion* at 0 K and is equal to the constant-pressure "heat of combustion" at 0 K.

For octane, the enthalpy of combustion per kilogram of fuel at 0 K, $h_{RP}{}^0$ (see Example 12) is constant over the range from 0 to 100% of theoretical fuel and is given by

$$h_{RP}{}^0 = h_P{}^0 - h_R{}^0 = -44{,}948 \text{ kJ/kg of octane},$$

where $h_R{}^0$ and $h_P{}^0$ denote the enthalpy at 0 K of the reactants and of the products, respectively, per kilogram of octane.

Values of the enthalpy per kilogram of *reactants* on the reactants base must therefore be augmented by the negative of $h_{RP}{}^0$ multiplied by the fraction of a kilogram of octane in each kilogram of reactants; that is, by

$$-\frac{w_f}{w_a + w_f} h_{RP}{}^0,$$

where $w_f$ and $w_a$ denote the mass rate of flow of fuel and air, respectively. Hence the enthalpy of the fuel-air mixture before combustion at state 3, reckoned from the products base, is given by

$$h_3 = 465.09 + 0.0216 \times 44{,}948 = 1435.97 \text{ kJ/kg}.$$

Since the enthalpies at states 3 and 4 are equal,

$$h_4 = 1435.97 \text{ kJ/kg},$$

where $h_4$ is reckoned from the products base state. Then

$$\bar{h}_4 = 1435.97 \times 28.879 = 41{,}469.4 \text{ J/g-mole},$$

where 28.879 is the molecular weight of the mixture found by interpolation from Table 9.

This value of $h_4$ and a table of enthalpies for products of combustion for 300% of theoretical air will yield the temperature at state 4 at the exit of the combustion chamber, because there is no change in chemical aggregation for such a mixture between the products base state and state 4. Since, however, a table for products of combustion for 300% of theoretical air or 33.33% of theoretical fuel is not available, linear interpolation, with respect to the percentage of theoretical fuel, between Tables 3 and 5 yields a value for the temperature as follows:

From Table 3, for 25% of theoretical fuel, $T_4 = 1304.9$ K.

From Table 5, for 50% of theoretical fuel, $T_4 = 1281.9$ K.

Hence, for 33.33% of theoretical fuel, we get

$$T_4 = 1304.9 + \frac{33.33 - 25}{50 - 25}(1281.9 - 1304.9)$$

$$= 1297.2 \text{ K}.$$

Similarly, we get for the relative pressure at state 4,

$$p_{r4} = 376.63 + \frac{33.33 - 25}{50 - 25}(391.10 - 376.63)$$

$$= 381.5.$$

Expansion: For isentropic expansion from 5 atm to 1 atm, we find the relative pressure at state 5 from that at state 4.

$$p_{r5} = p_{r4} \times \tfrac{1}{5} = 76.29.$$

The temperature and enthalpy at state 5 are found by linear interpolation, with respect to the percentage of theoretical fuel, between Tables 3 and 5. Thus for a value of $p_{r5}$ of 76.29, we get:

From Table 3, for 25% of theoretical fuel, $T_5 = 886.9$ K.

From Table 5, for 50% of theoretical fuel, $T_5 = 871.3$ K.

Hence, for 33.33% of theoretical fuel,

$$T_5 = 886.9 + \frac{33.33 - 25}{50 - 25}(871.3 - 886.9)$$

$$= 881.7 \text{ K}.$$

A similar calculation for $h_5$ yields

$$h_5 = 936.36 \text{ kJ/kg}.$$

The shaft work of the turbine per kilogram of products is now

$$W_t = 1435.97 - 936.36 = 499.61 \text{ kJ/kg}.$$

Performance: The shaft work of the compressor per kilogram of products is (from Table K)

$$W_c = \frac{w_a}{w_a + w_f}(h_2 - h_1) = 0.9784 \times 175.77$$

$$= 171.97 \text{ kJ/kg}.$$

The net work of the gas turbine per kilogram of products is

$$W_n = 499.61 - 171.97 = 327.64 \text{ kJ/kg},$$

since the work of compressing the liquid fuel is negligible. The heating value of the fuel supplied may be taken to be

$$-\frac{w_f}{w_a + w_f}h_{RP} = -0.0216 \times (-44{,}782)$$

$$= 967.3 \text{ kJ/kg},$$

where 44,782 is an arbitrary selected heat of combustion for octane at 25 °C. The efficiency of the gas turbine is then

$$\eta = \frac{327.64}{967.3} = 0.339.$$

The mixing and combustion processes may be combined and analyzed as a single process between sections 1 and 2 on the one hand, and section 4 on the other. Then the values for section 3 need not be determined.

If the products of combustion were treated as air in this example, the calculations for combustion and expansion would have been greatly simplified. Table 1 would have been used in place of Tables 3 and 5; no conversions to molar quantities and no interpolation between these tables would have been necessary. The temperature found at state 4 would have been 1334 K, an error of 37 degrees, and the efficiency would have been 0.348 an error of about 3%.

## Example 15 Diesel Engine

Air in a cylinder at 101.325 kPa (1 atm) and 350 K is compressed isentropically to one-tenth its initial volume. As the piston moves outward, liquid octane at 300 K is introduced and burned at such a rate as to maintain constant pressure. The total mass of octane introduced is one third the theoretical mass for combustion in the air in the cylinder. The products are expanded isentropically to the initial volume, at which point the exhaust valve opens and the pressure falls to 101.325 kPa. Find the states before and after combustion, at the opening of the exhaust valve, and at the return to atmospheric pressure.

*Solution.*

From Table 1 we obtain the data for state 1 given in Table L. To identify state 2, for which the entropy is the same as for state 1, we note that for state 2 the relative volume must be one tenth as large as for state 1.

For the adiabatic constant-pressure process between states 2 and 3 the enthalpy at state 3 is identical with the sum of the enthalpies of air in state 2 and fuel at 300 K. For each kilogram of air at 2, one third of the theoretical fuel or 0.0221 kg of fuel is introduced. Thus, each kilogram of products at 3 is formed from 0.0216 kg of fuel and 0.9784 kg of air. The enthalpy of this combination at 3 is then

$$h_3 = 0.9784(863.51) + 0.0216(44{,}948 - 38)$$

$$= 1814.91 \text{ kJ/kg,}$$

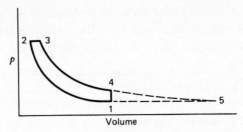

**Figure 2**

as reckoned from the products base (see Example 14). The molecular weight of this mixture, which is 28.879 and is obtained from Table 9, is then used to calculate the molar enthalpy.

The temperature at state 3 is obtained from the value of $\bar{h}_3$ by means of linear interpolation between Tables 3 and 5 (see Example 14), as follows: From Table 3, for 25% of theoretical fuel, $T_3 = 1608.8$ K. From Table 5, for 50% of theoretical fuel, $T_3 = 1577.7$ K. Hence, for 33.33% of theoretical fuel, we get

$$T_3 = 1608.8 + \frac{33.33 - 25}{50 - 25}(1577.7 - 1608.8)$$

$$= 1598.4 \text{ K.}$$

The internal energy, relative pressure, and relative volume at state 4 are calculated similarly by interpolation between Tables 3 and 5. The specific volume is determined with the aid of the equation of state

$$pv = RT.$$

The values are summarized in Table L.

## Table L   States in the Diesel Engine, Figure 2

| State | Table Number | T K | h kJ / kg | $\bar{h}$ J / g-mole | $p_r$ | u kJ / kg |
|-------|-------|-------|-------|-------|-------|-------|
| 1 | 1 | 350 | 350.73 | 10,159.6 | 2.3689 | 250.27 |
| 2 | 1 | 837.5 | 863.51 | 25,013.2 | 56.681 | 623.12 |
| 3 | 3, 5 | 1598.4 | 1814.91 | 52,412.5 | 950.43 | 1354.96 |
| 4 | 3, 5 | 970.9 | 1040.69 | 30,054.0 | 112.84 | 761.15 |
| 5 | 3, 5 | 747.9 | 783.51 | 22,696.9 | 39.72 | 568.21 |

| State | Table Number | $\bar{u}$ J / g-mole | $v_r$ | p kPa | $\hat{v}$ m³ / kg | Mass kg |
|-------|-------|-------|-------|-------|-------|-------|
| 1 | 1 | 7,249.5 | 42.407 | 101.3 | 0.9915 | 1 |
| 2 | 1 | 18,049.9 | 4.241 | 2424.7 | 0.0992 | 1 |
| 3 | 3, 5 | 39,130.3 | 0.013997 | 2424.7 | 0.1940 | 1.0221 |
| 4 | 3, 5 | 21,981.2 | 0.071541 | 288.2 | 0.9915 | 1.0221 |
| 5 | 3, 5 | 16,409.2 | 0.156863 | 101.3 | 2.1719 | 1.0221 |

(203)

Since states 3 and 4 lie on the same isentropic, state 4 may be identified from the relative volume, thus

$$\frac{v_{r4}}{v_{r3}} = \frac{\hat{v}_4}{\hat{v}_3},$$

where $\hat{v}$ denotes the volume per kilogram of air compressed and $\hat{v}_4$ is identical with $\hat{v}_1$.

The temperature $T_4$ is obtained from $v_{r4}$ by means of linear interpolation between Tables 3 and 5, as for $T_3$ above.

To identify state 5, which exists in the cylinder when atmospheric pressure is restored, we assume that the expansion of residual gases in the cylinder during exhaust is isentropic. The relative pressure at state 5 is found from the relation

$$\frac{p_{r5}}{p_{r3}} = \frac{p_5}{p_3}.$$

From Tables 3 and 5 we obtain by interpolation the temperature and relative volume and, through the equation of state, the specific volume.

The net indicated work is given by the algebraic sum of the areas under the curves 12, 23, and 34, thus

$$W = u_1 - u_2 + p_2(\hat{v}_3 - \hat{v}_2) + \frac{m_3}{m_2}(u_3 - u_4).$$

Substituting values from Table L, we get

$$W = (250.27 - 623.12) + 2424.7(0.1940 - 0.0992)$$
$$+ 1.0221(1354.96 - 761.15)$$
$$= 463.9 \text{ kJ/kg of air}$$

and (as in Example 14)

$$\eta = \frac{463.9}{0.0221 \times 44,782} = 0.469.$$

If the products of combustion were treated as air in this example, the temperature found at state 4 would have been 978 K, an error of 7 degrees, and the efficiency would have been 0.479 an error of about 2%.

### Example 16  Turbojet

A turbojet is to be designed for a speed of 650 km/hr at an altitude of 10,000 m (see Figure 3). Air is diffused isentropically from a relative velocity of 650 km/hr at section 1, the entrance to the diffuser, to zero velocity at section 2, the exit of the diffuser. The air is then compressed isentropically in steady flow through a fourfold increase in pressure from section 2 to section 3. Liquid octane at 300 K is introduced into the stream of compressed air at such a rate that the resultant homogeneous mixture at section 4 contains 25% of theoretical fuel. The octane burns completely at constant pressure, and the products of combustion leave the combustion chamber at section 5. Next, the products expand isentropically through a turbine to that pressure at section 6 which results in equality of power output from the turbine and power input to the compressor. Finally, the products expand reversibly and adiabatically through a nozzle to atmospheric pressure at section 7.

Calculate the efficiency of the power plant.

*Solution.*

Diffusion: The air enters the diffuser with a velocity of 650 km/hr or 180.6 m/sec, a temperature of

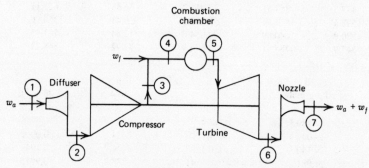

**Figure 3.**  Turbojet

(204)

### Table M  States in the Turbojet, Figure 3

| State | Table Number | T K | h kJ / kg | h̄ J / g-mole | $p_r$ | p kPa | V m / sec |
|---|---|---|---|---|---|---|---|
| 1 | 1, 60 | 223.1 | 223.33 | 6469.2 | 0.4906 | 26.42 | 180.6 |
| 2 | 1 | 239.4 | 239.64 | 6941.6 | 0.6273 | 33.78 | 0 |
| 3 | 1 | 355.8 | 356.55 | 10,328.1 | 2.5092 | 135.12 | 0 |
| 4 | 3, 5 | 343.4 | 350.12 | 10,118.8 |  | 135.12 | 0 |
| 5 | 3 | 1011.9 | 1082.06 | 31,272.2 | 130.04 | 135.12 | 0 |
| 6 | 3 | 852.9 | 897.43 | 25,942.7 | 65.30 | 67.85 | 0 |
| 7 | 3 | 669.1 | 692.02 | 19,999.9 | 25.43 | 26.42 | 641.0 |

223.15 K, and a pressure of 26.42 kPa. For zero exit velocity, application of the First Law results in

$$h_2 = h_1 + \frac{V_1^2}{2} = 223.33 + \frac{(180.6)^2}{(2)(1000)}$$

$$= 239.64 \text{ kJ/kg},$$

where the numerical value of $h_1$ is obtained from Table 1 at $T_1$. Table 1 at $h_2$ now gives $T_2 = 239.42$ K, $p_{r2} = 0.6273$, and $p_2 = 33.78$ kPa.

These values are summarized in Table M.

Compression: The work of isentropic compression in steady flow is the increase in enthalpy across the compressor if the velocities are small. State 3 at the exit of the compressor is identified through the relative pressure (Table 1),

$$p_{r3} = 4p_{r2} = 4 \times 0.6273 = 2.5092.$$

Hence $T_3$ is 355.78 K and $h_3$ is 356.55 kJ/kg of air. The work of compression is given by

$$W_c = h_3 - h_2 = 117.13 \text{ kJ/kg of air}.$$

Mixing and combustion: The same technique is used here as in Example 14, on the gas turbine. The results are given in Table M.

Turbine expansion: The equality of compressor power and turbine power is given by

$$h_3 - h_2 = \left(1 + \frac{w_f}{w_a}\right)(h_5 - h_6),$$

where $w_f$ and $w_a$ denote the mass rate of flow of fuel and air, respectively, and $h_6$ is the only unknown.

Nozzle expansion: For steady flow through the nozzle,

$$V_7^2 = 2(h_6 - h_7).$$

Since the nozzle expansion is isentropic, the relative pressure (from Table 3) at the exit of the nozzle determines state 7 as follows:

$$p_{r7} = p_{r6}\frac{p_7}{p_6} = 65.30\frac{26.42}{67.85} = 25.43$$

$$T_7 = 669.1 \text{ K}, \qquad h_7 = 692.02 \text{ kJ/kg}.$$

Hence the velocity $V_7$ leaving the nozzle is

$$V_7 = \sqrt{(2)(1000)(897.43 - 692.02)} = 640.95 \text{ m/sec}.$$

Performance: The thrust for a mass flow of one kilogram of air per second is given by the momentum equation in the form

$$\frac{T}{w_a} = \left(1 + \frac{w_f}{w_a}\right)V_7 - V_1.$$

The propulsive work for a mass flow of one kilogram of air per second is

$$\frac{W}{w_a} = \frac{T}{w_a}V_1.$$

The propulsive efficiency, the ratio of the propulsive work to the heating value of the fuel burned in the same period of time, is given by

$$\eta = \frac{\dfrac{W}{w_a}}{\dfrac{w_f}{w_a}(-h_{RP})} = \frac{V_1\left[\left(1 + \dfrac{w_f}{w_a}\right)V_7 - V_1\right]}{-\dfrac{w_f}{w_a}h_{RP}},$$

where $h_{RP}$ is the enthalpy of combustion at 25 °C. Hence

$$\eta = \frac{180.6\left(\dfrac{641.0}{0.9837} - 180.6\right)}{\dfrac{0.01628}{0.9837}(44{,}782)(1000)}$$

$$\eta = 0.115.$$

### Example 17    Isentropic Flow

Consider the steady, isentropic, one-dimensional flow of air at supersonic velocities. At a given section of the stream, the Mach number is 2.5, the stagnation temperature is 300 K, and the static pressure is 50 kPa.

Find stagnation pressure, static pressure, temperature, density, velocity, and mass velocity at sections of the stream where the Mach number is 2.5, 1.2, and 1.0.

*Solution.*   From Table 30, for a Mach number of 2.5, we get

$$\frac{p}{p_0} = 0.05853, \qquad \frac{T}{T_0} = 0.44444, \qquad \frac{\rho}{\rho_0} = 0.13169,$$

$$M^* = 1.8258.$$

From these and given values for $p$ and $T_0$, values for $p_0$ and $T$ are found. Either density, $\rho$ or $\rho_0$, may be found from the general relation

$$\rho = \frac{p}{RT},$$

and the other found from the value of $\rho/\rho_0$ above.
The velocity of sound is

$$a = \sqrt{kRT} = 20.05\sqrt{T} = 231.5 \text{ m/sec.}$$

The velocity is

$$V = Ma$$

and the mass velocity

$$G = \rho V.$$

Since Table 30 gives values of $p/p_0$, the pressure at any other Mach number $M$ may be found from $p/p_0$ and the previously calculated value of $p_0$.
The resultant values are summarized in Table N.

**Table N**

| M | 2.5 | 1.2 | 1.0 |
|---|---|---|---|
| p (kPa) | 50.0 | 352.3 | 451.3 |
| T (K) | 133.3 | 232.9 | 250.0 |
| $p_0$ (kPa) | 854.3 | 854.3 | 854.3 |
| $T_0$ (K) | 300.0 | 300.0 | 300.0 |
| $\rho$ (kg / m³) | 1.3065 | 5.2695 | 6.2893 |
| V (m / sec) | 578.79 | 367.18 | 317.02 |
| G (kg / m²sec) | 756.2 | 1934.9 | 1993.8 |

### Example 18    Rayleigh Line

Consider the steady, one-dimensional supersonic flow of air which is heated in the absence of friction and of area change. At one section of the stream, the Mach number is 2.5, the stagnation temperature is 300 K, and the static pressure is 50 kPa.

Find stagnation pressure, stagnation temperature, static pressure, temperature, density, velocity, and mass velocity at sections of the stream where the Mach number is 1.2 and 1.0.

*Solution.*   (*a*) The values of the quantities called for are given in Table N of Example 17 for a Mach number of 2.5. Table 36 gives the pressure, temperature, density, and velocity, each as a ratio to the corresponding quantity at a Mach number of unity on the same Rayleigh line. The values for a Mach number of unity can therefore be calculated. Thus,

$$p^* = \frac{p}{\dfrac{p}{p^*}} = \frac{50}{0.24616} = 203.1 \text{ kPa,}$$

and, similarly,

$$T^* = 352.0 \text{ K,} \qquad \rho^* = 2.0102 \text{ kg/m}^3,$$

$$V^* = 376.20 \text{ m/sec.}$$

For other values of the Mach number the ratios $p/p^*$, etc., now permit calculation of $p$, etc. The results are presented in Table O.

**Table O**

| M | 2.5 | 1.2 | 1.0 |
|---|---|---|---|
| p (kPa) | 50.0 | 161.6 | 203.1 |
| T (K) | 133.3 | 320.97 | 352.0 |
| $p_0$ (kPa) | 854.3 | 392.0 | 384.5 |
| $T_0$ (K) | 300.0 | 413.5 | 422.5 |
| $\rho$ (kg / m³) | 1.3065 | 1.7542 | 2.0102 |
| V (m / sec) | 578.79 | 431.09 | 376.20 |
| G (kg / m²sec) | 756.2 | 756.2 | 756.2 |

## Example 19 Fanno Line

Consider the steady, one-dimensional adiabatic flow of air at constant area with friction. At a given section of the stream, the Mach number is 2.5, the stagnation temperature is 300 K and the static pressure is 50 kPa.

(a) Find stagnation pressure, stagnation temperature, static pressure, temperature, density, velocity, and mass velocity at sections of the stream where the Mach number is 1.2 and 1.0.

(b) Find the distance between two sections of Mach number 2.5 and 1.2, respectively, if the flow is through a circular duct with a diameter of 3 cm and with an average friction coefficient of 0.003.

*Solution.* (a) The procedure parallels that of the preceding example. Values are first found for a Mach number of unity on the same Fanno line ( $p^*$, $T_0^*$, etc.) from the values for a Mach number of 2.5 and the ratios of Table 42. These with the values from Table 42 for any other given value of the Mach number yield the desired quantities. The results are presented in Table P.

(b) Table 42 gives values of the quantity $4fL_{max}/D$, where $L_{max}$ denotes the length of duct from Mach number $M$ to Mach number unity. The difference between the values of $L_{max}$ for two different values of Mach number is, therefore, the length of duct between the two sections corresponding to the two values of Mach number. Thus, for constant friction coefficient $f$,

$$ L = \frac{D}{4f}\left[\left(\frac{4fL_{max}}{D}\right)'' - \left(\frac{4fL_{max}}{D}\right)'\right]. $$

Thus, for a constant friction coefficient $f$ of 0.003, and a diameter of 3 cm the distance between two sections of Mach number 2.5 and 1.2, respectively, is

### Table P

| M | 2.5 | 1.2 | 1.0 |
|---|---|---|---|
| p (kPa) | 50.0 | 137.7 | 171.2 |
| T (K) | 133.3 | 232.9 | 250.0 |
| $p_0$ (kPa) | 854.3 | 333.9 | 324.0 |
| $T_0$ (K) | 300.0 | 300.0 | 300.0 |
| $\rho$ (kg / m$^3$) | 1.3065 | 2.0595 | 2.3853 |
| V (m / sec) | 578.79 | 367.18 | 317.02 |
| G (kg / m$^2$sec) | 756.2 | 756.2 | 756.2 |

given by

$$ L = \frac{D}{4f}\left[\left(\frac{4fL_{max}}{D}\right)_{2.5} - \left(\frac{4fL_{max}}{D}\right)_{1.2}\right] $$

$$ = \frac{3}{4 \times 0.003}(0.43197 - 0.03364) = 99.6 \text{ cm}. $$

## Example 20 Normal Shock

A steady flow of air passes through a one-dimensional normal shock. At the upstream section, the Mach number is 2.5, the stagnation temperature is 300 K, and the static pressure is 50 kPa.

Find the properties of the air at the upstream and downstream sections of the shock.

*Solution.* The properties at a Mach number of 2.5 are given in Table N of Example 17. These together with the quantities of Table 48, by obvious methods, give the results presented in Table Q.

### Table Q

| Section | Upstream | Downstream |
|---|---|---|
| M | 2.50 | 0.51299 |
| p (kPa) | 50.0 | 356.3 |
| T (K) | 133.3 | 284.9 |
| $p_0$ (kPa) | 854.3 | 426.3 |
| $T_0$ (K) | 300.0 | 300.0 |
| $\rho$ (kg / m$^3$) | 1.3065 | 4.3571 |
| V (m / sec) | 578.79 | 173.56 |
| G (kg / m$^2$sec) | 756.2 | 756.2 |

## Example 21 Two-Dimensional Shock

A steady parallel stream of air approaches a sharp corner with an angle of 20°, as shown in Figure 4. The air has a pressure of 150 kPa, a temperature of 300 K, and a Mach number of 2.0. For both weak and strong shocks, find the downstream Mach number and pressure, the ratio of downstream to upstream mass velocity, and the angle the shock makes with the original direction of flow.

*Solution.* Two solutions are possible, depending on the nature of the shock; both are given below. Let x and y refer to conditions upstream and downstream of the shock, respectively.

From Table 56, for $\omega' = 20°$ and $M_x = 2.0$, we get two values of $\alpha'$.

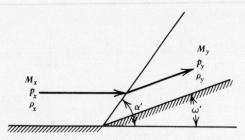

**Figure.4.** Two-dimensional shock

From Table 57, for $\omega' = 20°$ and the two values of $\alpha'$, we get two values of $M_y$.

From Tables 58 and 59, we get the corresponding values for $p_y/p_x$ and $\rho_y/\rho_x$.

The ratio of mass velocities downstream to upstream is given by

$$\frac{G_y}{G_x} = \frac{\rho_y}{\rho_x} \cdot \frac{V_y}{V_x} = \frac{\rho_y}{\rho_x} \cdot \frac{M_y}{M_x} \cdot \sqrt{\frac{T_y}{T_x}},$$

where

$$\frac{T_y}{T_x} = \frac{p_y/p_x}{\rho_y/\rho_x}.$$

The resulting values are given in Table R.

**Table R**

| n. | Upstream Section | Downstream Section | |
|---|---|---|---|
| | | Weak Shock | Strong Shock |
| M | 2.0 | 1.210 | 0.7282 |
| p (kPa) | 150.0 | 426.5 | 623.6 |
| T (K) | 300.0 | 417.7 | 488.3 |
| $\rho$ (kg / m³) | 1.7420 | 3.5572 | 4.4486 |
| $\alpha'$ (deg) | 0 | 53.44 | 74.25 |
| G / Gx | 1 | 1.458 | 1.186 |
| V (m / sec) | 21.96 | 15.68 | 10.20 |

(208)

# Acknowledgments

Tables 30 to 53 were reprinted from Meteor Report 14, entitled "The Mechanics and Thermodynamics of Steady One-Dimensional Gas Flow with Tables for Numerical Solutions," by A. H. Shapiro, W. R. Hawthorne, and G. M. Edelman, December 1, 1947. This report was prepared in the Gas Turbine Laboratory of the Massachusetts Institute of Technology for Project Meteor under contract with the Office of Naval Research of the United States Navy, and was reprinted by permission of the director of the project and of the authors. For this edition, the data in these tables have been rechecked and corrected where necessary.

Tables 55 to 59 are reprinted from Bumblebee Report 26, entitled "The Theory and Practice of Two-Dimensional Supersonic Pressure Calculations," by N. Edmonson, F. D. Murnaghan, and R. M. Snow, December 1945. This report was prepared in the Applied Physics Laboratory, The Johns Hopkins University, under contract with the Bureau of Ordnance of the United States Navy, and is reprinted here by permission of the director of the laboratory.

Personnel of the Research Division of the United Aircraft Corporation provided assistance in the preparation of Tables 26 through 29.

# Bibliography

1. J. O. Hirschfelder, C. F. Curtiss, and R. B. Bird, *Molecular Theory of Gases and Liquids*, John Wiley & Sons, Inc., New York, 1954.

2. J. H. Keenan, *Thermodynamics*, John Wiley & Sons, Inc., New York, 1941, p. 96.

3. J. E. Mayer and M. G. Mayer, *Statistical Mechanics*, John Wiley & Sons, Inc., New York, 1940; F. D. Rossini *Chemical Thermodynamics*, John Wiley & Sons, Inc., New York, 1950; and many others.

4. F. C. Andrews, *Equilibrium Statistical Mechanics*, John Wiley & Sons, Inc., New York, 1975.

5. G. Herzberg, *Spectra of Diatomic Molecules*, D. Van Nostrand Co., Inc., New York, 1950.

6. J. Hilsenrath, C. W. Beckett, W. S. Benedict, L. Fano, H. J. Hoge, J. F. Masi, R. L. Nuttall, Y. S. Touloukian, and H. W. Wolley, NBS Circular 564, National Bureau of Standards, Washington, D.C., 1955.

7. B. LeNeindre and B. Vodar, *Experimental Thermodynamics*, Vol. II, Butterworth and Co., Ltd., London, 1975.

8. Y. S. Touloukian and T. Makita, *Thermophysical Properties of Matter*, Vol. 6, IFI/Plenum Data Corporation, New York, 1970.

9. R. L. Dommett, RAE-TN-GW 429, 1–39, 1956 (AD 115386).

10. H. K. Kallman, RM 442, 1–44, 1950 (AD 103216).

11. E. R. Cohen and B. N. Taylor, *J. Phys. Chem. Ref. Data*, **2**, 663 (1973).

12. N. N. Greenwood, *Pure Appl. Chem.*, **37**, 4 (1974).

13. B. Rosen, *Spectroscopic Data of Diatomic Molecules*, Pergamon Press, London, 1970.

14. C. E. Moore, *Atomic Energy Levels*, Vol. I, National Bureau of Standards, Washington, D.C., 1949.

15. F. Din, *Thermodynamic Functions of Gases*, Vol. 2, Butterworths Scientific Publications, London, 1956.

16. N. B. Vargaftik, *Tables on the Thermophysical Properties of Liquids and Gases*, 2nd Ed., John Wiley & Sons, Inc., New York, 1975.

17. G. C. Maitland and E. B. Smith, *J. Chem. Eng. Data*, **17**, 150 (1972).

18. Y. S. Touloukian, S. C. Saxena, and P. Hestermans, *Thermophysical Properties of Matter*, Vol. 11, IFI/Plenum Data Corporation, New York, 1975.

19. Y. S. Touloukian, T. E. Liley, and S. C. Saxena, *Thermophysical Properties of Matter*, Vol. 3, IFI/Plenum Data Corporation, New York, 1970.

20. J. Kaye, paper presented at Annual Meeting of A.S.M.E., December 1947.

21. "Selected Values of Properties of Hydrocarbons and Related Compounds," American Petroleum Institute Research Project 44, Thermodynamics Research Center, Texas A & M University, College Station, Texas 77843 (loose-leaf data sheets, extant 1977).

22. Report of the CODATA Task Group on Key Values for Thermodynamics, J. D. Cox, Chairman, CODATA Bulletin No. 17, January 1976.

23. D. W. Scott, *J. Chem. Phys.*, **60**, 3144 (1974).

24. H. W. Woolley, R. B. Scott, and F. G. Brickwedde, *J. Res. Natl. Bur. Std.*, **41** 379 (1948).

25. L. Haar, Private communication, 1978.

26. D. R. Stull and H. Prophet, ed., *JANAF Thermochemical Tables*, 2nd Edition, NSRDS-NBS 37, U.S. Government Printing Office, Washington, D.C., 1971; M. W. Chase, private communication, the Dow Chemical Company, Midland, Michigan, 1977.

27. "Selected Values of Properties of Chemical Compounds," Thermodynamics Research Center Data Project, Thermodynamics Research Center, Texas A & M University, College Station, Texas 77843 (loose-leaf data sheets, extant 1978).

28. L. V. Gurvich, G. A. Khachkuruzov, V. A. Medvedev, I. V. Veyts, G. A. Bergman, V. S. Yungman, N. P. Rtishcheva, L. F. Kuratova, G. N. Yurkov, A. A. Kane, B. F. Yudin, B. I. Brounshteyn, V. F. Baybuz, V. A. Kvlividze, Ye. A Prozorovskiy, B. A. Vorobýev, *Thermodynamic Properties of Chemical Substances*, Academy of Sciences, U.S.S.R., Moscow, 1962.

29. S. Angus and B. Armstrong, *International Thermodynamic Tables of the Fluid State*, Vol. 1, Pergamon Press, New York, 1971.

30. S. Angus, B. Armstrong, and K. M. deReuck, *International Thermodynamic Tables of the Fluid State*, Vol. 3, Pergamon Press, New York, 1976.

31. V. J. Johnson, "Cryogens and Gases," Publication No. STP 537, American Society for Testing and Materials, Philadelphia, Pa., 1973.

32. G. M. Wilson, R. G. Clark, and F. L. Hyman, *Applied Thermodynamics*, American Chemical Society, Washington, D.C., 1968.

33. F. Din, *Thermodynamic Functions of Gases*, Vol. 1, Butterworths Scientific Publications, London, 1956.

34. F. Din, *Thermodynamic Functions of Gases*, Vol. 2, Butterworths Scientific Publications, London, 1956.

35. F. Din, *Thermodynamic Functions of Gases*, Vol. 3, Butterworths Scientific Publications, London, 1961.

36. A. A. Vasserman, Ya. Z. Kazavchinskii, and V. A. Rabinovich, *Thermophysical Properties of Air and Air Components*, Israel Program for Scientific Translation, Jerusalem, 1971 (translated from Russian).

37. A. H. Shapiro and W. R. Hawthorne, *J. Applied Mech.*, **14**, A317–337 (1947).

38. A. H. Shapiro and G. M. Edelman, *J. Applied Mech.*, **14** A154–162 (1947).

39. N. Edmonson, F. D. Murnaghan, and R. M. Snow, Bumblebee Report 26 (December 1945).

40. W. E. Forsythe, Ed., *Smithsonian Physical Tables*, Smithsonian Institution, Washington, D.C., 1969, 347.

41. E. R. Cohen and B. N. Taylor, *J. Phys. Chem. Ref. Data*, **2**, 663 (1973).

42. "The International System of Units (SI)," U.S. National Bureau of Standards Special Publication 330, U.S. Government Printing Office, Washington, D.C., 1977.